Natural Bio-active Compounds

Mohd Sayeed Akhtar
Mallappa Kumara Swamy
Uma Rani Sinniah
Editors

Natural Bio-active Compounds

Volume 1: Production and Applications

Springer

Editors
Mohd Sayeed Akhtar
Department of Botany
Gandhi Faiz-e-Aam College
Shahjahanpur, Uttar Pradesh, India

Mallappa Kumara Swamy
Department of Biotechnology
East West First Grade College of Science
Bengaluru, Karnataka, India

Uma Rani Sinniah
Department of Crop Science
Universiti Putra Malaysia
Serdang, Selangor, Malaysia

ISBN 978-981-13-7153-0 ISBN 978-981-13-7154-7 (eBook)
https://doi.org/10.1007/978-981-13-7154-7

This Springer imprint is published by the registered company Springer Nature Singapore Pte Ltd.
The registered company address is: 152 Beach Road, #21-01/04 Gateway East, Singapore 189721, Singapore

This book is dedicated to

Abu Ali Ibn Sina (980–1037 AD)

A great physician, scholar, philosopher, astronomer, writer of medieval times, and father of early modern medicine.

Foreword

Bio-active compounds produced from natural sources, such as plants, fungi, lichens, etc., exhibit superior chemo-diversity and possess several pharmacological significances. Some of the major classes of bio-active compounds include phenolics, alkaloids, tannins, saponins, lignin, glycosides, and terpenoids. The discovery of such unique compounds has inspired many scientific communities to explore their potential applications in various fields including agriculture and biomedicine. For instance, plant metabolites are utilized to manufacture eco-friendly biopesticides and as drug sources in medicine. Due to numerous health-promoting properties, these phytocompounds are widely used by humans as a source of medication since from ancient times to the modern world. The assessment of natural bio-active compounds for their wide-ranging therapeutic potential has led to the discovery of many drug leads in recent times. Natural products research has become a trust area among scientists aimed toward understanding the chemistry, analytical methodologies, biosynthetic mechanisms, and pharmacological activities of several natural compounds. In recent times, the natural product-based medicine is considered as the most suitable and safe to be used as an alternative medicine. In this regard, there is an unprecedented task to fulfill the increasing demand for natural metabolites by flavor and fragrance, food, and pharmaceutical industries. Thus, many natural resources are being explored to produce and accomplish the demand for natural bio-active compounds.

The present book entitled *Natural Bio-active Compounds: Volume 1 – Production and Applications* includes 22 chapters contributed by academicians, scientists, and researchers from different parts of the globe. In Chap. 1, Brazilian author provides a holistic point of view to the current strategies adopted to screen and produce novel endophyte-derived bio-active metabolites, while Chap. 2 by Indian and German authors discusses the unmapped repository of endophytic natural products. In Chap. 3, Braga Adelaide et al., describe the production of polyphenols by microbes, while in Chap. 4, Malaysian authors have discussed on the progress and advances made in the research of endolichenic fungi. Also, they have highlighted on the emerging biotechnological approaches in exploring endolichenic fungi. Chapter 5, by Anand Shyamlal Gupta, mentions the chemistry, medicinal importance, isolation, and strategic approaches for the purification of glycosides from natural sources. Likewise, in Chap. 6, Sridhar et al. describe the treatment of obesity by natural products-based pancreatic lipase inhibitors, and Chap. 7 by Saboon et al. describes the applications of natural compounds extracted from medicinal plants. Chapter 8, by Indian authors, narrates about the sources of seed oils, their methods of extractions, and bioactivity, while Chap. 9 by Desam Nagarjuna Reddy provides the comprehensive information on the specific chemical compounds occurring in essential oils and their medical applications and economic importance. In Chap. 10, an overview on the present scenario and future aspects of cellulose hydrogels and their applications is discussed by Pal et al., a group of Indian scientists. Chapter 11 by Mexican authors gives a detailed account on the current strategies for the production of plant secondary metabolites in a continuous and reliable manner, especially the influences of elicitors and eustressors on the production of plant secondary metabolites, while in Chap. 12, a collaborative work by Malaysian and Thailand researchers discusses the existing approaches in the management of colorectal cancer by targeting *KRAS* proto-oncogene. Similarly, Chap. 13 by Ansari and Akhtar explains the new insights on the recent progress of flavonoids as effective candidates in cancer therapeutics and prevention. Chapter 14 by Khairulmazmi and Tijjani highlights the uses and profiling of bio-active compounds of *Moringa oleifera,* their mode of action, and prospects in commercial biopesticides for agricultural applications. Subsequently, Chap. 15 focuses on the natural compound of genus *Brassica* and their therapeutic activities, while Chap. 16 by Kirubakari et al., entails the prospects of higher plants as antimicrobial agents. Chapters 17 and 18, by Indian authors, describe the phytochemistry and pharmacological properties of neem tree-derived bio-active compounds and the role of plant secondary metabolites acting on different targets for treating diabetes. In the next chapter, Javid et al., beautifully describe the pharmacological activities of *Leptadenia pyrotechnica.* Chapter 20 by Indian authors summarizes the therapeutic efficacy of garlic and its bio-active organosulfur compounds against risk factor-mediated atherosclerotic cardiovascular diseases, while Chap. 21, by Pakistani authors, provides comprehensive information on nutritive and pharmacological properties of *Physalis peruviana.* In the last chapter, Brazilian authors discuss on the content and chemical composition of the essential oil of *Baccharis milleflora* and their biological significances.

Understanding about various natural bio-active compounds is very much required in order to promote the drug discovery research and to complement the medical world by novel drug molecules with superior bioactivities. I believe this book surely provides updated information on the production and application of natural bio-active compounds to graduate and undergraduate students, teachers, industry persons, and healthcare professionals involved in natural product and therapeutic research areas. I congratulate the editorial board members, Dr. Mohd Sayeed Akhtar, Mallappa Kumara Swamy, and Uma Rani Sinniah, and all contributing authors for bringing the collection of their noble piece of work and also for the grand success of this book.

Abdul Shukor Juraimi

Professor, Department of Crop Science
Faculty of Agriculture
Universiti Putra Malaysia
Seri Kembangan, Malaysia

Preface

Secondary metabolites are the unique faction of compounds, produced by plants to protect itself against the various biotic and abiotic factors (diseases, pests, pathogens, herbivores, environmental stresses, etc.). Secondary metabolites do not influence the primary metabolic activities such as growth and reproduction of plants. The major classes include phenolics, alkaloids, tannins, saponins, lignin, glycosides, and terpenoids. Some of these compounds have become an integral part of plant-microbe interactions toward adapting to environmental irregularities. They regulate symbiosis, induce seed germination, and show allelopathic effect, i.e., inhibit other competing plant species in their environment. Moreover, these compounds induce adverse physiological activities, such as reduced digestive efficiency, reproductive failure, neurological problems, gangrene, goiter, and even death and also possess high toxicity. The discovery of such unique compounds has inspired many scientific communities to explore their potential applications in various fields including agriculture and biomedicine. For instance, plant secondary metabolites are utilized to manufacture eco-friendly biopesticides and as drug sources in medicine. Due to numerous health-promoting properties, these compounds are widely used as a source of medication since ancient times. The assessment of plant secondary metabolites for their wide-ranging therapeutic potential has led to the discovery of many drug leads in recent times. Therefore, this field of research has become a reliance area for researchers interested to explore the chemistry, analytical methodologies, biosynthetic mechanisms, and pharmacological activities of plant secondary metabolites.

The use of natural bio-active compounds and their products is considered as most suitable and safe to be used as an alternative medicine. Thus, there is an unprecedented task to fulfill the increasing demand for plant secondary metabolites by flavor and fragrance, food, and pharmaceutical industries. However, their supply has become one of the major constraints as their large-scale cultivation is very limited. Moreover, it is difficult to obtain a constant quantity of compounds from the cultivated plants as their yield fluctuates due to several factors including genotypic variations, geography, edaphic conditions, and harvesting and processing methods. In addition, medicinal plants have become endangered due to ruthless harvesting in nature. Alternatively, the plant tissue culture approaches can be well explored to produce secondary metabolites without practicing the conventional agriculture requiring more land space. In vitro cell and tissue cultures require less space and are grown under a controlled lab conditions and hence offer advantages of producing the desired compounds continuously without affecting their biosynthesis and quality. Furthermore, these cultures can be scaled up to produce metabolites in very large bioreactors, and also, using genetically engineered cells/tissues, novel products can be obtained. The proper knowledge and exploration of these in vitro approaches could provide an optional source to produce plant secondary metabolites from many medicinal plants in large scale.

Natural Bio-active Compounds: Volume 1 – Production and Applications is a very timely effort in this direction. This book volume with 22 contributions from the authors of Australia, Brazil, India, Malaysia, Mexico, Nigeria, Pakistan, Portugal, Saudi Arabia, and Thailand discusses the production and applications of natural bioactive compounds isolated from plants as well as microbial endophytes. Moreover, chemistry, pharmacological properties, and biotechnological approaches against various human diseases are also well discussed. This book will be a valuable resource for researchers to work toward identifying and characterizing new bioactive agents from a diversified flora and to enable the discovery of novel therapeutic leads in the near future against various diseases, and also for the graduate and undergraduate students, teachers, industry persons, and healthcare professionals involved in natural product and therapeutic research areas.

We are highly grateful to all our contributors for readily accepting our invitation and for sharing their knowledge. Further, we greatly appreciate their commitment in composing the chapters and enduring editorial suggestions to finally produce this venture. We are also thankful to Professor Abdul Shukor Juraimi for his suggestion and writing the foreword for this volume. We also thank the team of Springer International, especially Dr. Mamta Kapila and Raagaipriya Chandrasekaran for their generous cooperation at every stage of the publication.

Shahjahanpur, Uttar Pradesh, India — Mohd Sayeed Akhtar
Bengaluru, Karnataka, India — Mallappa Kumara Swamy
Serdang, Selangor, Malaysia — Uma Rani Sinniah

About the Book

This book has comprehensively reviewed latest information on various aspects of natural bioactive compounds produced from microbes, plants, and algae. It provides detailed information on several classes of phytocompounds including phenolics, alkaloids, tannins, glycosides, etc. and also discusses on their potential applications in various fields including agriculture and biomedicine. The health-promoting properties of these natural resources and their phytocompounds as detailed in the traditional medicine are detailed in this book with recent practical proofs and documentations with a special focus on their safety issues. Topics related to medicinal plants such as ethnopharmacology, phytochemistry, extraction methods, challenges in medicinal plants cultivation, toxicological effects, clinical studies, mode of action, potential biomolecular interactions, advancements in secondary metabolites production, targeted therapy, newly identified potential natural compounds, and novel drug discovery strategies including computational approaches are discussed in detail. Furthermore, various sources of natural products and their therapeutic applications will benefit to explore to overcome the current deficit in the supply of bioactive natural compounds. Overall, this book is a valuable resource for researchers to work toward identifying and characterizing new bioactive agents from a diversified flora, and to enable the discovery of novel therapeutic leads in the near future against various human ailments. This book is useful to industries, researchers, subject experts, and students working in multidisciplinary areas such as medicinal chemistry, pharmacology, biochemistry, and other topics related to drug discovery research.

Contents

About the Editors

Dr. Mohd Sayeed Akhtar (PhD) is an Assistant Professor at Gandhi Faiz-e-Aam College, Shahjahanpur, UP, India. He received his PhD from Aligarh Muslim University (AMU), India, in 2008, prior to conducting postdoctoral research at the Botanical Institute, University of Basel (BIB), Switzerland (2008–2010), and at Chonbuk National University (CBNU), Republic of Korea, in 2011. He was an Assistant Professor at Jimma University, Ethiopia (2011–2014), and a Fellow Researcher at the Institute of Tropical Agriculture, Universiti Putra Malaysia (UPM) (2014–2015). Dr. Akhtar has more than 15 years of research and 10 years of teaching experience in soil microbiology, applied microbiology, environmental microbiology, molecular biology, plant pathology, and plant nanobiotechnology. He has authored or coauthored more than 100 articles in peer-reviewed journals, conference proceedings, and book chapters and has edited 10 books with international publishers. Further, he serves as an Editorial Board Member and Reviewer for several high-impact international journals. His current research is focused on rhizospheric plant-microbe interactions and molecular biotechnology, bioremediation, biomineralization, nano-fertilizers, and nanobiotechnology.

Dr. Mallappa Kumara Swamy (PhD) is a Professor and Head of the Department of Biotechnology at East West First Grade College of Science, Bengaluru, Karnataka, India. He completed his postdoctoral research at the Department of Crop Science, Faculty of Agriculture, Universiti Putra Malaysia (UPM), Serdang, Selangor, Malaysia. Before that, he had worked as an Associate Professor and Head of the Department of Biotechnology, Padmashree Institute of Management and Sciences, Bangalore University, Bengaluru, India. He received his PhD (Biotechnology) from Acharya Nagarjuna University, Guntur, India, in 2013. He has more than 15 years of teaching and research experience in the fields of plant biotechnology, secondary metabolite production, phytochemistry, and bio-active studies. Dr. Swamy has authored 80 research publications in peer-reviewed journals and 16 book chapters. Recently, he has edited four books published by Springer Nature Singapore Pte Ltd., Singapore, one book by CRC Press LCC, USA and one book by Studium Press Pvt. Ltd., India. He also serves as an Editorial Board Member and Reviewer for several high-impact factor international journals. His current research is focused on cell and tissue culture technology for bio-active compound production and on nanotechnology for medical applications.

Dr. Uma Rani Sinniah (PhD) is a Professor at the Department of Crop Science, Faculty of Agriculture, Universiti Putra Malaysia (UPM). She has more than 35 years of teaching and research experience in the fields of plant biology, seed technology, cryobiology, germplasm conservation, and plant secondary metabolites and has successfully completed several projects supported by various national and international funding bodies. She has published more than 95 research articles in peer-reviewed journals and 8 book chapters. Her current research is focused on seed science and technology, cryopreservation, plant cell and tissue culture, and secondary metabolite production.

Production and Application of Novel Bio-active Compounds by Endophytic Microbes

1

Julio Alves Cardoso Filho

Contents

Abstract

Traditionally, endophytes are microorganisms that inhabit plant tissues, establishing an association with their hosts for most or all of their life without causing any apparent damage. Recently, researchers have shown an increased interest in the potential of endophytes to produce bio-active compounds with activity against numerous human, animal, and plant diseases. The determination of these bio-active molecules and their modes of action are technically challenging. Thus,

J. A. C. Filho (✉)
Center of Agricultural Sciences, Sector of Plant Pathology, Federal University of Alagoas, Rio Largo, AL, Brazil
e-mail: julioalvescardosofilho@yahoo.com.br

M. S. Akhtar et al. (eds.), *Natural Bio-active Compounds*,
https://doi.org/10.1007/978-981-13-7154-7_1

the aim of this chapter is to provide a holistic point of view to the current strategies adopted for screening and production of novel endophyte-derived metabolites, such as terpenoids, alkaloids, phenylpropanoids, aliphatic compounds, polyketides, and peptides, and to present their commercial applications in the medicinal, pharmaceutical, agricultural food, and cosmetic industries.

Keywords
Biotechnology industries · Pharmacology drugs · Polyketides · Nonribosomal peptides · Secondary metabolites

1.1 Introduction

An endophyte is defined as "any organism that forms latent asymptomatic infections within healthy tissue systems either in intercellular spaces (apoplasts) or inside the cells (symplasts) of plants" (Porras-Alfaro and Bayman 2011; Specian et al. 2012; Perotto et al. 2013; Chaparro et al. 2014; Dutta et al. 2014; Farrar et al. 2014; Akhtar et al. 2015; Hardoim et al. 2015; Kaul et al. 2016; Sengupta et al. 2017). The endophytic communities have been divided into different subgroups, including "mutualism, commensalism and parasitism", which are related to their host plants (Hardoim et al. 2012; Andreote et al. 2014; Nisa et al. 2015). The recent reviews highlight types of endophytic associations, such as algal endophytes (Sarasan et al. 2017), endophytic insect pathogenic fungi (Barelli et al. 2016; Moonjely et al. 2016; Behie et al. 2017), endophyte fungi (Vasundhara et al. 2016; Knapp et al. 2018; Akhtar and Panwar 2011; Akhtar et al. 2011; Akhtar et al. 2015; Swamy et al. 2016a, b), prokaryotic endophytes (Hollensteiner et al. 2018), endophytic actinobacteria (Álvarez-Pérez et al. 2017), plant growth-promoting bacteria (PGPB) (Akhtar and Siddiqui 2010; Akhtar and Azam 2014; Olanrewaju et al. 2017), plant growth-promoting fungi (PGPF) (Hossain et al. 2017), nematophagous endophytic fungi (Vidal-Diez de Ulzurrun and Hsueh 2018), mycorrhizas symbiosis (Berruti et al. 2016; Filho et al. 2017; Mills et al. 2018), actinorhizal symbiosis (Franche et al. 2016), and rhizobia symbiosis (Checcucci et al. 2017; Naveed et al. 2017). Most endophytes are unculturable (Liaqat and Eltem 2016); therefore, the analysis of their diversity and the molecular basis of their interactions with the plant are revealed by using molecular approaches (Kaul et al. 2016).

Endophytic microbes can drive the host plant demography (Saikkonen et al. 2016), shape plant communities (Yahr et al. 2016), guide the community structure and biodiversity of the aggregated organisms (Edwards et al. 2017), and have an impact on the phenotype and epigenome of their associated plants (Vannier et al. 2015). The field of drug discovery renewed our interest in endophytes microbes (Thatoi et al. 2013; Azevedo 2014; Lacava and Azevedo 2014; Kusari et al. 2015; Rukshana and Tamilselvi 2016; Sebastianes et al. 2017; Strobel 2018). Endophytes can synthesize homologous bio-active and structurally diverse secondary metabolites (SMs), such as alkaloids, benzopyranones, chinones, flavonoids, phenolic acids, quinones, steroids, peptides, terpenoids, tetralones, cytochalasines, quinols, xathones, chinones, isocumarines, and benzopyranones, that mimic the structure and function of their host compounds (Cragg and Newman 2013; Higginbotham et al. 2013; Bhardwaj and Agrawal 2014; Zhang et al. 2014a, b; Stierle and Stierle 2015; Chen et al. 2016; Newman and Cragg 2016; Agrawal et al. 2017; Sarasan et al. 2017; Deshmukh et al. 2018). The SMs are small molecules, which act as a defense compound under abiotic (e.g., acidity, drought, and salinity) or biotic stress conditions (e.g., parasitic symbiosis), or act as a signaling molecule during biotic interactions between organisms in their ecological niches (Wisecaver et al. 2014; Knox and Keller 2015). SMs are usually synthesized by mevalonic acid (Bian et al. 2017), methylerythritol 4-phosphate (Banerjee and Sharkey 2014), shikimate-chorismate (Tohge et al. 2013), polyketide (PKs), and nonribosomal polyketide (Harvey et al. 2015; Amoutzias et al. 2016; Martinez-Klimova et al. 2017). The PKs are biosynthesized through modular polyketide synthases (PKSs) type I from acetate and propionate building blocks (Knox and Keller 2015; Ray and Moore 2016; Vesth et al. 2016). The PKSs are classified into type I (Gallo et al. 2013), type II, and type III based on their structure and biochemistry (Yuzawa et al. 2016, 2017; Parvez et al. 2018). The polyketide-derived drugs include several anticancer drugs (epothilone; taxol or paclitaxel), antibiotics (erythromycin), insecticides (spinosyn A), and antifungals (amphotericin B) (Miller et al. 2008; Cane 2010; Osswald et al. 2014; Finzel and Burkart 2016). Moreover, the hybrid PKS–NRPS (HPN) are involved in peptide toxins biosynthesis, such as pectenotoxin and enuazonic acid and destruxins (Wu et al. 2013; Zhao et al. 2015). NRPs are produced by nonribosomal peptide synthetases (NRPSs). NRPSs use proteinogenic and nonproteinogenic amino acids (DNA non-encoded amino acids) as building blocks for the peptide chain assembly (Felnagle et al. 2008). NRPS are modular enzymes with multiple domains, namely acetylation, condensation, and thioesterase (Ayuso-Sacido and Genilloud 2005). Some necrotrophic phytopathogenic fungi produce phytotoxins host-specific toxins (HSTs) and non-host-specific toxins (non-HSTs) (Pusztahelyi et al. 2015) by the activity of polyketide synthases (PKSs) and nonribosomal peptide synthetases (NRPSs) (Scharf et al. 2014). These phytotoxins act as virulence factors (Vurro et al. 2018). Terpenoids are derived from the isomeric C-5 isoprenoid chain precursors isopentenyl and dimethylallyl diphosphate (IPP and DMAPP), which are synthesized from acetyl-CoA through the mevalonate pathway (MVA) that occurs among eukaryotes (Alberti et al. 2017) and in some prokaryotes (Lombard and Moreira 2011). The more common terpenoid fungi-derived products are mycotoxins, antibiotics, antitumor compounds, and phytohormones (Deka et al. 2017). The shikimate pathway occurs in microorganisms (e.g., oomycetes, ciliates, diatoms, ascomycetes, basidiomycetes and zygomycetes), plants, and green and red algae, but not in animals (Richards et al. 2006), and the end product is chorismate, the precursor for important SMs (e.g., flavors, fragrances, pharmaceuticals and food additives) (Tohge et al. 2013). MEP is also called the non-mevalonate or Rohmer pathway (Rohmer 1999). It occurs in algae, many eubacteria, and apicomplexan parasites, but not in archaea and animals (Eisenreich et al. 2004). MEP pathway enzymes (e.g., methyl erythritol phosphate) are effective targets for novel biosynthesis of antimalarial, antibacterial, and herbicidal agents (Matsue et al. 2010). Undoubtedly, the database of drugs and others web tools (e.g., standardizing APIs) enable users to compare endophytes genomes and their resulting SMs for their utilities and make further contributions, and they allow the discovery and sustainable production of these desirable bio-active secondary metabolites (Khater et al. 2016; Vesth et al. 2016). Considering these facts, endophytes are assuredly considered as a source of novel chemically bio-active compounds and a promissory reservoir for drug discovery (Brakhage 2013; Weber and Kim 2016).

1.2 Natural Products in Drug Discovery: Current Status and Future Perspectives

Natural products (NPs) is a holistic term for low molecular weight compounds and their derivatives isolated from plants, animals, minerals, and microbes (Nicoletti and Fiorentino 2015; Newman and Cragg 2016; Zhang et al. 2016). Historically, empirical research into natural products dates back to 1550 BC, and the scientific period began in the 1950s (Dias et al. 2012; David et al. 2014). Approximately 29.5% of FDA (Drug and Device Information From the Food and Drug Administration–FDA) delivered drugs are the derivatives of NPs (Gu et al. 2013). Despite the importance of NPs, research aimed at exploiting NPs and their derivatives drastically decreased during the past 30 years (Newman and Cragg 2016; de La Torre and Albericio 2018). A possible explanation for this decline can be attributed to the shift in technology used for drug discovery (Chang and Kwon 2016). However, in recent years, low drug productivity has renewed the focus on natural products and their derivatives as drug-discovery sources (Cragg and Newman 2013; Pawar et al. 2017). NPs research still has a good approach as drug (e.g., cancer chemotherapeutic and chemopreventive agents) candidates with applications in agriculture, medicine, and the biopharmaceutical industry (Beutler 2013; Booker et al. 2015). The framework of any drug discovery from the concept or idea to market consists of some basic steps, including disease selection, drug discovery, target discovery, database mining, target validation, structure-based drug design, fragment-based lead discovery, quantitative structure–activity relationships, measuring pharmacological activity (in vitro and in vivo studies), target selection, lead optimization (e.g., compound identification, discovery toxicology), lead validation (e.g., considerations for optimizing absorption, distribution, metabolism, excretion, and toxicity), trial evaluation in preclinical stage for first in human studies, approved clinical trials (e.g., approved or cleared by FDA), drug manufacturing, and available for sale (Sarker and Nahar 2012a, b; Booker et al. 2015; Patridge et al. 2016). On average, only one of 5000–10,000 of the new synthetic molecules in development approved for clinical trials becomes a manufacturing and commercial pharmaceutical drug because their toxicity is discovered in the clinical phases, inducing their rejection, and they are subsequently discarded (Ruiz-Torres et al. 2017). In principle, the bioprospecting for novel compounds from natural products and their derivatives as drug discovery sources is a critical and expensive step (scientific framework for drug development is ~$350 million) for current biopharmaceutical research (Mishra et al. 2017), and drug delivery requires a preparation and evaluation period of 12 years (Kesselheim et al. 2017). Natural crude extracts or fractions (e.g., plant, animal or microbe) usually occur as a complex mixture of unknown compounds with several types of polarities, and their separation processes are a painful challenge for their bioprospecting, isolation, structural identification, and biochemical characterization (Sacan et al. 2012; Sarker and Nahar 2012a, b). However, in the twenty-first century progress in virtual screening (Wright and Sieber 2016), combinatorial chemical techniques (Liu et al. 2017), high-throughput screening (HTS) platforms (Paytubi et al. 2017), rational drug design (Cozza 2017), high throughput chemistry (Shevlin 2017), and nanoprobes (Lyu and Pu 2017; Li et al. 2018) have provided useful tools for bioprospecting large compound libraries in a cost-effective manner (Owen et al. 2017) to discover drugs based on target-based screening (Lionta et al. 2014). These new innovative models of drug research and development for drug innovation (Shaw 2017; de La Torre and Albericio 2018) helped to build a new Golden Age of natural products drug discovery (Shen 2015; Liu and Wang 2017). We have briefly discussed the adoption and development of novel analytical technologies applied to the new drug discovery strategies and exploiting the NPs and their derivatives in the following sections.

1.2.1 High-Throughput Screening (HTS) for Unknown Natural Compound Detection

HTS technology, considered the "Rosetta stone" for drug design to the pharmaceutical industry, allows the identification of several hundred thousand synthetic compounds (e.g., in vitro assays) in several different types of libraries (e.g., combinatorial chemistry, genomics, protein, and peptide libraries). HTS technology can identify biological targets of interest (Paytubi et al. 2017), reduce the costs of drug development (Roy 2018), and has been conducted on microarrays cell-based assays (Nierode et al. 2016). In HTS libraries, the target molecules (e.g., small molecules, polymers, and antibodies) are arrayed on microarrays by robotic spotting technology (Kolluri et al. 2018) or soft lithography technology (Hong et al. 2017). A typical HTS system can screen 10,000 compounds per day, and ultra high-throughput screening (UHTS) can even conduct 100,000 assays per day in the commercial development of new drugs (Szymański et al. 2012). HTS technology has allowed (in vivo) the virtual screening of known molecules stored in public chemical databases [(PubChem (http://pubchem.ncbi.nlm.nih.gov), ChemSpider (http://www.chemspider.com), and ChEMBL (https://www.ebi.ac.uk/chembl). This technology has reduced animal test numbers and increased the biopharmaceutical industries drug discovery programs (Nowotka et al. 2017; Matsui et al. 2017). On the basis of products and services, the global HTS market is expected to reach USD 18.83 billion by 2021 (https://www.marketsandmarkets.com). The 3D cell-based HTS assays [e.g., microwell platform (Vrij et al. 2016), and microfluidic device (Edmondson et al. 2014; Chi et al. 2016)] provide a useful platform for discovery of NPs derivatives

(Ryan et al. 2016). These technologies improve in vitro the predictability and accuracy of drug screening (Sabhachandani et al. 2016). The National Institute of Health (NIH) Roadmap created the molecular Libraries Screening Center Network (MLSCN) (Huryn and Cosford 2007) as part of the Molecular Libraries Initiative (MLI) with the purpose of facilitating access to use of chemical probes and small-molecule tools for basic research that will interrogate novel biochemical pathways (Austin et al. 2004). The NIH National Center for Advancing Translational Sciences (NCATS) Chemical Genomics Center (NCGC) (www.ncgc.nih.gov), in the United States of America (USA), is another HTS center (Kaiser 2011) combining state-of-the-art technology with the best scientific minds in academia (Huggett 2016), government, and industry (Cox 2018). The NCGC focus is to translate the discoveries of the Human Genome Project into biology, specifically on new targets and untreatable diseases (Huang et al. 2011), by using industrial-scale HTS assays, informatics, and chemistry (Thomas et al. 2009). NCGC provides a more rapid development of research tools, better diagnostic methods, and disease treatments (Howe et al. 2015; Hu and Bajorath 2017) in the era of medicinal chemistry big data (Ekins et al. 2017). Recently, the HTS infrastructure, the European Lead Factory (ELF; https://www.europeanleadfactory.eu) project (Karawajczyk et al. 2017), some pharmaceutical companies (Bayer, AstraZeneca, UCB, Lundbeck, Sanofi, Merck), and their proprietary in-house compounds collections, joined to create a chemical space expansion for collaborative lead generation and drug discovery (Karawajczyk et al. 2015). The ELF project created the Joint European Compound Library (Besnard et al. 2015), an HTS library that is engaged in synthetic and or medicinal chemistry with 321,000 compounds (Karawajczyk et al. 2017) and linked to a cloud-based informatics system, the ELF Honest Data Broker (Paillard et al. 2016). Nowadays, some chemistry focused academic groups (Max-Planck Institute of Molecular Physiology, Germany; http://www.syncom.nl) and small and medium enterprises (Syncom, The Netherlands; http://www.mpi-dortmund.mpg.de/74682/Kumar) cooperate with ELF (Karawajczyk et al. 2015). Undoubtedly, academic and industrial ELF consortium partners offer a practical device to search for new synthetic and non-synthetic drugs discovery in chemical space (Shanks et al. 2015; Paillard et al. 2016), reducing the drug development framework, and helping the marketing of the drug (Bucci-Rechtweg 2017).

1.2.2 Hyphenated Techniques in Natural Products Analysis

NPs and their derivatives (e.g., crude extracts or fractions) are a mixture of unknown compounds with many types of chemical polarities, and their separation, screening, identification, and characterization are laborious processes (Sarker and Nahar 2012a, b). In order to obtain the structural information of the unknown compounds present in a crude sample, hyphenated systems usually create a multidimensional data set (e.g., chromatographic and spectroscopic data) for online identification and dereplication applications (Brusotti et al. 2014; Ibekwe and Ameh 2015). In these methods, the separation techniques [e.g., liquid chromatography (LC), high-performance liquid chromatography (HPLC), capillary electrophoresis (EC) or gas chromatography (GC)] are coupled to an online spectroscopic detection technology [e.g., infrared (IR), Fourier-transform infrared (FTIR), photodiode array (PDA), ultraviolet-visible (UV-vis) absorbance, or fluorescence emission, mass (MS), or NMR spectroscopy] (Patel et al. 2014). As a result, innumerable modern hyphenated techniques (e.g., CE-MS, GC-MS, LC-MS, LC-PDA, and LC-NMR) were developed (Yu et al. 2016) that allowed combining the better aspects of chromatographic and spectral methods, to build more powerful integrated systems (e.g., LC-PDA-MS, LC-MS-MS, LC-NMR-MS, and LC-PDA-NMR-MS) for isolation and analysis of NPs and their derivatives (Patel et al. 2012).

1.2.3 Dereplication of Natural Products Analysis

Dereplication approaches combine the use of chromatographic (e.g., TLC and HPLC) and spectroscopic techniques (e.g., UV–vis and IR) with NP databases (DBs) bioprospecting (Chervin et al. 2017; Pérez-Victoria et al. 2016; Prabhu et al. 2015). Nowadays, LD-based ion sources are used for pretreated samples in MS analyses of NPs and untreated samples (native form) in secondary ion mass spectrometry (SIMS) (Bhardwaj and Hanley 2014). In microbial NPs screening, the re-isolation of known compounds (Pérez-Victoria et al. 2016) makes this phase more laborious (Nielsen and Larsen 2015; Chervin et al. 2017). An ultra-performance liquid chromatography photodiode array high- resolution in tandem mass spectrometric (UPLC-PDA-HRMS-MS/MS) technique can be used for dereplication of fungal secondary metabolites in crude culture extracts (Tawfike et al. 2013), which limits the occurrence of false positives (Kildgaard et al. 2014; Wolfender et al. 2015). The quadrupole-type system, ion trap, time of flight (TOF), and Orbitrap (as MS analyzers) are also satisfactorily used for dereplication systems (Kildgaard et al. 2014), increasing the selectively and accurately applied to the dereplication strategies. The use of a ChemSpider (ACD-Structure Elucidator) hosted by the Royal Society of Chemistry is another dereplication strategy applied for NPs and their derivatives (Elyashberg et al. 2009). The molecular networking (MN) and in silico fragmentation tools (Sacan et al. 2012) provide new product strategies for dereplication approaches of secondary metabolites applied in NPs research (Allard et al. 2016; Masimirembwa and Thelingwani 2012).

1.2.4 Chemical Derivatization Strategies in Natural Products Analysis

Chemical derivatization strategies represent a more recent approach for the separation of lipids from complex NPs mixtures, including isomeric lipids (Jiang et al. 2017). In LC-MS, chemical derivatizations are commonly used to increase the MS ionization efficiency and selectivity, facilitate structure elucidation, and improve the chromatographic separation (Qi et al. 2014). In liquid chromatography-mass spectrometry (LC-MS), chemical derivatizations are required to reduce the polarities of the functional groups, improve their separation by chromatographic methods (TLC, LC, HPLC, and GC), and facilitate structure elucidation. Chemical derivatizations are applied in order to adapt the physicochemical properties of NPs derivatives products, generate derivatives, allow the synthesis of active molecular probes by conjugation of reporter tags (Robles and Romo 2014), and enable the structure-activity relationship (SAR), quantitative structure-activity relationships (QSAR), and molecular docking studies (Abdulfatai et al. 2017; Afifi et al. 2017).

1.3 Target-Based Drug Discovery (TBDD)

Nowadays, the pharmaceutical industry TBDD needs to select targets with reduced attrition rates in randomized trials (e.g., good laboratory practice stage of toxicology testing triggers) due to lack of differentiated efficacy (Chaparro et al. 2018). In this context, the traditional TBDD methods have been conducted by affinity chromatography, radiolabeling, and cell-based affinity tagging procedures (Azad and Wright 2012; Sakamoto et al. 2012). However, the current TBDD methods include both target-based (reverse pharmacology) and phenotypic target-based screening (target de-convolution) (Lee and Bogyo 2013; Vaidya 2014; Simoes-Pires et al. 2014; Jung and Kwon 2015; Nijman 2015; Arulsamy et al. 2016; Glenn and Croston 2017; Haasen et al. 2017). TBDD strategies are based on the interactions between a target and its phenotype (biological tractability) as well as an ability to modulate that phenotype using a chemical probe with a specific target (Garbaccio and Parmee 2016). The chemical probes are usually small-molecules used as molecular and biochemical modulators of a protein's function applied to TBDD to improve the target validation (Hajimahdi and Zarghi 2016). The current technologies applied to TBDD focusing on chemistry and phenotypic target-based screening and a summary of the potentiality, reliability, and limitations of these methods are discussed below.

1.3.1 Chemistry of Target-Based Screening

The chemistry-based screening target (e.g., cell-based target) presents four features for target validation based on the use of chemical probes, the exposure at the site of action, target engagement and selectivity, expression of functional pharmacology, and proof of phenotype perturbation (Bunnage et al. 2013; Garbaccio and Parmee 2016). These methods can help to optimize drugs use for pharmacogenomics-based personalized medicine (Mirsadeghi and Larijani 2017).

1.3.1.1 Chemical Genomics

Chemogenomics is an emerging research field that combines genomics, chemistry, and computational sciences for the rapid validation of new targeted therapeutics compounds, where a specific molecular target has their biological function modulated by a small molecule (Jones and Bunnage 2017; Rakers et al. 2018). The emergence of chemogenomics is due to the increasing number of in-house bioactivity databases (Lipinski et al. 2015) available both in commercial (GoSTAR: https://www.gostardb.com/) and public open databases (ChEMBL: https://www.ebi.ac.uk/chembl/) (Gaulton et al. 2017; Nowotka et al. 2017). The chemogenomics knowledge-based strategies provide a rational prediction of drug target gene interactions owing to the basic information of the computational design of target-directed combinatorial libraries found in the chemical space (Rakers et al. 2018). Nowadays, chemogenomics is considered as a viable alternative to some in silico approaches, such as docking, structure-based drug or ligand-based virtual screening strategies (Lionta et al. 2014). Currently, pharmacometabonomics is used to predict drug metabolism, pharmacokinetics, safety, and efficacy. It is complementary to pharmacogenomic and pharmacoproteomics or pharmacometabonomics (Shabaruddin et al. 2015; Everett 2016) and is therefore considered a new tool for personalized medicine (Everett 2015). However, the cost of pharmacogenomic assays continues to be very expensive to incorporate into standard health-care (Pink et al. 2014; Altar et al. 2015).

1.3.1.2 Chemical Proteomics

Chemical proteomics emerged on the West Coast, notably in Cravatt's lab at the Skaggs Institute of Chemical Biology (Adam et al. 2002) and Bogyo's lab; then at Celera in South San Francisco (Zanders 2012). Chemical proteomics is a field of chemical biology focused on the interaction between engineered small molecular probes (chemical probes) and proteome (Medina-Cleghorn and Nomura 2014). Chemical proteomics strategies usually combine phenotypic screening with target identification screening for novel drug targets (Medina-Cleghorn et al. 2015; Counihan et al. 2017; Piazza et al. 2018). In other words, chemical proteomics uses the design of small molecule probes to understand protein function (Cravatt et al. 2008), based on their action mode on protein expression and posttranslational modifications on the proteome-level in target cells or tissues of interest (Yu et al. 2016), and can identify small molecule targets in complex biological samples (Futamura et al. 2013). Chemical proteomics can help with the selection and validation of targets (Bantscheff and Drewes 2012; Liu and Guo 2014). The most usual applications of chemical probes are biological tractability (establishes the interaction between a target and its phenotype) and chemical tractability (ability to modulate its phenotype by a small molecule or chemical probe) (Garbaccio and Parmee 2016). In this context, these chemical probes provide a better understanding of pharmacokinetic or pharmacodynamic models by maximizing target identification and avoiding biases during target validation (Bunnage et al. 2013) resulting in the delivery of novel therapeutics (Garbaccio and Parmee 2016). Nowadays, the use of quantitative proteomics methodologies, such as protein profiling (Chen et al. 2017), compound-centric chemical proteomics (Wright and Sieber 2016), and drug affinity responsive target stability (Pai et al. 2015), DNA, RNA, protein or cell microarrays (Li 2016; Rothbauer et al. 2016), and microfluidic cell-chips (Carey et al. 2018), provides new insights into chemical proteomics for developing therapeutic agents (Pan et al. 2016; Olivon et al. 2017).

1.3.2 Phenotypic-Based Screening

Phenotypic-based screening (PBS) is also known as the neoclassic pharma strategy (Lee and Berg 2013). It exhibited some advantages over chemical target-based methods for bio-active compounds identification owing to continuous negative results in the pharmaceutical industry (Priest and Erdemli 2014; Walker et al. 2015; Moffat et al. 2017). PBS uses unbiased phenotypic assays to find large molecules with the ability to alter a specific phenotype of cells (cell proliferation), tissues or animals into HTS chemical space libraries (Ayotte and La Plante 2017). Recently, PBS has gained renewed importance in discovering first-in-class or best-in-class medicines (Lexchin 2014a, b; Lexchin 2016; Moffat et al. 2017). PBS assays approach key aspects of the physiological process like cell-cell interactions and signal transduction pathways (Ayotte and La Plante 2017; Isgut et al. 2018). They usually depend on the cell-based phenotypic assay (e.g., RNAi, reporter gene assay, CRISPR/Cas9 system, Reverse Phase Protein Arrays (RPPAs), cell viability (e.g. MTS, alamar blue, Annexin V-FITC flow cytometry assay), signaling pathway (e.g., GPCR, nuclear receptor, MAPK/ERK), disease-related phenotypic assay (e.g., neurodegenerative diseases, such as Alzheimer's and Parkinson's disease), and more recently network-based phenotype mapping (Fang 2015; Moerke and Fallahi-Sichani 2016; Moffat et al. 2017). Nowadays, automated microscope-based screening [e.g., high content screening (HCS), high content imaging (HCI), or image cytometry (IC)] is ideally suited for screening multi-targeted agents and drug combinations (Dolman et al. 2018; Verjans et al. 2018). The methods combine the molecular information, biological relevance, and patient data to increase the productivity of discovering first-in-class (Drawnel et al. 2017). PBS methods enable a new challenge to screen and focus drug combinations based on polypharmacology strategies (Isgut et al. 2018). However, for a robust PBS implementation, it will be necessary to build more sophisticated systems biology databases, such as the Human Microbiome Project (Bauer and Thiele 2018), Genome-Scale Metabolic Models (GEMs) (Rejc et al. 2017), the Human Metabolic Atlas (HMA; http://www.metabolicatlas.org) (Bauer and Thiele 2018), and Kyoto Encyclopedia of Genes and Genomes (KEGG; http://www.genome.jp/kegg/) (Hadadi et al. 2016). As a result, the information acquired can be incorporated and used in standard health-care and clinical practice (Pink et al. 2014; Altar et al. 2015). In 2010, the BioAssay Ontology (BAO; http://bioassayontology.org) project was created (Visser et al. 2011). This Project was developed to standardize the Minimum Information About a Cellular Assay (MIACA; http://sourceforge.net/projects/miaca) (Brazma et al. 2001), or to set up the ontologies (Zander Balderud et al. 2015). In terms of ontology resources, some databases storing information about HTS for the human cell lines (e.g., genotype-to-phenotype relationships) and anticancer drugs can be localized in the Genomics of Drug Sensitivity (GDSC; www.cancerrxgene.org) (Yang et al. 2013), Cellular Microscopy Phenotype Ontology (CMPO; www.ebi.ac.uk/cmpo (Jupp et al. 2016), and Human Phenotype Ontology (HPO; www.human-phenotype-ontology.github.io) (Köhler et al. 2017). These approaches provide the generation/extraction of derived ontologies (or perspectives) and analyze the activities of compounds for identification of artifacts technology (Mandavilli et al. 2018; Wang et al. 2017).

1.4 Endophytic Microorganisms as a Source of Bio-active Compounds

Bio-active compounds are produced by soil-borne endophytes fungi (e.g., *Trichoderma* spp., *Talaromyces* spp.) (Pusztahelyi et al. 2016; Zhai et al. 2016), soil-borne endophytes bacteria (e.g., *Pseudomonas putida*) (Santoyo et al. 2016; Frank et al. 2017; Honeker et al. 2017), and plant–symbiont nonpathogenic associations (e.g., *Daldinia eschscholtzii*, *Hypoxylon rickii*, and *Pestalotiopsis fici*) (Macías-Rubalcava and Sánchez-Fernández 2017; Helaly et al. 2018). These compounds have a critical role in plant resistance to biotic (Chadha et al. 2015; Vesterlund et al. 2011) and abiotic stress factors (Choudhary 2012), which benefits the host survival in return (Maheshwari et al. 2017). However, this is a complex phenomenon, and the mechanisms of this protection (e.g., deter herbivores by producing toxic alkaloids) process are poorly understood (Arora and Ramawat 2017). Recently, Newman and Cragg (2016) disclosed a list of all FDA approved drugs from 1981 to 2014, and over 35% of these chemotherapeutic candidates are produced by microbes and/or endophytes; therefore, the field of natural product research should be significantly expanded (Martinez-Klimova et al. 2017; Mazzoli et al. 2017; Gao et al. 2016; Uzma et al. 2018). Mayer et al. (2013) reported 102 marine natural products of animals, algae, fungi, and bacterial origin having antibacterial, antifungal, antiprotozoal, antituberculosis, and antiviral activities. Out of these, 68 exhibited a huge biodiversity of receptors and their molecular targets. This study received global technical support research from chemists and pharmacologists based in several countries, including Australia, Belgium, Brazil, Canada, China, Colombia, Cuba, Egypt, Fiji, France, Germany, Indonesia, Israel, Italy, Japan, Luxemburg, Malaysia, Mexico, the Netherlands, New Caledonia, New Zealand, Norway, Panama, Papua New Guinea, Philippines, South Africa, South Korea, Singapore, Spain, and Switzerland. This study culminated FDA-approved pharmaceuticals and 11 compounds in phase I, II, and III of clinical development. A considerable amount of literature has been published on the microbial richness of marine endophytes, and it highlights the correlation with their antimicrobial bio-active compounds and habitat complexity (Rédou et al. 2015; Sarasan et al. 2017; Deshmukh et al. 2018). In this context, deep-sea fungi like *Aspergillus fumigatus* (48X3-P3-P1), *A. terreus* (1H3-S0-P1), *Eurotium herbariorum* CB-33, *Fusarium oxysporum* (1H3-P0-P1, 4H1-P0-P1, and 4H1-P3-P3), *Penicillium bialowiezense* (CB-5, CB-7, and CB-8), *P. chrysogenum* (2H5-M3-P2-(3), CB-11, CB-17, and CB-24), *Penicillium* sp. (CB-16), and *Oidiodendron griseum* (CB-36) were isolated from 2000 m below the seafloor, and their antimicrobial screening revealed 33% of bio-active compounds (using 16 microbial targets) against pathogenic bacteria and fungi (Rédou et al. 2015). The deep subseafloor, and the others marine ecosystems, currently represent the last frontier of an untapped reservoir of novel bio-active molecules. However, the knowledge about the endophytes microbes and their SMs is still limited, and several aspects (e.g., SMs ecological functions and the plant-endophyte relationship) need to be studied and understood (Jia et al. 2016; Negreiros de Carvalho et al. 2016).

1.4.1 Endophytic Bio-active Compounds Production by Synthetic Biology and Metabolic Engineering

The endophytes bio-active compounds extraction, purification, chemical analysis, and new chemical bio-active entities (NCBEs) identification (especially those of high added value), in minimum commercial quantity or to synthesize chemically in industrial amounts (Khatri et al. 2017; Singh et al. 2017;), are commonly very tedious and challenging processes (Kadir et al. 2013). Therefore, engineering microbes based on next-generation sequencing methodologies (NGS) for the production of SMs and NCBEs characterization are emerging as advantageous alternative methodologies (Turner et al. 2018). Nevertheless, the biggest global pharma companies, such as Pfizer Inc., Merck & Co, Jonnson & Jonnson, F. Hoffmann-La Roche AG, and Sanofi, drastically reduced their industrial R&D investment in NPs during the past 30 years due to high rediscovery rates of NCBEs and the lack of innovative screening approaches (Katz and Baltz 2016).

The biosynthesis of SMs are co-regulated by gene clusters (BGCs) on a single genetic locus (Wallwey and Li 2011; Brakhage 2013; Cacho et al. 2015; Smanski et al. 2016; Vesth et al. 2016), localized on subtelomeric regions in the chromosome (Knox and Keller 2015), and around a synthase gene (Andersen et al. 2013). The "silent" or "cryptic" gene clusters (Zn(II)2Cys6, Cys2His2, and basic region-leucine zipper (bZIP) family are transcriptional regulators) provide endophytes microbes species (endophyte fungi) precise temporal and spatial control over the SMs expression and probably helps the intra, and inter-kingdom, horizontal cluster transfer (Hong et al. 2013; Ortiz et al. 2013; Knox and Keller 2015). Moreover, one of the major challenges in NP discovery is that only a tiny fraction of those BGCs have been characterized to date, partially due to the fact that they are transcriptionally silenced or do not express in totality in their native hosts (due to tight regulation) under standard laboratory conditions (Bamisile et al. 2018). In order to activate those BGCs [e.g., *cadA, mttA, mfs*) of *A. Niger* itaconic acid (IA)], and their cryptic pathways, many innovative approaches are used, such as one strain many compounds (Blumhoff et al. 2013; Hewage et al. 2014), co-culture (Kamdem et al. 2018), and microbial biotransformation (Donova 2017). The most common approaches employed for gene clusters activation in fungal endophytes include gene deletions (Andersen et al. 2013), modulation in epigenetic mechanisms (Deepika et al. 2016), proteomic (Brakhage and Schroeckh 2011), genome mining (van der Voort et al. 2015), and heterologous hosts for cloning and expression of fungal metabolites (Anyaogu and Mortensen 2015). The identification of loss of *aflR* expression (LaeA), a global regulator of SMs in *Aspergillus* spp. (Zhao et al. 2017), and many bioinformatic algorithm tools, such as Secondary Metabolite Analysis Shell (antiSMASH) (Weber et al. 2015), the Secondary Metabolite Unknown Region Finder (SMURF) (Khaldi et al. 2010; www.jcvi.org/smurf), and motif-independent de novo detection algorithm (MIDDAS-M) (Umemura et al. 2013) for SMB (www.secondarymetabolites.org) gene clusters, allow the prediction of super-clusters containing genes for more than one SM (Wiemann et al. 2013). More recently, synthetic biology and metabolic engineering were employed to activate those "silent" gene clusters, provide the insertion of targeted mutations through biosynthetic metabolic pathways in heterologous hosts, express the target mutant enzyme catalysts, and aimed to produce larger amounts of commercially and industrially high-value bio-active compounds (Boruta and Bizukojc 2017), such as biopigments applied to many industrial activities, e.g., food, textile, cosmetic, and pharmaceuticals (Narsing et al. 2017). Synthetic biology (SynBio) is related to design rationalizing of biological systems by applying the key concepts of engineering (Fletcher et al. 2016). These engineering concepts include the abstraction, modularity, and use of standard interfaces (Bhatia and Densmore 2013), such as Pigeon, a graphical user interface, clearly valuable for any design enterprise (Chelliah et al. 2013). Metabolic engineering is directly linked to the use of the systems-level for design, optimization of cellular metabolism, and the gene regulatory metabolic networks on a genome scale (Khatri et al. 2017). In general, there are three major steps in metabolic engineering for production of small molecule drugs, including pathway discovery, pathway assembly in the recombinant host (homologous or heterologous recombination), and pathway optimization (Zhang et al. 2015). Metabolic engineering and synthetic biology (SynBio) approaches have revolutionized many fields of biotechnology (e.g., renewable biofuel production) making possible the insertion of non-native and de novo biochemical pathways to generate high-value drop-in bio-active chemical compounds (Chiarabelli et al. 2013; Salehi et al. 2017). Consolidation of synthetic biology and metabolic engineering approaches doubles the knowledge of the biosynthetic pathways and organisms for which molecular tools (CRISPR/Cas9 or genome-scale metabolic models (GSMM) are available and optimized to any extent (Zhou et al. 2016; Harrington et al. 2017). Model organisms, such as *Saccharomyces cerevisiae* and *Escherichia coli*, remain widely used host strains for industrial production due to their robust and desirable traits (Lee et al. 2012).

1.4.1.1 Endophytic as Microbial Cell Biofactories of Bio-active Compounds

A brief overview of some endophytes as live platforms able to host and sustain the purposeful DNA designs as biofactory producers of bio-active metabolites obtained through synthetic biology and metabolic engineering is discussed below.

1.4.1.1.1 *Aspergillus* Species

There is a large volume of published studies describing the role of *Aspergillus*, which consists of over 340 taxonomically species, such as *A. fumigatus*, *A. flavus*, *A. niger*, *A. parasiticus*, *A. nidulans*, and *A. terreus*, recognized as cell factories for human, agricultural, and biotechnological applications (Meyer et al. 2015; Park et al. 2017). Heterologous expression was used to access the genes cryptic clusters (GCC) in *A. nidulans*. It is a practical and effective system for amplifying GCC from a target fungus (*A. terreus*) by placing them under control of a regulatable promoter (Bhetariya et al. 2011; Netzker et al. 2015) based on *LaeA* a regulator gene for SMs that allows the transfer and expression of the asperfuranone biosynthetic pathway (Chiang et al. 2013) into *A. nidulans*, and eliminates unwanted toxins (e.g., carcinogenic mycotoxins aflatoxins from *A. flavus*, and sterigmatocystin from *A. nidulans*) and allergic agents (e.g., allergic bronchopulmonary aspergillosis from *A. fumigatus*) (Park et al. 2017). Some *Aspergillus* metabolites and their applications include citric acid as a food additive, fumagillin as an antimicrobial agent, lovastatin as a hypolipidemic agent, and succinic acid as a flavor additive and detergent (Tobert 2003; Fallon et al. 2011; Sorensen et al. 2011; Sanchez et al. 2012). The genomes of *Aspergillus* species have now been sequenced, and genome-editing techniques have been rapidly developed for filamentous fungi (Teotia et al. 2016; de Vries et al. 2017). A putative gene cluster (*gedC* and *gedR*) that encodes the geodin production in *A. terreus* was transferred by heterologous reconstitution (genetic toolbox; USER cloning and USER fusion protocols) into *A. nidulans,* and the penicillin cluster (*pcbAB*, *pcbC*, and *pende*) of *P. chrysogenum* was rewired and expressed from a polycistronic gene cluster under control of a single xylose-inducible promoter in *A. nidulans*. The recent strategies for heterologous expression of fungal biosynthetic pathways in *Aspergilli* were reviewed by Anyaogu and Mortensen (2015). This genus presents an extensive metabolic engineering toolbox (Richter et al. 2014; Gressler et al. 2015); hence, *Aspergillus* can be employed as a multi-purpose biofactory for large scale production of bio-active secondary compounds and in development of strategies for converting biomass to bioenergy (Raghavendra et al. 2016).

1.4.1.1.2 *Penicillium chrysogenum*

Penicillium chrysogenum is a filamentous fungus used as an industrial producer of β-lactam antibiotics, such as penicillins and cephalosporins (van den Berg 2011). The penicillin biosynthesis is encoded by three genes [acvA (pcbAB), ipnA (pcbC), and aatA (pende)] (Terfehr et al. 2017). The biosynthesis of β-lactam in *P. chrysogenum* is regulated by a *LaeA* protein (a global regulator) and the Velvet complex proteins (VelA, VelB, VelC, and VosA) (Martín 2017). In 2008, the complete genome sequence of *P. chrysogenum* was elucidated (Martín 2017), unmasking genetic secrets of the industrial penicillin producer (van den Berg 2011). Recently, the metabolic engineering and synthetic biology approach resulted in the description of transcription factor *CreA* responsible for carbon repression (Cepeda-García et al. 2014). This finding opens the possibility of utilizing it to improve the industrial production of this antibiotic in others filamentous fungus (e.g., *Acremonium chrysogenum*) (Terfehr et al. 2017) and to use more sustainable methods for the fermentative production of unnatural antibiotics and related compounds (Salo et al. 2015). The classical strain *P. chrysogenum* has multiple copies of the penicillin biosynthesis cluster (pBC) encoded by three key enzymes: δ-(l-α-aminoadipyl)-L-cysteinyl-D-valine synthetase (ACVS), isopenicillin N synthase (IPNS), and isopenicillin N acyltransferase (IAT) (Nijland et al. 2010). Much of the literature since the mid-1990s emphasizes the applications of the *P. chrysogenum* genetic engineered strain for β-lactam antibiotics production (Martinez-Klimova et al. 2017). In 2016, we celebrated a historical medical framework for the 75th anniversary of the first medical systemic administration of penicillin in humans (Lobanovska and Pilla 2017). In 2001, the European Surveillance of Antimicrobial Consumption (ESAC) reviewed the antimicrobial resistance (AMR) strains based on the inappropriate prescription and administration of an antibiotic therapy (Fleming-Dutra et al. 2016; Nhung et al. 2017). On the other hand, the extensive livestock and agricultural use of antibiotics (about 63,000 to over 240,000 tons of annual global antibiotic use) contribute to antibiotic resistance (Osman et al. 2018). In this context, metabolic engineering and synthetic biology approaches were applied to *P. chrysogenum* (Weber et al. 2012). The overexpression of isopenicillin N Acyltransferase in *P. Chrysogenum* (Veiga et al. 2012), engineering of β-oxidation for improved semi-synthetic cephalosporin biosynthesis, can be used to overcome the current multidrug resistance (MDR) and extended spectrum beta lactamase (ESBL) or extended-spectrum cephalosporins (ESCs) caused by ESC-R *E. coli* and ESC-R *Salmonella* spp., resistant strains (Shrestha et al. 2017), due to plasmid-encoded AmpC β-lactamases (pAmpC) (mainly CMY-2) and CTX-M extended-spectrum β-lactamases (ESBLs) in Gram-negative (Trott 2013), which are considered major global health problems (Dandachi et al. 2018).

1.4.1.1.3 *Saccharomyces* Species

Yeast has been continuously used for the production of high-value small and large molecules, such as alcohols, acids, hydrocarbons, and proteins (Shi and Zhao 2017). Several others yeast species have been used in biomass production (*Trichosporon* spp.), food processing (*Kluyveromyces* spp.), feed nutrition (*Ogataea polymorpha*), degreasing and bioremediation (*Geotrichum candidum*), therapeutic and detergent (*T. fermentum*), and pharmaceutical (*Rhodotorula* spp.) (Johnson 2013a, b) industries. Recent advances in synthetic biology and metabolic engineering for yeasts (e.g., *S. cerevisiae* and *S. pastorianus*) presented new tools, such as genetic engineering toolbox, libraries of synthetic promoters [(e.g., prototrophic markers (*ADE1, HIS2, LEU2, AURA3*) and drug resistance markers (*CUP1, SFA1, ble, kan*)], ribosome binding sites, degradation tags, transcription terminators, plasmids, riboregulators, riboswitches, and more limited CRISPR/Cas9 genome editing (Lian et al. 2018). *S. cerevisiae* was engineered by synthetic biology tools (11.3 kbp *NRPS* gene *pcbAB* and the *NRPS* activator gene *npgA*) associated with long-read DNA sequencing (cytosolic synthesis of amino-adipyl-cysteinyl-valine (ACV)) to produce and secrete a β-lactam NP (benzyl penicillin) against *Streptococcus pyogenes* (Awan et al. 2017). This work opened up the use of baker's yeast (as standard chassis organism) to the rational engineering of NPs derived antibiotics. Recently, SWITCH, a dynamic CRISPR/Cas9 tool, was used for genome engineering and metabolic pathway control of genomic loci (*bts1, ypl062W, yjl064w, rox1*, and *erg9*) for cell factory construction in *S. cerevisiae* to mevalonate production (Jakočiūnas et al. 2016). The production of fuels and chemicals from xylose by engineered *S. cerevisiae* under industrial fermentation conditions to improve the bioconversion of xylose to ethanol (pentose metabolism) was reported to consolidate the key platform for future biorefineries (d'Espaux et al. 2017). Two cytochrome P450 monooxygenases from *Fusarium oxysporum* (FoCYP), FoCYP539A7 and FoCYP655C2, were cloned and heterologously expressed in an engineered *S. cerevisiae* mutant (the acyl-CoA oxidase enzyme and the β-oxidation pathway were inactivated) to provide the production of industrially valuable ω-hydroxy fatty acids (Durairaj et al.

2015). Membrane-anchored cytochrome P450 enzymes (CYPs) induced in *P. pastoris* through RAD52 over-expression optimizing the whole-cell biotransformation cultivation conditions (pH value) improved (at five-fold) the trans-nootkatol production, when compared with the initial strain (Wriessnegger et al. 2016). Metabolic engineering of *P. pastoris* (in cell culture in bioreactor cultivations) presents high-level production of the sesquiterpenoid (+)-nootkatone from simple carbon sources (Wriessnegger et al. 2016). The engineering strategy of the plant alkaloid pathway applied to *S. cerevisiae* resulted in the biosynthesis of protoberberine, protopine, and benzophenanthridine alkaloids (Trenchard and Smolke 2015). Two small molecule drugs, amorphadiene by mevalonate pathway (Baadhe et al. 2014), amorphadiene synthase, and a novel cytochrome P450 monooxygenase (YCF-AD1), and opioids (Galanie et al. 2015), were produced in engineered *S. cerevisiae* by using exogenous genes aiming to obtain synthetic antimalarial drugs and malaria vaccines. The overexpression of O-methyltransferase leads to improved vanillin (coding for O-methyltransferase, hsOMT) production only when complemented with model-guided network engineering (Brochado and Patil 2013). These examples of *Saccharomyces* species' metabolic engineering for food (Nevoigt 2008), beverage, and industrial biotechnology (Jansen et al. 2017), bioethanol (Johnson 2013a, b; Sànchez Nogué and Karhumaa 2015), and bulk and fine chemicals from renewable feedstocks consolidate their popularity as a production organism in industrial ("white") biotechnology (or low-cost production) (Nevoigt 2008). Recently, Lopes and Rocha (2017) presented a historical point of view of the genome-scale metabolic models (GSMMs) of yeast species. In that review, they updated the information of adaptive laboratory evolution strategies (e.g., genome-scale metabolic model, metabolism, constraint-based modeling, metabolic engineering, and yeast-cell factories) for metabolic engineering systems, and included the future perspectives for the yeast research field.

1.4.1.1.4 *Corynebacterium glutamicum*

C. glutamicum is a fast-growing, anaerobic Gram-positive, non-sporulating, non-motile, saprophytic actinomycete (Nakamura et al. 2007). It was metabolically engineered to produce 4-hydroxybenzoic acid (4-HBA) using the shikimate pathway overproduction by a growth arrested bioprocess (Kitade et al. 2018). This eco-friendly bioproduction of 4-HBA from biomass resources is a desirable, sustainable, and environment-friendly natural process (Chae et al. 2017; Park et al. 2018). A growth-arrested bioprocess using *C. glutamicum* has been successfully used for the production of biofuels, organic acids, antibiotics, and amino acids (Bückle-Vallant et al. 2014; Rohles et al. 2016). Recently, the RecET-assisted CRISPR-Cas9 genome editing method was used to regulate the metabolic network for the synthesis of bio-based products (e.g., shikimate and derived aromatic compounds) from renewable biomass (e.g., renewable feedstocks) using *Corynebacterium* species as cell factories (Kogure et al. 2016; Wang et al. 2018a). The above-mentioned confirms that *C. glutamicum* can be used as a platform for production of hydroxybenzoic acids, building blocks for the production of plastics, cosmetics, pharmaceuticals, food and feed supplements (Kallscheuer and Marienhagen 2018), and biotechnological production for value-added aromatic compounds of shikimate and derived aromatic compounds (Averesch and Krömer 2018).

1.4.1.1.5 *Escherichia coli*

Studies of SynBio and metabolic engineering have confirmed the effectiveness of engineered *E. coli* cell factories for the production of bulk chemicals (e.g., 3-hydroxy propionic acid and alcohols) from non-food renewable resources (e.g., lignocellulosic biomass), and they are environment-friendly when compared to the petrochemical route (Wang et al. 2007; Chen et al. 2013; Liu et al. 2016). Engineered *E. coli* as a powerful host for producing type I PKS erythromycin antibiotic (Peiru et al. 2005) is a real success story (Pfeifer et al. 2001; Zhang et al. 2008, 2010). Bacterial aromatic type II polyketide synthases are essential to human and animal life due to the biosynthesis of therapeutic compounds, including front-line antibiotics (Wang et al. 2007; Barajas et al. 2017), and anticancer drugs (Golinska et al. 2015). This bacterium has recently been engineered to produce taxadiene, an important precursor for taxol (important anticancer compound) synthesis, using a metabolic engineering approach to balance the taxadiene biosynthesis pathway (Soliman and Tang 2015). The shikimic acid (AS) production in *E. coli* from classical to metabolic engineering strategies was applied to improve its production (Martínez et al. 2015) and also gained attention in the pharmaceutical industry (Escalante et al. 2010; Cui et al. 2014) because it is used as a precursor for manufacturing oseltamivir phosphate (OSF) (Orozovic et al. 2014; Gupta et al. 2015), an antiviral inhibitor of neuraminidase, with applications for the influenza viruses (influenza A and B, the avian influenza virus H5N1, and the human influenza virus H1N1). The synthesis of pyrogallol was obtained using genetically engineered *E. coli* by the biosynthetic shikimate pathway (Wang et al. 2018b). Pyrogallol has broad applications in food and pharmaceutical industries (Martínez et al. 2015; Ozturk 2015).

1.4.1.1.6 *Pseudomonas* Species

Recently, *Pseudomonas putida* has emerged as a modular chassis (Loeschcke and Thies 2015; Nikel 2016). It is usually used as a cellular host used as a recipient in engineered biological systems in synthetic biology (Belda et al. 2016), with direct biotechnological applications in the fourth industrial revolution for sustainable manufacturing of fine and bulk chemicals (Dombrowski and Wagner 2014). A study by Kudjo (2007) nicely describes the biotechnological role of the *Pseudomonas* species associated with heavy metals (Cu, Cr, Cd, Pb, Ni, U, and Zn) and xenobiotic compounds bioremediation. Heavy metals contamination (even at low concentrations) negatively affects the diversity and the activity of agronomically important soil microbial communities (Guo et al. 2017). Bioremediation is an eco-friendly method that uses microorganisms or their enzymes to promote degradation and removal of contaminants from the environment (Kuroda and Ueda 2010; Rhodes 2014). In this context, a *P. putida* was genetically engineered (dehalogenase gene (dhaA31) into genome by dhlA promoter) for 1,2,3-trichloropropane (a toxin and carcinogen used in the paint industry) (Samin and Janssen 2012). It is a xenobiotic chlorinated compound of high chemical stability (Dvorak et al. 2014) and a carbon source for microbial growth under aerobic conditions (Samin and Janssen 2012). Moreover, its biosurfactant rhamnolipid biosynthesis was obtained through vitreoscilla hemoglobin gene (vgb) engineering in *P. aeruginisa* (recombinant strain, PaJC), aiming to increase biosurfactant production (Kahraman and Erenler 2012). These microbial biosurfactants have great industrial impact because of their biodegradability, low toxicity, high specific activity, and antimicrobial activity against a wide range of pathogenic mocrobes (Sridhar et al. 2015; Basit et al. 2018; Liwarska-Bizukojc et al. 2018; Nurfarahin et al. 2018).

1.4.1.1.7 *Streptomyces* Species

Streptomyces species is a soil-dwelling mycelial bacteria that forms sporulating aerial branches and belongs to the rhizospheric microbial communities (Chater et al. 2010). *Streptomyces* have a great biological, industrial, and clinical significance (Bekker et al. 2014) due to their ability to produce many bio-active compounds, ~2400 unique compounds (e.g., antibiotics, antifungal agents, food preservatives, immuno-suppressors, anthelmintic, and antitumor drugs) (Abdelmohsen et al. 2015; Niu et al. 2016; Dinesh et al. 2017). In this context, ~2500 papers published in the past 10 years were recovered by the PubMed database (https://www.ncbi.nlm.nih.gov/pubmed) for this richest known source of antibiotics. These findings highlight the importance of using *Streptomyces* species as a biocatalytic tool for transforming it into microbial cell factories with several biotechnological applications, such as green chemical transformations and biopharmaceutical and biofuel production (Spasic et al. 2018).

1.5 Nanotechnology Applied to Endophytic Bio-active Compounds Production: Potential and Limitations

Recently, nanotechnology has received attention at both the academic and industrial levels (Fraceto et al. 2016; Baker et al. 2017; Duhan et al. 2017; Kim et al. 2018). This context shows the importance of the biogenic synthesis of nanoparticles by endophytes microbes, including onco-therapeutic agents (Men et al. 2014; Nam et al. 2016; Conte et al. 2017), nutraceutical agents (Wang et al. 2014), delivery systems for cosmetics and dermal pharmaceuticals (Ganesan and Choi 2016; Lucia 2017; Kaul et al. 2018), crop protection agents (Pereira et al. 2014; Rao and Paria 2013; Patel et al. 2014), antimicrobial agents (Franci et al. 2015; Oktar et al. 2015; Dong et al. 2017), and delivery systems for antiinflammatory drugs (Serpe et al. 2013). Agri-nanotechnology uses polymeric nanoparticles to coat biofertilizer (e.g., AM fungi, Rhizobia) preparations to yield formulations (Panpatte et al. 2016; Servin et al. 2016). Nano carbon, nano alumino-silicate, mesoporous silica nanoparticles, nano-emulsions, and nano silver (Thul and Sarangi 2015; Fan et al. 2014) are used in agriculture for precision farming (Kim et al. 2018; Duhan et al. 2017). However, the questions regarding the toxicity of nanoparticles in food production still remains unknown (Maurer-Jones et al. 2013). Realistic scientific research to access the nanotoxicity of engineered nanomaterials to terrestrial and agricultural plant species has begun to answer these questions (Kah 2015; Iavicoli et al. 2017). However, the studies indicated the low to modest phytotoxicity of engineered nanoparticles (ENP) (nanoEHS) in terrestrial plant species (Gardea-Torresdey et al. 2014; Ma et al. 2015). In fact, these are the new class of emerging contaminants, and the current data set is insufficient to address the actual risk (Servin et al. 2016).

1.6 Conclusions and Future Prospects

Several studies have attempted to assess the impact of endophytes secondary metabolites and their derivatives as drug-discovery bio-active compounds, such as polyketides and peptides, where the microbes are appointed as the "holy grail" for most of the global pharma companies. Discovering and identifying the target molecules are the most challenging and time-consuming steps to resolve the persisting problems. In this regard, the metabolic pathways of the bio-active compounds encode the silent cryptic biosynthetic gene clusters (BGCs) and play a major role in complex regulation and poor gene expression under laboratory conditions. Nevertheless, many contemporary and versatile approaches based on multi-omics technologies (genomics, epigenomics, transcriptomics, proteomics, metabolomics and microbiomics), synthetic biology, and bio-engineering strategies will provide the "Rosetta Stone" for the development of real microbial cell factories by the identification of target genes and the manipulation of biosynthetic pathways to enhance the sustainable production of overlooked bio-active secondary metabolites, with applications in biofuels, biopharmaceuticals, fragrances, and food flavors industries.

Disclosure Statement The author declares that no conflicts of interest are present.

References

Abdelmohsen UR, Grkovic T, Balasubramanian S, Kamel MS, Quinn RJ, Hentschel U (2015) Elicitation of secondary metabolism in actinomycetes. Biotechnol Adv 33:798–811. https://doi.org/10.1016/j.biotechadv.2015.06.003

Abdulfatai U, Uzairu A, Uba S (2017) Quantitative structure-activity relationship and molecular docking studies of a series of quinazolinonyl analogues as inhibitors of gamma amino butyric acid aminotransferase. J Adv Res 8:33–43. https://doi.org/10.1016/j.jare.2016.10.004

Adam GC, Sorensen EJ, Cravatt BF (2002) Chemical strategies for functional proteomics. Mol Cell Proteomics 1:781–790

Afifi TH, Okasha RM, Ahmed HEA, Ilaš J, Saleh T, Abd-El-Aziz AS (2017) Structure-activity relationships and molecular docking studies of chromene and chromene based azo chromophores: a novel series of potent antimicrobial and anticancer agents. EXCLI J 16:868–902

Agrawal S, Acharya D, Adholeya A, Barrow CJ, Deshmukh SK (2017) Nonribosomal peptides from marine microbes and their antimicrobial and anticancer potential. Front Pharmacol 8:828. https://doi.org/10.3389/fphar.2017.00828

Akhtar MS, Azam T (2014) Effect of PGPR and antagonistic fungi on the growth, enzyme activity and fusarium root-rot of pea. Arch Phytopathol Plant Protect 47:138–148

Akhtar MS, Panwar J (2011) Arbuscular mycorrhizal fungi and opportunistic fungi: efficient root symbionts for the management of plant parasitic nematodes. Adv Sci Eng Med 3:165–175

Akhtar MS, Panwar J, Abdullah SNA, Swamy MK, Ashkani S (2015) Biocontrol of plant parasitic nematodes by fungi: efficacy and control strategies. In: Meghvanshi MK, Varma A (eds) Organic amendments and soil suppressiveness in plant disease management. Springer, Cham, pp 219–247. https://doi.org/10.1007/978-3-319-23075-7_11

Akhtar MS, Siddiqui ZA (2010) Role of plant growth promoting rhizobacteria in biocontrol of plant diseases and sustainable agriculture. In: Maheshwari DK (ed) Plant growth and health promoting bacteria, microbiology monographs, vol 18. Springer, Berlin, pp 157–196. https://doi.org/10.1007/978-3-642-13612-2_7

Akhtar MS, Siddiqui ZA, Wiemken A (2011) Arbuscular mycorrhizal fungi and rhizobium to control plant fungal diseases. In: Lichtfouse E (ed) Alternative farming systems, biotechnology, drought stress and ecological fertilisation. Sustainable agriculture reviews, vol 6. Springer, Dordrecht, pp 263–292. https://doi.org/10.1007/978-94-007-0186-1_9

Alberti F, Khairudin K, Venegas ER, Davies JA, Hayes PM, Willis LC, Bailey AM, Foster GD (2017) Heterologous expression reveals the biosynthesis of the antibiotic pleuromutilin and generates bioactive semi-synthetic derivatives. Nat Commun 8:1831. https://doi.org/10.1038/s41467-017-01659-1

Allard PM, Péresse T, Bisson J, Gindro K, Marcourt L, Pham VC, Roussi F, Litaudon M, Wolfender JL (2016) Integration of molecular networking and in-silico ms/ms fragmentation for natural products dereplication. Anal Chem 88:3317–3323. https://doi.org/10.1021/acs.analchem.5b04804

Altar CA, Carhart J, Allen JD, Hall-Flavin D, Winner J, Dechairo B (2015) Clinical utility of combinatorial pharmacogenomics-guided antidepressant therapy: evidence from three clinical studies. Mol Neuropsychiatry 1:145–155. https://doi.org/10.1159/000430915

Álvarez-Pérez JM, González-García S, Cobos R, Olego MÁ, Ibañez A, Díez-Galán A, Garzón-Jimeno E, Coque JJR (2017) Use of endophytic and rhizosphere actinobacteria from grapevine plants to reduce nursery fungal graft infections that lead to young grapevine decline. Appl Environ Microbiol 83:24. https://doi.org/10.1128/AEM.01564-17

Amoutzias GD, Chaliotis A, Mossialos D (2016) Discovery Strategies of bioactive compounds synthesized by nonribosomal peptide synthetases and type-i polyketide synthases derived from marine microbiomes. Marine Drugs 14:80. https://doi.org/10.3390/md14040080

Andersen MR, Nielsen JB, Klitgaard A, Petersen LM, Zachariasen M, Hansen TJ, Blicher LH, Gotfredsen CH, Larsen TO, Nielsen KF, Mortensen UH (2013) Accurate prediction of secondary metabolite gene clusters in filamentous fungi. Proc Natl Acad Sci U S A 110:E99–E107

Andreote FD, Gumiere T, Durrer A (2014) Exploring interactions of plant microbiomes. Sci Agric 71:528–539

Anyaogu DC, Mortensen UH (2015) Heterologous production of fungal secondary metabolites in *Aspergilli*. Front Microbiol 6:77. https://doi.org/10.3389/fmicb.2015.00077

Arora J, Ramawat KG (2017) An introduction to endophytes. In: Maheshwari DK (ed) Endophytes: biology and biotechnology. Sustainable development and biodiversity, vol 15. Springer, Cham, pp 1–23. https://doi.org/10.1007/978-3-319-66541-2_1

Arulsamy A, Kumari Y, Shaikh MF (2016) Reverse pharmacology: fast track path of drug discovery. Pharm Pharmacol Int J 4:77. https://doi.org/10.15406/ppij.2016.04.00077

Austin CP, Brady LS, Insel TR, Collins FS (2004) NIH molecular libraries initiative. Science 306:1138–1139

Averesch NJH, Krömer JO (2018) Metabolic engineering of the shikimate pathway for production of aromatics and derived compounds-present and future strain construction strategies. Front Bioeng Biotechnol 32. https://doi.org/10.3389/fbioe.2018.00032

Awan AR, Blount BA, Bell DJ, Ho JCH, McKiernan RM, Tom Ellis T (2017) biosynthesis of the antibiotic nonribosomal peptide Penicillin in baker's yeast. Nat Commun 8:15202. https://doi.org/10.1038/ncomms15202

Ayotte Y, La Plante S (2017) Resurgence of phenotypic screening for discovering drug leads. EC Microbiology 3:80–82

Ayuso-Sacido A, Genilloud O (2005) New PCR primers for the screening of NRPS and PKS-I systems in actinomycetes: detection and distribution of these biosynthetic gene sequences in major taxonomic groups. Microb Ecol 49:10–24

Azad MA, Wright GD (2012) Determining the mode of action of bioactive compounds. Bioorg Med Chem 20:1929–1939

Azevedo JL (2014) Endophytic fungi from Brazilian tropical hosts and their biotechnological applications. In: Kharwar RN, Upadhyay R, Dubey N, Raghuwansh R (eds) Microbial diversity and biotechnology in food security. Springer, New Delhi, pp 17–22. https://doi.org/10.1007/978-81-322-1801-2_2

Baadhe RR, Mekala NK, Rao Parcha S, Prameela Devi Y (2014) Optimization of amorphadiene production in engineered yeast by response surface methodology. 3Biotech 4:317–324

Baker S, Volova T, Prudnikova SV, Satish S, Prasad MNN (2017) Nanoagroparticles emerging trends and future prospect in modern agriculture system. Environ Toxicol Pharmacol 53:10–17

Bamisile BS, Dash CK, Akutse KS, Keppanan R, Wang L (2018) Fungal endophytes: beyond herbivore management. Front Microbiol 9:544. https://doi.org/10.3389/fmicb.2018.00544

Banerjee A, Sharkey TD (2014) Methylerythritol 4-phosphate (MEP) pathway metabolic regulation. Nat Prod Rep 31:1043–1055

Bantscheff M, Drewes G (2012) Chemoproteomic approaches to drug target identification and drug profiling. Bioorg Med Chem 20:1973–1978

Barajas JF, Blake-Hedges JM, Bailey CB, Curran S, Keasling JD (2017) Engineered polyketides: synergy between protein and host level engineering. Synth Syst Biotechnol 2:147–166

Barelli L, Moonjely S, Behie SW, Bidochka MJ (2016) Fungi with multifunctional lifestyles: endophytic insect pathogenic fungi. Plant Mol Biol 90:657–664

Basit M, Rasool MH, Naqvi SAR, Waseem M, Aslam B (2018) Biosurfactants production potential of native strains of *Bacillus cereus* and their antimicrobial, cytotoxic and antioxidant activities. Pak J Pharm Sci 31:251–256

Bauer E, Thiele I (2018) From network analysis to functional metabolic modeling of the human gut microbiota. mSystems 3:e00209–e00217. https://doi.org/10.1128/mSystems.00209-17

Behie SW, Moreira CC, Sementchoukova BL, Zelisko PM, Bidochka MJ (2017) Carbon translocation from a plant to an insect-pathogenic endophytic fungus. Nat Commun 8:14245. https://doi.org/10.1038/ncomms14245

Bekker V, Dodd A, Brady D, Rumbold K (2014) Tools for metabolic engineering in Streptomyces. Bioengineered 5:293–299

Belda E, van Heck RGA, López-Sánchez MJ, Cruveiller S, Barbe V, Fraser C, Klenk HP, Petersen J, Morgat A, Nikel PI, Vallenet D, Rouy Z, Sekowska A, Martins dos Santos VAP, de Lorenzo V, Danchin A, Médigue C (2016) The revisited genome of *Pseudomonas putida* KT2440 enlightens its value as a robust metabolic chassis. Environ Microbiol 18:3403–3424. https://doi.org/10.1111/1462-2920.13230

Berruti A, Lumini E, Balestrini R, Bianciotto V (2016) Arbuscular mycorrhizal fungi as natural biofertilizers: let's benefit from past successes. Front Microbiol 6:1559. https://doi.org/10.3389/fmicb.2015.01559

Besnard J, Jones PS, Hopkins AL, Pannifer AD (2015) The joint European compound library: boosting precompetitive research. Drug Discov Today 20:181–186

Beutler JA (2013) Natural products as tools for discovering new cancer targets. In: Koehn F (ed) Natural products and cancer drug discovery. Cancer drug discovery and development. Springer, New York, pp 213–237. https://doi.org/10.1007/978-1-4614-4654-5_9

Bhardwaj A, Agrawal P (2014) A review fungal endophytes: as a storehouse of bioactive compound. World. J Pharm Pharm Sci 3:228–237

Bhardwaj C, Hanley L (2014) Ion sources for mass spectrometric identification and imaging of molecular species. Nat Prod Rep 31:756–767

Bhatia S, Densmore D (2013) Pigeon: a design visualizer for synthetic biology. ACS Synth Biol 2:348–350

Bhetariya PJ, Madan T, Basir SF, Varma A, Usha SP (2011) Allergens/antigens, toxins and polyketides of important *Aspergillus* species. Indian J Clin Biochem 26:104–119

Bian G, Yuan Y, Tao H, Shi X, Zhong X, Han Y, Fu S, Fang C, Deng Z, Liu T (2017) Production of taxadiene by engineering of mevalonate pathway in *Escherichia coli* and endophytic fungus *Alternaria alternata* TPF6. Biotechnol J 12:1600697. https://doi.org/10.1002/biot.201600697

Blumhoff ML, Steiger MG, Mattanovich D, Sauer M (2013) Targeting enzymes to the right compartment: metabolic engineering for itaconic acid production by *Aspergillus niger*. Metab Eng 19:26–32

Booker A, Johnston D, Heinrich M (2015) Value chains of herbal medicines–ethnopharmacological and analytical challenges in a globalizing world. In: Mukherjee PK (ed) Evidence-based validation of herbal medicine. Elsevier, Amsterdam, pp 29–44

Boruta T, Bizukojc M (2017) Production of lovastatin and itaconic acid by *Aspergillus terreus*: a comparative perspective. World J Microbiol Biotechnol 33:34. https://doi.org/10.1007/s11274-017-2206-9

Brakhage AA (2013) Regulation of fungal secondary metabolism. Nat Rev Microbiol 11:21–32

Brakhage AA, Schroeckh V (2011) Fungal secondary metabolites-strategies to activate silent gene clusters. Fungal Genet Biol 48:15–22

Brazma A, Hingamp P, Quackenbush J, Sherlock G, Spellman P, Stoeckert C, Aach J, Ansorge W, Ball CA, Causton HC, Gaasterland T, Glenisson P, Holstege FC, Kim IF, Markowitz V, Matese JC, Parkinson H, Robinson A, Sarkans U, Schulze-Kremer S, Stewart J, Taylor R, Vilo J, Vingron M (2001) Minimum information about a microarray experiment (MIAME)-toward standards for microarray data. Nat Genet 29:365–371

Brochado AR, Patil KR (2013) Overexpression of O-methyltransferase leads to improved vanillin production in baker's yeast only when complemented with model-guided network engineering. Biotechnol Bioeng 110:656–659

Brusotti G, Cesari I, Dentamaro A, Caccialanza G, Massolini G (2014) Isolation and characterization of bioactive compounds from plant resources: the role of analysis in the ethnopharmacological approach. J Pharm Biomed Anal 87:218–228

Bucci-Rechtweg C (2017) Enhancing the pediatric drug development framework to deliver better pediatric therapies tomorrow. Clin Ther 39:1920–1932

Bückle-Vallant V, Krause F, Messerschmidt S, Eikmanns B (2014) Metabolic engineering of *Corynebacterium glutamicum* for 2-ketoisocaproate production. Appl Microbiol Biotechnol 98:297–311

Bunnage ME, Chekler ELP, Jones LH (2013) Target validation using chemical probes. Nat Chem Biol 9:195–199

Cacho RA, Tang Y, Chooi YH (2015) Next-generation sequencing approach for connecting secondary metabolites to biosynthetic gene clusters in fungi. Front Microbiol 5:774. https://doi.org/10.3389/fmicb.2014.00774

Cane DE (2010) Programming of erythromycin biosynthesis by a modular polyketide Synthase. J Biol Chem 285:27517–27523

Carey TR, Cotner KL, Li B, Sohn LL (2018) Developments in label-free microfluidic methods for single-cell analysis and sorting. Rev Nanomed Nanobiotechnol 24:e1529. https://doi.org/10.1002/wnan.1529

Cepeda-García C, Domínguez-Santos R, García-Rico RO, García-Estrada C, Cajiao A, Fierro F, Martín JF (2014) Direct involvement of the CreA transcription factor in penicillin biosynthesis and expression of the pcbAB gene in *Penicillium chrysogenum*. Appl Microbiol Biotechnol 98:7113–7124

Chadha N, Mishra M, Rajpal K, Bajaj R, Choudhary DK, Varma A (2015) An ecological role of fungal endophytes to ameliorate plants under biotic stress. Arch Microbiol 197:869–881. https://doi.org/10.1007/s00203-015-1130-3

Chae TU, Choi SY, Kim JW, Ko YS, Lee SY (2017) Recent advances in systems metabolic engineering tools and strategies. Curr Opin Biotechnol 47:67–82

Chang J, Kwon HJ (2016) Discovery of novel drug targets and their functions using phenotypic screening of natural products. J Ind Microbiol Biotechnol 43:221–231

Chaparro JM, Badri DV, Vivanco JM (2014) Rhizosphere microbiome assemblage is affected by plant development. ISME J 8:790–803

Chaparro MJ, Calderón F, Castañeda P, Fernández-Alvaro E, Gabarró R, Gamo FJ, Gómez-Lorenzo MG, Martín J, Fernández E (2018) Efforts aimed to reduce attrition in antimalarial drug discovery: a systematic evaluation of the current antimalarial targets portfolio. ACS Infect Dis 4:568–576

Chater KF, Biró S, Lee KJ, Palmer T, Schrempf H (2010) The complex extracellular biology of Streptomyces. FEMS Microbiol Rev 34:171–198

Checcucci A, DiCenzo GC, Bazzicalupo M, Mengoni A (2017) Trade, diplomacy, and warfare: the quest for elite *Rhizobia* inoculant strains. Front Microbiol 8:2207. https://doi.org/10.3389/fmicb.2017.02207

Chelliah V, Laibe C, Novere NL (2013) Biomodels database: a repository of mathematical models of biological processes. Methods Mol Biol 1021:189–199

Chen L, Zhang QY, Jia M, Ming QL, Yue W, Rahman K, Qin LP, Han T (2016) Endophytic fungi with antitumor activities: their occurrence and anticancer compounds. Crit Rev Microbiol 42:454–473

Chen X, Wong YK, Wang J, Zhang J, Lee YM, Shen HM, Lin Q, Hua ZC (2017) Target identification with quantitative activity based protein profiling (ABPP). Proteomics 17:1600212. https://doi.org/10.1002/pmic.201600212

Chen X, Zhou L, Tian K, Kumar A, Singh S, Prior BA, Wang Z (2013) Metabolic engineering of *Escherichia coli*: a sustainable industrial platform for bio-based chemical production. Biotechnol Adv 31:1200–1223

Chervin J, Stierhof M, Tong MH, Peace D, Hansen KO, Urgast DS, Andersen JH, Yu Y, Ebel R, Kyeremeh K, Paget V, Cimpan G, Wyk AV, Deng H, Jaspars M, Tabudravu JN (2017) Targeted dereplication of microbial natural products by high-resolution MS and predicted LC retention time. J Nat Prod 80:1370–1377

Chi CW, Ahmed AR, Dereli-Korkut Z, Wang S (2016) Microfluidic cell chips for high-throughput drug screening. Bioanalysis 8:921–937

Chiang YM, Oakley CE, Ahuja M, Entwistle R, Schultz A, Chang SL, Sung CT, Wang CCC, Oakley BR (2013) An efficient system for heterologous expression of secondary metabolite genes in *Aspergillus nidulans*. J Am Chem Soc 135:7720–7731

Chiarabelli C, Stano P, Luisi PL (2013) Chemical synthetic biology: a mini-review. Front Microbiol 4:285. https://doi.org/10.3389/fmicb.2013.00285

Choudhary DK (2012) Microbial rescue to plant under habitat-imposed abiotic and biotic stresses. Appl Microbiol Biotechnol 96:1137–1155

Conte R, Marturano V, Peluso G, Calarco A, Cerruti P (2017) Recent advances in nanoparticle-mediated delivery of anti-inflammatory phytocompounds. Int J Mol Sci 18:E709. https://doi.org/10.3390/ijms18040709

Counihan JL, Duckering M, Dalvie E, Ku WM, Bateman LA, Fisher KJ, Nomura DK (2017) Chemoproteomic profiling of acetanilide herbicides reveals their role in inhibiting fatty acid oxidation. ACS Chem Biol 12:635–642. https://doi.org/10.1021/acschembio.6b01001

Cox KL (2018) Implications of the reversal of net neutrality rules and next steps to protect the open internet. Res Libr Issues 293:12–25

Cozza G (2017) The development of CK2 inhibitors: from traditional pharmacology to *in silico* rational drug design. Pharmaceuticals 10:26. https://doi.org/10.3390/ph10010026

Cragg GM, Newman DJ (2013) Natural products: a continuing source of novel drug leads. Biochim Biophys Acta 1830:3670–3695

Cravatt BF, Wright AT, Kozarich JW (2008) Activity-based protein profiling: from enzyme chemistry to proteomic chemistry. Annu Rev Biochem 77:383–414

Croston GE (2017) The utility of target-based discovery. Expert Opin Drug Discovery 12:427–429

Cui YY, Ling C, Zhang YY, Huang J, Liu JZ (2014) Production of shikimic acid from *Escherichia coli* through chemically inducible chromosomal evolution and cofactor metabolic engineering. Microb Cell Factories 13:21. https://doi.org/10.1186/1475-2859-13-21

Dandachi I, Sokhn ES, Dahdouh EA, Azar E, El-Bazzal B, Rolain JM, Daoud Z (2018) Prevalence and characterization of multidrug resistant gram negative *Bacilli* isolated from Lebanese poultry: a nationwide study. Front Microbiol 9:550. https://doi.org/10.3389/fmicb.2018.00550

David B, Wolfender JL, Dias DA (2014) The pharmaceutical industry and natural products: historical status and new trends. Phytochem Rev 14:299–315

de la Torre BG, Albericio F (2018) The pharmaceutical industry in 2017. an analysis of fda drug approvals from the perspective of molecules. Molecules 23:533. https://doi.org/10.3390/molecules23030533

de Vries RP, Riley R, Wiebenga A, Aguilar-Osorio G, Amillis S, Uchima CA, Grigoriev IV (2017) Comparative genomics reveals high biological diversity and specific adaptations in the industrially and medically important fungal genus *Aspergillus*. Genome Biol 18:28. https://doi.org/10.1186/s13059-017-1151-0

Deepika VB, Murali TS, Satyamoorthy K (2016) Modulation of genetic clusters for synthesis of bioactive molecules in fungal endophytes: a review. Microbiol Res 182:125–140

Deka D, Tayung K, Jha DK (2017) Harnessing fungal endophytes for plant and human health. In: Maheshwari D (ed) Endophytes: biology and biotechnology. Sustainable development and biodiversity, vol 15. Springer, Cham, pp 59–98. https://doi.org/10.1007/978-3-319-66541-2_4

Deshmukh SK, Prakash V, Ranjan N (2018) Marine fungi: a source of potential anticancer compounds. Front Microbiol 8:2536. https://doi.org/10.3389/fmicb.2017.02536

d'Espaux L, Ghosh A, Runguphan W, Wehrs M, Xu F, Konzock O, Dev I, Nhan M, Gin J, Reider Apel A, Petzold CJ, Singh S, Simmons BA, Mukhopadhyay A, García Martín H, Keasling JD (2017) Engineering high-level production of fatty alcohols by *Saccharomyces cerevisiae* from lignocellulosic feedstocks. Metab Eng 42:115–125. https://doi.org/10.1016/j.ymben.2017.06.004

Dias DA, Urban S, Roessner U (2012) A historical overview of natural products in drug discovery. Metabolites 2:303–336. https://doi.org/10.3390/metabo2020303

Dinesh R, Srinivasan V, TES AM, Srambikkal H (2017) Endophytic actinobacteria: diversity, secondary metabolism and mechanisms to unsilence biosynthetic gene clusters. Crit Rev Microbiol 43:546–566. https://doi.org/10.1080/1040841X.2016.1270895

Dolman NJ, Samson BA, Chambers KM, Janes MS, Mandavilli BS (2018) Tools to measure autophagy using high content imaging and analysis. In: Johnston P, Trask O (eds) High content screening. Methods in molecular biology, vol 1683. Humana Press, New York, pp 59–71. https://doi.org/10.1007/978-1-4939-7357-6_5

Dombrowski U, Wagner T (2014) Mental strain as field of action in the 4th industrial revolution. Procedia CIRP 17:100–105. https://doi.org/10.1016/j.procir.2014.01.077

Dong ZY, Narsing Rao MP, Xiao M, Wang HF, Hozzein WN, Chen W, Li WJ (2017) Antibacterial activity of silver nanoparticles against *Staphylococcus warneri* synthesized using endophytic bacteria by photo-irradiation. Front Microbiol 8:1090. https://doi.org/10.3389/fmicb.2017.01090

Donova MV (2017) Steroid bioconversions. In: Barredo JL, Herráiz I (eds) Microbial steroids. Methods in molecular biology, vol 1645. Humana Press, New York, pp 1–13. https://doi.org/10.1007/978-1-4939-7183-1_1

Drawnel FM, Zhang JD, Küng E, Aoyama N, Benmansour F, Araujo Del Rosario A, Jensen Zoffmann S, Delobel F, Prummer M, Weibel F, Carlson C, Anson B, Iacone R, Certa U, Singer T, Ebeling M, Prunotto M (2017) Molecular phenotyping combines molecular information, biological relevance, and patient data to improve productivity of early drug discovery. Cell Chem Biol 24:624–634. https://doi.org/10.1016/j.chembiol.2017.03.016

Duhan JS, Kumar R, Kumar N, Kaur P, Nehra K, Duhan S (2017) Nanotechnology: the new perspective in precision agriculture. Biotechnol Rep 15:11–23

Durairaj P, Malla S, Nadarajan SP, Lee PG, Jung E, Park HH, Kim BG, Yun H (2015) Fungal cytochrome P450 monooxygenases of *Fusarium oxysporum* for the synthesis of ω-hydroxy fatty acids in engineered *Saccharomyces cerevisiae*. Microb Cell Factories 14:45. https://doi.org/10.1186/s12934-015-0228-2

Dutta D, Chandra K, Gogoi PR, Dutta P (2014) Endophytes: exploitation as a tool in plant protection. Braz Arch Biol Technol 57:621–629

Dvorak P, Bidmanova S, Damborsky J, Prokop Z (2014) Immobilized synthetic pathway for biodegradation of toxic recalcitrant pollutant 1,2,3-trichloropropane. Environ Sci Technol 48:6859–6866

Edmondson R, Broglie JJ, Adcock AF, Yang LJ (2014) Three-dimensional cell culture systems and their applications in drug discovery and cell-based biosensors. Assay Drug Dev Technol 12:207–218

Edwards JE, Forster RJ, Callaghan TM, Dollhofer V, Dagar SS, Cheng Y, Chang J, Kittelmann S, Fliegerova K, Puniya AK, Henske JK, Gilmore SP, O'Malley MA, Griffith GW, Smidt H (2017) PCR and omics based techniques to study the diversity, ecology and biology of anaerobic fungi: insights, challenges and opportunities. Front Microbiol 8:1657. https://doi.org/10.3389/fmicb.2017.01657

Eisenreich W, Bacher A, Arigoni D, Rohdich F (2004) Biosynthesis of isoprenoids via the non-mevalonate pathway. Cell Mol Life Sci 61:1401–1426

Ekins S, Spektor AC, Clark AM, Dole K, Bunin BA (2017) Collaborative drug discovery for more medicines for tuberculosis (MM4TB). Drug Discov Today 22:555–565

Elyashberg M, Blinov K, Molodtsov S, Smurnyy Y, Williams AJ, Churanova T (2009) Computer-assisted methods for molecular structure elucidation: realizing a spectroscopist's dream. J Cheminf 1:3. https://doi.org/10.1186/1758-2946-1-3

Escalante A, Calderón R, Valdivia A, de Anda R, Hernández G, Ramírez OT, Gosset G, Bolívar F (2010) Metabolic engineering for the production of shikimic acid in an evolved *Escherichia coli* strain lacking the phosphoenolpyruvate: carbohydrate phosphotransferase system. Microb Cell Factories 9:21. https://doi.org/10.1186/1475-2859-9-21

Everett JR (2015) Pharmacometabonomics in humans: a new tool for personalized medicine. Pharmacogenomics 16:737–754. https://doi.org/10.2217/pgs.15.20

Everett JR (2016) From metabonomics to pharmacometabonomics: the role of metabolic profiling in personalized medicine. Front Pharmacol 7:297. https://doi.org/10.3389/fphar.2016.00297

Fallon JP, Reeves EP, Kavanagh K (2011) The *Aspergillus fumigatus* toxin fumagillin suppresses the immune response of *Galleria mellonella* larvae by inhibiting the action of haemocytes. Microbiology 157:1481–1488

Fan R, Huang YC, Grusak MA, Huang CP, Sherrier DJ (2014) Effects of nano-TiO_2 on the agronomically-relevant rhizobium-legume symbiosis. Sci Total Environ 466:503–512. https://doi.org/10.1016/j.scitotenv.2013.07.032

Fang Y (2015) Combining label-free cell phenotypic profiling with computational approaches for novel drug discovery. Expert Opin Drug Discovery 10:331–343

Farrar K, Bryant D, Cope-Selby N (2014) Understanding and engineering beneficial plant-microbe interactions: plant growth promotion in energy crops. Plant Biotechnol J 12:1193–1206

Felnagle EA, Jackson EE, Chan YA, Podevels AM, Berti AD, McMahon MD, Thomas MG (2008) Nonribosomal peptide synthetases involved in the production of medically relevant natural products. Mol Pharm 5:191–211

Filho JAC, Sobrinho RR, Pascholati SF (2017) Arbuscular mycorrhizal symbiosis and its role in plant nutrition in sustainable agriculture. In: Meena V, Mishra P, Bisht J, Pattanayak A (eds) Agriculturally important microbes for sustainable agriculture. Springer, Singapore, pp 129–164. https://doi.org/10.1007/978-981-10-5343-6_5

Finzel K, Burkart MD (2016) Traffic control in modular polyketide synthases. ACS Cent Sci 2:9–11

Fleming-Dutra KE, Hersh AL, Shapiro DJ, Bartoces M, Enns EA, File TM Jr, Finkelstein JA, Gerber JS, Hyun DY, Linder JA, Lynfield R, Margolis DJ, May LS, Merenstein D, Metlay JP, Newland JG, Piccirillo JF, Roberts RM, Sanchez GV, Suda KJ, Thomas A, Woo TM, Zetts RM, Hicks LA (2016) Prevalence of inappropriate antibiotic prescriptions among US ambulatory care visits, 2010-2011. JAMA 315:1864–1873. https://doi.org/10.1001/jama.2016.4151

Fletcher E, Krivoruchko A, Nielsen J (2016) Industrial systems biology and its impact on synthetic biology of yeast cell factories. Biotechnol Bioeng 113:1164–1170

Fraceto LF, Grillo R, de Medeiros GA, Scognamiglio V, Rea G, Bartolucci C (2016) Nanotechnology in agriculture: which innovation potential does it have. Front Environ Sci 4:20. https://doi.org/10.3389/fenvs.2016.00020

Franche C, Normand P, Pawlowski K, Tisa LS, Bogusz D (2016) An update on research on *Frankia* and actinorhizal plants on the occasion of the 18th meeting of the frankia-actinorhizal plants symbiosis. Symbiosis 70:1. https://doi.org/10.1007/s13199-016-0431-x

Franci G, Falanga A, Galdiero S, Palomba L, Rai M, Morelli G, Galdiero M (2015) Silver nanoparticles as potential antibacterial agents. Molecules 20:8856–8874

Frank AC, Saldierna Guzmán JP, Shay JE (2017) Transmission of bacterial endophytes. microorganisms. Microorganisms 5:70. https://doi.org/10.3390/microorganisms5040070

Futamura Y, Muroi M, Osada H (2013) Target identification of small molecules based on chemical biology approaches. Mol BioSyst 9:897–914. https://doi.org/10.1039/c2mb25468a

Galanie S, Thodey K, Trenchard IJ, Interrante MF, Smolke CD (2015) Complete biosynthesis of opioids in yeast. Science 349:1095–1100

Gallo A, Ferrara M, Perrone G (2013) Phylogenetic study of polyketide synthases and nonribosomal peptide synthetases involved in the biosynthesis of mycotoxins. Toxins 5:717–742

Ganesan P, Choi DK (2016) Current application of phytocompound based nano-cosmeceuticals for beauty and skin therapy. Int J Nanomedicine 11:1987–2007

Gao XW, Liu HX, Sun ZH, Chen YC, Tan YZ, Zhang WM (2016) Secondary metabolites from the deep-sea derived fungus *Acaromyces ingoldii* FS121. Molecules:371. https://doi.org/10.3390/molecules21040371

Garbaccio RM, Parmee ER (2016) The impact of chemical probes in drug discovery: a pharmaceutical industry perspective. Cell Chem Biol 23:10–17

Gardea-Torresdey JL, Rico CM, White JC (2014) Trophic transfer, transformation, and impact of engineered nanomaterials in terrestrial environments. Environ Sci Technol 48:2526–2540

Gaulton A, Hersey A, Nowotka M, Bento AP, Chambers J, Mendez D, Mutowo P, Atkinson F, Bellis LJ, Cibrián-Uhalte E, Davies M, Dedman N, Karlsson A, Magariños MP, Overington JP, Papadatos G, Smit I, Leach AR (2017) The ChEMBL database in 2017. Nucleic Acids Res 45:D945–D954

Golinska P, Wypij M, Agarkar G, Rathod D, Dahm H, Rai M (2015) Endophytic actinobacteria of medicinal plants: diversity and bioactivity. Antonie Van Leeuwenhoek 108:267–289

Gressler M, Hortschansky P, Geib E, Brock M (2015) A new high-performance heterologous fungal expression system based on regulatory elements from the *Aspergillus terreus* terrein gene cluster. Front Microbiol 6:184. https://doi.org/10.3389/fmicb.2015.00184

Gu J, Gui Y, Chen L, Yuan G, Lu HZ, Xu X (2013) Use of natural products as chemical library for drug discovery and network pharmacology. PLoS One 8:e62839. https://doi.org/10.1371/journal.pone.0062839

Guo H, Nasir M, Lv J, Dai Y, Gao J (2017) Understanding the variation of microbial community in heavy metals contaminated soil using high throughput sequencing. Ecotoxicol Environ Saf 144:300–306

Gupta YK, Meenu M, Mohan P (2015) The Tamiflu fiasco and lessons learnt. Indian J Pharm 47:11–16

Haasen D, Schopfer U, Antczak C, Guy C, Fuchs F, Selzer P (2017) How phenotypic screening influenced drug discovery: lessons from five years of practice. Assay Drug Dev Technol 15:239–246

Hadadi N, Hafner J, Shajkofci A, Zisaki A, Hatzimanikatis V (2016) ATLAS of biochemistry: a repository of all possible biochemical reactions for synthetic biology and metabolic engineering studies. ACS Synth Biol 5:1155–1166

Hajimahdi Z, Zarghi A (2016) Progress in HIV-1 integrase inhibitors: a review of their chemical structure diversity. Iran J Pharm Res 15:595–628

Hardoim PR, Hardoim CCP, van Overbeek LS, van Elsas JD (2012) Dynamics of seed-borne rice endophytes on early plant growth stages. PLoS One 7:e30438. https://doi.org/10.1371/journal.pone.0030438

Hardoim PR, van Overbeek LS, Berg G, Pirttilä AM, Compant S, Campisano A, Doring M, Sessitsch A (2015) The hidden world within plants: ecological and evolutionary considerations for defining functioning of microbial endophytes. Microbiol Mol Biol Rev 79:293–320

Harrington LB, Doxzen KW, Ma E, Liu JJ, Knott GJ, Edraki A, Garcia B, Amrani N, Chen JS, Cofsky JC, Kranzusch PJ, Sontheimer EJ, Davidson AR, Maxwell KL, Doudna JA (2017) A broad spectrum inhibitor of CRISPR-Cas9. Cell 170:1224–1233

Harvey AL, Edrada-Ebel R, Quinn RJ (2015) The re-emergence of natural products for drug discovery in the genomics era. Nat Rev Drug Discov 14:111–129

Helaly SE, Thongbai B, Stadler M (2018) Diversity of biologically active secondary metabolites from endophytic and saprotrophic fungi of the ascomycete order Xylariales. Nat Prod Rep 35:993–1014

Hewage RT, Aree T, Mahidol C, Ruchirawat S, Kittakoop P (2014) One strain-many compounds (OSMAC) method for production of polyketides, azaphilones, and an isochromanone using the endophytic fungus *Dothideomycete* sp. Phytochemistry 8:87–94

Higginbotham SJ, Arnold AE, Ibañez A, Spadafora C, Coley PD, Kursar TA (2013) Bioactivity of fungal endophytes as a function of endophyte taxonomy and the taxonomy and distribution of their host plants. PLoS One 8:e73192. https://doi.org/10.1371/journal.pone.0073192

Hollensteiner J, Poehlein A, Granzow S, Liesegang H, Daniel R, Vidal S, Wemheuer F (2018) Draft genome sequence of the endophyte *Bacillus mycoides* strain GM5LP isolated from *Lolium perene*. Genome Announc 6:e01517–e01517. https://doi.org/10.1128/genomeA

Honeker LK, Neilson JW, Root RA, Gil-Loaiza J, Chorover J, Maier RM (2017) Bacterial rhizoplane colonization patterns of *Buchloe dactyloides* growing in metalliferous mine tailings reflect plant status and biogeochemical conditions. Microb Ecol 74:853–867

Hong HJ, Koom WS, Koh WG (2017) Cell microarray technologies for high-throughput cell-based biosensors. Sensors 17, 1293. https://doi.org/10.3390/s17061293

Hong SY, Roze LV, Linz JE (2013) Oxidative stress-related transcription factors in the regulation of secondary metabolism. Toxins 5:683–702

Hossain MM, Sultana F, Islam S (2017) Plant growth-promoting fungi (PGPF): phytostimulation and induced systemic resistance. In: Singh D, Singh H, Prabha R (eds) Plant-microbe interactions in agro-ecological perspectives. Springer, Singapore, pp 135–191

Howe EA, de Souza A, Lahr DL, Chatwin S, Montgomery P, Alexander BR, Nguyen DT, Cruz Y, Stonich DA, Walzer G, Rose JT, Picard SC, Liu Z, Rose JN, Xiang X, Asiedu J, Durkin D, Levine J, Yang JJ, Schürer SC, Braisted JC, Southall N, Southern MR, Chung TD, Brudz S, Tanega C, Schreiber SL, Bittker JA, Guha R, Clemons PA (2015) Bioassay research database (BARD): chemical biology and probe-development enabled by structured metadata and result types. Nucleic Acids Res 43:D1163–D1170. https://doi.org/10.1093/nar/gku1244

Hu Y, Bajorath J (2017) Entering the big data era in medicinal chemistry: molecular promiscuity analysis revisited. Future Sci OA 3:FSO179. https://doi.org/10.4155/fsoa-2017-0001

Huang R, Southall N, Wang Y, Yasgar A, Shinn P, Jadhav A, Nguyen DT, Austin CP (2011) The NCGC Pharmaceutical collection: a comprehensive resource of clinically approved drugs enabling repurposing and chemical genomics. Sci Transl Med 3:80. https://doi.org/10.1126/scitranslmed.3001862

Huggett B (2016) Academic partnerships 2015. Nat Biotechnol 34:372

Huryn DM, Cosford NDP (2007) The Molecular libraries screening center network (MLSCN): identifying chemical probes of biological systems. Ann Rep Med Chem 42:401–416. https://doi.org/10.1016/S0065-7743(07)42026-7

Iavicoli I, Leso V, Beezhold DH, Shvedova AA (2017) Nanotechnology in agriculture: opportunities, toxicological implications, and occupational risks. Toxicol Appl Pharmacol 329:96–111.https://doi.org/10.1016/j.taap.2017.05.025

Ibekwe NN, Ameh JS (2015) Hyphenated techniques in liquid chromatography as current trends in natural products analysis. Int Res J Pure Appl Chem 7:132–149

Isgut M, Rao M, Yang C, Subrahmanyam V, Rida PCG, Aneja R (2018) Application of combination high throughput phenotypic screening and target identification methods for the discovery of natural product-based combination drugs. Med Res Rev 38:504–524

Jakočiūnas T, Jensen MK, Keasling JD (2016) CRISPR/Cas9 advances engineering of microbial cell factories. Metab Eng 34:44–59. https://doi.org/10.1016/j.ymben.2015.12.003
Jansen MLA, Bracher JM, Papapetridis I, Verhoeven MD, de Bruijn H, de Waal PP, van Maris AJA, Klaassen P, Pronk JT (2017) *Saccharomyces cerevisiae* strains for second-generation ethanol production: from academic exploration to industrial implementation. FEMS Yeast Res 17:fox044. https://doi.org/10.1093/femsyr/fox044
Jia M, Chen L, Xin HL, Zheng CJ, Rahman K, Han T, Qin LP (2016) A friendly relationship between endophytic fungi and medicinal plants: a systematic review. Front Microbiol 7:906. https://doi.org/10.3389/fmicb.2016.00906
Jiang R, Jiao Y, Zhang P, Liu Y, Wang X, Huang Y, Zhang Z, Xu F (2017) Twin derivatization strategy for high-coverage quantification of free fatty acids by liquid chromatography–Tandem mass spectrometry. Anal Chem 89:12223–12230
Johnson EA (2013a) Biotechnology of non-*Saccharomyces* yeasts-the ascomycetes. Appl Microbiol Biotechnol 97:503–517. https://doi.org/10.1007/s00253-012-4497-y
Johnson EA (2013b) Biotechnology of non-*Saccharomyces* yeasts-the basidiomycetes. Appl Microbiol Biotechnol 97:7563–7577. https://doi.org/10.1007/s00253-013-5046-z
Jones LH, Bunnage ME (2017) Applications of chemogenomic library screening in drug discovery. Nat Rev Drug Discov 16:285–296
Jung HJ, Kwon HJ (2015) Target deconvolution of bioactive small molecules: the heart of chemical biology and drug discovery. Arch Pharm Res 38:1627–1641
Jupp S, Malone J, Burdett T, Heriche JK, Williams E, Ellenberg J, Parkinson H, Rustici G (2016) The cellular microscopy phenotype ontology. J Biomed Semant 7:28. https://doi.org/10.1186/s13326-016-0074-0
Kadir F, Ives P, Luitjens A, van Corven E (2013) Production and rurification of recombinant proteins. In: Crommelin D, Sindelar R, Meibohm B (eds) Pharmaceutical biotechnology. Springer, New York, pp 47–67
Kah M (2015) Nanopesticides and nanofertilizers: emerging contaminants or opportunities for risk mitigation? Front Chem 3:64. https://doi.org/10.3389/fchem.2015.00064
Kahraman H, Erenler SO (2012) Rhamnolipid production by *Pseudomonas aeruginosa* engineered with the Vitreoscilla hemoglobin gene. Prikl Biokhim Mikrobiol 48:212–217
Kaiser J (2011) National institutes of health. Drug-screening program looking for a home. Science 334:299. https://doi.org/10.1126/science.334.6054.299
Kallscheuer N, Marienhagen J (2018) *Corynebacterium glutamicum* as platform for the production of hydroxybenzoic acids. Microb Cell Factories 17:70. https://doi.org/10.1186/s12934-018-0923-x
Kamdem RST, Wang H, Wafo P, Ebrahim W, Özkaya FC, Makhloufi G, Janiak C, Sureechatchaiyan P, Kassack MU, Lin W, Liu Z, Proksch P (2018) Induction of new metabolites from the endophytic fungus *Bionectria* sp. through bacterial co-culture. Fitoterapia 124:132–136
Karawajczyk A, Giordanetto F, Benningshof J, Hamza D, Kalliokoski T, Pouwer K, Morgentin R, Nelson A, Müller G, Piechot A, Tzalis D (2015) Expansion of chemical space for collaborative lead generation and drug discovery: the European lead factory perspective. Drug Discov Today 20:1310–1316
Karawajczyk A, Orrling KM, de Vlieger JSB, Rijnders T, Tzalis D (2017) The European Lead Factory: a blueprint for public–private partnerships in early drug discovery. Front Med 3:75. https://doi.org/10.3389/fmed.2016.00075
Katz L, Baltz RH (2016) Natural product discovery: past, present, and future. J Ind Microbiol Biotechnol 43:155–176
Kaul S, Gulati N, Verma D, Mukherjee S, Nagaich U (2018) Role of nanotechnology in cosmeceuticals: a review of recent advances. J Pharm 27:3420204. https://doi.org/10.1155/2018/3420204
Kaul S, Sharma T, Dhar MK (2016) Omics tools for better understanding the plant-endophyte interactions. Front Plant Sci 7:955. https://doi.org/10.3389/fpls.2016.00955
Kesselheim AS, Sinha MS, Avorn J (2017) Determinants of market exclusivity for prescription drugs in the United States. JAMA Intern Med 177:1658–1664

Khaldi N, Seifuddin FT, Turner G, Haft D, Nierman WC, Wolfe KH, Fedorova ND (2010) SMURF: genomic mapping of fungal secondary metabolite clusters. Fungal Genet Biol 47:736–741

Khater S, Anand S, Mohanty D (2016) *In silico* methods for linking genes and secondary metabolites: the way forward. Synth Syst Biotechnol 1:80–88. https://doi.org/10.1016/j.synbio.2016.03.001

Khatri S, Saini RV, Chhillar AK (2017) Molecular farming approach towards bioactive compounds. In: Kalia V, Saini A (eds) Metabolic engineering for bioactive compounds. Springer, Singapore, pp 49–72

Kildgaard S, Mansson M, Dosen I, Klitgaard A, Frisvad JC, Larsen TO, Nielsen KF (2014) dereplication of bioactive secondary metabolites from marine-derived fungi by UHPLC-DAD-QTOFMS and a MS/HRMS library. Mar Drugs 12:3681–3705

Kim DY, Kadam A, Shinde S, Saratale RG, Patra J, Ghodake G (2018) Recent developments in nanotechnology transforming the agricultural sector: a transition replete with opportunities. J Sci Food Agric 98:849–864

Kitade Y, Hashimoto R, Suda M, Hiraga K, Inui M (2018) The production of 4-Hydroxybenzoic acid using metabolically engineered *Corynebacterium glutamicum*. Appl Environ Microbiol 84:e02587–e02517. https://doi.org/10.1128/AEM.02587-17

Knapp DG, Németh JB, Barry K, Hainaut M, Henrissat B, Johnson J, Kuo A, Lim JHP, Lipzen A, Nolan M, Ohm RA, Tamás L, Grigoriev IV, Spatafora JW, Nagy LG, Kovács GM (2018) Comparative genomics provides insights into the lifestyle and reveals functional heterogeneity of dark septate endophytic fungi. Sci Rep 8:6321. https://doi.org/10.1038/s41598-018-24686-4

Knox B, Keller N (2015) Key players in the regulation of fungal secondary metabolism. In: Zeilinger S, Martín JF, García-Estrada C (eds) Biosynthesis and molecular genetics of fungal secondary metabolites, Fungal Biology, vol 2. Springer, New York, pp 13–28

Kogure T, Kubota T, Suda M, Hiraga K, Inui M (2016) Metabolic engineering of *Corynebacterium glutamicum* for shikimate overproduction by growth-arrested cell reaction. Metab Eng 38:204–216

Köhler S, Vasilevsky NA, Engelstad M Foster E, McMurry J, Aymé S, Baynam G, Bello SM, Boerkoel CF, Boycott KM, Brudno M, Buske OJ, Chinnery PF, Cipriani V, Connell LE, Dawkins HJ, DeMare LE, Devereau AD, de Vries BB, Firth HV, Freson K, Greene D, Hamosh A, Helbig I, Hum C, Jähn JA, James R, Krause R F Laulederkind SJ, Lochmüller H, Lyon GJ, Ogishima S, Olry A, Ouwehand WH, Pontikos N, Rath A, Schaefer F, Scott RH, Segal M, Sergouniotis PI, Sever R, Smith CL, Straub V, Thompson R, Turner C, Turro E, Veltman MW, Vulliamy T, Yu J, von Ziegenweidt J, Zankl A, Züchner S, Zemojtel T, Jacobsen JO, Groza T, Smedley D, Mungall CJ, Haendel M, Robinson PN (2017) The human phenotype ontology in 2017. Nucleic Acids Res 45:D865-D876

Kolluri N, Klapperich CM, Cabodi M (2018) Towards lab-on-a-chip diagnostics for malaria elimination. Lab Chip 18:75–94

Kudjo DE (2007) Phytoremediation: the state of rhizosphere engineering for accelerated rhizodegradation of xenobiotic contaminants. J Chem Technol Biotechnol 82:228–232

Kuroda K, Ueda M (2010) Engineering of microorganisms towards recovery of rare metal ions: mini review. Appl Microbiol Biotechnol 87:53–60

Kusari P, Kusari S, Spiteller M, Kayser O (2015) Implications of endophyte-plant crosstalk in light of quorum responses for plant biotechnology. Appl Microbiol Biotechnol 99:5383–5390

Lacava PT, Azevedo JL (2014) Biological control of insect-pest and diseases by endophytes. In: Vijay VC, Alan CG (eds) Advances in endophytic research advances. Springer, New Delhi, pp 231–243

Lee J, Bogyo M (2013) Target deconvolution techniques in modern phenotypic profiling. Curr Opin Chem Biol 17:118–126. https://doi.org/10.1016/j.cbpa.2012.12.022

Lee JA, Berg EL (2013) Neoclassic drug discovery: the case for lead generation using phenotypic and functional approaches. J Biomol Screen 18:1143–1155. https://doi.org/10.1177/1087057113506118

Lee JW, Na D, Park JM, Lee J, Choi S, Lee SY (2012) Systems metabolic engineering of microorganisms for natural and non-natural chemicals. Nat Chem Biol 8:536–546

Lexchin J (2014a) How safe are new drugs. Market withdrawal of drugs approved in Canada between 1990 and 2009. Open Med 8:e14–e19

Lexchin J (2014b) Post-market safety warnings for drugs approved in Canada under the notice of compliance with conditions policy. Braz J Clin Pharmacol 79:847–859

Lexchin J (2016) How safe and innovative are first-in-class drugs approved by health Canada: a cohort study. Health Policy 12:65–75

Li J, Rao J, Pu K (2018) Recent progress on semiconducting polymer nanoparticles for molecular imaging and cancer phototherapy. Biomaterials 155:217–235

Li PCH (2016) Overview of microarray technology. In: Li P, Sedighi A, Wang L (eds) Microarray technology. Methods in molecular biology, vol 1368. Humana Press, New York, pp 3–4

Lian J, Mishra S, Zhao H (2018) Recent advances in metabolic engineering of *Saccharomyces cerevisiae*: new tools and their applications. Metab Eng (Online). https://doi.org/10.1016/j.ymben.2018.04.011

Liaqat F, Eltem R (2016) Identification and characterization of endophytic bacteria isolated from *in vitro* cultures of peach and pear rootstocks. 3Biotech 6:120. https://doi.org/10.1007/s13205-016-0442-6

Lionta E, Spyrou G, Vassilatis DK, Cournia Z (2014) Structure-based virtual screening for drug discovery: principles, applications and recent advances. Curr Top Med Chem 14:1923–1938

Lipinski CA, Litterman NK, Southan C, Williams AJ, Clark AM, Ekins S (2015) Parallel worlds of public and commercial bioactive chemistry data: miniperspective. J Med Chem 58:2068–2076

Liu P, Zhu X, Tan Z, Zhang X, Ma Y (2016) Construction of *Escherichia Coli* cell factories for production of organic acids and alcohols. Adv Biochem Eng Biotechnol 155:107–140

Liu R, Li X, Lam KS (2017) Combinatorial chemistry in drug discovery. Curr Opin Chem Biol 38:117–126

Liu XH, Wang T (2017) Natural products remain an important source for drug development in the post–genomic era. MOJ Bioequiv Availab 3:134–135

Liu Y, Guo M (2014) Chemical proteomic strategies for the discovery and development of anticancer drugs. Proteomics 14:399–411

Liwarska-Bizukojc E, Olejnik D, Delbeke EIP, Van Geem KM, Stevens CV (2018) Evaluation of biological properties and fate in the environment of a new class of biosurfactants. Chemosphere 200:561–568

Lobanovska M, Pilla G (2017) Penicillin's discovery and antibiotic resistance: lessons for the future. Yale J Biol Med 90:135–145

Loeschcke A, Thies S (2015) *Pseudomonas putida*-a versatile host for the production of natural products. Appl Microbiol Biotechnol 99:6197–6214

Lombard J, Moreira D (2011) Origins and early evolution of the mevalonate pathway of isoprenoid biosynthesis in the three domains of life. Mol Biol Evol 28:87–99

Lopes H, Rocha I (2017) Genome-scale modeling of yeast: chronology, applications and critical perspectives. FEMS Yeast Res (Online). https://doi.org/10.1093/femsyr/fox050

Lucia M (2017) Lipid-based nanoparticles as carriers for dermal delivery of antioxidants. Curr Drug Metab 18:469–480

Lyu Y, Pu K (2017) Recent advances of activatable molecular probes based on semiconducting polymer nanoparticles in sensing and imaging. Adv Sci 4:1600481. https://doi.org/10.1002/advs.201600481

Ma C, White JC, Parkash Dhanker O, Xing B (2015) Metal-based nanotoxicity and detoxification pathways in higher plants. Environ Sci Technol 49:7109–7122

Macías-Rubalcava ML, Sánchez-Fernández RE (2017) Secondary metabolites of endophytic *Xylaria* species with potential applications in medicine and agriculture. World J Microbiol Biotechnol 33:15. https://doi.org/10.1007/s11274-016-2174-5

Maheshwari R, Varahalarao V, Ramars A, Murty US (2017) Endophytic fungi as novel resources of natural therapeutics. Braz Arch Biol Technol 60:e17160542. https://doi.org/10.1590/1678-4324-2017160542

Mandavilli BS, Aggeler RJ, Chambers KM (2018) Tools to measure cell health and cytotoxicity using high content imaging and analysis. In: Johnston P, Trask O (eds) High content screening. Methods in molecular biology, vol 1683. Humana Press, New York, pp 33–46

Martín JF (2017) Key role of *LaeA* and velvet complex proteins on expression of β-lactam and PR-toxin genes in *Penicillium chrysogenum*: cross-talk regulation of secondary metabolite pathways. J Ind Microbiol Biotechnol 44:525–535

Martínez JA, Bolívar F, Escalante A (2015) Shikimic acid production in *Escherichia coli*: from classical metabolic engineering strategies to omics applied to improve its production. Front Bioeng Biotechnol 3:145. https://doi.org/10.3389/fbioe.2015.00145

Martinez-Klimova E, Rodríguez-Peña K, Sánchez S (2017) Endophytes as sources of antibiotics. Biochem Pharmacol 134:1–17

Masimirembwa C, Thelingwani R (2012) Application of *in silico*, *in vitro* and *in vivo* ADMET/PK platforms in drug discovery. In: Chibale K, Davies-Coleman M, Masimirembwa C (eds) Drug discovery in Africa. Springer, Berlin, pp 151–191

Matsue Y, Mizuno H, Tomita T, Asami T, Nishiyama M, Kuzuyama T (2010) The herbicide ketoclomazone inhibits 1-deoxy-D-xylulose 5-phosphate synthase in the 2-C-methyl-D-erythritol 4-phosphate pathway and shows antibacterial activity against *Haemophilus influenzae*. J Antibiot 63:583–588

Matsui JK, Primer DN, Molander GA (2017) Metal-free C–H alkylation of heteroarenes with alkyltrifluoroborates: a general protocol for 1°, 2° and 3° alkylation. Chem Sci 8:3512–3522

Maurer-Jones MA, Gunsolus I, Murphy CJ, Haynes CL (2013) Toxicity of engineered nanoparticles in the environment. Anal Chem 6:3036–3049

Mayer AMS, Rodríguez AD, Taglialatela-Scafati O, Fusetani N (2013) Marine Pharmacology in 2009–2011: marine compounds with antibacterial, antidiabetic, antifungal, anti-inflammatory, antiprotozoal, antituberculosis, and antiviral activities; affecting the immune and nervous systems, and other miscellaneous mechanisms of action. Marine Drugs 11:2510–2573

Mazzoli R, Riedel K, Pessione E (2017) Bioactive compounds from microbes. Front Microbiol 8:392. https://doi.org/10.3389/fmicb.2017.00392

Medina-Cleghorn D, Bateman LA, Ford B, Heslin A, Fisher KJ, Dalvie ED, Nomura DK (2015) Mapping proteome-wide targets of environmental chemicals using reactivity-based chemoproteomic platforms. Chem Biol 22:1394–1405

Medina-Cleghorn D, Nomura DK (2014) Exploring metabolic pat hways and regulation through functional chemoproteomic and metabolomic platforms. Chem Biol 21:1171–1184

Men K, Duan X, Wei XW, Gou ML, Huang MJ, Chen LJ, Qian ZY, Wei YQ (2014) Nanoparticle-delivered quercetin for cancer therapy. Anti Cancer Agents Med Chem 14:826–832

Meyer V, Fiedler M, Nitsche B, King R (2015) The cell factory aspergillus enters the big data era: opportunities and challenges for optimising product formation. Adv Biochem Eng Biotechnol 149:91–132

Miller K, Neilan B, Sze DM (2008) Development of Taxol and other endophyte produced anti-cancer agents. Recent Pat Anticancer Drug Discov 3:14–19

Mills BJW, Batterman AS, Field KJ (2018) Nutrient acquisition by symbiotic fungi governs Palaeozoic climate transition. Philos Trans R Soc 373:20160503. https://doi.org/10.1098/rstb.2016.0503

Mirsadeghi S, Larijani B (2017) Personalized medicine: pharmacogenomics and drug development. Acta Med Iran 55:150–165

Mishra M, Sidoti D, Avvari GV, Mannaru P, Ayala DFM, Pattipati KR, Kleinman DL (2017) A context-driven framework for proactive decision support with applications. IEEE Access 5:12475. https://doi.org/10.1109/ACCESS.2017.2707091

Moerke N, Fallahi-Sichani M (2016) Reverse phase protein arrays for compound profiling. Curr Protoc Chem Biol 8:179–196

Moffat JG, Vincent F, Lee JA, Eder J, Prunotto M (2017) Opportunities and challenges in phenotypic drug discovery: an industry perspective. Nat Rev Drug Discov 16:531–543

Moonjely S, Barelli L, Bidochka MK (2016) Insect pathogenic fungi as endophytes. Adv Genet 94:107–135

Nakamura J, Hirano S, Ito H, Wachi M (2007) Mutations of the *Corynebacterium glutamicum* NCgl1221 gene, encoding a mechanosensitive channel homolog, induce L-glutamic acid production. Appl Environ Microbiol 73:4491–4498

Nam JS, Sharma AR, Nguyen LT, Chakraborty C, Sharma G, Lee SS (2016) Application of bioactive quercetin in oncotherapy: from nutrition to nanomedicine. Molecules 21:E108. https://doi.org/10.3390/molecules21010108

Narsing Rao MP, Xiao M, Li WJ (2017) Fungal and bacterial pigments: secondary metabolites with wide applications. Front Microbiol 8:1113. https://doi.org/10.3389/fmicb.2017.01113

Naveed M, Hussain MB, Mehboob I, Zahir ZA (2017) Rhizobial amelioration of drought stress in legumes. In: Zaidi A, Khan M, Musarrat J (eds) Microbes for legume improvement. Springer, Cham, pp 341–365

Negreiros de Carvalho PL, Silva Ede O, Chagas-Paula DA, Hortolan Luiz JH, Ikegaki M (2016) Importance and implications of the production of phenolic secondary metabolites by endophytic fungi: a mini-review. Mini-Rev Med Chem 16:259–271

Netzker T, Fischer J, Weber J, Mattern DJ, König CC, Valiante V, Schroeckh V, Brakhage AA (2015) Microbial communication leading to the activation of silent fungal secondary metabolite gene clusters. Front Microbiol 6:299. https://doi.org/10.3389/fmicb.2015.00299

Nevoigt E (2008) Progress in metabolic engineering of *Saccharomyces cerevisiae*. Microbiol Mol Biol Rev 72:379–412

Newman DJ, Cragg GM (2016) Natural products as sources of new drugs from 1981 to 2014. J Nat Prod 79:629–661

Nhung NT, Chansiripornchai N, Carrique-Mas JJ (2017) Antimicrobial resistance in bacterial poultry pathogens: a Review. Front Vet Sci 4:126. https://doi.org/10.3389/fvets.2017.00126

Nicoletti R, Fiorentino A (2015) Plant bioactive metabolites and drugs produced by endophytic fungi of *Spermatophyta*. Agriculture 5:918–970

Nielsen KF, Larsen TO (2015) The importance of mass spectrometric dereplication in fungal secondary metabolite analysis. Front Microbiol 6:71. https://doi.org/10.3389/fmicb.2015.00071

Nierode G, Kwon PS, Dordick JS, Kwon S-J (2016) Cell-based assay design for high-content screening of drug candidates. J Microbiol Biotechnol 26:213–225

Nijland JG, Ebbendorf B, Woszczynska M, Boer R, Bovenberg RAL, Driessen AJM (2010) Nonlinear biosynthetic gene cluster dose effect on *Penicillin* production by *Penicillium chrysogenum*. Appl Environ Microbiol 76:7109–7115

Nijman SMB (2015) Functional genomics to uncover drug mechanism of action. Nat Chem Biol 11:942–948

Nikel PI (2016) Systems and synthetic biology approaches for metabolic engineering of *Pseudomonas putida*. In: Castro-Sowinski S (ed) Microbial models: from environmental to industrial sustainability. Microorganisms for sustainability, vol 1. Springer, Singapore, pp 3–22

Nisa H, Kamili AN, Nawchoo IA, Shafi S, Shameem N, Bandh SA (2015) Fungal endophytes as a prolific source of phytochemicals and other bioactive natural products: a review. Microb Pathog 82:50–59

Niu G, Chater KF, Tian Y, Zhang J, Tan H (2016) Specialised metabolites regulating antibiotic biosynthesis in *Streptomyces* spp. FEMS Microbiol Rev 40:554–573

Nowotka MM, Gaulton A, Mendez D, Bento AP, Hersey A, Leach A (2017) Using ChEMBL web services for building applications and data processing workflows relevant to drug disc. Expert Opin Drug Discovery 12(8):757–767

Nurfarahin AH, Mohamed MS, Phang LY (2018) Culture medium development for microbial-derived surfactants production-an overview. Molecules 23:E1049. https://doi.org/10.3390/molecules23051049

Oktar FN, Yetmez M, Ficai D, Ficai A, Dumitru F, Pica A (2015) Molecular mechanism and targets of the antimicrobial activity of metal nanoparticles. Curr Top Med Chem 15:1583–1588

Olanrewaju OS, Glick BR, Babalola OO (2017) Mechanisms of action of plant growth promoting bacteria. World J Microbiol Biotechnol 33:197. https://doi.org/10.1007/s11274-017-2364-9

Olivon F, Allard PM, Koval A, Righi D, Genta-Jouve G, Neyts J, Apel C, Pannecouque C, Nothias LF, Cachet X, Marcourt L, Roussi F, Katanaev VL, Touboul D, Wolfender JL, Litaudon M

(2017) Bioactive natural products prioritization using massive multi-informational molecular networks. ACS Chem Biol 12:2644–2651

Orozovic G, Orozovic K, Järhult JD, Olsen B (2014) Study of Oseltamivir and Zanamivir resistance-related mutations in influenza viruses isolated from wild Mallards in Sweden. PLoS One 9:e89306. https://doi.org/10.1371/journal.pone.0089306

Ortiz SC, Trienens M, Rohlfs M (2013) Induced fungal resistance to insect grazing: reciprocal fitness consequences and fungal gene expression in the *Drosophila-Aspergillus* model system. PLoS One 8:e74951

Osman KM, Kappell AD, Elhadidy M, ElMougy F, El-Ghany WAA, Orabi A, Mubarak AS, Dawoud TM, Hemeg HA, Moussa IMI, Hessain AM, Yousef HMY (2018) Poultry hatcheries as potential reservoirs for antimicrobial-resistant *Escherichia coli*: a risk to public health and food safety. Sci Rep 8:5859. https://doi.org/10.1038/s41598-018-23962-7

Osswald C, Zipf G, Schmidt G, Maier J, Bernauer HS, Müller R, Wenzel SC (2014) Modular construction of a functional artificial epothilone polyketide pathway. ACS Synth Biol 3:759–772. https://doi.org/10.1021/sb300080t

Owen L, Pennington B, Fischer A, Jeong K (2017) The cost-effectiveness of public health interventions examined by NICE from 2011 to 2016. J Public Health 18:1–10

Ozturk SSB (2015) Acethylcholinesterase inhibitory potential and antioxidante properties of pyrogallol. J Enzym Inhib Med Chem 30:761–766

Pai MY, Lomenick B, Hwang H, Schiestl R, McBride W, Loo JÁ, Huang J (2015) Drug affinity responsive target stability (DARTS) for small molecule target identification. Methods Mol Biol 1263:287–298

Paillard G, Cochrane P, Jones PS, van Hoorn WP, Caracoti A, van Vlijmen H, Pannifer AD (2016) The ELF honest data broker: informatics enabling public–private collaboration in a precompetitive arena. Drug Discov Today 21:97–102

Pan S, Zhang H, Wang C, Yao SC, Yao SQ (2016) Target identification of natural products and bioactive compounds using affinity-based probes. Nat Prod Rep 33:612–620

Panpatte DG, Jhala YK, Shelat HN, Vyas RV (2016) Nanoparticles: the next generation technology for sustainable agriculture. In: Singh D, Singh H, Prabha R (eds) Microbial inoculants in sustainable agricultural productivity. Springer, New Delhi, pp 289–300

Park HS, Jun SC, Han KH, Hong SB, Yu JH (2017) Diversity, application, and synthetic biology of industrially important *Aspergillus* fungi. Adv Appl Microbiol 100:161–202

Park J, Shin H, Lee SM, Um Y, Woo HM (2018) RNA-guided single/double gene repressions in *Corynebacterium glutamicum* using an efficient CRISPR interference and its application to industrial strain. Microb Cell Factories 17:4. https://doi.org/10.1186/s12934-017-0843-1

Parvez A, Giri S, Giri GR, Kumari M, Bisht R, Saxena P (2018) Novel type III polyketide synthases biosynthesize methylated polyketides in *Mycobacterium marinum*. Sci Rep 8:6529. https://doi.org/10.1038/s41598-018-24980-1

Patel DA, Patel AC, Nolan WC, Zhang Y, Holtzman MJ (2012) High throughput screening for small molecule enhancers of the interferon signaling pathway to drive next-generation antiviral drug discovery. PLoS One 7:e36594. https://doi.org/10.1371/journal.pone.0036594

Patel N, Desai P, Patel N, Jha A, Gautam HK (2014) Agronanotechnology for plant fungal disease management: a review. Int J Curr Microbiol App Sci 3:71–84

Patridge E, Gareiss P, Kinch MS, Hoyer D (2016) An analysis of FDA-approved drugs: natural products and their derivatives. Drug Discov Today 21:204–207

Pawar SV, Ho JC, Yadav GD, Yadav VG (2017) The impending renaissance in discovery and development of natural products. Curr Top Med Chem 17:251–267

Paytubi S, de La Cruz M, Tormo JR, Martín J, González I, González-Menendez V, Genilloud O, Reyes F, Vicente F, Madrid C, Balsalobre C (2017) A High-throughput screening platform of microbial natural products for the discovery of molecules with antibiofilm properties against *Salmonella*. Front Microbiol 8:326. https://doi.org/10.3389/fmicb.2017.00326

Peiru S, Menzella HG, Rodriguez E, Carney J, Gramajo H (2005) Production of the potent antibacterial polyketide Erythromycin C in *Escherichia coli*. Appl Environ Microbiol 71:2539–2547

Pereira AES, Grillo R, Mello NFS, Rosa AH, Fraceto LF (2014) Application of poly (epsilon-caprolactone) nanoparticles containing atrazine herbicide as an alternative technique to control weeds and reduce damage to the environment. J Hazard Mater 268:207–215. https://doi.org/10.1016/j.jhazmat.2014.01.025
Pérez-Victoria I, Martín J, Reyes F (2016) Combined LC/UV/MS and NMR strategies for the dereplication of marine natural products. Planta Med 82:857–871
Perotto S, Angelini P, Bianciotto V, Bonfante P, Girlanda M, Kull T, Mello A, Pecoraro L, Perini C, Persiani MA, Saitta A, Sarrocco S, Vannacci G, Venanzoni R, Venturella G, Selosse MA (2013) Interactions of fungi with other organisms. Plant Biosyst 147:208–218
Pfeifer BA, Admiraal SJ, Gramajo H, Cane DE, Khosla C (2001) Biosynthesis of complex polyketides in a metabolically engineered strain of *E. coli*. Science 291:1790e2
Piazza I, Kochanowski K, Cappelletti V, Fuhrer T, Noor E, Sauer U, Picotti P (2018) A map of protein-metabolite interactions reveals principles of chemical communication. Cell 172:358–372.e23. https://doi.org/10.1016/j.cell.2017.12.006
Pink J, Pirmohamed M, Lane S, Hughes DA (2014) Cost-effectiveness of pharmacogenetics guided warfarin therapy vs alternative anticoagulation in atrial fibrillation. Clin Pharmacol Ther 95:199–207
Porras-Alfaro A, Bayman P (2011) Hidden fungi, emergent properties: endophytes and microbiomes. Annu Rev Phytopathol 49:291–315
Prabhu G, Agarwal S, Sharma V, Madurkar SM, Munshi P, Singh S, Sen S (2015) A natural product based DOS library of hybrid systems. Eur J Med Chem 95:41–48
Priest BT, Erdemli G (2014) Phenotypic screening in the 21st century. Front Pharmacol 5:264. https://doi.org/10.3389/fphar.2014.00264
Pusztahelyi T, Holb IJ, Pócsi I (2015) Secondary metabolites in fungus-plant interactions. Front Plant Sci 6:573. https://doi.org/10.3389/fpls.2015.00573
Pusztahelyi T, Holb IJ, Pócsi I (2016) Plant-fungal interactions: special secondary metabolites of the biotrophic, necrotrophic, and other specific interactions. In: Mérillon JM, Ramawat K (eds) Fungal Metabolites. Reference series in phytochemistry. Springer, Cham, pp 1–58
Qi BL, Liu P, Wang QY, Cai WJ, Yuan BF, Feng YQ (2014) Derivatization for liquid chromatography mass spectrometry. TrAC Trends Anal Chem 59:121–132
Raghavendra MP, Nayaka SC, Gupta VK (2016) Microbial enzymes for conversion of biomass to bioenergy. In: Gupta V (ed) Microbial enzymes in bioconversions of biomass. Biofuel and biorefinery technologies, vol 3. Springer, Cham, pp 1–26
Rakers C, Najnin RA, Polash AH, Takeda S, Brown JB (2018) Chemogenomic active learning's domain of applicability on small, sparse qHTS matrices: a study using cytochrome P450 and nuclear hormone receptor families. Chem Med Chem 13:511–521. https://doi.org/10.1002/cmdc.201700677
Rao KJ, Paria S (2013) Use of sulfur nanoparticles as a green pesticide on *Fusarium solani* and *Venturia inaequalis* phytopathogens. RSC Adv 3:10471–10478. https://doi.org/10.1039/c3ra40500a
Ray L, Moore BS (2016) Recent advances in the biosynthesis of unusual polyketide synthase substrates. Nat Prod Rep 33:150–161. https://doi.org/10.1039/c5np00112a
Rédou V, Navarri M, Meslet-Cladière L, Barbier G, Burgaud G (2015) Species richness and adaptation of marine fungi from deep-subseafloor sediments. Appl Environ Microbiol 81:3571–3583. https://doi.org/10.1128/AEM.04064-14
Rejc Ž, Magdevska L, Tršelič T, Osolin T, Vodopivec R, Mraz J, Pavliha E, Zimic N, Cvitanović T, Rozman D, Moškon M, Mraz M (2017) Computational modelling of genome-scale metabolic networks and its application to CHO cell cultures. Comput Biol Med 88:150–160
Rhodes CJ (2014) Mycoremediation (bioremediation with fungi)-growing mushrooms to clean the earth. Chem Speciat Bioavailab 26:196–198. https://doi.org/10.3184/095422914X14047407349335
Richards TA, Dacks JB, Campbell SA, Blanchard JL, Foster PG, McLeod R, Roberts CW (2006) Evolutionary origins of the eukaryotic shikimate pathway: gene fusions, horizontal gene transfer, and endosymbiotic replacements. Eukaryot Cell 5:1517–1531

Richter L, Wanka F, Boecker S, Storm D, Kurt T, Vural Ö, Süßmuth R, Meyer V (2014) Engineering of *Aspergillus niger* for the production of secondary metabolites. Fungal Biol Biotechnol 14:1–4

Robles O, Romo D (2014) Chemo-and site-selective derivatizations of natural products enabling biological studies. Nat Prod Rep 31:318–334

Rohles CM, Gießelmann G, Kohlstedt M, Wittmann C, Becker J (2016) Systems metabolic engineering of *Corynebacterium glutamicum* for the production of the carbon-5 platform chemicals 5-aminovalerate and glutarate. Microb Cell Factories 15:154. https://doi.org/10.1186/s12934-016-0553-0

Rohmer M (1999) The discovery of a mevalonate-independent pathway for isoprenoid biosynthesis in bacteria, algae and higher plants. Nat Prod Rep 16:565–574

Rothbauer M, Charwat V, Ertl P (2016) Cell microarrays for biomedical applications. In: Li P, Sedighi A, Wang L (eds) Microarray Technology. Methods in molecular biology, vol 1368. Humana Press, New York, pp 273–291

Roy A (2018) Early probe and drug discovery in academia: a minireview. High Throughput 7:4. https://doi.org/10.3390/ht7010004

Ruiz-Torres V, Encinar JA, Herranz-López M, Pérez-Sánchez A, Galiano V, Barrajón-Catalán E, Micol V (2017) An updated review on marine anticancer compounds: the use of virtual screening for the discovery of small-molecule cancer drugs. Molecules 22:E1037. https://doi.org/10.3390/molecules22071037

Rukshana BS, Tamilselvi KS (2016) Endophytes are plant helpers: an overview. Int J Curr Microbiol App Sci 5:424–436

Ryan SL, Barid AM, Vaz G, Urguhart AJ, Senge M, Richard DJ, O'Byme KJ, Davies AM (2016) Drug discovery approaches utilizing three-dimensional cell culture. Assay Drug Dev Technol 14:19–28

Sabhachandani P, Motwani V, Cohen N, Sarkar S, Torchilin V, Konry T (2016) Generation and functional assessment of 3D multicellular spheroids in droplet based microfluidics platform. Lab Chip 16:497–505

Sacan A, Ekins S, Kortagere S (2012) Applications and limitations of *in silico* models in drug discovery. In: Larson R (ed) Bioinformatics and drug discovery. Methods in molecular biology (Methods and Protocols), vol 910. Humana Press, Totowa, pp 87–124

Saikkonen K, Young CA, Helander M, Schardl CL (2016) Endophytic Epichloë species and their grass hosts: from evolution to applications. Plant Mol Biol 90:665–675. https://doi.org/10.1007/s11103-015-0399-6

Sakamoto S, Hatakeyama M, Ito T, Handa H (2012) Tools and methodologies capable of isolating and identifying a target molecule for a bioactive compound. Bioorg Med Chem 20:1990–2001

Salehi B, Ayatollahi SA, Segura-Carretero A, Kobarfard F, Contreras MDM, Faizi M, Sharifi-Rad M, Tabatabai SA, Sharifi-Rad J (2017) Bioactive chemical compounds in *Eremurus persicus* (Joub. & Spach) Boiss. essential oil and their health implications. Cell Mol Biol 63:1–7. https://doi.org/10.14715/cmb/2017.63.9.1

Salo OV, Ries M, Medema MH, Lankhorst PP, Vreeken RJ, Bovenberg RAL, Driessen AJM (2015) Genomic mutational analysis of the impact of the classical strain improvement program on β–lactam producing *Penicillium chrysogenum*. BMC Genomics 16:937. https://doi.org/10.1186/s12864-015-2154-4

Samin G, Janssen DB (2012) Transformation and biodegradation of 1,2,3-trichloropropane (TCP). Environ Sci Pollut Res Int 19:3067–3078

Sanchez JF, Somoza AD, Keller NP, CCC W (2012) Advances in *Aspergillus* secondary metabolite research in the post-genomic era. Nat Prod Rep 29:351–371

Sànchez Nogué V, Karhumaa K (2015) Xylose fermentation as a challenge for commercialization of lignocellulosic fuels and chemicals. Biotechnol Lett 37:761–772

Santoyo G, Moreno-Hagelsieb G, Orozco-Mosqueda Mdel C, Glick BR (2016) Plant growth-promoting bacterial endophytes. Microbiol Res 183:92–99

Sarasan M, Puthumana J, Job N, Han J, Lee JS, Philip R (2017) Marine algicolous endophytic fungi-A promising drug resource of the era. J Microbiol Biotechnol 27:1039–1052

Sarker SD, Nahar L (2012a) An introduction to natural products isolation. In: Sarker S, Nahar L (eds) Natural products isolation. Methods in molecular biology (Methods and Protocols), vol 864. Humana Press, New York, pp 1–25. https://doi.org/10.1007/978-1-61779-624-1_1

Sarker SD, Nahar L (2012b) Hyphenated techniques and their applications in natural products analysis. In: Sarker S, Nahar L (eds) Natural products isolation. Methods in molecular biology (Methods and Protocols), vol 864. Humana Press, New York, pp 301–340. https://doi.org/10.1007/978-1-61779-624-1_12

Scharf DH, Heinekamp T, Brakhage AA (2014) Human and plant fungal pathogens: the role of secondary metabolites. PLoS Pathog 10:e1003859. https://doi.org/10.1371/journal.ppat.1003859

Sebastianes FLS, de Azevedo JL, Lacava PT (2017) Diversity and biotechnological potential of endophytic microorganisms associated with tropical mangrove forests. In: de Azevedo J, Quecine M (eds) Diversity and benefits of microorganisms from the tropics. Springer, Cham, pp 37–56. https://doi.org/10.1007/978-3-319-55804-2_3

Sengupta S, Ganguli S, Singh PK (2017) Metagenome analysis of the root endophytic microbial community of Indian rice (*Oriza sativa* L). Genome Data 12:41–43

Serpe L, Canaparo R, Foglietta F, Zara GP (2013) Innovative formulations for the controlled and site-specific delivery of antiinflammatory drugs. Curr Pharm Des 19:7219–7236

Servin A, Elmer W, Mukherjee A, Torre-Roche R, Hamdi H, White JC, Bindraban P, Dimkpa C (2016) A review of the use of engineered nanomaterials to suppress plant disease and enhance crop yield. J Nanopart Res 17:92. https://doi.org/10.1007/s11051-015-2907-7

Shabaruddin FH, Fleeman ND, Payne K (2015) Economic evaluations of personalized medicine: existing challenges and current developments. Pharmaco Econ Personal Med 8:115–126

Shanks E, Ketteler R, Ebner D (2015) Academic drug discovery within the United Kingdom: a reassessment. Nat Rev Drug Discov 14:510–513

Shaw DL (2017) Is open science the future of drug development. Yale J Biol Med 90:147–151

Shen B (2015) A new golden age of natural products drug discovery. Cell 163:1297–1300

Shevlin M (2017) Practical high-throughput experimentation for chemists. ACS Med Chem Lett 8:601–607

Shi S, Zhao H (2017) Metabolic engineering of oleaginous yeasts for production of fuels and chemicals. Front Microbiol 8:2185. https://doi.org/10.3389/fmicb.2017.02185

Shrestha A, Bajracharya AM, Subedi H, Turha RS, Kafle S, Sharma S, Neupane S, Chaudhary DK (2017) Multi-drug resistance and extended spectrum beta lactamase producing gram negative bacteria from chicken meat in Bharatpur metropolitan, Nepal. BMC Res Notes 10:574. https://doi.org/10.1186/s13104-017-2917-x

Simoes-Pires C, Hostettmann K, Haouala A, Cuendet M, Falquet J, Graz B, Christen P (2014) Reverse pharmacology for developing an anti-malarial phytomedicine. The example of *Argemone mexicana*. Int J Parasitol Drugs Drug Resist 4:338–346

Singh M, Kumar A, Singh R, Pandey KD (2017) Endophytic bacteria: a new source of bioactive compounds. 3Biotech 7:315. https://doi.org/10.1007/s13205-017-0942-z

Smanski MJ, Zhou H, Claesen J, Shen B, Fischbach MA, Voigt CA (2016) Synthetic biology to access and expand nature's chemical diversity. Nat Rev Microbiol 14:135–149

Soliman S, Tang Y (2015) Natural and engineered production of taxadiene with taxadiene synthase. Biotechnol Bioeng 112:229–235. https://doi.org/10.1002/bit.25468

Sorensen A, Lubeck PS, Lubeck M, Nielsen KF, Ahring BK, Teller PJ, Frisvad JC (2011) *Aspergillus saccharolyticus* sp. nov., a black *Aspergillus* species isolated in Denmark. Int J Syst Evol Microbiol 61:3077–3083

Spasic J, Mandic M, Djokic L, Nikodinovic-Runic J (2018) Streptomyces spp. in the biocatalysis toolbox. Appl Microbiol Biotechnol 102:3513–3536

Specian V, Sarragiotto MH, Pamphile JA, Clemente E (2012) Chemical characterization of bioactive compounds from the endophytic fungus *Diaporthe helianthus* isolated from *Luehea divaricata*. Braz J Microbiol 43:1174–1182

Sridhar B, Karthik R, Pushpam AC, Vanitha MC (2015) Production and purification of biosurfactants from *Saccharomyces cerevisiae* and *Pseudomonas aeruginosa* and its application on fruit salads. Int J Adv Res Eng Technol 6:97–104

Stierle AA, Stierle DB (2015) Bioactive secondary metabolites produced by the fungal endophytes of conifers. Nat Prod Commun 10:1671–1682

Strobel G (2018) The emergence of endophytic microbes and their biological promise. J Fungi 4:57. https://doi.org/10.3390/jof4020057

Swamy MK, Akthar MS, Sinniah UR (2016a) Root exudates and their molecular interactions with rhizospheric microbes. In: Hakeem KR, and Akhtar MS (eds), Plant, soil and microbes: mechanisms and molecular interactions. Springer Cham, pp 59–77

Swamy MK, Akthar MS, Sinniah UR (2016b) Response of PGPR and AM Fungi toward growth and secondary metabolite production in medicinal and aromatic plants. In: Hakeem KR, Akhtar MS (eds) Plant, soil and microbes: mechanisms and molecular interactions. Springer, Cham, pp 147–170

Szymański P, Markowicz M, Mikiciuk-Olasik E (2012) Adaptation of high-throughput screening in drug discovery-toxicological screening tests. Int J Mol Sci 13:427–52. https://doi.org/10.3390/ijms13010427

Tawfike AF, Viegelmann C, Edrada-Ebel R (2013) Metabolomics and dereplication strategies in natural products. In: Roessner U, Dias D (eds) Metabolomics tools for natural product discovery. Methods in molecular biology (Methods and Protocols), vol 1055. Humana Press, Totowa, pp 227–244

Teotia P, Kumar M, Varma A, Kumar V (2016) Molecular tools for strain improvement in *Aspergillus*. In: Gupta VG (ed) New and future developments in microbial biotechnology and bioengineering. Elsvier, Amsterdam, pp 119–124

Terfehr D, Dahlmann TA, Kück U (2017) Transcriptome analysis of the two unrelated fungal β-lactam producers *Acremonium chrysogenum* and *Penicillium chrysogenum*: velvet-regulated genes are major targets during conventional strain improvement programs. BMC Genomics 18:272. https://doi.org/10.1186/s12864-017-3663-0

Thatoi H, Behera BC, Mishra RR, Dutta SK (2013) Biodiversity and biotechnological potential of microorganisms from mangrove ecosystems: a review. Ann Microbiol 63:1–19. https://doi.org/10.1007/s13213-012-0442-7

Thomas CJ1, Auld DS, Huang R, Huang W, Jadhav A, Johnson RL, Leister W, Maloney DJ, Marugan JJ, Michael S, Simeonov A, Southall N, Xia M, Zheng W, Inglese J, Austin CP (2009) The pilot phase of the NIH chemical genomics center. Curr Top Med Chem 9:1181–1193

Thul ST, Sarangi BK (2015) Implications of nanotechnology on plant productivity and its rhizospheric environment. In: Siddiqui MH, Al-Whaibi M, Mohammad F (eds) Nanotechnology and plant sciences. Springer, Cham, pp 37–53. https://doi.org/10.1007/978-3-319-14502-0_3

Tobert JA (2003) Lovastatin and beyond: the history of the HMG-CoA reductase inhibitors. Nat Rev Drug Discov 2:517–526

Tohge T, Watanabe M, Hoef-gen R, Fernie AR (2013) Shikimate and phenylalanine biosynthesis in the green lineage. Front Plant Sci 4:62. https://doi.org/10.3389/fpls.2013.00062

Trenchard IJ, Smolke CD (2015) Engineering strategies for the fermentative production of plant alkaloids in yeast. Metab Eng 30:96–104

Trott D (2013) β-Lactam resistance in gram-negative pathogens isolated from animals. Curr Pharm Des 19:239–249

Turner TL, Kim H, Kong II, Liu JJ, Zhang GC, Jin YS (2018) Engineering and evolution of *Saccharomyces cerevisiae* to produce biofuels and chemicals. Adv Biochem Eng Biotechnol 162:175–215

Umemura M, Koike H, Nagano N, Ishii T, Kawano J, Yamane N, Kozone I, Horimoto K, Kazuo S, Asai K, Yu J, Bennett JW, Machida M (2013) MIDDAS-M: Motif-independent de novo detection of secondary metabolite gene clusters through the integration of genome sequencing and transcriptome data. PLoS One 8:e84028. https://doi.org/10.1371/journal.pone.0084028

Uzma F, Mohan CD, Hashem A, Konappa NM, Rangappa S, Kamath PV, Singh BP, Mudili V, Gupta VK, Siddaiah CN, Chowdappa S, Alqarawi AA, Abdallah EF (2018) Endophytic fungi-alternative sources of cytotoxic compounds: a review. Front Pharmacol 9:309. https://doi.org/10.3389/fphar.2018.00309

Vaidya ADB (2014) Reverse pharmacology-a paradigm shift for drug discovery and development. Curr Res Drug Discov 1:39–44

van den Berg MA (2011) Impact of the *Penicillium chrysogenum* genome on industrial production of metabolites. Appl Microbiol Biotechnol 92:45–53

van der Voort M, Meijer HJG, Schmidt Y, Watrous J, Dekkers E, Mendes R, Dorrestein PC, Gross H, Raaijmakers JM (2015) Genome mining and metabolic profiling of the rhizosphere bacterium *Pseudomonas* sp. SH-C52 for antimicrobial compounds. Front Microbiol 6:693. https://doi.org/10.3389/fmicb.2015.00693

Vannier N, Mony C, Bittebière AK, Vandenkoornhuyse P (2015) Epigenetic mechanisms and microbiota as a toolbox for plant phenotypic adjustment to environment. Front Plant Sci 6:1159. https://doi.org/10.3389/fpls.2015.01159

Vasundhara M, Kumar A, Reddy MS (2016) Molecular approaches to screen bioactive compounds from endophytic fungi. Front Microbiol 7:1774. https://doi.org/10.3389/fmicb.2016.01774

Veiga T, Gombert AK, Landes N, Verhoeven MD, Kiel JA, Krikken AM, Nijland JG, Touw H, Luttik MA, van der Toorn JC, Driessen AJ, Bovenberg RA, van den Berg MA, van der Klei IJ, Pronk JT, Daran JM (2012) Metabolic engineering of β-oxidation in *Penicillium chrysogenum* for improved semi-synthetic cephalosporin biosynthesis. Metab Eng 14:437–448. https://doi.org/10.1016/j.ymben.2012.02.004

Verjans ET, Doijen J, Luyten W, Landuyt B, Schoofs L (2018) Three-dimensional cell culture models for anticancer drug screening: worth the effort. J Cell Physiol 233:2993–3003

Vesterlund SR, Helander M, Faeth SH, Hyvönen T, Saikkonen K (2011) Environmental conditions and host plant origin override endophyte effects on invertebrate communities. Fungal Divers 47:109–118

Vesth TC, Brandl J, Andersen MR (2016) Fun gene cluster S: predicting fungal gene clusters from genome and transcriptome data. Synth Syst Biotechnol 1:122–129

Vidal-Diez de Ulzurrun G, Hsueh YP (2018) Predator-prey interactions of nematode-trapping fungi and nematodes: both sides of the coin. Appl Microbiol Biotechnol 102:3939. https://doi.org/10.1007/s00253-018-8897-5

Visser U, Abeyruwan S, Vempati U, Smith RP, Lemmon V, Schürer SC (2011) BioAssay Ontology (BAO): a semantic description of bioassays and high-throughput screening results. BMC Bioinformatics 12:257. https://doi.org/10.1186/1471-2105-12-257

Vrij EJ, Espinoza S, Heilig M, Kolew A, Schneider M, van Blitterswijk CA, Truckenmüller RK, Rivron NC (2016) 3D high throughput screening and profiling of embryoid bodies in thermoformed microwell plates. Lab Chip 16:734–742

Vurro M, Boari A, Casella F, Zonno MC (2018) Fungal phytotoxins in sustainable weed management. Curr Med Chem 25:268–286

Walker A, Turner S, Johnson R (2015) Pharma and biotech valuations: divergent perspectives. Bus Dev Licens J 22:10–13

Wallwey C, Li SM (2011) Ergot alkaloids: structure diversity, biosynthetic gene clusters and functional proof of biosynthetic genes. Nat Prod Rep 28:496–510

Wang B, Hu Q, Zhang Y, Shi R, Chai X, Liu Z, Shang X, Zhang Y, Wen T (2018a) A RecET-assisted CRISPR-Cas9 genome editing in *Corynebacterium glutamicum*. Microb Cell Factories 17:63. https://doi.org/10.1186/s12934-018-0910-2

Wang J, Shen X, Yuan Q, Yan Y (2018b) Microbial synthesis of pyrogallol using genetically engineered *Escherichia coli*. Metab Eng 45:134–141

Wang L, Dash S, Ng CY, Maranas CD (2017) A review of computational tools for design and reconstruction of metabolic pathways. Synth Syst Biotechnol 2:243–252

Wang S, Su R, Nie S, Sun M, Zhang J, Wu D, Moustaid-Moussa N (2014) Application of nanotechnology in improving bioavailability and bioactivity of diet-derived phytochemicals. J Nutr Biochem 25:363–376

Wang Y, Boghigian BA, Pfeifer BA (2007) Improving heterologous polyketide production in *Escherichia coli* by overexpression of an S-adenosylmethionine synthetase gene. Appl Environ Microbiol 77:367–373

Weber SS, Polli F, Boer R, Bovenberg RAL, Driessen AJM (2012) Increased penicillin production in *Penicillium chrysogenum* production strains via balanced over expression of isopenicillin N acyltransferase. Appl Environ Microbiol 78:7107–7113

Weber T, Blin K, Duddela S, Krug D, Kim HU, Bruccoleri R, Lee SY, Fischbach MA, Müller R, Wohlleben W, Breitling R, Takano E, Medema MH (2015) AntiSMASH 3.0-a comprehensive resource for the genome mining of biosynthetic gene clusters. Nucleic Acids Res 43:W237–W243

Weber T, Kim HU (2016) The secondary metabolite bioinformatics portal: computational tools to facilitate synthetic biology of secondary metabolite production. Synth Syst Biotechnol 1:69–79

Wiemann P, Guo CJ, Palmer JM, Sekonyela R, Wang CCC, Keller NP (2013) Prototype of an intertwined secondary-metabolite supercluster. Proc Natl Acad Sci U S A 110:17065–17070

Wisecaver JH, Slot JC, Rokas A (2014) The evolution of fungal metabolic pathways. PLoS Genet 10:e1004816. https://doi.org/10.1371/journal.pgen.1004816

Wolfender JL, Marti G, Thomas A, Bertrand S (2015) Current approaches and challenges for the metabolite profiling of complex natural extracts. J Chromatogr A 1382:136–164

Wriessnegger T, Moser S, Emmerstorfer-Augustin A, Leitner E, Müller M, Kaluzna I, Schürmann M, Mink D, Pichler H (2016) Enhancing cytochrome P450-mediated conversions in *P. pastoris* through RAD52 over-expression and optimizing the cultivation conditions. Fungal Genet Biol 89:114–125. https://doi.org/10.1016/j.fgb.2016.02.004

Wright MH, Sieber SA (2016) Chemical proteomics approaches for identifying the cellular targets of natural products. Nat Prod Rep 33:681–708

Wu CC, Chen TH, Liu BL, Wu LC, Chen YC, Tzeng YM, Hsu SL (2013) Destruxin B isolated from entomopathogenic fungus *Metarhizium anisopliae* induces apoptosis via a Bcl-2 family-dependent mitochondrial pathway in human nonsmall cell lung cancer cells. Evidence-Based Compl Altern Med 2013:548929. https://doi.org/10.1155/2013/54892

Yahr R, Schoch CL, Dentinger BTM (2016) Scaling up discovery of hidden diversity in fungi: impacts of barcoding approaches. Philos Trans R Soc 371:20150336. https://doi.org/10.1098/rstb.2015.0336

Yang W, Soares J, Greninger P, Edelman EJ, Lightfoot H, Forbes S, Bindal N, Beare D, Smith JA, Thompson IR, Ramaswamy S, Futreal PA, Haber DA, Stratton MR, Benes C, McDermott U, Garnett MJ (2013) Genomics of drug sensitivity in cancer (GDSC): a resource for therapeutic biomarker discovery in cancer cells. Nucleic Acids Res 41:D955–D961

Yu G, Liang W, Liu J, Meng D, Wei L, Chai T, Cai Y (2016) Proteomic analysis of differential expression of cellular proteins in response to Avian H9N2 virus infection of A549 Cells. Front Microbiol 7:1962. https://doi.org/10.3389/fmicb.2016.01962

Yuzawa S, Keasling JD, Katz L (2016) Insights into polyketide biosynthesis gained from repurposing antibiotic-producing polyketide synthases to produce fuels and chemicals. J Antibiot 69:494–499

Yuzawa S, Keasling JD, Katz L (2017) Bio-based production of fuels and industrial chemicals by repurposing antibiotic-producing type I modular polyketide synthases: opportunities and challenges. J Antibiot 70:378–385

Zander Balderud L, Murray D, Larsson N, Vempati U, Schürer SC, Bjäreland M, Engkvist O (2015) Using the Bioassay ontology for analyzing high-throughput screening data. J Biomol Screen 20:402–415

Zanders ED (2012) Overview of chemical genomics and proteomics. In: Zanders E (ed) Chemical genomics and proteomics. Methods in molecular biology (Methods and Protocols), vol 800. Humana Press, New York, pp 3–10

Zhai MM, Li J, Jiang CX, Shi YP, Di DL, Crews P, Wu QX (2016) The bioactive secondary metabolites from *Talaromyces* species. Nat Prod Bioprospect 6:1–24

Zhang H, Boghigian BA, Pfeifer BA (2010) Investigating the role of native propionyl-CoA and methylmalonyl-CoA metabolism on heterologous polyketide production in *Escherichia coli*. Biotechnol Bioeng 105:567–e73

Zhang HW, Ying C, Bai XL (2014a) Advancement in endophytic microbes from medicinal plants. Int J Pharm Sci Res 5:1589–1600

Zhang P, Li X, Wang BG (2016) Secondary metabolites from the marine algal-derived endophytic fungi: chemical diversity and biological activity. Planta Med 82:832–842

Zhang S, Zhao X, Tao Y, Lou C (2015) A novel approach for metabolic pathway optimization: oligo-linker mediated assembly (OLMA) method. J Biol Eng 9:23. https://doi.org/10.1186/s13036-015-0021-0

Zhang W, Li Y, Tang Y (2008) Engineered biosynthesis of bacterial aromatic polyketides in *Escherichia coli*. Proc Natl Acad Sci U S A 105:20683–20688

Zhang XY, Tang GL, Xu XY, Nong XH, Qi SH (2014b) Insights into deep-sea sediment fungal communities from the east Indian ocean using targeted environmental sequencing combined with traditional cultivation. PLoS One 9:e109118. https://doi.org/10.1371/journal.pone.0109118

Zhao K, Shao B, Yang D, Li F, Zhu J (2015) Natural occurrence of *Alternaria* toxins in wheat based products and their dietary exposure in China. PLoS One 10:e0132019. https://doi.org/10.1371/journal.pone.0132019

Zhao X, Spraker JE, Bok JW, Velk T, He Z-M, Keller NP (2017) A cellular fusion cascade regulated by *LaeA* is required for sclerotial development in *Aspergillus flavus*. Front Microbiol 8:1925. https://doi.org/10.3389/fmicb.2017.01925

Zhou W, Wang Y, Lu A, Zhang G (2016) Systems pharmacology in small molecular drug discovery. Int J Mol Sci 17:246. https://doi.org/10.3390/ijms17020246

Endophytes: The Unmapped Repository for Natural Products

2

Indira P. Sarethy, Nidhi Srivastava, and Sharadwata Pan

Contents

I. P. Sarethy · N. Srivastava
Department of Biotechnology, Jaypee Institute of Information Technology, Noida, India

S. Pan (✉)
School of Life Sciences Weihenstephan, Technical University of Munich, Freising, Germany
e-mail: sharadwata.pan@tum.de

M. S. Akhtar et al. (eds.), *Natural Bio-active Compounds*,
https://doi.org/10.1007/978-981-13-7154-7_2

Abstract

Endophytes are the microorganisms present within the living tissues of plants. It has been suggested that every plant harbours at least one to two endophytes. However, very few plants have been characterized for their endophytic population, in comparison to their known diversity. A search in PubMed using the keyword 'endophyte' shows that there has been a considerable increase in the number of publications focusing on endophytes, i.e., 32 in the 2000s to ~500 in 2017 and roughly 200 in the first few months of 2018. The trend suggests increasing interest in endophytes as sources of novel metabolites. The pointers in early studies had shown the presence of novel natural compounds produced by endophytes. Indeed, the therapeutic molecules in many plants have been proposed to be produced by endophytes and not the host plants themselves. These molecules have the potential to serve as added resources in the desperate search for bio-active compounds which can combat various diseases and syndromes prevalent today and which are fast-losing effective therapeutics. Their presence in plants growing in diverse habitats adds to their potential for chemodiversity. Modern omics technologies, involving next-generation sequencing, metagenomics and metatranscriptomics, have shown a promise in better understanding of plant-endophyte relationships and can play a significant role in establishing the biosynthetic potential of endophytes. Therefore, bioprospecting for endophytes constitutes an attractive area of research. Thus, the current chapter provides a comprehensive account of these microorganisms as they correlate to various habitats, their role in 'benefit-sharing' with their hosts and the recent technologies which have unveiled their involvement in various aspects of their host plants' lives.

Keywords

Bio-active compounds · Endophytes · Natural products · Novel metabolites · Omics

2.1 Introduction

Many groups of microorganisms interact with plants and can be associated with the outside of the plant (root zone or colonizing the external parts), in which case they are termed rhizospheric, or they can be inside the plant, in a symbiotic harmonious relationship (Dudeja et al. 2012). These microorganisms have a complex interaction system with the host plants, contributing to growth promotion activities, inhibiting pathogenic attacks and facilitating seed germination and nutrient uptake (Hurek et al. 2002; Ryan et al. 2008). Endophytes, microorganisms growing inside the plant body, are increasingly being studied for their significant role in enhancing plant growth, improving the nutrition profile and in defence (Chen et al. 1995; Hallmann et al. 2006). The knowledge that the endophytes are the potentiators of bio-active metabolite production has triggered active interest in involving high-throughput processes in understanding and elucidating the role of these microorganisms. The excellent reviews by Strobel and Daisy (2003) and Strobel et al. (2004) provide in-depth information on the various types of bio-active metabolites obtained from endophytes. Trujillo et al. (2015) have shown that actinobacteria, a group hitherto known as prolific producers of bio-active metabolites, are also major contributors of the plant-microorganism interaction systems.

A look at Fig. 2.1 shows that scientists have evinced growing interest in endophytes, leading to increased volumes of publication outputs (a PubMed survey), involving endophytes. Interestingly, the genomics aspects of endophyte studies have started picking up speed from 2005 onwards, coinciding with the establishment of next-generation sequencing technologies. Furthermore, the studies on endophytes related to bio-active compound production have also kept pace with genomics studies. Keeping in mind the immense potential of endophytes in agriculture, medicine and industry, the current chapter provides a comprehensive account of these microorganisms as they correlate to various habitats, their role in 'benefit-sharing' with their hosts and the recent technologies which have unveiled their involvement in various aspects of their host plants' lives.

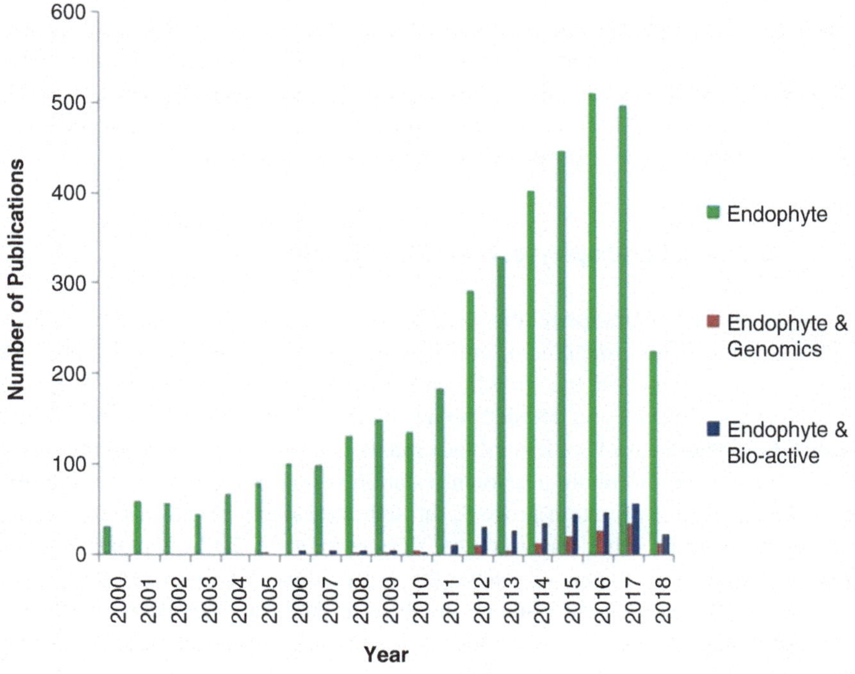

Fig. 2.1 The number of publications related to endophytes as obtained from PubMed

2.2 Evolutionary Progress of Endophytes

Unlike previous accounts of developmental antiquity of endophytes, which predominantly focused on fungi, it is imperative to include the growth and evolution of plant-bacteria relationships, considering the fact that the internal physiological parts of the plant body may as well be proportionately cohabitated by bacterial colonies (Hardoim et al. 2015). For elaborate discussions on the developed track record of endophytes, readers may refer to recent texts (Hardoim et al. 2015; Arora and Ramawat 2017). There are diverse viewpoints over the exact evolutionary timeline of the endophytes. Paleontological evidences indicate a timeline of around 400 million years and earlier for the connotation between endophytes and terrestrial plants (Arora and Ramawat 2017). Alternatively, the first ever evidences of such associations have been traced back to the Ordovician period, with a more specific epoch of 460 million years (Bidartondo et al. 2011). In spite of this ambiguity, the chronological sequence of evolution of endophytes is clear during the development process. While the plants transformed their primary habitats from aquatic to land, harsh environmental conditions such as high proportion of carbon dioxide, temperature instabilities, insufficient soil nutrients, etc. have facilitated a beneficial relationship between the endophytes and their hosts, providing a platform for the hosts to thrive the conditions (Bonfante and Selosse 2010). Substantiations of adaptation of the endophytes to the microenvironment of the plants exist, through genetic alterations and synthesizing metabolic products during the process (Arora and Ramawat 2017). The extensive and systematic evolutionary progression of the endophytes are aptly reflected in their rich diversity, transcending several kinds of habitats and a wide variety of host plants (Sun and Guo 2012). A series of useful references on fungal and bacterial endophytes and their corresponding host plants have been collated by Arora and Ramawat (2017). In the below mentioned sections, the fungal and bacterial associations with their hosts have been discussed in details.

2.2.1 Bacterial Endophyte: Host Plant Evolution

The mechanism of symbiotic relationship between plant and bacterial species, especially from the genus *Rhizobium*, and leguminous plants has been the principal focus to contemplate on the growth and development of bacterial endophytes (Hardoim et al. 2015). For a detailed review on the manifestations of such bacterial colonization onto the roots of leguminous plants, see the review by Kondorosi et al. (2013). Except the morphological manifestation of the rhizobia that are primarily linked to the nitrogen fixation process, through the establishment of root nodules, there seem to be no other physiological development in the host plant, which is mostly devoid of any otherwise noteworthy expressions (Malfanova et al. 2013). It is imperative that several rhizobia are involved in the process, especially affecting monocots such as sugarcane, rice and maize and the contribution of such nitrogen assimilation to the full nitrogen reservoir in case of nonleguminous plant species remains debatable (Reinhold-Hurek and Hurek 2011; Hardoim et al. 2015). Irrespective of such arguments, the extent of nitrogen fertilization and type of the host plant are the most crucial factors which need to be investigated to comprehend the evolutionary progress of the bacterial endophytes (Hoeksema et al. 2010). The route of penetration through the rhizosphere is the most accepted outlook on the origin for the bacterial endophytes (Arora and Ramawat 2017). At this point, the bacterial endophytes have been linked to produce a wide range of enzymes including pectinase and cellulose, which facilitate the disintegration of the cell wall and smoothen the entry route (Wang and Dai 2011). Their gradual migration from roots to the plant interior depends on several factors, and only a handful can survive the process to ultimately progress to the xylem tissues (Arora and Ramawat 2017).

2.2.2 Fungal Endophyte: Host Plant Evolution

The understanding of evolution of fungal endophyte and plant symbiosis is mainly based on two perspectives: horizontal and vertical communications. While the vertical transmission of endophytes targets a transfer amongst generations through seeds, horizontal transmission refers to the same, however, mediated via soil or deteriorated plant remains (Herrera et al. 2016). According to Bidartondo et al. (2011), the first line of communications between the ancient group of land plants in the Ordovician epoch and fungal endophytes was orchestrated by the fungal genus *Glomeromycota*. It is true that different classes of fungi have been known to be in the interaction mode with higher plant families. However, the indications corresponding to the genetic associations between arbuscular mycorrhizal fungi (AMF) in the plants with nodular beneficial characteristics point towards a coevolution of fungal endophytic populations in conjugation with their hosts (Hardoim et al. 2015). This mutual preferentialism expedited even more complex degrees of growth and development for the fungal endophytes and their hosts, provided multiple perpetrators took part in the process (Frey-Klett et al. 2011). The aforementioned primeval plant-AMF symbiotic evidences point towards a subsequent and wider evolutionary progress for the later generations of endophytes and hosts (Hardoim et al. 2015). However, Brundrett (2002) commented on several contextual aspects, including presence or absence of robust communication between plant and AMF initially; evolution of exo- and endodermis, tissues that favoured foreign fungal cohabitation; and a gradual but steady development of arbuscules during the process. It was hinted that even more severe forms of plant-fungal relations developed with the evolutionary growth, which resulted in plant species that completely utilize their fungal equivalents throughout the communication process (Hardoim et al. 2015).

2.3 Give and Take Relationship Between the Host and Endophytes

The beneficial relationship between the host plants and the endophytes has been documented in detail in recent reviews (Rudgers et al. 2010; Hardoim et al. 2015; Jia et al. 2016; Pinto-Carbó et al. 2018; Rho et al. 2018a). There is a great deal of advantages conferred by the endophytic fungi on their hosts, by enhancing the secretion of plant growth regulators such as auxins and cytokinins, which grossly impact their general growth and physiological development (Jia et al. 2016). A range of endophytic fungi has demonstrated benefits including enhanced mineral and nutrient assimilation, seed sprouting and manifestation of distinct enzymatic actions including expression of genetic factors (Behie et al. 2012; Jia et al. 2016). There are multitude of factors and outlooks based on which general observations have been made and inferred to reach rational conclusions. Past and recent studies, both original research articles and reviews, suffice to that, including insightful discussions on the underlying mechanisms and relevant applications. Undoubtedly, benefits reach both ends, i.e., hosts and endophytes, which have facilitated their progressive evolution, as outlined in the previous section. In this section, we relook at the aspect with selected perspectives.

2.3.1 Biotic and Abiotic Stress Control

A range of defence reactions attributed to endophytic fungi to their corresponding host plants with significant pharmaceutical properties have been collated recently (Jia et al. 2016). The authors have attributed a variety of stress sustenance methods to endophytic fungi including salt and drought stress, insect-repelling features, protection from foreign pathogens as well as tolerance against heavy metals and heat tolerance. The endophytic fungi adopt various dissimilar mechanisms while countering these biotic and abiotic stresses, thereby conferring shielding actions to the host plants. For instance, while the fortification against salt stress has been linked to the enhanced solvable protein amount (Liu et al. 2011), the guard during drought conditions is attributed to the enhanced nutrient assimilation and modification of host metabolic activities (Meng and He 2011). Endophytic fungi are known to synthesize insect-specific toxic compounds which help their host plants to repel insects (Gange et al. 2012). The stress mediated by thermal fluctuations and heat is mainly countered by endophytic actions that help moderate the absorption of amino acids (Khan et al. 2013). By forbidding the migration of electron from the quinone receiver QA to QB, host plants are able to fight the ill effects of metal ions such as Pb^{2+} (Li and Zhang 2015). Endophytic fungi have also been known to offer biotic stress control by secreting toxic pathogens that prove to be fatal for certain disease-causing fungi (Yang et al. 2012; Wang et al. 2012). Protection against abiotic stresses is mainly linked with the ability of the endophytic fungi to synthesize antioxidant products (Jia et al. 2016). Very recently, Rho et al. (2018a) have reported meta-analysis on alleviation of a range of stress types in a wide range of plants by endophytic actions. Their outcomes highlight the superiority of the endophytes, without any noticeable host-endophyte selectivity, in countering saline stress, nitrogen insufficiency and famine.

2.3.2 Modification of Group Configuration

Although several studies and reports are available targeting symbiotic relationship between host plants and endophytes on case by case accounts, the effects of endophytic associations with their hosts on a community level are hitherto less explored. Rudgers et al. (2010) mitigated this lacuna with a detailed and prolonged study on the conformation of the community of fescue grass belonging to the genus, *Lolium*, as a consequence of alteration of the genotypes of the foliar fungus belonging to the genus, *Neotyphodium*. The authors concluded that the variations in the genotypes of the endophytes may demonstrate flowing consequences on the constitution of the host groups. In addition to its preservation and organization worth, the work also demonstrates direct influence of endophyte genotype modifications on their host groups. Lately, Paungfoo-Lonhienne et al. (2015) outlined a wide variety (~15) of fungal genera which do not depend on culture environments, by examining the influence of the quantity of nitrogen-containing manures on the fungal community harbouring soil and root environments in sugarcanes. The authors reported that the constitution of the fungal group is greatly altered as a function of the fertilizer amount and that with increasing concentration, although it does not affect the fungal taxonomic assortment to a great extent. Additionally, there lies a direct, negative impact on the environment, including evidences in favour of manifestation of pathogenic fungal strains. Recently, Glynou et al. (2017) have linked a higher degree of ecological acclimatization and distribution for the root endophytic fungi to their genotypic multiplicity. The authors reported that environmental selection has little to do with this well-organized dissemination trend and that the resident group congregation is more dependent on specific biotic communications.

2.3.3 Foliage Nodular Interdependence

A significant number of review article and studies have appeared till date, targeting root nodule endophytic fungi and their beneficial relationship with the host plants, as described previously. However, careful profiling of bacterial endophytes and detailed reports on their symbiotic accounts on the leaf nodules are scarce. As outlined earlier, much of the recent attention have been centred around the root nodules and advantages on host plants through fungal endophyte settlement. However, nodules may as well be associated with leaves, which have been discovered in both mono- (family, Dioscoreaceae) and dicotyledonous (families, Primulaceae and Rubiaceae) plants, where a symbiotic association is developed with the bacterial endophytes in the form of lifelong colonization and synthesis of secondary products with active biological characteristics (Lemaire et al. 2012a; Carlier et al. 2017; Pinto-Carbó et al. 2018). The distinctiveness of the symbiont and the communication approach have also been studied, with studies reporting a taxonomic affiliation for the bacteria to the genus *Burkholderia*, and a ‘vertical’ communication procedure for majority of general host-regulated leaf nodule associations, thereby demonstrating an enhanced host specificity by the perpetrating bacterial species (Lemaire et al. 2011, 2012b; Carlier and Eberl 2012; Carlier et al. 2016; Pinto-Carbó et al. 2016). A variety of communication routes and symbiotic means have been reported between bacterial endophytes and the role of foliar nodes in the host plants including plant hormone secretions, iron absorption, bio-active metabolite synthesis, scavenging of free radicals and general host identification and interactions. For a detailed collation of recent and useful references, see the review by Pinto-Carbó et al. (2018). According to the authors, leaf nodule symbiosis is poised to be a plush foundation for a wide array of solicitations in sustainable agriculture and biopharmaceutical domains.

2.4 Habitat versus Stress versus Endophytes

Infectious diseases and multidrug-resistant pathogens account for half of the global mortality (Burnham et al. 2018). This has necessitated the urgent need to counter this high mortality by new drug discovery. Natural products have provided the best skeletons for drug molecules. Understudied habitats, which can potentially contribute to novel bio-active metabolites, are being screened in large numbers. Endophytes represent one such niche habitat. In turn, the habitats where the hosts of the endophytes live also contribute significantly to the type of products and functioning of the ecosystem as a whole. Endophytic diversity plays a key role in managing nutritional requirements of associated host plants and providing resistance to biotic and abiotic stress. The survival of the host can be extended in extreme environments like hot and cold deserts or those with nutrient limitations. The resident endophytes influence the host plant metabolic system by several complex mechanisms such as induction of particular genes of host under stress conditions, bio-active metabolite production, plant growth promotion via production of phytohormones, root morphology modulation, facilitation of nutrient uptake, regulation of osmotic pressure and physiological/functional modulation in transpiration (Fang et al. 2013; Verma et al. 2017).

Endophytes that facilitate growth of the host plant have been experimentally shown to offer protection against various extremes of stresses such as flooding, drought, temperature, salinity, pathogen attacks, nematode infection and metal and organic chemical contaminants (Glick 2015). Abiotic stresses are responsible for the imbalance in the water content of the host plant. Endophytes maintain the desired osmotic potential by increasing cell matrix potential and facilitating retention of the water content in the soil and in the host plant. Production of the reactive oxygen species (ROS) is the primary response of the host plant towards stress, leading to an oxidative burst. Endophytes bring the debilitating effects of the oxidative burst under control (Lata et al. 2018). The enzyme 1-aminocyclopropane-1-carboxylate (ACC) deaminase has been shown to play a vital role in stress alleviation. Increasing ethylene levels directly contribute to stress effects. ACC exuded by plant cells is taken up by the endophytes and cleaved into α-ketobutyrate and ammonia, bypassing its conversion into ethylene within plant cells, resulting in less exposure to ethylene and thereby growth inhibition (Vurukonda et al. 2016).

Phosphate is also necessary for the plant for its metabolism. Phosphate is present in insoluble form and remains attached to the organic group. Endophytic bacteria or fungi lower the pH of the soil by releasing organic phosphates, or they solubilize inorganic phosphates by the production of phytases, C-P lyases and phosphonatases (Oteino et al. 2015). In this section, we provide a brief outline of the various habitats where endophytes have been shown to contribute significantly to host survival.

2.4.1 Desert

Desert is a region of barren land, wherein crop productivity is diminished due to water deficit, salinity, high temperature, nutrient deficiency and undesired levels of soil pH (Li et al. 2017). Plants capable of growing in such locations have high chances of harbouring endophytes which help them withstand these adverse conditions and support in managing/ alleviating the stress. For instance, dark septate endophytes (DSE), isolated from *Gymnocarpos przewalskii*, were inoculated into the root of a xerophytic plant *Ammopiptanthus mongolicus*. Amongst these DSE, two strains (*Phialophora* sp. and *Embellisia chlamydospora*) were shown to enhance calcium and potassium contents. Other strains (*Knufia* sp. and *Leptosphaeria* sp.) were involved in elevating the biomass of *A. mongolicus* (Li et al. 2018) for better survival. Gonzalez-Teuber et al. (2017) reported around hundred fungal endophytes symbiotic in the root of quinoa plants (*Chenopodium quinoa*) from the salt lakes of the Atacama Desert. Many were identified as belonging to *Penicillium*, *Phoma* and *Fusarium* sp. The isolates were found to enhance plant growth and made their host tolerant to abiotic stress.

2.4.2 Freshwater

Freshwater (lake, pond, river or a wetland) is characterized by having a low concentration of salt, unlike oceans. Plants growing along running waters are conferred with unique properties by their endophytes. They may help their host by secreting plant growth-promoting substances. For instance, You et al. (2015) reported a novel endophyte *Aspergillus clavatus* Y2H0002, isolated from the roots of *Nymphoides peltata* obtained from Dalsung wetland. The culture filtrate of this endophyte, when applied on Waito-c rice seedlings, showed growth-promoting activities mediated by gibberellins (GAs, GA1, GA3 and GA4). Rho et al. (2018b) reported that members of endophytes of family *Salicaceae*, when inoculated on rice plants, demonstrated plant growth-promoting activity as well as regulated water stress by modulating stomatal behaviour.

2.4.3 Mountains

Mountains refer to high-elevation areas of Earth with a cold harsh climate and with daily shifts in temperatures which are major constraints to plant growth. Endophytes assist their host plants in resisting unfavourable conditions. For instance, Chowdhury et al. (2017) reported nearly 2000 bacterial endophytes obtained from mountain-cultivated *Panax ginseng* grown in 24 different locations of the Republic of Korea. Out of these 252 bacteria were further analysed for their properties. More than 70% of the isolates showed siderophore production, 47% were associated with phosphate solubilisation, 67% produced IAA-like indole derivatives, 13% showed hydrogen cyanide (HCN) production and 40% demonstrated ß-glucosidase activity.

2.4.4 Grasslands

Grasslands generally are present in tropical/temperate climates. Loss of biodiversity in these habitats occurs due to biotic stress (herbivory and insect-pest infections). Endophytes provide protection to their host populations. For instance, *Achnatherum inebrians* (drunken horse grass) in three sites of Northwest China flourished due to its toxic endophyte (*Epichloë gansuensis*) in comparison with non-endophytic host populations (Yao et al. 2015). Endophytes have also been reported to help their hosts by making them stress-resistant. For instance, Wang et al. (2018) demonstrated that the endophyte *E. gansuensis* made its host, grass (*A. inebrians*), resistant to biotic and abiotic stresses under nitrogen-deficient environment. *E. gansuensis* modulated nitrogen reductase and nitrite reductase, enzymes involved in nitrogen metabolism. Fresh and dry weight of the endophyte-infected leaves was found to be higher than that of noninfected plant leaves. Similar enhancement was also observed with NO_3^-, NH^{4+}, N and P content, glutamine synthetase activity, N accumulation, N utilization and N uptake efficiency.

2.5 Bio-active Metabolites: From Host or Endophyte?

Strobel and Daisy (2003) have shown in their review that one or more endophytes always reside in the more than 400,000 plants in the world. Endophyte fraction is still being estimated, and it is suggested that unexplored areas could represent a vast plethora of microbial diversity, who could be the key players in the ecosystem functioning. Furthermore, the endophytes represent an understudied class of microorganisms. Having shown the promise of producing novel bio-active metabolites, in many cases being the actual producers rather than their respective hosts, high-throughput screening mechanisms are expected to speed up the process of characterizing and identifying novel metabolites. Endophytes are considered as 'chemical synthesizers' (Chutulo and Chalannavar 2018) with many showing excellent therapeutic potential. As seen from Fig. 2.2, various antimicrobial, antiparasitic, anticancer, antioxidants, immunosuppressants and other compounds belonging to multifunctional groups such as alkaloids,

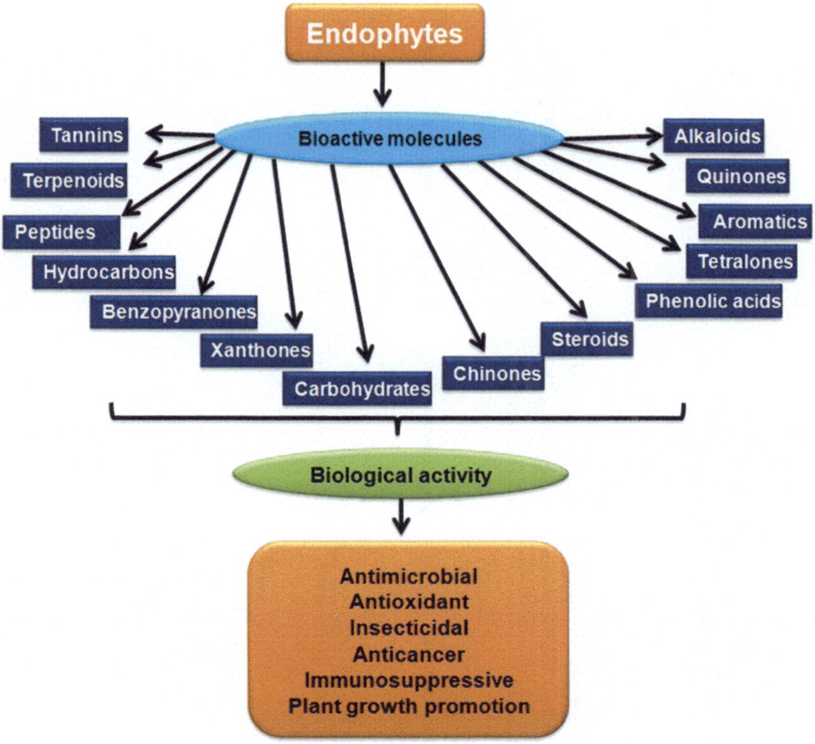

Fig. 2.2 Various bio-active metabolites produced by endophytes

phenolic acids, quinones, steroids, saponins, tannins, terpenoids, tetralones, xanthones, benzopyranones and chinones have been documented from endophytes (Gouda et al. 2016). Antibiosis, competition for nutrients and niches and induced systemic resistance (ISR) are some of the possible mechanisms of biocontrol adopted by the endophytes (Malfanova et al. 2013).

Bio-active compounds from endophytes are increasingly being characterized, and many have shown promising antimicrobial, anticancer, antiparasitic, cytotoxic, insecticidal and antioxidant activities (Table 2.1). Many reviews have provided detailed accounts of bio-active compounds from endophytes, and readers are requested to refer to these (Pimentel et al. 2011; Gouda et al. 2016; Singh et al. 2017). It is interesting to note that many plants with medicinal or therapeutic values are being relooked at in the context of whether it is the host plant itself or its constituent endophyte that is responsible for the bio-active metabolite production. A comprehensive list of the recently identified and characterized bio-active compounds isolated from endophytes has been summarized (Table 2.1).

Table 2.1 Bio-active compounds with therapeutic potential produced by endophytes

Bio-active compounds	Structural moiety	Activity	Endophyte	Host	References
Xylarphthalide A	Phthalide derivative	Antibacterial	*Xylaria* sp. GDG-102	*Sophoraton chinensis*	Zheng et al. (2018)
3-(Sec-butyl)-6-ethyl-4,5-dihydroxy-2-methoxy-6-methylcyclohex-2-enone	Diazole derivative	Antimicrobial	*Phaeophleos poravochysiae*	*Vochysia divergens*	Savi et al. (2018)
Nemanifuranones A	Polyketide	Antimicrobial	*Nemania serpens* (Pers.)	*Vitis vinifera*	Ibrahim et al. (2017)
Rb_1, Rd and 20(S)-Rg_3 Ginsenoside Re	Triterpenoidsaponins	Antimicrobial	*Fusarium* sp. PN8 *Aspergillus* sp. PN17	*Panax notoginseng*	Jin et al. (2017)
(2E,5E)-Phenyltetradeca-2,5-dienoate	Phenolic ester, fatty acid	Antibacterial	*Pseudomonas aeruginosa* strain UICC B-40	*Neesia altissima*	Pratiwi et al. (2017)
(1) 2-Methyl-3-nonyl prodiginine (2) Bis (2-ethylhexyl) phthalate, (3) Preaustinoid A	(1) Pyrrolyldipyrromethene core skeleton (2) Phthalide derivative (3) Meroterpene	Antimicrobial and anticancer	*Epicoccumnigrum*	*Ferula sumbul*	Perveen et al. (2017)
(1) 3′-Hydroxy-5-methoxy-3,4-methylenedioxybiphenyl (2) 3′-Hydroxy-5,5′-dimethoxy-3,4-methylenedioxybiphenyl	Biphenyls	Antibacterial, antioxidant and anticancer	*Streptomyces* sp.	*Boesenbergia rotunda*	Taechowisan et al. (2017)

(1) (R)-2,3-dihydro-2,5-dihydroxy-2-methylchromen-4-one (2) (2R, 4S)-2,3-dihydro-2-methyl-benzopyran-4,5-diol (3) (R)-3-Methoxyl-1-(2,6-(4) dihydroxy phenyl)-butan-1-one, (5) 7-O-α-D-ribosyl-5-hydroxy-2-methyl-4H-chromen-4-one (6) 7-O-α-D-Ribosyl-2,3-dihydro-5-hydroxy-2-methyl-chromen-4-one, daldinium A	Benzopyran derivatives	Antimicrobial activities and glucose uptake stimulating activities	*Daldinia eschscholzii*	*Dendrobium chrysotoxum*	Hu et al. (2017)
(1) Cyclo-(L-Val-L-Pro), (2) Cyclo-(L-Leu-L-Pro), (3) Cyclo-(L-Phe-L-Pro), (4) Cyclo-(L-Val-L-Phe), (5) N-(7-Hydroxy-6-methyl-octyl)-acetamide	Diketopiperazine (DKP) compounds	Antibacterial and cytotoxic	*Streptomyces* sp. UK 25	*Zingiber spectabile*	Alshaibani et al. (2017)
Butyrolactone I and Butyrolactone V	4-Carbon lactone	Antibacterial, antitumor and anti-leishmanicidal activities	*Aspergillus terreus*-F7	*Hyptis suaveolens*	da Silva et al. (2017)
(1) 3,6,9-Trihydroxy-7-methoxy-4,4-dimethyl-3,4-dihydro-1H-benzo[g] isochromene-5,10-dione, (2) Fusarubin (3) 3-O-Methylfusarubin (4) Javanicin	Naphthoquinone derivatives	Antibacterial and anti-mycobacterial	*Fusarium solani*	*Glycyrrhiza glabra*	Shah et al. (2017)

(continued)

Table 2.1 (continued)

Bio-active compounds	Structural moiety	Activity	Endophyte	Host	References
Pestalactone A-C (3)	Isocoumarin	Antifungal	*Pestalotiopsis* sp.	*Photinia fraseri*	Song et al. (2017)
(1) Cryptosporiopsin (2) 5-Hydroxycryptosporiopsin (3) (+)-Cryptosporiopsinol (4) Mellein	(1), (2), (3), Cyclopentene carboxylic acid derivative (4) Dihydroisocoumarin	Antifungal	*Pezicula sporulosa*	*Picea rubens* and *P. mariana*	McMullin et al. (2017)
Phomopoxides A-G (1-7)	Polyoxygenated cyclohexenoids	Cytotoxic and antifungal	*Phomopsis* sp. YE3250	*Paeonia delavayi*	Huang et al. (2018)
5-Hydroxy-8-methoxy-4-phenylisoquinolin-1(2H)-one	Isoquinolone alkaloid	Antifungal	*Penicillium* sp. R22	*Nerium indicum*	Ma et al. (2017)
3-Hydroxy-5-methoxyhex-5-ene-2,4-dione	Ketone derivative	Anticandidal activity	*Diaporthe* sp. ED2	*Orthosiphon stamieus*	Tong et al. (2017)
Terrein (4, 5-Dihydroxy-3-(1-propenyl)-2-cyclopenten-1-one)	Cyclopenten derivative	Antifungal and anticancer	*Aspergillus terreus* JAS-2	*Achyranthes aspera*	Goutam et al. (2017)
Paclitaxel	Alkaloid	Anticancer	*Phomamedicaginis*	*T. wallichiana var. mairei*	Zaiyou et al. (2017)
Peniproline A	Pyrrolidone derivative	Anticancer	*Penicillium decumbens* CP-4	–	Wang et al. (2017)
(−)-(10E,15S)-6-Chloro-10(11)-dehydrocurvularin, (−)-(10E,15S)-10(11)-dehydrocurvularin	Lactone derivative	Anticancer	*Alternaria* sp. AST0039	*Astragalus lentiginosus*	Bashyal et al. (2017)
Cladosporol A	Dimeric tetralone derivative	Anticancer	*Cladosporium cladosporioides*	*Datura innoxia*	Koul et al. (2017)
7-epi-10-Deacetyltaxol	Alkaloid	Anticancer	*Pestalotiopsis microspora*	*Taxodium mucronatum*	Subban et al. (2017)
Camptothecin	Monoterpeneindole alkaloid	Anticancer	*Fusarium solani*	*Camptotheca acuminate*	Ran et al. (2017)

(1'Z)-Dechloromycorrhizin A	Chlorinated para-quinone	Nematicidal activities	*Lachnum pygmaeum*	*Picea rubens* and *P. mariana*	McMullin et al. (2017)
Prodigiosin	Tripyrrole derivative	Anticancer, antimalarial, antibacterial, antifungal, antiproliferative and immunosuppressive activities	*Serratia marcescens*	*Beta vulgaris* L	Khanam and Chandra (2018)
(1) Terreic acid (2) 6-Methylsalicylic acid	(1) Quinone derivative (2) Salicylic acid derivative		*Pseudocercospora* sp. ESL 02	*Elaeocarpus sylvestris*	Prihantini and Tachibana (2017)
(1) Gallic acid (2) Rutin (3) Phlorizin (4) 2,4-di-tert-Butylphenol (5) 2,6-di-tert-Butyl hydroquinone	(1) Phenolic acid derivative (2) Flavonoid (3) Flavonoid (4) Quinone	Antioxidant	*Fusarium sp.*	*Fritillaria unibracteata* var. *wabuensis*	Pan et al. (2017)

2.6 Bioremediation Potential of Endophytes

Bioremediation is the process of removal of pollutants such as heavy metals, volatile organic compounds, greenhouse gases, crude oil, hydrocarbons and radionuclides from the biosphere utilizing microbial metabolism. Various characteristics of the contaminant (mobility, solubility, degradability and bioavailability) influence the bioremediation mechanism. Microbial growth and metabolism are critical factors that influence the biodegradation of contaminants (Stępniewska and Kuźniar 2013). Many studies have substantiated that endophytes can speed up degradation process by interacting with host plants (Govarthanan et al. 2016; Krishnamurthy and Naik 2017). This section succinctly shows the role of recently published data on endophytes, which have been characterized to play a role in the degradation of contaminants (Table 2.2). For details of other endophytes documented in bioremediation processes, readers are requested to refer to the articles by Krishnamurthy and Naik (2017), Govarthanan et al. (2016) and Stępniewska and Kuźniar (2013). More widely known for their growth-promoting activities, endophytes play an important role in nutrient cycling and soil quality improvement in turn, promoting growth of their respective host plants by nitrogen fixation, phosphate solubilisation, secreting phytohormones and other molecules such as siderophores and biosurfactants (Hassan 2017; Kandel et al. 2017). Many endophytes are multifarious, facilitating plant growth promotion along with bioremediation activities. An overview of such types of endophytes has been summarized in tabular form (Table 2.3).

2.7 Metaomics in Endophyte Studies

Conventional cultivable approaches provided the initial information on existence of a microcosm within plant bodies' unique niche habitat. Over a period of time, realization dawned that the cultivable fraction is a mere 1% of the actual microbial world. With the fact being applicable equally to the endophytic population, studies started focusing on realizing the uncultivable/environmental fraction. The bio-active compounds produced by endophytes have been documented to be of different types, depending on the location of the endophyte inside the host and, indeed, the geographical location of the host. Chutulo and Chalannavar (2018) have documented the entire endophyte profile of the medicinally important neem (*Azadirachta indica*) and have comprehensively documented the varied microbial populations in different parts of the plant body system and the diverse kinds of bio-active metabolites obtained from them. It is indicative that such a rich medical repository has not been studied by modern technologies (Fig. 2.3). There is probably a rich population of endophytes waiting to be discovered.

Technological advances and increasingly lower costs of high-throughput next-generation sequencing techniques have resulted in fine in-depth detailing of the microbiome constituents inside the plant body. Starting with comparative genomics using conventional Sanger sequencing techniques and moving on to next-generation sequencing (NGS) involving metagenomics, metaproteomics and

Table 2.2 Contribution of endophytes in bioremediation

Contaminants	Structural moiety	Degradation products	Endophyte	Host	References
Triclosan (TCS)	Polychloro phenoxy phenol	2-Chlorohydroquinone, 2, 4-dichloropheno, and hydroquinone	*Penicillium oxalicum* B4	*Artemisia annua*	Tian et al. (2018)
Phenanthrene	Polycyclic aromatic hydrocarbons (PAHs)	Tricarboxylic acid (TCA) cycle intermediates	*Pseudomonas* sp. Ph6-gfp	*Brassica chinensis*	Sun et al. (2018)
di-n-Butyl-phthalate (DBP)	Phthalic acid esters (PAEs)	Mono butyl-phthalate, phthalic acid and other intermediates	*Bacillus megaterium* strain YJB3	*Canna indica*	Feng et al. (2018)
Chlorpyrifos (CP)	Organophosphate	O,O-diethyl O-3,5,6-trichloropyridinol	*Sphingomonas* sp. (strain HJY)	*Allium tuberosum*	Feng et al. (2017a)
Trichloroethylene (TCE)	Halocarbon	Release of chloride ions	*Enterobacter* sp. strain PDN3	*Poplulus* sp.	Doty et al. (2017)
Benzene Phenols	Hydrocarbons	Carbon, hydrogen	*Achromobacter* sp. (AIEB-7), *Pseudomonas* sp. (AIEB-4) and *Alcaligenes* sp. (AIEB-6) *Bacillus* sp. (AIEB-1), *Enterobacter* sp. (AIEB-3) and *Acinetobacter* sp. (AIEB-2)	*Cannabis sativa*	Iqbal et al. (2018)
Pyrene degradation in *Triticumaestivum* L.	Polycyclic aromatic hydrocarbon degrading	Tricarboxylic acid (TCA) cycle intermediates	*Serratia* sp. PW7	*Plantago asiatica*	Zhu et al. (2017)

Table 2.3 Endophyte exhibiting plant growth promotion activity concomitant with bio-/phytoremediation

Contaminants	Degradation mechanism	Plant growth promoter molecules	Endophytes	Host	References
Petroleum degradation	Biotransformation	Siderophores, phosphate solubilization, 1-aminocyclopropane-1-carboxylate deaminase, nitrogen fixation and indole-3-acetic acid production as well as biosurfactant production	*Streptomyces* sp.	*Helianthemum lippii, Zygophyllum album, Bassia mauricata*	Baoune et al. (2018)
Heavy metals	Biotransformation and/or accumulation	IAA, ACC deaminase and solubilize phosphate	*Mucor* sp. MHR-7	*Brassica campestris*	Zahoor et al. (2017)
Chlorpyrifos	Detoxification	Indole acetic acid and siderophore production, secretion of phosphate solubilization and 1-aminocyclopropane-1-carboxylate deaminase	*Pseudomonas aeruginosa* strain RRA, *Bacillus megaterium* strain RRB, *Sphingobacterium siyangensis* strain RSA, *Stenotrophomona spavanii* strain RSB and *Curtobacterium plantarum* strain RSC	*Oryza sativa*	Feng et al. (2017b)
Tannery effluent	Bioremediation and detoxification	Phytoharmones production	*Pantoea stewartii* ASI11, *Microbacterium arborescens* HU33 and *Enterobacter* sp. HU38	*Leptochloa fusca*	Ashraf et al. (2018)
Heavy metals	Removal of heavy metals	Nutrient uptake enhancement, siderophores and other growth promoters	*Phialocephala fortinii* and *R. veluwensis*	*Clethra barbinervis*	Yamaji et al. (2016)
Cadmium	Phyto-stabilization/ immobilization of heavy metals	Indole-3-acetic acid (IAA)	Fungal strains, RSF-4L and RSF-6L	*Solanum nigrum*	Khan et al. (2017)

Arsenic	Detoxification through phytochelatincomplexation	Siderophore and IAA and ACC deaminase	*Ensifera dhaerens* strain 91R, *Rhizobium herbae* strain 32E,*Variovorax paradoxus* strain 28EY and *Phyllobacterium myrsinacearum* strain 28EW	*Betula celtiberica*	Mesa et al. (2017)
Cadmium and lead	Accumulation	Indole-3-acetic acid (IAA) and siderophore	*Fusarium* sp. CBRF44	*Brassica napus*	Shi et al. (2017)
Cadmium and lead	Accumulation	Phosphate-solubilizing activities	*Penicillium* sp. CBRF65	*Brassica napus*	Shi et al. (2017)

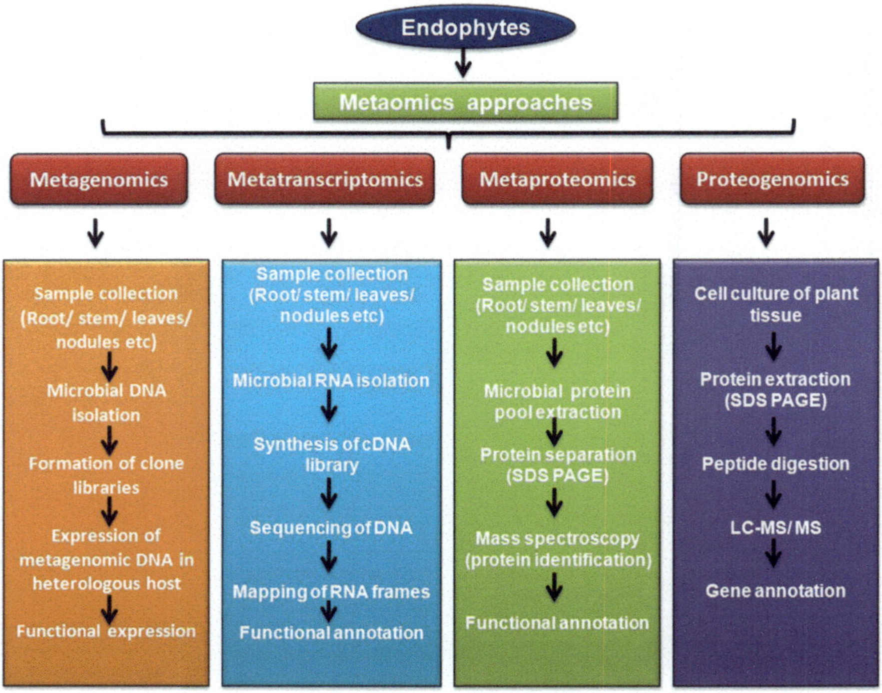

Fig. 2.3 An overview of the various 'omics' approaches

metatranscriptomics, the studies have come a long way (Kaul et al. 2016). There is a considerable amount of data being unveiled of the endophytes and the role they play in their complex relationships with their hosts. Microorganisms hitherto not known to be endophytes have been deduced to be colonizing the interiors of the plant body. In the following sections, we highlight the dovetailing of evolving sequencing technology platforms with that of increasing knowledge output on endophytes.

2.7.1 Comparative Genomics/Whole-Genome Sequencing

Santoyo et al. (2016) have provided a critical review of whole-genome sequencing studies of endophytes that have resulted in elucidation of various mechanisms by which endophytes promote growth of their host plants. Whole-genome sequencing studies have shown existence of genes conferring traits related to bio-active metabolite production (fusaric acid resistance proteins), supporting the host plant in growth-promoting activities, signaling (N-acyl homoserine lactone synthases) and surface attachment (hyperadherence factors).

Amongst many other studies and some of which are briefly summarized in Table 2.2, Remali et al. (2017) have provided the whole-genome sequence of *Streptomyces kebangsaanesis* SUK12, a novel endophyte colonizing the ethnomedicinal plant *Portulaca oleracea*, and characterized for production of phenazine class of antibiotics. Whole-genome sequencing of *Methylobacterium extorquens* DSM13060, an endophyte of pine, revealed its potential for producing antimicrobial defensin-like peptides, based on computational and prediction models (Tejesvi et al. 2016). The gene was further cloned and the peptide produced in *E. coli*. It was shown to exhibit antimicrobial activity against *Staphylococcus aureus* and *Bacillus subtilis*. This approach was different in that initial extensive computational analysis was used to predict potential antimicrobial peptides, in silico tools were used to predict activity of the putative peptides and their similarity to other antimicrobial peptides was confirmed using appropriate software. It was after this that the gene sequence was cloned and expressed successfully. The MB1533 defensin-like peptide was the first reported from *Methylobacterium* as also from bacterial endophytes. Amongst the various extensively studied bacterial endophytes, *Rhodotorula graminis* WP1 was the first yeast endophyte, whose genome is sequenced (Firrincieli et al. 2015). Colonizing the stems of *Populus trichocarpa*, its genome sequence revealed the presence of genes involved in the synthesis of (R)-acetoin and (R, R)-2,3-butandiol, which are known to provide resistance to phytopathogens and also promote plant growth. The authors have also speculated that this yeast colonizes in a filamentous form, with putative gene sequences indicating the same, although under culture conditions, the yeast did not show a filamentous morphology. Antitoxin systems (comprising of multi-antimicrobial extrusion protein (MATE), multidrug resistance efflux transporters (EmrE) and putative proteins with a multidrug resistance protein domain) potentially strengthen the endophyte against other competing ones.

Similarly, whole-genome sequencing provides the confidence in analysing entire genomes; their utility cannot be underestimated in comparison with NGS-based studies, which tend to provide more fragmented information. NGS studies have gained momentum in recent times, offering information data running into gigabytes, but suffer from certain challenges and limitations, which whole-genome sequencing and comparative genomics studies address. In a similar manner, the challenge of whole-genome sequencing lies in obtaining cultivable microorganisms successfully, which, as mentioned earlier, cover 1% of the existing population.

2.7.2 Metagenomics

Metagenomics, the pre-eminent technology focusing on sequencing of environmental communities, has provided interesting inputs on the roles endophytes play. Many studies have provided a profile of the endophytic microbiome in various types of plants – those with medicinal and therapeutic properties, important crop plants and those subjected to various abiotic and biotic stresses. Sengupta et al. (2017) profiling the endophyte community of rice, the major food crop of Asia, documented that members of *Bacillus* dominated the root system, probably contributing to nitrogen fixation. Mashiane et al. (2017) showed that the endophytic community considerably differed amongst genetically modified Bt maize and its non-modified parental line, at the pre- and post-flowering stages. Members of *Gammaproteobacteria* dominated in Bt maize, while *Alphaproteobacteria* and *Actinobacteria* were the major communities in non-Bt maize phyllospheres. There were also differences in abundance of beneficial endophytes, indicating that genetic modification causes a significant effect on endophytic communities.

An interesting study by Tian et al. (2015) focused on roots of nematode-infected tomato plants pre- and postinfection. Differences in microbial communities of both rhizosphere and endophyte were observed. Furthermore, a group of bacteria – *Rhodocyclales*, *Sphingobacteriales*, *Rhizobiales*, *Enterobacteriales*, *Flavobacteriales* and *Burkholderiales* – were enriched in the root gall, suggesting their potential role in the infection process. Functional metagenomics showed that this enriched bacterial population contained genes coding for proteins involved in degradation of polysaccharides, carbohydrate and protein metabolism and nitrogen fixation. The root-knot microbiome contained genes involved in nitrogen fixation and assimilation, providing the clue that this group of endophytes offers a mechanism to provide nitrogen to the nematodes via the host supply. *Actinobacteria*, comprising of *Streptomycetales* and *Micromonosporales*, predominated in healthy roots. In contrast to the widely held view that endophytes are largely beneficial to their hosts, this study provided a new perspective, indicating that endophyte populations may be actively contributing to nematode infections.

In an elegant experiment, Sánchez-López et al. (2018) have exploited metagenomics data mining to understand the seed endophyte microbiome of *Crotalaria pumila*, an annual herbaceous tropical plant species which is known to tolerate a wide variety of environmental stresses. The core microbiome of *C. pumila* seeds showed the presence of genes related to nitrogen fixation, photosynthesis and methanol metabolism. Colonization by a representative *Methylobacterium* sp. (strain Cp3) was also studied using Sanger sequencing and confocal microscopy to understand the mode of colonization. *Methylobacterium* was postulated to be the major endophyte conferring metal tolerance. The endophyte could utilize the methanol released by plant cells as an additional substrate for metabolism. It could also tolerate metal stress and produce IAA. Furthermore, this organism was present across three generations of the host plant, thus providing a new perspective on endophyte colonization in harsh environments.

Some difficulties encountered while carrying out metagenomic analyses is the necessity to avoid the co-isolation of host genome. Elsebai et al. (2014) have developed a method wherein this possibility was avoided and a novel gene having putative antibacterial activity was discovered presumed to originate from the fungal endophyte of *Empetrum nigrum* L. The same research group also obtained a defensin (endopiceasin) from *Picea glauca* EST libraries, which was postulated to have originated from its fungal endophyte (Mygind et al. 2005).

2.7.3 Metaproteomics/Metaproteogenomic / Metatranscriptomics

Genome-based information has made rapid strides and provided considerable data on the vast uncultivable fraction of microorganisms inhabiting the biosphere. Nevertheless, the extensive analysis required to assemble the data and derive productive information in terms of function of the microbial community requires proteomic tools. Metaproteomics involves the large scale of proteins expressed by microbial communities (Maron et al. 2007). The metaproteome can serve as the missing link between metagenomic data and the functional profile of microbial communities. Metatranscriptomics provides the expression profile of the gene dataset (Aguiar-Pulido et al. 2016). Metaproteogenomics provides a combined approach utilizing both metagenomics and metaproteomics and thereby unveils a larger protein dataset than metaproteomics alone. A combination approach involving metatranscriptomics also can still further enlarge ecosystem functioning details.

Metaproteomics has gained attraction largely due to improved, fast and high-throughput technologies such as peptide ionization in mass spectrometry and efficient data analysis using bioinformatics tools (Maron et al. 2007). Hence, database comparison of the obtained ionization patterns and correlating with their corresponding gene sequences provide the finer details of niche-specific role of the community. Knief et al. (2012) characterized the phyllosphere and rhizosphere community of rice plants using a metaproteogenomics approach. More than 4000 proteins were identified, and differences in protein types were found. Those involved in methanogenesis and methanotrophy predominated in the rhizosphere, while methylotrophic proteins were obtained in the phyllosphere. While this study does not focus on endophytes, it does provide an indication of differential protein patterns in and around the ecosystem of plant. Metaproteomics and proteogenomics studies are largely lacking with respect to endophytes due to which large-scale proteome datasets are not available. However, it is anticipated that more publications will shortly become available, considering that the 'omics' platform is widely available to scientists globally.

2.8 Conclusions and Future Prospects

This chapter provides a comprehensive account of endophytes and the critical role they play in a wide variety of activities. It is probable that many more activities may be attributable to endophytes; once more metagenomic, metaproteomic and proteogenomic studies become available. It is imperative to understand that plants were once considered as producers of therapeutic molecules, and today there is a dawning realization that these molecules originate from the constituent endophytes. In a similar way, endophytes were considered to provide protection to their hosts from pathogens, and yet, a recent study has shown that endophytes facilitate debilitating nematode establishment too. 'Omics' studies are still in their infancy, with respect to endophytes. The unseen microbiome within host plants is yet to be deciphered comprehensively. Therefore, more promise is expected from endophytes in the coming years.

References

Aguiar-Pulido V, Huang W, Suarez-Ulloa V, Cickovski T, Mathee K, Narasimhan G (2016) Metagenomics, metatranscriptomics, and metabolomics approaches for microbiome analysis. Evol Bioinform Online 12:5–16

Alshaibani MM, Zin NM, Jalil J, Sidik NM, Ahmad SJ, Kamal N, Edrada-Ebel R (2017) Isolation, purification, and characterization of five active diketopiperazine derivatives from endophytic *Streptomyces* SUK 25 with antimicrobial and cytotoxic activities. J Microbiol Biotechnol 27:1249–1256

Arora J, Ramawat KG (2017) An introduction to endophytes. In: Maheshwari D (ed) Endophytes: biology and biotechnology, sustainable development and biodiversity, vol 15. Springer, Cham, pp 1–23

Ashraf S, Afzal M, Naveed M, Shahid M, Ahmad Zahir Z (2018) Endophytic bacteria enhance remediation of tannery effluent in constructed wetlands vegetated with *Leptochloa fusca*. Int J Phytoremed 20:121–128

Baoune H, Ould El Hadj-Khelil A, Pucci G, Sineli P, Loucif L, Polti MA (2018) Petroleum degradation by endophytic *Streptomyces* spp. isolated from plants grown in contaminated soil of southern Algeria. Ecotoxicol Environ Saf 147:602–609

Bashyal BP, Wijeratne EM, Tillotson J, Arnold AE, Chapman E, Gunatilaka AA (2017) Chlorinated dehydrocurvularins and alterperylenepoxide a from *alternaria* sp. ast0039, a fungal endophyte of *Astragalus lentiginosus*. J Nat Prod 80:427–433

Behie SW, Zelisko PM, Bidochka MJ (2012) Endophytic insect parasitic fungi translocate nitrogen directly from insects to plants. Science 336:1576–1577

Bidartondo MI, Read DJ, Trappe JM, Merckx V, Ligrone R, Duckett JG (2011) The dawn of symbiosis between plants and fungi. Biol Lett 7:574–577

Bonfante P, Selosse MA (2010) A glimpse into the past of land plants and of their mycorrhizal affairs: from fossils to evo-devo. New Phytol 186:267–270

Brundrett MC (2002) Co-evolution of roots and mycorrhizas of land plants. New Phytol 154:275–304

Burnham JP, Olsen MA, Stwalley D, Kwon JH, Babcock HM, Kollef MH (2018) Infectious diseases consultation reduces 30-day and 1-year all-cause mortality for multidrug-resistant organism infections. Open Forum Infect Dis 5:ofy026. https://doi.org/10.1093/ofid/ofy026

Carlier AL, Eberl L (2012) The eroded genome of a *Psychotria* leaf symbiont: hypotheses about lifestyle and interactions with its plant host. Environ Microbiol 14:2757–2769

Carlier A, Fehr L, Pinto-Carbó M, Schäberle T, Reher R, Dessein S, König G, Eberl L (2016) The genome analysis of *Candidatus Burkholderia crenata* reveals that secondary metabolism may be a key function of the *Ardisia crenata* leaf nodule symbiosis. Environ Microbiol 18:2507–2522

Carlier A, Cnockaert M, Fehr L, Vandamme P, Eberl L (2017) Draft genome and description of *Orrella dioscoreae* gen. nov. sp. nov., a new species of *Alcaligenaceae* isolated from leaf acumens of *Dioscorea sansibarensis*. Syst Appl Microbiol 40:11–21

Chen C, Bauske E, Musson G, Rodriguezkabana R, Kloepper J (1995) Biological control of *Fusarium* wilt on cotton by use of endophytic bacteria. Biol Control 5:83–91

Chowdhury EK, Jeon J, Rim SO, Park YW, Lee SK, Bae H (2017) Composition, diversity and bioactivity of culturable bacterial endophytes in mountain-cultivated ginseng in Korea. Sci Rep 7:10098. https://doi.org/10.1038/s41598-017-10280-7

Chutulo EC, Chalannavar RK (2018) Endophytic mycoflora and their bioactive compounds from *Azadirachta indica*: a comprehensive review. J Fungi 4:42. https://doi.org/10.3390/jof4020042

da Silva IP, Brissow E, Kellner Filho LC, Senabio J, de Siqueira KA, Vandresen Filho S, Damasceno JL, Mendes SA, Tavares DC, Magalhaes LG, Junior PA, Januario AH, Soares MA (2017) Bioactive compounds of *Aspergillus terreus*-F7, an endophytic fungus from *Hyptis suaveolens* (L.) Poit. World J Microbiol Biotechnol 33:62. https://doi.org/10.1007/s11274-017-2228-3

Doty SL, Freeman JL, Cohu CM, Burken JG, Firrincieli A, Simon A, Khan Z, Isebrands JG, Lukas J, Blaylock MJ (2017) Enhanced degradation of TCE on a superfund site using endophyte-assisted poplar tree phytoremediation. Environ Sci Technol 51:10050–10058

Dudeja SS, Giri R, Saini R, Suneja-Madan P, Kothe E (2012) Interaction of endophytic microbes with legumes. J Basic Microbiol 52:248–260

Elsebai MF, Tejesvi MV, Pirttila AM (2014) Endophytes as a novel source of bioactive new structures. In: Verma V, Gange A (eds) Advances in endophytic research. Springer, New Delhi, pp 191–202

Fang W, Yang L, Zhu X, Zeng L, Li X (2013) Seasonal and habitat dependent variations in culturable endophytes of *Camellia sinensis*. J Plant Pathol Microb 4:169. https://doi.org/10.4172/2157-7471.1000169

Feng F, Li Y, Ge J, Chen J, Jiang W, He S, Liu X, Yu X (2017a) Degradation of chlorpyrifos by an endophytic bacterium of the *Sphingomonas genus* (strain HJY) isolated from Chinese chives (*Allium tuberosum*). J Environ Sci Health B 52:736–744

Feng F, Ge J, Li Y, He S, Zhong J, Liu X, Yu X (2017b) Enhanced degradation of chlorpyrifos in rice (*Oryza sativa* L.) by five strains of endophytic bacteria and their plant growth promotional ability. Chemosphere 184:505–513

Feng NX, Yu J, Mo CH, Zhao HM, Li YW, Wu BX, Cai QY, Li H, Zhou DM, Wong MH (2018) Biodegradation of di-n-butyl phthalate (DBP) by a novel endophytic *Bacillus megaterium* strain YJB3. Sci Total Environ 616:117–127

Firrincieli A, Otillar R, Salamov A, Schmutz J, Khan Z, Redman RS, Fleck ND, Lindquist E, Grigoriev IV, Doty SL (2015) Genome sequence of the plant growth promoting endophytic yeast *Rhodotorula graminis* WP1. Front Microbiol 6:978. https://doi.org/10.3389/fmicb.2015.00978

Frey-Klett P, Burlinson P, Deveau A, Barret M, Tarkka M, Sarniguet A (2011) Bacterial-fungal interactions: hyphens between agricultural, clinical, environmental, and food microbiologists. Microbiol Mol Biol Rev 75:583–609

Gange AC, Eschen R, Wearn JA, Thawer A, Sutton BC (2012) Differential effects of foliar endophytic fungi on insect herbivores attacking a herbaceous plant. Oecologia 168:1023–1031

Glick BR (2015) Beneficial plant-bacterial interactions. Springer, Switzerland, p 243

Glynou K, Ali T, Haghi Kia S, Thines M, Maciá-Vicente JG (2017) Genotypic diversity in root-endophytic fungi reflects efficient dispersal and environmental adaptation. Mol Ecol 26:4618–4630

Gonzalez-Teuber M, Vilo C, Bascunan-Godoy L (2017) Molecular characterization of endophytic fungi associated with the roots of *Chenopodium quinoa* inhabiting the Atacama Desert, Chile. Genom Data 11:109–112

Gouda S, Das G, Sen SK, Shin HS, Patra (2016) Endophytes: a treasure house of bioactive compounds of medicinal importance. Front Microbiol 7:1538. https://doi.org/10.3389/fmicb.2016.01538

Goutam J, Sharma G, Tiwari VK, Mishra A, Kharwar RN, Ramaraj V, Koch B (2017) Isolation and characterization of "terrein" an antimicrobial and antitumor compound from endophytic fungus *Aspergillus terreus* (JAS-2) associated from *Achyranthes aspera* Varanasi, India. Front Microbiol 8:1334

Govarthanan M, Mythili R, Selvankumar T, Kamala-Kannan S, Rajasekar A, Chang Y-C (2016) Bioremediation of heavy metals using an endophytic bacterium *Paenibacillus* sp. RM isolated from the roots of *Tridax procumbens*. 3Biotech 6:242

Hallmann J, Berg G, Schulz B (2006) Isolation procedures for endophytic microorganism. In: Schulz BE, Boyle CC, Sieber T (eds) Microbial root endophytes, vol 9. Springer, Berlin/Heidelberg, pp 299–319

Hardoim PR, Van Overbeek LS, Berg G, Pirttilä AM, Compant S, Campisano A, Döring M, Sessitsch A (2015) The hidden world within plants: ecological and evolutionary considerations for defining functioning of microbial endophytes. Microbiol Mol Biol Rev 79:293–320

Hassan SE (2017) Plant growth-promoting activities for bacterial and fungal endophytes isolated from medicinal plant of *Teucrium polium* L. J Adv Res 8:687–695

Herrera SD, Grossi C, Zawoznik M, Groppaa MD (2016) Wheat seeds harbour bacterial endophytes with potential as plant growth promoters and biocontrol agents of *Fusarium graminearum*. Microbiol Res 186:37–43

Hoeksema JD, Chaudhary VB, Gehring CA, Johnson NC, Karst J, Koide RT, Pringle A, Zabinski C, Bever JD, Moore JC, Wilson GWT, Klironomos JN, Umbanhowar J (2010) A meta-analysis of context-dependency in plant response to inoculation with mycorrhizal fungi. Ecol Lett 13:394–407

Hu M, Yang XQ, Zhou Q-Y, Li SQ, Wang BY, Ruan BH, Yang YB, Zhang ZX, Zhou H, Ding ZT (2017) Benzopyran derivatives from endophytic *Daldinia eschscholzii* JC-15 in *Dendrobium chrysotoxum* and their bioactivities. Nat Prod Res (Online). https://doi.org/10.1080/14786419.2017.1419236

Huang R, Jiang BG, Li XN, Wang YT, Liu SS, Zheng KX, He J, Wu SH (2018) Polyoxygenated cyclohexenoids with promising α-glycosidase inhibitory activity produced by *Phomopsis* sp. YE3250, an endophytic fungus derived from *Paeonia delavayi*. J Agric Food Chem 66:1140–1146

Hurek T, Handley LL, Reinhold-Hurek B, Piche Y (2002) *Azoarcus* grass endophytes contribute fixed nitrogen to the plant in an unculturable state. Mol Plant Microbe Interact 15:233–242

Ibrahim A, Sorensen D, Jenkins HA, Ejim L, Capretta A, Sumarah MW (2017) Epoxynemanione A, nemanifuranones A-F, and nemanilactones A-C, from *Nemania serpens*, an endophytic fungus isolated from *Riesling grapevines*. Phytochemistry 140:16–26

Iqbal A, Arshad M, Hashmi I, Karthikeyan R, Gentry TJ, Schwab AP (2018) Biodegradation of phenol and benzene by endophytic bacterial strains isolated from refinery wastewater-fed *Cannabis sativa*. Environ Technol 39:1705–1714

Jia M, Chen L, Xin HL, Zheng CJ, Rahman K, Han T, Qin LP (2016) A friendly relationship between endophytic fungi and medicinal plants: a systematic review. Front Microbiol 7:906

Jin Z, Gao L, Zhang L, Liu T, Yu F, Zhang Z, Wang B (2017) Antimicrobial activity of saponins produced by two novel endophytic fungi from *Panax notoginseng*. Nat Prod Res 31:2700–2703

Kandel SL, Joubert PM, Doty SL (2017) Bacterial endophyte colonization and distribution within plants. Microorganisms 5:E77. https://doi.org/10.3390/microorganisms5040077

Kaul S, Sharma TK, Dhar M (2016) "Omics" tools for better understanding the plant-endophyte interactions. Front Plant Sci 7:955

Khan AL, Kang SM, Dhakal KH, Hussain J, Adnan M, Kim JG, Lee IJ (2013) Flavonoids and amino acid regulation in *Capsicum annuum* L. by endophytic fungi under different heat stress regimes. Sci Hortic 155:1–7

Khan AR, Ullah I, Waqas M, Park GS, Khan AL, Hong SJ, Ullah R, Jung BK, Park CE, Ur-Rehman S, Lee IJ, Shin JH (2017) Host plant growth promotion and cadmium detoxification in *Solanum nigrum*, mediated by endophytic fungi. Ecotoxicol Environ Saf 136:180–188

Khanam B, Chandra R (2018) Comparative analysis of prodigiosin isolated from endophyte *Serratia marcescens*. Lett Appl Microbiol 66:194–201

Knief C, Delmotte N, Chaffron S, Stark M, Innerebner G, Wassmann R, von Mering C, Vorholt JA (2012) Metaproteogenomic analysis of microbial communities in the phyllosphere and rhizosphere of rice. ISME J 6:1378–1390

Kondorosi E, Mergaert P, Kereszt A (2013) A paradigm for endosymbiotic life: cell differentiation of Rhizobium bacteria provoked by hostplant factors. Annu Rev Microbiol 67:611–628

Koul M, Kumar A, Deshidi R, Sharma V, Singh RD, Singh J, Sharma PR, Shah BA, Jaglan S, Singh S (2017) Cladosporol A triggers apoptosis sensitivity by ROS-mediated autophagic flux in human breast cancer cells. BMC Cell Biol 18:26. https://doi.org/10.1186/s12860-017-0141-0

Krishnamurthy YL, Naik BS (2017) Endophytic fungi bioremediation. In: Maheshwari DK, Annapurna K (eds) Endophytes in crop productivity and protection. Springer, Cham, pp 47–60

Lata R, Chowdhury S, Gond SK Jr, White JF (2018) Induction of abiotic stress tolerance in plants by endophytic microbes. Lett Appl Microbiol 66:268–276

Lemaire B, Robbrecht E, van Wyk B, Van Oevelen S, Verstraete B, Prinsen E, Smets E, Dessein S (2011) Identification, origin, and evolution of leaf nodulating symbionts of *Sericanthe* (Rubiaceae). J Microbiol 49:935–941

Lemaire B, Van Oevelen S, De Block P, Verstraete B, Smets E, Prinsen E, Dessein S (2012a) Identification of the bacterial endosymbionts in leaf nodules of *Pavetta* (Rubiaceae). Int J Syst Evol Microbiol 62:202–209

Lemaire B, Janssens S, Smets E, Dessein S (2012b) Endosymbiont transmission mode in bacterial leaf nodulation as revealed by a population genetic study of *Psychotria leptophylla*. Appl Environ Microbiol 78:284–287

Li X, Zhang L (2015) Endophytic infection alleviates Pb^{2+} stress effects on photosystem II functioning of *Oryza sativa* leaves. J Hazard Mater 295:79–85

Li Y, Cheng C, An D (2017) Characterisation of endophytic bacteria from a desert plant *Lepidium perfoliatum* L. Plant Protect Sci 53:32–43

Li X, He X, Hou L, Ren Y, Wang S, Su F (2018) Dark septate endophytes isolated from a xerophyte plant promote the growth of *Ammopiptanthus mongolicus* under drought condition. Sci Rep 8:7896

Liu XZ, Song WL, Zhang K, Ye YC, Dai CC (2011) Effects of two kinds of endophytic fungi infection on water stress of seedlings of *Chrysanthemum morifolium*. Acta Hortic Sin 38:335–342

Ma YM, Qiao K, Kong Y, Li MY, Guo LX, Miao Z, Fan C (2017) A new isoquinolone alkaloid from an endophytic fungus R22 of *Nerium indicum*. Nat Prod Res 31:951–958

Malfanova N, Lugtenberg B, Berg G (2013) Bacterial endophytes: who and where, and what are they doing there. In: de Bruijn FJ (ed) Molecular microbial ecology of the rhizosphere. Wiley-Blackwell, Hoboken, pp 15–37

Maron PA, Ranjard L, Mougel C, Lemanceau P (2007) Metaproteomics: a new approach for studying functional microbial ecology. Microb Ecol 53:486–493

Mashiane RA, Ezeokoli OT, Adeleke RA, Bezuidenhout CC (2017) Metagenomic analyses of bacterial endophytes associated with the phyllosphere of a Bt maize cultivar and its isogenic parental line from South Africa. World J Microbiol Biotechnol 33:80. https://doi.org/10.1007/s11274-017-2249-y

McMullin DR, Green BD, Prince NC, Tanney JB, Miller JD (2017) Natural products of *Picea* endophytes from the Acadian forest. J Nat Prod 80:1475–1483

Meng JJ, He XL (2011) Effects of AM fungi on growth and nutritional contents of *Salvia miltiorrhiza* Bge. under drought stress. J Agric Univ Hebei 34:51–55

Mesa V, Navazas A, González-Gil R, González A, Weyens N, Lauga B, Gallego JL, Sánchez J, Peláez AI (2017) Use of endophytic and rhizosphere bacteria to improve phytoremediation of arsenic-contaminated industrial soils by autochthonous *Betula celtiberica*. Appl Environ Microbiol 83:e03411–e03416

Mygind PH, Fischer RL, Schnorr KM, Hansen MT, Sönksen CP, Ludvigsen S, Raventós D, Buskov S, Christensen B, De Maria L, Taboureau O, Yaver D, Elvig-Jørgensen SG, Sørensen MV, Christensen BE, Kjaerulff S, Frimodt-Moller N, Lehrer RI, Zasloff M, Kristensen HH (2005) Plectasin is a peptide antibiotic with therapeutic potential from a saprophytic fungus. Nature 437:975–980

Oteino N, Lally RD, Kiwanuka S, Lloyd A, Ryan D, Germaine KJ, Dowling DN (2015) Plant growth promotion induced by phosphate solubilizing endophytic Pseudomonas isolates. Front Microbiol 6:745

Pan F, Tian-Jiao Su, Shi-Mei Cai, Wei Wu (2017) Fungal endophyte-derived *Fritillaria unibracteata* var. wabuensis: diversity, antioxidant capacities *in vitro* and relations to phenolic, flavonoid or saponin compounds. Sci Rep 7:42008

Paungfoo-Lonhienne C, Yeoh YK, Kasinadhuni NR, Lonhienne TG, Robinson N, Hugenholtz P, Ragan MA, Schmidt S (2015) Nitrogen fertilizer dose alters fungal communities in sugarcane soil and rhizosphere. Sci Rep 5:8678

Perveen I, Raza MA, Iqbal T, Naz I, Sehar S, Ahmed S (2017) Isolation of anticancer and antimicrobial metabolites from *Epicoccum nigrum*; endophyte of *Ferula sumbul*. Microb Pathog 110:214–224

Pimentel MR, Molina G, Dionísio AP, Maróstica MR, Pastore GM (2011) The use of endophytes to obtain bioactive compounds and their application in biotransformation process. Biotechnol Res Int Vol 2011:576286. https://doi.org/10.4061/2011/576286

Pinto-Carbó M, Sieber S, Dessein S, Wicker T, Verstraete B, Gademann K, Eberl L, Carlier A (2016) Evidence of horizontal gene transfer between obligate leaf nodule symbionts. ISME J 10:2092–2105

Pinto-Carbó M, Gademann K, Eberl L, Carlier A (2018) Leaf nodule symbiosis: function and transmission of obligate bacterial endophytes. Curr Opin Plant Biol 44:23–31

Pratiwi RH, Hidayat I, Hanafi M, Mangunwardoyo W (2017) Antibacterial compound produced by *Pseudomonas aeruginosa* strain UICC B-40, an endophytic bacterium isolated from *Neesia altissima*. J Microbiol 55:289–295

Prihantini AI, Tachibana S (2017) Antioxidant compounds produced by *Pseudocercospora* sp. ESL 02, an endophytic fungus isolated from *Elaeocarpus sylvestris*. Asian Pac J Trop Biomed 7:110–115

Ran X, Zhang G, Li S, Wang J (2017) Characterization and antitumor activity of camptothecin from endophytic fungus *Fusarium solani* isolated from *Camptotheca acuminate*. Afr Health Sci 17:566–574

Reinhold-Hurek B, Hurek T (2011) Living inside plants: bacterial endophytes. Curr Opin Plant Biol 14:435–443

Remali J, Loke K-K, Ng CL, Aizat WM, Tiong J, Zin NM (2017) Whole-genome shotgun sequence of phenazine-producing endophytic *Streptomyces kebangsaanensis* SUK12. Genom Data 13:7–10

Rho H, Hsieh M, Kandel SL, Cantillo J, Doty SL, Kim SH (2018a) Do endophytes promote growth of host plants under stress. A meta-analysis on plant stress mitigation by endophytes. Microb Ecol 75:407–418

Rho H, Van Epps V, Wegley N, Doty SL, Kim SH (2018b) Salicaceae endophytes modulate stomatal behavior and increase water use efficiency in rice. Front Plant Sci 9:188

Rudgers JA, Fischer S, Clay K (2010) Managing plant symbiosis: fungal endophyte genotype alters plant community composition. J Appl Ecol 47:468–477

Ryan RP, Germaine K, Franks A, Ryan DJ (2008) Bacterial endophytes: recent developments and applications. FEMS Microbiol 278:1–9

Sanchez-Lopez AS, Pintelon I, Stevens V, Imperato V, Timmermans JP, Gonzalez-Chávez C, Carrillo-Gonzalez R, Van Hamme J, Vangronsveld J, Thijs S (2018) Seed endophyte microbiome of *Crotalaria pumila* unpeeled: identification of plant-beneficial Methylobacteria. Int J Mol Sci 19:E291. https://doi.org/10.3390/ijms19010291

Santoyo G, Moreno-Hagelsieb G, del Carmen Orozco-Mosqueda M, Glick BR (2016) Plant growth-promoting bacterial endophytes. Microbiol Res 183:92–99

Savi DC, Shaaban KA, Gos FMWR, Ponomareva LV, Thorson JS, Glienke C, Rohr J (2018) *Phaeophleospora vochysiae* Savi and *Glienke* sp. nov. isolated from *Vochysia divergens* found in the Pantanal, Brazil, produces bioactive secondary metabolites. Sci Rep 8:3122. https://doi.org/10.1038/s41598-018-21400-2

Sengupta S, Ganguli S, Singh PK (2017) Metagenome analysis of the root endophytic microbial community of Indian rice (*O. sativa* L.). Genomics Data 12:41–43

Shah A, Rather MA, Hassan QP, Aga MA, Mushtaq S, Shah AM, Hussain A, Baba SA, Ahmad Z (2017) Discovery of anti-microbial and anti-tubercular molecules from *Fusarium solani*: an endophyte of *Glycyrrhiza glabra*. J Appl Microbiol 122:1168–1176

Shi Y, Xie H, Cao L, Zhang R, Xu Z, Wang Z, Deng Z (2017) Effects of Cd- and Pb-resistant endophytic fungi on growth and phytoextraction of *Brassica napus* in metal-contaminated soils. Environ Sci Pollut Res Int 24:417–426

Singh M, Kumar A, Singh R, Pandey KD (2017) Endophytic bacteria: a new source of bioactive compounds. 3Biotech 7:315. https://doi.org/10.1007/s13205-017-0942-z

Song RY, Wang XB, Yin GP, Liu RH, Kong LY, Yang MH (2017) Isocoumarin derivatives from the endophytic fungus, *Pestalotiopsis* sp. Fitoterapia 122:115–118

Stępniewska Z, Kuźniar A (2013) Endophytic microorganisms--promising applications in bioremediation of greenhouse gases. Appl Microbiol Biotechnol 97:9589–9596

Strobel G, Daisy B (2003) Bioprospecting for microbial endophytes and their natural products. Microbiol Mol Biol Rev 67:491–502

Strobel G, Daisy B, Castillo U, Harper J (2004) Natural products from endophytic microorganisms. J Nat Prod 67:257–268

Subban K, Singh S, Subramani R, Johnpaul M, Chelliah J (2017) Fungal 7-epi-10-deacetyltaxol produced by an endophytic *Pestalotiopsis microspora* induces apoptosis in human hepatocellular carcinoma cell line (HepG2). BMC Compl Altern Med 17:504. https://doi.org/10.1186/s12906-017-1993-8

Sun X, Guo LD (2012) Endophytic fungal diversity: review of traditional and molecular techniques. Mycology 3:65–76

Sun K, Habteselassie MY, Liu J, Li S, Gao Y (2018) Subcellular distribution and biotransformation of phenanthrene in pakchoi after inoculation with endophytic *Pseudomonas* sp. as probed using HRMS coupled with isotope-labeling. Environ Pollut 237:858–867

Taechowisan T, Chaisaeng S, Phutdhawong WS (2017) Antibacterial, antioxidant and anticancer activities of biphenyls from *Streptomyces* sp. BO-07: an endophyte in *Boesenbergia rotunda* (L.) Mansf A. Food Agric Immunol 28:1330–1346

Tejesvi MV, Andersen B, Antcheva N, Brinch KS, Koskimäki JJ, Kristensen HH, Tossi A, Pirttilä AM (2016) MB1533 is a defensin-like antimicrobial peptide from the intracellular meristem endophyte of scots pine *Methylobacterium extorquens* DSM13060. J Microb Biochem Technol 8:1. https://doi.org/10.4172/1948-5948.1000252

Tian B-Y, Cao Y, Zhang K-Q (2015) Metagenomic insights into communities, functions of endophytes, and their associates with infection by root-knot nematode, *Meloidogyne incognita*, in tomato roots. Sci Rep 5:17087

Tian H, Ma YJ, Li WY, Wang JW (2018) Efficient degradation of triclosan by an endophytic fungus *Penicillium oxalicum* B4. Environ Sci Pollut Res Int 25:8963–8975

Tong WY, Leong CR, Tan WN, Khairuddean M, Zakaria L, Ibrahim D (2017) Endophytic *Diaporthe* sp. ED2 produces a novel anti-candidal ketone derivative. J Microbiol Biotechnol 27:1065–1070

Trujillo ME, Riesco R, Benito P, Carro L (2015) Endophytic actinobacteria and the interaction of micromonospora and nitrogen fixing plants. Front Microbiol 6:1341. https://doi.org/10.3389/fmicb.2015.01341

Verma SK, Gond SK, Mishra A, Sharma VK, Kumar J, Singh DK, Kumar A, Kharwar RN (2017) Fungal endophytes representing diverse habitats and their role in plant protection. In: Satyanarayana T, Deshmukh S, Johri BN (eds) Developments in fungal biology and applied mycology. Springer, Singapore, pp 135–157

Vurukonda SSKP, Vardharajula S, Shrivastava M, Ali SZ (2016) Enhancement of drought stress tolerance in crops by plant growth promoting rhizobacteria. Microbiol Res 184:13–24

Wang Y, Dai CC (2011) Endophytes: a potential resource for biosynthesis, biotransformation, and biodegradation. Ann Microbiol 61:207–215

Wang Y, Xu L, Ren W, Zhao D, Zhu Y, Wu X (2012) Bioactive metabolites from *Chaetomium globosum* L18, an endophytic fungus in the medicinal plant *Curcuma wenyujin*. Phytomedicine 19:364–368

Wang X, Li J, Yu S, Ye L, Feng M, Li J (2017) Peniproline A, a new 1-phenylamino-2-pyrrolidone metabolite from the endophytic fungus *Penicillium decumbens* CP-4. Nat Prod Res 31:1772–1777

Wang J, Nan Z, Christensen MJ, Zhang X, Tian P, Zhang Z, Niu X, Gao P, Chen T, Ma L (2018) Effect of *Epichloë gansuensis* endophyte on the nitrogen metabolism, nitrogen use efficiency, and stoichiometry of *Achnatherum inebrians* under nitrogen limitation. J Agric Food Chem 66:4022–4031

Yamaji K, Watanabe Y, Masuya H, Shigeto A, Yui H, Haruma T (2016) Root fungal endophytes enhance heavy-metal stress tolerance of *Clethra barbinervis* growing naturally at mining sites via growth enhancement, promotion of nutrient uptake and decrease of heavy-metal concentration. PLoS One 11:e0169089

Yang NY, Jiang S, Shang EX, Tang YP, Duan JA (2012) A new phenyl pentanamine alkaloid produced by an endophyte *Bacillus subtilis* isolated from *Angelica sinensis*. J Chem Res 36:647–647

Yao X, Christensen MJ, Bao G, Zhang C, Li X, Li C, Nan Z (2015) A toxic endophyte-infected grass helps reverse degradation and loss of biodiversity of over-grazed grasslands in northwest China. Sci Rep 5:18527

You YH, Kwak TW, Kang SM, Lee MC, Kim JG (2015) *Aspergillus clavatus* Y2H0002 as a new endophytic fungal strain producing gibberellins isolated from *Nymphoides peltata* in fresh water. Mycobiol 43:87–91

Zahoor M, Irshad M, Rahman H, Qasim M, Afridi SG, Qadir M, Hussain A (2017) Alleviation of heavy metal toxicity and phytostimulation of *Brassica campestris* L. by endophytic *Mucor* sp. MHR-7. Ecotoxicol Environ Saf 142:139–149

Zaiyou J, Li M, Xiqiao H (2017) An endophytic fungus efficiently producing paclitaxel isolated from *Taxus wallichiana* var. mairei 96:e7406

Zheng N, Yao F, Liang X, Liu Q, Xu W, Liang Y, Liu X, Li J, Yang R (2018) A new phthalide from the endophytic fungus *Xylaria* sp. GDG-102. Nat Prod Res 32:755–760

Zhu X, Wang W, Crowley DE, Sun K, Hao S, Waigi MG, Gao Y (2017) The endophytic bacterium *Serratia* sp. PW7 degrades pyrene in wheat. Environ Sci Pollut Res Int 24:6648–6656

Microbial Hosts as a Promising Platform for Polyphenol Production

3

Adelaide Braga, Isabel Rocha, and Nuno Faria

Contents

Abstract

Plants synthesize a variety of different secondary metabolites, such as polyphenols, terpenoids, alkaloids, etc., with pharmaceutical and nutraceutical importance. Polyphenols have shown numerous health benefits with rare side effects. However, the extraction of these compounds from natural sources cannot meet the increasing consumer demand for natural products, and its purification is often difficult, making the overall process too expensive. In contrast, microbial pro-

A. Braga (✉) · N. Faria
Centre of Biological Engineering, University of Minho, Braga, Portugal
e-mail: abraga@deb.uminho.pt

I. Rocha
Centre of Biological Engineering, University of Minho, Braga, Portugal

Instituto de Tecnologia Química e Biológica, Universidade Nova de Lisboa, Oeiras, Portugal

M. S. Akhtar et al. (eds.), *Natural Bio-active Compounds*,
https://doi.org/10.1007/978-981-13-7154-7_3

duction of polyphenols is a powerful alternative to produce natural products in large amounts, in an environmentally sustainable way. Nevertheless, plant-derived polyphenols are not naturally produced by microorganisms and therefore require the integration of the heterologous pathway from plants through genetic engineering techniques. In the present chapter, the recent advances in microbial production of plant-derived polyphenols, emphasizing on flavonoids, anthocyanins, curcuminoids, and stilbenes, have been summarized. In addition, different strategies used to increase the product yield, and the production processes are also highlighted.

Keywords
Anthocyanin · Curcuminoids · Flavonoids · Stilbenes · Microorganisms

3.1 Introduction

The adaptation of plants to their surrounding environment is strongly facilitated by their ability to produce and use certain secondary metabolites. These compounds are not directly associated with plant growth and/or propagation. Instead, they are involved in the plant protection against UV radiation, oxidative stress, and defense against pathogens or herbivores and play a crucial role in allelopathy and tritrophic interactions (Jovanov et al. 2017). Based on their appealing characteristics, plant secondary metabolites have been used since the ancient times in traditional medicine. Currently, they are an important source of valuable compounds for food additives, pharmaceutics, and fine chemicals (Hussain et al. 2012). Based on their chemical structure and the pathways by which they are synthesized, secondary metabolites can be divided into three chemical distinct groups, terpenoids, alkaloids, and polyphenols (Hussain et al. 2012), and nowadays, more than 200,000 different secondary metabolites are known (Yonekura-Sakakibara and Saito 2009).

Polyphenols are produced through the shikimate/phenylpropanoid pathway. They are made of two or more aromatic rings attached with at least two phenolic hydroxyl groups (Fig. 3.1) (Sytar Oksana et al. 2012; Lin et al. 2016). These compounds have numerous applications in food industries as colorants, fragrances, and flavoring agents (van Summeren-Wesenhagen and Marienhagen 2013; Milke et al. 2018). Nowadays, many studies suggest that polyphenols possess health-protecting effects against cardiovascular diseases, cancer, diabetes, and Alzheimer disease, attracting a great deal of research interest (Dudnik et al. 2018). The consumer preference, reflecting the recent trends toward a "healthy lifestyle", lead to an increasing demand for polyphenols and their market grows globally. The polyphenol market size has been rising and is expected to continue for the next years. In 2015, it reached a global value of US$ 757 million (Allied Market Research 2017). The increasing application of polyphenols in food, beverages, pharmaceutical, and cosmetic industry should drive it even further (Allied Market Research 2017), reaching a global

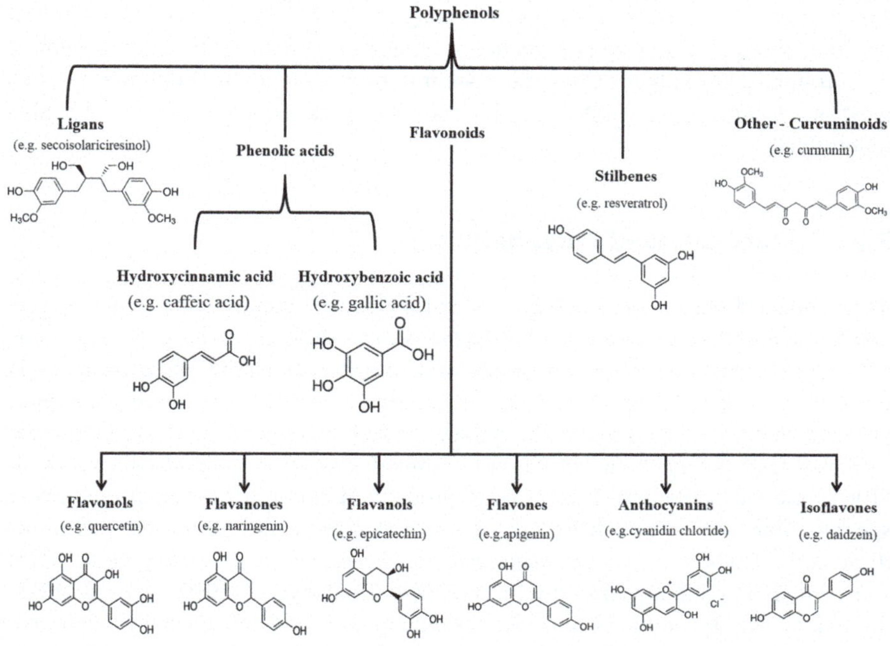

Fig. 3.1 Classification and structure of the main polyphenol classes

consumption of 33,880 tons and a market value of US$ 1.33 billion in 2024, according to Grand View Research (2016) report. This increasing demand makes it necessary to find new, more efficient and sustainable production processes, able to provide the required quantities to satisfy the market needs.

Polyphenols can be extracted from plants, but such processes have many disadvantages, such as low concentration of the product of interest, seasonal variation, risk of plant disease, stability of the compound, and trade restrictions (Atanasov et al. 2015; Jovanov et al. 2017). Alternatively to the extraction from plant, the chemical synthesis represents an attractive technology for their production. However, these technologies usually imply the use of harsh conditions (toxic catalysts, high pressure and temperature, among others) and tend to produce a mixture of similar molecules that are difficult to purify (Bicas et al. 2016).

Considering the disadvantages of chemical production, i.e., the quality of the product, health and environmental issues, and the lack of capacity of natural production at the industrial scale to meet the market needs, addressing an alternative way for polyphenol production through low-cost and environmentally friendly processes became crucial. Based on this, microorganisms have emerged as attractive platforms for their biosynthesis (Bicas et al. 2016). Microorganisms can grow with high growth rates and achieve high biomass yields in scalable cultivation and production processes, but they do not naturally produce polyphenols. In order to make them do that, it is necessary to functionally integrate heterologous pathway from plants that naturally produce the polyphenol in question (Chouhan et al. 2017;

Milke et al. 2018). Thus, in the present chapter, the recent advances in microbial production of plant-derived polyphenols, emphasizing on flavonoids, anthocyanins, curcuminoids, and stilbenes, have been summarized. In addition, different strategies used to increase the product yield and the production processes are also highlighted.

3.2 Plant-Derived Polyphenols

The number of phenol rings and the structural elements that bind these rings to one another define the classification of polyphenols into different groups. The most relevant are the phenolic acids, flavonoids, and stilbenes. In nature, these compounds exist in free forms; however, they are frequently conjugated with another polyphenol or with organic acids and carbohydrates, which make it difficult and inefficient to isolate and purify them via the conventional extraction methods. Ogata et al. (1967) firstly described the heterologous production of polyphenols in *Rhodotorula* species. Since then, there have been a lot of studies regarding the production of these compounds in microorganisms, namely, *Escherichia coli* (Jeong et al. 2015; Liu et al. 2016) and *Saccharomyces cerevisiae* (Sydor et al. 2010; Li et al. 2015, 2016) and more recently in *Corynebacterium glutamicum* (Kallscheuer et al. 2016a, b) and *Lactococcus lactis* (Dudnik et al. 2018).

Flavonoids are the largest family of polyphenols and are composed of 15-carbon atoms with two aromatic rings connected through a 3-carbon bridge (Ahmed et al. 2017). They are produced in fruit skin as a response against biotic and abiotic stress such as microbial invasions and environmental stress, among others (Treutter 2006). Flavonoids have been classified into six subgroups, including flavanones (e.g., hesperetin and naringenin); flavones (e.g., luteolin and apigenin); isoflavones (e.g., genistein and daidzein); flavonols (e.g., quercetin and kaempferol); flavan-3-ols (e.g., catechin and epicatechin); and anthocyanidins (e.g., cyanidin and delphinidin) (Trantas et al. 2015).

Anthocyanins are flavonoids, usually found in plant flowers and fruits, which impart color (red, blue, and purple) to them in order to attract pollinators and protect them from excessive sunlight (Chouhan et al. 2017; Zha and Koffas 2017a). They are the glycoside (bonded to a sugar moiety) form of anthocyanidins and are characterized by a hydroxyl group in position 3 and a C-ring fully unsaturated (Zha and Koffas 2017b). For humans, their interest arises from their antioxidant properties (Tsuda 2012). The most common anthocyanidins are cyanidin (red), peonidin (pink), malvidin (reddish purple), pelargonidin (orange red), delphinidin (bluish purple), and petunidin (purple) (Khoo et al. 2017).

Curcuminoids are diarylheptanoids (C6-C7-C6), isolated from the rhizome of turmeric (*Curcuma longa Linn.*), being the curcumin its major component (Rodrigues et al. 2015b). These compounds have been used in traditional medicine due to several therapeutic properties, as well as a food additive (Sood and Nagpal 2013; Amalraj et al. 2017; Hewlings and Kalman 2017). Stilbenes are polyphenols with C6-C2-C6 structure, acting as phytoalexins (they are produced by plants when they are attacked by bacteria, fungi, and viruses). These compounds are normally present in soya, peanuts, wine, and grapes. The most common example of stilbenes is resveratrol (Katsuyama et al. 2007b; Kiselev 2011; Ahmed et al. 2017). Stilbenes can be further decorated resulting in pinostilbene (3-methoxy-4′,5-dihydroxy-trans-stilbene), pterostilbene (trans-3,5-dimethoxy-4-hydroxystilbene), piceid, and resveratroloside (Kallscheuer et al. 2017; Wang et al. 2015a; Zhang et al. 2009).

3.3 Metabolic Pathways Leading to Polyphenol Production

Polyphenols are products of the shikimate/phenylpropanoid pathways. The amino acids L-phenylalanine (L-Phe) and L-tyrosine (Tyr) are the major precursors for their synthesis (Light et al. 2012; Milke et al. 2018). L-Tyr biosynthesis starts with the conversion of chorismate (CHO) to prephenate (PHA) by the enzyme chorismate mutase (CM). The enzyme prephenate dehydrogenase (PDH) catalyzes the conversion of the intermediate PHA to 4-hydroxyphenylpyruvate (HPP) that is transaminated to L-Try (Schenck and Maeda 2018). The biosynthetic pathway for L-Phe biosynthesis also starts with the conversion of CHO to PHA followed by a reaction yielding phenylpyruvate (PPY) catalyzed by the enzyme prephenate dehydratase (PDT) (Vargas-Tah and Gosset 2015) (Fig. 3.2a).

Thereafter, both amino acids are deaminated to the phenylpropanoids cinnamic acid and *p*-coumaric acid, respectively, in a non-oxidative manner, by the activity of phenylalanine ammonia lyase (PAL) (MacDonald and D'Cunha 2007) and tyrosine ammonia lyases (TAL) (Nishiyama et al. 2010) (Fig. 3.2b). The enzyme P450 monooxygenase cinnamate-4-hydroxylase (C4H) can oxidize the cinnamic acid yielding the phenylpropanoid *p*-coumaric acid (Jendresen et al. 2015). For the synthesis of stilbenes and flavonoids, the phenylpropanoids are converted to the phenylpropanoid CoA thioesters by the action of the 4-coumarate-CoA ligase (4CL) enzyme. In the next step, two different types of III polyketide synthases (stilbene synthases, STS, or chalcone synthases, CHS) catalyze the formation of a tetraketide intermediate consuming the phenylpropanoid CoA thioesters and three molecules of malonyl-CoA, yielding a stilbene (catalyzed by the enzyme STS) or chalcone (catalyzed by the enzyme CHS) (Tropf et al. 1994; Austin et al. 2004). Chalcones are then converted to their isomer naringenin by chalcone isomerase (CHI). Naringenin represents the most important precursor molecule for almost all flavonoids and can be further converted to flavanonols (dihydroflavonols), flavonols, and anthocyanidins by hydroxylation (e.g., flavanone 3-hydroxylase (F3H), flavonoid 3′-hydroxylase (F3′H), and flavonoid 3′,5′-hydroxylase (F3′5′H)) and reduction by dihydroflavonol 4-reductase (DFR), followed by the oxidation from anthocyanidin synthase (ANS) to generate the anthocyanidins. They are further glycosylated by flavonoid glucosyltransferases (FGTs), giving the anthocyanins (Zha and Koffas 2017a, b) (Fig 3.2b).

For curcuminoid biosynthesis, 4CL converts *p*-coumaric acid to *p*-coumaroyl-CoA, and *p*-coumaroyl shikimate transferase (CST), *p*-coumaroyl 5-O-shikimate 3=-hydroxylase (CS3=H), and caffeoyl-CoAO-methyltransferase (CCoAOMT)

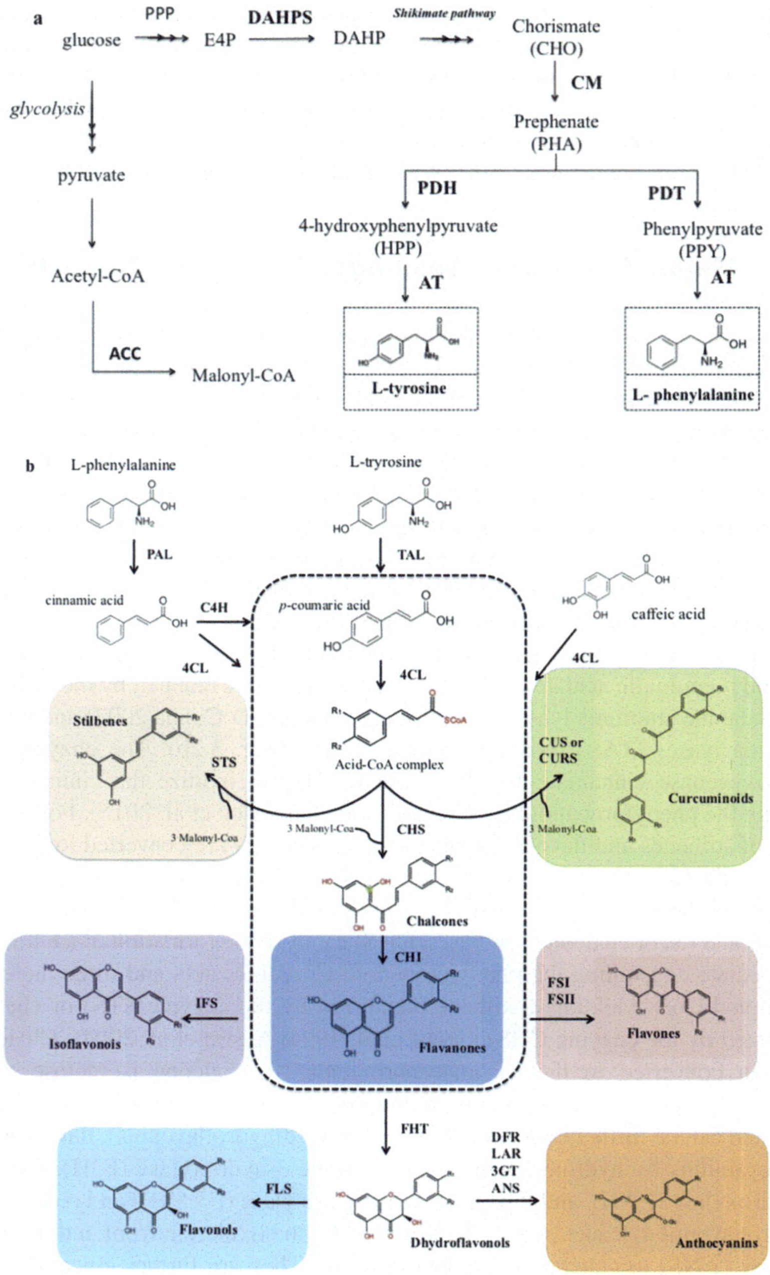

Fig. 3.2 Schematic overview of biosynthetic pathways for stilbenes, flavonoids, and curcuminoids synthesis: (**a**) Biosynthesis of aromatic amino acids by the shikimate pathway (adapted with permission from Milke et al. (2018), (**b**) biosynthesis of polyphenols starting from aromatic amino acids. E4P erythrose 4-phosphate, PPP pentose phosphate pathway, DAHPS DAHP synthase, DAHP 3-deoxy-D-arabinoheptulosonate 7-phosphate, ACC acetyl-CoA carboxylase, CM chorismate mutase, PDT

convert it to feruloyl-CoA. Both phenylpropanoid CoA thioesters are then converted to diketide-CoAs by diketide-CoA synthase (DCS), by condensation with malonyl-CoA. The enzyme curcumin synthases (CURSs) and curcuminoid synthase (CUS) catalyze the formation of curcuminoids by condensing the diketide-CoAs with coumaroyl-CoA and feruloyl-CoA (Rodrigues et al. 2015b; Wang et al. 2015a; Fang et al. 2017) (Fig. 3.2b). In the next steps, a variety of polyphenols can be generated by the action of decorating enzymes (such as acyltransferases and glycosyltransferases) (D'Auria 2006), with glycosyl, methyl, acetyl, or other acyl groups (Harborne and Mabry 2013). The decoration of polyphenols increases its stability, bioavailability, and water solubility (Tsao 2010).

3.4 Polyphenol Production in Engineered Microorganisms

Use of microorganisms to produce a wide variety of compounds has long been described (Du et al. 2013). However, polyphenols are not naturally produced by microorganisms, and the integration of the heterologous pathway from plants is essential (Dudnik et al. 2018). One of the most important factors in the microbial polyphenol production is the selection of the host organism. In the last years, several studies reported the successful expression of the heterologous biosynthetic pathway for polyphenol production in different microorganisms such as *E. coli*, *L. lactis*, *S. cerevisiae*, *S. venezuelae*, and *C. glutamicum*, among others (Wang et al. 2011a; Liu et al. 2016; Kallscheuer et al. 2016b, 2017). The main bottlenecks for microbial polyphenol production are the precursor availability and low activity of the key enzymes in the heterologous hosts. This can be overcome through the supplementation of the production medium with the main precursors, such as L-Phe, L-Try, *p*-coumaric acid, or cinnamic acid, improving polyphenol production through an increase in the amount of malonyl-CoA. The ability of *E. coli* strains to use *p*-coumaric acid and L-Tyr to produce 1308 and 35 mg/l resveratrol, respectively, was described by Lim et al. (2011) and Wu et al. (2013b). For polyphenol production directly from glucose (de novo synthesis), the pentose phosphate pathway and the shikimate pathway were additionally engineered to provide sufficient amounts of L-Try or L-Phe (Mei et al. 2015).

Another issue in this bioprocess is the low intracellular pool of malonyl-CoA. However, the addition of certain compounds to the culture medium, such as cerulenin, is a strategy commonly used. This inhibitor of fatty acid synthesis improves the level of malonyl-CoA and consequently the polyphenol titers (Yang

Fig. 3.2 (continued) prephenate dehydratase, PDH prephenate dehydrogenase, AT aminotransferase, TAL tyrosine ammonia lyase, PAL phenylalanine ammonia lyase, 4CL 4-coumarate, CoA ligase, C4H, cinnamate 4-hydroxylase, STS stilbene synthase, CHS chalcone synthase, CHI chalcone isomerase, IFS 2-hydroxyisoflavanone synthase, FLS flavonol synthase, DFR dihydroflavonol reductase, ANS anthocyanidin synthase, 3GT anthocyanidin 3-glycosyltransferase, FSI soluble flavone synthase, FSII membrane-bound flavone synthase, FHT flavanone 3b-hydroxylase, LAR leucocyanidin reductase, CUS curcuminoid synthase, CURS curcumin synthase

et al. 2015). Nevertheless, the unfeasible price of cerulenin for industrial applications (Santos et al. 2011; de Fouchécour et al. 2018) and its inhibitory effect in cell growth are the main restrictions for using it in large-scale fermentation (Lim et al. 2011; van Summeren-Wesenhagen and Marienhagen 2015; Milke et al. 2018). In fact, the medium supplementation with precursors or cerulenin increases the process costs, and its scale-up for industrial scale is unattractive. In order to overcome these limitations, the use of genetic engineering techniques becomes mandatory (Chityala and Nandana 2017). The production titer of some polyphenols obtained through metabolic engineering in microorganisms, in the past 8 years, has been presented in tabular form (Table 3.1).

3.5 Flavonoids

Flavonoids consist of a large group of poplyphenolic compounds that are responsible for the pigments that color most flower, fruits, and seeds (Shashank and Pandey 2013). The core structure of flavonoid consists of a 15-carbon scaffold made of two benzene rings (A and B) linked via a heterocyclic pyran ring (C). They can be divided into a variety of subclasses such as flavan-3-ols (e.g., catechin and epicatechin), flavones (e.g., luteolin and apigenin), flavonols (e.g., quercetin and kaempferol), flavanones (e.g., hesperetin and naringenin), anthocyanins (e.g., cyanidin and delphinidin), and isoflavonoids (e.g., genistein and daidzein), according to the substitution patterns of the central C ring (Shashank and Pandey 2013). These polyphenols have attracted much attention due to their properties and applications such as antiviral, antibacterial, anti-obesity, and anticancer activities (Wang et al. 2009; Si et al. 2010). Flavonoids are produced through the phenylpropanoid pathway (Fig. 3.2b). Nevertheless, further reactions catalyzed by different enzymes can modify the basic flavonoid backbone to produce a diversity of flavonoids and their derivatives. These enzymes include members of the cytochrome P450 hydroxylase requiring the activity of an NADPH-cytochrome P450 reductase (CPR), NADPH-dependent reductase, 2-oxoglutarate-dependent dioxygenase (ODDs), O-methyltransferase (OMT), acyl and glycosyltransferase (UGT) families (Choi et al. 2012; Koirala et al. 2016).

3.5.1 Microbial Production of Flavonoids

The interest in microbial flavonoid production arisen in 2003, and since then a lot of studies reported the synthesis of different flavonoids in engineered hosts (Wang et al. 2011a; Pandey et al. 2015; Trantas et al. 2015; Jones et al. 2017; Delmulle et al. 2018). Since the major intermediates in flavonoid biosynthesis are flavanones, such as naringenin, much research has focused on their production in microorganisms (Koopman et al. 2012; Wu et al. 2014; Milke et al. 2018) and different approaches developed and applied to increase its production.

Table 3.1 Production titers of some polyphenols by metabolic engineering different microorganisms (published after 2010)

Compound	Microorganism	Heterologous enzymes (Sources)	Substrate/precursor	Titer (mg/l)	References
Resveratrol	*E. coli* BW27784	4CL (*Arabidopsis thaliana*)	*p*-Coumaric acid	0.160	Afonso et al. (2015)
		STS (*Arachis hypogaea*)			
	E. coli BW27784	4CL (*A. thaliana*)	*p*-Coumaric acid	404	Lim et al. (2011)
		STS (*A. hypogaea*)			
		4CL (*A. thaliana*)		1380	
		STS (*V. vinifera*)			
		4CL (*Petroselinum crispum*)		142	
		STS (*A. hypogaea*)			
		4CL (*P. crispum*)		610	
		STS (*V. vinifera*)			
		4CL (*A. thaliana*)	*p*-Coumaric acid	2340	
		STS (*V. vinifera*)			
	E. coli C41 (DE3)	TAL (*Saccharothrix espanaensis*)	*p*-Coumaric acid	1.4	Choi et al. (2011)
		4CL (*Streptomyces coelicolor*)			
		STS (*Arachis hypogaea*)			
	E. coli BW25113 (DE3)	TAL (*Rhodotorula glutinis*)	Glucose	4.6	Liu et al. (2016)
		4CL (*P. crispum*)			
		STS (*V. vinifera*)			
	E. coli BL21 (DE3)	TAL (*R. glutinis*)	L-tyrosine	35.02	Wu et al. (2013b)
		4CL (*P. crispum*)			
		STS (*V. vinifera*)			
		matB and *matC*) (*Rhizobium trifolii*)			

(continued)

Table 3.1 (continued)

Compound	Microorganism	Heterologous enzymes (Sources)	Substrate/precursor	Titer (mg/l)	References
	E. coli BL21(*DE3*)	*tyyA* and *aroG*—integrated in the *E. coli* genome	Glucose	304.5	Wu et al. (2017b)
		TAL (*Trichosporon cutaneum*)			
		4CL (*P. crispum*)			
		STS (*V. vinifera*)			
		matB and *matC* (*R.* trifolii)			
		*fab*F, *fab*B, *fab*I, *fab*D repressed			
	E.coli W	TKT (*E. coli*)	Glycerol	22.58	Camacho-Zaragoza et al. (2016)
	(pheA-)Rg	*aroG* (*E. coli*)			
		TAL (*R. glutinis*)			
		Gene *pheA* was deleted			
	E.coli W-Vv	4CL (*S. coelicolor*)			
		STS (*V. vinifera*)			
	L. lactis	TAL, 4CL, STS, ACC (from different sources)	L-tyrosine	0.45–1.37	Gaspar et al. (2016)
	C. glutamicum DelAro^3	STS (*A. hypogaea*)	*p*-Coumaric acid	12	Kallscheuer et al. (2016b)
		4CL (*P. crispum*)	*p*-Coumaric acid + cerulenin	158	
	C. glutamicum DelAro4	*aroH* (*E. coli*)	Glucose	12	
		TAL(*F. johnsoniae*)	Glucose + cerulenin	59	
		4CL(*Petroselinum*)	Glucose (40 gL^{-1})	4	Braga et al. (2018a)
		STS(*A. hypogaea*)	Glucose (80 gL^{-1})	12	
			Glucose (fed-batch)	7	

	Industrial yeast	4CL1 (*A. thaliana*)	*p*-Coumaric acid	391	Sydor et al. (2010)
		STS (*V. vinifera*)			
	S. cerevisiae W303-1A	4CL1 (*A. thaliana*)	*p*-Coumaric acid	3.1	Shin et al. (2011b)
		STS (*A. hypogaea*)			
	S. cerevisiae WAT11	TAL (*R. sphaeroides*)	Tyrosine	1.9	Wang et al. (2011b)
		4CL::STS, 4CL1 (*A. thaliana*)-STS (*V. vinifera*) fusion enzyme			
	S. cerevisiae WAT11	4CL1 (*A. thaliana*)	*p*-Coumaric acid	14.4	Wang and Yu (2012)
		STS (*V. vinifera*)			
	S. cerevisiae CEN. PK102-5B	4CL1 (*A. thaliana*)	Glucose	415.65	Li et al. (2015)
		STS (*V. vinifera*)			
		TAL (*H. aurantiacus*)			
	S. cerevisiae CEN. PK102-5B	4CL1 (*A. thaliana*)	Ethanol	531.41	Li et al. (2015)
		STS (*V. vinifera*)			
		TAL (*H. aurantiacus*)			
Naringenin	*E. coli* BL21 (DE3)	4CL (*P. crispum*)	*p*-Coumaric acid	474	Xu et al. (2011)
		CHS (*Petunia x hybrida*)			
		CHI (*M. sativa*)			
		ACC (*P. luminescens*)			
		PGK, PDH (*E. coli*)			
	S. cerevisiae IMX198	PAL, C4H, CPR, 4CL, CHS, CHI (*A. thaliana*)	Glucose	109	Koopman et al. (2012)
		TAL (*R. capsulatus*)			
		ARO4G2265 (*S. cerevisiae*)			

(continued)

Table 3.1 (continued)

Compound	Microorganism	Heterologous enzymes (Sources)	Substrate/precursor	Titer (mg/l)	References
7-O-methyl aromadendrin	*E. coli* BL21 (DE3)	4CL (*P. crispum*)	*p*-Coumaric acid	30	Malla et al. (2012)
		CHS (*P. hybrid*)			
		CHI (*M. sativa*)			
Pinocembrin	*E. coli* BL21 (DE3)	PAL (*R. glutinis*)	Glucose	40	Wu et al. (2013a)
		4CL (*P. crispum*)			
		CHS (*P. hybrid*)			
		CHI (*M. sativa*)			
	E. coli BL21 (DE3)	PAL (*R. glutinis*)	Glucose	528.8	Wu et al. (2016)
		4CL (*P. crispum*)			
Eriodictyol	*E. coli* BL21 (DE3)	PAL (*R. glutinis*)	Tyrosine	107	Zhu et al. (2014)
		4CL (*P. crispum*)			
		CHS (*P. hybrid*)			
		CHI (*M. sativa*)			
		F3'H (*Gerbera hybrida*)			
		CPR (*Catharanthus roseus*)			
Curcumin	*E. coli K-12 MG1655(DE3)*	DCS (*Curcuma longa*)	Caffeic acid	70	Rodrigues et al. (2015a)
		CURS1 (*Curcuma longa*)			
		4CL1 (*A. thaliana*)			
Curcuminoids (bisdemethoxy curcumin and dicinnamoyl methane)	*E. coli B-CU10*	PAL (*A. thaliana*)	Glucose	4.63 + 6.95	Kim et al. (2017)
		TAL (*Saccharothrix espanaensis*)			
		4CL (*Oryza sativa*)			
		CUS (*O. sativa*)			
Cyanidin 3-O-glucoside	*E. coli* BL21 (DE3)	ANS (*M. domestica*)	Catechin	350	Lim et al. (2015)
		F3GT (*P. hybrida*)			

Apigenin glucoside	*E. coli* BL21 (DE3)	OleD GT (*S. antibioticus*)	Apigenin	4.67	Choi et al. (2012)
Hypolaetin	*E. coli* BL21 (DE3)	Sam5 (*S. espanaensis*)	Luteolin	88	Lee et al. (2014)
		OsP450 reductase (*O. sativa*)			
Kaempferol	*E. coli* BL21 (DE3)	GbFLS (*G. biloba*)	Dihydrokaempferol	n.e.	Xu et al. (2012)
			Naringenin	n.e.	
Kaempferol-3-o-glucoside	*E. coli* BL21 (DE3)	F3H and FLS1(*A. thaliana*)	Naringenin	109.3	Malla et al. (2013)
		UGT78K1 (*Glycine max*)			
		Phosphoglucomutase (*Nocardia farcinica*)			
		UTP-glucose-1-phosphate uridylyltransferase (*E. coli*)			

The first attempts for microbial production of flavonoids relied on medium supplementation with phenylpropanoic acid precursors (Hwang et al. 2003; Yan et al. 2005b). Watts et al. (2004) constructed an *E. coli* harboring PAL from *Rhodobacter capsulatus* and 4CL and CHS from *A. thaliana* and produced 20.8 mg/l of naringenin from phenylpropanoic acids. Similarly, Jiang et al. (2005) also expressed PAL from the red yeast *R. toruloides*, 4CL from the plant *A. thaliana*, and CHS from the plant *Hypericum androsaemum* in *S. cerevisiae* to produce naringenin (7 mg/l) and pinocembrin (0.8 mg/l) using *p*-coumaric acid, ferulic acid, and caffeic acid as substrates. Recently, Kallscheuer et al. (2016b) evaluated the CHS and CHI enzymes originating from petunia (*Petunia hybrida*) for flavonoid production in *C. glutamicum* that also carries the *4cl* gene of *P. crispum*. This strain accumulates 35.2 mg/l of naringenin from 5 mM *p*-coumaric acid and 37.4 mg/l of eriodictyol from 5 mM caffeic acid, in the presence of 25 mM cerulenin. Another approach that has also been described is the flavonoid production from the amino acid precursors L-Phe and L-Try (Yan et al. 2005b; Trantas et al. 2009; Stahlhut et al. 2015; Lyu et al. 2017). Hwang et al. (2003) assembled the *RrPAL* from *Rhodotorula rubra*, *ScCCL* from *Streptomyces coelicolor* A3(2), and *GeCHS* from *Glycyrrhizae chinata* in *E. coli* for pinocembrin and naringenin from L-Phe and L-Tyr, respectively. Soon after, a similar strategy was attempted in *S. cerevisiae* by Ro and Douglas (2004), and Ralston et al. (2005) also introduced partial flavonoid and isoflavonoid biosynthetic pathways in *S. cerevisiae*.

In the past few years, efforts have been arising for de novo synthesis of flavonoids starting from cheap and renewable carbon sources. The enzymes PAL/TAL, 4CL, CHS, and CHI were harbored in *E. coli* (Santos et al. 2011; Wu et al. 2013a) and *S. cerevisiae* (Koopman et al. 2012). About 29 mg/l of naringenin and 40 mg/l of pinocembrin were obtained from glucose in *E. coli* strains, whereas naringenin production from glucose in *S. cerevisiae* was 109 mg/l. Kallscheuer et al. (2016b) deregulated the shikimate pathway and introduced a heterologous TAL from *Flavobacterium johnsoniae* in *C. glutamicum* enabling a production of 32 mg/l naringenin from glucose. These titers can be further increased with medium supplementation with cerulenin (84 mg/l naringenin obtained by Santos et al. (2011)). More recently, Rodriguez et al. (2017) engineered *S. cerevisiae* for de novo production of naringenin, liquiritigenin, kaempferol, resokaempferol, quercetin, and fisetin from glucose, with good production titers for kaempferol (26.57 mg/l) and quercetin (20.38 mg/l). This is also the first report of de novo biosynthesis of resokaempferol and fisetin in yeasts. Duan et al. (2017) also assembled a FLS from *Populus deltoides* for kaempferol production in *S. cerevisiae*; however, the titer achieved was lower (6.97 mg/l) than the ones previously described by Rodriguez et al. (2017). Nevertheless, the authors also tested other strategies to increase the kaempferol production, such as the overexpression of acetyl-CoA biosynthetic pathway and *p*-coumarate supplementation, combined with a fed-batch process allowing a production of 66.29 mg/l.

This is a clear hint that one of the issues in microbial flavonoid production is the low level of free malonyl-CoA (Zha et al. 2009) and several efforts have been carried out to increase the intracellular malonyl-CoA pools and consequently the flavonoid titers (Fowler et al. 2009; Zha et al. 2009). To increase the malonyl-CoA pool, different strategies have been implemented, such as overexpression of ACC complex genes (Leonard et al. 2007) and knockout of acetate kinase (*ackA*) and acetaldehyde dehydrogenase (*adhE*) (Zha et al. 2009). Strategies at the genome level were also applied to optimize the flavanone production in *E. coli* (Fowler et al. 2009). Another strategy was described by Leonard et al. (2008) in an *E. coli* strain. The overexpression of malonyl-CoA synthetase (*matB*) and malonate carrier protein (*matC*) in a culture medium supplemented with malonate and cerulenin increased the flavavone production to 710 mg/l. Meanwhile, other flavonoids as genistein, kaempferol, and quercetin were produced by feeding naringenin to engineered yeast cells (Trantas et al. 2009). Also, the medium supplementation with *p*-coumaric acid allowed the production of kaempferol and quercetin. Fisetin was produced from L-Try (Leonard et al. 2006; Santos et al. 2011; Stahlhut et al. 2015). Leonard et al. (2005) reported the construction of *S. cerevisiae* strains able to produce different flavones (chrysin, apigenin, and luteolin) and intermediate flavanones from phenylpropanoid acids. Further expression of flavone synthase (FSI) derived from parsley allowed the production of 2–3 mg/l of apigenin and luteolin. The same group further tested the production of flavone in an *E. coli* that was previously modified for flavanone production, achieving a production of 0.4 mg/l of apigenin, 10 μg/l of luteolin, and 0.2 mg/l of genkwanin (Leonard et al. 2006).

Lee et al. (2014) reported the production of other flavone, hypolaetin (88 mg/l), from luteolin in *Saccharothrix espanaensis*. The production of isoflavones, such as genistein, was also described by Leonard and Koffas (2007) in *E. coli* and by Trantas et al. (2009) in *S. cerevisiae*. The production of 5-deoxyflavanone (flavanones present in leguminous plants) like liquiritigenin, isoliquiritigenin, 7-hydroxyflavanone, butin, and butein was described by Yan et al. (2007) in *E. coli* and *S. cerevisiae*, with titers ranging from 0.5 to 17 mg/l. Co-culture strategies have recently been highlighted as an interesting and alternative strategy to reduce the host metabolic burden and increase the flavonoid titers. Jones et al. (2016) used this strategy with an *E. coli-E. coli* co-culture for the production of flavan-3-ols (40.7 mg/l). The production of naringenin from D-xylose (an alternative carbon source) was described by Zhang et al. (2017), with a co-culture system of *S. cerevisiae* and *E. coli*, attaining a production of 21.16 mg/l. Recently, the production of apigetrin from glucose (an apigenin-7-O-β-d-glucopyranoside of apigenin) in an *E. coli-E. coli* co-culture system was described by Thuan et al. (2018b). After the process scale-up, a yield of 16.6 mg/l was achieved.

Another interesting approach in microbial production of polyphenols is the ability to use microorganisms to produce unusual molecules, in this case flavonoids that are not natively produced in plants. Even though plants do not produce these compounds, they seem very promising due to their biological activity. Katsuyama et al. (2007b) used a recombinant *E. coli* strain to produce natural and unnatural flavonoids from different precursors, allowing the production of 14 flavanones, 13 flavones, and 8 flavonols. Moreover, different flavonoids have been generated by the action of various decorating enzymes. The production of methylated flavanones from glucose (sakuranetin, ponciretin) and kaempferol 3-O-rhamnoside in *E. coli* has been described by Kim et al. (2013b) and Yang et al. (2014) with a production titer of 42.5 mg/l of ponciretin and 40.1 mg/l of sakuranetin, respectively, and 57 mg/l of kaempferol 3-O-rhamnoside.The glycosylation of kaempferol and quercetin by a rhamnose flavonol glycosyltransferase allowed the production of the corresponding 3-O-rhamnosides at concentrations of 150 and 200 mg/l, respectively (Kim et al. 2009). Xia and Eiteman (2017) engineered *E. coli* strains to generate high concentrations of quercetin glucosides from quercetin. After the process scale-up to bioreactor, a quercetin glucoside titer of 3.9 g/l was achieved. The glycosylation of different isoflavonoids, such as genistein, daidzein, or formononetin, has also been successfully applied, leading the production of daidzein, genistin, or ononinin (Li et al. 2014; Pandey et al. 2014). The production of sophoricoside, a 4′ glucoside of genistein, was achieved in *E. coli* (Ruby et al. 2014). The production of unnatural flavonoid glycosides was also reported (Joe et al. 2010; Kim et al. 2010, 2013a; Yoon et al. 2012).

3.6 Anthocyanins

Anthocyanins are water-soluble pigmented flavonoids, produced by terrestrial plants. They are widely applicable to pharmaceutical products, food processing, cosmetic manufacturing, and solar cell development. Purple, red, and blue pigments extracted from fruits, vegetables, and flowers are used as natural food colorants and dyes (Khoo et al. 2017). Their production only occurs under stress conditions (Tsuda 2012) or infection by pathogens (Bakowska-Barczak 2005). Some anthocyanins have also been used in traditional medicine to treat various diseases like cardiovascular diseases, cancer, neurodegenerative diseases, obesity, and diabetes (Wallace and Monica 2013; Mierziak et al. 2014). In plants, anthocyanins are stored in vacuoles after their production, and a diversity of colors with different stabilities can be observed depending on the pH (He et al. 2010). As anthocyanins are very unstable at basic and neutral pH, plants use different strategies to stabilize them. The most common strategies consist of using structural decorations, lowering the pH, and copigmentation in vacuoles (Passeri et al. 2016; Kallam et al. 2017).

3.6.1 Microbial Production of Anthocyanins

The first report concerning the microbial anthocyanin biosynthesis was done by Yan et al. (2005a). An *E. coli* strain cloned with the genes of F3H and ANS from *Malus domestica*, DFR from *Anthurium andraeanum*, and flavonoid 3-O-glucosyltransferase (F3GT) from *Petunia hybrid* produced 6 mg/l cyanidin 3-O-glucoside and 5.6 mg/l pelargonidin 3-O-glucoside, using naringenin and eriodictyol as precursors. However, the titers obtained were very low, and different strategies have been further employed to enhance the anthocyanin productivity. Enzyme screening and selection is an important way to increase the anthocyanin production. Yan et al. (2008) screened ANS from different plant sources (*A. majus*,

Gerbera hybrida, *P. hybrida*, and *M. domestica*), and the maximum cyanidin production in *E. coli* was observed with ANS from *P. hybrid*. However, the production titers are still low since the heterologous expression of the enzymes from plants in prokaryotic cells is still a challenge. One alternative to overcome this issue is the fused expression of multiple enzymes in successive steps to simulate the enzyme complex that probably exists in plants. The cyanidin 3-O-glucoside titer was increased by fusing F3GT from *A. thaliana* with the N-terminus of ANS from *P. hybrid* with a pentapeptide linker (Yan et al. 2008).

In fact, balancing the cofactor and/or co-substrates supply for electron transfer and enzyme activation/stabilization are significant challenges for the efficient biosynthesis of anthocyanins (Turnbull et al. 2004; Yan et al. 2008). UDP-glucose is an essential cofactor for the glycosylation of some anthocyanins. Its adequate supply is essential for their efficient production. The overexpression of genes responsible for its biosynthesis and simultaneously blocking the competitive UDP-glucose consumption pathways are commonly used as a strategy. This was described by Leonard et al. (2008) and Yan et al. (2008), and in both studies an increase in cyanidin 3-O-glucoside production has been observed. Other important co-substrates for anthocyanin overproduction are the sodium ascorbate and S-adenosyl-L-methionine, necessary for anthocyanin methylation (Yan et al. 2008; Cress et al. 2017).

Apart from the issues previously described, another obstacle in anthocyanin biosynthesis is the product toxicity to cells. Lim et al. (2015) identified a transporter protein responsible for the transportation of catechin and cyanidin 3-O-glucoside in *E. coli*, and an anthocyanin titer of 350 mg/l was observed, which is the highest production level reported to date. An interesting strategy that could be tested is the introduction of plant transporters in microbial hosts. However, until now, these transporters have not been tested in bacterial strains. Nevertheless, the culture process parameters also need to be optimized in order to achieve an efficient anthocyanin production. Plants produce and stabilize anthocyanins through pH adjustments. Stabilizing them in producing hosts is difficult because yeast and bacteria usually grow around a pH of 7. A two-step cultivation strategy was proposed by Yan et al. (2008) to overcome this problem, and with the proposed strategy, it was possible to achieve a cyaniding 3-O-glucoside titer of 38.9 mg/l. Other factors are also important in microbial anthocyanin production, such as dissolved oxygen, temperature, and substrate feeding (Lim et al. 2015; Zhao et al. 2015; Zha and Koffas 2017b).

Efforts have been made in recent years to increase the anthocyanin titers obtained with microbial hosts; however, many difficulties still remain to be overcome. In fact, although it was possible to increase the production of cyanidin-3-O-glucoside up to 350 mg/l, it was necessary to supplement the medium with the precursor flavan-3-ol, and its direct production from flavonoids never exceeded 2.07 mg/l (Yan et al. 2005a, 2008; Lim et al. 2015; Cress et al. 2017). This may be related to the metabolic burden of expressing an extended pathway, as previously discussed. Recently, Jones et al. (2017) described the complete biosynthesis of pelargonidin-3-O-glucoside using a four-strain *E. coli* polyculture collectively expressing 15 heterologous genes, via de novo synthesis from glucose, achieving a production of 9.5 mg/l. Until 2017 the heterologous synthesis of anthocyanins has only been accomplished in *E. coli*. However, *S. cerevisiae* presents some unique advantages over *E. coli* for the design and construction of a biosynthetic pathway for the production of anthocyanins: it has a food-grade status (GRAS organism) and has intracellular compartments similar to those of plants, and the P450 enzymes would be more adequately expressed in an eukaryotic organism (Sahdev et al. 2008). Hence, Eichenberger et al. (2018) firstly described de novo production of pelargonidin-3-O-glucoside, cyanidin-3-O-glucoside, and delphinidin-3-O-glucoside in *S. cerevisiae*. This study represents an important step toward sustainable industrial production of anthocyanins since describes for the first time de novo biosynthesis of anthocyanins in yeast and for the first time in a single microorganism.

3.7 Curcuminoids

Curcuminoids such as curcumin are isolated from the rhizome of turmeric (*Curcuma longa* L.) and are responsible for its yellow color (Palve and Nayak 2012). However, other two related curcuminoids as demethoxycurcumin and bisdemethoxycurcumin were also present in the turmeric (Amalraj et al. 2017). Nowadays the commercially available curcumin is isolated from the rhizome of *C. longa* and contains a mixture of curcumins (77%), demethoxycurcumin (8 %), and bisdemethoxycurcumin (5%). The term curcumin is normally used to represent all the three curcuminoids found in the turmeric extract (Hewlings and Kalman 2017). Several studies have demonstrated health-enhancing properties of curcuminoids including antimicrobial, neuroprotective, antioxidant, anti-inflammatory, anticancer, cardioprotective, and radioprotective effects, among others (Shin et al. 2011a; Amalraj et al. 2017).

3.7.1 Microbial Production of Curcuminoids

The first work that report the heterologous production of curcuminoids in *E. coli* dates back to 2008 (Katsuyama et al. 2008). The authors used the enzymes from the phenylpropanoid pathway (PAL from *Rhodotorula rubra* and 4CL from *Lithospermum erythrorhizon*) and CUS from *Oryza sativa* for curcuminoid biosynthesis. The engineered strain produced a curcuminoid titer of around 35.3 mg/l, in the presence of L-Try and/or L-Phe. Further, they also tested the curcuminoid production by directly supplying phenylpropanoid acids (*p*-coumaric acid, cinnamic acid, and ferulic acid). This *E. coli* strain carrying only 4CL, CUS, and ACC from *C. glutamicum* (overexpressed to increase the intracellular pool of malonyl-CoA) was able to produce around 100 mg/l of curcuminoids. A similar study was reported by Katsuyama et al. (2010). They described the production of 15 asymmetric curcuminoids through the simultaneous addition of two different unnatural carboxylic acids analogous to *p*-coumaric acid (the precursors), using an *E. coli* harboring CUS, ACC, and 4CL. Wang et al. (2015a) managed to co-express five enzymes including TAL, 4CL, C3H, caffeic acid 3-O-methyltransferase (COMT), and CUS in *E. coli*. This resulted in the total biosynthesis of the curcumin, with a titer of

0.6 mg/l. By removal of COMT, the authors demonstrated the rational design of new molecules, generating two novel molecules, 8- and 8-OH unmethylated, that leads to the production of the curcumin analog dicaffeoylmethane.

Rodrigues et al. (2015a) reported the curcuminoid production through ferulic and caffeic acid. An *E. coli* strain harboring with genes for DCS and CURS1, isolated from *C. longa*, and the gene for 4CL isolated from *A. thaliana* was able to produce 70 mg/l of curcumin from ferulic acid. Moreover, curcuminoids were produced from L-Try through the caffeic acid pathway. TAL from *R. glutinis* and 4-coumarate 3-hydroxylase (C3H) from *S. espanaensis* were selected to produce caffeic acid. This study reported an alternative pathway for curcumin production from L-Try through the production of caffeic acid as an intermediate. A recent study from Kim et al. (2017) reported the constructions of a recombinant *E. coli* strain harboring PAL from *A. thaliana* or TAL from *Saccharothrix espanaensis*, together with 4CL from *O. sativa* and CUS also from *O. sativa*. They observed that introducing genes coding for tyrosine-specific TAL and phenylalanine-specific PAL, respectively, it was possible to synthesize a specific curcuminoid (bisdemethoxycurcumin and dicinnamoylmethane). These results stressed out the importance of this enzyme to direct the pathway to a specific curcuminoid, if desired.

A significant progress has been achieved in the microbial production of curcuminoids in recent years; however, the established methodologies are still prohibitive for process scale-up. Generally, to achieve a high biomass production and a suitable protein production level, a first step of growth in lysogeny broth (LB) is performed. As soon as the cells reach the exponential growth phase, they are harvested and transferred to M9 modified minimal salt medium. The curcuminoid production begins by adding the substrates (amino acids or ferulic acid). Other studies also reported the LB supplementation with glucose after protein expression (Wang et al. 2013; Wang et al. 2015a). This strategy is suitable for laboratory scale fermentations, but its scale-up is laborious and expensive. Couto et al. (2017) described a promising strategy for curcumin production in *E. coli*. The authors studied the effect of different culture conditions in curcumin production, and they observed that one-step cultivation in terrific broth can be a very interesting alternative medium to produce curcumin, with a titer of 817.7 mM. This is the highest concentration of curcumin reported so far in a heterologous organism.

Another drawback in this process is the toxicity of the precursor (caffeic acid, *p*-coumaric acid, and ferulic acid) to the cells (Zhang and Stephanopoulos 2013). Alternative strategies should be considered such as the addition of lower substrate concentrations at the beginning of the experiment and further additions following their consumption rate (stepwise fed-batch) (Huang et al. 2013). One of the major bottlenecks for curcumin production, as other polyphenols, in engineered microorganisms is the low intracellular availability of malonyl-CoA. Significant efforts have been devoted to the development of different approaches to increase the intracellular pool of malonyl-CoA (Fowler et al. 2009; Xu et al. 2011).

The dependence on precursor feeding is also an issue in this process, due to the high market price of these compounds (Santos et al. 2011). Nevertheless, the development of engineered strains able to convert cheaper substrates, such as glucose, may solve this problem, making it possible to produce curcuminoids via de novo synthesis (Rodriguez et al. 2014; Santos et al. 2012). Furthermore, the use of residues and wastes as an alternative medium is also an interesting approach. Rice bran pitch that is rich in ferulic acid was used for curcumin production. Starting with 11 mg of ferulic acid extracted from 500 mg of rice bran pitch, the *E. coli* strain engineered by Katsuyama et al. (2008) harboring 4CL, CUS, and ACC genes produced 57 mg/l of curcumin. A co-culture strategy was recently described by Fang et al. (2017) with two different *E. coli* strains, the *E. coli* rpoA14 (DE3), used to biosynthesize *p*-coumaric acid from glucose, and the *E. coli* BL21 Star™ (DE3), which produces the curcuminoids from *p*-coumaric. This strategy allows the production of 6.3 mg/l of curcuminoids in 22 h. Many attempts have been made to implement the heterologous biosynthesis of curcuminoids; however, to the author's knowledge, until now the heterologous synthesis of curcuminoids has only been achieved in *E. coli* strains.

3.8 Stilbenes

Stilbenes are characterized by the presence of a 1,2-diphenylethylene nucleus and can be naturally found in berries, grapes, and peanuts and in the constituents of other plants. In plants, they are produced as a defense mechanism against outside stresses like infections and UV radiation (Kiselev 2011; Sytar Oksana et al. 2012; Reinisalo et al. 2015). A huge interest has risen in this class of polyphenols due to their health-enhancing properties, including anticancer, antiaging, antiatherogenic, anti-inflammatory, and antioxidant activities (Sharma et al. 2007; Quideau et al. 2011; Atanasov et al. 2015). Resveratrol (3,5,4-trihydroxy-transstilbene) is the stilbene most extensively studied, due to its large spectrum of biological activities (De Filippis et al. 2017). An increasing interest in its production rose after the discovery that resveratrol might be one of the factors responsible for the low incidence of cardiovascular diseases in the French population despite having a fat-rich diet (also known as "French paradox") (Yang et al. 2014). As other polyphenols, stilbenes can be further decorated by acylation (hydrangeic acid), glycosylation (piceid, resveratroloside), or O-methylation (pinostilbene, pterostilbene) (Kim et al. 2002; Fabris et al. 2008; Jeong et al. 2015; Wang et al. 2015b). Over the last years, this class of polyphenols has received considerable interest, and several studies have reported its production in microbial hosts.

3.8.1 Microbial Production of Stilbenes

Engineering microbial host for stilbene production might represent an interesting alternative for its production in large quantities. Since resveratrol is one of the stilbenes mostly investigated and engineered in microorganisms, this topic will focus on the recent advances in its microbial production. In recent years, microbial production of resveratrol was achieved in yeast and bacteria by the introduction of the heterologous pathway from plants and engineering of the microbial host metabolism toward increased production (Beekwilder et al. 2006; Yang et al. 2015). Several studies have reported the successful expression of the heterologous biosynthetic pathway for resveratrol production in bacteria, such as *E. coli*, *L. lactis*, and *C. glutamicum*, and in yeasts *S. cerevisiae* and *Yarrowia lipolytica* (Becker et al. 2003; Beekwilder et al. 2006; Huang et al. 2006; Zhang et al. 2006; Donnez et al. 2009; Trantas et al. 2009; Sydor et al. 2010; Kallscheuer et al. 2016b; Dudnik et al. 2018).

The first works concerning resveratrol production in *E. coli* are based on the biotransformation of its precursors like L-Tyr and *p*-coumaric acid. Watts et al. (2006) firstly described the resveratrol production in an engineered *E. coli* strain BW27784 harboring 4CL and STS genes from different plants and achieved a resveratrol production of 104.5 mg/l, with the supplementation of *p*-coumaric acid. Other studies also reported the resveratrol production from the precursors of phenylpropanoid pathway and/or from phenylpropanoid acids; however, the resveratrol titers obtained were very low (< 100 mg/l) (Watts et al. 2006; Beekwilder et al. 2006; Wu et al. 2013b; Zhang et al. 2015). Lim et al. (2011) achieved a resveratrol production of 2.39 g/l with an *E. coli* strain BW27784 harboring 4CL from *A. thaliana* and STS from *V. vinifera*, in the presence of 15 mM *p*-coumaric acid and cerulenin.

The first report on microbial production of resveratrol in *S. cerevisiae* was described by Beekwilder et al. (2006). The genes 4CL2 from *Nicotiana tabacum* cv. Samsun and STS from *V. vinifera* were integrated in the genome of *S. cerevisiae* CEN.PK113-3B, achieving a resveratrol production of 5.8 mg/l from 820.8 mg/l *p*-coumaric acid. Different approaches were further developed achieving a maximum resveratrol titer of 391 mg/l obtained with an engineered industrial Brazilian *S. cerevisiae* strain that overexpressed STS and 4CL1 genes with medium supplementation with 15 mM *p*-coumaric acid (Sydor et al. 2010).

Besides the commonly used hosts *E. coli* and *S. cerevisiae*, other industrially relevant bacteria and yeast were used as a chassis for resveratrol production. Gaspar et al. (2016) and Kallscheuer et al. (2016b) demonstrated that *L. lactis* and *C. glutamicum*, respectively, are able to produce resveratrol, showing the assembly and functional expression of synthetic pathways for resveratrol production. Another bacterium described for resveratrol production is *S. venezuelae* (Park et al. 2009). Huang et al. (2006) also reported the resveratrol production in the nonconventional yeast *Y. lipolytica* (ATCC 20362 strain).

The resveratrol production can be improved by exploring alternative enzymes from other sources that allow higher resveratrol yield or more specific enzymes that allow the production of the desired compound with fewer by-products. Also, protein engineering and mutagenesis have been applied to improve resveratrol production in microorganisms (Zhang et al. 2006, 2015; Wang et al. 2011b; Wang and Yu 2012; Wu et al. 2013b). One of the major bottlenecks for resveratrol production in engineered microorganisms, as other polyphenols, is the low intracellular availability of malonyl-CoA, and an improvement in its production can be achieved by redirecting more malonyl-CoA into the resveratrol biosynthetic pathway (Katsuyama et al. 2007a; Lim et al. 2011; Choi et al. 2011; Bhan et al. 2013; Wu et al. 2013b; Yang et al. 2015).

Lim et al. (2011) achieved a resveratrol production of 2.3 g/l by a two-step biotransformation from *p*-coumaric acid in presence of cerulenin, with an *E. coli* strain. Kallscheuer et al. (2016b) also engineered a *C. glutamicum* strain for resveratrol obtaining a titer of 158 mg/l from 5 mM of *p*-coumaric acid in presence of 25 mM cerulenin. Nevertheless, these approaches are not feasible for large-scale fermentations. Alternative strategies, such as rerouting native metabolic flows, using stoichiometric modeling to improve malonyl-CoA availability and expression of heterologous genes coding for a malonyl-CoA synthetase and a malonate importer protein, were successfully attempted (Zha et al. 2009; Yang et al. 2015; Wu et al. 2017b).

The production of phenylpropenoic acids as cinnamic acid or *p*-coumaric acid could be increased through the heterologous expression of PAL or TAL genes (Huang et al. 2013; Zhang and Stephanopoulos 2013). Trantas et al. (2009) cloned a *S. cerevisiae* strain with PAL, C4H, 4CL, RS, and CPR genes. This strain was able to produce 0.29 mg/l of resveratrol from L-Phe. A similar strategy was further described by Shin et al. (2012) in a *S. cerevisiae* carrying PAL from *R. toruloides*, C4H and 4CL1 genes from *A. thaliana*, and the STS gene from *A. hypogaea*, reaching a resveratrol titer of 5.8 mg/l from L-Try. This strategy has also been described for *E. coli* (Katsuyama et al. 2007a; Wu et al. 2013b; Wang et al. 2015a).

Since L-Try and L-Phe are native amino acids that can be overproduced in some microorganisms, de novo production of resveratrol from simple carbon sources is also an interesting alternative, since it allows the use of cheap and renewable substrates. Li et al. (2015) described de novo production of resveratrol from glucose or ethanol in *S. cerevisiae* in fed-batch fermentation, resulting in a resveratrol titer of 415.65 and 531.41 mg/l from glucose or ethanol, respectively. Recently, a site-specific integration strategy was utilized to chromosomally insert resveratrol biosynthetic pathway in *E. coli*, but it only produced a titer of 4.6 mg/l of resveratrol from glucose (Liu et al. 2016). Soon after, Wu et al. (2017a) achieved a resveratrol production of 304.5 mg/l in *E. coli* from glucose, using a TAL from *Trichosporon cutaneum* (TcTAL). Nevertheless, the resveratrol titer achieved is still lower than the one reported in *S. cerevisiae* (Li et al. 2015). Kallscheuer et al. (2016b) also constructed a *C. glutamicum* strain able to produce 60 mg/l resveratrol from glucose, after deregulation of the shikimate pathway and introduction of a heterologous TAL from *F. johnsoniae*. The resveratrol production with this strain was further optimized at bioreactor scale by Braga et al. (2018a). They observed that an increase in glucose concentration from 40 to 80 g/l leads to an increase in the resveratrol titer from 4 to 12 mg/l (Braga et al. 2018a); however, the concentration attained was lower than the ones described by Kallscheuer et al. (2016b), in shake flaks. The impact of oxygen on resveratrol biosynthesis and stability was further discussed by Braga et al. (2018a). They observed that high oxygen concentrations in the bioreactor affected negatively the resveratrol titers with *C. glutamicum*, since they noticed that the resveratrol concentration decreased significantly after reaching a maximum product concentration and also that the resveratrol production in bioreactor is lower than the ones obtained in shake flask. To overcome this issue, an *in situ* product removal strategy was applied by Braga et al. (2018b) for resveratrol production with *C. glutamicum* using Amberlite XAD-7HP as adsorbent. They observed an increase (from 75% to 95%) in the amount of extracellular resveratrol produced. With this strategy the potential problems with the toxicity of resveratrol to the cells and undesired oxidation were avoided.

Co-culture strategies have recently been described by Camacho-Zaragoza et al. (2016) for resveratrol production. The authors firstly described a co-culture strategy with two *E. coli* strains for the resveratrol production from glycerol, attaining a final titer of 22.6 mg/l. In order to enhance resveratrol stability, solubility, and uptake into human cells, it must be protected from light, oxygen, and harsh pH conditions. This is achieved through the use of decorating enzymes (Rimando et al. 2002; Chao et al. 2010; Fulda 2010). Consequently, other stilbenes like pinosylvin, piceatannol, as well as methylated and glucosylated resveratrol have been produced in microbial hosts. A brief overview of the recent research achievements on its biosynthesis will also been presented.

The production of O-methylated derivatives of resveratrol pinostilbene and pterostilbene from L-Try has been described by Katsuyama et al. (2007a) in an *E. coli* strain expressing the pinosylvin methyltransferase (PMT) homologue from *Oryza sativa*. They attained a production of 18 and 5.8 mg/l of pinostilbene and pterostilbene, respectively. However, Kang et al. (2014) first reported the production of the methylated resveratrol compounds bis-methyl and tri-methyl resveratrol (3,4′-dimethoxy-5-hydroxystilbene and 3,5,4′-trimethoxystilbene) in an *E. coli* culture without precursor feeding in the culture. Soon after, Jeong et al. (2014) reported the production of 34 mg/l of pinostilbene from 1 mM resveratrol in *E. coli* expressing the resveratrol O-methyltransferase gene from *Vitis riparia*. The same group described the production of pinostilbene from *p*-coumaric acid in *E. coli* through co-expression of multiple enzymes (CCL, STS, ROMT), achieving a maximum titer of 2.6 mg/l (Jeong et al. 2015). Wang et al. (2015b) also expressed ROMT from *V. vinifera* in *E. coli* and *S. cerevisiae* allowing a pterostilbene titer of 50 mg/l and 2.2 mg/l, respectively, from *p*-coumaric acid. Recently Heo et al. (2017) demonstrated de novo synthesis of pterostilbene in a tyrosine overproducing *E. coli* strain, reaching a titer of 33.6 mg/l. *C. glutamicum* was also engineered to produce pterostilbene from *p*-coumaric acid, achieving a titer of 42 mg/l (Kallscheuer et al. 2017).

The microbial production of pinosylvin, a resveratrol analogue of polyketide stilbenoid, has also been reported. Great efforts have been made to enable its production via de novo synthesis. With co-expression of PAL, 4CL, and STS in *E. coli*, Wang et al. (2015a) achieved de novo biosynthesis of pinosylvin from glucose with a titer of 13.3 mg/l. The addition of cerulenin allowed the production of 70 mg/l pinosylvin from glucose in *E. coli* (van Summeren-Wesenhagen and Marienhagen 2015). More recently, Liang et al. (2016) produced 47.40 mg/l of pinosylvin from glycerol, using CRISPRi to inactivate the malonyl-CoA consumption pathway in order to increase its availability. Nevertheless, the highest pinosylvin titer obtained from glucose, without any precursor supplementation, was described by Wu et al. (2017a). The hydroxylation of resveratrol enables the production of piceatannol, and various monooxygenases were reported to efficiently perform this reaction (Lee et al. 2012; Furuya and Kino 2014). However, its de novo production from glucose was also reported by Wang et al. (2015b) attaining a production of 21.5 mg/l.

The expression of glucosyltransferases from different sources allowed the production of resveratrol glucoside derivatives, like piceid (resveratrol-3-O-glucoside) and resveratroloside (resveratrol-O-glucoside) in *E. coli* (Ozaki et al. 2012; Choi et al. 2014). However, Thuan et al. (2018a) described a different strategy for resveratrol glucoside production. They used a co-culture with two *E. coli* strains, one expressing two enzymes that convert *p*-coumaric acid into resveratrol and another expressing glucosyltransferase to convert the resveratrol into its glucosidated forms: polydatin and resveratroloside.

3.9 Conclusions and Future Prospects

Plant secondary metabolites, such as stilbenes, flavonoids, anthocyanins, and curcuminoids, have shown several health benefits, and efforts have been made in order to increase the production titers obtained with microbial hosts. In fact, some polyphenols (e.g., stilbenes and flavonoids) are produced at gram-scale from inexpensive carbon sources; however this is still a challenge for more complex structures. The major drawbacks in this process are the low activity of the enzymes from plants in heterologous hosts and the deficient supply of precursor molecules by the microbial metabolism. Nevertheless, we believe that using synthetic biology approaches and metabolic engineering tools combined with process engineering and optimization, the polyphenol titers and process yields obtained can be further improved, making the microbial production of polyphenols competitive and economically feasible at industrial scale.

Acknowledgments We would like to thank the European Union Framework Program 7 "BacHBerry" (www.bachberry.eu), Project No. FP7- 613793 for financial support, the Portuguese Foundation for Science and Technology (FCT) under the scope of the strategic funding of UID/BIO/04469 unit, COMPETE 2020 (POCI-01-0145-FEDER-006684), and BiotecNorte operation (NORTE-01-0145-FEDER-000004) funded by the European Regional Development Fund under the scope of Norte2020–Programa Operacional Regional do Norte.

References

Afonso MS, Ferreira S, Domingues FC, Silva F (2015) Resveratrol production in bioreactor: assessment of cell physiological states and plasmid segregational stability. Biotechnol Rep 5:7–13

Ahmed E, Arshad M, Zakriyya Khan M, Shoaib Amjad M, Mehreen Sadaf H, Riaz I, Sabir S, Ahmad N, Ejaz Ahmed C (2017) Secondary metabolites and their multidimensional prospective in plant life. J Pharmacogn Phytochem 205:205–214

Allied Market Research (2017) https://www.alliedmarketresearch.com/polyphenol-market. Accessed 20 May 2018

Amalraj A, Pius A, Gopi S, Gopi S (2017) Biological activities of curcuminoids, other biomolecules from turmeric and their derivatives – a review. J Tradit Compl Med 7:205–233

Atanasov AG, Waltenberger B, Eva-Maria Pferschy-Wenzig TL, Wawrosch C, Uhrin P, Temml V, Wang L, Schwaiger S, Heiss EH, Rollinger JM, Schuster D, Breuss JM, Bochkov V, Mihovilovic MD, Kopp B, Bauer R, Dirscha VM, Stuppner H (2015) Discovery and resupply of pharmacologically active plant-derived natural products: a review. Biotechnol Adv 33:1582–1614

Austin MB, Bowman ME, Ferrer JL, Schröder J, Noel JP (2004) An aldol switch discovered in stilbene synthases mediates cyclization specificity of type III polyketide synthases. Chem Biol 11:1179–1194

Bakowska-Barczak A (2005) Acylated anthocyanins as stable, natural food colorants – a review. Polish J Food Nutr Sci 1455:107–116

Becker JVW, Armstrong GO, Van Der Merwe MJ, Lambrechts MG, Vivier MA, Pretorius IS (2003) Metabolic engineering of *Saccharomyces cerevisiae* for the synthesis of the wine-related antioxidant resveratrol. FEMS Yeast Res 4:79–85

Beekwilder J, Wolswinkel R, Jonker H, Hall R, De Rie Vos CH, Bovy A (2006) Production of resveratrol in recombinant microorganisms. Appl Environ Microbiol 72:5670–5672

Bhan N, Xu P, Khalidi O, Koffas MAG (2013) Redirecting carbon flux into malonyl-CoA to improve resveratrol titers: proof of concept for genetic interventions predicted by OptForce computational framework. Chem Eng Sci 103:109–114

Bicas JL, Maróstica Jr MR, Pastore GM (2016) Biotechnological production of natural ingredients for food industry. Bentham Science Publishers, pp 1–492

Braga A, Oliveira J, Silva R, Ferreira P, Rocha I, Kallscheuer N, Marienhagen J, Faria N (2018a) Impact of the cultivation strategy on resveratrol production from glucose in engineered *Corynebacterium glutamicum*. J Biotechnol 265:70–75

Braga A, Silva M, Oliveira J, Silva AR, Ferreira P, Ottens M, Rocha I, Faria N (2018b) An adsorptive bioprocess for production and recovery of resveratrol with *Corynebacterium glutamicum*. J Chem Technol Biotechnol 93:1661–1668

Camacho-Zaragoza JM, Hernández-Chávez G, Moreno-Avitia F, Ramírez-Iñiguez R, Martínez A, Bolívar F, Gosset G (2016) Engineering of a microbial coculture of *Escherichia coli* strains for the biosynthesis of resveratrol. Microb Cell Fact 15:163. https://doi.org/10.1186/s12934-016-0562-z

Chao J, Li H, Cheng KW, Yu MS, Chang RCC, Wang M (2010) Protective effects of pinostilbene, a resveratrol methylated derivative, against 6-hydroxydopamine-induced neurotoxicity in SH-SY5Y cells. J Nutr Biochem 21:482–489

Chityala S, Nandana V (2017) Biotechnology of commercial microbial products. In: Singh D, Singh HPR (eds) Plant-microbe interactions in agro-ecological perspectives. Springer, Singapore

Choi O, Wu CZ, Kang SY, Ahn JS, Uhm TB, Hong YS (2011) Biosynthesis of plant-specific phenylpropanoids by construction of an artificial biosynthetic pathway in *Escherichia coli*. J Indian Microbiol Biotechnol 38:1657–1665

Choi SH, Ryu M, Yoon YJ, Kim DM, Lee EY (2012) Glycosylation of various flavonoids by recombinant oleandomycin glycosyltransferase from *Streptomyces antibioticus* in batch and repeated batch modes. Biotechnol Lett 34:499–505

Choi O, Lee JK, Kang SY, Pandey RP, Sohng JK, Ahn JS, Hong YS (2014) Construction of artificial biosynthetic pathways for resveratrol glucoside derivatives. J Microbiol Biotechnol 24:614–618

Chouhan S, Sharma K, Zha J, Guleria S, Koffas MAG (2017) Recent advances in the recombinant biosynthesis of polyphenols. Front Microbiol 8:1–16

Couto MR, Rodrigues JL, Rodrigues LR (2017) Optimization of fermentation conditions for the production of curcumin by engineered *Escherichia coli*. J Royal Soc Interface 14:20170470. https://doi.org/10.1098/rsif.2017.0470

Cress BF, Leitz QD, Kim DC, Amore TD, Suzuki JY, Linhardt RJ, Koffas MAG (2017) *CRISPRi*-mediated metabolic engineering of *E. coli* for O-methylated anthocyanin production. Microb Cell Fact 16:1–14. https://doi.org/10.1186/s12934-016-0623-3

D'Auria JC (2006) Acyltransferases in plants: a good time to be BAHD. Curr Opin Plant Biol 9:331–340

De Filippis B, Ammazzalorso A, Fantacuzzi M, Giampietro L, Maccallini C, Amoroso R (2017) Anticancer activity of stilbene-based derivatives. Chem Med Chem 12:558–570

de Fouchécour F, Sánchez-Castañeda A-K, Saulou-Bérion C, Spinnler HÉ (2018) Process engineering for microbial production of 3-hydroxypropionic acid. Biotechnol Adv 36:1207–1222

Delmulle T, de Maeseneire SL, de Mey M (2018) Challenges in the microbial production of flavonoids. Phytochem Rev 17:229–247
Donnez D, Jeandet P, Clément C, Courot E (2009) Bioproduction of resveratrol and stilbene derivatives by plant cells and microorganisms. Trends Biotechnol 27:706–713
Du J, Shao Z, Zhao H (2013) Engineering microbial factories for synthesis of value-added products. J Indian Microbiol Biotechnol 38:873–890
Duan L, Ding W, Liu X, Cheng X, Cai J, Hua E, Jiang H (2017) Biosynthesis and engineering of kaempferol in *Saccharomyces cerevisiae*. Microb Cell Fact 16:1–10. https://doi.org/10.1186/s12934-017-0774-x
Dudnik A, Almeida AF, Andrade R, Avila B, Bañados P, Barbay D, Bassard JE, Benkoulouche M, Bott M, Braga A, Breitel D, Brennan R, Bulteau L, Chanforan C, Costa I, Costa RS, Doostmohammadi M, Faria N, Feng C, Fernandes A, Ferreira P, Ferro R, Foito A, Freitag S, Garcia G, Gaspar P, Godinho-Pereira J, Hamberger B, Hartmann A, Heider H, Jardim C, Julien-Laferriere A, Kallscheuer N, Kerbe W, Kuipers OP, Li S, Love N, Marchetti-Spaccamela A, Marienhagen J, Martin C, Mary A, Mazurek V, Meinhart C, Sevillano DM, Menezes R, Naesby M, Nørholm MHH, Okkels FT, Oliveira J, Ottens M, Parrot D, Pei L, Rocha I, Rosado-Ramos R, Rousseau C, Sagot MF, Dos Santos CN, Schmidt M, Shelenga T, Shepherd L, Silva AR, da Silva MH, Simon O, Stahlhut SG, Solopova A, Sorokin A, Stewart D, Stougie L, Su S, Thole V, Tikhonova O, Trick M, Vain P, Veríssimo A, Vila-Santa A, Vinga S, Vogt M, Wang L, Wang L, Wei W, Youssef S, Neves AR, Forster J (2018) BacHBerry: BACterial hosts for production of bioactive phenolics from bERRY fruits. Phytochem Rev 17:291–326
Eichenberger M, Hansson A, Fischer D, Dürr L, Naesby M (2018) *De novo* biosynthesis of anthocyanins in *Saccharomyces cerevisiae*. FEMS Yeast Res 18:foy046. https://doi.org/10.1093/femsyr/foy046
Fabris S, Momo F, Ravagnan G, Stevanato R (2008) Antioxidant properties of resveratrol and piceid on lipid peroxidation in micelles and monolamellar liposomes. Biophys Chem 135:76–83
Fang Z, Jones JA, Zhou J, Koffas MAG (2017) Engineering *Escherichia coli* co-cultures for production of curcuminoids from glucose. Biotechnol J 13:e1700576. https://doi.org/10.1002/biot.201700576
Fowler ZL, Gikandi WW, Koffas MAG (2009) Increased malonyl coenzyme A biosynthesis by tuning the *Escherichia coli* metabolic network and its application to flavanone production. Appl Environ Microbiol 75:5831–5839
Fulda S (2010) Resveratrol and derivatives for the prevention and treatment of cancer. Drug Discov Today 15:757–765
Furuya T, Kino K (2014) Regioselective synthesis of piceatannol from resveratrol: catalysis by two-component flavin-dependent monooxygenase HpaBC in whole cells. Tetrahedron Lett 55:2853–2855
Gaspar P, Dudnik A, Neves AR, Föster J (2016) Engineering *Lactococcus lactis* for stilbene production. In: Abstract from 28th International Conference on Polyphenols 2016. Vienna, Austria
Grand View Research 2016 (2016) Grand view research. https://www.grandview research.com/press-release/global-polyphenols-market. Accessed 18 May 2018
Harborne JB, Mabry TJ (2013) The flavonoids: advances in research. Springer Science, Business Media LLC, New York/Philadelphia
He F, Mu L, Yan GL, Liang NN, Pan QH, Wang J, Reeves MJ, Duan CQ (2010) Biosynthesis of anthocyanins and their regulation in colored grapes. Molecules 15:9057–9091
Heo KT, Kang SY, Hong YS (2017) *De novo* biosynthesis of pterostilbene in an *Escherichia coli* strain using a new resveratrol O-methyltransferase from *Arabidopsis*. Microb Cell Fact 16:30. https://doi.org/10.1186/s12934-017-0644-6
Hewlings S, Kalman D (2017) Curcumin: a review of its effects on human health. Foods 6:92. https://doi.org/10.3390/foods6100092
Huang LL, Xue Z, Zhu QQ (2006) Method for the production of resveratrol in a recombinant oleaginous microorganism. US Patent No. 11/436160
Huang Q, Lin Y, Yan Y (2013) Caffeic acid production enhancement by engineering a phenylalanine over-producing *Escherichia coli* strain. Biotechnol Bioeng 110:3188–3196

Hussain MS, Fareed S, Ansari S, Rahman MA, Ahmad IZ, Saeed M (2012) Current approaches toward production of secondary plant metabolites. J Pharm Bioallied Sci 4:10–20

Hwang EI, Hwang EI, Ohnishi Y, Ohnishi Y, Horinouchi S, Horinouchi S (2003) Production of plant-specific flavanones by *Escherichia coli* containing an artificial gene cluster. Appl Environ Microbiol 69:2699–2706

Jendresen CB, Stahlhut SG, Li M, Gaspar P, Siedler S, Förster J, Maury J, Borodina I, Nielsen AT (2015) Highly active and specific tyrosine ammonia-lyases from diverse origins enable enhanced production of aromatic compounds in bacteria and *Saccharomyces cerevisiae*. Appl Environ Microbiol 81:4458–4476

Jeong YJ, An CH, Woo SG, Jeong HJ, Kim YM, Park SJ, Yoon BD, Kim CY (2014) Production of pinostilbene compounds by the expression of resveratrol O-methyltransferase genes in *Escherichia coli*. Enzyme Microb Technol 54:8–14

Jeong YJ, Woo SG, An CH, Jeong HJ, Hong YS, Kim Y-M, Ryu YB, Rho MC, Lee WS, Kim CY (2015) Metabolic engineering for resveratrol derivative biosynthesis in *Escherichia coli*. Mol Cells 38:318–326

Jiang H, Wood KV, Morgan JA (2005) Metabolic engineering of the phenylpropanoid pathway in *Saccharomyces cerevisiae*. Appl Environ Microbiol 71:2962–2969

Joe EJ, Kim BG, An BC, Chong Y, Ahn JH (2010) Engineering of flavonoid O-methyltransferase for a novel regioselectivity. Mol Cells 30:137–141

Jones JA, Vernacchio VR, Sinkoe AL, Collins SM, Ibrahim MHA, Lachance DM, Hahn J, Koffas MAG (2016) Experimental and computational optimization of an *Escherichia coli* co-culture for the efficient production of flavonoids. Metab Eng 35:55–63

Jones JA, Vernacchio VR, Collins SM, Shirke AN, Xiu Y, Englaender JA, Cress BF, McCutcheon CC, Linhardt RJ, Gross RA, Koffas MAG (2017) Complete biosynthesis of anthocyanins using *E. coli* polycultures. mBio 8:e00621-17. https://doi.org/10.1128/mBio.00621-17

Jovanov A, Petrov P, Đorđev V, Zdun CG, Šavikin K, Bugarski B (2017) Polyphenols extraction from plant sources. Lek Sirovine 37:45–49

Kallam K, Appelhagen I, Luo J, Albert N, Zhang H, Deroles S, Hill L, Findlay K, Andersen ØM, Davies K, Martin C (2017) Aromatic decoration determines the formation of anthocyanic vacuolar inclusions. Curr Biol 27:945–957

Kallscheuer N, Vogt M, Kappelmann J, Krumbach K, Noack S, Bott M, Marienhagen J (2016a) Identification of the phd gene cluster responsible for phenylpropanoid utilization in *Corynebacterium glutamicum*. Appl Microbiol Biotechnol 100:1871–1881

Kallscheuer N, Vogt M, Stenzel A, Gätgens J, Bott M, Marienhagen J (2016b) Construction of a *Corynebacterium glutamicum* platform strain for the production of stilbenes and (2S)-flavanones. Metab Eng 38:47–55

Kallscheuer N, Vogt M, Bott M, Marienhagen J (2017) Functional expression of plant-derived O-methyltransferase, flavanone 3-hydroxylase, and flavonol synthase in *Corynebacterium glutamicum* for production of pterostilbene, kaempferol, and quercetin. J Biotechnol 258:190–196

Kang S, Lee JK, Choi O, Kim CY, Jang J, Hwang BY, Hong YS (2014) Biosynthesis of methylated resveratrol analogs through the construction of an artificial biosynthetic pathway in *E. coli*. BMC Biotechnol 14:67. https://doi.org/10.1186/1472-6750-14-67

Katsuyama Y, Funa N, Horinouchi S (2007a) Precursor-directed biosynthesis of stilbene methyl ethers in *Escherichia coli*. Biotechnol J 2:1286–1293

Katsuyama Y, Funa N, Miyahisa I, Horinouchi S (2007b) Synthesis of unnatural flavonoids and stilbenes by exploiting the plant biosynthetic pathway in *Escherichia coli*. Chem Biol 14:613–621

Katsuyama Y, Matsuzawa M, Funa N, Horinouchi S (2008) Production of curcuminoids by *Escherichia coli* carrying an artificial biosynthesis pathway. Microbiology 154:2620–2628

Katsuyama Y, Hirose Y, Funa N, Onhnishi Y, Horinouchi S (2010) Precursor-directed biosynthesis of curcumin analogs in *Escherichia coli*. Biosci Biotechnol Biochem 74:641–645

Khoo HE, Azlan A, Tang ST, Lim SM (2017) Anthocyanidins and anthocyanins: colored pigments as food, pharmaceutical ingredients, and the potential health benefits. Food Nutr Res 61:1361779. https://doi.org/10.1080/16546628.2017.1361779

Kim HJ, Chang EJ, Cho SH, Chung SK, Park HD, Choi SW (2002) Antioxidative activity of resveratrol and its derivatives isolated from seeds of *Paeonia lactiflora*. Biosci Biotechnol Biochem 66:1990–1993

Kim DH, Kim BG, Jung NR, Ahn JH (2009) Production of genistein from naringenin using *Escherichia coli* containing isoflavone synthase-cytochrome P450 reductase fusion protein. J Microbiol Biotechnol 19:1612–1616

Kim BG, Joe EJ, Ahn JH (2010) Molecular characterization of flavonol synthase from poplar and its application to the synthesis of 3-O-methylkaempferol. Biotechnol Lett 32:579–584

Kim HJ, Kim BG, Ahn JH (2013a) Regioselective synthesis of flavonoid bisglycosides using *Escherichia coli* harboring two glycosyltransferases. Appl Microbiol Biotechnol 97:5275–5282

Kim MJ, Kim BG, Ahn JH (2013b) Biosynthesis of bioactive O-methylated flavonoids in *Escherichia coli*. Appl Microbiol Biotechnol 97:7195–7204

Kim EJ, Cha MN, Kim BG, Ahn JH (2017) Production of curcuminoids in engineered *escherichia coli*. J Microbiol Biotechnol 27:975–982

Kiselev KV (2011) Perspectives for production and application of resveratrol. Appl Microbiol Biotechnol 90:417–425

Koirala N, Thuan NH, Ghimire GP, Van Thang D, Sohng JK (2016) Methylation of flavonoids: chemical structures, bioactivities, progress and perspectives for biotechnological production. Enzyme Microb Technol 86:103–116

Koopman F, Beekwilder J, Crimi B, Van Houwelingen A, Hall RD, Bosch D, Van Maris AJA, Pronk JT, Daran JM (2012) *De novo* production of the flavonoid naringenin in engineered *Saccharomyces cerevisiae*. Microb Cell Fact 11:155. https://doi.org/10.1186/1475-2859-11-155

Lee N, Kim EJ, Kim BG (2012) Regioselective hydroxylation of trans -resveratrol via inhibition of tyrosinase from *streptomyces avermitilis* MA4680. ACS Chem Biol 7:1687–1692

Lee H, Kim BG, Ahn JH (2014) Production of bioactive hydroxyflavones by using monooxygenase from *Saccharothrix espanaensis*. J Biotechnol 176:11–17

Leonard E, Koffas MAG (2007) Engineering of artificial plant cytochrome P450 enzymes for synthesis of isoflavones by *Escherichia coli*. Appl Environ Microbiol 73:7246–7251

Leonard E, Yan Y, Lim KH, Koffas M (2005) Investigation of two distinct flavone synthases for plant-specific flavone biosynthesis in *Saccharomyces cerevisiae*. Appl Environ Microbiol 71:8241–8248

Leonard E, Chemler J, Hong K (2006) Expression of a soluble flavone synthase allows the biosynthesis of phytoestrogen derivatives in *Escherichia coli*. Appl Environ Microbiol 70:85–91

Leonard E, Lim K, Saw P, Koffas MAG (2007) Engineering central metabolic pathways for high-level flavonoid production in *Escherichia coli*. Appl Environ Microbiol 73:3877–3886

Leonard E, Yan Y, Fowler ZL, Li Z, Lim CG, Lim KH, Koffas MAG (2008) Strain improvement of recombinant *Escherichia coli* for efficient production of plant flavonoids. Mol Pharm 5:257–265

Li J, Li Z, Li C, Gou J, Zhang Y (2014) Molecular cloning and characterization of an isoflavone 7-O-glucosyltransferase from *Pueraria lobata*. Plant Cell Rep 33:1173–1185

Li M, Kildegaard KR, Chen Y, Rodriguez A, Borodina I, Nielsen J (2015) *De novo* production of resveratrol from glucose or ethanol by engineered *Saccharomyces cerevisiae*. Metab Eng 32:1–11. https://doi.org/10.1016/j.ymben.2015.08.007

Li M, Schneider K, Kristensen M, Borodina I, Nielsen J (2016) Engineering yeast for high-level production of stilbenoid antioxidants. Sci Rep 6:1–8. https://doi.org/10.1038/srep36827

Liang JL, Guo LQ, Lin JF, He ZQ, Cai FJ, Chen JF (2016) A novel process for obtaining pinosylvin using combinatorial bioengineering in *Escherichia coli*. World J Microbiol Biotechnol 32:102. https://doi.org/10.1007/s11274-016-2062-z

Light SH, Halavaty AS, Minasov G, Shuvalova L, Anderson WF (2012) Structural analysis of a 3-deoxy-D-arabino-heptulosonate 7-phosphate synthase with an N-terminal chorismate mutase-like regulatory domain. Protein Sci 21:887–895

Lim CG, Fowler ZL, Hueller T, Schaffer S, Koffas MAG (2011) High-yield resveratrol production in engineered *Escherichia coli*. Appl Environ Microbiol 77:3451–3460

Lim CG, Wong L, Bhan N, Dvora H, Xu P, Venkiteswaran S, Koffas MAG (2015) Development of a recombinant *Escherichia coli* strain for overproduction of the plant pigment anthocyanin. Appl Environ Microbiol 81:6276–6284

Lin D, Xiao M, Zhao J, Li Z, Xing B, Li X, Kong M, Li L, Zhang Q, Liu Y, Chen H, Qin W, Wu H, Chen S (2016) An overview of plant phenolic compounds and their importance in human nutrition and management of type 2 diabetes. Molecules 21:E1374. https://doi.org/10.3390/molecules21101374

Liu X, Lin J, Hu H, Zhou B, Zhu B (2016) De novo biosynthesis of resveratrol by site-specific integration of heterologous genes in *Escherichia coli*. FEMS Microbiol Lett 363:1–5. https://doi.org/10.1093/femsle/fnw061

Lyu X, Ng KR, Lee JL, Mark R, Chen WN (2017) Enhancement of naringenin biosynthesis from tyrosine by metabolic engineering of *Saccharomyces cerevisiae*. J Agric Food Chem 65:6638–6646

MacDonald MJ, D'Cunha GB (2007) A modern view of phenylalanine ammonia lyase. Biochem Cell Biol 85:273–282

Malla S, Koffas MAG, Kazlauskas RJ, Kim BG (2012) Production of 7-O-Methyl aromadendrin, a medicinally valuable flavonoid, in *Escherichia coli*. Appl Environ Microbiol 78:684–694

Malla S, Pandey RP, Kim BG, Sohng JK (2013) Regiospecific modifications of naringenin for astragalin production in *Escherichia coli*. Biotechnol Bioeng 110:2525–2535

Mei YZ, Liu RX, Wang DP, Wang X, Dai CC (2015) Biocatalysis and biotransformation of resveratrol in microorganisms. Biotechnol Lett 37:9–18

Mierziak J, Kostyn K, Kulma A (2014) Flavonoids as important molecules of plant interactions with the environment. Molecules 19:16240–16265

Milke L, Aschenbrenner J, Marienhagen J, Kallscheuer N (2018) Production of plant-derived polyphenols in microorganisms: current state and perspectives. Appl Microbiol Biotechnol 102:1575–1585

Nishiyama Y, Yun CS, Matsuda F, Sasaki T, Saito K, Tozawa Y (2010) Expression of bacterial tyrosine ammonia-lyase creates a novel p-coumaric acid pathway in the biosynthesis of phenylpropanoids in *Arabidopsis*. Planta 232:209–218

Ogata K, Uchiyama K, Yamada H (1967) Metabolism of aromatic amino acid in microorganisms: part 1. Formation of cinnamic acid from phenylalanine. Agric Biol Chem 31:200–206

Ozaki S, Imai H, Iwakiri T, Sato T, Shimoda K, Nakayama T, Hamada H (2012) Regioselective glucosidation of trans-resveratrol in *Escherichia coli* expressing glucosyltransferase from *Phytolacca americana*. Biotechnol Lett 34:475–481

Palve Y, Nayak P (2012) Curcumin: a wonder anticancer drug. Int J Pharm Biomed Sci 3:60–69

Pandey RP, Parajuli P, Koirala N, Lee JH, Park Y II, Sohng JK (2014) Glucosylation of isoflavonoids in engineered *Escherichia coli*. Mol Cells 37:172–177

Pandey RP, Parajuli P, Koffas MAG, Sohng JK (2015) Microbial production of natural and non-natural flavonoids: pathway engineering, directed evolution and systems/synthetic biology. Biotechnol Adv 34:634–662

Park SR, Yoon JA, Paik JH, Park JW, Jung WS, Ban YH, Kim EJ, Yoo YJ, Han AR, Yoon YJ (2009) Engineering of plant-specific phenylpropanoids biosynthesis in *Streptomyces venezuelae*. J Biotechnol 141:181–188

Passeri V, Koes R, Quattrocchio FM (2016) New challenges for the design of high value plant products: stabilization of Anthocyanins in plant vacuoles. Front Plant Sci 7:1–9. https://doi.org/10.3389/fpls.2016.00153

Quideau S, Deffieux D, Douat-Casassus C, Pouységu L (2011) Plant polyphenols: chemical properties, biological activities, and synthesis. Angew Chemie-Int Ed 50:586–621

Ralston L, Subramanian S, Matsuno M, Yu O (2005) Partial reconstruction of flavonoid and isoflavonoid biosynthesis in yeast using soybean type I and type II chalcone isomerases. Plant Physiol 137:1375–1388

Reinisalo M, Kårlund A, Koskela A, Kaarniranta K, Karjalainen RO (2015) Polyphenol stilbenes: molecular mechanisms of defence against oxidative stress and aging-related diseases. Oxid Med Cell Longev 2015:340520. https://doi.org/10.1155/2015/340520

Rimando AM, Cuendet M, Desmarchelier C, Mehta RG, Pezzuto JM, Duke SO (2002) Cancer chemopreventive and antioxidant activities of pterostilbene, a naturally occurring analogue of resveratrol. J Agric Food Chem 50:3453–3457

Ro D, Douglas CJ (2004) Reconstitution of the entry point of plant phenylpropanoid metabolism in yeast (*Saccharomyces cerevisiae*). J Biol Chem 279:2600–2607

Rodrigues JL, Araújo RG, Prather KLJ, Kluskens LD, Rodrigues LR (2015a) Production of curcuminoids from tyrosine by a metabolically engineered *Escherichia coli* using caffeic acid as an intermediate. Biotechnol J 10:599–609

Rodrigues JL, Prather KLJ, Kluskens LD, Rodrigues LR (2015b) Heterologous Production of Curcuminoids. Microbiol Mol Biol Rev 79:39–60

Rodriguez A, Martínez JA, Flores N, Escalante A, Gosset G, Bolivar F (2014) Engineering *Escherichia coli* to overproduce aromatic amino acids and derived compounds. Microb Cell Fact 13:1–15. https://doi.org/10.1186/s12934-014-0126-z

Rodriguez A, Strucko T, Stahlhut SG, Kristensen M, Svenssen DK, Forster J, Nielsen J, Borodina I (2017) Metabolic engineering of yeast for fermentative production of flavonoids. Bioresour Technol 245:1645–1654

Ruby S, Kumar RJ, Vishwakarma RK, Singh S, Khan BM (2014) Molecular cloning and characterization of genistein 4′-O-glucoside specific glycosyltransferase from *Bacopa monniera*. Mol Biol Rep 41:4675–4688

Sahdev S, Khattar SK, Saini KS (2008) Production of active eukaryotic proteins through bacterial expression systems: a review of the existing biotechnology strategies. Mol Cell Biochem 307:249–264

Santos CNS, Koffas M, Stephanopoulos G (2011) Optimization of a heterologous pathway for the production of flavonoids from glucose. Metab Eng 13:392–400

Santos CNS, Xiao W, Stephanopoulos G (2012) Rational, combinatorial, and genomic approaches for engineering L-tyrosine production in *Escherichia coli*. Proc Natl Acad Sci USA 109:13538–13543

Schenck CA, Maeda HA (2018) Tyrosine biosynthesis, metabolism, and catabolism in plants. Phytochemistry 149:82–102

Sharma S, Kulkarni SK, Chopra K (2007) Effect of resveratrol, a polyphenolic phytoalexin, on thermal hyperalgesia in a mouse model of diabetic neuropathic pain. Fundam Clin Pharmacol 21:89–94

Shashank K, Pandey AK (2013) Chemistry and biological activities of flavonoids. Sci World J 2013:533–548

Shin SK, Ha TY, McGregor RA, Choi MS (2011a) Long-term curcumin administration protects against atherosclerosis via hepatic regulation of lipoprotein cholesterol metabolism. Mol Nutr Food Res 55:1829–1840

Shin SY, Han NS, Park YC, Kim MD, Seo JH (2011b) Production of resveratrol from *p*-coumaric acid in recombinant *Saccharomyces cerevisiae* expressing 4-coumarate: Coenzyme A ligase and stilbene synthase genes. Enzyme Microb Technol 48:48–53

Shin SY, Jung SM, Kim MD, Han NS, Seo JH (2012) Production of resveratrol from tyrosine in metabolically engineered *Saccharomyces cerevisiae*. Enzyme Microb Technol 51:211–216

Si HY, Li DP, Wang TM, Zhang HL, Ren FY, Xu ZG, Zhao YY (2010) Improving the anti-tumor effect of genistein with a biocompatible superparamagnetic drug delivery system. J Nanosci Nanotechnol 10:2325–2331

Sood S, Nagpal M (2013) Role of curcumin in systemic and oral health: an overview. J Nat Sci Biol Med 4:3–7

Stahlhut SG, Siedler S, Malla S, Harrison SJ, Maury J, Neves AR, Forster J (2015) Assembly of a novel biosynthetic pathway for production of the plant flavonoid fisetin in *Escherichia coli*. Metab Eng 31:84–93

Sydor T, Schaffer S, Boles E (2010) Considerable increase in resveratrol production by recombinant industrial yeast strains with use of rich medium. Appl Environ Microbiol 76:3361–3363

Sytar Oksana, Marian B, Mahendra R, Bo SH (2012) Plant phenolic compounds for food, pharmaceutical and cosmetics production. J Med Plants Res 6:2526–2539

Thuan NH, Trung NT, Cuong NX, Van Cuong D, Van Quyen D, Malla S (2018a) *Escherichia coli* modular coculture system for resveratrol glucosides production. World J Microbiol Biotechnol 34:75. https://doi.org/10.1007/s11274-018-2458-z
Thuan NH, Chaudhary AK, Van Cuong D, Cuong NX (2018b) Engineering co-culture system for production of apigetrin in *Escherichia coli*. J Ind Microbiol Biotechnol 45:175–185
Trantas E, Panopoulos N, Ververidis F (2009) Metabolic engineering of the complete pathway leading to heterologous biosynthesis of various flavonoids and stilbenoids in *Saccharomyces cerevisiae*. Metab Eng 11:355–366
Trantas EA, Koffas MAG, Xu P, Ververidis F (2015) When plants produce not enough or at all: metabolic engineering of flavonoids in microbial hosts. Front Plant Sci 6:7. https://doi.org/10.3389/fpls.2015.00007
Treutter D (2006) Significance of flavonoids in plant resistance: a review. Environ Chem Lett 4:147–157
Tropf S, Lanz T, Rensing SA, Schröder J, Schröder G (1994) Evidence that stilbene synthases have developed from chalcone synthases several times in the course of evolution. J Mol Evol 38:610–618
Tsao R (2010) Chemistry and biochemistry of dietary polyphenols. Nutrients 2:1231–1246
Tsuda T (2012) Dietary anthocyanin-rich plants: biochemical basis and recent progress in health benefits studies. Mol Nutr Food Res 56:159–170
Turnbull JJ, Nakajima JI, Welford RWD, Yamazaki M, Saito K, Schofield CJ (2004) Mechanistic studies on three 2-oxoglutarate-dependent oxygenases of flavonoid biosynthesis: anthocyanidin synthase, flavonol synthase, and flavanone 3β-hydroxylase. J Biol Chem 279:1206–1216
van Summeren-Wesenhagen PV, Marienhagen J (2013) Putting bugs to the blush: metabolic engineering for phenylpropanoid-derived products in microorganisms. Bioeng 4:355–362
van Summeren-Wesenhagen PV, Marienhagen J (2015) Metabolic engineering of *Escherichia coli* for the synthesis of the plant polyphenol pinosylvin. Appl Environ Microbiol 81:840–849
Vargas-Tah A, Gosset G (2015) Production of cinnamic and p-hydroxycinnamic acids in engineered microbes. Front Bioeng Biotechnol 3:1–10. https://doi.org/10.3389/fbioe.2015.00116
Wallace TC, Monica GM (2013) Anthocyanins in health and disease. Taylor C. CRC Press, Boca Raton
Wang Y, Yu O (2012) Synthetic scaffolds increased resveratrol biosynthesis in engineered yeast cells. J Biotechnol 157:258–260
Wang BF, Wang JS, Lu JF, Kao TH, Chen BH (2009) Antiproliferation effect and mechanism of prostate cancer cell lines as affected by isoflavones from soybean cake. J Agric Food Chem 57:2221–2232
Wang Y, Chen S, Yu O (2011a) Metabolic engineering of flavonoids in plants and microorganisms. Appl Microbiol Biotechnol 91:949–956
Wang Y, Halls C, Zhang J, Matsuno M, Zhang Y, Yu O (2011b) Stepwise increase of resveratrol biosynthesis in yeast *Saccharomyces cerevisiae* by metabolic engineering. Metab Eng 13:455–463
Wang S, Zhang S, Zhou T, Zeng J, Zhan J (2013) Design and application of an *in vivo* reporter assay for phenylalanine ammonia-lyase. Appl Microbiol Biotechnol 97:7877–7885
Wang S, Zhang S, Xiao A, Rasmussen M, Skidmore C, Zhan J (2015a) Metabolic engineering of *Escherichia coli* for the biosynthesis of various phenylpropanoid derivatives. Metab Eng 29:153–159
Wang Y, Bhuiya MW, Zhou R, Yu O (2015b) Pterostilbene production by microorganisms expressing resveratrol O-methyltransferase. Ann Microbiol 65:817–826
Watts KT, Lee PC, Schmidt-dannert C (2004) Exploring recombinant flavonoid biosynthesis in metabolically engineered *Escherichia coli*. Chembiochem 55108:500–507
Watts KT, Lee PC, Schmidt-Dannert C (2006) Biosynthesis of plant-specific stilbene polyketides in metabolically engineered *Escherichia coli*. BMC Biotechnol 6:1–12. https://doi.org/10.1186/1472-6750-6-22
Wu J, Du G, Zhou J, Chen J (2013a) Metabolic engineering of *Escherichia coli* for (2 S)-pinocembrin production from glucose by a modular metabolic strategy. Metab Eng 16:48–55

Wu J, Liu P, Fan Y, Bao H, Du G, Zhou J, Chen J (2013b) Multivariate modular metabolic engineering of *Escherichia coli* to produce resveratrol from l-tyrosine. J Biotechnol 167:404–411
Wu J, Du G, Zhou J, Chen J (2014) Systems metabolic engineering of microorganisms to achieve large-scale production of flavonoid scaffolds. J Biotechnol 188:72–80
Wu J, Zhang X, Zhou J, Dong M (2016) Efficient biosynthesis of (2S)-pinocembrin from D-glucose by integrating engineering central metabolic pathways with a pH-shift control strategy. Bioresour Technol 218:999–1007
Wu J, Zhang X, Zhu Y, Tan Q, He J, Dong M (2017a) Rational modular design of metabolic network for efficient production of plant polyphenol pinosylvin. Sci Rep 7:1–15. https://doi.org/10.1038/s41598-017-01700-9
Wu J, Zhou P, Zhang X, Dong M (2017b) Efficient *de novo* synthesis of resveratrol by metabolically engineered *Escherichia coli*. J Ind Microbiol Biotechnol 44:1083–1095. https://doi.org/10.1007/s10295-017-1937-9
Xia T, Eiteman MA (2017) Quercetin glucoside production by engineered *Escherichia coli*. Appl Biochem Biotechnol 182:1358–1370
Xu P, Ranganathan S, Fowler ZL, Maranas CD, Koffas MAG (2011) Genome-scale metabolic network modeling results in minimal interventions that cooperatively force carbon flux towards malonyl-CoA. Metab Eng 13:578–587
Xu F, Li L, Zhang W, Cheng H, Sun N, Cheng S, Wang Y (2012) Isolation, characterization, and function analysis of a flavonol synthase gene from *Ginkgo biloba*. Mol Biol Rep 39:2285–2296
Yan Y, Chemler J, Huang L, Martens S, Koffas MAG (2005a) Metabolic engineering of anthocyanin biosynthesis in *Escherichia coli*. Appl Environ Microbiol 71:3617–3623
Yan Y, Kohli A, Koffas MAG (2005b) Biosynthesis of natural flavanones in *Saccharomyces cerevisiae*. Appl Environ Microbiol 71:5610–5613
Yan Y, Huang L, Koffas MAG (2007) Biosynthesis of 5-deoxyflavanones in microorganisms. Biotechnol J 2:1250–1262
Yan Y, Li Z, Koffas MAG (2008) High-yield anthocyanin biosynthesis in engineered Escherichia coli. Biotechnol Bioeng 100:126–140
Yang SM, Han SH, Kim BG, Ahn JH (2014) Production of kaempferol 3-O-rhamnoside from glucose using engineered *Escherichia coli*. J Ind Microbiol Biotechnol 41:1311–1318
Yang Y, Lin Y, Li L, Linhardt RJ, Yan Y (2015) Regulating malonyl-CoA metabolism via synthetic antisense RNAs for enhanced biosynthesis of natural products. Metab Eng 29:217–226
Yonekura-Sakakibara K, Saito K (2009) Functional genomics for plant natural product biosynthesis. Nat Prod Rep 26:1466. https://doi.org/10.1039/b817077k
Yoon J-A, Kim B-G, Lee WJ, Lim Y, Chong Y, Ahn J-H (2012) Production of a novel Quercetin glycoside through metabolic engineering of *Escherichia coli*. Appl Environ Microbiol 78:4256–4262
Zha J, Koffas MAG (2017a) Anthocyanin production in engineered microorganisms. In: Schwab W, Lange B, Wüst M (eds) Biotechnology of natural products. Springer, Cham, pp 81–97
Zha J, Koffas MAG (2017b) Production of anthocyanins in metabolically engineered microorganisms: current status and perspectives. Synth Syst Biotechnol 2:259–266
Zha W, Rubin-Pitel SB, Shao Z, Zhao H (2009) Improving cellular malonyl-CoA level in *Escherichia coli* via metabolic engineering. Metab Eng 11:192–198
Zhang H, Stephanopoulos G (2013) Engineering *E. coli* for caffeic acid biosynthesis from renewable sugars. Appl Microbiol Biotechnol 97:3333–3341
Zhang Y, Li SZ, Li J, Pan X, Cahoon RE, Jaworski JG, Wang X, Jez JM, Chen F, Yu O (2006) Using unnatural protein fusions to engineer resveratrol biosynthesis in yeast and mammalian cells. J Am Chem Soc 128:13030–13031
Zhang H, Matsuda H, Yamashita C, Nakamura S, Yoshikawa M (2009) Hydrangeic acid from the processed leaves of *Hydrangea macrophylla var. thunbergii* as a new type of anti-diabetic compound. Eur J Pharmacol 606:255–261
Zhang E, Guo X, Meng Z, Wang J, Sun J, Yao X, Xun H (2015) Construction, expression, and characterization of *Arabidopsis thaliana* 4CL and *Arachis hypogaea* RS fusion gene 4CL::RS in *Escherichia coli*. World J Microbiol Biotechnol 31:1379–1385

Zhang W, Liu H, Li X, Liu D, Dong XT, Li FF, Wang EX, Li BZ, Yuan YJ (2017) Production of naringenin from D-xylose with co-culture of *E. coli* and *S. cerevisiae*. Eng Life Sci 17:1021–1029

Zhao S, Jones JA, Lachance DM, Bhan N, Khalidi O, Venkataraman S, Wang Z, Koffas MAG (2015) Improvement of catechin production in *Escherichia coli* through combinatorial metabolic engineering. Metab Eng 28:43–53

Zhu S, Wu J, Du G, Zhou J, Chen J (2014) Efficient synthesis of eriodictyol from L-tyrosine in *Escherichia coli*. Appl Environ Microbiol 80:3072–3080

Endolichenic Fungi from Common Lichens as New Sources for Valuable Bio-active Compounds

4

Krystle Angelique Aguda Santiago
and Adeline Su Yien Ting

Contents

Abstract

Lichens are known to produce bio-active compounds. However, their natural slow growth limited progress in researching lead compounds from lichens for drug development. In recent years, a rare group of fungi were found to grow in the lichens, identified as endolichenic fungi (EF). The discovery of this group of fungi is relatively new but has attracted notable attention attributed to their phar-

K. A. A. Santiago · A. S. Y. Ting (✉)
School of Science, Monash University Malaysia,
Bandar Sunway, Selangor, Darul Ehsan, Malaysia
e-mail: adeline.ting@monash.edu; adelsuyien@yahoo.com

M. S. Akhtar et al. (eds.), *Natural Bio-active Compounds*,
https://doi.org/10.1007/978-981-13-7154-7_4

macological potential (antimicrobial, antiviral, antioxidant, and antitumor). Interestingly, these endolichenic fungi produce bio-active compounds that may be distinct from those produced by their host lichens. With an estimate of 20,000 known lichens reported worldwide, there are tremendous opportunities to discover valuable endolichenic fungi. Thus, the aim of present chapter is to provide an overview on the progress and advances made in the research of endolichenic fungi. Also, it discusses on the emerging biotechnological approaches to explore endolichenic fungi.

Keywords
Bio-active compounds · Biodiversity · Biotechnology · Conservation · Lichens

4.1 Introduction

Lichens are a specialized group of symbionts, formed from the association between a fungus (the mycobiont) and a photosynthetic partner (the phycobiont/photobiont). The common mycobiont is a fungus from the phylum *Ascomycota*, while the photobiont is either a green alga or cyanobacterium (Nash 2008). Subsequent studies revealed the possible presence of yeast within the lichen thalli as the third partner in the symbiosis (Spribille et al. 2016). This supports the theory that a lichen thallus may serve as a consortium of several microorganisms such as fungi and bacteria, with either destructive or little effects on both the mycobiont and photobiont (Honneger et al. 2013). The presence of consortium of microorganisms in the lichen thallus was further verified in recent years, when another group of fungi (with lineages to *Pezizomycotina*, phylum *Ascomycota*) was discovered. This group of fungi is described as the endolichenic fungi and has close association with the green algal photobiont (Arnold et al. 2009). The discovery of endolichenic fungi was accidental, as attempts to isolate the lichen mycobiont into pure cultures revealed their growth from the thallus sample (Kellogg and Raja 2017). Endolichenic fungi are distinct from the lichen mycobiont and lichenicolous fungi, as they are found exclusively within the interiors of lichens. Unlike lichenicolous fungi, endolichenic fungi do not cause symptoms on the lichen thalli (Tripathi and Joshi 2015; Wang et al. 2016). The nature of endolichenic fungi is therefore described to be similar to plant endophytic fungi, which resides within the internal tissues of host plants without causing any symptoms to the host (U'ren et al. 2010; Kellogg and Raja 2017). Nevertheless, endolichenic fungal assemblages were notably distinct from endophytes of vascular plants, with the exception of endophytic fungi inhabiting mosses (U'ren et al. 2012).

The discovery of endolichenic fungi and their interaction with other organisms through endolichenism is regarded as one of the greatest milestones in understanding the major ecological transitions of microbial interactions (i.e., pathogenicity, endophytism, symbiosis) and ecological distinctiveness of *Ascomycetes* (Arnold et al. 2009; U'ren et al. 2012). Arnold et al. (2009) hypothesized that endolichenism played a key role in the evolution of endophytism in the phylum *Ascomycota*, suggesting that endolichenic fungi represent the starting point to the evolution of most endophytic fungi existing in nature. Endolichenism suggests that lichens are "cradles" of fungal diversification, as lichen thallus could potentially harbor various "missing fungi" (Arnold et al. 2009). "Missing fungi" are those undiscovered and undescribed fungi among the hypothesized 5.1 million fungal species existing on the planet (Blackwell 2011), which are anticipated to exist in tropical forests, unexplored habitats, and lost or hidden species (Hawksworth and Rossman 1997). Over the years, several species of endolichenic fungi have been reported, which include species of *Aspergillus*, *Chaetothyriales*, *Neurospora*, *Nodulisporium*, *Phaeosphaeria*, *Penicillium*, *Ulocladium*, and *Xylaria*. These endolichenic fungi are studied for their potential and bioactivity in antibacterial, antifungal, antioxidant, and cytotoxic activities.

The interest in bio-active compounds produced by endolichenic fungi stems from the fact that the compounds are relatively distinct from those produced by the mycobiont of the lichen (Kellogg and Raja 2017; Singh et al. 2017). Their unique and diverse chemical structures provide new lead compounds or molecules for drug development. To date, approximately 10% of the known endolichenic fungi have been studied for their medicinal properties (Singh et al. 2017). Most of these studies, however, lacked the complete structural identification of the targeted bio-active chemical constituents. As a consequence, none of these compounds reached the final stages of drug discovery for therapeutic purposes (Singh et al. 2017). To facilitate discoveries on compounds from endolichenic fungi, the "one strain, many compounds" (OSMAC) approach is adopted (Wijeratne et al. 2010; Wang et al. 2013a; Padhi et al. 2017; Yuan et al. 2016). The utilization of "OMICS"-based techniques is also beneficial, as it will provide a clearer and broader understanding of the compounds produced by these microorganisms. These approaches will be discussed in the next few sections, as part of the biotechnological approaches to further harness bio-active compounds produced by endolichenic fungi and their potential for future commercialization.

With the increasing interest in endolichenic fungi, lichens are at risk of overcollection and overexploitation. Hence, it is imperative to consider its conservation to enable sustainable exploration of endolichenic fungi for their bio-active compounds, particularly when lichens are slow-growing (Shukla et al. 2014). Several conservation strategies and forest management practices can be implemented to minimize the threats of lichen extinction and ensuring the conservation of lichen biodiversity (Shukla et al. 2014). These include standardized preparation of lichen inventory in a given area, provision of protocol for lichen sampling, and strategies to minimize environmental disturbances. Generally, these strategies aim to protect the lichen species and to conserve their habitats (Scheidegger and Worth 2009). Thus, the aim of present chapter is to provide an overview on the progress and advances made in the research of endolichenic fungi and also discuss the emerging biotechnological approaches in exploring endolichenic fungi.

4.2 Origin, Biodiversity, and Distribution of Endolichenic Fungi

The evolutionary origins of endolichenic fungi are not well understood, presumably attributed to the relatively low number of lichens sampled and the limited studies distinguishing fungi growing on lichen surfaces from those occurring within lichen thalli (Arnold et al. 2009; Kellogg and Raja 2017). However, since lichens are thought to exist prior to the existence of plants, Arnold et al. (2009) hypothesized that lichens could have possibly harbored fungi (i.e., the endolichenic fungi), which are presumably ancestors to the plant-associated endophytic fungi (plant endophytes). Their investigation recommended that endolichenic fungi could have played an important role in the evolution of plant endophytes (Arnold et al. 2009). Both endolichenic fungi and endophytes share similar traits, occurring within another living organism without causing any negative effects (U'ren et al. 2010; Kellogg and Raja 2017). More importantly, endolichenism was postulated as the precursor to major ecological transitions such as symbiosis, pathogenicity, and endophytism (Fig. 4.1). Endolichenic fungi are therefore suggested to serve as "evolutionary incubator" for microbial transitions to other associations, coining the term "cradles of fungal diversification" (Arnold et al. 2009). The endolichenic fungi, however, remained independent of lichenized ancestors and lichen-forming fungi, suggesting that they are taxonomically and ecologically distinct from mycobionts and lichenicolous fungi (Arnold et al. 2009).

Despite the close similarities of endolichenic fungi with endophytes, endolichenic fungi are largely distinct from plant endophytes with the only exception of endophytes occurring within mosses (U'ren et al. 2010, 2012). Endolichenic fungi are also distinct from other fungi in the environment. Peršoh and Rambold (2012) determined that endolichenic fungi from the lichen *Letharietum vulpinae* formed a cluster independent from soil-inhabiting and rock-colonizing fungi. In a more recent study, Padhi et al. (2017) isolated *Aspergillus tubingensis* from the lichen *Parmelia caperata*. This fungus morphologically resembles *A. niger* and was initially thought to be similar to other *Aspergillus* species, which are opportunistic human pathogens and endophytes. However, a variation in the ITS2 region of *A. tubingensis* proved that the isolate is distinct from known *Aspergillus* species. U'ren et al. (2010) further revealed that endolichenic fungi are not incidental or saprotrophic fungi entrapped by the lichen thalli, hence explaining their distinct species clusters.

Since, the first isolation of endolichenic fungi in 1990 (Petrini et al. 1990), studies on diversity assessments have revealed species-rich endolichenic fungal

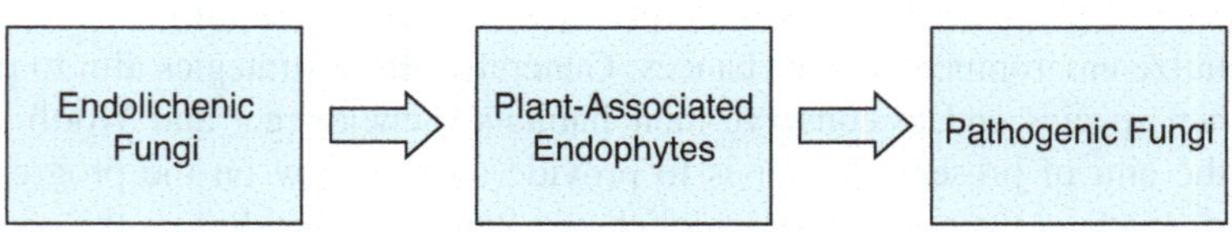

Fig. 4.1 The major ecological transitions explaining the evolutionary origins in the phylum *Ascomycota*

assemblages from different types of lichens from a wide range of ecosystems (Girlanda et al. 1997; Zhang et al. 2015; Wang et al. 2016). For example, fruticose lichens *Cladonia* and *Stereocaulon* collected from the forested mountain range in Germany yielded 62 species of endolichenic fungi (Petrini et al. 1990). These include species of *Acremonium*, *Alternaria*, *Botrytis*, *Cladosporium*, *Colletotrichum*, *Fusarium*, *Geotrichum*, *Paecilomyces*, *Penicillium*, *Phoma*, and *Trichoderma*. The foliose lichen *Parmelia taractica* from the coniferous forest in Italy yielded 95 endolichenic fungal taxa, which included species of *Alternaria*, *Cladosporium*, *Fusarium*, *Mucor*, *Paecilomyces*, *Penicillium*, *Phoma*, *Tolypocladium*, *Trichoderma*, and *Ulocladium* (Girlanda et al. 1997). The foliose lichens *Parmotrema* sp. and *Pseudocyphellaria* sp. and the fruticose lichen *Usnea* sp., from the montane forest of Sri Lanka, yielded 23 species of endolichenic fungi, which include species of *Acremonium*, *Aspergillus*, *Chrysosporium*, *Cladosporium*, *Curvularia*, *Fusarium*, *Nigrospora*, *Periconia*, and *Phoma* (Kannangara et al. 2009).

These studies on endolichenic fungal species sampled from various lichens across diverse environments revealed their ubiquitous nature and distribution. Arnold et al. (2009) recovered more than 60 isolates from the arctic, boreal, temperate, and tropical regions within the USA. Similarly, another extensive study in Arizona, USA, reported more than 500 endolichenic fungal isolates, which were successfully recovered from microsites of different altitude, temperature, rainfall, and vegetation (U'ren et al. 2010). Even the most pristine regions of the world, such as the alpine regions and the Arctic, harbored endolichenic fungal communities. Twenty-five isolates representing 20 genera were reported from lichens in Himalaya (Tripathi and Joshi 2015) and a high diversity (over 300 fungal operational taxonomic units, OTUs) of endolichenic fungi was reported in Svalbard, High Arctic (Zhang et al. 2015). Interestingly, endolichenic fungi were also reported to be present in fossilized lichen *Chlorolichenomycites salopensis* (Honneger et al. 2013). The great number of endolichenic fungi isolated from different types of lichens (including fossilized lichens) from different parts of the world clearly shows the wide distribution of these cryptic organisms.

The interaction of endolichenic fungi with lichens, as well as factors permitting coexistence, is still unknown although it has been hypothesized that the age and chemical composition of lichen may influence the composition of endolichenic fungi (Suryanarayanan and Thirunavukkarasu 2017). Other possible factors influencing distribution and diversity of endolichenic fungi include the biotic factors (lichen mycobiont, interaction with host) and abiotic factors (topography, climate, geographical distance) (U'ren et al. 2010, 2012; Chagnon et al. 2015). The biotic factors play a major role in determining the endolichenic communities in lichens. The lichen mycobiont selects the compatible photobiont when forming a lichen association (Piercey-Normore and Deduke 2011), consequently influencing the diversity of endolichenic fungi as endolichenic fungi are closely related to the lichen photobiont (Arnold et al. 2009). In addition, the existence of endolichenic fungi in lichens may also be attributed to their interaction with host. Chagnon et al. (2015) postulated that the endolichenic fungi are more nested and connected relative to endophytic networks. This suggested that plants are more selective in choosing their fungal partners than lichens. As a consequence, higher abundance and diversity of endolichenic fungi is usually observed, compared to plant endophytes in a given site (Chagnon et al. 2015). In addition, abiotic factors such as topography, climate, and geographical distance also play an important role in shaping the diversity of endolichenic fungi. For example, prolonged exposure of lichen thalli to rainfall is somewhat correlated to the increase in the diversity of endolichenic fungi. This is attributed to enhance photosynthetic rates of the lichen photobionts due to the presence of dew (U'ren et al. 2012). U'ren et al. (2012) also highlighted the effects of geographical distance. It was observed that species composition of endolichenic fungal communities from lichens collected within neighboring places were more similar than with those collected from distant sites. In brief, the endolichenic fungi are a phylogenetic disparate group compared to existing mycobionts and lichenicolous fungi and other fungal species from various habitats. They can be found across diverse habitats. They exist in lichens and derive nutrients from and benefit from protection conferred by the host lichen (Kellogg and Raja 2017). In return, the endolichenic fungi produce various bio-active compounds that may confer benefits to their lichen host (Kellogg and Raja 2017). These bio-active compounds may also benefit humans and are now explored to potentiate their value.

4.3 Bio-active Compounds from Endolichenic Fungi

Endolichenic fungi produce a wide array of bio-active compounds. These include polyketides, alkaloids, terpenoids, steroids, and cyclic peptides (Gao et al. 2016). The chemical investigation of endolichenic fungi has become well-known 17 years after the first paper on endolichenic fungal diversity was published (Paranagama et al. 2007). This has led to the characterization and isolation of over 176 bio-active compounds, in which 104 of the compounds represent novel chemical structures (Gao et al. 2016) and exhibited several biological activities such as antioxidant, antiviral, antibacterial, antifungal, and cytotoxic activities, with the latter two more commonly pursued (Table 4.1; Fig. 4.2). Other activities such as inhibition of acetylcholinesterase and $A\beta_{42}$ aggregation activity (anti-Alzheimer's disease), nematicidal, insecticidal, and promotion of root elongation were also reported. Interestingly, the bio-active compounds produced by endolichenic fungi were somewhat different from those produced by the lichen mycobiont (Kellogg and Raja 2017; Singh et al. 2017).

4.3.1 Antioxidant Compounds

Bio-active compounds from endolichenic fungi have been reported to have antioxidant activities. Samanthi et al. (2015a) isolated two novel polyketides from the endolichenic fungus *Penicillium citrinum* inhabiting the lichen *Parmotrema* sp. In their study, the antioxidant activity of one of the polyketides was comparable to the antioxidant standard butylated hydroxytoluene (BHT). In a more recent study,

Table 4.1 Examples of endolichenic fungi and their bio-active compounds produced and valuable properties. Data is sampled from discoveries reported from 2007 to early 2018

Endolichenic fungi	Lichen host(s)	Major bio-active compounds	Bioactivities	References
Alternaria alternata	*Usnea aciculifera*	(+)-(2S,3S,4aS)-altenuene, (−)-(2S,3S,4aR)-isoaltenuene	Antiviral	He et al. (2012)
Apiospora montagnei	*Cladonia* sp.	Libertellenone L, 23-*O*-acetyl-*N*-hydroxyapiosporamide, 8-hydroxy-3-hydroxymethyl-9-oxo-9H-xanthene-1-carboxylic acid methyl ether, arthrinin A, arthrinin B, myrocin A, libertellenone G	Cytotoxic	Wang et al. (2017)
Aspergillus tubingensis	*Parmelia caperata*	–	Antimicrobial	Padhi et al. (2017)
Aspergillus versicolor	*Lobaria retigera*	8-*O*-methylversicolorin A, 8-*O*-methylversicolorin B, 8-*O*-methylaverythin, 1′-*O*-ethyl-6,8-di-*O*-methylaverantin	Cytotoxic	Dou et al. (2014)
Aspergillus versicolor	–	diorcinol D	Antifungal	Li et al. (2015a, b)
Aspergillus sp.	*Parmelia* sp.	–	Antimicrobial	Padhi and Tayung (2015)
Biatriospora sp.	–	Biatriosporin D	Antifungal	Zhou et al. (2016b) and Zhang et al. (2017)
Broomella sp.	*Parmotrema* sp.	–	Antifungal	Kannangara et al. (2009)
Chaetothyriales sp.	*Umbilicaria* sp.	Chaetothyrins A–C	Antifungal, cytotoxic	Zhou et al. (2016a)
Chaetomium globosum	*Evernia strumnepalense*	Chaetoglobosin Y, chaetoglobosin E, isochaetoglobosin D, chaetoglobosin G	Cytotoxic	Zheng et al. (2014)
Chrysosporium sp., *Cladosporium* sp.	*Pseudocyphellaria* sp., *Parmotrema* sp., *Usnea* sp.	–	Antifungal	Kannangara et al. (2009)

(continued)

Table 4.1 (continued)

Endolichenic fungi	Lichen host(s)	Major bio-active compounds	Bioactivities	References
Coniochaeta sp.	*Xanthoria mandschurica*	Conioxepinol A–D, coniofurol A, conioxanthone A, coniothiepinols A and B, coniothienol A	Antibacterial, cytotoxic	Wang et al. (2010b)
Corynespora sp.	*Usnea cavernosa*	Corynesporol, 1-hydroxydehydroherbarin, herbarin	Cytotoxic	Paranagama et al. (2007)
Cryptosporiopsis diversispora	–	–	Antifungal	Hwang et al. (2011)
Curvularia trifolii	*Usnea* sp.	Macrocyclic lactone, macrocyclic ketone	Anti-inflammatory, antioxidant, cytotoxic	Samanthi et al. (2015b)
Curvularia sp.	*Pseudocyphellaria* sp., *Usnea* sp.	–	Antifungal	Kannangara et al. (2009)
Cytospora sp.	*Parmelia* sp.	–	Antimicrobial	Padhi and Tayung (2015)
Floricola striata	*Umbilicaria* sp.	Floricolins A–J, betulinan C, BTH-II0204-207, betulinan A, betulinan B	Antifungal	Li et al. (2016)
Fusarium sp.	*Parmelia* sp.	-	Antimicrobial	Padhi and Tayung(2015)
Geopyxis aff. *majalis*	*Pseudevernia intensa*	Geopyxin A–D	Cytotoxic	Wijeratne et al. (2012)
Geopyxis sp.	*Pseudevernia intensa*	Geopyxin A, E, and F	Cytotoxic	Wijeratne et al. (2012)
Lecythophora sp.	*Parmotrema tinctorum*, *Cladonia evansii*	Oxaspirol A, B, C, and D	ATPase activity (cytotoxic)	Wijeratne et al. (2016)
Myxotrichum sp.	*Cetraria islandica*	Myxotritones A–C	Root elongation	Yuan et al. (2016)
		Myxodiol A, myxotrichin A–C	Antifungal, cytotoxic	Yuan et al. (2013)
Neurospora terricola	*Evernia strumcirrhatum*	Terricollenes A-C, terricollene	Cytotoxic	Zhang et al. (2009)

Nigrospora sphaerica	*Parmelinella wallichiana*	(+)-(2S,3S,4aS)-altenuene, (−)-(2S,3S,4aR)-isoaltenuene	Antiviral	He et al. (2012)
Nigrospora sp.	*Usnea* sp.	-	Antifungal	Kannangara et al. (2009)
Nodulisporium sp.	*Everniastrum* sp.	Nodulisporipyrones A–D	Antimicrobial	Zhao et al. (2015)
		Nodulisporisteriods A–B, demethoxyviridin, inoterpene B	Anti-Aβ_{42} aggregation activity (anti-Alzheimer's disease)	Zheng et al. (2013)
Ophiosphaerella korrae	*Physcia caesia*	Ophiosphaerellins A–I, ophiosphaerekorrins A–B	Acetylcholinesterase inhibitor (anti-Alzheimer's disease)	Li et al. (2018)
Penicillium citrinum	*Parmotrema* sp.	Polyketides	Antioxidant	Samanthi et al. (2015a)
Penicillium pinophilum	*Pseudocyphellaria* sp.	unidentified compounds (3)	Anti-inflammatory, insecticidal	Cooray et al. (2017)
Penicillium sp.	*Parmelia* sp., *Pseudocyphellaria* sp.	–	Antimicrobial	Kannangara et al. (2009); Padhi and Tayung (2015)
Periconia sp.	*Parmelia* sp.	Pericocins A–D	Antimicrobial	Wu et al. (2015a)
		Pericolactines A–C	–	Wu et al. (2015b)
Pestalotiopsis sp.	*Clavaria* sp.	Ambuic acid, six ambuic acid derivatives, torreyanic acid	Antimicrobial	Ding et al. (2009)
Phaeosphaeria sp.	*Heterodermia obscurata*	Phaeosphaerins A–F	Cytotoxic	Li et al. (2012)
Phialocephala fortinii	*Parmelia* sp.	Six spirobisnaphthalenes, four perylenequinones, five naphthalenone (structures 1-4,7-13)	Antifungal	Xie et al. (2016)
Phialophora sp.	*Cetrelia braunsiana*	(+)-(2S,3S,4aS)-altenuene, (−)-(2S,3S,4aR)-isoaltenuene	Antiviral	He et al. (2012)

(continued)

Table 4.1 (continued)

Endolichenic fungi	Lichen host(s)	Major bio-active compounds	Bioactivities	References
Phialophora sp.	*Cladonia ochrochlora*	Xinshengin, phialophoriol, altenusin	–	Ye et al. (2013)
Phoma sp.	*Pseudocyphellaria* sp.	–	Antifungal	Kannangara et al. (2009)
Phomopsis sp.	*Parmelia* sp.	–	Antimicrobial	Padhi and Tayung (2015)
Pleosporales sp.	–	Cucurbitacins A–E, benzocoumarins	–	Jiao et al. (2015)
Preussia africana	*Ramalina calicaris*	Preussochromone A, preussochromones B–F	Cytotoxic	Zhang et al. (2012)
Scopulariopsis sp.	*Cladonia gracilis*	1-(4′-hydroxy-3′5′-dimethoxy-phenyl)-1,8-dimethoxynaphthalen-2(*1H*)-one, 1,8-dimethoxynaphthalen-2-ol	-	Yang et al. (2012)
Thielavia microspora	–	–	Antifungal	Hwang et al. (2011)
Tolypocladium cylindrosporum	*Lethariella zahlbruckneri*	Tolypocladenols A1, A2, and B, tolypyrodinone A	Cytotoxic	Li et al. (2015a, b)
Trichoderma sp.	*Parmelia* sp.	–	Antimicrobial	Padhi and Tayung (2015)
Ulocladium sp.	*Everniastrum* sp.	Tricycloalternarenes F–H, ophiobolins P–T	Antimicrobial, cytotoxic	Wang et al. (2013a,b)
Xylaria grammica	*Menagazzia* sp.	Grammicin	Nematicidal	Kim et al. (2018)
Xylaria sp.	*Leptogium saturninum*	Two cyclic pentapeptides, blazein, ganodesterone, ergosterin	Antifungal	Wu et al. (2011)
–	*Parmotrema austrosinense*	(3*R*)-5-Hydroxymellein	Antioxidant	Zhao et al. (2017)

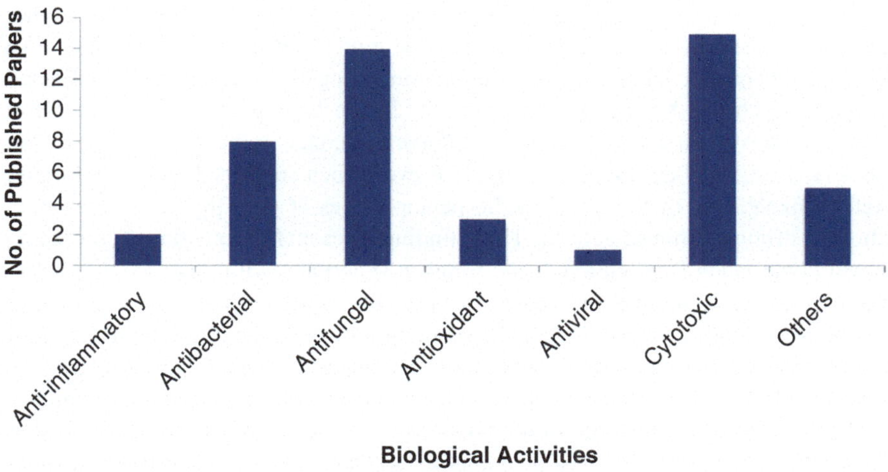

Fig. 4.2 Number of paper published on the chemical investigations of bio-active compounds from endolichenic fungi from 2007 to early 2018

(3*R*)-5-hydroxymellein was isolated from an unidentified endolichenic fungus inhabiting *P. austrosinense* (Zhao et al. 2017). This compound exhibited antioxidant activity, comparable with those of the commercial antioxidants, ascorbic acid, and butyl hydroxyl anisole. This (3*R*)-5-hydroxymellein compound serves as protectant for human keratinocytes when exposed to UV rays (280–315 nm).

4.3.2 Antimicrobial Compounds

Bio-active compounds from endolichenic fungi were also tested for antibacterial and antifungal activities. Wang et al. (2010a) isolated the first naturally occurring thiepinol and thienol from the endolichenic fungus *Coniochaeta* sp. These polyketides showed significant inhibition against the Gram-positive bacteria *Enterococcus faecium* and *E. faecalis*. Similarly, Ding et al. (2009) reported that the novel compounds, such as ambuic acid and ambuic derivatives, and torreyanic acid analogue were isolated from the endolichenic fungus *Pestalotiopsis* sp. inhabiting the lichen *Clavaria* sp. Among these polyketides, ambuic acid and one of its derivatives showed moderate inhibition against the Gram-positive bacterium *Staphylococcus aureus* with IC_{50} values of 43.9 and 27.8 μM, respectively. However, no inhibition was observed against Gram-negative bacteria and fungi, including yeast. On the contrary, diorcinol D, a diphenyl ethyl derivative isolated from the endolichenic fungus *Aspergillus versicolor*, showed fungicidal activities against the yeast *C. albicans* (Li et al. 2015a, b). Diorcinol D disrupted the cell membrane, resulting in high-osmotic pressure stress, thereby stimulating the production of reactive oxygen species (ROS) as a stress response. The elevated level of ROS exceeded the cell's threshold, leading to cell death.

Some of the bio-active compounds produced by endolichenic fungi may require synergism with other existing compounds to render antimicrobial activities. Wu et al. (2011) revealed this using ten proline-containing cyclopentapeptides extracted from endolichenic *Xylaria* sp. inhabiting the lichen *Leptogium saturninum*. These compounds were first tested against *C. albicans* at concentrations of 100 μg/ml but showed no signs of antifungal activity. However, when combined with 0.004 μg/ml ketoconazole, one of the novel cyclic pentapeptides at 6.25 μg/ml concentration showed strong antifungal activity. These findings present the potential of combining novel compounds from endolichenic fungi in synergism with other existing compounds to render desired outcomes. In brief, the antimicrobial activities of bio-active compounds produced by endolichenic fungi were generally effective against Gram-positive bacteria, yeasts, and other fungi, but not against Gram-negative bacteria (Table 4.2). Nevertheless, there is still justification to expound the antimicrobial properties of compounds from endolichenic fungi, as they have been found to respond to synergism with existing compounds. There is room for further investigations and more discoveries into the novel compounds.

4.3.3 Cytotoxic Compounds

The first evaluation of cytotoxic activities from bio-active compounds produced by endolichenic fungi was by Paranagama et al. (2007). Nine compounds, which include novel heptaketides (corynesporol and 1-hydroxydehydroherbarin), herbarin, and herbarin derivatives, were isolated from the endolichenic fungus *Corynespora* sp. BA-10763 from the fruticose lichen *Usnea cavernosa.* The compounds were tested for their cytotoxicity and inhibition toward cell migration of human metastatic breast (MDA-MB-231) and prostate (PC-3M) cancer cells. The compound dehydroherbarin was found to be most effective, significantly inhibiting the migration of PC-3M and MDA-MB-231 cells with the application of 5 μM of dehydroherbarin. Nevertheless, the efficiency of dehydroherbarin was limited to inhibition of cell migration and not cytotoxicity. To address this limitation, structural modification to the bio-active compounds (e.g., acetylation of corynesporol and reduction of herbarin) has been attempted to generate derivatives that may possibly be cytotoxic. Structural modifications have to be attempted in a cautious manner as this could lead to either the complete loss or enhance cytotoxic activities. In a study by Wang et al. (2017), the acetylation and aromatization to the pyridine alkaloids isolated from the endolichenic fungus *Apiospora montagnei* from the lichen *Cladonia* sp. led to a complete loss of cytotoxicity. However, with the addition of hydroxyl group, a tenfold increase in cytotoxicity was detected. Similarly, structural modification resulting in the lack of methyl substituent at C-5 of isocoumarin from *A. montagnei* enhanced cytotoxic activity (Wang et al. 2017). Zhang et al. (2009) evaluated the cytotoxic activity of the endolichenic fungi *Neurospora terricola* from the foliose lichen *Everniastrum cirrhatum*. Several novel polyketides were discovered: terricollenes A–C, terricolyne, 1-*O*-methylterricolyne, and 1-*O*-acetylterricolyne. However, only terricollene A and C and 1-*O*-methylterricolyne

Table 4.2 Identified bio-active compounds from endolichenic fungi exhibiting antimicrobial activities (published after 2009)

Compounds	Microbe	IC_{50}/MIC/MIC_{80}	References
Ambuic acid	*Staphylococcus aureus* (ATCC 6538)	IC_{50}: 43.9 μM	Ding et al. (2009)
Ambuic acid derivative ($C_{21}H_{28}O_7$)	*Staphylococcus aureus*(ATCC 6538)	IC_{50}: 27.8 μM	Ding et al. (2009)
Coniothiepinol A	*Enterococcus faecium*	IC_{50}: 3.93 ± 0.18 μg/mL	Wang et al. (2010a)
	Enterococcus faecalis	IC_{50}: 11.51 ± 0.45 μg/mL	Wang et al. (2010a)
Cyclo(*N*-methyl-$_L$-Phe-$_L$-Val-$_D$-Ile-$_L$-Leu-$_L$-Pro)	*Candida albicans*	MIC: 6.25 μg/mL (in synergism)	Wu et al. (2011)
Tricycloalternarene 1b	Bacille *Calmette-Guerrin*	MIC: 125 μg/mL	Wang et al. (2013a)
Ophiobolin P	Methicillin-resistant *Staphylococcus aureus*	MIC: 62.5 μM	Wang et al. (2013b)
	Bacillus subtilis	MIC: 31.3 μM	Wang et al. (2013b)
Ophiobolin P	Bacille *Calmette-Guerrin*	MIC: >250 μM	Wang et al. (2013b)
Ophiobolin T	Methicillin-resistant *S. aureus*	MIC: 31.3 μM	Wang et al. (2013b)
	Bacillus subtilis	MIC: 15.6 μM	Wang et al. (2013b)
	Bacille *Calmette-Guerrin*	MIC: 31.3 μM	Wang et al. (2013b)
Diorcinol D	*Candida albicans*	MIC_{80}: 8 mg/L	Li et al. (2015a, b)
	Candida krusei	MIC_{80}: 32 mg/L	Li et al. (2015a, b)
	Candida tropicalis	MIC_{80}: 16 mg/L	Li et al. (2015a, b)
	Candida glabrata	MIC_{80}: 32 mg/L	Li et al. (2015a, b)
	Candida parapsilosis	MIC_{80}: 16 mg/L	Li et al. (2015a, b)
Nodulisporipyrone A	*Candida albicans*	MIC: 500 μg/mL	Zhao et al. (2015)
	Aspergillus niger	MIC: 31 μg/mL	Zhao et al. (2015)
Nodulisporipyrone B	*Candida albicans*	MIC: 250 μg/mL	Zhao et al. (2015)
	Aspergillus niger	MIC: 31 μg/mL	Zhao et al. (2015)
Nodulisporipyrone C	*Candida albicans*	MIC: 250 μg/mL	Zhao et al. (2015)
	Aspergillus niger	MIC: 31 μg/mL	Zhao et al. (2015)

(continued)

Table 4.2 (continued)

Compounds	Microbe	IC_{50}/MIC/MIC_{80}	References
Nodulisporipyrone D	*Candida albicans*	MIC: 250 µg/mL	Zhao et al. (2015)
	Aspergillus niger	MIC: 31 µg/mL	Zhao et al. (2015)
Floricolin C	*Candida albicans*	MIC: 8 µg/mL	Li et al. (2016)
Chaetothyrin A	*Candida albicans*	MIC: >240 µM	Zhou et al. (2016a)
Chaetothyrin B	*Candida albicans*	MIC: >240 µM	Zhou et al. (2016a)
Chaetothyrin C	*Candida albicans*	MIC: >240 µM	Zhou et al. (2016a)

exhibited modest cytotoxicity against the human tumor cells HeLa and MCF-7, with IC_{50} values ranging from 53.3 to 92.6 µM and 59.2 µM, respectively. Another endolichenic fungi, *Ulocladium* sp., inhabiting the same lichen species (*Everniastrum* sp.) produced tricycloalternarene (terpenoid) with strong cytotoxicity against the same human tumor cells (IC_{50}value of 8.58 µM) (Wang et al. 2013a). *Ulocladium* sp. also produced ophiobolin T and 6-epi-ophiobolin G, terpenoid compounds with strong cytotoxic activities against HepG2 (human hepatocellular liver carcinoma), with IC_{50} values of 0.24 and 0.37 µM, respectively (Wang et al. 2013b). In addition, the endolichenic *Coniochaeta* sp. inhabiting the lichen *Xanthoria mandschurica* also produce new polyketides effective against four human tumor cell lines, which include HepG2, HeLa (cervical epithelium), A549 (human lung carcinoma), and MDA-MB-231 (human breast adenocarcinoma) (Wang et al. 2010b). The compounds conioxepinol B and conioxepinol D showed moderate IC_{50} values against HeLa cells (36.2 µM) and A6549 (40.9 µM) and MDA-MB-231 (41.4 µM), respectively.

Endolichenic fungi have also been discovered to produce pigments with biological activities known as phototoxins/photosensitizers (Zhou and Liu 2010). One of the most interesting groups of phototoxins is the perylenequinones, which are known to absorb and transform light energy to generate extremely cytotoxic reactive oxygen species (Daub et al. 2005). These toxins localize in different parts of the cell and can cause detrimental effects such as cell death (Zhou and Liu 2010). Li et al. (2012) identified 12 polyketides produced by *Phaeosphaeria* sp. from the foliose lichen *Heterodermia obscurata.* The polyketides are phaeosphaerins A, B, C, D, E, and F; hypocrellins A and C; elsinochromes A, B, and C; and calphostin D. Six of the phototoxins are novel (phaeosphaerins A to F), and they showed inhibition against three human prostate cancer cells by accumulating within the lysosomes of the tumor cells and causing cell death. The IC_{50} values ranged from 2.2 to 25 µM. In addition, phaeosphaerin C and hypocrellin A were found to respond positively to light, increasing their cytotoxicity upon light irradiation. Another interesting cytotoxic compound of natural occurrence was thiopyranchromenone (preussochromone A). This compound, together with five other novel chromone derivatives, was isolated from the endolichenic fungus *Preussia africana* from the lichen *Ramalina calicaris* (Zhang et al. 2012). These compounds were tested against four human tumor cell lines, with preussochromone A and C showing significant cytotoxic effects against lung carcinoma epithelial cells (IC_{50} values of 8.34 and 5.75 μM, respectively). Preussochromone A is the first naturally occurring thiopyranchromenone isolated, which carries with it the 3,4-dihydrothiopyrano[2,3-b] chromen-5(2H)-one skeleton. This polyketide has never been isolated from natural products, as was only previously found in synthetic compounds (Zhang et al. 2012). This discovery exalts endolichenic fungi to a new level of importance as producers of beneficial compounds.

Several other natural *ent*-kaurane diterpenes and their modified derivatives/analogues have also been found to have anticancer activity. Wijeratne et al. (2012) isolated these compounds from the endolichenic fungi *Geopyxis* aff. *majalis* and *Geopyxis* sp., found in the foliose lichen *Pseudevernia intensa*. All compounds were subjected to cytotoxicity assay against five cancer cell lines. Geopyxin B (IC_{50} 2.10–6.32 μM) and 1-*O*-acetylmethylgeopyxin A (IC_{50} 0.39–1.58 μM), in natural *ent*-kaurane and methyl ester analogues, respectively, were observed to have high cytotoxic activities. Wijeratne et al. (2016) also isolated novel oxaspirols A, B, C, and D from the endolichenic fungus *Lecythophora* sp. from two lichen species, *Parmotrema tinctorum* (oxaspirols B, C, and D) and *Cladonia evansii* (oxaspirols A, B, and C). Oxaspirol B was found to be active, showing specific p97 ATPase inhibitory activity.

4.3.4 Other Valuable Bio-active Compounds

In addition to cytotoxic, antimicrobial, and antioxidant properties, the endolichenic fungi have also been reported to produce other beneficial compounds with profound biological activities. This includes antiviral, anti-Alzheimer's disease, insecticidal, and nematicidal activities. The compounds that are responsible for these properties vary and are produced by the diverse species of endolichenic fungi. For example, the endolichenic fungus *Nigrospora sphaerica* found in the lichen *Cetrelia braunsiana* produces heptaketides alternariol and alternariol-9-methyl ether, which showed antiviral activity at IC_{50} values of 13.5 and 23.5 μM, respectively (He et al. 2012). On the other hand, the endolichenic fungus *Ophiosphaerella korrae* from the lichen *Physcia caesia* produces polyketide-derived compounds with inhibitory effect toward acetylcholinesterase (AChE) (anti-Alzheimer's disease). These compounds include ophiosphaerellin A to I and ophiosphaerekorrins A and B, with ophiosphaerellin C exhibiting the strongest AChE activity with application at 1.25 μg (Li et al. 2018). Similarly, Zheng et al. (2013) isolated nodulisporisteroid A and B, alongside with demethoxyviridin and inoterpene B from the endolichenic fungus *Nodulisporium* sp. These compounds showed inhibition toward aggregation of $A\beta_{42}$ (a small peptide involved in the development of Alzheimer's disease). In their investigation, demethoxyviridin displayed anti-$A\beta_{42}$ aggregation activity, with IC_{50} value of 13.4 μM.

The insecticidal and nematicidal activities are valuable for application in crop protection. Kim et al. (2018) isolated the bio-active compound grammicin from the endolichenic fungus *Xylaria grammica* from the foliose lichen *Menegazzia* sp., which showed strong nematicidal activity against the parasitic nematode *Meloidogyne incognita*. In their study, grammicin was most effective toward the second-stage juveniles and nematode eggs. This bio-active compound was also prepared into a powder-type formulation and was applied on pot and field experiments to suppress the development of root-knot nematode disease in tomato and melon plants. In brief, bio-active compounds produced by endolichenic fungi offer a wide range of applications. Several of these compounds have novel chemical structures, offering beneficial applications. With the emerging biotechnological tools, bioprospecting of these compounds can be explored for the utilization and benefit of mankind.

4.4 Biotechnological Approaches to Harness Bio-active Compounds from Endolichenic Fungi

The study of endolichenic fungi and their bio-active compounds are based on biotechnological approaches. These approaches are adopted to achieve optimum conditions or best practices to isolate and establish cultures of endolichenic fungi, to produce sufficient bio-active compounds via fermentation, and to screen for their valuable bio-active compounds via the "one strain, many compounds" (OSMAC) approach.

4.4.1 Culturing and Isolation Techniques for Endolichenic Fungi

In the first isolation approach, no chemical surface sterilization was utilized (Petrini et al. 1990). Instead, rinsing of thalli in sterile tap water with agitation and mesh filtration was performed. The lichen thalli were then cut into small pieces and plated onto 2% Malt Extract Agar (MEA) (2% malt extract, 0.4% yeast extract, 2% agar). The authors were able to isolate 506 fungal taxa. Girlanda et al. (1997) adopted a similar approach but introduced the use of surface disinfectants such as H_2O_2. This gave rise to 117 fungal taxa. Comparatively, it was found that incorporation of H_2O_2 in the surface sterilization technique yielded lesser endolichenic fungi. On the contrary, Suryanarayanan et al. (2005) modified the disinfectants and used ethanol and sodium hypochlorite (NaOCl) instead of H_2O_2. They successfully recovered 242 isolates belonging to 21 genera. It is subsequently noted that to achieve optimum isolation conditions, the type and concentration of disinfectants as well as the exposure time is to be factored. In general, fewer endolichenic fungi will be recovered with prolonged exposure of the lichen thalli in 0.5% NaOCl (Arnold et al. 2009; Kellogg and Raja 2017). It was also discovered that most researchers advocate the use of 2% MEA to recover high numbers of endolichenic fungi (Kannangara et al. 2009; Arnold et al. 2009; Peršoh and Rambold 2012; U'ren et al. 2012; Chagnon et al. 2015; Vinayaka et al. 2016). With improvements to isolation and culture establishment approaches, advent molecular tools (pyrosequencing, metagenomics) can be further introduced to allow the study of non-culturable or fastidious species of endolichenic fungi (He and Zhang 2012; U'ren et al. 2014).

4.4.2 Submerged and Solid-State Fermentation for Production of Bio-active Compounds

Fermentation is a critical process to induce the production of fungal bio-active compounds. There are two types of fermentation: (i) submerged fermentation (SmF) and (ii) solid-state fermentation (SSF). SmF is defined as a fermentation process utilizing liquid substrates such as molasses and broths (Subramaniyam and Vimala 2012). SSF, on the other hand, is a fermentation process that involves inert solid matrix/natural substrate (wheat bran, rice, rice straw, hay, fruit and vegetable waste, paper pulp) and is conducted in the absence or near absence of free water (Bhargav et al. 2008; Singhania et al. 2009; Subramaniyam and Vimala 2012). Chemical investigations on endolichenic fungi utilized both types, although SSF (He et al. 2012; Dou et al. 2014; Jiao et al. 2015; Zhao et al. 2015; Li et al. 2016; Wang et al. 2017) is more common than SmF (Paranagama et al. 2007; Wang et al. 2010b; Wijeratne et al. 2012; Padhi et al. 2017). Although both types of fermentation methods can be used, filamentous fungi such as endolichenic fungi are most ideally adapted for SSF (Krishna 2005; Singhania et al. 2009). The bio-active compounds produced by SSF are more stable and produced in higher quantities than SmF (Subramaniyam and Vimala 2012). Jiao et al. (2015) discovered that the bio-active compounds produced by the endolichenic fungi *Pleosporales* sp. differed when cultured in SSF and SmF. SSF yielded the compounds cucurbitacins D and E, 3,10-dihydroxy-4,8-dimethyl-6-methylbenzocoumarin, 3,8,10-trihydroxy-4-methoxy-6 methyl benzocoumarin, and 2,5-dimethoxy-3,6-bis(4-methoxy phenyl)-1,4-benzoquinone, while SmF yielded the compounds cucurbitacins A, B, and C, (5*R*)-5-hydroxy-2,3-dimethylcyclohex-2-en-1-one, dankasterone A, and (17*R*)-4-hydroxy-17-methylincisterol.

4.4.3 Screening for Bio-active Compounds: The OSMAC Approach

"One strain, many compounds" or OSMAC is a simple strategy that aims to increase the number of bio-active compounds available to a target organism by altering the fermentation parameters such as media composition, pH, temperature, aeration, and addition of enzymes (Kusari et al. 2012; Kellogg and Raja 2017). This approach has been adopted to explore the chemical diversity of bio-active compounds produced by endolichenic fungi. OSMAC approach was applied to several studies. Wang et al. (2013a) used this method to determine bio-active compounds produced by the endolichenic fungus *Ulocladium* sp. isolated from the lichen *Everniastrum* sp. The fungus was inoculated into two different culture media, glucose malt yeast extract medium and Czapek's medium, to derive the various bio-active compounds that were produced in response to the media. Similarly, Padhi and Tayung (2015) evaluated the effects of different culture media as well as their incubation period and antimicrobial potential of bio-active compounds produced. Wang et al. (2013b) further utilized the OSMAC method with adjustments to the media composition for the fermentation of the endolichenic fungus *Ulocladium* sp. As a result, five new sesterterpenes (ophiobolins P–T) with strong cytotoxicity were isolated. With OSMAC approach, the metabolic pathway of a microorganism is not restricted, thus achieving a more diverse metabolic profile that may confer stronger bioactivities.

4.5 Commercialization Potential of Bio-active Compounds from Endolichenic Fungi

Endolichenic fungi produce unique and diverse bio-active compounds, with potential for development into commercial valuable products. Commercialization of these products is still untapped and has so much more room for exploration as currently no drugs have been developed from endolichenic fungi. The most extensive works conducted by Zhang et al. (2017) and Kim et al. (2018) under in vivo assessments. Zhang et al. (2017) found the antifungal biatriosporin D isolated from *Biatriospora* sp., which was effective in controlling *C. albicans*. This was evaluated in a nematode model *Caenorhabditis elegans*, where the treatment of biatriosporin D inhibited the transformation of *C. albicans* to the infectious hyphal form. This propelled the possible use of biatriosporin D as a commercial antifungal compound for the treatment of yeast infection (*C. albicans*). In another study, Kim et al. (2018) used the compound grammicin, isolated from the endolichenic fungus *X. grammica* to suppress the development of root-knot disease caused by nematodes on tomato and melon. They suggested the feasibility of grammicin as a marketable control agent against plant-parasitic nematodes. Thus, it is evident that bio-active compounds isolated from endolichenic fungi have the potential to be used for the commercialization purposes.

4.6 Conservational Importances of Lichen and Their Endolichenic Fungi

As the demands for novel bio-active compounds increases, there will be increasing explorations on this niche group of endolichenic fungi. Consequently, there is a possibility that lichens will soon be exhausted rapidly. Since, lichens are slow-growers (Shukla et al. 2014). Therefore, their conservation is crucial to allow bioprospecting of endolichenic fungi from lichens in a more sustainable manner especially for biomonitoring and biodiversity assessment. For instance, standardized protocols for lichen sampling and data management are designed to accurately monitor and determine patterns of distribution and diversity at larger geographical scales (Shukla et al. 2014). However, the chemical investigations provide the accurate and consistent lichen sample. Collection of all lichens present in one area is not recommended because this may lead to the loss of biodiversity. There should also be more awareness on the nature of lichens and their beneficial value to humans. Conservation practices can help drive this important message. There should also be more strategies to upkeep the forests and reduce pollution as lichens are significantly affected by deforestation activities and pollution. Lichens are poikilohydrous in nature, having their water status passively depending on the surrounding environment (Nash 2008). As such, the growth of lichens is highly susceptible to numerous pollutants and other environmental disturbances (Seaward 2008; Shukla et al. 2014). When lichen growth is affected, they slowly die, gradually affecting the availability of endolichenic fungi for research explorations.

4.7 Conclusions and Future Prospects

Endolichenic fungi and their various bio-active compounds have opened the new area of bioprospecting valuable compounds from a relatively uncommon group of fungi. Still, there is much that can be explored from endolichenic fungi, leading to drug discovery and other beneficial applications. In addition, in depth studies on the association of endolichenic fungi could be performed to understand their ecological roles and the effects of the bio-active compounds produced to the host itself. With the emerging biotechnological tools and improvements to the culturing techniques, compounds from endolichenic fungi can be elucidated to explore their bio-active potential in near future for the discovery of novel drugs.

References

Arnold AE, Miadlikowska J, Higgins KL, Sarvate SD, Gugger P, Way A, Hofstetter V, Kauff F, Lutzoni F (2009) A phylogenetic estimation of trophic transition networks for Ascomycetous fungi: are lichens cradles of symbiotrophic fungal diversification. Syst Biol 58:283–297

Bhargav S, Panda BP, Ali M, Javed S (2008) Solid-state fermentation: an overview. CABEQ 22:49–70

Blackwell M (2011) The fungi: 1,2,3…5.1 million species. Am J Bot 98:426–438

Chagnon PL, U'ren JM, Miadlikowska J, Lutzoni F, Arnold AE (2015) Interaction type influences ecological network structure more than local abiotic conditions: evidence from endophytic and endolichenic fungi at a continental scale. Oecologia 180:181–191

Cooray MPM, Gobika T, Paranagama PA (2017) Insecticidal properties of secondary metabolites produced by an endolichenic fungi *Penicillium pinophilum* against stored grain insect pest *Callosobruchus maculatus*. In: Proceedings of the 2nd international conference on bioscience and biotechnology, vol 2, pp 107–115. (ISSN 2513-2695 online). https://doi.org/10.17501/biotech.2017.2110

Daub ME, Sonia H, Chung KR (2005) Photoactivated perylenequinone toxins in fungal pathogenesis of plants. FEMS Microbiol Lett 252:197–206

Ding G, Li Y, Fu S, Liu S, Wei J, Che Y (2009) Ambuic acid and torreyanic acid derivatives from the endolichenic fungus *Pestalotiopsis* sp. J Nat Prod 72:182–186

Dou Y, Wang X, Jiang D, Wang H, Jiao Y, Lou H, Wang X (2014) Metabolites from *Aspergillus versicolor*, an endolichenic fungus from the lichen *Lobaria retigera*. Drug Discov Ther 8:84–88

Gao H, Zou J, Li J, Zhao H (2016) Endolichenic fungi: a potential treasure trove for discovery of special structures and bioactive compounds. In: Studies in natural products chemistry, vol 48. Elsevier, Dordrecht, pp 347–397

Girlanda M, Isocrono D, Bianco C, Luppi-Mosca AM (1997) Two foliose lichens as microfungal ecological niches. Mycologia 89:531–536

Hawksworth DL, Rossman AY (1997) Where are all the undescribed fungi. Phytopathology 87:888–891

He Y, Zhang Z (2012) Diversity of organism in the *Usnea longissima* lichen. Afr J Microbiol Res 6:4797–4804

He JW, Chen GD, Gao H, Yang F, Li XX, Peng T, Guo LD, Yao XS (2012) Heptaketides with antiviral activity from three endolichenic fungal strains *Nigrospora* sp., *Alternaria* sp. and *Phialophora* sp. Fitoterapia 83:1087–1091

Honneger R, Axe L, Edwards D (2013) Bacterial epibionts and endolichenic actinobacteria and fungi in the lower devonian lichen *Chlorolichenomycites salopensis*. Fungal Biol 117:512–518

Hwang HG, Kim YN, Baik KS, Choi SK (2011) Isolation and characteristics of endolichenic fungi producing antifungal compound. Korean J Microbiol 47:97–101

Jiao Y, Li G, Wang HY, Liu J, Li XB, Zhang LL, Zhao ZT, Lou HX (2015) New metabolites from endolichenic fungus *Pleosporales* sp. Chem Biodivers 12:1095–1104

Kannangara BTSDP, Rajapaksha RSCG, Paranagama PA (2009) Nature and bioactivities of endolichenic fungi in *Pseudocyphellaria* sp., *Parmotrema* sp. and *Usnea* sp. at Hakgala montane forest in Sri Lanka. Lett Appl Microbiol 48:203–209

Kellogg JJ, Raja HA (2017) Endolichenic fungi L., a new source of rich bioactive secondary metabolites on the horizon. Phytochem Rev 16:271–293

Kim TY, Jang JY, Yu NH, Chi WJ, Bae CH, Yeo JH, Park AR, Hur JS, Park HW, Park JY (2018) Nematicidal activity of grammicin produced by *Xylaria grammica* KCTC 13121BP against *Meloidogyne incognita*. Pest Manag Sci 74:384–391

Krishna C (2005) Solid-state fermentation systems-an overview. Crit Rev Biotechnol 25:1–30

Kusari S, Hertweck C, Spiteller M (2012) Chemical ecology of endophytic fungi: origins of secondary metabolites. Chem Biol 19:792–798

Li G, Wang H, Zhu R, Sun L, Wang L, Li M, Li Y, Liu Y, Zhao Z, Lou H (2012) Phaeosphaerins A–F, cytotoxic perylenequinones from an endolichenic fungus, *Phaeosphaeria* sp. J Nat Prod 75:142–147

Li XB, Li L, Zhu RX, Li W, Chang WQ, Zhang LL, Wang XN, Zhao ZT, Lou HX (2015a) Tetramic acids and pyridone alkaloids from the endolichenic fungus *Tolypocladium cylindrosporum*. J Nat Prod 78:2155–2160

Li Y, Chang W, Zhang M, Li X, Jiao Y, Lou H (2015b) Diorcinol D exerts fungicidal action against *Candida albicans* through cytoplasm membrane destruction and ROS accumulation. PLoS One 10:e0128693. https://doi.org/10.1371/journal.pone.0128693

Li W, Gao W, Zhang M, Li YL, Li L, Li XB, Chang WQ, Zhao ZT, Lou HX (2016) *p*-terphenyl derivatives from the endolichenic fungus *Floricola striata*. J Nat Prod 79:2188–2194

Li Y, Zhu R, Zhang J, Xie F, Wang X, Xu K, Qiao Y, Zhao Z, Lou H (2018) Ophiosphaerellins A–I, polyketide-derived compounds from the endolichenic fungus *Ophiosphaerella korrae*. ACS Omega 3:176–180

Nash TH III (2008) Introduction. In: Nash TH (ed) Lichen biology. Cambridge University Press, Edinburgh/Cambridge, pp 1–8

Padhi S, Tayung K (2015) *In vitro* antimicrobial potentials of endolichenic fungi isolated from thalli of *Parmelia* lichen against some human pathogens. Beni-Suef Univ J Basic Appl Sci 4:299–306

Padhi S, Das D, Panja S, Tayung K (2017) Molecular characterization and antimicrobial activity of an endolichenic fungus, *Aspergillus* sp. isolated from *Parmelia caperata* of Similipal Biosphere Reserve, India. Interdiscip Sci Comput Life Sci 9:237–246

Paranagama PA, Wijeratne EMK, Burns AM, Marron MT, Gunatilaka MK, Arnold AE, Gunatilaka AAL (2007) Heptaketides from *Corynespora* sp. inhabiting the cavern beard lichen, *Usnea cavernosa*: first report of metabolites of an endolichenic fungus. J Nat Prod 70:1700–1705

Peršoh D, Rambold G (2012) Lichen-associated fungi of the *Letharietum vulpinae*. Mycol Progress 11:753–760

Petrini O, Hake U, Dreyfuss MM (1990) An analysis of fungal communities isolated from fruticose lichens. Mycologia 82:444–451

Piercey-Normore MD, Deduke C (2011) Fungal farmers or algal escorts: lichen adaptation from the algal perspective. Mol Ecol 20:3708–3710

Samanthi KAU, Wickramaarachchi S, Wijeratni EMK, Paranagama PA (2015a) Two new antioxidant active polyketides from *Penicillium citrinum*, an endolichenic fungus isolated from *Parmotrema* species in Sri Lanka. J Nat Sci Found Sri Lanka 43:119–126

Samanthi KAU, Wickramaarachchi S, Wijeratni EMK, Paranagama PA (2015b) Two new bioactive polyketides from *Curvularia trifolii*, an endolichenic fungus isolated from *Usnea* sp. in Sri Lanka. J Nat Sci Found Sri Lanka 43:217–224

Scheidegger C, Worth S (2009) Conservation strategies for lichens: Insights from population biology. Fungal Biol Rev 23:55–66

Seaward MRD (2008) Environmental role of lichens. In: Nash TH III (ed) Lichen biology. Cambridge University Press, Edinburgh/Cambridge, pp 274–298

Shukla V, Upreti DK, Bajpai R (2014) Lichens to biomonitor the environment. Management and conservational approaches. Springer, New Delhi, pp 171–178

Singh BN, Upreti DK, Gupta VK, Dai XF, Jiang Y (2017) Endolichenic fungi: a hidden reservoir of next generation biopharmaceuticals. Trends Biotechnol 35:808–813

Singhania RR, Patel AK, Soccol CR, Pandey A (2009) Recent advances in solid-state fermentation. Biochem Eng J 44:13–18

Spribille T, Tuovinen V, Resl P, Vanderpool D, Wolinski H, Aime MC, Schneider K, Stabentheiner E, Toome-Heller M, Thor G, Mayrhofer H, Johannesson H, McCutcheon JP (2016) Basidiomycete yeasts in the cortex of ascomycete macrolichens. Science 353:488–492

Subramaniyam R, Vimala R (2012) Solid-state and submerged fermentation for the production of bioactive substances: a comparative study. Int J Sci Nat 3:480–486

Suryanarayanan TS, Thirunavukkarasu N (2017) Endolichenic fungi: the lesser known fungal associates of lichens. Mycology 8:189–196

Suryanarayanan TS, Thirunavukkarasu N, Hariharan GN, Balaji P (2005) Occurrence of non-obligate microfungi inside lichen thalli. Sydowia 57:120–130

Tripathi M, Joshi Y (2015) Endolichenic fungi in Kumaun Himalaya: a case study. In: Upreti DK, Divakar PK, Shukla V, Bajpai R (eds) Recent advances in lichenology. Springer, New Delhi, pp 111–120

U'ren JM, Lutzoni F, Miadlikowska J, Arnold AE (2010) Community analysis reveals close affinities between endophytic and endolichenic fungi in mosses and lichens. Microb Ecol 60:340–353

U'ren JM, Lutzone F, Miadlikowska J, Laetsch AD, Arnold AE (2012) Host and geographic structure of endophytic and endolichenic fungi at a continental scale. Am J Bot 99:898–914

U'ren JM, Riddle JM, Monacell JT, Carbone I, Miadlikowska J, Arnold AE (2014) Tissue storage and primer selection influence pyrosequencing-based inferences of diversity and community composition of endolichenic and endophytic fungi. Mol Ecol Resour 14:1032–1048

Vinayaka KS, Krishnamurthy YL, Banakar S, Prashith Kekuda TR (2016) Association and variation of endophytic fungi among some macrolichens in central western ghats, Southern India. Int J Curr Microbiol App Sci 5:115–124

Wang Y, Niu S, Liu S, Guo L, Che Y (2010a) The first naturally occurring thiepinols and thienol from an endolichenic fungus *Coniochaeta* sp. Org Lett 12:5081–5083

Wang Y, Zheng Z, Liu S, Zhang H, Li E, Guo L, Che Y (2010b) Oxepinochromenones, furochromenone, and their putative precursors from the endolichenic fungus *Coniochaeta* sp. J Nat Prod 73:920–924

Wang QX, Bao L, Yang XL, Guo H, Ren B, Guo LD, Song FH, Wang WZ, Liu HW, Zhang LX (2013a) Tricycloalternarenes F–H: three new mixed terpenoids produced by an endolichenic fungus *Ulocladium* sp. using OSMAC method. Fitoterapia 85:8–13

Wang QX, Bao L, Yang XL, Liu DL, Guo H, Dai HQ, Song FH, Zhang LX, Guo LD, Li SJ, Liu HW (2013b) Ophiobolins P–T, five new cytotoxic and antibacterial sesterterpenes from the endolichenic fungus *Ulocladium* sp. Fitoterapia 90:220–227

Wang Y, Zheng Y, Wang X, Wei X, Wei J (2016) Lichen-associated fungal community in *Hypogymnia hypotrypa* (Parmeliaceae:Ascomycota) affected by geographic distribution and altitude. Front Microbiol 7:1231. https://doi.org/10.3389/fmicb.2016.01231

Wang H, Umeokoli BO, Eze P, Heering C, Janiak C, Müller WEG, Orfali RS, Hartmann R, Dai H, Lin W, Liu Z, Proksch P (2017) Secondary metabolites of the lichen-associated fungus *Apiospora montagnei*. Tetrahedron Lett 58:1702–1705

Wijeratne EMK, Bashyal BP, Gunatilaka MK, Arnold AE, Gunatilaka AAL (2010) Maximizing chemical diversity of fungal metabolites: biogenetically related heptaketides of the endolichenic fungus *Corynespora* sp. J Nat Prod 73:1156–1159

Wijeratne EMK, Bashyal BP, Liu MX, Rocha DD, Gunaherath GMKB, U'ren JM, Gunatilaka MK, Arnold AE, Whitesell L, Gunatilaka AAL (2012) Geopyxins A–E, *ent*-kaurane diterpenoids from endolichenic fungal strains *Geopyxis* aff. *majalis* and *Geopyxis* sp. AZ0066: structure-activity relationships of geopyxins and their analogues. J Nat Prod 75:361–369

Wijeratne EMK, Gunaherath GMKB, Chapla VM, Tillotson J, de la Cruz F, Kang M, U'ren JM, Araujo AR, Arnold AE, Chapman E, Gunatilaka AAL (2016) Oxaspirol B with p97 inhibitory activity and other oxaspirols from *Lecythophora* sp. FL1375 and FL1031, endolichenic fungi inhabiting *Parmotrema tinctorum* and *Cladonia evansii*. J Nat Prod 79:340–352

Wu W, Dai H, Bao L, Ren B, Lu J, Luo Y, Guo L, Zhang L, Liu H (2011) Isolation and structural elucidation of proline-containing cyclopentapeptides from an endolichenic *Xylaria* sp. J Nat Prod 74:1303–1308

Wu YH, Xiao GK, Chen GD, Wang CX, Hu D, Lian YY, Lin F, Guo LD, Yao XS, Gao H (2015a) Pericocins A–D, new bioactive compounds from *Periconia* sp. Nat Prod Commun 10:2127–2130

Wu YH, Chen GD, He RR, Wang CX, Hu D, Wang GQ, Guo LD, Yao XS, Gao H (2015b) Pericolactines A–C, a new class of diterpenoid alkaloids with unusual tetracyclic skeleton. Sci Rep 5:17082. https://doi.org/10.1038/srep17082

Xie F, Chang W, Zhang M, Li Y, Li W, Shi H, Zheng S, Lou H (2016) Quinone derivatives isolated from the endolichenic fungus *Phialocephala fortinii* are Mdr1 modulators that combat azole resistance in *Candida albicans*. Sci Rep 6:33687. https://doi.org/10.1038/srep33687

Yang F, Chen GD, Gao H, Li XX, Wu Y, Guo LD, Yao XS (2012) Two new naphthalene derivatives from an endolichenic fungal strain *Scopulariopsis* sp. J Asian Nat Prod Res 14:1059–1063

Ye F, Chen GD, He JW, Li XX, Sun X, Guo LD, Li Y, Gao H (2013) Xinshengin, the first altenusin with tetracyclic skeleton core from *Phialophora* spp. Tetrahedron Lett 54:4551–4554

Yuan C, Wang HY, Wu CS, Jiao Y, Li M, Wang YY, Wang SQ, Zhao ZT, Lou HX (2013) Austdiol, fulvic acid and citromycetin derivatives from an endolichenic fungus, *Myxotrichum* sp. Phytochem Lett 6:662–666

Yuan C, Guo YH, Wang HY, Ma XJ, Jiang T, Zhao JL, Zou ZM, Ding G (2016) Allelopathic polyketides from an endolichenic fungus *Myxotrichum* sp. by using OSMAC strategy. Sci Rep 6:19350. https://doi.org/10.1038/srep19350

Zhang F, Liu S, Lu X, Guo L, Zhang H, Che Y (2009) Allenyl and alkynyl phenyl ethers from the endolichenic fungus *Neurospora terricola*. J Nat Prod 72:1782–1785

Zhang F, Li L, Niu S, Si Y, Guo L, Jiang X, Che Y (2012) A thiopyranchromenone and other chromone derivatives from an endolichenic fungus, *Preussia africana*. J Nat Prod 75:230–237

Zhang T, Wei XL, Zhang YQ, Liu HY, Yu LY (2015) Diversity and distribution of lichen-associated fungi in Ny-Ålesund Region (Svalbard, High Arctic) as revealed by 454 pyrosequencing. Sci Rep 5:14850

Zhang M, Chang W, Shi H, Zhou Y, Zheng S, Li Y, Li L, Lou H (2017) Biatriosporin D displays anti-virulence activity through decreasing the intracellular cAMP levels. Toxicol Appl Pharmacol 322:104–112

Zhao Q, Wang CX, Yu Y, Wang GQ, Zheng QC, Chen GD, Lian YY, Lin F, Guo LD, Gao H (2015) Nodulisporipyrones A–D, new bioactive α-pyrone derivatives from *Nodulisporium* sp. J Asian Nat Prod Res 17:567–575

Zhao L, Kim JC, Paik MJ, Lee W, Hur JS (2017) A multifunctional and possible skin UV protectant, (3*R*)-5-hydroxymellein, produced by an endolichenic fungus isolated from *Parmotrema austrosinense*. Molecules 22:E26. https://doi.org/10.3390/molecules22010026

Zheng QC, Chen GD, Kong MZ, Li GQ, Cui JY, Li XX, Wu ZY, Guo LD, Cen YZ, Zheng YZ, Gao H (2013) Nodulisporisteriods A and B, the first 3,4-*seco*-4-methyl-progesteroids from *Nodulisporium* sp. Steroids 78:896–901

Zheng QC, Kong MZ, Zhao Q, Chen GD, Tian HY, Li XX, Guo LD, Li J, Zheng YZ, Gao H (2014) Chaetoglobosin Y, a new cytochalasan from *Chaetomium globosum*. Fitoterapia 93:126–131

Zhou ZY, Liu JK (2010) Pigments of fungi (macromycetes). Nat Prod Rep 27:1531–1570

Zhou YH, Li XB, Zhang JZ, Li L, Zhang M, Chang W-Q, Wang X-N, Lou H-X (2016a) Three new drimane-type sesquiterpenoids, chaetothyrins A–C, from an endolichenic fungus *Chaetothyriales* sp. J Asian Nat Prod Res 18:409–414

Zhou YH, Zhang M, Zhu RX, Zhang JZ, Xie F, Li XB, Chang WQ, Wang XN, Zhao ZT, Lou HX (2016b) Heptaketides from an endolichenic fungus *Biatriospora* sp. and their antifungal activity. J Nat Prod 79:2149–2157

Strategic Approaches for the Purification of Glycosides from Natural Sources

5

Anand Shyamlal Gupta

Contents

Abstract

Glycosides from plant origin are broadly classified as into C-, S-, N- and O-glycosides on the basis of the linkage between glycone and aglycone. The chemistry of glycone and aglycone moiety plays a very critical role for the isolation and purification of these glycosides. Globally, glycosides have a great demand in various industrial sectors, such as pharmaceuticals and food and agro-based products. These glycosides can be isolated from various natural sources by

A. S. Gupta (✉)
Amity Institute of Biotechnology, Amity University, Mumbai, Maharashtra, India
e-mail: anandguptapharma@gmail.com

M. S. Akhtar et al. (eds.), *Natural Bio-active Compounds*,
https://doi.org/10.1007/978-981-13-7154-7_5

different extraction processes. Generally, the extract comprises of various structural analogs, which may be bioisosteres, isomers, pathway intermediates or degradation pathway metabolites. These moieties should be separated from the parent molecule for a desired pharmacological effect of glycosides. These structural analogs have a low separation factor, i.e., they are very difficult to separate due to their similar physico-chemical properties with the desired molecule of interest. The structural analogs of glycosides have a severe tendency to compete or interfere in the purification process to obtain the desired molecule. Different unit operations are employed for purifying a molecule from its impurities, but the chromatography provides more number of stages for purification as compared to other unit operations. The provision of the multiple stages makes the chromatography a selective tool for separation and purification of glycosides from their structural analogs along with other unit operations. In recent days, new techniques like simulated moving bed (SMB) for polishing stage have been developed to separate glycosides continuously with high purity. The present chapter deals with chemistry, medicinal importance, isolation and strategic approaches for the purification of glycosides from natural sources.

Keywords

Chromatography · Glycosides · Structural isomers · Solvent extraction · Tandem column

5.1 Introduction

Glycosides are the secondary metabolite produced in plant through various biosynthetic pathways and possess a variety of pharmacological and therapeutic applications. Treatment of various diseases and etiologic conditions by using these glycosides is an age-old practice. Glycosides consist of both sugar moiety (glycone) and non-sugar moiety (aglycone) (Kar 2003). The synthesis of sugars in plants is carried out through the photosynthesis and biochemical pathways (Fig. 5.1). However, there is no generalized or common biosynthetic pathway for aglycone moiety, as each class of glycosides has its specific route of biosynthesis. The most common pathways for the biosynthesis of various glycosides include shikimic acid and acetate pathways. These two pathways may be directly involved in the biosynthesis of glycosides, or they are produced via generation of some metabolic intermediates, such as amino acids. Glycosides are classified on the basis of their sugar moiety, linkage between glycone and aglycone, and on their therapeutic uses (Kar 2003; Bartnik and Facey 2017).

Globally, the glycoside has a great demand in various industrial sectors like food, pharmaceuticals, nutraceuticals and cosmetics (Zhang 2014). It can be isolated from various natural sources by adopting different extraction processes. Usually, the crude plant extracts comprise of various structural analogs, which may be bioisosteres, isomers, pathway intermediates or degradation pathway metabolites. These moieties should be separated from the parent molecule for a desired pharmacological effect of any secondary metabolites including glycosides (Edwards et al. 1976).

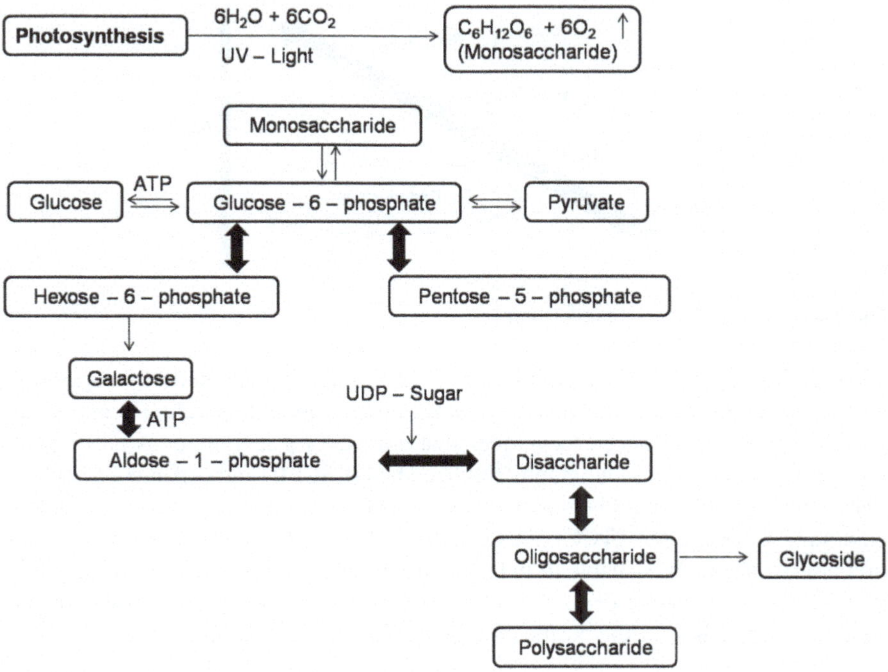

Fig. 5.1 The synthesis of sugar in plants is done by photosynthesis and biochemical pathway

In case of steviol glycosides which comprise of structural analogs like steviosides and rebaudosides, the taste of the steviol glycosides as a natural sweetener is highly affected on the basis of the steviosides (Kennelly 2003). The content of steviosides is very critical as it imparts a bitter taste in the various nutraceutical doses and zero calorie preparations in the food industries (Megeji et al. 2005).

The conventional approach for the purification involves the solid-liquid extraction (SLE) of glycosides from various parts of the plants, roots, rhizomes, barks, bulbs, flowers and leaves followed by other unit operations (Scholfield and Dutton 1955). The conventional approach of purification, involving various unit operations, seems to be compromised during the scale-up of the process. The outcome of these approaches also leads to a compromised yield and purity. As a conventional approach, the SLE can act as an isolating or recovery step followed by precipitation, chromatography and crystallization for the purification and polishing. The strategic approach for the purification at different stages as a general scheme should aim for recovery, isolation, purification and polishing (RIPP) (Fig. 5.2). The RIPP scheme advancement should not only increase the yield and purity at every step but also be effective in reducing the process cost for the purification or downstream processing (DSP) (Ghosh 2006). The strategic approach for the purification should have the correct sequence of the unit operations involved in the process. Also, it should include recent advancements, such as supercritical fluid extraction (SCFE) and aqueous two-phase separation (ATPS) methods, for a selective extraction in the purification process.

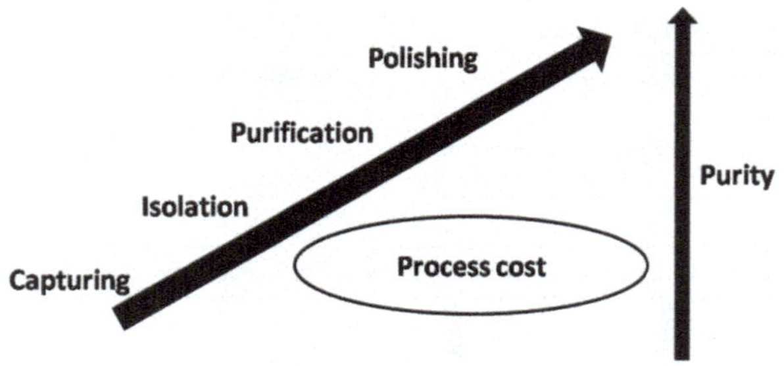

Fig. 5.2 General Scheme for purification

Also, the strategic process should be operated at a continuous mode for separating and purifying glycosidic molecules so that the process should be techno-economical. Different types of advanced techniques that are used for a continuous purification include liquid solid chromatography fluidized bed (LSCFB) and expanded bed chromatography (EBC) and polishing includes simulated moving bed (SMB) and tandem column chromatography (TCC) (Du et al. 2004). The strategically developed extraction process should be scalable from lab to pilot level for the commercialized production. The present chapter deals with chemistry, medicinal importance, isolation and strategic approaches for the purification of glycosides from natural sources.

5.1.1 Challenges for Purification of Glycoside Molecules from Natural Sources

The process for purification of glycosides and other secondary metabolites, such as antibiotic structural analogs, poses various challenges. Due to similar physicochemical properties and low separation factor, the purification of synthetic and semi-synthetic derivatives has challenges for designing and the development of a specific purification process (Edwards et al. 1976). Some of them are mentioned below:

1). The presence of structural analogs such as isomeric impurities, non-isomeric impurities and byproducts formed is a major threat to process development and its abundance in the feed, i.e. types and amounts of impurities (impurity profile).
2). The isomeric impurities present can be positional isomers, functional isomers, geometrical isomers and bioisosteres and are very difficult to separate; the non-isomeric impurities are pathway intermediates generated in biological or chemical synthesis which are comparably less difficult to separate, i.e. the level of difficulty (separation factor) (Yang et al. 2010).
3). The unit operations utilized for the separation of small molecules (Jenkins et al. 1996) from their different structural analogs must result with high selectivity and maximum recovery with uncompromised stability of the product as an outcome, i.e. selection and correct sequencing of unit operations and stability of molecule.
4). The developed process should be industrially efficient having easy scalability and reproducibility.

5). The finished product should meet the regulatory specifications for purity with the desired yield of process, i.e. techno-economical process.

This chapter will provide the basic fundamentals for the development of an intensive, integrative, cost-effective and strategic purification process for separation of secondary metabolites such as glycosides. It will also provide an overview to separate glycosides from their structural related impurities, bioisosteres, structural analogs and biosynthetic intermediates; to grasp the significance of correct sequencing and integration of unit operations like membrane, distillation, precipitation and/or crystallization to benefit overall manufacturing process; and also, to understand the process optimization and various critical parameters in different unit operations involved for process designing. The validation of the developed process with reproducibility as a key factor obtains the batch-to-batch desired yield and purity of the product.

5.2 Sources and Chemistry of Glycosides

Plants are the major sources for glycosides, and they are widely distributed in various parts of them as natural sources. On the basis of the content present in various parts, the specific part is used for commercial application by extractive process for isolation, e.g. Sennosides from *Senna* leaves and Shatavarin I, II, III and IV from Shatavari (*Asparagus racemosus*). In some cases, it is also found that glycosides are present in whole plants, e.g. Chirantin in Chirata (*Swertia chirata*); various sources and different parts or whole plants are given in detail (Table 5.1). One of the extensive ways of glycoside classification is based on the aglycone moiety present in structure along with pharmacological significance (Table 5.1) (Kar 2003). Glycosides are the naturally occurring molecules, consisting of two chemical moieties as the aglycone; as genin and the glycone; and as saccharide or sugars linked together, the sugar moiety is linked to the aglycone moiety by forming a glycosidic bond between them (Evans 2009). The different types of glycosidic linkage are illustrated below:

1) O-glycosides: the bond between the genin and glycone moiety involves oxygen atom as a glycosidic linkage, for example, Sennosides from *Senna* leaves (reaction 1).
2) C-glycosides: the bond between the genin and glycone moiety involves carbon atom as a glycosidic linkage, e.g. aloin from *Aloe*. The C-glycoside is comparatively stronger than other glycosides due to the involvement of C-C bond; high energetics are required for the hydrolysis of the C-C bonds (reaction 2).
3) S-glycosides: the bond between the genin and glycone moiety involves sulphur atom (thio) as a glycosidic linkage, e.g. sinigrin from black mustard (reaction 3).
4) N-glycosides: the bond between the genin and glycone moiety involves nitrogen atom as a glycosidic linkage, e.g. nucleosides present in plants and animals. The existence of the N-glycosides occurs in plants as in the form of cyanogenic glycosides (CN) or isocyanogenic (NC) rather than nitrogen atom. The common example of the CN glycosides is prunasin and amygdalin from almond (reaction 4).

Table 5.1 Natural sources of glycoside from different plant parts and their chemical constituents

Glycosides	Common name	Parts of plant	Major Chemical constituent
Anthracene	Curacao Aloe	Latex of leaves	Aloe – emodin
	Alexandria senna	Dried leaflets	Sennosides – A, B, C, D
	Indian Senna	Dried leaflets	Sennosides – A, B, C, D
	Cascara sagrada	Dried barks	Barbaloin, Deoxybarbaloin
	Frangula	Dried barks	Frangulin A, B
Phenol	Bearberry	Dried leaves	Arbutin, Methylarbutin
	Canadian wintergreen	Leaves	Gaultherin
	Poplar – willow	Barks	Salicin
	Populus	Leaves, barks	Populin
Steroidal	Digitalis	Dried leaves	Purpurea glycoside A, B, C Digitalin, Diginin
	European white Squill and red squill	Bulbs	Scillaren A, glucoscillaren A, Scillaridin A
	Indian squill	Bulbs	Scillaren A and B
	Strophanthus	Dried ripe seeds	K-strophanthoside, K-strophanthride β and cymarin
Flavonoids	Parsley	Seeds, leaves	Apiin, 7-glucoapigenin
	Diosmin	Dried leaves	Diosmetin
	Rutin	Seeds, leaves	Rutin, Quercetin
	Quercetin	Barks	Quercetin
	Hesperidin	Fruits and peels	Hesperidin
	Carthamin	Herbs	Carthamin
	Tephrosin	Leaves	Toxicarol
Coumarin and furanocoumarin	Horse chestnut	Fruit and barks	Aesculin
	Chicory plant	Flowers	Cichorin
	Daphne	Barks	Daphnin
	European ash	Barks	Fraxin
	Khellol	Seeds	Khellinin
	Psoralen	Dried ripe fruits	Psoralen
	Cantharide beetles	Dried insect	Cantharidin
Cyanogenetic	Bitter almond	Dried ripe kernels	Amygdalin
	Wild cherry bark	Dried barks	Prunasin
	Linseed	Ripe seeds	Linamarin
Thioglycosides	Black mustard	Dried ripe seeds	Sinigrin
	White mustard	Dried ripe seeds	Sinalbin

(continued)

Table 5.1 (continued)

Glycosides	Common name	Parts of plant	Major Chemical constituent
Saponin	Dioscorea	Dried tubers	Dioscin
	Solanum	Dried berries	Solasonine
	Shatavari	Dried roots	Shatavarin I, II, III and IV
	Ginseng	Dried roots	Ginsenoside, Oleanolic acid, Panaxadiol, panaxatriol
	Liquorice	Dried roots and rhizomes	Glycyrrhizic acid
	Senega	Dried roots	Senegin, Polygallic acid
	Bacopa	Fresh stems and leaves	Bacoside A and bacoside B, Asiatic acid and Brahmic acid
	Quillaja bark	Inner dried bark	Quillaic acid, Quillaiasapotoxin
Aldehyde	Vanilla	Unripe fruit, beans, pods	Glucovanillin, Vanillin alcohol
Bitter	Picrorhiza	Rhizomes	Picroside I, II and kutkoside
	Gentian	Dried rhizomes and roots	Gentopicrin, Gentisin, Gentianic acid
	Chirata	Dried plant	Amarogentin and chiratin
Miscellaneous	Aminoglycosides	*Streptomyces griseus, Streptomyces fradiae*	Streptomycin, Neomycin

5.3 Medicinal and Pharmaceutical Importance

Glycosides as phytochemicals are being utilized as therapeutics and medicine for prevention and treatment of various diseases from ancient ages to modern world in the form of crude extract and/or purified finished products. These glycosidic molecules can be classified on the basis of aglycone moiety present in structure along with their medicinal and pharmaceutical importance for different uses (Kar 2003; Evans 2009; Table 5.2).

5.4 Isolation and Separation Technique

These glycosides are biologically derived from natural sources, such as plants and various parts of the plants. The biologically derived molecules are differentiated as primary metabolites and secondary metabolites. The primary metabolites consist of proteins, lipids, nucleic acid, carbohydrate and vitamins; and secondary metabolite consists of terpenoids, flavonoids, phenolic, alkaloids, antibiotic and steroids (Irchhaiya et al. 2015). The biological synthesis leads to the formation of different structural analogs which are structurally similar to molecules of interest and act as impurities or side stream to the parent molecule.

Table 5.2 Medicinal and other uses of glycosides

Aglycones	Common name	Botanical name	Uses
Anthracene glycosides	Curacao Aloe	*Aloe barbadensis*	Laxatives and purgative
	Alexandria senna	*Cassia senna*	
	Indian Senna	*Cassia angustifolia*	
	Cascara sagrada	*Rhamnus purshiana*	
	Frangula	*Rhamnus frangula*	
Phenol glycosides	Bearberry	*Bergenia crassifolia*	Diuretic, antiseptic and treatment of urinary tract infection, analgesic
	Canadian wintergreen	*Gaultheria procumbens*	
	Poplar	*Populus nigra*	
	Willow	*Salix fragilis, Salix purpurea*	
	Populus	*Populus tremula*	
Steroid glycosides (Cardiac glycosides)	Digitalis	*Digitalis purpurea*	Cardiotonic, cardiac stimulant, expectorant in bronchitis and asthma, diuretics
		Digitalis lanata	
		Digitalis lutea	
		Digitalis thapsi	
	European white squill and Red squill	*Urginea maritima*	
	Indian squill	*Urginea indica*	
	Strophanthus	*Strophanthus hispidus Strophanthus kombe*	
Flavonoid glycosides	Parsley	*Petroselinum sativum*	Flavouring agent, Treatment of bleeding, industrial dyes, blood vessels fragility and treatment of cardiovascular related diseases
	Diosmin	*Barosma crenulata*	
	Rutin	*Fagopyrum esculentum*	
	Quercetin	*Quercus tinctoria*	
	Hesperidin	*Citrus sinensis*	
	Carthamin	*Carthamus tinctorius*	
	Tephrosin	*Tephrosia vogelii*	

(continued)

Table 5.2 (continued)

Aglycones	Common name	Botanical name	Uses
Coumarin and Furanocoumarin glycosides	Horse chestnut	*Aesculus hippocastanum*	Treatment of diarrhoea, as a febrifuge or antipyretic, bitter tonic and astringent
	Chicory plant	*Cichorium intybus*	
	Daphne	*Daphne mezereum*	
	European ash	*Fraxinus excelsior*	
	Khellol	*Eranthis hyemalis Ammivis naga*	
	Psoralen	*Psoralea corylifolia*	
	Cantharide beetles	*Cantharis vesicatoria*	
Cyanogenetic glycosides	Bitter almond	*Prunus amygdalus*	Skin lotions and liniments, treatment of scabies and skin diseases, preparation of I_2 ointments and cresol soap solution
	Wild cherry bark	*Prunus serotina*	
	Linseed	*Linum usitatissimum*	
Thioglycosides	Black mustard	*Brassica nigra*	Black mustard oil is been used for cooking, pickle; as condiment and spices; as a counter-irritant and rubefacient in plasters and poultices
	White mustard	*Brassica alba*	
Saponin glycosides	Dioscorea	*Dioscorea delitoidea*	Rheumatoid arthritis treatment, steroidal drugs, galactogogue, antioxytocic activity, general tonic, stimulant, carminative and diuretic activities, treatment of insomnia, anaemia, gastritis, diabetes and treatment for sexual impotence, taste masking, expectorant, sweetener and taste improviser in beverages, bronchitis, emetic, nerve tonic, cardiotonic, emulsifier and foaming agent
	Solanum	*Solanum khasianum*	
	Shatavari	*Asparagus racemosus*	
	Ginseng	*Panax ginseng*	
	Liquorice	*Glycyrrhiza glabra*	
	Senega	*Polygala senega*	
	Bacopa	*Bacopa monnieri*	
	Quillaja bark	*Quillaja saponaria*	
Aldehyde glycosides	Vanilla	*Vanilla planifolia*	Flavouring agents
Bitter glycosides	Picrorhiza	*Picrorhiza kurroa*	Bitter tonic, treatment of jaundice
	Gentian	*Gentiana lutea*	
	Chirata	*Swertia chirata*	
Miscellaneous	Streptomycin	*Streptomyces griseus*	Antibacterial

These structural analogs may be bioisosteres, isomers, pathway intermediates or degradation pathway metabolites or moieties which are to be separated from the parent molecule for desired pharmacological effect (Yang et al. 2010). The difference of one or more functional group, atom or chemical moiety in the structure forms structural analogs or structurally similar compounds with agonist or antagonist pharmacological activity depending on the structural activity relationship (SAR). The separations of these structural analogs are difficult and cumbersome. The different unit operations like extraction, precipitation, crystallization, distillation, drying, filtration, evaporation, sedimentation, centrifugation and chromatography are used for isolation, separation and purification of molecules. The developed separation process should have unit operation which provides multiple numbers of transfer units to purify a molecule from its structural analogs (Harrison et al. 2015).

These structural analogs have low separation factor, i.e. they are very difficult to separate due to their similar physico-chemical properties with the desired molecule of interest (Wei et al. 2010). The structural analogs are present in high or low abundance with severe tendency to compete or interfere in the purification process of the desired molecule. Different unit operations are employed for purifying a molecule from its impurities, but chromatography as a unit operation provides more number of stages for purification as compared with other unit operations. The provision of the multiple stages makes chromatography as a selective tool for the separation and purification of small molecules from their structural analogs along with other unit operations (Jenkins et al. 1996; Carta 2002). The basis for the generalized isolation and separation technique for any biomolecules including glycosides by chromatography consists of the two approaches, namely, thermodynamic-based and kinetic-based approach.

5.4.1 Thermodynamic-Based Approach

The purification process in chromatography involves thermodynamic interaction of the molecules between the adsorbent and glycoside. The interaction involved is based on parameters, such as particle size, surface area, resin matrix and counter ions, which are related to capacity factor (k) as a governing factor. The process involves different sequential steps, such as loading, washing, elution, regeneration and flushing (Carta 2002). The thermodynamic involved in extraction process is governed by the factor called partition coefficient (K), which is governed by the differential solubility of solutes in specific solvents (Scholfield and Dutton 1955). The importance lies in designing the washing mobile phase so that complete removal of impurities should be achieved with very low impact on recovery of the process. The designing of mobile phase is decided by the interaction between the absorbent and glycosidic bond and, also, the chemical properties of the impurities. The elution designed should result into the highest purity in single step of chromatographic separation which is not possible in all purification process development. The chromatography involved in this approach is expanded bed chromatography (EBC) and packed bed chromatography (PBC) with high porous and large bead size (Du et al. 2004).

5.4.2 Kinetic-Based Approach

The glycoside parent molecules and impurities are structural isomers, positional isomers and stereoisomers to one another, which have low degree of separation. The separation of such molecule can be done by using kinetic-based chromatography as a high resolving tool for various glycoside molecules and their isomers. The separation utilizes the differential migration rate of molecules in specific chromatographic conditions such as mobile phase flow rate, loading concentration, loading volume, particle size of adsorbent and bed height. The chromatography involved in this approach is simulated moving bed (SMB) and packed bed chromatography (PBC) with high to medium porosity and small bead size (Wang et al. 2017). The criticality lies in designing the precise mobile phase in which the differential migration rate can be generated between the isomers, so that the structural impurities can be removed and desired purity of the glycoside is attained. The designing of mobile phase is mainly dependent on the polarity of the mobile phase, pH and in some cases the conductivity. The mode of interaction between the absorbent and the chemical properties of the glycoside molecule and its impurities has a significant impact on migration rate of the molecules. The elution designed should result into the highest purity.

5.5 Development of Purification Strategy

The chemistry of glycone and aglycone moiety plays a very critical role for isolation and purification of these glycosides. These glycosides can be isolated from various natural sources by extraction process. The extract comprises of various structural analogs which may be bioisosteres, isomers, pathway intermediates or degradation pathway metabolites. These moieties should be separated from the parent molecule for desired pharmacological effect of glycosides (Ghosh 2006). The strategic development of purification process for active constituent can be multistep and tedious and generally combines various separation techniques depending on the solubility, pH and stability of the compounds to be separated (Ghosh 2006). The strategy for difficult separations involving glycosidic molecules and their structural analogs requires high-performance unit operation, e.g. chromatography with intensification and integration of process steps to increase purity, yield and, to reduce process time, production cost as well as associated capital expenditure (Carta 2002). The general strategy for purification involves four different stages:

1). Recovery: To process the bulk volume or large quantity for capturing the molecule of interest along with associated impurity for conversion into small volume and ease of processing.
2). Isolation: To improve the purity of the captured molecule by removing impurities and synergize the thermodynamic properties of the molecule in further developmental stages. In many cases, capturing and isolation can be achieved in same step.

3). Purification: To increase the purity of the molecule by using technique such as chromatography, precipitation and other unit operation resulting into removal of maximum impurities and enhanced purity with traces of structural similar analogs.
4). Polishing: Remove structural analogs and traces of impurities to meet the finished product specification by using controlled techniques such as crystallization or kinetic-based chromatography.

The degree of separation (R_s) and selectivity (α) is the important parameters for the separation of the structural analogs which plays a very important role in the process development and strategy design. The R_s and α have high impact on the kinetic mode of separation, whereas the capacity factor (k) governs basis of the thermodynamic mode in chromatographic separation. Also, partition coefficient (K) is the indicative parameter for the degree of hydrophobicity for extraction as well as in chromatographic separation (Jonas and De Planas 1974). The various strategic approaches can be achieved by utilizing different physico-chemical properties mentioned below.

5.5.1 Aglycone Charge

The charge, i.e. positive and/or negative on the aglycone, plays a critical role which can be used for the purification of various glycosides as one aspect for the development of the purification process. The charge development critically depends on the pH of the extract. The glycoside which may have a significant effect of pH includes cyanogenic glycoside, thiocyanate glycoside and steroidal-terpenoidal glycoside (Neuberger and Wilson 1971). The aglycone structure in above-mentioned glycosides imparts a specific charge on the molecule which can be used to separate the uncharged species or low (%) ionized species by the chromatographic separation. The pH-based strategic approach can be used for the purification of the glycoside by high-throughput separation technique which involves extraction followed by ion exchange chromatography (Du et al. 2012). This strategy eliminates the unit operations such as precipitation, membrane separation and decolorization prior to the polishing. In certain cases it may be observed that aglycone charge strategy may provide the result as partial purification; but this approach may provide the partial purified glycosidic molecule which can be easily polished to the desired high purity by crystallization, indicating with reduced unit operation in the process.

5.5.2 Aglycone Hydrophobicity

The degree of hydrophobicity has a significant impact on the strategic separation of the structural isomer, functional isomers, positional isomers and bioisosteres. The aglycone moiety and its analogs with a small difference of carbon atom have a significant difference in their hydrophobic index, which can be utilized for the separation of the structural analogs (Jonas and De Planas 1974). The differential functional groups, such as in halogens (X = Cl, Br, I, F), have a dramatic effect on the separation, as the change in hydrophobicity can be observed in the aglycone with chlorine and isomer without chlorine (i.e. dechlorinated). However, the separation of such isomeric aglycone has low separation factor but can be purified by kinetic-based chromatographic separation for the strategic purification process development and scale-up as a novel strategy (Du et al. 2012).

The presence of anthracene aglycone in aloe and emodin are functional isomers at C-3 and C-6 position due to which there is significant hydrophobic differences (Kar 2003; Bartnik and Facey 2017). This difference can be exploited for binding, separation and resolution based on the kinetic, i.e. the differential migration rate on the chromatographic column. The desired extract of *Aloe* can be directly implied on the hydrophobic interaction chromatography (HIC) in which adsorbent is of bigger diameter (>300μ) for capturing of the aloe and emodin from other isomers and colour impurities. The elution of the above column which is partially purified can be introduced on smaller diameters (50 to 120μ) which can be utilized for kinetic-based separation with elution of highest purity which can be further polished by crystallization. Similarly, the difference of hydrophobicity exists in emodin and emodin dianthrone which can be separated by HIC kinetic-based separation (Su and Ferguson 1973).

5.5.3 Aglycone Accessibility

The porosity and pore diameter of membrane and chromatographic adsorbent have a significant role for process development for purification of glycosides, e.g. apigenin glycosides like apiin (apigenin 7-O-apioglucoside), apigetrin (apigenin 7-glucoside), vitexin (apigenin 8-C-glucoside), isovitexin (apigenin 6-C-glucoside), rhoifolin (apigenin 7-O-neohesperidoside) and schaftoside (apigenin 6-C-glucoside 8-C-arabinoside) (Kar 2003). The molecular shape and size of the aglycone as well as glycone moiety can be exploited for purification development (Ghosh 2006). The spatial arrangement creates a differential accessibility of the molecules. The separation of the glycosidic bioisosteres with less sugar moiety can be separated, e.g. in apiin and rhoifolin (Kar 2003). The small difference in the molecular size in the oligomers, such as dimers and trimers of the glycone moiety, can be used for separation of the glycosides (Ge et al. 2017). The aglycone accessibility can also be applied in the separation of cis–trans isomeric configuration. The trans isomer has more stable confirmation in alkene-containing glycosides due to high molecular size as compared to the cis isomers, for example crocin as a glycoside in saffron having cis–trans isomer. The current strategy can be used in membrane-based chiral separation for the E–Z and R–S configuration (Dembitsky 2004). The molecular separation assisted with the enzyme-based dynamic kinetic resolution (DKR) for chiral separation of glycoside has been implied to improvise the enantiomeric excess of the chiral resolution, e.g. cascarosides A, B, C and D (Bartnik and Facey 2017). The DKR can be used for the extractive conversion which can be implied for the membrane-based separation and/or chromatography based on MWCO and/or accessibility (Su and Ferguson 1973).

5.5.4 Glycone Coordinate Chemistry

The difference in the glycone moiety (sugars) can be utilized as an important aspect for the purification of glycoside with metallic ions such as divalent species Ca^{++}, Mg^{++}, Pb^{++} and Zn^{++} and monovalent atomic species like Na^{+} and K^{+}. These interactions can be used in immobilized metal affinity chromatography (IMAC) where R_s of the glycoside molecule with their structural analogs which differed in their glycone, e.g. in the flavonoids hesperidin with rutinose and quercetin with rhamnose, can be separated and purified based on the interaction between metallic ion and glycone coordinate chemistry (Bartnik and Facey 2017). The components in the plant extract such as nucleic acids and colours can be removed by the utilization of the coordinate chemistry; but, there may be a non-specific interaction with protein present in the extract with histidine exposed (Axelrod 1965). Hence, the elution fraction from IMAC may contain protein and glycoside. Later in the process, the proteins can be removed by using salt-based precipitation or anti-solvent approach based on the solubility of the protein and glycoside present (Axelrod 1965). The separation on the basis of anomeric sugars in the glycosides can be utilized for purification from its structural analogs towards higher purity in the polishing step (Augestad and Berner 1954). The strategy for separation can also be developed by individual or combining the two approaches as explained below.

5.5.5 Tandem Approach

The sequential arrangement of same unit operation to enhance the purity after each unit operation is called tandem process. The tandem process can be used in membrane-based separation, chromatography and extraction. In the membrane-based separation of glycoside from the polyphenolic, colour and polysaccharides can be achieved by sequential arrangement of different MWCO membranes, e.g. microfiltration, ultrafiltration and nanofiltration (Chhaya et al. 2013). In the commercial-scale purification, the tedious separation of glycoside isomers based on the functional group or the structures can be performed by using chromatographic tandem approach, in which the binding and elution in the primary column (1°) is governed by thermodynamic approach and subsequent arrangement of the secondary column (2°) for polishing. The primary column (1°) and secondary column (2°) can be of same mode or different mode. In the same mode of separation, the column arranged may be of $HIC_1 - HIC_2$, $AEX_1 - AEX_2$, $CEX_1 - CEX_2$, whereas in a different mode of separation, column arranged may be $HIC_1 - AEX_2$, $AEX_1 - HIC_2$, $HIC_1 - CEX_2$, $CEX_1 - HIC_2$, $AEX_1 - CEX_2$, $CEX_1 - AEX_2$ as given (Fig. 5.3) (Mouly et al. 1998).

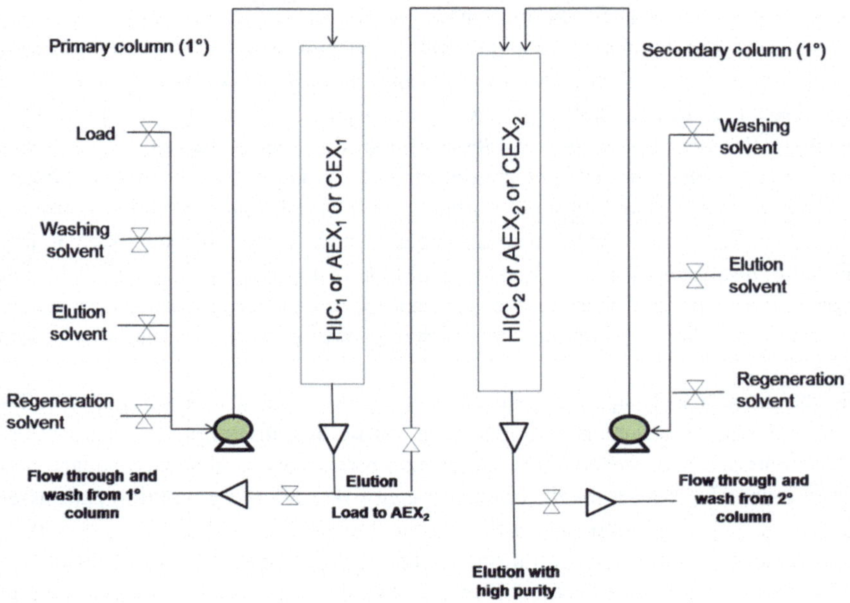

Fig. 5.3 Tandem column chromatography for glycoside purification

5.5.6 Kinetic Continuous Column Chromatography

Simulated moving bed (SMB) is a recent advance and emerging technique which has been utilized for the separation of the structural glycone isomers which are with very low degree of separation, i.e. R_s<0.4 or below. The technique is based on differential migration rates of isomers in varied zones of loading, washing, elution and regeneration in continuous cycle as the column travels in each zone. The current strategy can be utilized for separation of dilute stream with two components with highest purity indicating a super strategic polishing step. The SMB-based strategy yields higher productivity due to continuous process; but high cost of instrument and capital expenditure is the matter of concern. The processing aspect positively helps for scalability of diluted stream but limits itself for two component isomeric separation.

5.6 Scale-Up of the Purification Strategy

The scale-up from laboratory level to pilot plant level has 50–100 fold increase in the volume of purification set up, whereas scale-up from pilot plant to commercial scale is 10–50 fold. The scale-up of strategy involving adsorption and chromatography processes deals with changes in flow rates (FR) and in the dimensions of the column (d × h). The scale-up for industrial process is done on the basis of aspect ratio, defined as ratio of diameter (d) to height (h). The d/h value is maintained during the process development, but the ratio may not be feasible in scale-up to pilot and commercialization. In thermodynamic-based strategy, scale-up is done by increasing the bed diameter, and bed height is constant, where the objective of process is just for partial purification or capturing; and also, this strategy may not be applicable where kinetic base separation is required. In the kinetic separation strategy, migration rate of glycoside and their structural isomer is the critical parameter of the process (Du et al. 2012). The migration rate is kept constant by maintaining the linear velocity (u) and number of plates (N) from lab scale to pilot to commercialization with an aim that the resolution between the isomer does not vary in any scale. The objectives of thermodynamic and kinetic approach are usually different; therefore the scale-up of these two operational factors is to be considered separately for harmonious outcome of the process wrt purity, yield and economics (Carta 2002).

In the process scale-up, it is desirable to maintain a balance between thermodynamic and kinetic approach parameters; in thermodynamic approach, parameters such as particle size, surface area, pore size, resin matrix, ligand chemistry and counter ions, which are related to capacity factor (k), act as a governing parameter (Du et al. 2012). The kinetic separation parameters such as mobile phase flow rate, linear velocity, loading concentration, loading volume, particle size of adsorbent and bed height play a very critical role. The balance between thermodynamic and kinetic approach can be provided by Van Deemter's equation (5.1), which is widely used for process characterization and fundamental for scale-up of the chromatography expressed as

$$H = A + \frac{B}{u} + Cu \tag{5.1}$$

where "u" is the linear flow velocity (cm/min), "H" is the plate height of the column (cm) and "A", "B" and "C" are constants. "A" reflects the quality of the packing of the column and is independent of the linear flow velocity. "A" is small when the column is packed well and is homogeneous throughout its length. "B" is a measure of the band broadening due to longitudinal diffusion of the sample components along the edge of their respective bands as they travel across the column. It decreases with increasing linear flow velocity because the sample components spend less time undergoing diffusion inside the column. "C" includes contributions from the binding kinetics (adsorption/desorption) as well as the mass transfer of the sample components to and from the packing particles. In preparative chromatography, process is carried out at high flow velocity "u" in order to increase the outcome of the scale-up, which simplifies Eq. (5.1) in the form of Eq. (5.2):

$$H = Cu \tag{5.2}$$

The plate height "H" is equal to the length of the column (L) divided by the total number of plates (N); Equation (5.3) depicts that "H" is smaller for a more efficient column:

$$\frac{L}{N} = Cu \tag{5.3}$$

To maintain the degree of separation, the total number of plates (N) is to be kept constant, even though the bed height is changed; accordingly linear flow velocity (u) should be changed so that the term L/u remains constant and safeguards the column performance for desired resolution of structural analogs for kinetic separation and for purification and polishing.

In most thermodynamic-based separations of glycosides with the high load volume at pilot scale, the scale-up is done by increasing the column diameter rather than height to achieve the desired column volume with linear flow velocity (u) which is maintained so that residence time of the product is unchanged with similar purity profile and desired yield. It is also to be considered that the mobile phase in lab scale and pilot scale are similar in pH, composition, viscosity and conductivity. The change in diameter of adsorbent (d_p) is generally observed due to column pressure generated by small size adsorbent. The efficiency of the chromatographic column can be maintained by keeping the "N" constant based on the correlative term mentioned in Eq. (5.4).

$$N = \frac{L}{u.dp^2} \tag{5.4}$$

Based on Eq. (5.4) which indicates that in scale-up strategy the increased d_p can be utilized to reduce the chromatographic back pressure, the process can also be operated by reducing the linear velocity so that number of plates (N), i.e. the efficiency, can be maintained at different levels of scale-up process.

5.7 Conclusions and Future Prospects

The glycosides can be isolated from various natural sources by extraction process. The extract comprises of various structural analogs which may be bioisosteres, isomers, pathway intermediates or degradation pathway metabolites. These moieties should be separated from the parent molecule for desired pharmacological effect of glycosides. These structural analogs have low separation factor, i.e., they are very difficult to separate due to their similar physico-chemical properties with the desired molecule of interest. The structural analogs of glycosides have a severe tendency to compete or interfere in purification process of the desired molecule. The differences in physico-chemical aspects such as aglycone charge, accessibility, hydrophobicity and glycone coordinate chemistry can be utilized for separation and purification of glycosides strategically. The strategy for separation can also be developed by individual or combining the two approaches as tandem for thermodynamic and kinetic separation and SMB for continuous chromatography. The thermodynamic and kinetic process parameters should be optimized on the basis of scalability, reproducibility and techno-economic viability.

The various unit operations can be employed for purifying a molecule from its impurities; but the correct sequencing of the unit operations should be strategically arranged based on the RIPP scheme which should be scaled up to desired level with standard quality output. It is the fact that chromatography provides more number of stages for purification as compared with other unit operations. The provision of the multiple stages by high value of "N", i.e., number of plates, makes chromatography as a selective tool for separation and purification of glycosides from their structural analogs along with other unit operations. The differences in the hydrophobicity due to aglycone and glycone of the structural analogs and the glycosidic parent molecule can be utilized for separation by HIC. The similar approach of separation can be used for ion exchange process (AEX, CEX) in which the molecular differences arise on the basis of charge at specific pH for separation of isomers and structural analogs. In the process development, either batch or continuous mode separation, based on thermodynamic and/or kinetic process, various strategies can be developed and scaled up. In recently developed techniques, continuous separation of glycosides with high purity can be achieved by simulated moving bed (SMB) at the polishing stage.

Acknowledgments I am thankful to Dr. Neetin Desai, Head Of Institute, Amity University, Amity Institute of Biotechnology, Mumbai, for infrastructural support. I pay my gratitude and special thanks to Dr. Sandeep Pai for proof reading and my colleagues Dr. Piyush Kumar and Dr. Nilesh Wagh for their editorial help.

References

Augestad I, Berner E (1954) Chromatographic separation of anomeric glycosides. Acta Chem Scand 8:251–256

Axelrod B (1965) Mono-and oligosaccharides. In: Bonner J, Varner JE (eds) Plant Biochemistry. Academic Press, New York, pp 231–257

Bartnik M, Facey P (2017) Glycosides. In: Badal S, Delgoda R (eds) Pharmacognosy: fundamentals, applications and strategies. Academic Press, New York, pp 101–161

Carta G (2002) Scale-up and optimization in preparative chromatography: principles and biopharmaceutical applications. In: Rathore AS, Velayudhan A (eds) Chromatographic science series, vol 88. Marcel Dekker, Inc, New York, p 342

Chhaya, Majumdar G, De S (2013) Primary clarification of *Stevia* extract: a comparison between centrifugation and microfiltration. Sep Sci Technol 48:113–121

Dembitsky VM (2004) Chemistry and biodiversity of the biologically active natural glycosides. Chem Biodivers 1:673–781

Du Q, Chen P, Jerz G, Winterhalter P (2004) Preparative separation of flavonoid glycosides in leaves extract of *Ampelopsis grossedentata* using high-speed counter-current chromatography. J Chromatogr A 1040:147–149

Du H, Wang H, Yu J, Liang C, Ye W, Li P (2012) Enrichment and purification of total flavonoid c-glycosides from *Abrus mollis* extracts with macroporous resins. Ind Eng Chem Res 51:7349–7354

Edwards EI, Epton R, Marr G (1976) A new class of semi-synthetic antibiotics: ferrocenylpenicillins and cephalosporins. J Organomet Chem 107:351–357

Evans WC (2009) Trease and evans pharmacognosy, 16th edn. Saunders Ltd, New York

Ge L, Zhang W, Zhou G, Ma B, Mo Q, Chen Y, Wang Y (2017) Nine phenylethanoid glycosides from *Magnolia officinalis* var. *biloba* fruits and their protective effects against free radical-induced oxidative damage. Sci Rep 7:45342. https://doi.org/10.1038/srep45342

Ghosh R (2006) Principles of bioseparations engineering. World Scientific Publishing company, Co Pte Ltd, Singapore. (doi:10.1142/6183; assessed on July 7th 2018)

Harrison RG, Todd P, Rudge SR, Petrides DP (2015) Bioseparations science and engineering, Topics in chemical engineering. Oxford University Press, Oxford

Irchhaiya R, Kumar A, Yadav A, Gupta N, Kumar S, Gupta N, Kumar S, Yadav V, Prakash A, Gurjar H (2015) Metabolites in plants and its classification. World J Pharm Pharma Sci 4:287–305

Jenkins A, Kratochvil P, Stepto R, Suter U (1996) Glossary of basic terms in polymer science (IUPAC Recommendations 1996). Pure Appl Chem 68:2287–2311

Jonas H, De Planas GM (1974) Extraction and chromatographic purification of digitalis cardiac glycosides and their binding to plant pigments. Prep Biochem 4:411–434

Kar A (2003) Pharmacognosy and pharmacobiotechnology. New Age Inter'l, Pvt. Ltd Publishers, New Delhi

Kennelly EJ (2003) Sweet and non-sweet constituents of *Stevia rebaudiana*. In: Kinghorn AD (ed) Stevia-the Genus Stevia. CRC Press/Taylor and Francis, London, pp 81–99

Megeji N, Kumar J, Singh V, Kaul V, Ahuja PS (2005) Introducing *Stevia rebaudiana*, a natural zero-calorie sweetener. Curr Sci 88(5):801–804

Mouly P, Gaydou EM, Auffray A (1998) Simultaneous separation of flavanone glycosides and polymethoxylated flavones in citrus juices using liquid chromatography. J Chromatogr A 800:171–179

Neuberger A, Wilson B (1971) The separation of glycosides on a strongly basic ion-exchange resin: An interpretation in terms of acidity. Carbohydr Res 17:89–95

Scholfield CR, Dutton HJ (1955) Purification of phosphatides. Google Patents No. US3047597A

Su SCY, Ferguson N (1973) Extraction and separation of anthraquinone glycosides. J Pharm Sci 62:899–901

Wang D, Du N, Wen L, Zhu H, Liu F, Wang X, Du J, Li S (2017) An efficient method for the preparative isolation and purification of flavonoid glycosides and caffeoylquinic acid derivatives from leaves of *Lonicera japonica* Thunb. using high speed counter-current chromatography (HSCCC) and Prep-HPLC guided by DPPH-HPLC experiments. Molecules 22:229. https://doi.org/10.3390/molecules22020229

Wei M, Wang S, He Y, Fang Y, Chen Y (2010) Separation of structurally similar nocathiacin analogues by reversed phase chromatography. J Chromatogr A 1217:3038–3043

Yang H, Yan F, Wu D, Huo M, Li J, Cao Y, Jiang Y (2010) Recovery of phytosterols from waste residue of soybean oil deodorizer distillate. Bioresour Technol 101:1471–1476

Zhang J (2014) Global drug discovery outsourcing market gaining momentum (https://www.lifescienceleader.com/doc/global-drug-discovery-outsourcing-market-gaining-momentum-0001; assessed on June 28th 2018)

Natural Products-Based Pancreatic Lipase Inhibitors for Obesity Treatment

6

S. N. C. Sridhar, Ginson George, Aanchal Verma, and Atish Tulshiram Paul

Contents

Abstract

Obesity is a multifactorial metabolic disorder, majorly caused due to high-fat diet associated with decline in physical activity. Recent decades have seen a rapid upsurge in the obese population with over 650 million obese adults worldwide, as indicated by the recent WHO statistics. Further, a drastic change in the global food system that resulted in overconsumption of energy has transformed obesity into a global epidemic. Pancreatic lipase inhibition is one among the several targets explored for the treatment of obesity and is considered safe and effective due to the fact that the inhibitor does not require any systemic absorption. Orlistat, a potent pancreatic lipase inhibitor, had been clinically approved since 1998 for

S. N. C. Sridhar · G. George · A. Verma · A. T. Paul (✉)
Laboratory of Natural Product Chemistry, Department of Pharmacy, Birla Institute of Technology and Science, Pilani (BITS Pilani), Pilani, Rajasthan, India
e-mail: atish.paul@pilani.bits-pilani.ac.in

M. S. Akhtar et al. (eds.), *Natural Bio-active Compounds*,
https://doi.org/10.1007/978-981-13-7154-7_6

long-term treatment of obesity. However, recent FDA reports have cited severe adverse effects with long-term use of orlistat, including hepatotoxicity and acute pancreatitis, highlighting the urge for safer and effective antiobesity therapeutics. Natural products have been an effective source for the treatment of various disorders and diseases while producing lesser adverse effects. Therefore, the present chapter discusses the various natural products explored for their potential towards pancreatic lipase inhibition under different chemical classes. Furthermore, a preliminary structure-activity relationship has been discussed.

Keywords
Metabolic disorder · Obesity · Orlistat · Pancreatic lipase

6.1 Introduction

Obesity is a multifactorial metabolic disorder defined by excessive deposition of lipids that presents risk to the human health. Worldwide, obesity has tripled since 1975 with recent reports from the World Health Organization (WHO, 2008) indicating a rapid upsurge in obese population. As of 2016, over 650 million adults (aged above 18) were found obese, accounting to 13% of global population. While obesity has been prevalent in adults, recent decades have seen a phenomenal rise in adolescent and childhood obesity with 340 million adolescents (aged below 18) and 41 million children (aged under 5) being found overweight or obese. In 2017, the World Obesity Federation declared obesity as a chronic, relapsing progressive disease process (Bray et al. 2017). Obesity is mainly caused due to high-fat diet associated with decline in physical activity; however, there are other factors including genetic, social and cultural that share a minor role in causing obesity (Grundy 1998). Obesity is preventable and does not itself impose any risk to the health. A chronic condition, however, might lead to severe comorbid risks including insulin resistance, diabetes mellitus (Type II), hypertension, dyslipidaemia, coronary heart disease, sleep apnea, gall bladder disease, gout, osteoarthritis and certain cancers (Khaodhiar et al. 1999). With these associated comorbid conditions and the rapidly growing obese population, obesity ranks fifth among the global deaths (http://easo.org/education-portal/obesity-facts-figures/).

Overweight and obesity are preliminarily identified using the body mass index (BMI), while waist-hip ratio (WHR) is considered more effective in diagnosing obesity. As represented in Table 6.1, an individual with BMI ranging between 25 and 29.9 kg/m^2 is considered overweight, while a BMI greater than 30 kg/m^2 (or) WHR above 0.85 (for women) and 1.00 (for men) is considered obese. The current guideline suggested by the American Heart Association/American College of Cardiology/The Obesity Society (AHA/ACC/TOS) recommends a minimum loss of 500 kcal per day, through physical activity and diet modification to achieve significant weight loss. However, a negative response to lifestyle modification alone might necessitate either bariatric surgery and/or antiobesity pharmacotherapy as adjunctive strategies

Table 6.1 Parameters used to diagnose obesity

Category	BMI (kg/m^2)	WHR
Overweight	25–29.9	ND
Obesity	30–34.9 (class I)	> 0.85 (for women)
	35–39.9 (class II)	> 1.00 (for men)
	≥ 40 (class III)	

ND Not defined

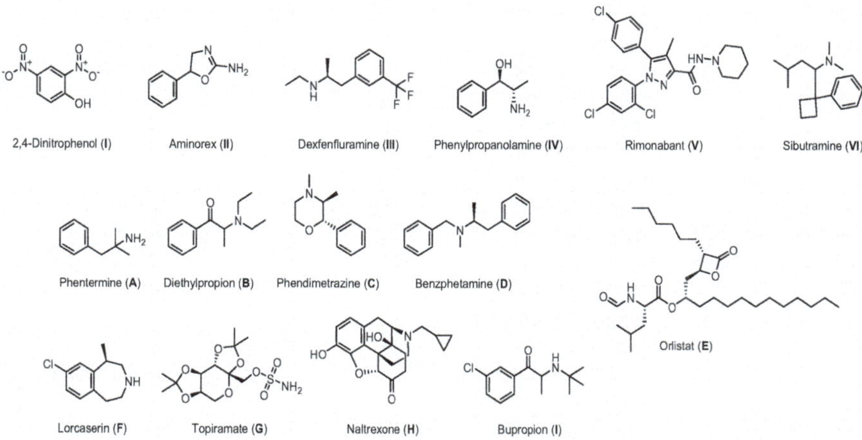

Fig. 6.1 List of withdrawn (**I-VI**) and approved drugs (**A-I**) for the treatment of obesity

to achieve significant weight loss (Patel 2015). Of these two adjuncts, bariatric surgery is recommended only to patients with BMI ≥ 40 kg/m^2 or BMI ≥ 35 kg/m^2 with comorbidities, while the antiobesity pharmacotherapy is recommended to the other classes of obese patients who do not qualify for surgery (https://asmbs.org/patients/who-is-a-candidate-for-bariatric-surgery). Since the 1930s, many drugs were approved for the treatment of obesity; however, most of these drugs were withdrawn due to their severe adverse effects during post marketing surveillance. Currently, ten drugs are clinically approved for the treatment of obesity (Fig. 6.1). These approved drugs are further classified under two main classes based on the duration of administration: short-term and long-term antiobesity pharmacotherapy (Manning et al. 2014; Haslam 2016). A concise list of the approved and withdrawn drugs used in the treatment of obesity is summarised in tabular form (Tables 6.2 and 6.3).

Apart from, new targets are being explored towards the treatment of obesity and can be sub-divided into three categories: (a) central targets and hormones, the regulation of which suppresses the appetite; (b) the peripheral hormones of the gastrointestinal (GI) tract, pancreas and adipose tissue, which aid in satiety or appetite suppression; and (c) the peripheral targets through which lipid metabolism is modulated (Fig. 6.2). The detailed information of these targets is studied by several past researchers (Chakrabarti 2009; Kadomatsu et al. 2011; Mauvais-Jarvis 2011; Colon-Gonzalez et al. 2013; Fani et al. 2014; Kimple et al. 2014).

Table 6.2 List of withdrawn antiobesity drugs

Drug	Mechanism of action	Year approved	Year withdrawn	Reasons for withdrawal
Dinitrophenol (**I**)	Uncoupler of oxidative phosphorylation in phospholipid bilayer	1933	1938	Dermatitis, agranulocytosis, visual impairment and death
Aminorex (**II**)	Anorectic stimulant	1965	1968	Pulmonary hypertension
Amphetamines (schedule II)	Stimulates CNS through norepinephrine release, increases resting energy expenditure and suppresses appetite	1945–1962	1971	Addiction, hypertension, myocardial toxicity
Fenfluramine and dexfenfluramine (**III**)	Serotoninergic agent, suppresses appetite	1973/1996	1997	Valvular heart disease
Phenylpropanolamine (**IV**)	Norepinephrine release inducer, suppresses appetite	1982	2000	Haemorrhagic stroke
Rimonabant (**V**)	Cannabinoid (CB_1) receptor antagonist, suppresses appetite	2006	2009	Mood disorders, suicidal ideation
Sibutramine (**VI**)	Monoamine reuptake inhibitor, suppresses appetite	1997	2010	Cardiovascular risks and stroke

Table 6.3 List of currently available antiobesity drugs

Drug	Year approved	Mechanism of action
Short-term		
Phentermine (**A**)	1959	Norepinephrine release inducers and appetite suppressors
Diethylpropion (**B**)	2011	
Phendimetrazine (**C**)	2010	
Benzphetamine (**D**)	2010	
Long-term		
Orlistat (**E**)	1999	Pancreatic lipase inhibitor
Lorcaserin (**F**)	2012	5-HT_{2C} receptor agonist
Phentermine/topiramate (**G**) extended release (ER)	2012	Sympathomimetic/increases GABA activity and modulates voltage-gated ion channels
Naltrexone (**H**) /bupropion (**I**)	2014	Opioid receptor antagonist/aminoketone antidepressant
Liraglutide	2014	GLP-1 receptor agonist

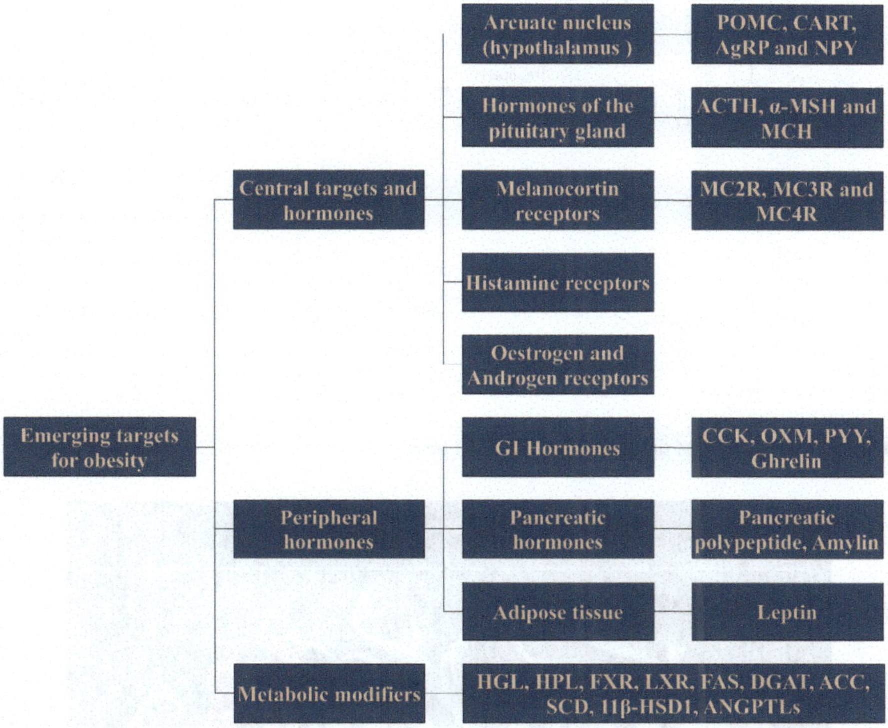

Fig. 6.2 Schematic representation of various targets being explored for the treatment of obesity. (*POMC* proopiomelanocortin, *CART* cocaine- and amphetamine-regulated transcript, *AgRP* agouti-related peptide, *NPY* neuropeptide Y, *ACTH* adrenocorticotropic hormone, *α-MSH* α-melanocyte-stimulating hormone, *MCH* melanin-concentrating hormone, *CCK* cholecystokinin, *OXM* oxyntomodulin, *PYY* peptide YY, *HGL* human gastric lipase, *HPL* human pancreatic lipase, *FXR* farnesoid X receptor, *LXR* liver X receptor, *FAS* fatty acid synthase, *DGAT* diglyceride acyltransferase, *ACC* acetyl-CoA carboxylase, *SCD* stearoyl-CoA desaturase, *11β-HSD1* 11β-hydroxysteroid dehydrogenase, *ANGPTL* angiopoietin-like proteins)

Lipases, or more appropriately the triacylglycerol lipases, are a family of digestive enzymes, classified under the serine hydrolases (EC 3.1.1.3), and include the lingual lipase, gastric lipase and the pancreatic lipase (PL). In humans, the lingual lipase possesses a negligible role in the digestion of lipids while a major part of the lipid digestion occurs in the stomach and duodenum. As represented in Fig. 6.3, the dietary triglycerides (TG) are first hydrolysed to diglycerides (DG) in the presence of gastric lipase in the stomach, releasing one free fatty acid (FA). The DG is further hydrolysed to monoglyceride (MG) and a free FA in the presence of PL. Of these two lipases, gastric lipase is primarily involved in the hydrolysis of short chain esters and does not exhibit prominent role in adult humans, while 60–70% of the lipid digestion occurs in the presence of PL. However, the fatty acids generated during the gastric lipolysis act as emulsifiers alongside the bile salts, for the digestion of long chain fatty esters in the duodenum (Bauer et al. 2005).

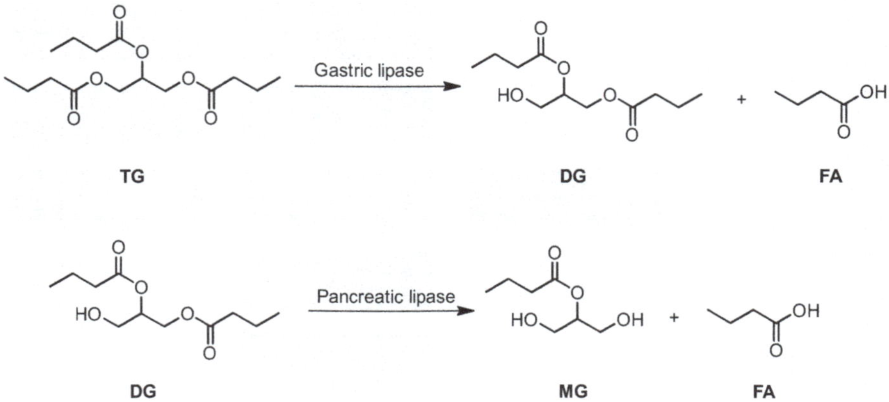

Fig. 6.3 Physiology of lipid digestion in the GI tract

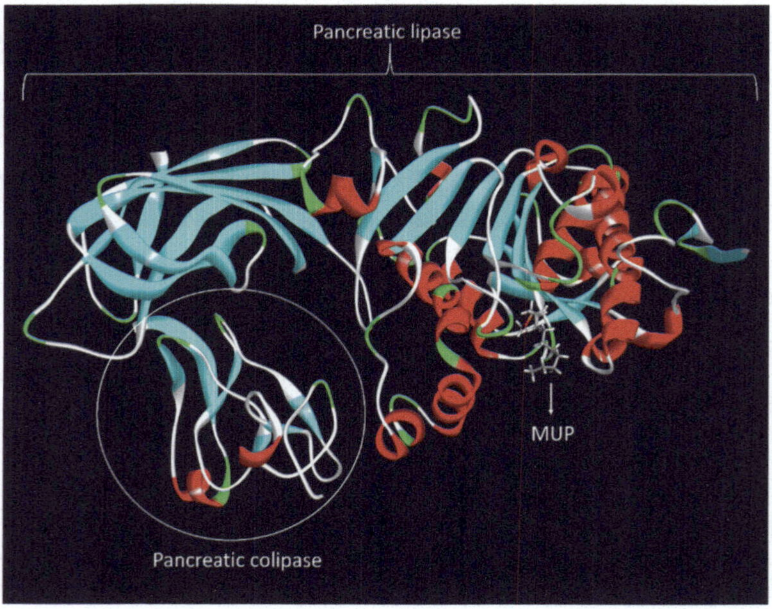

Fig. 6.4 Secondary structure of the human PL-colipase complex co-crystallised with methoxyundecyl phosphinic acid at the active site

6.1.1 Pancreatic Lipase and Its Crystal Structure

The human PL is encoded by the PNLIP gene located at 10q25.3 region of the chromosome and is secreted from the pancreatic exocrine, along with the other pancreatic enzymes (Davis et al. 1991; Palade et al. 2008). The crystal structure of the human PL is composed of 449 amino acids. As represented in Fig. 6.4, the larger

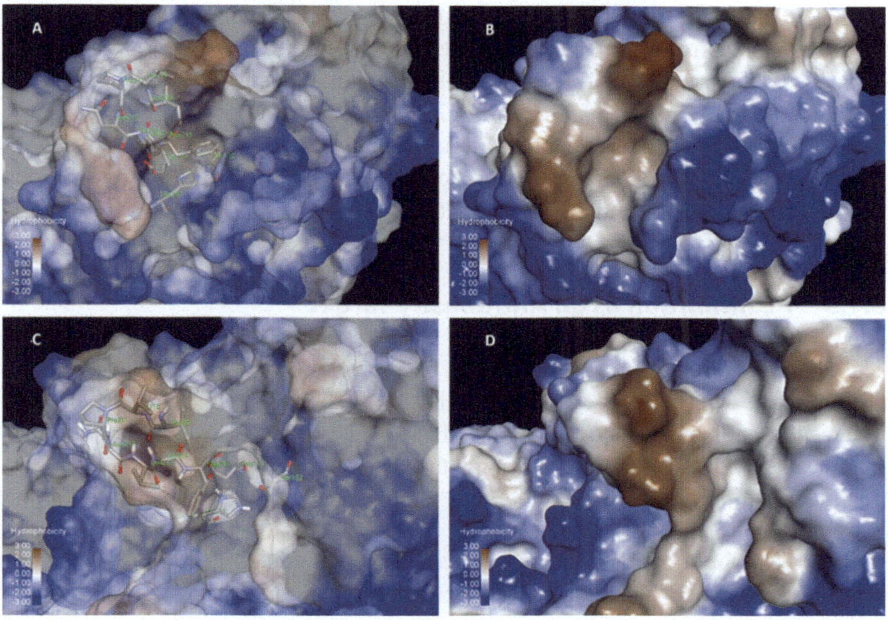

Fig. 6.5 Representation of closed (**a**, **b**) and open (**c**, **d**) lid forms of human PL. A and C represent transparent hydrophobic surface with active site amino acids; B and D include only hydrophobic surface clearly representing the closed and open lid conformations (in brown), respectively

protein chain constitutes the PL, while the smaller chain constitutes the pancreatic colipase, a small protein (with 85 amino acids) bound to the C-terminal of PL and is involved in the activation of the PL. The active site of the human PL comprises a catalytic triad containing Ser 152-Asp 176-His 263. This triad is highly restricted and is surrounded by the hydrophobic lid domain which consists of the amino acids Gly 76-Lys 80 and Leu 213-Met 217. During the inactivated phase, the triad is inaccessible and enclosed within the lid domain. However, the activation of the PL leads to conformational change in the lid domain, resulting in the opening of the active site. Accordingly, the human PL exists in two conformations: the closed lid (or the inactivated form) and the open lid (or the activated form) as represented in Fig. 6.5. The crystal structures of these two conformations were revealed through X-ray diffractions and are designated by the PDB codes, 1N8S and 1LPB, respectively (van Tilbeurgh et al. 1992; Egloff et al. 1995).

6.1.2 Activation of Pancreatic Lipase and Digestion of Lipids

The physiology of lipid digestion involves a series of events, starting with the formation of lipid micelles in the duodenum, in the presence of bile salts and the free fatty acids (released from the gastric lipolysis). This micelle formation allows the interfacial activation of PL (Chapus et al. 1976), facilitating hydrophobic interactions of the long alkyl chains of the lipids with the hydrophobic lid domain of the PL. This phenomenon results in conformational change of PL from the closed lid to the open lid form (van Tilbeurgh et al. 1993). The conformational change is further facilitated through a salt bridge formation by Arg256-Asp257 with Tyr267-Lys268 (Lowe 2002). The subsequent steps involve various biochemical reactions between the ester linkage of the triglyceride and the catalytic triad (Fig. 6.6), which results in the ester hydrolysis of the triglyceride (Kokkinou et al. 2012).

6.1.3 PL Inhibition as an Antiobesity Target

The above facts clearly indicate that PL is the primary enzyme involved in the digestion of dietary lipids. Consequently, PL inhibition would result in lipid indigestion and subsequent prevention of fat intake into the systemic circulation. Inhibition of PL is considered among the safe and effective strategies for the treatment of obesity, due to the fact that the target is peripheral and the inhibitor does not require any systemic absorption.

The PL inhibitory potential of a chemical molecule is determined through the use of an in vitro assay, which comprises the enzyme and a suitable substrate added into a pH-maintained buffer medium. Of the various sources of PL, viz. human, porcine, hog, rat and mice, porcine PL is majorly used in the PL inhibition assay, due to the fact that this enzyme has a high similarity index with the human PL (87% as indicated by the NCBI BLAST tool) and is highly economic to use. Likewise, various substrates, including triolein, tributyrin, olive oil, 4-nitrophenyl butyrate, 4-methylumbelliferyl oleate and 4-nitrophenyl palmitate, are used in PL inhibition assay. However, the enzyme turnover varies with variation in the substrate used (Roskoski 2007). Consequently, the potential of the chemical molecule to inhibit the PL varies with the substrate, resulting in heterogeneity in the reported IC_{50} or % inhibition. Henceforth, the PL inhibitory activity of the natural products determined using different assay procedures cannot be comparable; however, the PL inhibitory potential of the natural products discussed in this book chapter has been defined either as potent, potential, moderate or poor, by comparing their IC_{50} (or % inhibition) with that of the orlistat (standard) reported using similar assay procedure. For instance, if the IC_{50} of orlistat is x μM, then potent indicates $IC_{50} \leq 10x$; potential will be 10x–25x, while moderate and poor will be 25x–50x and > 50x, respectively. The present chapter discusses the various natural products explored for their potential towards pancreatic lipase inhibition under different chemical classes. Furthermore, a preliminary structure-activity relationship has also been discussed.

6.2 Natural Products-Based PL Inhibitors

Orlistat, a potent PL inhibitor, is one among the other drugs clinically approved for long-term treatment of obesity and has been reported for tolerable side effects including steatorrhea, oily stools and frequent or urgent bowel movements (Heck

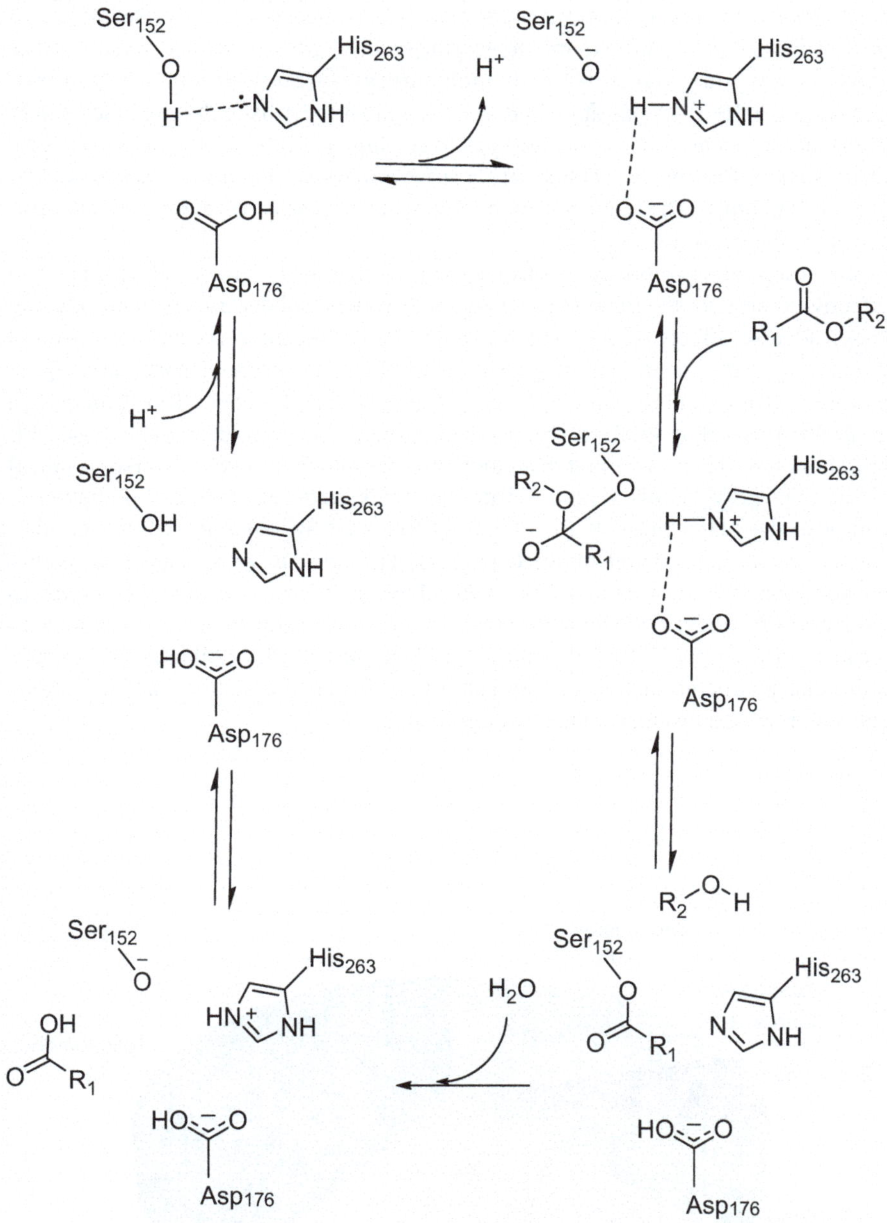

Fig. 6.6 Schematic representation of series of biochemical reactions at the catalytic triad of PL during ester hydrolysis

et al. 2000). However, recent reports from FDA cited severe adverse effects of orlistat including hepatotoxicity, acute pancreatitis, gall stones and renal injuries (http://www.fda.gov/safety/medwatch/safetyinformation/ucm215504.htm). Further, in 2010, FDA has approved a revised label for orlistat that included safety information about cases of severe liver injury (http://www.fda.gov/Drugs/DrugSafety/PostmarketDrugSafetyInformationforPatientsandProviders/ucm213038.htm). These events highlighted the necessity of safer and effective drugs for the treatment of obesity.

Natural products represent a vast reservoir of chemical entities and have been an effective source for the treatment of various disorders and diseases while producing lesser adverse effects (Cragg and Newman 2013). Recent years has seen a rapid projection in the identification of natural products as potential pancreatic lipase inhibitors (Birari and Bhutani 2007; de la Garza Hernández et al. 2011; Lunagariya et al. 2014; Seyedan et al. 2015). To date, around 750 natural products-based PL inhibitors have been identified that can be further classified under diverse chemical classes, viz. polyphenols, saponins, triterpenes, alkaloids, etc. Of these, polyphenols comprise the major class of PL inhibitors followed by saponins, while the other classes can be considered minor (Fig. 6.7). The present book chapter is mainly focussed on the discussion of the natural products that exhibited potential to moderate PL inhibition. The authors are suggested to refer the original articles to retrieve the complete list of natural products-based PL inhibitors. Further, a preliminary structure-activity relationship has also been discussed within various classes of natural products, wherever applicable.

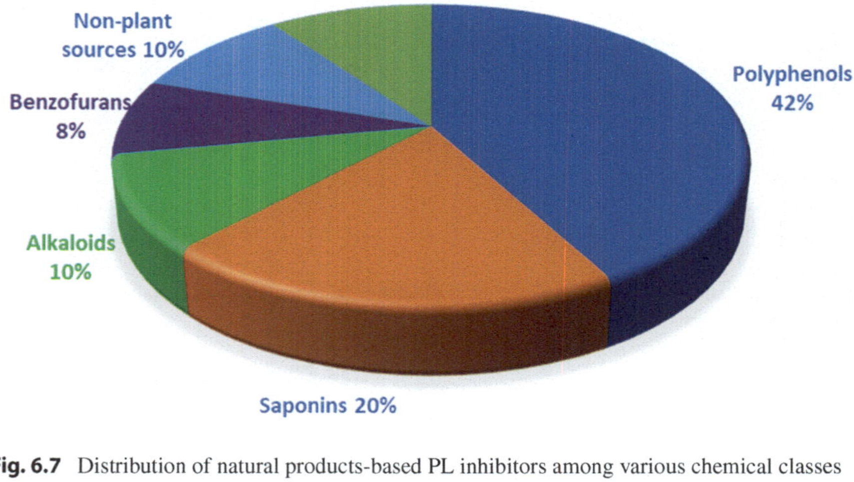

Fig. 6.7 Distribution of natural products-based PL inhibitors among various chemical classes

6.2.1 Polyphenols

Polyphenols represent a wide range of phytochemicals found largely in fruits, leaves and vegetables, and are generally involved in defence mechanism against ultraviolet radiation or aggression by pathogens. These compounds are well known for their multifaceted biological activities and are widely explored for their potential as anticarcinogenic, anti-inflammatory, antiobesity agents, etc. (Li et al. 2014; Srivastava and Kumar Mishra 2015). Within the area of pancreatic lipase inhibitors, polyphenols contribute to the most explored class with over 200 compounds reported to date (Buchholz and Melzig 2015). These polyphenols can be further classified into various subclasses, including flavonoids, phenolic acids, etc.

6.2.1.1 Flavonoids

Flavonoids consist a C6-C3-C6 structural backbone, in which the two C6 units (Ring A and Ring B) are phenolic in nature and are linked to a chromane ring (Ring C). These flavonoids are further divided into various classes depending on the hydroxylation pattern and variations in the chromane ring (Tsao 2010). A general representation of the structural backbones of various flavonoid classes is provided in Fig. 6.8.

6.2.1.2 Flavanols and Its Oligomers

Flavanols and its oligomers constitute the major class of PL inhibitory polyphenols, with a majority of flavanols identified from *Camellia sinensis* (Theaceae). In a study conducted by Nakai et al., around 50 flavanol derivatives were isolated from the leaves of *C. sinensis* and screened for PL inhibitory activity using 4-methylumbelliferyl oleate (4-MUO) (Nakai et al. 2005). These derivatives included the unsubstituted flavanols, their galloylated esters, and various flavanol-based dimers. The unsubstituted flavanols, viz. (+)-catechin (1), (−)-epicatechin (2), (+)-gallocatechin (3) and (−)-epigallocatechin (4), did not exhibit potential PL inhibitory activity ($IC_{50} > 20$ μM), with an exception for the 8-C-ascorbyl (−)-epigallocatechin (5) that exhibited a potential IC_{50} of 0.646 μM. However, the galloylated esters of these flavanols (6–12) exhibited a potential PL inhibitory activity with IC_{50} values less than 1 μM, highlighting the importance of gallate substitution (Fig. 6.9 and Table 6.4). Similar reports were identified by Ivanov et al., wherein (+)-catechin 3-*O*-gallate (13) and (+)-catechin 3,5-di-O-gallate (14), isolated from the aqueous ethanol extract of *Bergenia crassifolia* rhizomes (Saxifragaceae), exhibited potential inhibitory activity towards PL with IC_{50} of 4.52 and 0.706 μM, respectively (Ivanov et al. 2011). Furthermore, in two independent studies, epigallocatechin-3-*O*-gallate (10) and its analogs were determined to exhibit non-covalent interactions with the active site of PL (Wu et al. 2013; Wang et al. 2014).

The study conducted by Nakai et al. (2005) also reported several flavan-based dimers, listed under various subclasses (Fig. 6.10). Oolonghomobisflavan A (15) was the most active compound in the study with an IC_{50} of 0.048 μM, and the greater potential of this analog over other dimers was explained to the presence of the methylene bridge linking the 8,8′-positions of the two flavan units (Nakai et al. 2005).

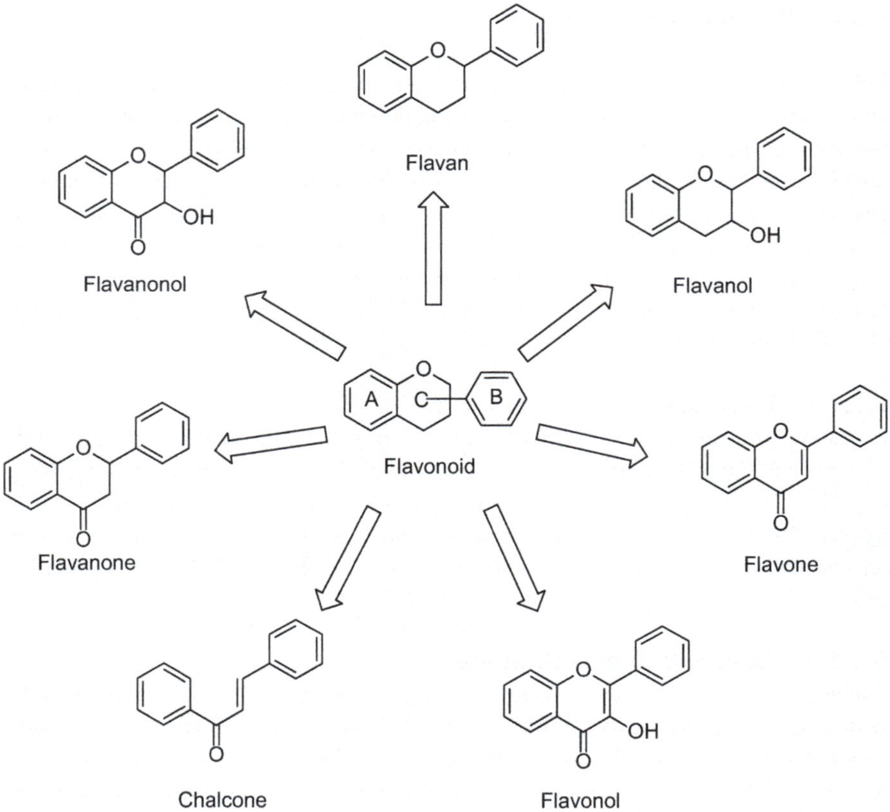

Fig. 6.8 A general representation of flavonoid backbone and its various classes

Further, a trimer of epigallocatechin (20) was also reported for PL inhibition; however, the compound possessed a lesser potential (IC_{50} value of 0.129 μM), compared to its dimeric analog, oolonghomobisflavan A. In another study conducted by Shannon et al., the PL inhibitory potential of theaflavin (21) and its galloyl esters was determined (Fig. 6.11). The study highlighted the prominent role of the galloyl ester and its location for greater potential of the galloylated esters (22–24) over their unsubstituted counterpart, theaflavin (21). Molecular modelling studies revealed that the hydroxyl groups of the galloyl ester interacted with the Asn 262, Asp 205 and His 263 through H-bonds, while its carbonyl group bonded with Cys 262, resulting in stable conformation of these molecules in the active site as well as in facilitating the protonation of His 263 (Glisan et al. 2017).

Various procyanidins isolated from the fruits of *Cassia nomame* (Fabaceae) were also screened for PL inhibitory potential using 4-MUO as substrate (Hatano et al. 1997). These procyanidins, however, exhibited a moderate PL inhibitory activity, with an exception for KA-2 (25) that possessed an IC_{50} of 5.5 μM (Fig. 6.12). Further various procyanidins from cocoa extracts with varying degree of polymerisation (n = 2–10) were examined for their in vitro PL inhibitory potential using

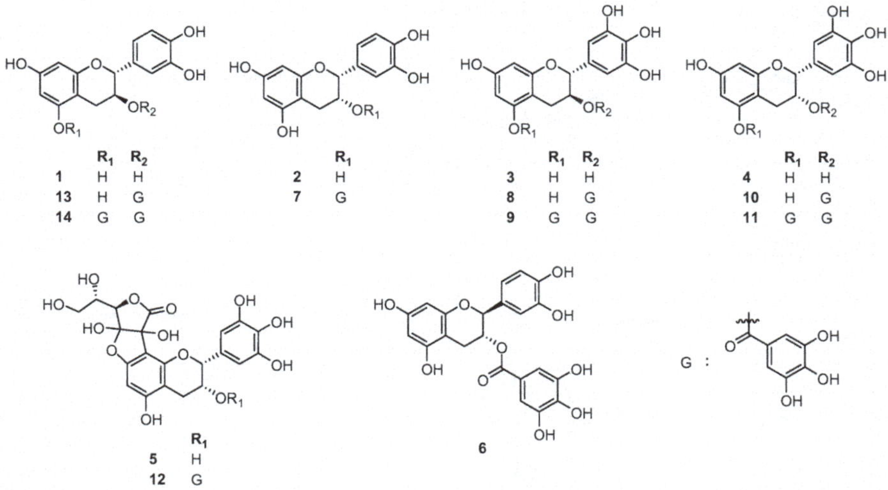

Fig. 6.9 Representation of various flavanols and their gallates from *C. sinensis* and *B. crassifolia* (G: galloyl)

Table 6.4 PL inhibitory activity of flavanol gallates reported from *C. sinensis*

Compound	IC_{50} (μM)
(−)-catechin 3-*O*-gallate (**6**)	0.543
(−)-epicatechin 3-*O*-gallate (**7**)	0.452
(−)-gallocatechin 3-*O*-gallate (**8**)	0.437
(−)-gallocatechin 3,5-di-*O*-gallate (**9**)	0.213
(−)-epigallocatechin 3-*O*-gallate (**10**)	0.349
(−)-epigallocatechin 3,5-di-*O*-gallate (**11**)	0.098
8-C-ascorbyl (−)-epigallocatechin 3-*O*-gallate (**12**)	0.791

4-NPB as substrate. Gu et al. (2011) reported that the PL inhibitory potential of these polymers increased with the degree of polymerisation. Enzyme kinetic studies revealed that these polymers exhibited a mixed (competitive and uncompetitive) type of inhibition on PL.

6.2.1.3 Flavones, Flavonols and Chalcones

Flavones and their hydroxy derivatives (flavonols) contribute to the second most explored PL inhibitory class of polyphenols. In a study conducted by Lee et al. (2010), luteolin (26) and its C-glycoside derivatives (27–31), isolated from the leaves of *Eremochloa ophiuroides* (Poaceae), were screened for PL inhibition assay using 4-NPB as substrate, wherein the glycosides exhibited poor activity with IC_{50} ranging from 18 to 50 μM. Luteolin (26), however, possessed negligible activity (Fig. 6.13). On contrary, Rahim et al. reported that the glycosides of apigenin were less active compared to their aglycone counterpart. However, neither apigenin nor its glycosides possessed PL inhibitory activity (Rahim et al. 2015).

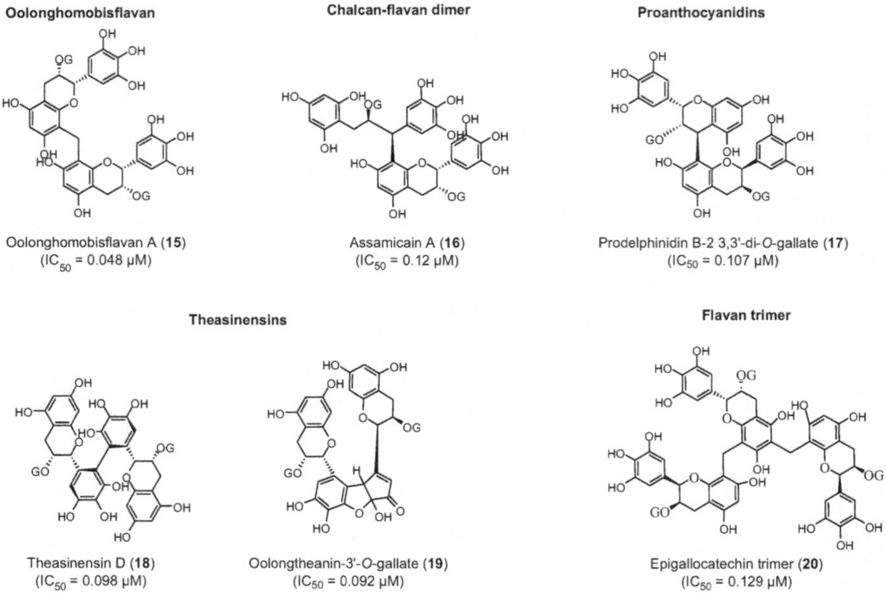

Fig. 6.10 Various classes of flavan oligomers and their compounds with potential PL inhibitory activity (G = galloyl)

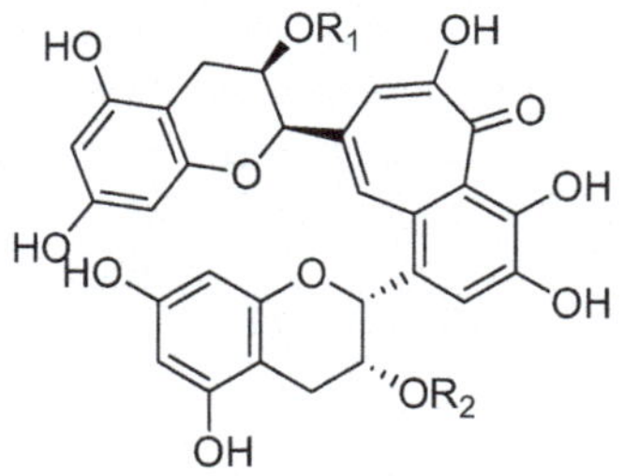

	R_1	R_2	IC_{50} (μM)
Theaflavin (**21**)	H	H	> 10
Theaflavin-3-*O*-gallate (**22**)	H	G	3.0
Theaflavin-3'-*O*-gallate (**23**)	G	H	4.2
Theaflavin-3,3'-di-*O*-gallate (**24**)	G	G	1.9

Fig. 6.11 PL inhibitory potential of theaflavin and its galloyl esters (G: galloyl)

Apart from, prenylated derivatives constitute another subclass of the flavones, which were reported for potential PL inhibitory activity. Examples include norartocarpin (32), brosimone I (33) and hypargyflavone A (34) isolated from the stems of *Artocarpus nitidus* and *A. hypargyreus* (Moraceae), respectively (Zhao et al. 2009; Yu et al. 2012), which possessed potent PL inhibitory activity (Fig. 6.14). Of the various flavonols, quercetin, kaempferol and their glycosides contribute to the

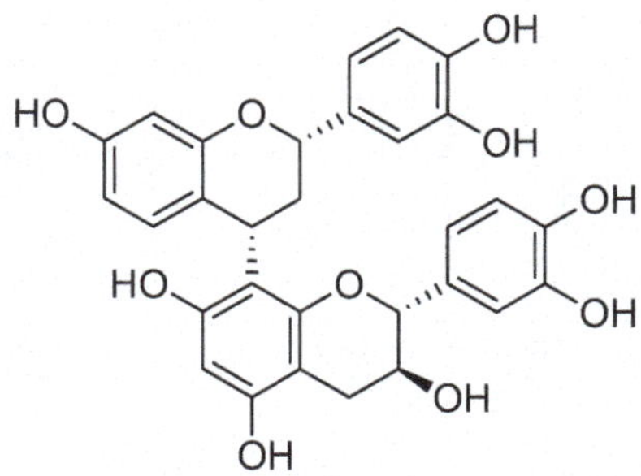

Fig. 6.12 Procyanidin KA-2 (**25**) from the fruits of *Cassia nomame*

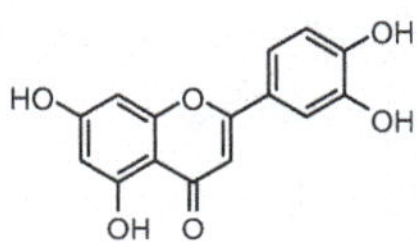

Luteolin (**26**)
(IC_{50} > 200 μM)

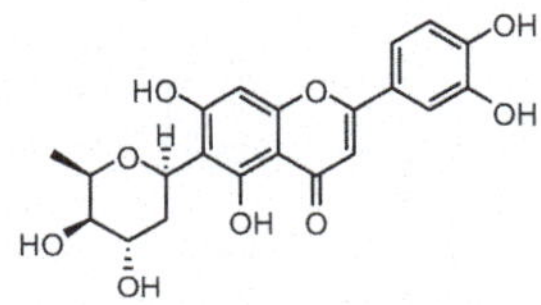

Luteolin 6-*C*-β-D-boivinopyranoside (**27**)
(IC_{50} = 50.5 μM)

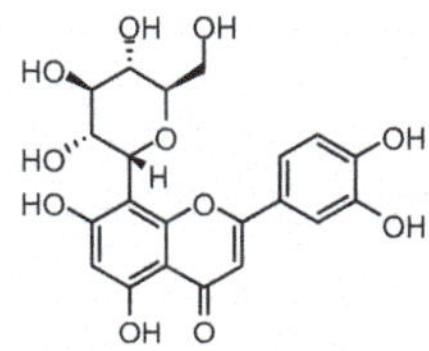

Orientin (**28**)
(IC_{50} = 31.6 μM)

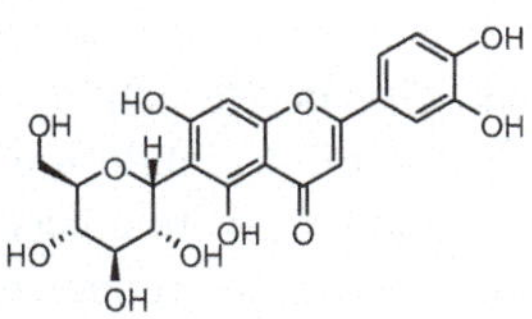

Isoorientin (**29**)
(IC_{50} = 44.6 μM)

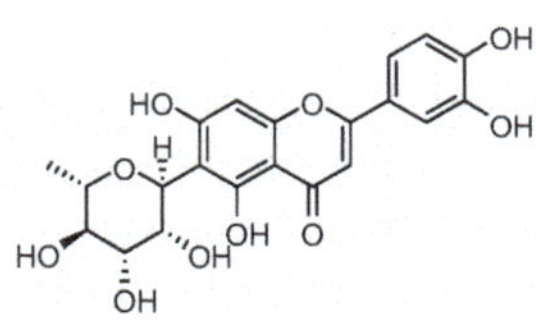

Derhamnosylmaysin (**30**)
(IC_{50} = 25.9 μM)

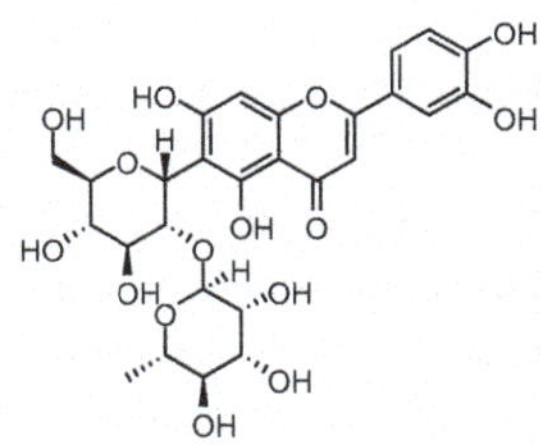

isoorientin 2-*O*-α-L-rhamnoside (**31**)
(IC_{50} = 18.5 μM)

Fig. 6.13 Luteolin and its *C*-glycosides from *Eremochloa ophiuroides*

majorly explored flavonol derivatives for PL inhibition. In a study conducted by Sergent et al., quercetin (35) and kaempferol (36) possessed IC_{50} of 21.5 and 13.4 μM, respectively (Fig. 6.15), in PL inhibition assay using 4-MUO as substrate (Sergent et al. 2012). However, the glycoside derivatives of these flavonols did not possess PL inhibitory activity (Sugimoto et al. 2009; Yuda et al. 2012).

Chalcones represent a special class of flavonoids that does not contain the chromane ring. Chalcones exhibited a similar pattern of activity as that of flavones, wherein morachalcone A (37), a prenylated chalcone from the leaves of *Morus alba* (Moraceae), possessed potential PL inhibitory activity with IC_{50} of 6.2 μM (Fig. 6.15). The unsubstituted chalcones, however, possessed poor PL inhibition (Jeong et al. 2015).

6.2.1.4 Flavanones and Flavanonols

A majority of the flavanones reported for PL inhibition were isolated from the root barks of *Cudrania tricuspidata* (Moraceae), of which two prenylated flavanones,

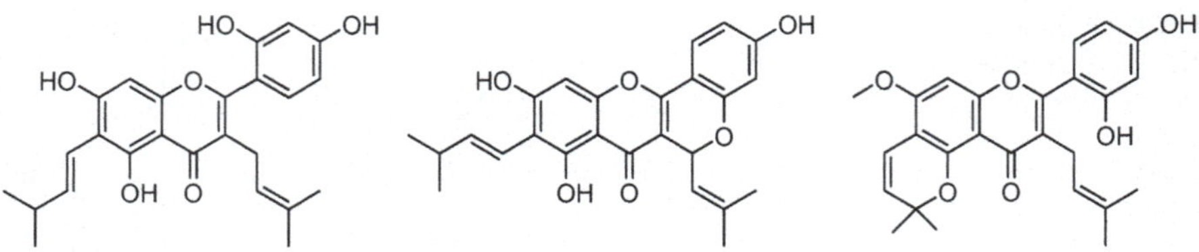

Norartocarpin (**32**) (IC_{50} = 1.8 μM)   Brosimone I (**33**) (IC_{50} = 3.4 μM)   Hypargyflavone A (**34**) (IC_{50} = 2.3 μM)

Fig. 6.14 Prenylated flavones from *Artocarpus sps* with potent PL inhibitory activity

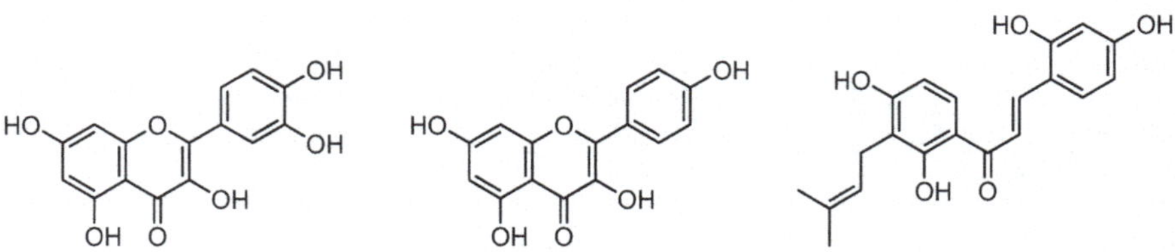

Quercetin (**35**)   Kaempferol (**36**)   Morachalcone A (**37**)

Fig. 6.15 Chemical structures of quercetin (**35**), kaempferol (**36**) and morachalcone A (**37**)

Cudraflavanones A and D (38 and 39), possessed potential to moderate activity. However, three other prenylated flavanones (40–42) exhibited poor PL inhibition (Fig. 6.16). A structural analysis of these flavanones indicated that C6 position was favourable for prenyl substitution (as seen in 38 and 39), while C8 position reduced the PL inhibitory activity (for 40, 41 and 42). Further, Cudraflavanone A (38) was reported to exhibit competitive nature of inhibition on PL, as identified through enzyme kinetic studies (Jo et al. 2015). Similar reports were identified with morusalnol A (43), isolated from root barks of *Morus alba*, that possessed potent IC_{50} value of 0.71 μM (Jeong et al. 2015).

Two flavanone glycosides, hesperidin (44) and neohesperidin (45), isolated from the peels of *Citrus unshiu* fruits (Rutaceae) exhibited poor PL inhibition with IC_{50} values of 52.4 and 75.3 μM (Kawaguchi et al. 1997). However, the unsubstituted flavanones, viz. liquiritigenin (46), naringenin (47) and isosakuranetin (48) (Fig. 6.17), did not possess PL inhibitory activity (Birari et al. 2011; Sergent et al. 2012; Jo et al. 2013). Further, (+)-taxifolin (49), epitaxifolin (50), aromadendrine (51) and 6-hydroxy aromadendrine (52) are the only flavanonols evaluated for PL inhibition assay (Ahn et al. 2013a; Jo et al. 2015); however, they did not possess any activity (Fig. 6.18).

6.2.1.5 Flavans

Flavans are among the least explored flavonoid classes studied for PL inhibition. To date, only five flavans were reported and include 7,4′-dihydroxyflavan (53) and 3′,7-dihydroxy-4′-methoxyflavan (54) from the stem barks of *Broussonetia kanzinoki* (Moraceae) and hispaglabridin A (55), glabridin (56) and its 4′-O-methoxy

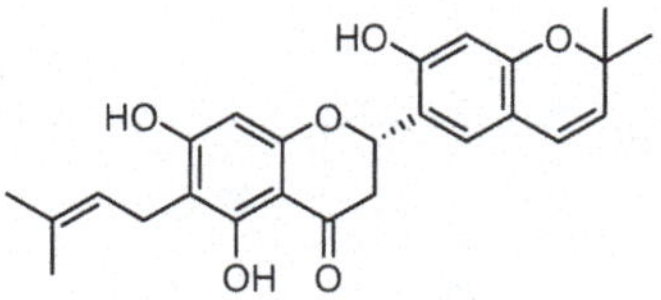

Cudraflavanone A (**38**)
(IC_{50} = 6.5 μM)

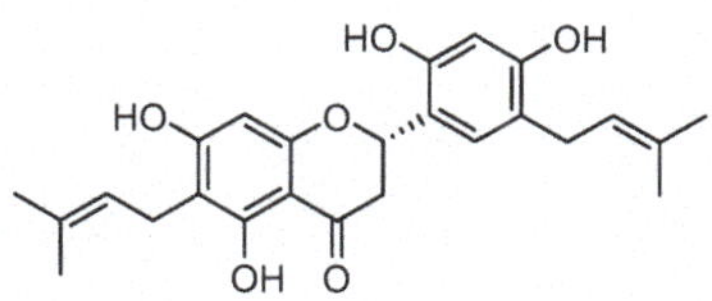

Cudraflavanone D (**39**)
(IC_{50} = 9.0 μM)

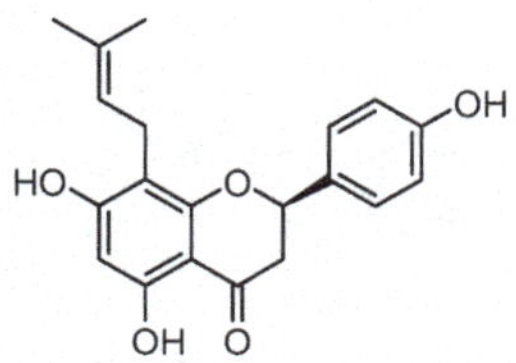

8-Prenylnaringenin (**40**)
(IC_{50} = 76.9 μM)

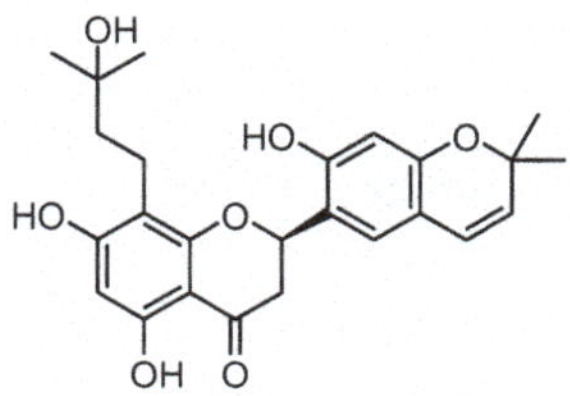

Compound **41**
(IC_{50} = 84.6 μM)

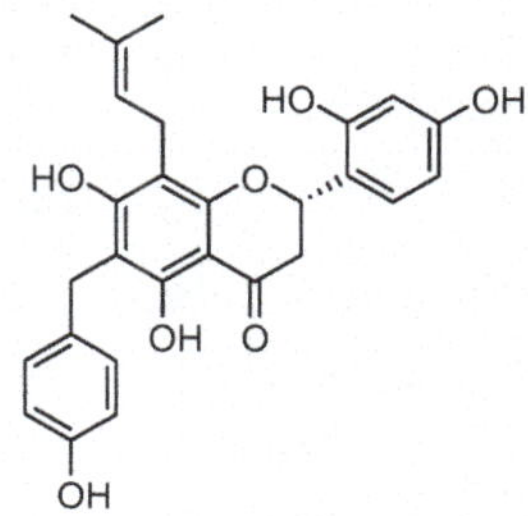

Cudracuspiflavanone A (**42**)
(IC_{50} = 54.8 μM)

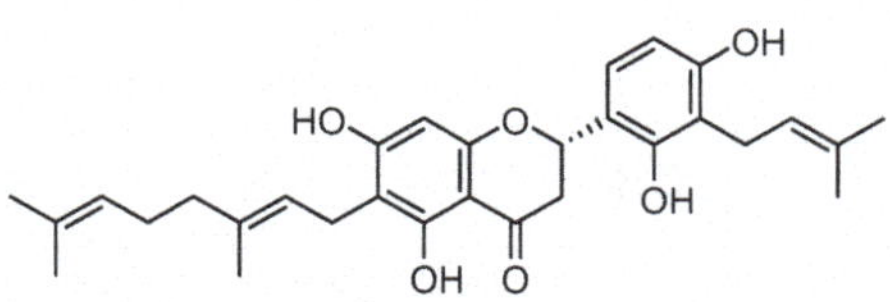

Morusalnol A (**43**)
(IC_{50} = 0.71 μM)

Fig. 6.16 Prenylated flavanones from *C. tricuspidata* and *M. alba* and their PL inhibitory activity

derivative (57), isolated from the roots of *Glycyrrhiza glabra* (Fabaceae) (Birari et al. 2011; Ahn et al. 2012). These flavans, however, exhibited poor PL inhibitory activity with the IC_{50} values ranging from 85.1 to 485.6 μM (Fig. 6.19). Apart from, isoflavones and anthocyanidins are the other two classes of flavonoids, which were least explored and did not possess PL inhibitory activity (Guo et al. 2009; Birari et al. 2011; You et al. 2011; Jo et al. 2015).

6.2.1.6 Phenolic Acids

Phenolic acids represent one of the earlier classes, explored for their role in PL inhibition. In a study conducted by Karamać et al., various derivatives of benzoic and cinnamic acids were screened for PL inhibition using 4-NPA substrate. However, none of the acids exerted greater than 40% inhibition at a final concentration of 10 μM (Karamać and Amarowicz 1996). Apart from, gallic acid and various galloylated glucose derivatives (gallotannins), isolated from *Galla rhois* (formed by

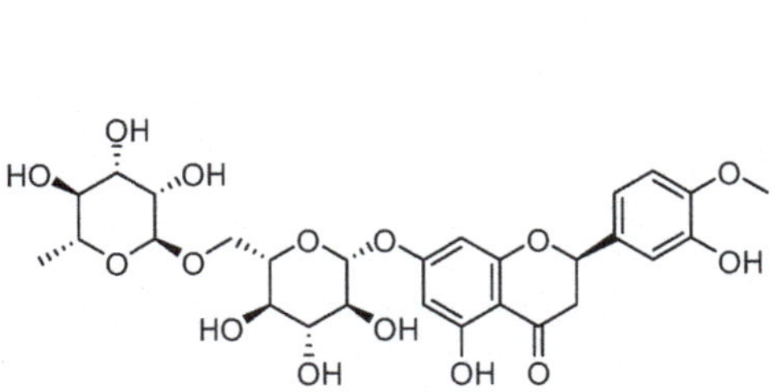

Hesperidin (**44**)
(IC_{50} = 52.4 μM)

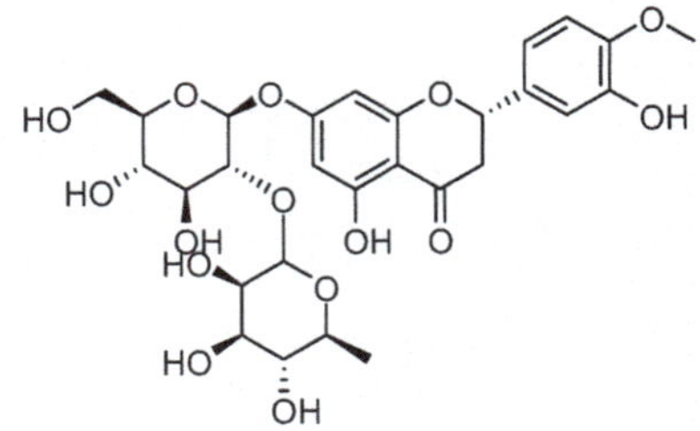

Neohesperidin (**45**)
(IC_{50} = 75.3 μM)

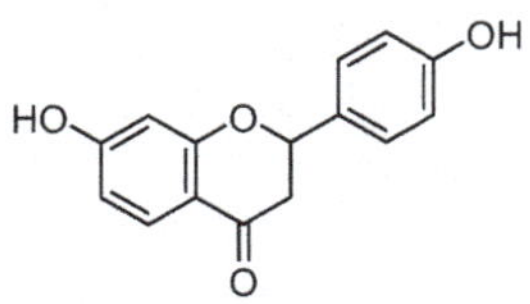

Liquiritigenin (**46**)

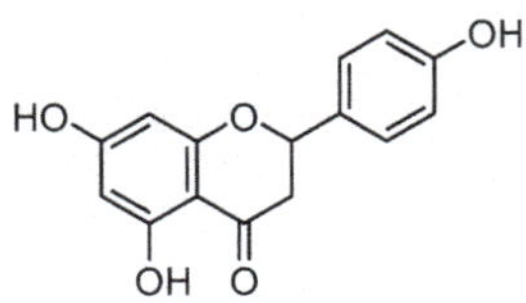

Naringenin (**47**)

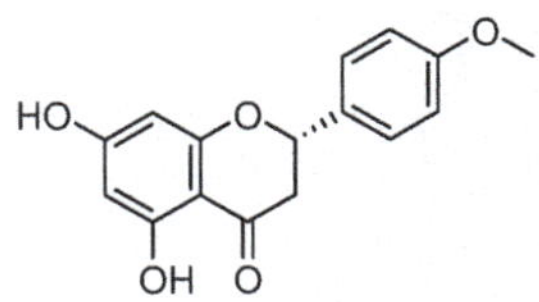

Isosakuranetin (**48**)

Fig. 6.17 Summary of various flavanone glycosides and simple flavanones reported for PL inhibitory activity

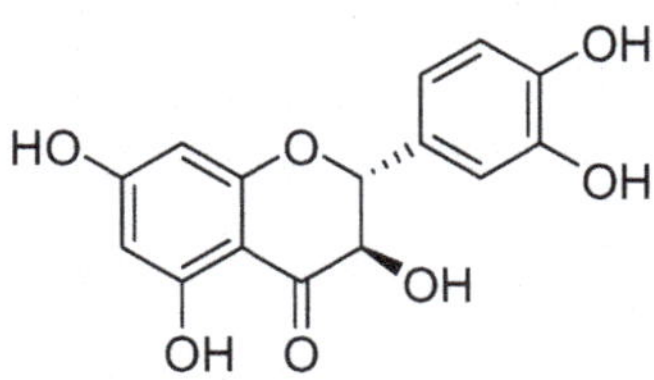

(+)-Taxifolin (**49**)

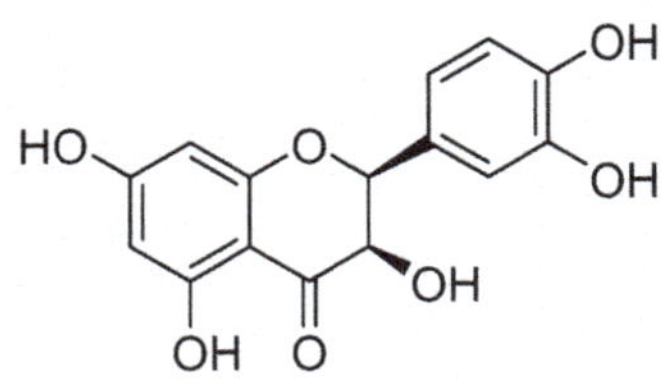

Epitaxifolin (**50**)

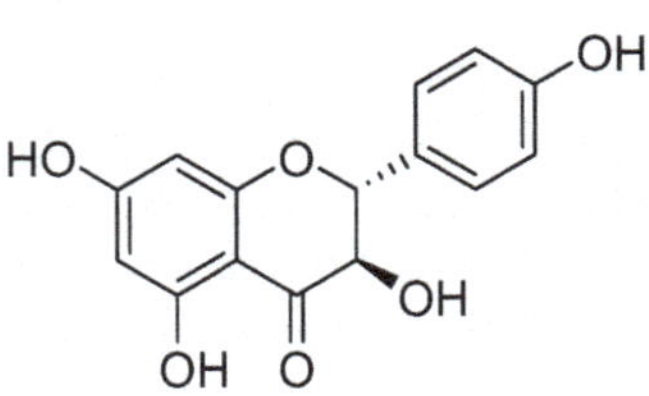

Aromadendrine (**51**)

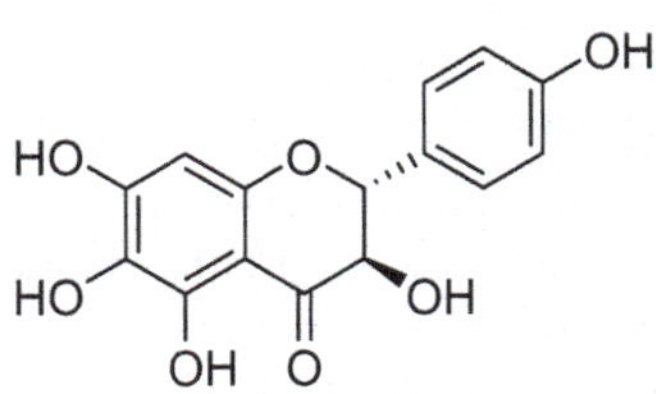

6-Hydroxy aromadendrine (**52**)

Fig. 6.18 Summary of various flavanonols reported for PL inhibitory activity

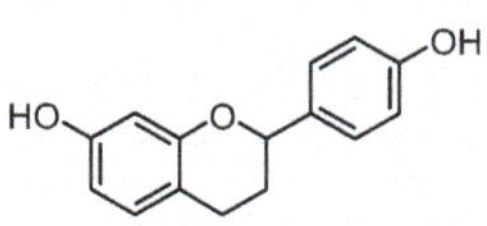

7,4'-Dihydroxy flavan (**53**)
(IC_{50} = 85.1 μM)

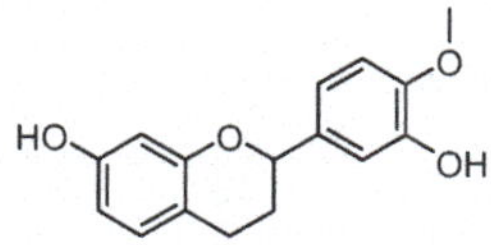

3',7-Dihydroxy-4'-methoxyflavan (**54**)
(IC_{50} > 100 μM)

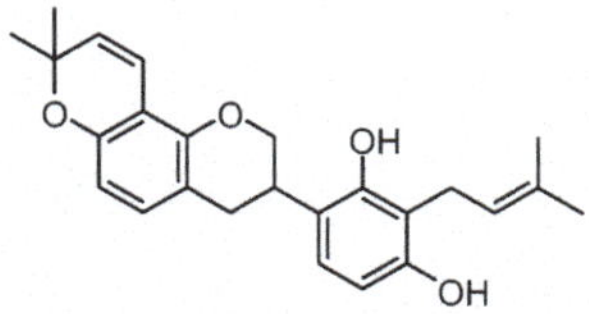

Hispaglabridin A (**55**)
(IC_{50} = ND)

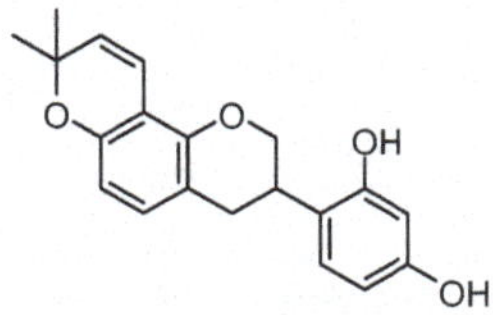

Glabridin (**56**)
(IC_{50} = 485.6 μM)

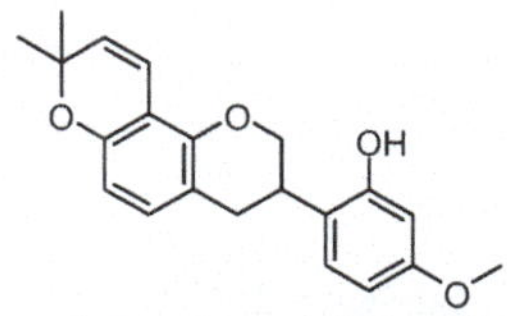

4'-*O*-methoxy glabridin (**57**)
(IC_{50} = 404.9 μM)

Fig. 6.19 Chemical structures of various flavans and their PL inhibitory activity (The IC_{50} of hispaglabridin A (**55**) was not determined (ND), but it exhibited PL inhibition by 43.8% at a concentration of 250 μg/ml)

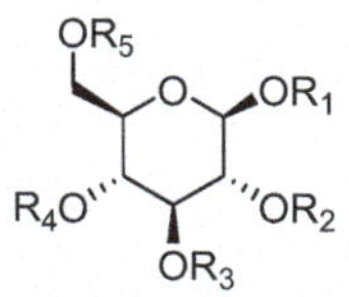

	R_1	R_2	R_3	R_4	R_5
58	G	G	H	G	G
59	G	G	G	G	G
60	G	diG	G	G	G

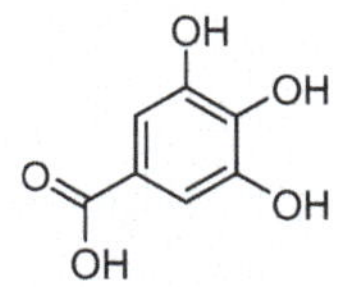

Gallic acid (**61**)

diG:

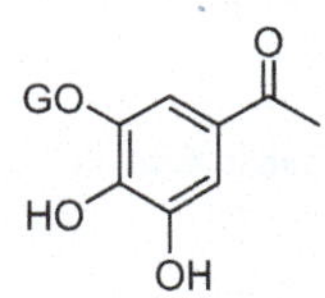

Fig. 6.20 Chemical structures of gallic acid and various gallotannins (G: galloyl)

aphids on the leaves of *Rhus javanica* (Anacardiaceae)), constitute the major components of phenolic acids screened for PL inhibition. A potential to moderate activity was exhibited by various gallotannins (58–60), and the activity was proportional to the number of galloyl substitutions (Fig. 6.20 and Table 6.5). Gallic acid (61), however, did not possess PL inhibitory activity (Kwon et al. 2013). In a study conducted by Narita et al. (2012), various caffeoylquinic and feruloylquinic acids, isolated from decaffeinated coffee beans extract, were screened for PL inhibition in the presence of olive oil as substrate. The dicaffeoyl acids (61–63) exhibited comparatively greater potential PL inhibitory activity, followed by the monocaffeoyl derivatives (64–66), while the feruloyl acids (67–69) exhibited poor PL inhibitory activity (Fig. 6.21 and Table 6.6).

Table 6.5 PL inhibitory activity of gallotannins representing inverse relation between activity and the number of galloyl units

Compound	IC_{50} (μM)	Number of galloyl units
58	23.2	4
59	15.9	5
60	3.5	6

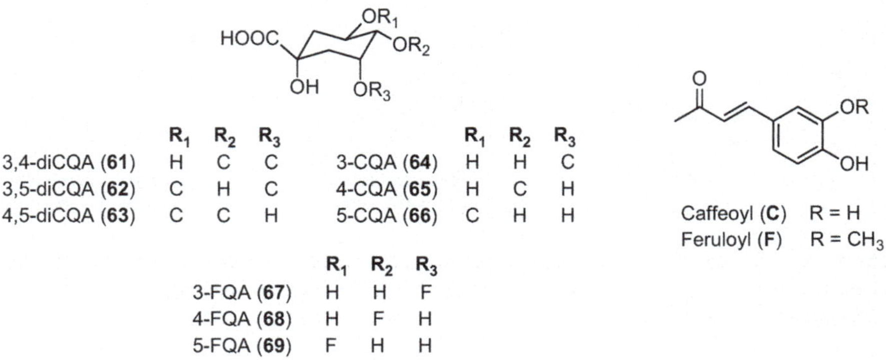

Fig. 6.21 Chlorogenic acids from coffee beans extract. (*CQA* caffeoylquinic acid, *FQA* feruloylquinic acid)

Table 6.6 PL inhibitory activity of various chlorogenic acids from decaffeinated coffee beans extract

Compound	IC_{50} (mM)	Compound	IC_{50} (mM)	Compound	IC_{50} (mM)
3,4-diCQA (**61**)	0.62	3-CQA (**64**)	3.09	3-FQA (**67**)	>8.00
3,5-diCQA (**62**)	0.75	4-CQA (**65**)	3.53	4-FQA (**68**)	>8.00
4,5-diCQA (**63**)	0.48	5-CQA (**66**)	3.19	5-FQA (**69**)	>8.00

6.2.1.7 Miscellaneous Polyphenols

Apart from flavonoids and phenolic acids, few stilbene derivatives (Fig. 6.22) and phloroglucinol derivatives (Fig. 6.23) were also evaluated for their PL inhibitory potential. Examples include resveratrol (70) and its *O*-glycosides, viz. *cis*- and *trans*-piceid (71, 72) from *Vitis vinifera* (Vitaceae) and a prenylated stilbene, morusibene A (73), from the root barks of *Morus alba* (Kim et al. 2014c; Ha et al. 2016). Few dihydrostilbenes from the rhizomes of *Dioscorea opposita* (Dioscoreaceae) were also reported, with compounds 74 and 75 that exhibited potent PL inhibitory activity (Yang et al. 2014). To summarise, polyphenols contributed to the most explored class of phytochemicals for their potential towards PL inhibition. In particular, flavanols and its various derivatives account to the majority of PL inhibitory polyphenols. Within the class of flavanols, the

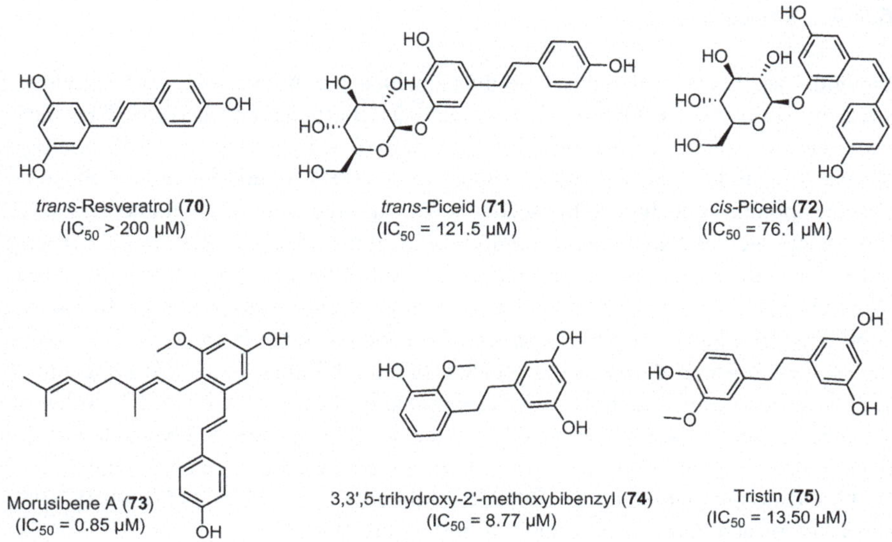

Fig. 6.22 Resveratrol and its derivatives (**70–72**) from *Vitis vinifera*, morusibene A (**73**) from *Morus alba* and dihydrostilbenes (**74–75**) from *D. opposita* and their PL inhibitory activity

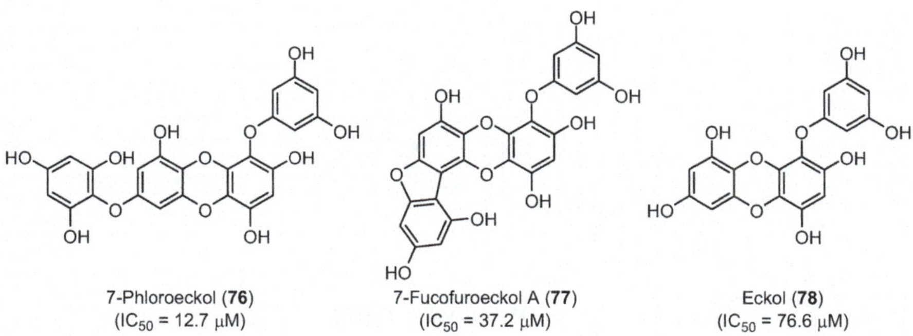

Fig. 6.23 7-Phloroeckol (**76**), fucofuroeckol A (**77**) and eckol (**78**) from *E. bicyclis* and their PL inhibitory activity

methylene-bridged dimers possessed potent PL inhibitory activity. Further, the number of galloyl units played a prominent role, with the unsubstituted flavanols that exhibited poor PL inhibition. Apart from the galloyl esters, the activity was also upraised by the prenyl substitution, as seen with various classes of polyphenols, viz. flavanones, chalcones and stilbenes. However, substitution of glycosides resulted in decreased PL inhibitory potential.

6.2.2 Saponins

Saponins are a class of natural products, which are nonvolatile, surface-active, structurally diverse and chemically referred to as triterpenes and steroids. They have been widely explored for various biological activities, including anti-inflammatory, antiparasitic and haemolytic effects (Sparg et al. 2004). Within the area of PL inhibition, saponins contribute to the second most explored class of phytochemicals. Of this, plants such as *Platycodon grandiflorum* (Campanulaceae), *Acanthopanax senticosus* (Araliaceae), *Ilex paraguariensis* (Aquifoliaceae) and *Sapindus rarak* (Sapindaceae) contribute to the major sources of PL inhibitory saponins. In a study conducted by Zhao et al. (2005), seven triterpenoidal saponins were isolated from the roots of *P. grandiflorum* and evaluated for PL inhibitory assay. Prosapogenin D (79) was found to be the most active compound in the series, with an IC_{50} value of 1.3 mM. In another study conducted by Xu et al. (2005), various glycoside derivatives of prosapogenin D from *P. grandiflorum* were evaluated, wherein platycodins A, C and deapioplatycodin D (80–82) inhibited PL by 96.7, 94.8 and 88.33%, respectively, at a final concentration of 500 μg/mL (Fig. 6.24).

Various triterpenoidal saponins from the fruits of *A. senticosus* were also evaluated for PL inhibition (Li et al. 2007). Silphioside F (83), copteroside B (84), hederagenin 3-*O*-β-D-glucuronopyranoside 6′-*O*-methyl ester (85) and gypsogenin

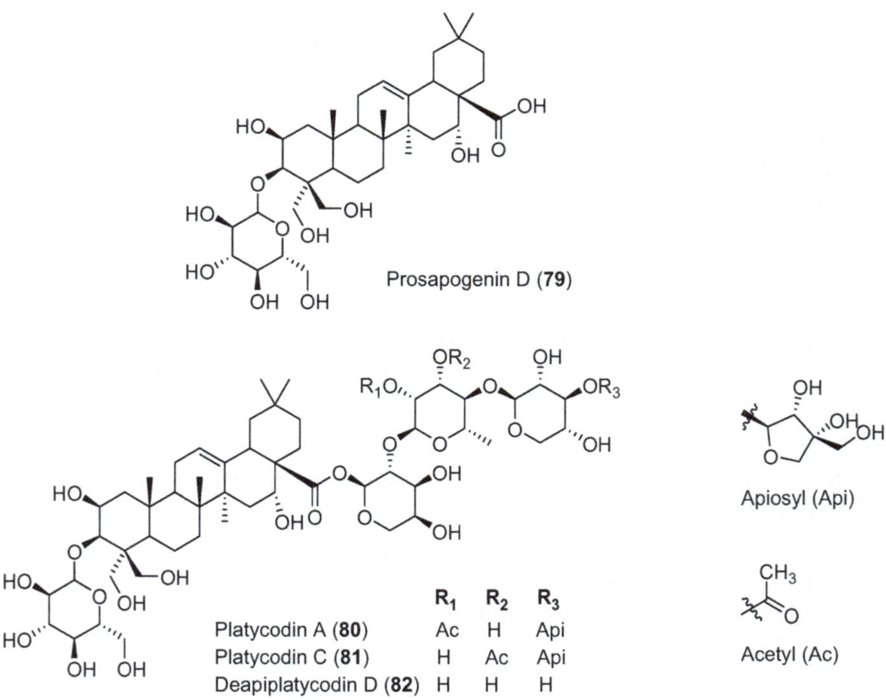

Fig. 6.24 Prosapogenin D (**79**) and its glycoside derivatives (**80–82**) from *P. grandiflorum*

Table 6.7 PL inhibitory activity of triterpenoidal saponins from fruits of *A. senticosus*

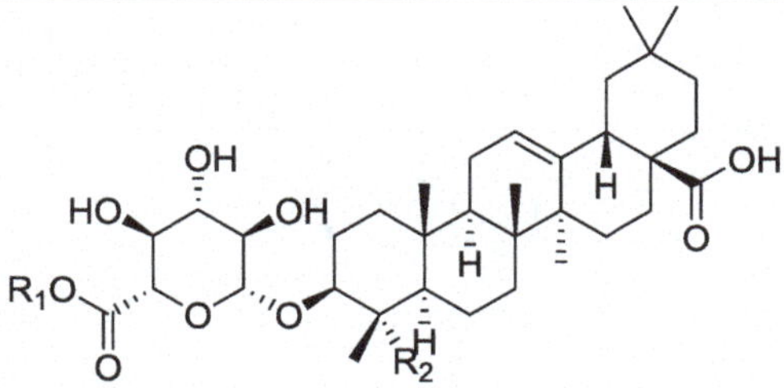

Compound	R_1	R_2	IC_{50} (mM)
Silphioside F (**83**)	-H	$-CH_3$	0.22
Copteroside B (**84**)	-H	-OH	0.25
Hederagenin 3-*O*-β-D-glucuronopyranoside 6'-*O*-methyl ester (**85**)	$-CH_3$	-OH	0.26
Gypsogenin 3-*O*-β-D-glucuronide (**86**)	-H	-CHO	0.29

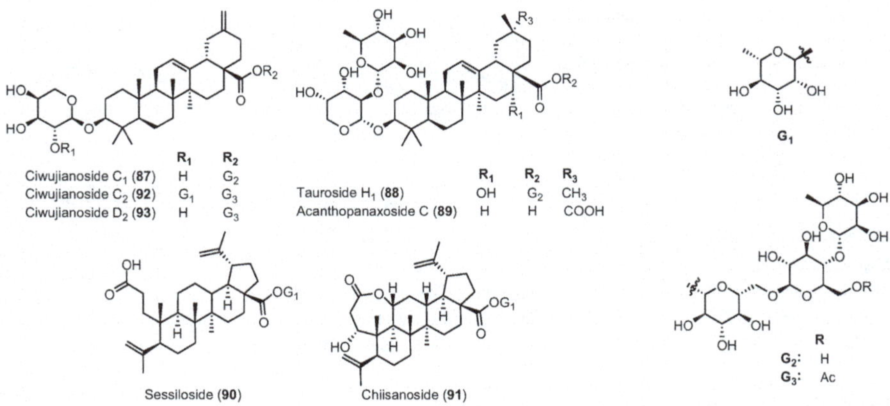

Fig. 6.25 Triterpenoidal saponins from the leaves of *A. senticosus* (Ac: acetyl)

3-*O*-β-D-glucuronide (86) exhibited negligible PL inhibition (Table 6.7). Further, in a study conducted by Jiang et al. (2006), 15 triterpenoidal saponins from the leaves of *A. senticosus* were evaluated for their PL inhibitory activity; while ciwujianoside C_1 (87), tauroside H_1 (88), acanthopanaxoside C (89), sessiloside (90) and chiisanoside (91) inhibited around 50% of the enzyme activity, ciwujianosides C_2 (92) and D_2 (93) enhanced the enzyme activity to more than 140% at a concentration of 1 mg/mL (Fig. 6.25).

In a study conducted by Morikawa et al. (2009), 21 saponins, including four novel oleanane-type triterpene oligoglycosides (94–97) and four sesquiterpene oligoglycosides, were isolated from the pericarps of *Sapindus rarak*, wherein the triterpenes (Fig. 6.26) exhibited poor PL inhibitory activity. The sesquiterpenes, however, were inactive against PL. Likewise, in another study conducted by Sugimoto et al. (2009), 21 saponins, including three novel triterpenes,

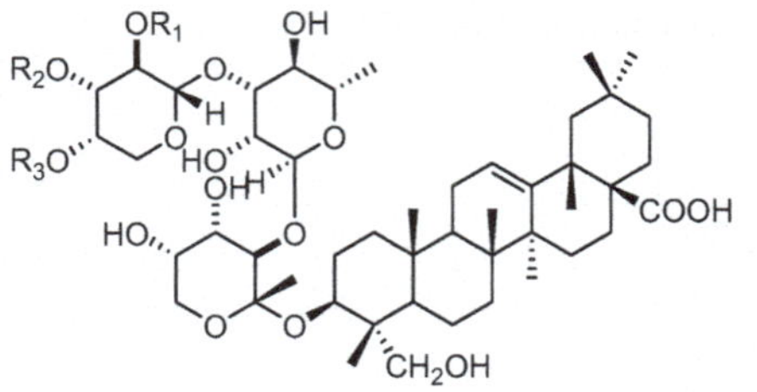

	R_1	R_2	R_3	IC_{50} (μM)
Rarasaponin I (**94**)	Ac	H	H	131
Rarasaponin II (**95**)	H	Ac	H	172
Rarasaponin III (**96**)	H	H	Ac	576

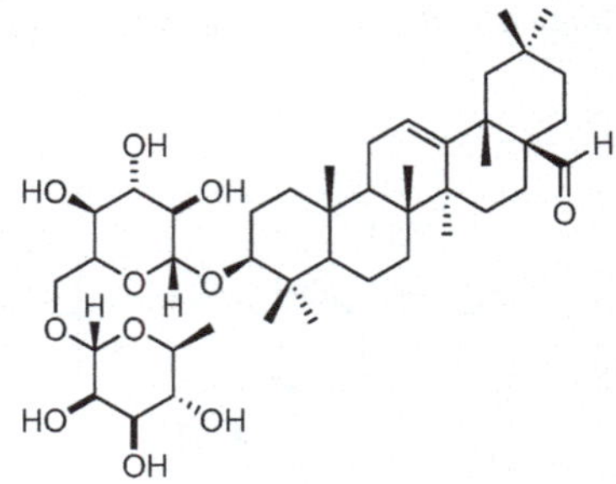

Raraoside A (**97**)
(IC_{50} = 151 μM)

Fig. 6.26 Rarasaponins (**94–96**) and raraoside A (**97**) from *Sapindus rarak* (Ac: acetyl)

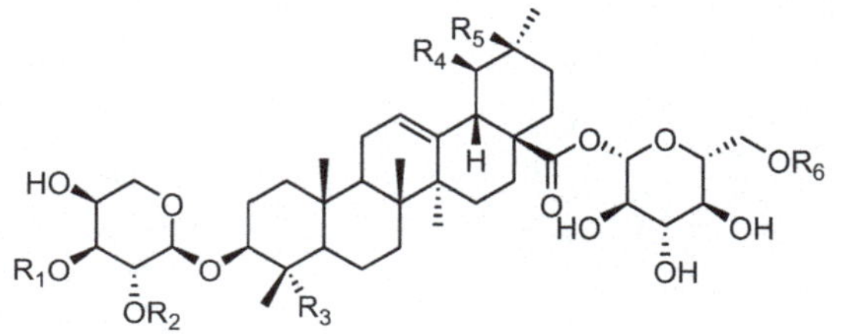

G_1:

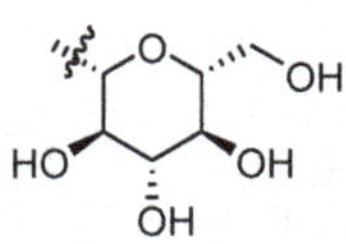

G_2:

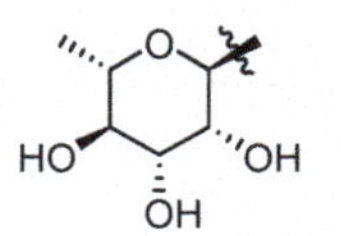

	R_1	R_2	R_3	R_4	R_5	R_6
Mateglycoside A (**98**)	G_1	G_2	CH_3	H	CH_3	G^1
Mateglycoside B (**99**)	G_1	H	CH_2-OH	CH_3	H	H
Mateglycoside C (**100**)	H	H	CH_2-OH	CH_3	H	H
Matesaponin A (**101**)	G_1	H	CH_3	CH_3	H	H
Nudicaucin C (**102**)	G_1	G_2	CH_3	H	CH_3	H

Fig. 6.27 Matesaponins from *Ilex paraguariensis*

mateglycosides A, B and C (98–100), were isolated from the leaves of *I. paraguariensis*. Matesaponin I (101) and nudicaucin C (102) exhibited moderate PL inhibitory activity (94% and 77%, respectively, at 100 μM); however, the mateglycosides (98–100) did not possess PL inhibitory activity (Fig. 6.27).

Apart from, various classes of saponins, viz. chakasaponins, gypsosaponins, perennisaponins and scabiosaponins, respectively, from *C. sinensis*, *Gypsophila oldhamiana* (Caryophyllaceae), *Bellis perennis* (Asteraceae) and *Scabiosa tschiliensis* (Caprifoliaceae), include the other major PL inhibitory saponins. In a study conducted by Zheng et al. (2007) three triterpene saponins, gypsosaponins A–C (103–105), from the roots of *G. oldhamiana* were subjected to PL inhibition assay, wherein gypsosaponins A and C exhibited poor PL inhibition (58.2 and 50.3% respectively, at 1 mg/mL), while gypsosaponin B exhibited 99.2% inhibition at this concentration (Fig. 6.28). However, gypsosaponin B (104) was considered an artefact by Zheng et al. (2007) as it was not detected in the extract during HPLC

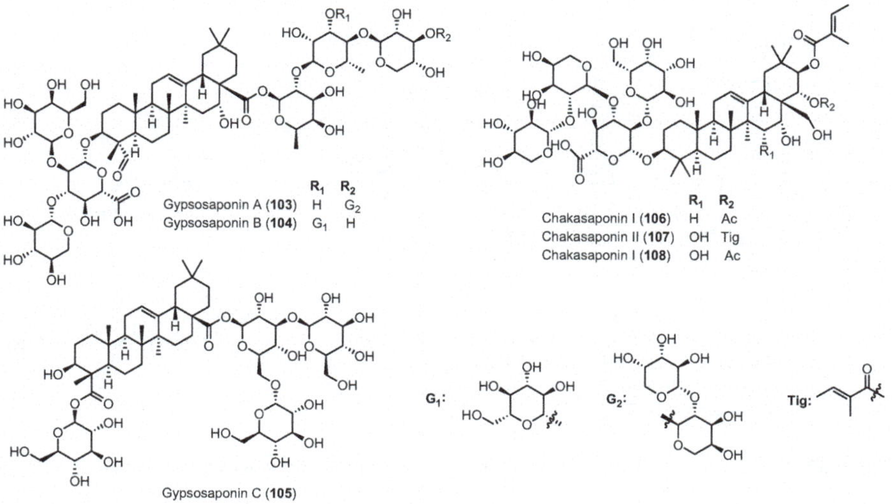

Fig. 6.28 Gypsosaponins and chakasaponins from *G. oldhamiana* and *C. sinensis*, respectively (*Ac* acetyl, *Tig* tigloyl)

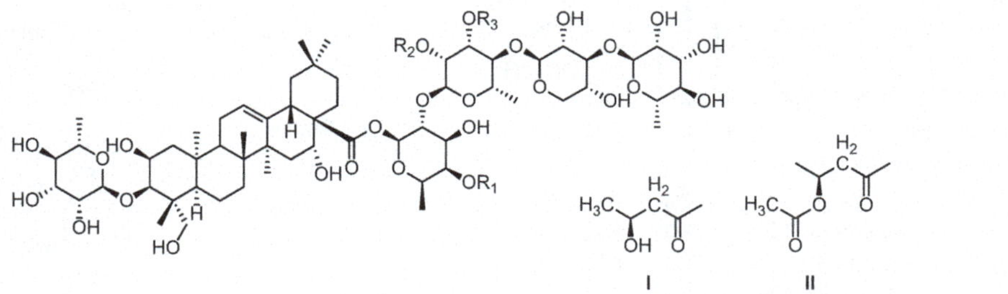

	R^1	R^2	R^3	IC_{50} (μM)
Perennisaponin G (**109**)	Ac	Ac	H	163
Perennisaponin H (**110**)	Ac	H	Ac	137
Perennisaponin I (**111**)	I	Ac	H	147
Perennisaponin J (**112**)	II	Ac	H	148
Perennisaponin K (**113**)	II	H	Ac	223
Perennisaponin L (**114**)	III	Ac	H	81.4
Perennisaponin M (**115**)	IV	Ac	H	195

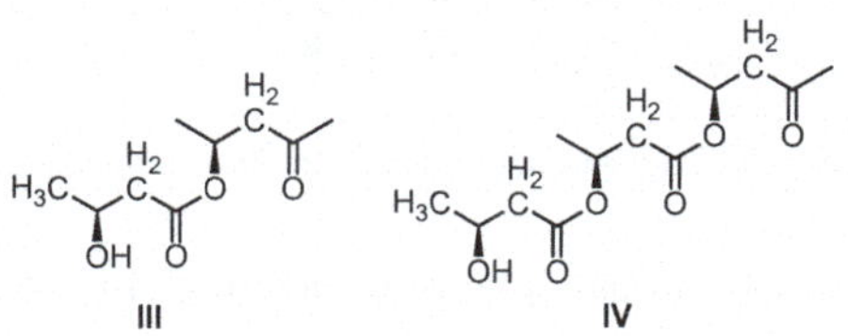

Fig. 6.29 Perennisaponins and their PL inhibitory activity from *Bellis perennis*. (*Ac* acetyl)

analysis. In another study, Yoshikawa et al. reported that the chakasaponins I–III (106–108), from the flower buds of *C. sinensis*, exhibited poor PL inhibition (Fig. 6.28); however, they exhibited potent acceleration of GI transit, when subjected to in vivo studies at a dose of 100 mg/kg p.o. in male ddY mice (Yoshikawa et al. 2009). In a study conducted by Morikawa et al. (2010) seven perennisaponins (109–115) from the flowers of *B. perennis* exhibited moderate PL inhibition (Fig. 6.29). Perennisaponin L (116) was the most active compound in the series, with an IC_{50} of 81.4 μM. Further, in a study conducted by Zheng et al. (2004), 13

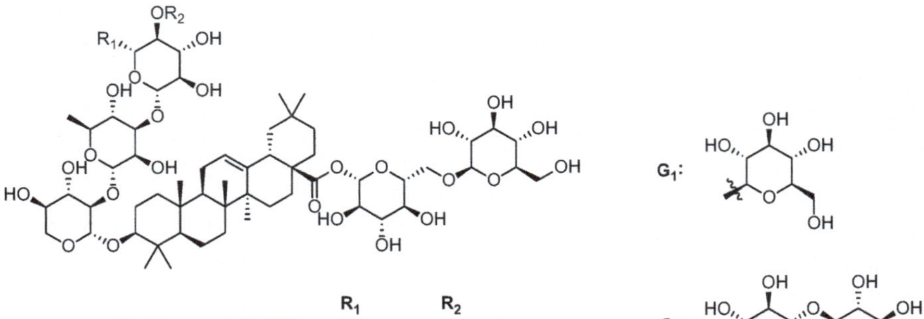

	R_1	R_2
Scabiosaponin E (**116**)	H	H
Scabiosaponin F (**117**)	CH_2-OH	H
Scabiosaponin G (**118**)	CH_2-OH	G_1
Hookeroside A (**119**)	H	G_1
Hookeroside B (**120**)	H	G_2

Fig. 6.30 Chemical structures of PL inhibitory scabiosaponins from *Scabiosa tschiliensis*

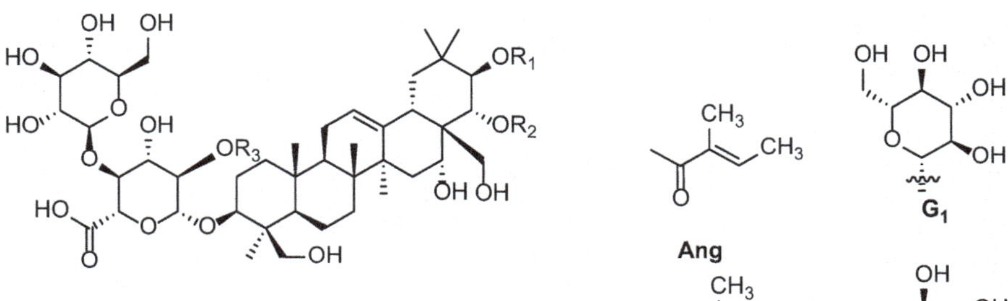

	R_1	R_2	R_3
Escin Ia (**121**)	Tig	Ac	G_1
Escin IIa (**122**)	Tig	Ac	G_2
Escin Ib (**123**)	Ang	Ac	G_1
Escin IIb (**124**)	Ang	Ac	G_2

Fig. 6.31 Escin derivatives from the seeds of *Aesculus turbinata*. (*Ac* acetyl)

triterpenoidal saponins, including 11 scabiosaponins and two hookerosides, were evaluated, wherein scabiosaponins E–G (116–118) and hookerosides A and B (119, 120) exhibited greater than 60% PL inhibition at a final concentration of 1 mg/mL (Fig. 6.30). Apart from, various escin derivatives from the edible seeds of *Aesculus turbinata* (Sapindaceae) were evaluated for PL inhibitory assay (Zheng et al. 2004). Kimura et al. reported that the escins (121–124) exhibited greater potential activity (with IC_{50} ranging from 25 to 50 μM), followed by the desacylescins (50–100 μM), while deacetylescins were poor PL inhibitors (> 100 μM). Further, an angeloyl (Ang) moiety at C-21 caused greater potency (as seen with escins Ib and IIb) compared to the tigloyl (Tig) moiety, as in escins Ia and IIa (Fig. 6.31).

Various other saponins that were explored for PL inhibition assay include ginsenosides and chikusetsusaponins from *Panax sps* (Araliaceae), dioscin derivatives from *Dioscorea nipponica* (Dioscoreaceae), sesquiterpenes from *Alisma orientale*

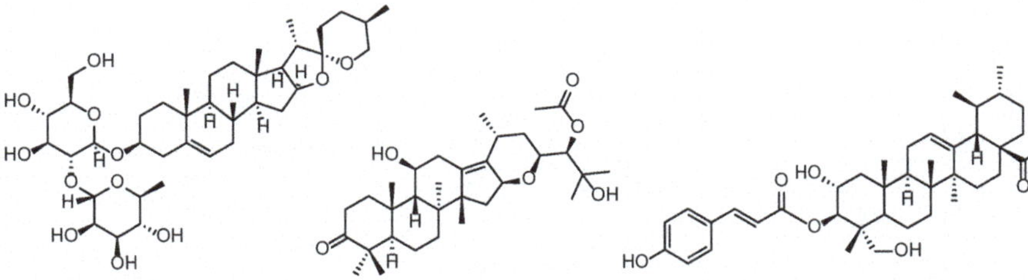

Prosapogenin A (**125**) (IC_{50} = 2.48 μM)

Alisol F24-acetate (**126**) (IC_{50} = 45.5 μM)

3-*O*-*trans*-*p*-coumaroyl actinidic acid (**127**) (IC_{50} = 14.95 μM)

Fig. 6.32 Saponins and their PL inhibitory activity from *D. nipponica*, *A.orientale* and *A. arguta*

(Alismataceae) and few triterpene acids from *Actinidia arguta* (Actinidiaceae). Of these, prosapogenin A (125), alisol F 24-acetate (126) and 3-*O*-*trans*-*p*-coumaroyl actinidic acid (127) from *D. nipponica*, *A.orientale* and *A. arguta*, respectively exhibited moderate to poor PL inhibitory activity (Fig. 6.32). Ginsenosides and chikusetsusaponins from *Panax sps*, however, did not possess PL inhibitory activity (Kwon et al. 2003; Han et al. 2005; Jang et al. 2008; Liu et al. 2008; Cang et al. 2017). Around 100 triterpene derivatives were explored to determine their PL inhibitory potential. These triterpenes majorly constituted an oleanane-type nucleus. While an appropriate structure-activity relationship could not be constructed, a preliminary analysis indicated the role of glycosides to impart PL inhibitory potency to this class of phytochemicals. Moreover, the activity varied with the number of glycoside units and their location. A similar phenomenon could be observed with the acetyl substitution, as seen with the perennisaponins.

6.2.3 Alkaloids

The term "alkaloid" was first proposed and defined by W. Meissner as *"plant-derived substance that reacts like alkalis"*. However, the modern definition of alkaloid was proposed by S. W. Pelletier as *"a cyclic organic compound containing nitrogen in a negative oxidation state which is of limited distribution among living organisms"* (Pelletier 1983). Alkaloids have been explored for a wide range of pharmacological activities, the major being their potential activity as anticarcinogenic compounds; however, very few alkaloids have been explored in the area of PL inhibition, with 40 alkaloids reported to date. Further, these alkaloids belonged to various subclasses, viz. pyrroles, benzylisoquinolines, carbazoles and bisindoles.

In a study reported by Kim et al. 17 pyrrole alkaloids from the fruits of *Morus alba* were evaluated for PL inhibition assay (Kim et al. 2014b). The *p*-hydroxybenzyl derivative (128) was the most active compound in the series; however, it exhibited a moderate PL inhibition (70% at 100 μM). Various benzylisoquinoline alkaloids from *Nelumbo nucifera* (Nelumbonaceae), *Berberis sps* and *Papaver somniferum* (Papaveraceae) were studied for their PL inhibitory potential. Liriodenine (129) from *N. nucifera* exhibited a moderate PL inhibition (45% at 100 μM), followed by

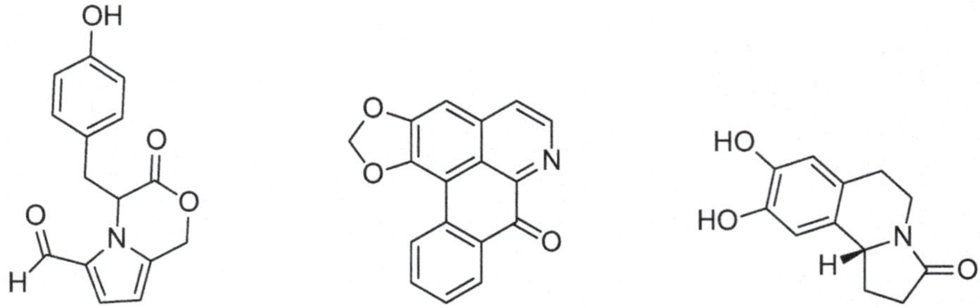

Compound **128** Liriodenine (**129**) Oleracein E (**130**)

Fig. 6.33 Alkaloids from *M. alba* and *N. nucifera*

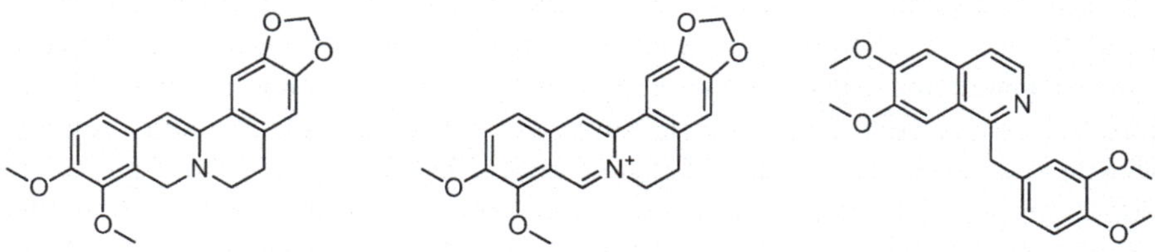

Dihydroberberine (**131**) (IC_{50} = 23.7 μM) Berberine (**132**) (IC_{50} = 314.5 μM) Papaverine (**133**) (IC_{50} = 106.6 μM)

Fig. 6.34 Isoquinoline alkaloids and their PL inhibitory activity from *Berberis* sp. and *P. somniferum*

oleracein E (130), that exhibited 40% inhibition. Nevertheless, liriodenine (129) possessed potential inhibitory activity against adipocyte differentiation (>80%) at this concentration (Fig. 6.33). However, other *Nelumbo* alkaloids possessed poor PL inhibitory activity (Ahn et al. 2013a).

In another study, Mohammad et al. reported that dihydroberberine (131) possessed potential PL inhibitory activity over its unsaturated derivative, berberine (132). Molecular docking studies of these alkaloids in the active site of PL indicated a similar binding pattern, exhibiting π-stacking with Phe 77 and Phe 215, as well as H-bond with Ser 152 and His 263. However, the potential activity of dihydroberberine was explained due to the absence of a permanent cationic centre on the nitrogen, leading to a higher binding affinity (Mohammad et al. 2013). Similar interactions were identified for papaverine (133) by Al-Masri et al. in the molecular modelling studies (Al-Masri 2013), while papaverine exhibited poor PL inhibition in the in vitro assay (Fig. 6.34).

Apart from, few carbazole alkaloids (134–137) have been reported from the leaves of *Murraya koenigii* (Rutaceae). While mahanimbine (134) that consisted a prenyl substitution exhibited comparatively greater potential towards PL inhibition (Fig. 6.35), the unprenylated alkaloids exhibited poor inhibitory activity (Birari et al. 2009). Further, in a study conducted by Sridhar et al., various synthetic derivatives

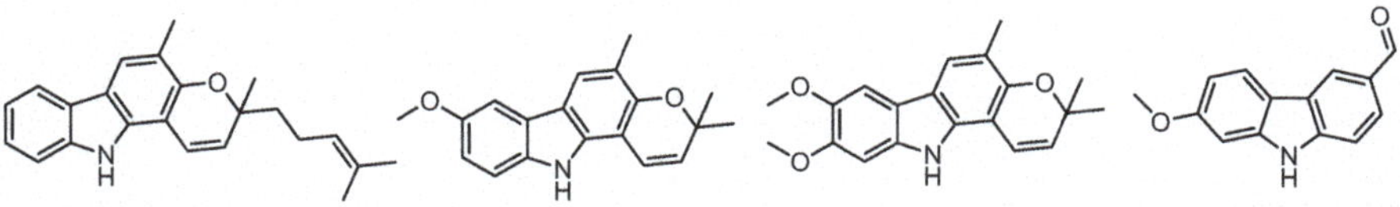

Mahanimbin (**134**) (IC_{50} = 17.9 μM)

Koenimbine (**135**) (IC_{50} = 168.6 μM)

Koenigicine (**136**) (IC_{50} = 428.6 μM)

Clausazoline K (**137**) (IC_{50} > 500 μM)

Fig. 6.35 Carbazole alkaloids from *M. koenigii* and their PL inhibitory activity

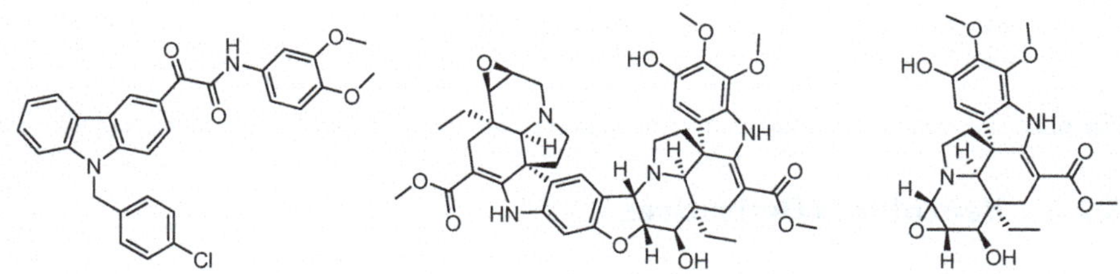

Synthetic carbazole derivative (**138**)

Conophylline (**139**)

Taberhanine (**140**)

Fig. 6.36 Synthetic carbazole derivative (**138**), conophylline (**139**) and taberhanine (**140**)

containing a carbazole nucleus fused with an α-ketoamide were evaluated for their PL inhibitory potential. The trimethoxy derivative with N-benzyl substitution (138) exhibited a potential PL inhibitory activity with IC_{50} of 6.31 μM (Fig. 6.36). Sridhar et al. reported that the carbonyl group of the ketoamide possibly exhibited a covalent bond interaction with Ser 152, while the carbazole involved in π-stacking with Phe 77, Phe 114 and Phe 215 of the lid domain, as analysed through molecular docking and dynamics simulation (Sridhar et al. 2017a). In another study by Sridhar et al., the PL inhibitory potency of conophylline (139), a bis-indole alkaloid from the leaves of *Tabernaemontana divaricata* (Apocynaceae), was reported (Sridhar et al. 2017b), wherein it exhibited a potent activity (IC_{50} = 3.31 μM). Molecular dynamics simulation of conophylline (139) highlighted that the dimeric extension stabilised the ligand through hydrophobic interactions that was not observed with its monomeric counterpart, taberhanine (140). To summarise, around 40 alkaloids were explored for their PL inhibitory potential. However, these alkaloids exhibited moderate to poor inhibitory activity with an exception for conophylline, mahanimbine and the synthetic carbazole derivative. In the presence of a hydrophobic extension, viz. prenyl unit (as in mahanimbine), the aryl wings (as in synthetic derivative) or dimeric extension (as in conophylline) stabilised these molecules through interactions with the lid domain, while the reactive carbonyl group (ester and α-ketoamide in conophylline and synthetic derivative, respectively) might have resulted in covalent bond formation with Ser152, causing greater potency of these alkaloids against PL.

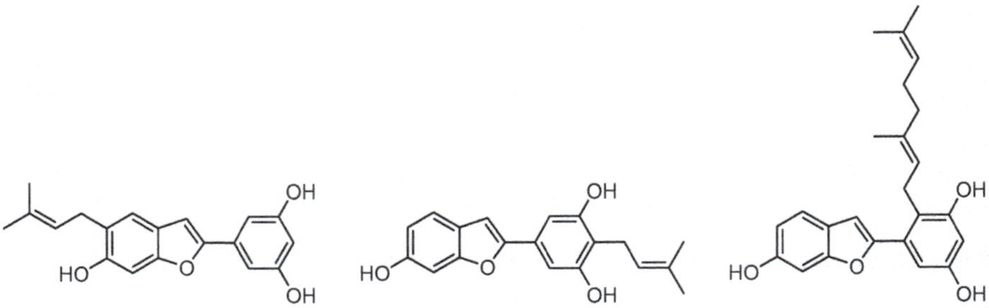

Moracin N (**141**) (IC_{50} = 29.7 μM)   Moracin C (**142**) (IC_{50} = 32 μM)   Albafuran A (**143**) (IC_{50} > 70 μM)

Fig. 6.37 Prenylated benzofurans from the leaves of *M. alba* and their PL inhibitory activity

6.2.4 Benzofuran Derivatives

Benzofurans are among the major group of phytochemicals, which contain an oxygen heterocycle nucleus, and are widely explored for various biological activities, viz. anticarcinogenic, antimicrobial, antiviral, anti-inflammatory, anti-Alzheimer's, etc. (Khanam and Shamsuzzaman 2015). However, they are less explored in the area of PL inhibition, similar to alkaloids. To date, around 40 benzofurans from various plants were explored, with the major source being *Morus alba* and *Shorea roxburghii* (Dipterocarpaceae). In two different studies conducted on the leaves and root bark of *M. alba*, 15 benzofuran derivatives were studied for their PL inhibitory potential (Jeong et al. 2015; Ha et al. 2016). The prenylated benzofurans, Moracin N (141) and Moracin C (142) from the leaves, exhibited greater potential over the unsubstituted benzofurans. Albafuran A (143), a prenylated benzofuran, however, did not possess PL inhibitory activity (Fig. 6.37). Similar reports were observed with the prenylated benzofurans from the root barks of *M. alba*; however, they exhibited poor PL inhibitory activity compared to the standard drug, orlistat (Fig. 6.38).

In a study conducted by Morikawa et al. (2012) various benzofuran derivatives from bark of *S. roxburghii* were studied for their PL inhibitory potential. While the dimeric benzofurans (148–149) exhibited comparatively potential activity, the monomeric derivatives (150–151) possessed moderate to poor PL inhibition (Fig. 6.39). Further, Wilsonol C (152), a benzofuran trimer from the roots of *Vitis vinifera*, also exhibited potent PL inhibition (Kim et al. 2014c). Apart from, artokinin (153), a prenylated benzofuran, and acernikol (154), a sesquilignan-substituted benzofuran from *Artocarpus nitidus* and *Fraxinus rhynchophylla* (Oleaceae), respectively (Zhao et al. 2009; Ahn et al. 2013b), are the other phytochemicals in this class reported for PL inhibitory activity (Fig. 6.40). To summarise, around 40 benzofuran derivatives were explored, wherein the prenylated benzofurans as well as the dimers possessed potential activity, clearly indicating the role of prenyl units in imparting greater potency to the natural products towards PL inhibition.

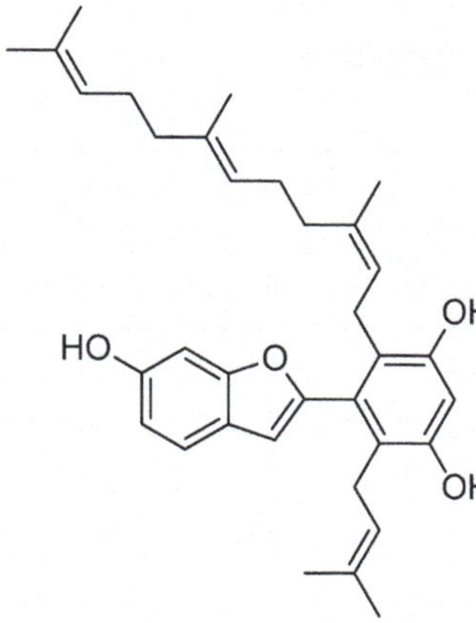

Morusalfuran B (**144**)
(IC_{50} = 0.39 μM)

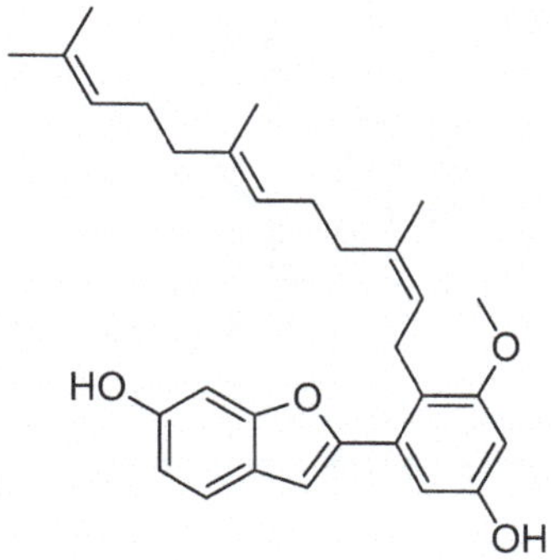

Morusalfuran C (**145**)
(IC_{50} = 0.57 μM)

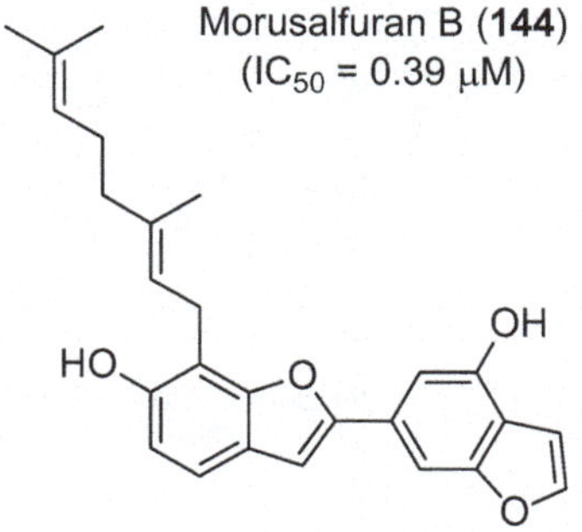

Morusalfuran F (**146**)
(IC_{50} = 0.48 μM)

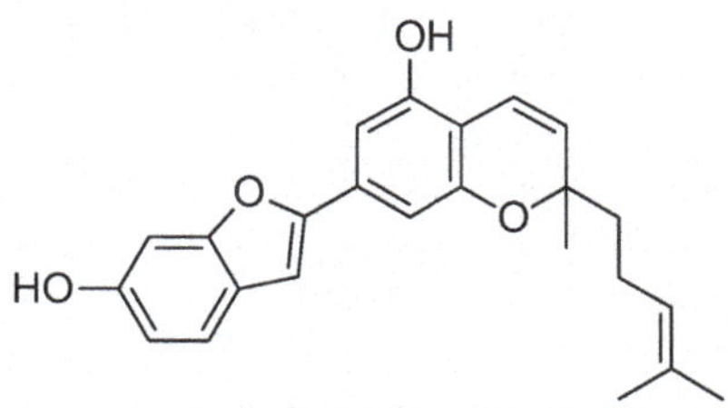

Morusalfuran G (**147**)
(IC_{50} = 0.76 μM)

Fig. 6.38 Prenylated benzofurans from the root barks of *M. alba* and their PL inhibitory activity

6.2.5 PL Inhibitors from Non-plant Sources

6.2.5.1 Monascus Pigments

Monascus pigments are a group of natural compounds, widely utilised as natural food colourants in East Asia and possess a range of biological activities including anti-mutagenic, antimicrobial activities and potential antiobesity characteristics. In two different studies conducted by Kim et al., over 50 *Monascus* pigments, derived with various amino acids, were extracted and evaluated for PL inhibitory activity (Kim et al. 2007a; Kim et al. 2007b). The L-Tyr ethyl ester derivative (158) was found to be the most active in the series, with a potential IC_{50} of 13.8 μM, alongside few other derivatives that exhibited moderate PL inhibitory activity (Fig. 6.41). Further, these pigments exhibited a non-competitive mode of inhibition on PL, as revealed through the enzyme kinetics.

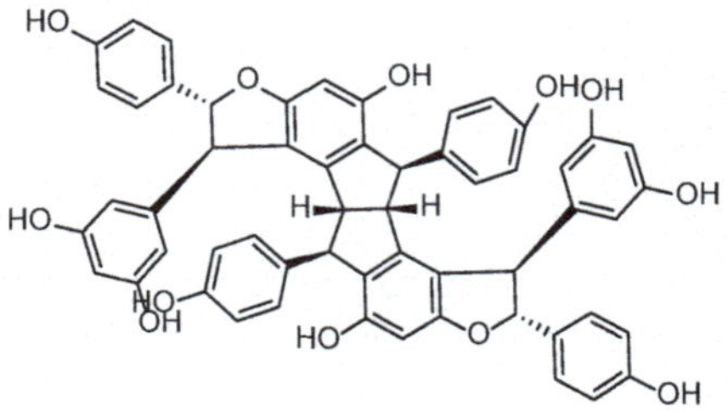

Ampelopsin H (**148**)
(IC_{50} = 9.3 μM)

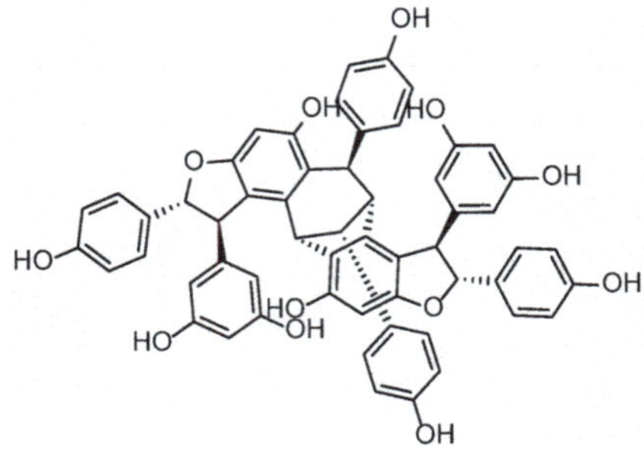

Vaticanol C (**149**)
(IC_{50} = 12.5 μM)

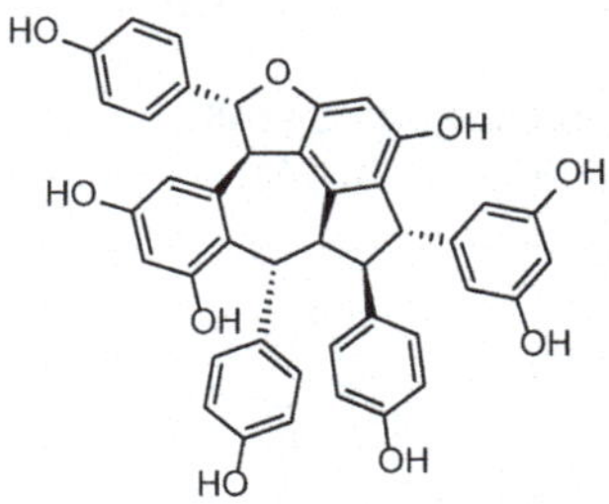

Vaticanol A (**150**)
(IC_{50} = 51.7 μM)

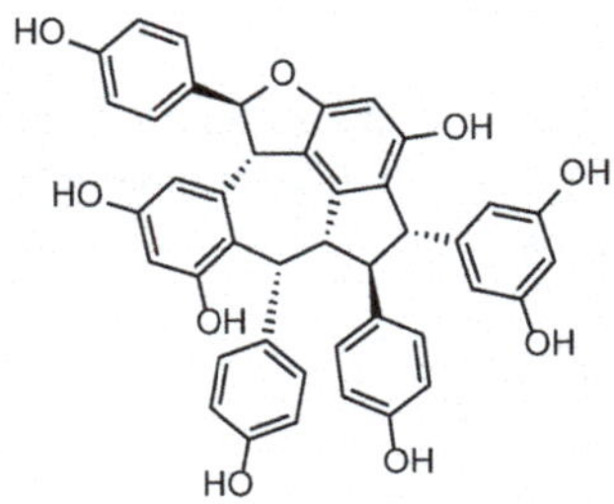

Vaticanol E (**151**)
(IC_{50} = 85.5 μM)

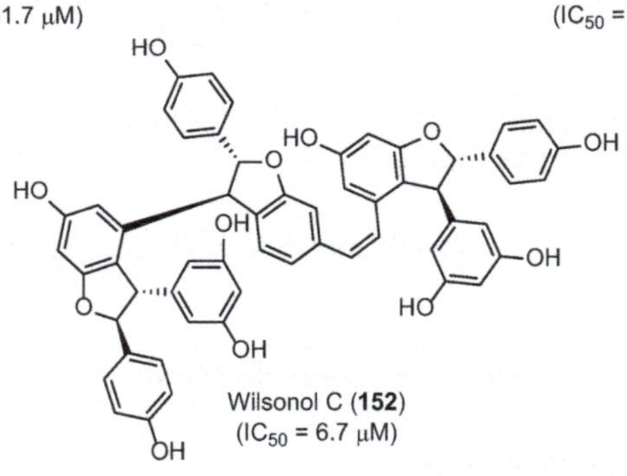

Wilsonol C (**152**)
(IC_{50} = 6.7 μM)

Fig. 6.39 Monomers, dimers and trimer of benzofurans and their PL inhibitory activity

6.2.5.2 Lactones and Their Derivatives

Lactones from fungi are the earliest class of molecules explored for their potential towards PL inhibition. Apart from the clinically approved drug, orlistat, various other fungal-derived lactones were evaluated since the 1980s. Among these, the major lactones include the ebelactones, panclicins and vibralactones obtained from various species of *Streptomyces* and *Boreostereum*. In a study conducted by Umezawa et al. (1980) two ebeleactones, A (160) and B (161), were isolated along with esterastin (162), from the MG7-Gl strain of actinomycetes (a strain closely related to

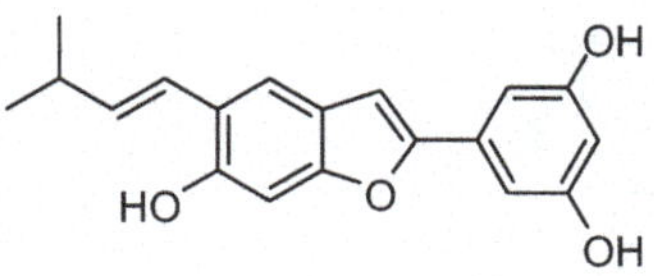

Artokinin (**153**)
(IC_{50} = 3.7 μM)

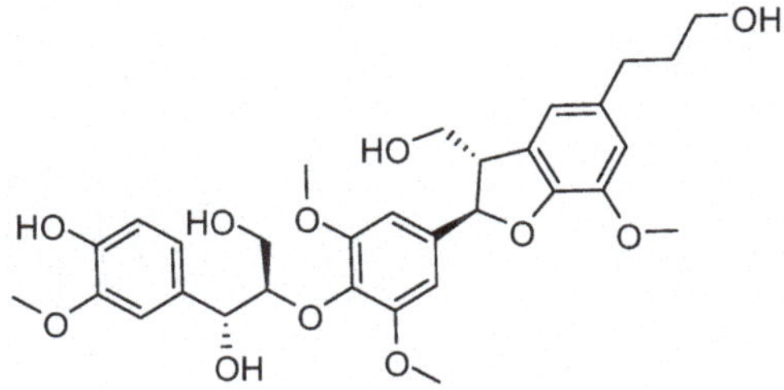

Acernikol (**154**)
(IC_{50} = ND)

Fig. 6.40 Benzofurans from *A. nitidus* and *F. rhynchophylla* (The IC_{50} of Acernikol was not determined (ND) but exhibited 40% inhibition at 170 μM)

Streptomyces aburaviensis). All the three lactones possessed a potent inhibitory activity against PL (Fig. 6.42). Esterastin (162), a closely related analog of orlistat, was the most active compound with an IC_{50} of 0.4 nM (Umezawa et al. 1980).

In another study, Mutoh et al. (1994) isolated five structural analogs of orlistat, panclicins A–E (163–167), from *Streptomyces sp.* NR 0619 and evaluated its PL inhibitory potential. Panclicin C (165) was the most active in the series with an IC_{50} of 0.62 μM, followed by panclicins D and E (Fig. 6.43). Further, all these panclicins exhibited an irreversible inhibition of PL. In a study conducted by Chen et al. (2016) vibralactone and its oxime derivatives from *Boreostereum vibrans*, a basidiomycete, were evaluated for their PL inhibitory potential. The oximes exhibited a greater potential over vibralactone (168), however, did not possess potential PL inhibitory activity comparable to that of orlistat. Vibralactoxime K (169) was the most active in the series with an IC_{50} of 11.1 μM (Fig. 6.4).

6.2.5.3 Methyl Xestospongoate and Its Derivatives

Xestospongia testudinaria (Petrosiidae), commonly known as the Chinese marine sponge, is a sessile marine filter feeder, of the phylum *Porifera*. Gong et al. (2016) reported that the methyl xestospongoate (170) obtained from this sponge possessed potent PL inhibitory activity (IC_{50} = 3.1 μM). Further studies resulted in the isolation of various related analogs with potential PL inhibitory activities (Liang et al. 2014). The most active compound in the series was xestospongenyne (171), which possessed a potent IC_{50} of 0.61 μM (Fig. 6.45).

6.2.6 Miscellaneous

Apart from various classes of PL inhibitors detailed above, various other minor classes (viz. xanthones, phthalides, nucleotides) and few miscellaneous phytochemicals were also explored for their PL inhibitory potential and are discussed below.

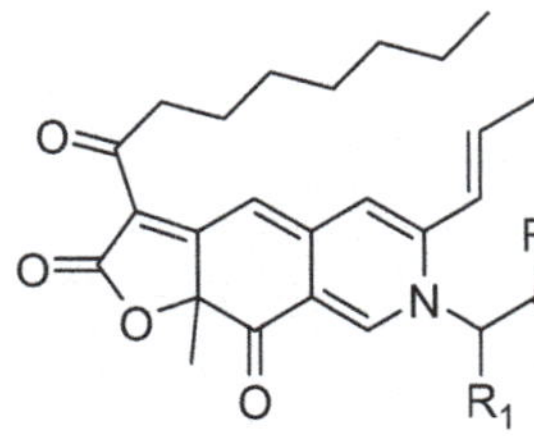

Compound	R_1	R_2	IC_{50} (μM)
H-Pen (**155**)		-OH	24.0
H-Cha (**156**)	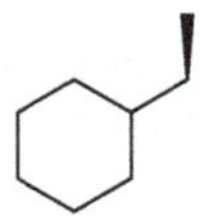	-OH	42.4
L-Phe-L-Phe (**157**)	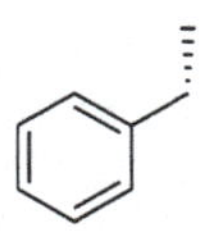	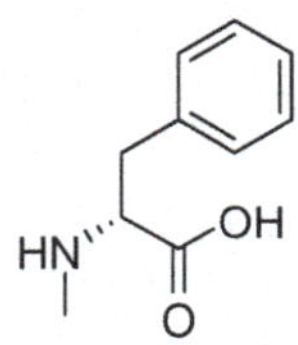	24.7
L-Tyr Ethyl ester (**158**)	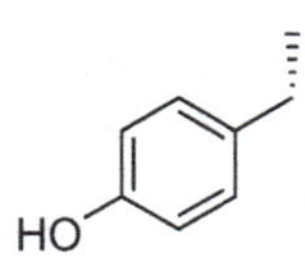	$-OC_2H_5$	13.8
L-Tyr-L-Tyr (**159**)	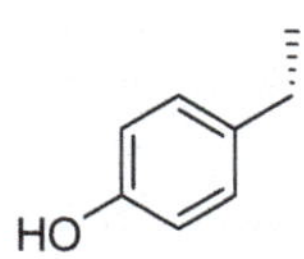	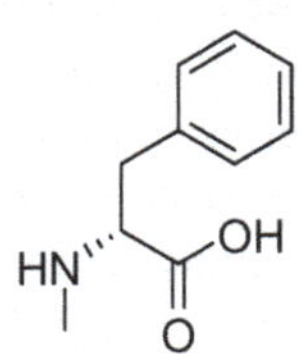	32.1

Fig. 6.41 Various *Monascus* pigments and their PL inhibitory activity

6.2.6.1 Xanthones

Xanthones, or more specifically *9H*-xanthen-9-ones, are a group of phytochemicals, which can be classified under the broad chemical class of oxygen heterocycles, alongside benzofurans and flavonoids. The only study that reported the PL inhibitory potential of xanthones was conducted by Chae et al. on the pericarps of *Garcinia mangostana* (Guttiferae) (Chae et al. 2016). α-Mangostin (172) possessed potent activity with IC_{50} of 5 μM, followed by γ-mangostin (173) and gartanin (174). Further, α-mangostin was found to exhibit a non-competitive nature of inhibition on PL (Table 6.8).

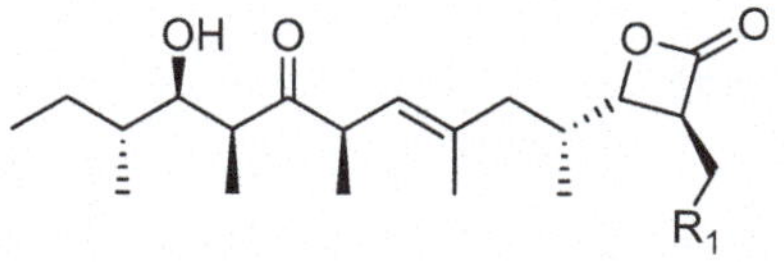

	R_1	IC_{50} (nM)
Ebelactone A (**160**)	H	8.9
Ebelactone B (**161**)	CH_3	2.3

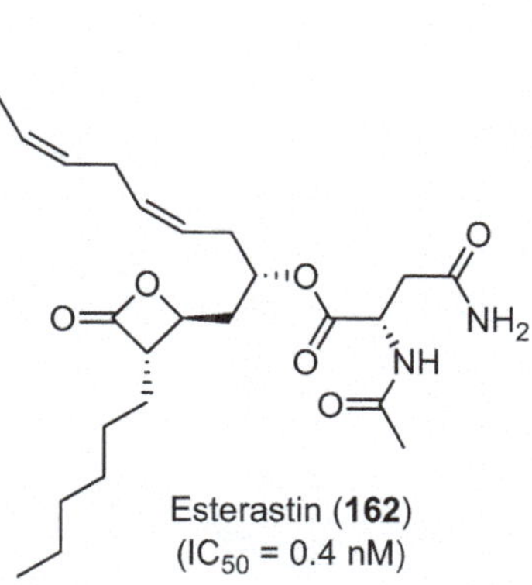

Esterastin (**162**)
(IC_{50} = 0.4 nM)

Fig. 6.42 PL inhibitory activities of ebelactones and esterastin

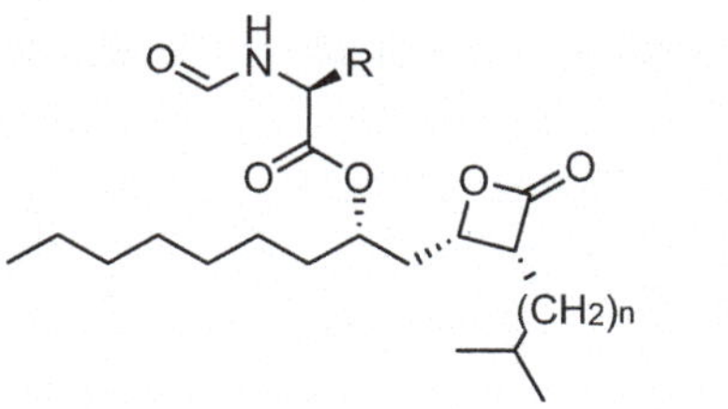

	R	n	IC_{50} (μM)
Panclicin A (**163**)	CH_3	7	2.9
Panclicin C (**165**)	H	7	0.62

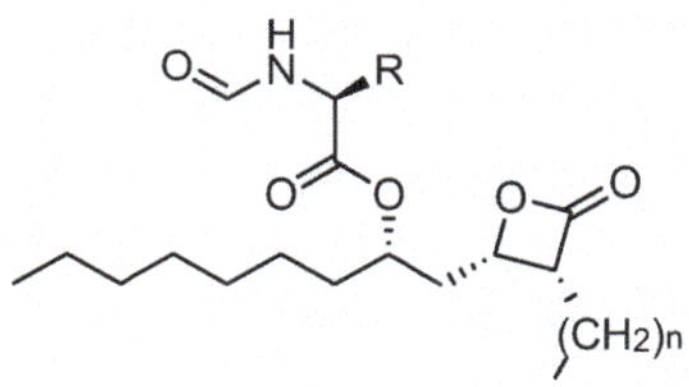

	R	n	IC_{50} (μM)
Panclicin B (**164**)	CH_3	9	2.6
Panclicin D (**166**)	H	9	0.66
Panclicin E (**167**)	H	11	0.89

Fig. 6.43 Panclicins and their PL inhibitory activities from *Streptomyces* sp. NR 0619

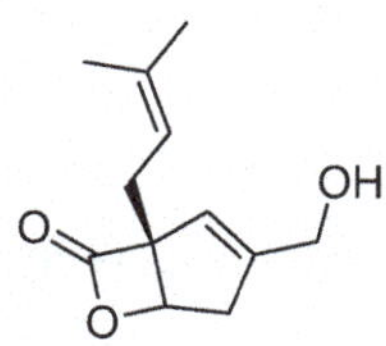

Vibralactone (**168**)
(IC_{50} = 48.7 μM)

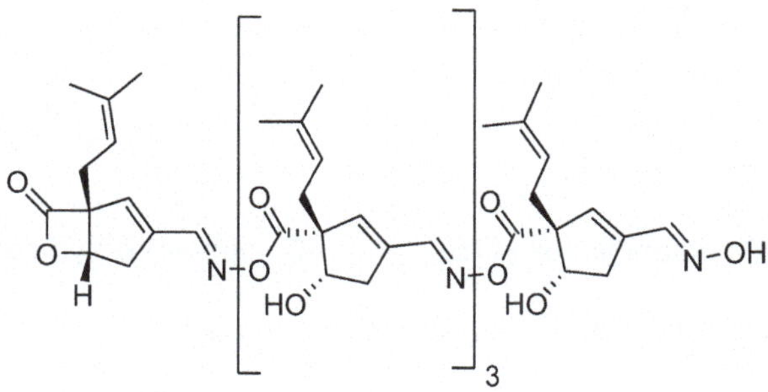

Vibralactoxime K (**169**)
(IC_{50} = 11.1 μM)

Fig. 6.44 Vibralactone (**168**) and its oxime derivative (**169**) from *B. vibrans* and their PL inhibitory activity

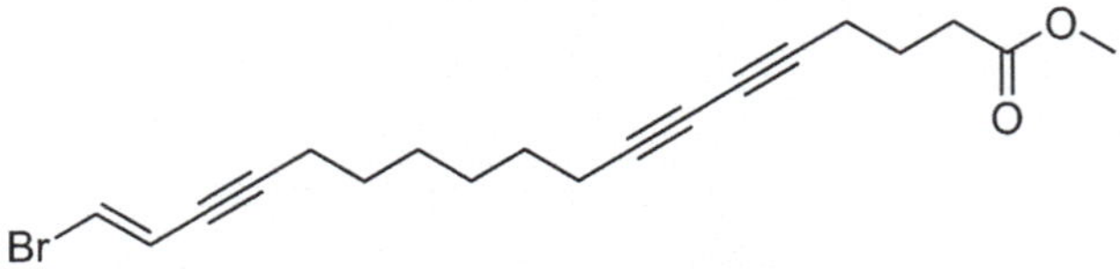

Methyl xestospongoate (**170**)

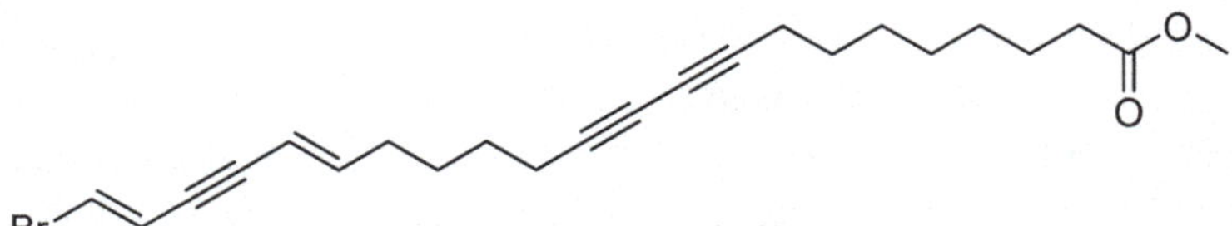

Xestospongenyne (**171**)

Fig. 6.45 Xestospongic acid derivatives from Chinese marine sponge

Table 6.8 Xanthones with potential PL inhibitory activity from the pericarps of *G. mangostana*

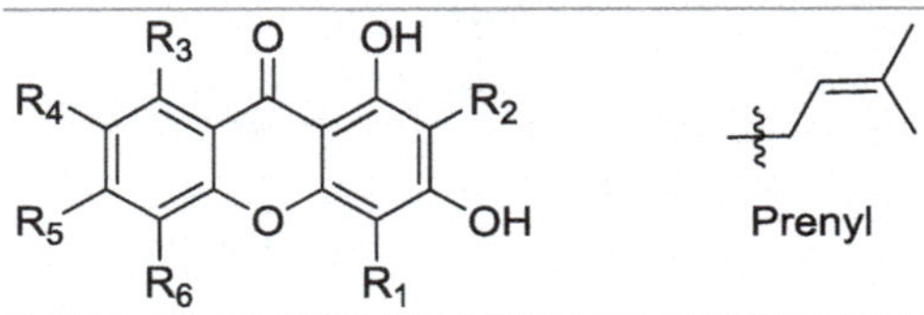

Compound	R_1	R_2	R_3	R_4	R_5	R_6	IC_{50} (μM)
α-Mangostin (**170**)	H	Prenyl	Prenyl	$-OCH_3$	-OH	-H	5
γ-Mangostin (**171**)	H	Prenyl	Prenyl	-OH	-OH	-H	10
Gartanin (**172**)	Prenyl	Prenyl	-OH	-H	-H	-OH	12

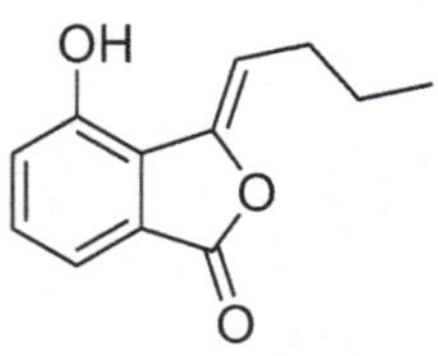

Senkyunolide B (**175**)

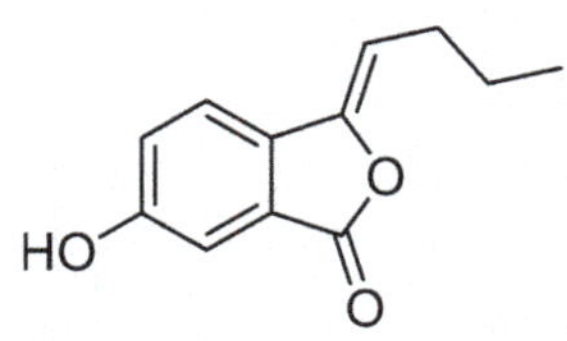

3-Butylidene-6-hydroxy-isobenzofuranone (**176**)

Fig. 6.46 Phthalide derivatives from *C. officinale*

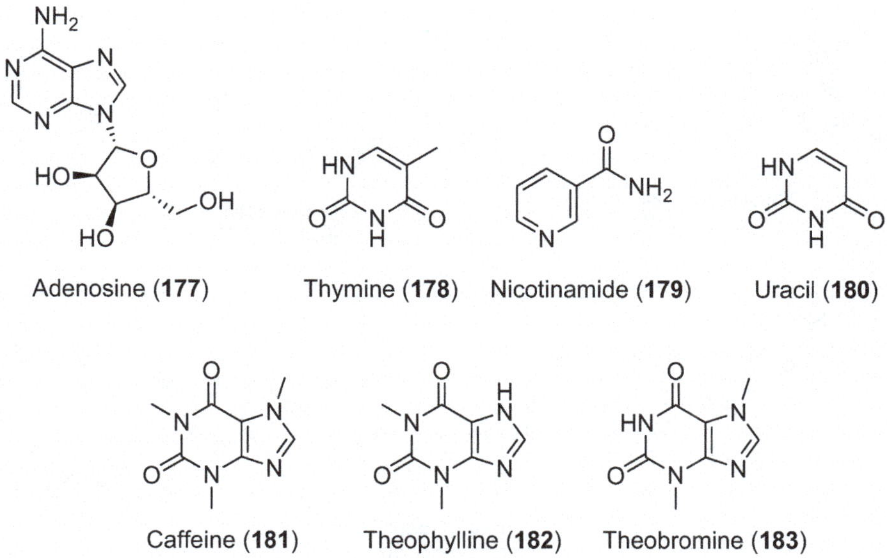

Fig. 6.47 Various nucleotide derivatives explored for their PL inhibitory activity

6.2.6.2 Phthalides

Phthalides are a group of phytochemicals that contain a five-membered oxygen heterocycle in the form of a lactone. In a study conducted by Mo et al. various phthalide derivatives from the rhizomes of *Cnidium officinale* (Umbelliferae) were evaluated for their PL inhibitory potential (Mo et al. 2016). Senkyunolide B (175) inhibited PL by 55% at 100 μM, followed by 3-butylidene-6-hydroxy-isobenzofuranone (176) that exhibited 40% PL inhibition at this concentration (Fig. 6.46).

6.2.6.3 Nucleotides

Various nucleotide derivatives including adenosine (177), thymine (178), nicotinamide (179) and uracil (180), isolated from *Cordyceps militaris* (Clavicipitaceae), were evaluated for PL inhibition assay; however, they did not possess activity (Fig. 6.47). Likewise, the methylxanthine analogs, viz. caffeine (181), theophylline (182) and theobromine (183), exhibited poor PL inhibition (Wikiera et al. 2012; Kim et al. 2014a). Apart from, various other natural products evaluated for PL inhibitory activity include cassiamin A (184) from *Cassia siamea* (Fabaceae) and crocetin (185) and crocin (186) from *Gardenia jasminoides* (Rubiaceae) (Bitou· et al. 1999; Lee et al. 2005; Kumar et al. 2013), which were reported to possess moderate to poor PL inhibitory activity (Fig. 6.48).

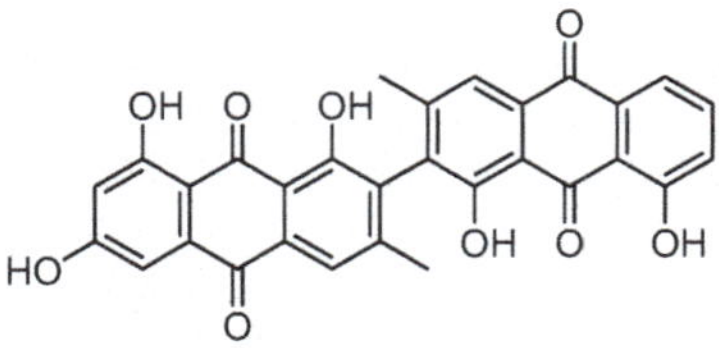

Cassiamin A (**184**)
(IC_{50} = 41.8 μM)

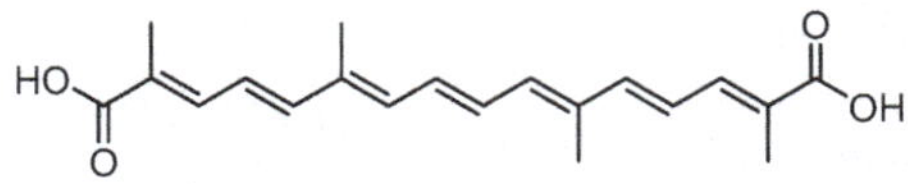

Crocetin (**185**)
(IC_{50} = 6.4 mM)

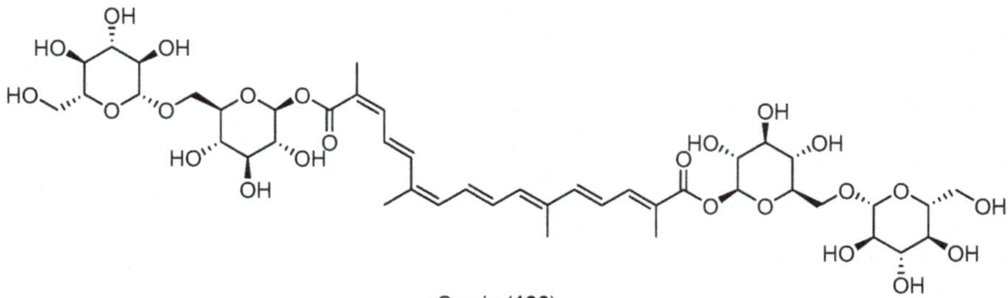

Crocin (**186**)
(IC_{50} = 2.76 mM)

Fig. 6.48 Miscellaneous phytochemicals and their PL inhibitory activity

6.3 Conclusion and Future Prospects

A significant number of natural products were evaluated for their PL inhibitory potential. However, very few number of natural products were analysed to understand their nature of inhibition and binding pattern, using enzyme kinetics and molecular modelling studies, respectively. Moreover, most of these natural products lack further preclinical and clinical studies that would determine their potential as antiobesity agents. These facts clearly highlight the necessity for further in silico and in vivo studies of these natural products, which would lead to more concrete structure-activity relationship for the development of better pharmacophores with better efficacies and lower adverse effects.

Acknowledgements The authors acknowledge the financial support received from DST-SERB (Grant. No. YSS/2014/000283) for the research work on pancreatic lipase inhibitors. Mr. S N C Sridhar thankfully acknowledges Birla Institute of Technology and Science, Pilani (BITS Pilani), Pilani campus and CSIR for providing fellowship (File No: 09/719(0088)/2018-EMR-I).

References

Ahn JH, Liu Q, Lee C, Ahn MJ, Yoo HS, Hwang BY, Lee MK (2012) A new pancreatic lipase inhibitor from *Broussonetia kanzinoki*. Bioorg Med Chem Lett 22:2760–2763

Ahn JH, Kim ES, Lee C, Kim S, Cho SH, Hwanga BY, Lee MK (2013a) Chemical constituents from *Nelumbo nucifera* leaves and their anti-obesity effects. Bioorg Med Chem Lett 23:3604–3608

Ahn JH, Shin E, Liu Q, Kim SB, Choi KM, Yoo HS, Hwang BY, Lee MK (2013b) Secoiridoids from the stem barks of *Fraxinus rhynchophylla* with pancreatic lipase inhibitory activity. Nat Prod Res 27:1132–1135

Al-Masri IM (2013) Pancreatic lipase inhibition by papaverine: investigation by simulated molecular docking and subsequent *in vitro* evaluation. Jordan J Pharm Sci 6:271–279

Bauer E, Jakob S, Mosenthin R (2005) Principles of physiology of lipid digestion. Asian Australas J Anim Sci 18:282–295

Birari RB, Bhutani KK (2007) Pancreatic lipase inhibitors from natural sources: unexplored potential. Drug Discov Today 12:879–889

Birari R, Roy SK, Singh A, Bhutani KK (2009) Pancreatic lipase inhibitory alkaloids of *Murraya koenigii* leaves. Nat Prod Commun 4:1089–1092

Birari RB, Gupta S, Mohan CG, Bhutani KK (2011) Antiobesity and lipid lowering effects of *Glycyrrhiza* chalcones: experimental and computational studies. Phytomedicine 18:795–801

Bitou N, Ninomiya M, Tsujita T, Okuda H (1999) Screening of lipase inhibitors from marine algae. Lipids 34:441–445

Bray GA, Kim KK, Wilding JPH (2017) Obesity: a chronic relapsing progressive disease process. A position statement of the world obesity federation. Obes Rev 18:715–723

Buchholz T, Melzig MF (2015) Polyphenolic compounds as pancreatic lipase inhibitors. Planta Med 81:771–783

Cang J, Wang C, Huo XK, Tian XG, Sun CP, Deng S, Zhang BJ, Zhang HL, Liu KX, Ma XC (2017) Sesquiterpenes and triterpenoids from the rhizomes of *Alisma orientalis* and their pancreatic lipase inhibitory activities. Phytochem Lett 19:83–88

Chae HS, Kim EY, Han L, Kim NR, Lam B, Paik JH, Yoon KD, Choi YH, Chin YW (2016) Xanthones with pancreatic lipase inhibitory activity from the pericarps of *Garcinia mangostana* L.(Guttiferae). Eur J Lipid Sci Technol 118:1416–1421

Chakrabarti R (2009) Pharmacotherapy of obesity: emerging drugs and targets. Expert Opin Ther Targets 13:195–207

Chapus C, Semeriva M, Bovier-Lapierre C, Desnuelle P (1976) Mechanism of pancreatic lipase action. 1. Interfacial activation of pancreatic lipase. Biochemistry 15:4980–4987

Chen HP, Zhao ZZ, Li ZH, Dong ZJ, Wei K, Bai X, Zhang L, Wen CN, Feng T, Liu JK (2016) Novel natural oximes and oxime esters with a vibralactone backbone from the basidiomycete *Boreostereum vibrans*. ChemistryOpen 5:142–149

Colon-Gonzalez F, Kim GW, Lin JE, Valentino MA, Waldman SA (2013) Obesity pharmacotherapy: what is next. Mol Asp Med 34:71–83

Cragg GM, Newman DJ (2013) Natural products: a continuing source of novel drug leads. Biochim Biophys Acta Gen Subj 1830:3670–3695

Davis RC, Diep A, Hunziker W, Klisak I, Mohandas T, Schotz MC, Sparkes RS, Lusis AJ (1991) Assignment of human pancreatic lipase gene (PNLIP) to chromosome 10q24-q26. Genomics 11:1164–1166

de la Garza Hernández AL, Milagro FI, Boque N, Campión J, Martínez JA (2011) Natural inhibitors of pancreatic lipase as new players in obesity treatment. Planta Med 77:773–785

Egloff MP, Marguet F, Buono G, Verger R, Cambillau C, van Tilbeurgh H (1995) The 2.46 Å. Resolution structure of the pancreatic lipase-colipase complex inhibited by a C11 alkyl phosphonate. Biochemistry 34:2751–2762

Fani L, Bak S, Delhanty P, van Rossum EF, van den Akker EL (2014) The melanocortin-4 receptor as target for obesity treatment: a systematic review of emerging pharmacological therapeutic options. Int J Obes 38:163–169

Glisan SL, Grove KA, Yennawar NH, Lambert JD (2017) Inhibition of pancreatic lipase by black tea theaflavins: comparative enzymology and *in silico* modeling studies. Food Chem 216:296–300

Gong JX, Wang HY, Yao LG, Li XW, Guo YW (2016) First total synthesis of the marine natural brominated polyunsaturated lipid xestospongenyne as a potent pancreatic lipase inhibitory agent. Synlett 27:391–394

Grundy SM (1998) Multifactorial causation of obesity: implications for prevention. Am J Clin Nutr 67:S563–S572
Gu Y, Hurst WJ, Stuart DA, Lambert JD (2011) Inhibition of key digestive enzymes by cocoa extracts and procyanidins. J Agric Food Chem 59:5305–5311
Guo Y, Wu G, Su X, Yang H, Zhang J (2009) Antiobesity action of a daidzein derivative on male obese mice induced by a high-fat diet. Nutr Res 29:656–663
Ha MT, Tran MH, Ah KJ, Jo KJ, Kim J, Kim WD, Cheon WJ, Woo MH, Ryu SH, Min BS (2016) Potential pancreatic lipase inhibitory activity of phenolic constituents from the root bark of *Morus alba* L. Bioorg Med Chem Lett 26:2788–2794
Han LK, Zheng YN, Yoshikawa M, Okuda H, Kimura Y (2005) Anti-obesity effects of chikusetsu-saponins isolated from *Panax japonicus* rhizomes. BMC Complement Altern Med 5:9
Haslam D (2016) Weight management in obesity-past and present. Int J Clin Pract 70:206–217
Hatano T, Yamashita A, Hashimoto T, Ito H, Kubo N, Yoshiyama M, Shimura S, Itoh Y, Okuda T, Yoshida T (1997) Flavan dimers with lipase inhibitory activity from *Cassia nomame*. Phytochemistry 46:893–900
Heck AM, Yanovski JA, Calis KA (2000) Orlistat, a new lipase inhibitor for the management of obesity. Pharmacother J Hum Pharmacol Drug Ther 20:270–279
Ivanov SA, Nomura K, Malfanov IL, Sklyar IV, Ptitsyn LR (2011) Isolation of a novel catechin from *Bergenia* rhizomes that has pronounced lipase-inhibiting and antioxidative properties. Fitoterapia 82:212–218
Jang DS, Lee GY, Kim J, Lee YM, Kim JM, Kim YS, Kim JS (2008) A new pancreatic lipase inhibitor isolated from the roots of *Actinidia arguta*. Arch Pharm Res 31:666–670
Jeong JY, Jo YH, Kim SB, Liu Q, Lee JW, Mo EJ, Lee KY, Hwang BY, Lee MK (2015) Pancreatic lipase inhibitory constituents from *Morus alba* leaves and optimization for extraction conditions. Bioorg Med Chem Lett 25:2269–2274
Jiang W, Li W, Han L, Liu L, Zhang Q, Zhang S, Nikaido T, Koike K (2006) Biologically active triterpenoid saponins from *Acanthopanax senticosus*. J Nat Prod 69:1577–1581
Jo YH, Kim SB, Ahn JH, Liu Q, Hwang BY, Lee MK (2013) Inhibitory activity of benzophenones from *Anemarrhena asphodeloides* on pancreatic lipase. Nat Prod Commun 8:481–483
Jo YH, Kim SB, Liu Q, Lee JW, Hwang BY, Lee MK (2015) Benzylated and prenylated flavonoids from the root barks of *Cudrania tricuspidata* with pancreatic lipase inhibitory activity. Bioorg Med Chem Lett 25:3455–3457
Kadomatsu T, Tabata M, Oike Y (2011) Angiopoietin-like proteins: emerging targets for treatment of obesity and related metabolic diseases. FEBS J 278:559–564
Karamać M, Amarowicz R (1996) Inhibition of pancreatic lipase by phenolic acids-examination *in vitro*. Z Naturforsch C 51:903–906
Kawaguchi K, Mizuno T, Aida K, Uchino K (1997) Hesperidin as an inhibitor of lipases from porcine pancreas and *Pseudomonas*. Biosci Biotechnol Biochem 61:102–104
Khanam H, Shamsuzzaman (2015) Bioactive benzofuran derivatives: a review. Eur J Med Chem 97:483–504
Khaodhiar L, McCowen KC, Blackburn GL (1999) Obesity and its comorbid conditions. Clin Cornerstone 2:17–31
Kim JH, Kim HJ, Kim C, Jung H, Kim YO, Ju JY, Chin CS (2007a) Development of lipase inhibitors from various derivatives of monascus pigment produced by Monascus fermentation. Food Chem 101:357–364
Kim JH, Kim HJ, Park HW, Youn SH, Choi DY, Shin CS (2007b) Development of inhibitors against lipase and α-glucosidase from derivatives of monascus pigment. FEMS Microbiol Lett 276:93–98
Kim SB, Ahn B, Kim M, Ji HJ, Shin SK, Hong IP, Kim CY, Hwang BY, Lee MK (2014a) Effect of *Cordyceps militaris* extract and active constituents on metabolic parameters of obesity induced by high-fat diet in C58BL/6J mice. J Ethnopharmacol 151:478–484
Kim SB, Chang BY, Hwang BY, Kim SY, Lee MK (2014b) Pyrrole alkaloids from the fruits of *Morus alba*. Bioorg Med Chem Lett 24:5656–5659

Kim YM, Lee EW, Eom SH, Kim TH (2014c) Pancreatic lipase inhibitory stilbenoids from the roots of *Vitis vinifera*. Int J Food Sci Nutr 65:97–100

Kimple ME, Neuman JC, Linnemann AK, Casey PJ (2014) Inhibitory G proteins and their receptors: emerging therapeutic targets for obesity and diabetes. Exp Mol Med 46:e102

Kokkinou M, Theodorou LG, Papamichael EM (2012) Aspects on the catalysis of lipase from porcine pancreas (type VI-s) in aqueous media: development of ion-pairs. Braz Arch Biol Technol 55:231–236

Kumar D, Karmase A, Jagtap S, Shekhar R, Bhutani KK (2013) Pancreatic lipase inhibitory activity of cassiamin A, a bianthraquinone from *Cassia siamea*. Nat Prod Commun 8:195–198

Kwon CS, Sohn HY, Kim SH, Kim JH, Son KH, Lee JS, Lim JK, Kim JS (2003) Anti-obesity effect of *Dioscorea nipponica* Makino with lipase-inhibitory activity in rodents. Biosci Biotechnol Biochem 67:1451–1456

Kwon OJ, Bae JS, Lee HY, Hwang JY, Lee EW, Ito H, Kim TH (2013) Pancreatic lipase inhibitory gallotannins from *galla rhois* with inhibitory effects on adipocyte differentiation in 3T3-L1 cells. Molecules 18:10629–10638

Lee IA, Lee JH, Baek NI, Kim DH (2005) Antihyperlipidemic effect of crocin isolated from the fructus of *Gardenia jasminoides* and its metabolite crocetin. Biol Pharm Bull 28:2106–2110

Lee EM, Lee SS, Chung BY, Cho JY, Lee IC, Ahn SR, Jang SJ, Kim TH (2010) Pancreatic lipase inhibition by C-glycosidic flavones isolated from *Eremochloa ophiuroides*. Molecules 15:8251–8259

Li F, Li W, Fu H, Zhang Q, Koike K (2007) Pancreatic lipase-inhibiting triterpenoid saponins from fruits of *Acanthopanax senticosus*. Chem Pharm Bull 55:1087–1089

Li AN, Li S, Zhang YJ, Xu XR, Chen YM, Li HB (2014) Resources and biological activities of natural polyphenols. Nutrients 6:6020–6047

Liang LF, Wang T, Cai YS, He WF, Sun P, Li YF, Huang Q, Taglialatela-Scafati O, Wang HY, Guo YW (2014) Brominated polyunsaturated lipids from the Chinese sponge *Xestospongia testudinaria* as a new class of pancreatic lipase inhibitors. Eur J Med Chem 79:290–297

Liu W, Zheng Y, Han L, Wang H, Saito M, Ling M, Kimura Y, Feng Y (2008) Saponins (Ginsenosides) from stems and leaves of *Panax quinquefolium* prevented high-fat diet-induced obesity in mice. Phytomedicine 15:1140–1145

Lowe ME (2002) The triglyceride lipases of the pancreas. J Lipid Res 43:2007–2016

Lunagariya NA, Patel NK, Jagtap SC, Bhutani KK (2014) Inhibitors of pancreatic lipase: state of the art and clinical perspectives. EXCLI J 13:897–921

Manning S, Pucci A, Finer N (2014) Pharmacotherapy for obesity: novel agents and paradigms. Ther Adv Chronic Dis 5:135–148

Mauvais-Jarvis F (2011) Estrogen and androgen receptors: regulators of fuel homeostasis and emerging targets for diabetes and obesity. Trends Endocrinol Metab 22:24–33

Mo EJ, Yang HJ, Jeong JY, Kim SB, Liu Q, Hwang BY, Lee MK (2016) Pancreatic lipase inhibitory phthalide derivatives from the rhizome of *Cnidium officinale*. Rec Nat Prod 10:148–153

Mohammad M, Al-masri IM, Issa A, Khdair A, Bustanji Y (2013) Inhibition of pancreatic lipase by berberine and dihydroberberine: an investigation by docking simulation and experimental validation. Med Chem Res 22:2273–2278

Morikawa T, Xie Y, Asao Y, Okamoto M, Yamashita C, Muraoka O, Matsuda H, Pongpiriyadacha Y, Yuan D, Yoshikawa M (2009) Oleanane-type triterpene oligoglycosides with pancreatic lipase inhibitory activity from the pericarps of *Sapindus rarak*. Phytochemistry 70:1166–1172

Morikawa T, Li X, Nishida E, Nakamura S, Ninomiya K, Matsuda H, Oda Y, Muraoka O, Yoshikawa M (2010) Medicinal flowers. Part 29. Acylated oleanane-type triterpene bisdesmosides: perennisaponins G, H, I, J, K, L, and M with pancreatic lipase inhibitory activity from the flowers of *Bellis perennis*. Helv Chim Acta 93:573–586

Morikawa T, Chaipech S, Matsuda H, Hamao M, Umeda Y, Sato H, Tamura H, Ninomiya K, Yoshikawa M, Pongpiriyadacha Y, Hayakawa T, Muraoka O (2012) Anti-hyperlipidemic constituents from the bark of *Shorea roxburghii*. J Nat Med 66:516–524

Mutoh M, Nakada N, Matsukuma S, Ohshima S, Yoshinari K, Watanabe J, Arisawa M (1994) Panclicins, novel pancreatic lipase inhibitors. J Antibiot 47:1369–1375

Nakai M, Fukui Y, Asami S, Toyoda-Ono Y, Iwashita T, Shibata H, Mitsunaga T, Hashimoto F, Kiso Y (2005) Inhibitory effects of oolong tea polyphenols on pancreatic lipase *in vitro*. J Agric Food Chem 53:4593–4598

Narita Y, Iwai K, Fukunaga T, Nakagiri O (2012) Inhibitory activity of chlorogenic acids in decaffeinated green coffee beans against porcine pancreas lipase and effect of a decaffeinated green coffee bean extract on an emulsion of olive oil. Biosci Biotechnol Biochem 76:2329–2331

Palade GE, Siekevitz P, Caro LG (2008) Structure, chemistry and function of the pancreatic exocrine cell. In: de Reuck A, Cameron M (eds) Ciba foundation symposium-the exocrine pancreas: normal and abnormal functions. Wiley, Chichester, pp 23–55

Patel D (2015) Pharmacotherapy for the management of obesity. Metab Exp 64:1376–1385

Pelletier SW (1983) The nature and definition of an alkaloid. Alkaloids Chem Biol Perspect 1:1–31

Rahim ATMA, Takahashi Y, Yamaki K (2015) Mode of pancreatic lipase inhibition activity *in vitro* by some flavonoids and non-flavonoid polyphenols. Food Res Int 75:289–294

Roskoski R (2007) Michaelis-Menten kinetics. In: xPharm: the comprehensive pharmacology reference. Elsevier, Amsterdam, pp 1–10

Sergent T, Vanderstraeten J, Winand J, Beguin P, Schneider YJ (2012) Phenolic compounds and plant extracts as potential natural anti-obesity substances. Food Chem 135:68–73

Seyedan A, Alshawsh MA, Alshagga MA, Koosha S, Mohamed Z (2015) Medicinal plants and their inhibitory activities against pancreatic lipase: a review. Evidence-Based Compl Altern Med 2015:973143. https://doi.org/10.1155/2015/973143

Sparg S, Light ME, Van Staden J (2004) Biological activities and distribution of plant saponins. J Ethnopharmacol 94:219–243

Sridhar SNC, Ginson G, Venkataramana Reddy PO, Tantak MP, Kumar D, Paul AT (2017a) Synthesis, evaluation and molecular modelling studies of 2-(carbazol-3-yl)-2-oxoacetamide analogues as a new class of potential pancreatic lipase inhibitors. Bioorg Med Chem 25:609–620

Sridhar SNC, Mutya S, Paul AT (2017b) Bis-indole alkaloids from *Tabernaemontana divaricata* as potent pancreatic lipase inhibitors: molecular modelling studies and experimental validation. Med Chem Res 26:1268–1278

Srivastava T, Kumar Mishra S (2015) Novel function of polyphenols in human health: a review. Res J Res J Phytochem 9:116–126

Sugimoto S, Nakamura S, Yamamoto S, Yamashita C, Oda Y, Matsuda H, Yoshikawa M (2009) Brazilian natural medicines. III. Structures of triterpene oligoglycosides and lipase inhibitors from mate, leaves of *Ilex paraguariensis*. Chem Pharm Bull 57:257–261

Tsao R (2010) Chemistry and biochemistry of dietary polyphenols. Nutrients 2:1231–1246

Umezawa H, Aoyagi T, Uotani K, Hamada M, Takeuchi T, Takahashi S (1980) Ebelactone, an inhibitor of esterase, produced by *Actinomycetes*. J Antibiot 33:1594–1596

van Tilbeurgh H, Sarda L, Verger R, Cambillau C (1992) Structure of the pancreatic lipase-procolipase complex. Nature 359:159–162

van Tilbeurgh H, Egloff MP, Martinez C, Rugani N, Verger R, Cambillau C (1993) Interfacial activation of the lipase-procolipase complex by mixed micelles revealed by X-ray crystallography. Nature 362:814–820

Wang S, Sun Z, Dong S, Liu Y, Liu Y (2014) Molecular interactions between (−)-epigallocatechin gallate analogs and pancreatic lipase. PLoS One 9:e111143

WHO (2008) Waist circumference and waist-hip ratio: report of a WHO expert consultation. World Heal Organ pp 8–11

Wikiera A, Mika M, Żyła K (2012) Methylxanthine drugs are human pancreatic lipase inhibitors. Polish J Food Nutr Sci 62:109–113

Wu X, He W, Yao L, Zhang H, Liu Z, Wang W, Ye Y, Cao J (2013) Characterization of binding interactions of (−)-epigallocatechin-3-gallate from green tea and lipase. J Agric Food Chem 61:8829–8835

Xu BJ, Han LK, Zheng YN, Lee JH, Sung CK (2005) *In vitro* inhibitory effect of triterpenoidal saponins from *Platycodi Radix* on pancreatic lipase. Arch Pharm Res 28:180–185

Yang MH, Chin YW, Yoon KD, Kim J (2014) Phenolic compounds with pancreatic lipase inhibitory activity from Korean yam (*Dioscorea opposita*). J Enzyme Inhib Med Chem 29:1–6

Yoshikawa M, Sugimoto S, Kato Y, Nakamura S, Wang T, Yamashita C, Matsuda H (2009) Acylated oleanane-type triterpene saponins with acceleration of gastrointestinal transit and inhibitory effect on pancreatic lipase from flower buds of Chinese tea plant (*Camellia sinensis*). Chem Biodivers 6:903–915

You Q, Chen F, Wang X, Luo PG, Jiang Y (2011) Inhibitory effects of muscadine anthocyanins on α-glucosidase and pancreatic lipase activities. J Agric Food Chem 59:9506–9511

Yu MH, Zhao T, Yan GR, Yang HX, Wang HY, Hou AJ (2012) New isoprenylated flavones and stilbene derivative from *Artocarpus hypargyreus*. Chem Biodivers 9:394–402

Yuda N, Tanaka M, Suzuki M, Asano Y, Ochi H, Iwatsuki K (2012) Polyphenols extracted from black tea (*Camellia sinensis*) residue by hot-compressed water and their inhibitory effect on pancreatic lipase *in vitro*. J Food Sci 77:H254–H261

Zhao HL, Sim JS, Shim SH, Ha YW, Kang SS, Kim YS (2005) Antiobese and hypolipidemic effects of platycodin saponins in diet-induced obese rats: evidences for lipase inhibition and calorie intake restriction. Int J Obes 29:983

Zhao T, Yan GR, Pan SL, Wang HY, Hou AJ (2009) New isoprenylated 2-arylbenzofurans and pancreatic lipase inhibitory constituents from *Artocarpus nitidus*. Chem Biodivers 6:2209–2216

Zheng Q, Koike K, Han LK, Okuda H, Nikaido T (2004) New biologically active triterpenoid saponins from *Scabiosa tschiliensis*. J Nat Prod 67:604–613

Zheng Q, Li W, Han L, Koike K (2007) Pancreatic lipase-inhibiting triterpenoid saponins from *Gypsophila oldhamiana*. Chem Pharm Bull 55:646–650

Natural Compounds Extracted from Medicinal Plants and Their Applications

7

Saboon, Sunbal Khalil Chaudhari, Sohaib Arshad, Muhammad Shoaib Amjad, and Mohd Sayeed Akhtar

Contents

Abstract

Plant natural products have played an important role in the lives of human beings for their use as a source of food and medicine. The medicinal properties in plants typically result from the different combinations of these natural compounds known as phytochemicals. Generally, these phytochemicals are classified into primary and secondary compounds. Primary compounds include chlorophyll,

Sunbal Khalil Chaudhari and Mohd Sayeed Akhtar have equally contributed for this chapter.

Saboon
Department of Botany, PMAS Arid Agriculture University, Rawalpindi, Pakistan

S. K. Chaudhari (✉)
Department of Botany, The University of Lahore, Sargodha, Pakistan
e-mail: sunbal.khalil@imbb.uol.edu.pk; khalilsunbal@yahoo.com

S. Arshad
Department of Plant Sciences, Quaid-i-Azam University, Islamabad, Pakistan

M. S. Amjad
Department of Botany, Women Univeristy of Azad Jammu & Kashmir, Bagh, Pakistan

M. S. Akhtar
Department of Botany, Gandhi Faiz-e-Aam College, Shahjahanpur, Uttar Pradesh, India
e-mail: sayeedbot@gmail.com

M. S. Akhtar et al. (eds.), *Natural Bio-active Compounds*,
https://doi.org/10.1007/978-981-13-7154-7_7

proteins, and sugars, while secondary compounds include terpenoids, alkaloids, flavonoids, and phenolic. Many fruits, vegetables, and herbs contain a great variety of phytochemical such as phenolic compounds (phenolic acids, flavonoids, quinones, coumarins, lignans, stilbenes, tannins), nitrogen compounds (alkaloids, amines, betalains), vitamins (A, C, D, E), and terpenes (including carotenoids). Different phytochemicals have different pharmacological activities, i.e., terpenoids have antibacterial, anti-inflammatory, anticancer, antimalarial, and antiviral activities. Alkaloids mostly have anesthetics properties. Phenolic compounds play important role in neutralizing free radicals. Flavonoids, one of the large groups of phenolic compounds, have various clinical properties such as anti-atherosclerotic, anti-inflammatory, antitumor, and antiviral. Due to this the phytochemicals are considered as a rich source of natural antioxidants and achieve an appreciable role in the development of modern drug for diseases, i.e., tumor, hepatic diseases, and arthritis. Apart from medicine, these natural compounds are also used as flavoring agents, fragrances, and functional additives by the cosmetic and pharmaceutical industries. Many of these well-known and traditionally used natural compounds extracted from tea, wine, fruit, vegetables, and spices are already being introduced commercially, both as medicine and additives in food supplements. Thus, natural compounds become an alternate health-care system to resolve the health problems of the world in today's era. Therefore, the aim of this chapter is to provide an overview on the various types of phytochemicals and their medicinal importance, which may be helpful for the researchers to design new drugs against different disorders.

Keywords

Alkaloids · Bio-active compounds · Flavonoids · Polyphenols · Medicinal plants

7.1 Introduction

Natural products from plants have been documented to possess diversified health benefits for thousands of years through traditional uses, such as Babylon around 2000 BC which gives instructions for preparation and administration of the medicinal herbs and Greek herbalist which listed 500 plants with medicinal properties. Romans also contributed to accumulative knowledge of medicinal plants by introducing more than 200 plants to Britain. In fact today modern scientific investigation increases the awareness for the use of natural products for health care. In the beginning of the twenty-first century, scientists have brought an interesting trend in pharmaceutical development: return to nature as a source of potential drugs (Georgiev 2013; Lanzotti 2014). Through this effort various phytochemicals, such as flavonoids, alkaloids, phenolic acids, anthocyanins, lignans, stilbenes, polysaccharides, carotenoids, essential oils, and certain vitamins (A, C, and E), have received an increased attention due to their important role in preventing and managing of dangerous diseases such as cancers, diabetes, Alzheimer's diseases, and cardiovascular diseases (Andrae-Marobela et al. 2013; Shao and Xiao 2013; Xiao et al. 2014; Xiao 2015; Xiao and Jiang 2015).

There are more than thousand known and many unknown phytochemicals. According to an estimate, about 4000 phytochemicals have been cataloged, and about 150 have been studied in detail (Meagher and Thomson 1999). These phytochemicals are present in different parts of the plants, i.e., leaves, roots, stem, flower, seed, and fruits (Costa et al. 1999). These phytochemicals provide a natural defense mechanism to plants against pathogens and environmental hazards, i.e., stress and pollution (Gibson et al. 1998). Although phytochemicals are not essential nutrients in plants, in scientific investigation most of these phytochemicals showed good biological activities such as antioxidant activity, antimicrobial activity, modulation of detoxification enzymes, stimulation of the immune system, decrease of platelet aggregation, and modulation of hormone metabolism and antitumor property. Because of these properties, many researchers have performed experiments to isolate and disclose the beneficial health effects of these natural compounds which resulted into the base of modern drugs we use today. Large numbers of phytochemicals are isolated from herbs, spices, fruits, vegetables, legumes, whole grains, nuts, and fungi. Therefore, the aim of this chapter is to provide an overview on the various types of phytochemicals, and their medicinal importance, which may be helpful for the researchers to design new drugs against different disorders.

7.2 Narrative of Different Natural Bio-active Compounds

The different types of natural bio-active compounds found in the plants are as follows:

7.2.1 Alkaloids

The term alkaloid was first coined by French scientist Carl Friedrich Wilhelm Meibner in 1819. The word "alkaloids" derives from "alkaline" which is used to describe any nitrogen-containing base (pH >7) (Mueller-Harvey and McAllan 1992). So, alkaloids are defined as the group of naturally occurring plant product that contains mostly basic nitrogen atoms. They have heterocyclic ring structure with nitrogen in a negative oxidation state (Pelletier 1983). This group also includes some related compounds with neutral (McNaught and Wilkinson 1997) and even weakly acidic properties (Manske 1965) due to presence of other elements like O, H, S, etc. The first medically useful alkaloid was morphine, isolated from plant *Papaver somniferum* in 1805 by German chemist Friedrich Sertürner (Heinrich 2013). Alkaloids are most diverse, effective, and medicinally important plant substances, bitter in taste, optically active, colorless, and crystalline or liquid at room temperature. Around 12,000 known structures of alkaloids represent one of the biggest groups of natural products. It was found that 20% of plant species contain alkaloids, but the major source of alkaloid is flowering plants.

Generally, alkaloids are extremely toxic because their main role is to ensure plant survival against microorganisms, insects, and herbivores. They protect plant by means of allelopathically active chemicals and marked therapeutic potential (Molyneux et al. 1996). Alkaloids extracted from plant have been used as ingredients in liquid medicine and poisons. These may be used for treating large number of ailments including snakebite, fever, and insanity. Pure form of alkaloids and their derivatives are used as medicinal agents all over the world due to their analgesic, antiseptic, antispasmodic, antimicrobial, and cardiovascular activities (Stary 1996). In humans, alkaloid mostly affects the nervous system, particularly the action of neurotransmitters like acetylcholine, epinephrine, norepinephrine, gamma-aminobutyric acid, dopamine, and serotonin. For instance, berberine is used in ophthalmics and sanguinarine in toothpastes an antiseptic (Cordell 1983).

7.2.1.1 Classification of Alkaloids

Many researchers have proposed different classifications for alkaloids on the basis of their biosynthetic precursor and heterocyclic ring system. One of the popular classifications that divide the whole class of compounds into three categories is as follows (Eagleson 1994):

- True alkaloids are the compounds which derive from amino acid and a heterocyclic ring with nitrogen, e.G., atropine, nicotine, etc.
- Proto-alkaloids are the compounds which contain nitrogen atom derived from an amino acid which is not a part of the heterocyclic ring, e.G., adrenaline, ephedrine, etc.
- Pseudo-alkaloids are the compounds that do not originate from amino acids, e.g., caffeine, theobromine, etc.

Due to vast diversity in structure, the alkaloids are further divided into different classes which include piperidine, indole, purine, imidazole, tropane, quinolizidine, isoquinoline, benzylisoquinoline, pyrrolizidine, and pyrrolidine. Indole alkaloids are characterized by the presence of serotonin with known 2000 compounds. The most explored pharmacological compounds are vincamine, vincristine, vinblastine, strychnine, ajmalicine, and ajmaline (Kainsa et al. 2012). Similarly, tropane alkaloids are derived from the amino acid ornithine. The important member of this class includes scopolamine, hyoscyamine, cocaine, and atropine having many valuable medicinal exercises (Ziegler and Facchini 2008).

Quinoline and isoquinoline are another important heterocyclic class of alkaloids formed by fusion of the benzene ring to the pyridine ring. Quinine is one of the important members of this class, used as medicine against malaria parasite *Plasmodium*. Other important members are camptothecin, echinopsine, homocamptothecin, chinidin, cinchonidine, folipdine, and dihydroquinine (Marella et al. 2013). On the other hand, isoquinoline alkaloids are the structural isomer of quinoline alkaloids. Many important alkaloids like narcotines, protopines, morphine, codeine, and thebaine belong to this class. These alkaloids have the potential to act as analgesic and narcotic drug, antitussives (codeine), and muscle relaxant and have antitumor properties associated with papaverine and noscapine, respectively. The alkaloid sanguinarine from this group has antimicrobial activity (Frick et al. 2005).

These classes of alkaloid are also known to exhibit biological activities like antihyperglycemic, antitumor, and antibacterial activity (Nassiri 2013). However, purine alkaloids are obtained from purine, and the important members of this class are caffeine, theobromine, theophylline, and aminophylline. They possess many important biological properties such as antioxidant, anti-inflammatory, antidiabetic, and hyperlipidemia (Herman and Herman 2013; Li et al. 2013). Similarly, piperidine alkaloids occur widely in the plant as well as animal kingdom. It is highly studied and about 700 alkaloids of this structural type are known. They have saturated heterocyclic ring. These compounds are known for their toxicity, but apart from the toxicity, these compounds also possess important pharmacological properties such as bactericidal, antihistaminic, antitumor, central nervous system stimulant and depressant, herbicidal, insecticidal, and fungicidal properties (Singh et al. 2012). The best known examples of this class are coniine, lobeline, and cynapine.

Pyridine alkaloids are similar to piperidine alkaloids except that their heterocyclic nitrogen-containing nucleus is unsaturated. The important examples of pyridine alkaloids are anabasine, nicotine, anatabine, and epibatidine. These alkaloids exhibited strong antimicrobial properties (Machado et al. 2012). Likewise, imidazole alkaloids are derived from amino acid L-histidine. The important and well-studied member of this class is pilocarpine obtained from *Pilocarpus jaborandi.* This alkaloid is valuable in ophthalmic practices such as glaucoma (Cronemberger et al. 2012). However, the presence of pyrrolizidine alkaloids are structurally consist of two five membered rings which share a common nitrogen. Senecionine, heliotrine, and clivorine are the common member of pyrrolizidine alkaloids. It is used in the plant defense against herbivores, possesses hepatotoxic properties, and also is an important compound for the treatment of diseases like cancer and diabetes (Majik and Tilve 2012). However, pyrrolidine alkaloids are derived from amino acids ornithine and lysine with addition of acetate/malonate units. Important members of this class are putrescine, hygrine, and cuscohygrine. Researches on these compound showed that they possess remarkable antibacterial, antifungal, and antitubercular properties (Parmar et al. 2012).

7.2.1.2 Application of Alkaloids

Alkaloids are important for the protection and survival of plant against microorganisms, insects, and some other animals feeding on plants. It could be used in dyes, spices, drugs, poisons, etc. However, it also has many important pharmacological activities for human welfare. For instance, the quinine is obtained from *Cinchona officinalis.* It is used for the treatment of malaria for a long time. Certain other alkaloids, such as allocrytopine, columbamine, dehydroocoteine, jatrorrhizine, norcorydine, thalifendine, ushinsunine, and bisbenzylisoquinoline, are also used as antimalarial drug against *Plasmodium falciparum* (Wright et al. 2000). Alkaloid sorbicillactone A, isolated from *Penicillium chrysogenum*, and coscinamide alkaloids exhibited cytopathic effects against HIV-1 virus (Bokesch et al. 2000; Bringmann et al. 2003; Lohombo-Ekomba et al. 2004). Similarly, indole class of alkaloids from *Eudistoma olivaceum* showed good antiviral activity against HSV-1, HSV-2, and vaccinia virus (Gul and Hamann 2005). However, dragmacidin alkaloids from *Spongosorites* sp. were reported to inhibit feline leukemia virus (Wright et al. 1992), while aporphine alkaloids from *Magnolia grandiflora* are effective against herpes simplex and poliovirus type 1, respectively. Alkaloids from fresh ripen fruit of *Emblica officinalis*, bisbenzylisoquinoline alkaloids (cycleanine and cocsoline) isolated from *Delphinium* spp., and imidazole derivatives showed effective bacterial activity against gram-positive and gram-negative pathogenic bacteria (De Luca 2006; Rahman et al. 2009). Aporphine alkaloids isolated from *Pseuduvaria setosa* were known to display antituberculosis activity against *Mycobacterium tuberculosis* (Wirasathien et al. 2006). However, phenanthridine is isolated from *Chelidonium majus*. Similarly, quinoline alkaloids (skimmianine, kokisaginine, mescaline) and bisbenzylisoquinoline alkaloids were also reported to exhibit antifungal activity against the clinical drug-resistant yeast and *Leucoagaricus gongylophorus* (Biavatti et al. 2002; Lohombo-Ekomba et al. 2004; Meng et al. 2009). Likewise, pyrrolizidine alkaloids (senecionine) exhibited antitrypanosomal activity (Nibret et al. 2009), while diterpenoid alkaloids from *Delphinium* spp. possess antifeedant activity against the different insect species infectious to plants (Gonzalez-Coloma et al. 1998).

Antitumor actions of alkaloids are very well explored, and their products are available for the treatment of lethal tumors; the dimeric indoles, vincristine, and vinblastine isolated from *Catharanthus roseus* are most commonly used for the patients suffering from leukemia and Hodgkin's disease. They inhibit the leukemia by causing the depolymerization of protein which form mitotic spindle in cell division as a result of this hindrance in cell division occurs and tumor formation reduces (Tari et al. 1986). Some other alkaloids, i.e., camptothecin, isoquinoline, and aporphine alkaloids from the tubers of *Stephania pierrei*, benzylisoquinoline alkaloids from *Stephania* spp., *Cyclea* spp., pyrrolizidine alkaloids (senecionine) and *Berberis curare* also have similar properties to cure leukemia (Goto et al. 1996; Angerhofer et al. 1999; Nibret et al. 2009). In another study, Yui et al. (2001) reported lycorine and lycoricidinol alkaloids which inhibit the TNF-α production by inhibiting the protein synthesis or by altering the cysteine/methionine incorporation into the macrophages. In some cases lycorine and its synthetic derivative induce cell cycle arrest (Lamoral-Theys et al. 2010).

Alkaloids also possess good antioxidant properties to scavenge free radical; the important member having the ability to scavenge free radical include quinolone from *Oryza sativa* cv. *Heugjinmi*, norditerpene, beta-carboline alkaloids, pyrrole alkaloid, berberine, canadine, anonaine, and antioquine and most of indole alkaloid (Chung and Woo 2001; Kolak et al. 2006; Correche et al. 2008). Some alkaloids have stimulant property such as caffeine, yohimbine, and nicotine. Alkaloids also act as muscle relaxant such as D-tubocurarine which have the ability to obstruct the acetylcholine receptor; some other examples of muscle relaxant have aporphine alkaloid (Das et al. 1997; Sotnikova et al. 1997). Morphine one of the oldest and important alkaloids is used as analgesic (Rao et al. 1978). Most of the indole alkaloids are used as antihypertensive. The important alkaloids quinidine and spareien are used to treat heart diseases as antiarrhythmic. Tropane alkaloids from *Atropa belladonna* is a well-known anticholinergic alkaloid. Thus, the alkaloids help mankind against certain life-threatening disease such as tumor and cardiac diseases, but certain alkaloids have shown reverse effects such as asphyxia, paralysis, or in some extreme condition patient death. So they must be used with prescription in small amount (Schmeller and Wink 1998; Buckingham 2010).

7.2.2 Polyphenols

Polyphenols are one of the largest classes of phytochemicals and the most commonly distributed in the plant kingdom. More than 8000 polyphenolic compounds have been identified in various plant species. Their main function is to protect plants against pathogens and environmental stresses (Beckman 2000). In food, polyphenols contribute toward the quality of food by giving oxidative stability, astringency, color, odor, flavor, bitterness, and nutrition (Cheynier 2005). The astringency, bitterness, and color present in red wine are due to the contribution of phenolics (Lesschaeve and Noble 2005). In dietary food, phenolics also serve as important oxygen reservoirs and act as substrates for browning reactions. Initially, polyphenols have been considered as anti-nutrients by different nutritionists, due to presence of tannins, which have adverse effects on health because they decrease the energy by decreasing activities of digestive enzymes, amino acid and protein availabilities, and mineral uptake (Salunkhe et al. 1982). But modern techniques and research showed that the food rich in polyphenols offered some protection against lethal diseases such as cancers, cardiovascular diseases, diabetes, osteoporosis, and neurodegenerative diseases (Arts and Hollman 2005; Graf et al. 2005).

Plant phenolic compounds arise from a common intermediate, phenylalanine, or its close precursor, shikimic acid. Phenol is considered as the simplest unit of this group. Primarily polyphenols occur in conjugated forms, with one or more sugar units linked to hydroxyl (-OH) groups. So in simple words, we can say that polyphenols are hydroxyl containing class of chemical compounds. In which hydroxyl group is bonded directly to an aromatic hydrocarbon. In polyphenol there is variation in the bonding of sugar unit to aromatic hydrocarbon, the sugar may be directly linked to an aromatic carbon. They may occur in association with other compounds, like organic acids, amines, lipids, and other phenols (Kondratyuk and Pezzuto 2004). The different associations of phenol rings with other structural elements give different characteristics to phenol rings due to which there function varies. So polyphenols are divided into different classes, i.e., phenolic acids (hydroxybenzoic acids and hydroxycinnamic acids), flavonoids (flavonols, flavones, flavanols, flavanones, isoflavones, proanthocyanidins), stilbenes, tannins (hydrolyzable and condensed tannins), and lignans (Spencer et al. 2008). In these groups flavonoids are the largest group of plant phenols and the most explored class (Dai and Mumper 2010).

7.2.2.1 Classification of Polyphenols

7.2.2.1.1 Phenolic Acids

The phenolic acids are the compounds which generally have carboxylic acid as their functional group. Naturally occurring phenolic acids contain two distinct carbon structures: such as hydroxycinnamic acid (C6C3) and hydroxybenzoic acid (C6C1) structures. Basic skeleton of these two structures is same, but due to difference in the position of hydroxyl and methoxyl groups on the benzene rings, their characteristics vary. Phenolic acids hardly exist in free form except the processed food majority of phenolic acid linked through ester, ether, or acetal bonds either to structural components of the plant, larger polyphenols, or smaller organic molecules (e.g., glucose, quinic acid). These linkages give rise to a vast range of their derivatives. Hydroxybenzoic acids are found in the free form as well as combined into esters of glycosides. Some of them are constituents of hydrolysable tannins (gallotannins and ellagitannins) which are compounds containing a central core of glucose or another polyol (Dai and Mumper 2010); gallotannins are mostly present in mangoes and ellagitannins in red fruit.

7.2.2.1.2 Flavonoids

Flavonoids comprise the most abundant, diverse, and well-studied group of polyphenols. More than 4000 flavonoids have been recognized which occur in different plant parts. Flavonoids are hydroxylated phenolic substances; their basic structure consists of two aromatic rings bound together by three linear carbon atoms. Their central chain usually forms a closed pyran ring with one of the benzene rings. The hydroxyl functional group is attached with C6-C3 unit. Most of the flavonoids are glycones (with attached sugar), but small amount of flavonoids occur in the form of aglycones (without sugar). Due to the variation in the type of heterocycle and arrangement and number of hydroxyl groups, presence of double bond, and the extent of alkylation and glycosylation, flavonoids are divided into six subclasses: flavonols, flavones, flavanones, flavanols, anthocyanins, and isoflavones (Kahkonen et al. 1999).

7.2.2.1.2.1 Flavonols

Flavonols are the most abundant flavonoids in foods and commonly present in all higher plants. Main members of this group are kaempferol, myricetin, and quercetin (Manach et al. 2005). Among them the most abundant is quercetin which is also well studied for their biological role. Flavonols differ from other groups due to position of hydroxyl functional group which is present at C3 position. The six-membered rings present in flavonol are known as pyrone. The associated sugar present in flavonols is mostly glucose or rhamnose, but other sugars may also be involved. Their concentration varies in different types of fruits and vegetable depending on plant type, growth, light, degree of ripeness, season, food preparation, and processing (Aherne and O'Brien 2002).

7.2.2.1.2.2 Flavones

In plants flavones are less common than flavonols. They are characterized by the presence of unsaturated carbon (a double bond between C2 and C3) in the flavan skeleton. It is also a six-membered pyrone ring condensed with the benzene ring. The flavones consist chiefly of glycosides of luteolin and apigenin. The important edible sources of flavones identified are parsley and celery (Manach et al. 2005). Cereals such as millet and wheat contain C-glycosides of flavones (Graefe et al. 2001). Large quantities of hydrophobic flavonoids are known as polymethoxylated flavones. These are found in the skin of citrus fruit such as tangeretin, nobiletin, and sinensetin (Nielsen et al. 2003).

7.2.2.1.2.3 Flavanones

It is present in high concentration in citrus fruit, in moderate concentration in tomatoes, and in some aromatic plants. These are saturated carbon chain with hydroxyl group attached to C3 atom. The six-membered ring present in flavanones is a dihydro-derivative of the pyrone ring. Flavanones are generally glycosylated by a disaccharide at position seven which imparts a bitter taste to fruit or in some cases less flavor due to presence of rutinose. Due to hydrolysis of glycoside flavanones, the nonsugar components are formed such as naringenin in grapefruit, hesperetin in oranges, and eriodictyol in lemons.

7.2.2.1.2.4 Flavanols

Flavanols exist in both monomer and the polymer form. Catechin is a monomer and found in many types of fruit (Manach et al. 2005). Green tea is a rich source of monomer, but black tea contains few monomers, and on fermentation they condense into theaflavins (dimers) and thearubigins (polymers). Structurally like flavanones, the flavanols contain a saturated carbon chain with a hydroxyl group in the C3 atom. The other types of flavanols contain epicatechin, gallocatechin, epigallocatechin, and epigallocatechin gallate present in some fruits, seeds, and tea (Wittig et al. 2001).

7.2.2.1.2.5 Isoflavones

Isoflavones contain pyran ring, in which the phenyl group is usually substituted at the C2 position of the pyran ring. In isoflavonoids the substitution is at C3 position. In isoflavones the basic skeleton of flavonoids is modified by aryl migration. They may be present as aglycones or glycosides, depending on the preparation of soya. In the human diet, soya and its processed products are the main source of isoflavones. The isoflavone content of soya varies greatly by geographic zone, growing conditions, and processing.

7.2.2.1.2.6 Anthocyanins

Anthocyanins are water-soluble glycosides/acylglycosides. They are derivatives of 2-phenylbenzopyrylium or flavylium salts. They are positively charged at acidic pH due to this equilibrium establish which result in formation of flavylium cation. They present ubiquitously in higher plants and give color to fruits.

7.2.2.1.3 Stilbenes

It belongs to a non-flavonoid class of phenolic compounds. The basic skeleton of stilbenes contains 14 carbon (C6-C2-C6) atoms with two phenyl rings connected by a carbon methylene bridge. In which one ring carries two hydroxyl groups, while the other ring is substituted by hydroxyl and methoxyl groups in different position. They may occur in free form or glycosylated forms as dimeric, trimeric, and polymeric stilbenes. Stilbenes originate from the phenylpropanoid pathway same as flavonoids, but they have a different structure because of polyketide portion which undergoes a different type of cyclization including loss of one carbon by decarboxylation. The occurrence of stilbenes is low in human diet. But their production in plant can be changed by transformation of a single gene, stilbene synthase. One of the most studied stilbenes is trans-resveratrol (3,4′,5-trihydroxystilbene) found largely in grapes. Studies show that resveratrol has anticarcinogenic effects (Zhu et al. 2004).

7.2.2.1.4 Tannins

Tannin word was originally coined by Seguin to describe the substances present in plant, which are responsible for tanning leather. Tannin is a descriptive term use for a group of polymeric phenolic substances. It is water-soluble and forms reversible and irreversible complexes with proteins, alkaloids, nucleic acids, polysaccharides, and minerals (Schofield et al. 2001). Their molecular weight ranges from 500 to 3000. They are capable of tanning leather and precipitating gelatin from solution. They possess the property known as astringency, and they are found in almost every plant part. On the basis of structural differences, tannins are divided into two major groups: hydrolysable tannins and condensed tannins (Mc-Leod 1974).

Hydrolyzable tannins contain a central core of polyhydric alcohol (glucose) and hydroxyl groups which are esterified by gallic acid. On the bases of esterification, hydrolysable tannins are further divided into three types which are gallotannins, ellagitannins (hexahydroxydiphenic acid), and complex tannins. Gallotannins are all those tannins in which galloyl units or their derivatives are bound to other diverse polyol, catechin, or triterpenoid units. On hydrolysis with acid, base, or enzyme, they yield glucose and gallic acid, e.g., tannic acid (Chinese tannin), *Hamamelis* tannin, Turkish tannin, *Acer* tannin, and *Tara* tannin. Ellagitannins are also hydrolyzable tannins. They are characterized by the coupling of two galloyl units to each other without a glycosidically linked catechin unit. They undergo lactonization to produce ellagic acid, e.g., corilagin obtained from *Caesalpinia coriaria* and *Terminalia chebula*, chebulinic acid, and chebulagic acid. Complex tannins are those tannins in which the catechin unit is bound glycosidically to a gallotannin or an ellagitannin unit, e.g., acutissimin. Condensed tannins are the second class of tannins; they are structurally more complex than hydrolyzable tannins; their complete structures are not completely disclosed. They are mainly considered as the polymerized products of flavan-3-ols and flavan-3,4-diols or sometimes the mixture of both. They are formed by linkage of C4 of one catechin with C8 or C6 of the next monomeric catechin. They are widely distributed in fruits, vegetables, forage, plants, cocoa, red wine, and certain food grains, such as sorghum, finger millets, and legume.

7.2.2.1.5 Lignans

Lignans are composed of *p*-coumaryl alcohol, hydroxycinnamic alcohols, sinapyl alcohol, and coniferyl alcohol. Lignans comprise a whole class of compounds with a similar basic skeleton (C6-C3) and is mostly present in free form (Willför et al. 2006). They are formed by stereoselective coupling of two hydroxycinnamic alcohol units at the central atoms of their side chain. Lignans were observed in more than 70 plant families (Saleem et al. 2005). In addition, they occur in many plant foods like oil seeds, cereals, fruits, and vegetables. During metabolism lignin is converted into enterodiol and enterolactone by the intestinal microflora (Heinonen et al. 2001). In lignan the podophyllotoxins are strong cytotoxic; it has the same mechanism of action as that of indole alkaloids vinblastine and vincristine; it inhibits cell division by binding to tubulin and preventing its polymerization. Its other derivatives (etoposide, etopophos, and teniposide) are mild cytotoxic.

7.2.2.2 Application of Polyphenols

Phenolic acids are important because of their pharmacological activities such as antimicrobial, cytotoxicity, anti-inflammatory, and antitumor activities. In addition to these properties, the flavonoids act as powerful antioxidants scavenging free radicals to protect the human body from dangerous diseases, and this property is dependent on the attachment and number of hydroxyl group. Moreover, flavonoids also possess enzyme inhibition activity, antimicrobial activity, anti-allergic activity, estrogenic activity, and vascular activity (Atmani et al. 2009). However, resveratrol, a stilbene found in many food sources, is considered to be beneficial for health. The stilbene also showed reduction in heart diseases by inhibition of oxidation of LDL cholesterol and platelet aggregation. Resveratrol one of the important stilbenes increases longevity by activation of sirtuins, NAD^+-dependent protein deacetylases involved in aging, which respond to oxidative stress and are induced by a low-calorie diet. Resveratrol also increases the lifespans of baker yeasts and fruit flies (Wood et al. 2004). It also possesses an antioxidant and anti-infammatory and anti-tumorigenic activity (Surh et al. 1999) and also increases the proliferation of estrogen-dependent T47D breast cells. In some cases it acts as a phytoestrogen receptor (Gehm et al. 1997).

Similarly, tannins are used as astringents, diuretics, hemostatic, anti-inflammatory, antioxidant, and antiseptic and for tumors and diarrhea. Instead of health benefits, tannins are used in dye industry to make textile dyes and in production of ink. In food industry tannins are used to clarify wine, beer, and fruit juices and also used as nutritional antioxidant and also used as coagulants in rubber production (Gyamfi and Aniya 2002). Palavy and Priscilla (2006) reported their significances in lethal disease like AIDS and cancers. However, lignans are used as treatment for warts, but in large quantity they are too toxic for systemic application. They also possess some cytotoxic properties, which lead to the synthesis of their semisynthetic derivatives such as 4′-demethylpodophyllotoxin. It prevents the cancer by stabilization of topoisomerase-DNA complexes. Podophyllotoxins are one of the potent cytotoxic compounds, while etoposide and its prodrug teniposide are also used to treat small-cell lung cancer and testicular cancer with certain lymphomas in combination with other drugs. Teniposide is used as therapy for childhood acute lymphocytic leukemia, secoisolariciresinol, an alkaloid is used as phytoestrogens.

7.3 Conclusions and Future Prospects

The use of natural bio-active compounds for the cure of various diseases is now globally accepted and becoming popular as an alternative system of medicine. These natural bio-active compounds may serve as a potential candidate for the search and designing of new drugs against the different kinds of curable and lethal diseases. In future, more researches on the molecular modeling and novel drug designing are desired to find the exact cause and cure against the deadliest diseases.

References

Aherne SA, O'Brien NM (2002) Dietary flavonols: chemistry, food content, and metabolism. Nutrition 18:75–81

Andrae-Marobela K, Ghislain FW, Okatch H, Majinda R (2013) Polyphenols: a diverse class of multi-target anti-HIV-1 agents. Curr Drug Metab 7:392

Angerhofer CK, Guinaudeau H, Wongpanich V, Pezzuto JM, Cordell GA (1999) Antiplasmodial and cytotoxic activity of natural bisbenzylisoquinoline alkaloids. J Nat Prod 62:59–66

Arts ICW, Hollman PCH (2005) Polyphenols and disease risk in epidemiologic studies. Am J Clin Nutr 81:317–325

Atmani D, Nassima C, Dina A, Meriem B, Nadjet D, Hania B (2009) Flavonoids in human health: from structure to biological activity. Curr Nutr Food Sci 5:225–237

Beckman CH (2000) Phenolic-storing cells: keys to programmed cell death and periderm formation in wilt disease resistance and in general defence responses in plants. Physiol Mol Plant Pathol 57:101–110

Biavatti MW, Vieira PC, da Silva MFGF, Fernandes JB, Victor SR, Pagnocca FC, Albuquerque S, Caracelli I, Zukerman-Schpector J (2002) Biological activity of quinoline alkaloids from *Raulinoa echinata* and X-ray structure of flindersiamine. J Braz Chem Soc 13:66–70

Bokesch HR, Pannell LK, McKee TC, Boyd MR (2000) Coscinamides A, B and C, three new bis-indole alkaloids from the marine sponge *Cosinoderma* sp. Tetrahedron Lett 41:6305–6308

Bringmann G, Lang G, Muhlbacher J, Schaumann K, Steffens S, Rytik PG, Hentschel U, Morschhäuser J, Müller WE (2003) Sorbicillactone a: a structurally unprecedented bioactive novel-type alkaloid from a sponge-derived fungus. Prog Mol Subcell Biol 37:231–253

Buckingham J (2010) Dictionary of alkaloids, 1st edn. CRC Press, Boca Raton

Cheynier V (2005) Polyphenols in foods are more complex than often thought. Am J Clin Nutri 81:S223–S229

Chung HS, Woo WS (2001) A quinolone alkaloid with antioxidant activity from the aleurone layer of anthocyanin-pigmented rice. J Nat Prod 64:1579–1580

Cordell GA (1983) Introduction to alkaloids: a biogenic approach. Wiley-Interscience, New York, p 1055

Correche ER, Andujar SA, Kurdelas RR, Gomez Lechon MJ, Freile ML, Enriz RD (2008) Antioxidant and cytotoxic activities of canadine: biological effects and structural aspects. Bioorg Med Chem 16:3641–3651

Costa MA, Zia ZQ, Davin LB, Lewis NG (1999) Toward engineering the metabolic pathways of cancer-preventing lignans in cereal grains and other crops. In: Romeo JT (ed) Recent advances in phytochemistry, phytochemicals in human health protection, nutrition, and plant defense, vol 33. Kluwer Academic, New York, pp 67–87

Cronemberger S, Calixto N, Moraes MN, Castro ID, Lana PC, Loredo AF (2012) Efficiency of one drop of 2% pilocarpine to reverse the intraocular pressure peak at 6:00 A.M. in early glaucoma. Vision Pan. Am J Ophthalmol 11:14–16

Dai J, Mumper R (2010) Plant phenolics: extraction, analysis and their antioxidant and anticancer properties. Molecules 15:7313–7352

Das M, Vedasiromoni JR, Chauhan SPS, Ganguly DK (1997) Effect of green tea (*Camellia sinensis*) extract on the rat diaphragm. J Ethnopharmacol 57:197–201

De Luca L (2006) Naturally occurring and synthetic imidazoles: their chemistry and their biological activities. Curr Med Chem 13:1–23

Eagleson M (1994) Concise encyclopedia chemistry, 1st edn. Walter de Gruyter, Berlin

Frick S, Kramell R, Schmidt J, Fist AJ, Kutchan TM (2005) Comparative qualitative and quantitative determination of alkaloids in narcotic and condiment *Papaver somniferum*. J Nat Prod 68:666–673

Gehm BD, McAndrews JM, Pei YC, Jameson JL (1997) Resveratrol, a polyphenolic compound found in grapes and wine, is an agonist for the estrogen receptor. Proc Natl Acad Sci U S A 94:14138–14143

Georgiev MI (2013) Coming back to nature: plants as a vital source of pharmaceutically important metabolites-part II A. Curr Med Chem 20:975

Gibson EL, Wardel J, Watts CJ (1998) Fruit and vegetable consumption, nutritional knowledge and beliefs in mothers and children. Appetite 31:205–228

Gonzalez-Coloma A, Guadano A, Gutierrez C, Cabrera R, de La Pena E, de La Fuente G et al (1998) Antifeedant delphinium diterpenoid alkaloids. Structure-activity relationships. J Agric Food Chem 46:286–290

Goto S, Okutomi T, Suma Y, Kera J, Soma G, Takeuchi S (1996) Induction of tumor necrosis factor by a camptothecin derivative, irinotecan, in mice and human mononuclear cells. Anticancer Res 16:2507–2511

Graefe EU, Wittig J, Mueller S (2001) Pharmacokinetics and bioavailability of quercetin glycosides in humans. J Clin Pharmacol 41:492–499

Graf BA, Milbury PE, Blumberg JB (2005) Flavonols, flavones, flavanones and human health: epidemiological evidence. J Med Food 8:281–290

Gul W, Hamann MT (2005) Indole alkaloid marine natural products: an established source of cancer drug leads with considerable promise for control of parasitic, neurological and other diseases. Life Sci 78:442–453

Gyamfi MA, Aniya Y (2002) Antioxidant properties of Thonningianin A, isolated from the African medicinal herb, Thonningia sanguinea. Biochem Pharmacol 63:1725–1737

Heinonen S, Nurmi T, Liukkonen K (2001) *In vitro* metabolism of plant lignans: new precursors of mammalian lignans enterolactone and enterodiol. J Agric Food Chem 49:3178–3186

Heinrich M (2013) Ethnopharmacology and drug discovery. In: Reedijk J (ed) Elsevier reference module in chemistry, molecular sciences and chemical engineering. Elsevier, Oxford

Herman A, Herman AP (2013) Caffeine's mechanisms of action and its cosmetic use. Skin Pharmacol Physiol 26:8–14

Kahkonen MP, Hopia AI, Vuorela HJ, Rauha JP, Pihlaja K, Kujala TS, Heinonen M (1999) Antioxidant activity of plant extracts containing phenolic compounds. J Agric Food Chem 47:3954–3962

Kainsa S, Kumar P, Rani P (2012) Medicinal plants of Asian origin having anticancer potential: short review. Asian J Biomed Pharm Sci 2:1–7

Kolak U, Ozturk M, Ozgokce F, Ulubelen A (2006) Norditerpene alkaloids from *Delphinium linearilobum* and antioxidant activity. Phytochemistry 67:2170–2175

Kondratyuk TP, Pezzuto JM (2004) Natural product polyphenols of relevance to human health. Pharm Biol 42:46–63

Lamoral-Theys D, Decaestecker C, Mathieu V, Dubois J, Kornienko A, Kiss R, Evidente A, Pottier L (2010) Lycorine and its derivatives for anticancer drug design. Mini-Rev Med Chem 10:41–50

Lanzotti V (2014) Drugs based on natural compounds: recent achievements and future perspectives. Phytochem Rev 13:725–726

Lesschaeve I, Noble AC (2005) Polyphenols: factors influencing their sensory properties and their effects on food and beverage preferences. Am J Clin Nutri 81:S330–S335

Li S, Lo CY, Pan MH, Lai CS, Ho CT (2013) Black tea: chemical analysis and stability. Food Funct 4:10–18

Lohombo-Ekomba ML, Okusa PN, Penge O, Kabongo C, Choudhary MI, Kasende OE (2004) Antibacterial, antifungal, antiplasmodial, and cytotoxic activities of *Albertisia villosa*. J Ethnopharmacol 93:331–335

Machado PA, Hilario FF, Carvalho LO, Silveira MLT, Alves RB, Freitas RP, Coimbra ES (2012) Effect of 3-alkylpyridine marine alkaloid analogues in *Leishmania* species related to American cutaneous *Leishmaniasis*. Chem Biol Drug Res 80:745–751

Majik MS, Tilve SG (2012) Pyrrolizidine alkaloids pyrrolams A-D: survey on synthetic efforts, biological activities and studies on their stability. Synthesis 44:2373–2681

Manach C, Williamson G, Morand C, Scalbert A, Remesy C (2005) Bioavailability and bioefficacy of polyphenols in humans. I. Review of 97 bioavailability studies. Am J Clin Nutr 81:S230–S242

Manske RHF (1965) The alkaloids. Chemistry and physiology. Academic Press, New York

Marella A, Tanwar OP, Saha R, Ali MR, Srivastava S, Akhter M et al (2013) Quinoline: a versatile heterocyclic. Saudi Pharm J 21:1–12

Mc-Leod MN (1974) Plant tannins: their role in forage quality. Nutr Abstr Rev 44:803–812

McNaught D, Wilkinson A (1997) IUPAC, compendium of chemical terminology. Blackwell Scientific Publications, Oxford

Meagher E, Thomson C (1999) Vitamin and mineral therapy. In: Morrison G, Hark L (eds) Medical nutrition and disease, 2nd edn. Blackwell Science Inc, Malden, Massachusetts, pp 33–58

Meng F, Zuo G, Hao X, Wang G, Xiao H, Zhang J, Xu G (2009) Antifungal activity of the benzo[c] phenanthridine alkaloids from *Chelidonium majus* Linn. Against resistant clinical yeast isolates. J Ethnopharmacol 125:494–496

Molyneux RJ, Nash RJ, Asano N (1996) Alkaloids. In: Pelletier SW (ed) Chemical and biological perspectives, vol 11. Pergamon, Oxford

Mueller-Harvey I, McAllan AB (1992) Tannins. Their biochemistry and nutritional properties. In: Morrison IM (ed) Advances in plant cell biochemistry and biotechnology, vol 1. JAI Press Ltd, London, pp 151–217

Nassiri M (2013) Simple, one-pot, and three-component coupling reactions of azaarenes (phenanthridine, isoquinoline, and quinoline), with acetylenic esters involving methyl propiolate or ethyl propiolate in the presence of nh-heterocyclic or 1,3-dicarbonyl compounds. Synth Commun 43:157–168

Nibret E, Sporer F, Asres K, Wink M (2009) Antitrypanosomal and cytotoxic activities of pyrrolizidine alkaloid-producing plants of Ethiopia. J Pharm Pharmacol 61:801–808

Nielsen IL, Dragsted LO, Ravn-Haren G, Freese R, Rasmussen SE (2003) Absorption and excretion of black currant anthocyanins in humans and Watanabe heritable hyperlipidemic rabbits. J Agric Food Chem 51:2813–2820

Palavy K, Priscilla MD (2006) Standardisation of selected Indian medicinal herbal raw material containing polyphenols as major constituents. J Pharmacol Sci 68:506–509

Parmar NJ, Pansuriya BR, Barad HA, Kant R, Gupta VK (2012) An improved microwave assisted one-pot synthesis, and biological investigations of some novel aryldiacenyl chromeno fused pyrrolidines. Bioorg Med Chem Lett 22:4075–4079

Pelletier SW (1983) The nature and definition of an alkaloids. In: Alkaloids: chemical and biological perspectives. Wiley, New York

Rahman S, Akbor MM, Howlader A, Jabbar A (2009) Antimicrobial and cytotoxic activity of the alkaloids of Amlaki (*Emblica officinalis*). Pak J Biol Sci 12:1152–1155

Rao RVK, Ali N, Reddy MN (1978) Occurrence of both sapogenins and alkaloid lycorine in *Curculigo orchioides*. Indian J Pharm Sci 40:104–105
Saleem M, Kim HJ, Ali MS, Lee YS (2005) An update on bioactive plant lignans. Nat Prod Rep 22:696–716
Salunkhe DK, Jadhav SJ, Kadam SS, Chavan JK (1982) Chemical, biochemical, and biological significance of polyphenols in cereals and legumes. Crit Rev Food Sci Nutr 17:277–305
Schmeller T, Wink M (1998) Utilization of alkaloids in modern medicine. Springer, Heidelberg, pp 435–459
Schofield P, Mbugua DM, Pell AN (2001) Analysis of condensed tannins: a review. Anim Feed Sci Technol 91:21–40
Shao R, Xiao JB (2013) Natural products for treatment of Alzheimer's disease and relative diseases: understanding their mechanism of action. Curr Neuropharmacol 11:367
Singh AK, Chawla R, Rai A, Yadav LDS (2012) NHC-catalysed diastereoselective synthesis of multi-functionalised piperidines *via* cascade reaction of enals with azalactones. Chem Commun 48:3766–3788
Sotnikova R, Kettmann V, Kostalova D, Taborska E (1997) Relaxant properties of some aporphine alkaloids from *Mahonia aquifolium*. Methods Find Exp Clin Pharmacol 19:589–597
Spencer JP, Abd El Mohsen MM, Minihane AM, Mathers JC (2008) Biomarkers of the intake of dietary polyphenols: strengths, limitations and application in nutrition research. Br J Nutr 99:12–22
Stary F (1996) The natural guide to medicinal herbs and plants. Barnes & Noble Inc, New York
Surh YJ, Hurh YJ, Kang JY, Lee E, Kong G, Lee SJ (1999) Resveratrol, an antioxidant present in red wine, induces apoptosis in human promyelocytic leukemia (HL60) cells. Cancer Lett 140:1–10
Tari C, Fournier N, Briand C, Ducet G, Crevat A (1986) Action of vinca alkaloides on calcium movements through mitochondrial membrane. Pharmacol Res Commun 18:519–528
Willför SM, Smeds AI, Holmbom BR (2006) Chromatographic analysis of lignans. J Chromatogr A 1012:64–77
Wirasathien L, Boonarkart C, Pengsuparp T, Suttisri R (2006) Biological activities of alkaloids from *Pseuduvaria setosa*. Pharm Biol 44:274–278
Wittig J, Herderich M, Graefe EU, Veit M (2001) Identification of quercetin glucuronides in human plasma by high-performance liquid chromatography-tandem mass spectrometry. J Chromatogr B 753:237–243
Wood JG, Rogina B, Lavu S, Howitz KT, Helfand SL, Tatar M, Sinclair DA (2004) Sirtuin activators mimic caloric restriction and delay ageing in metazoans. Nature 430:686–689
Wright AE, Pomponi SA, Cross SS, McCarthy P (1992) A new bis (indole) alkaloid from a deep-water marine sponge of the genus *Spongosorites*. J Organomet Chem 57:4772–4775
Wright CW, Marshall SJ, Russell PF, Anderson MM, Phillipson JD, Kirby GC, Warhurst DC, Schiff PL (2000) *In vitro* antiplasmodial, antiamoebic, and cytotoxic activities of some monomeric isoquinoline alkaloids. J Nat Prod 63:1638–1640
Xiao JB (2015) Natural polyphenols and diabetes: understanding their mechanism of action. Curr Med Chem 22:2
Xiao JB, Jiang HX (2015) A review on the structure-function relationship aspect of polysaccharides from tea materials. Crit Rev Food Nutr 55:930–938
Xiao JB, Muzashvili TS, Georgiev MI (2014) Advance on biotechnology for glycosylation of high-value flavonoids. Biotechnol Adv 32:1145–1156
Yui S, Mikami M, Mimaki Y, Sashida Y, Yamazaki M (2001) Inhibition effect of Amaryllidaceae alkaloids, Lycorine and Lycoricidinol on macrophage TNF-α production. Yakugaku Zasshi 121:167–171
Zhu YJ, Agbayani R, Jackson MC, Tang CS, Moore PH (2004) Expression of the grapevine stilbene synthase gene VST1 in papaya provides increased resistance against diseases caused by *Phytophthora palmivora*. Planta 220:241–250
Ziegler J, Facchini PJ (2008) Alkaloid biosynthesis: metabolism and trafficking. Annu Rev Plant Biol 59:735–769

Seed Oils as a Source of Natural Bio-active Compounds

8

Lalduhsanga Pachuau, Chanam Melody Devi, Abhinab Goswami, Supriya Sahu, and Rajat Subhra Dutta

Contents

L. Pachuau (✉)
Department of Pharmaceutical Sciences, Assam University, Silchar, Assam, India
e-mail: aduhapc@gmail.com

C. M. Devi · A. Goswami · R. S. Dutta
Department of Pharmacy, Regional Institute of Paramedical and Nursing Sciences, Aizawl, Mizoram, India

S. Sahu
Department of Pharmaceutical Sciences, Dibrugarh University, Dibrugarh, Assam, India

M. S. Akhtar et al. (eds.), *Natural Bio-active Compounds*,
https://doi.org/10.1007/978-981-13-7154-7_8

Abstract

Seed oils have enormous potential applications in different industries. They have been increasingly demonstrated to be a viable and renewable source of healthy dietary fatty acids and other bio-active compounds. Seed oils exhibit great multiplicity in their fatty acids, tocopherols, phospholipids, sphingolipids, and sterols compositions depending on the plant species. Essential fatty acids, such as omega-3 and omega-6 PUFAs are derived from seed oils, and extensive investigations on their potential use as an alternative to petroleum products in different industries are also being carried out. This chapter examines the sources of seed oils, their methods of extraction, and characterization along with their bioactivity. The botanical sources of some important seed oils along with their reported bio-active constituents are listed out. It was observed that different methods of extractions and the extraction conditions have influence on the yield and quality of oils. Various methods to monitor the quality and profiling of the seed oil and the beneficial and health-promoting activities of phytochemicals along with their cosmetic applications are also highlighted.

Keywords

Bioactivities · Cosmetics · Nutrition · Phospholipids · Seed oils

8.1 Introduction

Seeds are often considered as residues of agriculture-based industries (Veronezi and Jorge 2012). However, studies conducted over the years have shown that seeds of various plants contain several bio-active components. As evidenced by recent PubMed publication search, there has been increasing enthusiasm toward investigations on the multi-bioactivities and health benefits of these seed oils (Fig. 8.1). The seed oils extracted from various plants, such as pumpkin, bitter melon, Kalahari melon, kenaf, roselle, hemp, *Eruca*, *Alseodaphne andersonii*, *Eucommia ulmoides*, *Garcinia xanthochymus*, and others, are rich sources of polyunsaturated fatty acids and antioxidants, such as tocols, bioflavonoids, and phytosterols (Nyam et al. 2009a; Khoobchandani et al. 2010, 2011; Adams et al. 2012; Pieszka et al. 2013; Paz et al. 2014, Chaliha et al. 2017; Zhang et al. 2018). These seed oils are extracted by various methods including mechanical process and chemical or solvent methods as well as through supercritical fluid extraction method. Each method has its own advantages and limitations which are discussed in detail in the following sections. The mechanical screw press method is widely used for extraction of seed oils due to its simplicity and ease of operation, while solvent extraction is known for their completeness of extraction (Ogunniyi 2006; Bhuiya et al. 2015; Yusuf et al. 2017).

The physicochemical characterization of the extracted seed oils is necessary to monitor the quality of the oil and assure their safety. A relationship exists between fatty acids, tocopherols, phospholipids, triacylglycerol, sphingolipids, and sterols composition of the oil with its functional properties and applications (Matthaus 2012).

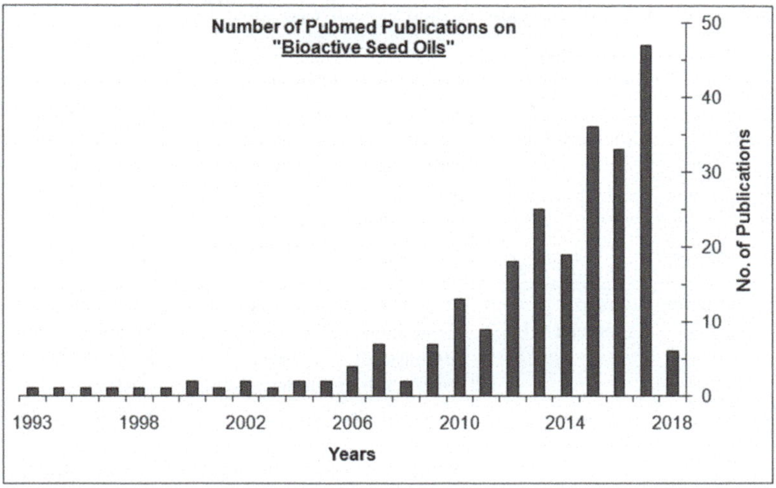

Fig. 8.1 Number of PubMed publications on bio-active seed oils

A wide range of bioactivities exhibited by the seed oils include antioxidant, antimicrobial, and antiproliferative properties, which have been attributed mainly to their fatty acid components. The current chapter discusses on the sources, methods of extraction of seed oils, and their characterization methods and bioactivities. The application of seed oils in the cosmetic industry is also highlighted.

8.2 Sources and Extraction of Methods

8.2.1 Sources

Seed oils have been extracted from various sources. Plant seeds obtained from fruits, tubers, and other vegetative parts of the food and nonfood crops have been widely explored for extraction. Recent trend also indicates that increased attention has been given toward the utilization of underutilized agricultural products as well as the by-products and wastes from food processing industries as sources of seed oils (Nyam et al. 2009b). Table 8.1 provides various sources of seed oils along with their bio-active components.

8.2.2 Methods of Extraction

8.2.2.1 Mechanical Screw Press Extraction Method

Screw press oil extraction is widely used in commercial and industrial seed oil extraction, as it provides safety and simplicity. The whole process can easily be maintained and operated by semiskilled supervisors and can be adapted quickly as per the needs to improve the efficiency of the oil recovery (Singh and Bargale 2000; Faugno et al. 2016). Industrial screw press extractors are available in wide range of

Table 8.1 Some sources of seed oils and their bio-active compounds

Plant name	Common name (family)	Bio-active compounds	References
Amaranthus cruentus	Amaranth (Amaranthaceae)	α-, β-, γ-, δ-tocopherols, 24-methylenecholesterol, campesterol, stigmasterol, β-sitosterol, Δ5-avenasterol, Δ7-avenasterol, Δ7-stigmasterol, α-spinasterol, citrostandienol, cycloartenol	Czaplicki et al. (2011) and Ogrodowska et al. (2012, 2014)
Annona muricata	Soursop (Annonaceae)	Fatty acids, α-, γ-tocopherols, carotenoids, campesterol, stigmasterol, β-sitosterol, epicatechin, p-coumaric acid	Navaratne and Subasinghe (2014) and Silva and Jorge (2016)
Arabidopsis thaliana	Arabidopsis (Brassicaceae)	α-, γ- tocopherols	Kothari (2013)
Arctium lappa	Burdock (Asteraceae)	Arctin, arctigenin, arctignan E	Kothari (2013)
Borago officinalis	Borage (Boraginaceae)	γ-, δ-tocopherols, β-sitosterol, Δ5-avenasterol, gramisterol, citrostandienol, cycloartenol	Czaplicki et al. (2011)
Butea parviflora	Bastard teak (Fabaceae)	Oleic acid, linoleic acid, palmitic acid, behenic acid	Kaki et al. (2016)
Camellia sinensis	Tea (Theaceae)	Saponin, oleiferasaponin A_1	Czaplicki et al. (2011), Kothari (2013) and Wang et al. (2017)
Cannabis sativa	Hemp (Cannabaceae)	Palmitic acid, stearic acid, oleic acid, linoleic acid, γ-linolenic acid, α-linolenic acid, eicosenoic acid, behenic acid	Leizer et al. (2000), Paz et al. (2014) and Porto et al. (2015)
Carica papaya	Papaya (Caricaceae)	Fatty acids, α-, β-tocopherols, carotenoids, campesterol, stigmasterol, β-sitosterol,	Silva and Jorge (2016)
Citrullus lanatus	Watermelon (Cucurbitaceae)	Lycopene, β-carotene, vitamins (B, C, and E), minerals (K, Mg, Ca, and Fe), amino acid (citrulline), phenolics	Choudhary et al. (2015), Navaratne and Subasinghe (2014), Nyam et al. (2009b, 2011) and Salanta et al. (2015)
Citrus medica	Citron (Rutaceae)	Fatty acids, α-tocopherols, campesterol, β-sitosterol, p-coumaric acid, salicylic acid, quercetin	Silva and Jorge (2016)
Citrus sinensis	Orange (Rutaceae)	Fatty acids, α-tocopherols, carotenoids, campesterol, β-sitosterol, salicylic acid	Silva and Jorge (2016)
Combretum kraussii	Forest bushwillow (Combretaceae)	Combretastatin B5	Kothari (2013)

(continued)

Table 8.1 (continued)

Plant name	Common name (family)	Bio-active compounds	References
Cucumis melo var. *inodorus*	Melon (Cucurbitaceae)	Fatty acids, α-, γ-tocopherols, carotenoids, β-sitosterol	Silva and Jorge (2016)
Cucurbita moschata	Pumpkin (Cucurbitaceae)	Fatty acids, α-, γ-tocopherols, carotenoids, stigmasterol, β-sitosterol, stigmastanol, salicylic acid	Bardaa et al. (2016a), Czaplicki et al. (2011), Hrabovski et al. (2012), Nyam et al. (2009b), Patel (2013), Saavedra et al. (2013) and Veronezi and Jorge (2012)
Eruca sativa	Arugula (Brassicaceae)	Allyl isothiocyanate, phenylethyl isothiocyanate, sulforaphane	Bansal et al. (2015), Gulfraz et al. (2011), Khoobchandani et al. (2011) and Sanad and Mabrouk (2016)
Fortunella margarita Swingle	Kumquat (Rutaceae)	Fatty acids, carotenoids, campesterol, stigmasterol, β-sitosterol, p-coumaric acid, salicylic acid, quercetin	Silva and Jorge (2016)
Fragaria × *ananassa* Duchesne	Strawberry (Rosaceae)	Fatty acids, γ, α-tocopherols, carotenoids, β-sitosterol, stigmastanol, caffeic acid,	Silva and Jorge (2016)
Hancornia speciosa var. *pubescens*	Mangaba (Apocynaceae)	Fatty acids, α-tocopherols, carotenoids, β-sitosterol, stigmastanol	Silva and Jorge (2016)
Helianthus	Sunflower (Asteraceae)	Sesquiterpene lactone, diterpene, flavonoids	Stoia and Oancea (2013)
Hibiscus cannabinus	Kenaf (Malvaceae)	Vitamin E, β-sitosterol, alpha-linolenic acid (ALA)	Nyam et al. (2012, 2013, 2015) and Yoshime et al. (2016)
Hibiscus sabdariffa	Roselle (Malvaceae)	Linoleic/oleic acid, fatty acids, β-sitosterol, campesterol, δ-5-avenasterol, cholesterol	Al-Okbi et al. (2017a), Dhar et al. (2015) and Nyam et al. (2009a, 2013, 2015)
Juglans regia	Walnut (Juglandaceae)	α-, β-, γ-, δ-tocopherols, campesterol, stigmasterol, β-sitosterol, Δ5- avenasterol, citrostandienol, cycloartenol	Czaplicki et al. (2011)
Linum usitatissimum	Linseed (Linaceae)	α-, γ-, δ-tocopherols, campesterol, stigmasterol, β-sitosterol, Δ5- avenasterol, cycloartenol, 24-methylene cycloartenol.	Czaplicki et al. (2011)

(continued)

Table 8.1 (continued)

Plant name	Common name (family)	Bio-active compounds	References
Mangifera indica L.	Mango (Anacardiaceae)	Fatty acids, α-tocopherols, stigmasterol, β-sitosterol, gallic acid, epicatechin, salicylic acid, quercetin	Silva and Jorge (2016)
Momordica charantia L.	Bitter melon (Cucurbitaceae)	Calcium, magnesium, phosphorous, potassium, iron, zinc, ascorbic acid, tocopherols, folate, catechin, epicatechin, gallic acid, saponins, peptides, alkaloids, polyunsaturated fatty acids (PUFAs)	Dandawate et al. (2016), Kai et al. (2014), Padmashree et al. (2011), Saini et al. (2017) and Yoshime et al. (2016)
Momordica cochinchinensis	Gac (Cucurbitaceae)	Lycopene	Kothari (2013)
Oenothera spp.	Evening primrose (Onagraceae)	α-, γ-, δ-tocopherols, campesterol, stigmasterol, β-sitosterol, Δ5- avenasterol	Czaplicki et al. (2011)
Olea europaea	Olive (Oleaceae)	Protein, fats, phenols	Rodriguez et al. (2008)
Papaver somniferum	Poppy (Papaveraceae)	α-, γ-tocopherols, campesterol, stigmasterol, β-sitosterol, Δ5- avenasterol	Czaplicki et al. (2011)
Passiflora edulis Sims	Passion fruit (Passifloraceae)	Fatty acids, β, γ, α-tocopherols, carotenoids, campesterol, stigmasterol, β-sitosterol, stigmastanol, caffeic acid, p-coumaric acid, salicylic acid	Silva and Jorge (2016)
Persea americana	Avocado (Lauraceae)	Glycolipids, phospholipids, furoic acid, abscisic acid	Dabas et al. (2013)
Phaleria macrocarpa	God's crown (Thymelaeaceae)	Mahkoside A, dodecanoic acid, palmitic acid, des-acetyl flavicordin-A, flavicordin-A, flavicordin-D, flavicordin-A, glucoside, ethyl stearate, lignans, sucrose	Easmin et al. (2015)
Phaseolus vulgaris	Green beans (Fabaceae)	Antifungal peptide (vulgarinin)	Kothari (2013)
Phoenix dactylifera	Date (Arecaceae)	Palmitic acid, linoleic acid, lauric acid, myristic acid, stearic acid, p-hydroxybenzoic acid, protocatechuic acid, m-coumaric acid, gallic acid, potassium, sodium, calcium, magnesium, iron, manganese, zinc, copper	Farsi and Lee (2011)
Plukenetia volubilis	Sacha inchi (Eurphobiaceae)	Linolenic acid	Romero et al. (2009)

(continued)

Table 8.1 (continued)

Plant name	Common name (family)	Bio-active compounds	References
Punica granatum	Pomegranate (Lythraceae)	Phytosterols, tocopherols	Durdevic et al. (2018); Khoddami and Roberts (2015)
Psidium guajava	Guava (Myrtaceae)	Fatty acids, α-, β-, γ-tocopherols, carotenoids, β-sitosterol, stigmastanol, p-coumaric acid, salicylic acid, quercetin	Silva and Jorge (2016)
Rubus idaeus	Red raspberry (Rosaceae)	Fatty acids, carotenoid, tocopherols, phenolic content	Parry et al. (2005)
Rubus sp.	Marion berry	Fatty acids, carotenoid, tocopherols, phenolic contents	Parry et al. (2005)
Rubus ursinus × *idaeus*	Boysenberry (Rosaceae)	Fatty acids, carotenoid, tocopherols, phenolic contents	Parry et al. (2005)
Sesamum indicum	Sesame (Pedaliaceae)	γ-tocopherols, campesterol, stigmasterol, β-sitosterol, Δ5- avenasterol, Δ7-avenasterol, sesamin	Czaplicki et al. (2011) and Kothari (2013)
Solanum lycopersicum	Tomato (Solanaceae)	Fatty acids, γ- tocopherols, carotenoids, β-sitosterol, stigmastanol, p-coumaric acid, salicylic acid	Silva and Jorge (2016)
Thymus vulgaris	Thyme (Lamiaceae)	Linoleic acid, oleic acid, stearic acid, palmitic acid, γ-tocopherol	Assiri et al. (2016)
Vitis labrusca	Grape (Vitaceae)	Fatty acids, α-, β-, γ-tocopherols, carotenoids, campesterol, stigmasterol, β-sitosterol, catechin, caffeic acid	Freitas et al. (2017), Kothari (2013), Silva and Jorge (2016) and Teixeira et al. (2014)
Vaccinium corymbosum	Blueberry (Ericaceae)	Fatty acids, carotenoid, tocopherols, phenolic content	Parry et al. (2005)

capacities ranging from 40 to 1000 kg/h (Singh and Bargale 2000). To mechanical press the seeds for extraction of oil, seeds are crushed first and then warmed using steam-jacketed vessel to reduce the moisture content. Crushed seeds are then loaded into hydraulic presses for mechanical expression and extraction of the oil (Ogunniyi 2006). The pre-pressing procedure of crushing and cooking is essential for various reasons. Breaking down the seeds weakens the oil cell walls, and heating completes the breakdown of the oil cells while also lowering the viscosity of the oil (Khan and Hanna 1983). The operating conditions of the mechanical press are highly critical in determining the yield of the oil from the source. The major factors that govern efficient extraction of oils by this method include pressure, temperature, and moisture content (Savoire et al. 2013). In many cases, mechanical pressing alone results in the removal of only about 45% of the oil present necessitating recovery of the remaining oil by solvent extraction method (Ogunniyi 2006; Stanisavljevic et al. 2009). Variation in the method, such as introduction of cold-pressing has been

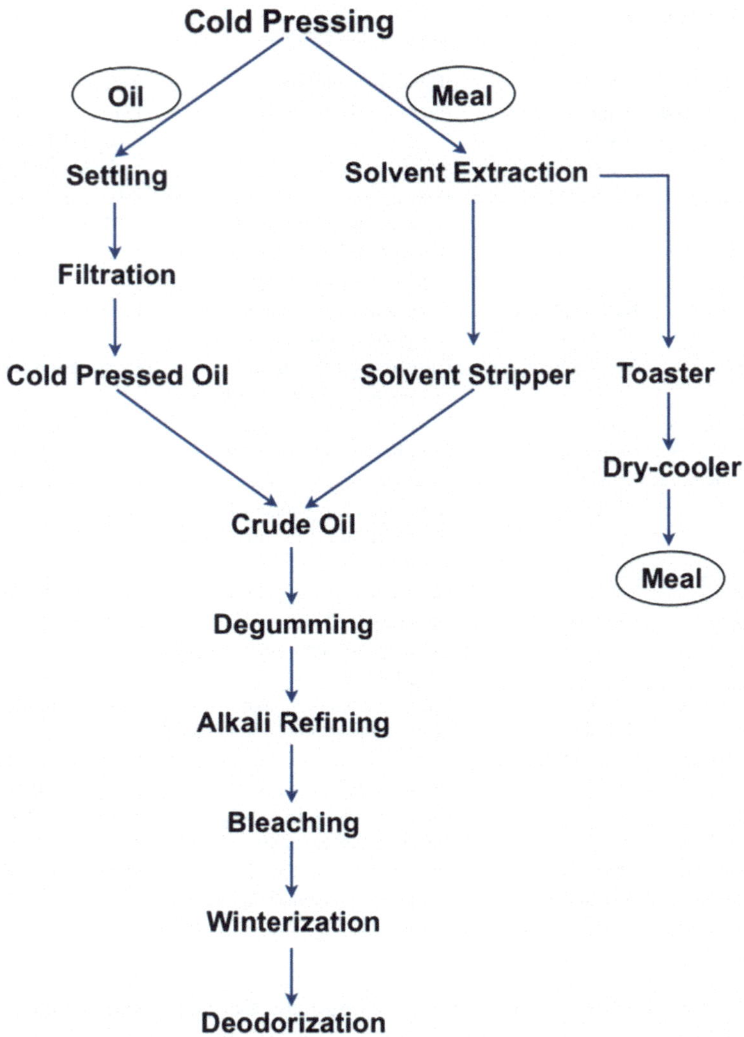

Fig. 8.2 Cold-pressed flaxseed oil processing. (Reproduced with permission from Shim et al. 2015, Copyright©2015 Elsevier Ltd)

successfully applied as an alternative to the more traditional pressing method (Shim et al. 2015). The different steps in cold pressing and refining are shown in Fig. 8.2.

8.2.2.2 Solvent Extraction Method

Solvent extraction is an old method of extraction that has been effectively used as an alternative to the hydraulic press in the extraction of seed oils. It involves the use of low-boiling-point extracting solvent which dissolved the oil and then evaporated to leave behind the extracted seed oils. This method can be used to extract almost 100% of the seed oil, and the refined extracted oil has better keeping property compared to the expressed oil (Swisher and Fiero 1932). Wide ranges of solvents are available to choose, and these solvents can be effectively removed and recovered. Conventional solvent extraction methods include percolation, maceration, heating under reflux, and Soxhlet extraction (Easmin et al. 2015). Soxhlet apparatus is the most commonly used solvent extraction method, and extraction may be carried out between 40 and 70 °C (Nyam et al. 2009a; Nyam et al. 2012; Porto et al. 2015). Organic solvents such as chloroform, methanol, n-hexane, heptanes, and petroleum ether are commonly used, and solvents are removed under reduced pressure using a rotary evaporator. The major drawback of this method is that the process can be time consuming and the organic solvents used in the extraction are hazardous to health and environment. Moreover, complete removal of the solvent is difficult; as a result, solvent with high purity and extraction selectivity is required for efficient extraction (Easmin et al. 2015).

8.2.2.2.1 Ultrasound-Assisted Solvent Extraction (UASE)

Both nondestructive low-intensity ultrasound and the more disruptive high-intensity ultrasound are finding increasing use in food industries for various purposes (McClements, 1995). High-intensity ultrasound-assisted solvent extraction is one of the methods used to improve the conventional solvent extraction process. This process significantly reduces the consumption of the organic extraction solvent and the extraction time while increasing the recovery of the targeted compounds (Gil-Chavez et al. 2013). The better efficiency of extraction achieved with ultrasonic technique is due to the phenomenon called as acoustic cavitation (Gil-Chaves et al. 2013; Easmin et al. 2015). This ultrasound-induced cavitation resulted in disruption of the biological cell walls and increased solvent accessibility, thereby facilitating the release of their contents (Dolatowski et al. 2007). By this method efficient extraction can be achieved within just 5 min, while conventional solvent extraction took 24 h (Sora and Villamiel 2010).

8.2.2.2.2 Microwave-Assisted Solvent Extraction (MASE)

Microwave-assisted solvent extraction has become an attractive alternative to the conventional solvent extraction method. Microwaves are a form of electromagnetic radiation ranging in wavelengths from 1 mm to 1 m, which when absorbed are converted into thermal energy (Zhang et al. 2011). During MASE, absorption of the wave resulted in heating of the moisture inside the cells which then evaporate and create a high pressure on the cell wall (Gil-Chavez et al. 2013). This process disrupts the physical integrity of the cellular structure, improves porosity, and allows the extracting solvent to penetrate biological matrix readily to dissolve out the desired compound. All these processes require only a fraction of the time taken by conventional solvent extraction method. Studies have shown that the optimized 5 min MASE resulted in better yield of pomegranate seed oils than 8 h each of Soxhlet and cold extraction process (Cavdar et al. 2017). In addition, oils derived from MASE also exhibit better physicochemical properties, total phenolic content, and antioxidant activity.

8.2.2.2.3 Supercritical Fluid Extraction (SFE)

Extraction of the desired components with a solvent at a temperature and pressure exceeding the critical point for the solvent is called supercritical fluid extraction (SFE). SFE is a novel method of solvent extraction that is fast, efficient, and safe and that is gaining popularity. Compared to the conventional solvent extraction methods, SFE avoids the use of harmful and toxic solvents and instead use solvents such as carbon dioxide which are environment-friendly. SFE extends the possibility of extracting products free of residual solvents, and by fine-tuning the supercritical fluid temperature and pressure, the solubility and selectivity of fatty acids can also be achieved (Perez et al. 2015). SFE technology has been successfully applied in the extraction of various seed oils including grape seeds (Prado et al. 2012; Perez et al. 2015).

8.2.3 Processing and Refining of Edible Oils

Crude oils obtained through mechanical or solvent extraction methods may contain undesirable components that may affect the safety and stability of the oil. Therefore, refining the crude oil prior to their availability for consumption is a necessary step to attain high-quality edible oils. Refining crude oils have been achieved through physical and chemical processing methods. Some of the deleterious substances present in crude oil include oxidation substances, free fatty acids, phospholipids, pigments, and metal salts (Cmolik and Pokorny 2000). The refining procedure aims to remove these substances and keep the essential components intact while minimizing the oil loss in the process (Vaisali et al. 2014). Traditional refining process for edible oils includes degumming, neutralization/deacidification, bleaching, and deodorization (Cvengros 1995). The first step in edible oil refining is the degumming process that removes phospholipids, trace metals, and mucilaginous materials (Vaisali et al. 2014). This is accomplished through chemical method using water and acid or enzymatic method or membrane degumming. Deacidification followed the degumming process which is also a critical step in the refining of seed oils as rancid flavor and oxidation are expedited by the presence of free fatty acids (Chaiyasit et al. 2007). Chemical and physical methods are the two most common deacidification processes in vegetable oil refining. In chemical method, the free fatty acids are removed by washing crude oils with sodium hydroxide or sodium carbonate solution, hence called as alkali refining. An alternative method, called as minimal refining that replaces NaOH with $Ca(OH)_2$, MgO, and Na_2SiO_3, has also been developed (Ghazani et al. 2013). An important step in the physical deacidification is the use of superheated steam under low pressure and at temperatures higher than 220 °C during which free fatty acids and undesirable volatiles are removed (Cmolik and Pokorny 2000). Color compounds, products of oxidation, trace metals, and remaining phospholipids are removed during bleaching step by utilizing bleaching clays or charcoal (Vaisali et al. 2014). Generally, deodorization by steam distillation completes the refining process, and a high-quality, light-colored refined

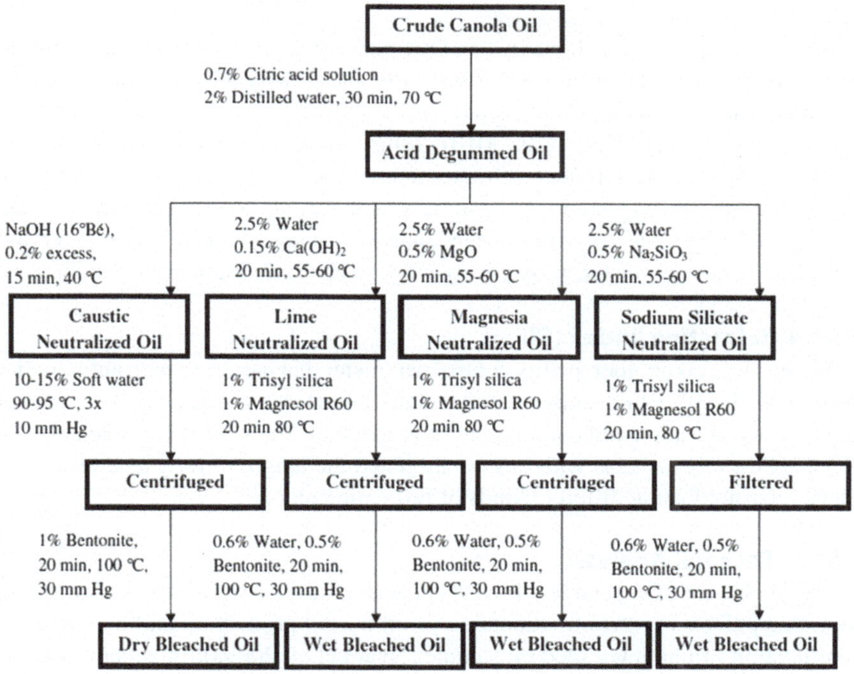

Fig. 8.3 Flow sheet showing processing of crude canola oil by traditional and minimal refining methods. (Reproduced with permission from Ghazani et al. 2013, Copyright©2013 AOCS)

oil which is free from objectionable odor is obtained. A chart showing refining process of crude canola oil is depicted in Fig. 8.3.

8.3 Characterization of Seed Oils

8.3.1 Physical Characterization

8.3.1.1 Color, Odor, and Taste

Edible oils have their characteristic color, odor, and taste. The color of the oil could be an indication of the quality. The characteristic odor and taste of the oil are also consequence of their composition and specific method of extraction (Markovic and Bastic 1976).

8.3.1.2 Specific Gravity

Specific gravity is an important characteristic of edible seed oils as it relates to the density of the oil. It is the ratio of the density of the test substance to the density of the reference substance at a given temperature and pressure. According to the US Pharmacopeia (2005), the reference liquid is water, and measurements are conducted at 25 °C. Specific gravity is measured by following Method I (Pycnometer) or Method II (Oscillating transducer density meter).

8.3.1.3 Viscosity

There is a specific relationship between fatty acid composition of vegetable oils and their rheology. It has been shown that viscosity and density decreases with increase in unsaturation while high saturation and polymerization leads to increase in viscosity and density (Kim et al. 2010). Temperature also influences density and viscosity as oils become lighter at higher temperature. Oils with low viscosity and density are highly suggested for consumers (Zahir et al. 2017). The viscosity and rheology of vegetable oils have been analyzed using Ostwald viscometer and controlled-stress rheometer, respectively (Kim et al. 2010; Zahir et al. 2017).

8.3.1.4 Refractive Iindex (RI)

Refractive index is another quality control parameter that increases with autoxidation (Arya et al. 1969). RI depends on molecular weights, fatty acid chain lengths, and degree of unsaturation and conjugation (Sadoudi and Ahmed 2017). In general, the RI values for oils increase with unsaturation and the length of fatty acid chains. RI can be measured using different kinds of refractometers.

8.3.1.5 Thermal Analysis

Thermal behavior of vegetable oils has been analyzed by using differential scanning calorimetry (DSC) control fractionation of oil during production and may assist in identification of unknown seed oil sample (Nyam et al. 2009a). In general, the oils with high saturated fatty acid content exhibit DSC melting and crystallization profiles at higher temperature.

8.3.1.6 Smoke, Fire, and Flash Point

The smoke, fire, and flash points of vegetable oils are measured to determine the thermal stability of oils as they indicate the temperature of decomposition, ignition, and continued combustion, respectively (Morgan 1942). They are related to the free fatty acid contents of the oil, and the lower the free fatty acid content or the shorter the chain, the lower are the smoke, fire, and flash points (Wang 2011).

8.3.2 Chemical Characterization

8.3.2.1 Iodine Value

The degree of unsaturation of fats and oils can be expressed in terms of its iodine value. It is the number of grams of iodine that reacts with 100 g of the oil (Amri 2011). The higher the iodine value, the more are the C=C bonds in the fat or oil, and the more are the chances for rancidity. Iodine value is one of the important quality control and standardization parameters for vegetable oils. Wijs solution method is the most commonly followed method for the determination of iodine value (Simurdiak et al. 2016).

8.3.2.2 Acid Value

Acid value is defined as the number of milligram of potassium hydroxide required to neutralize free acids in 1 g of the oil (Evans 1996). It is an important quality indicator of vegetable oils. Decomposition of glycerides in oils leads to liberation of free fatty acids, thus increasing the acid value. An increase or rise in acid value ergo indicates rancidification of the oil (Evans 1996; Shah and Seth 2010). Acid-base titration method is mostly followed in determination of acid value; however, several novel techniques for acid value determination have been developed (Kardash and Tur'yan 2005).

8.3.2.3 Peroxide Value

Peroxide value is a test for oxidative rancidity; higher peroxide value of seed oils indicates rancidity (Shah and Seth 2010). During oxidative rancidity, peroxides are formed by the oxidation of double bonds in unsaturated fatty acids resulting in the production of aldehydes, ketones, and other low molecular weight acid (Baiao and Lara 2005). Therefore, peroxide value is one of the most important parameters that indicates the quality of vegetable oils (Nyam et al. 2009a).

8.3.2.4 Saponification Value

Saponification value is the quantity in milligram of potassium hydroxide required to neutralize the free fatty acids and to saponify the esters present in 1 g of the oil (Indian Pharmacopoeia 2007). It measures the average molecular weight of all the fatty acids present and oils with shorter triglyceride chains that exhibit higher saponification value (Baiao and Lara 2005; Shah and Seth 2010). It is usually determined by titrimetric methods.

8.3.2.5 Unsaponifiable Matter

Unsaponifiable matter consists of compounds such as sterols that are extracted from the remaining of the saponification process through the use of solvents such as diethyl ether (Evans 1996; Shah and Seth 2010). This unsaponifiable matter may exhibit beneficial effects against different disease conditions (Dabas et al. 2013).

8.3.2.6 Fatty Acid Composition

Determination of fatty acids composition is essential for complete characterization of seed oils. Profiling the fatty acids of a lipid is important as the functional and therapeutic properties of lipids are determined by their fatty acid composition, and proper profiling of fatty acids also makes it possible to determine adulterants in seed oils (Sun et al. 2015). Gas chromatography (GC) coupled with mass spectrometry (MS) or with other detectors has been the most common method used for analysis of fatty acids components in lipids (Watanabe et al. 1973; Nyam et al. 2009a; Sun et al. 2015). However, since GC analysis requires derivatization of the fatty acids into their volatile forms such as methyl esters, a highly sensitive liquid chromatography (LC)-MS method that bypasses the need for derivatization has also been developed to analyze and profile the fatty acids in lipids (Bromke et al. 2015).

8.4 Bioactivities of Seed Oils

8.4.1 Nutraceutical and Antioxidant Activity of Seed Oils

Seeds and seed oils have been important part of food ingredients since ancient times. Studies have shown that seeds are promising sources of lipids, proteins, and ash (Veronezi and Jorge 2012). Lipid profiling of seed oils has unveiled beneficial antioxidant phytochemicals such as tocopherols, carotenoids, phenolic, and polyphenolic compounds, including the special fatty acid α-linolenic acid (Parry et al. 2005). Cold-pressed seed oils obtained from black caraway, carrot, cranberry, and hemp were found to exhibit strong free radical scavenging activities and also inhibit human LDL oxidation (Yu et al. 2005). The antioxidant activity and oxidative stability of marketed cold-pressed seeds and fruit oils from macadamia, avocado, sesame, safflower, pumpkin, rose hip, Linola, flaxseed, walnut, hempseed, poppy, and milk thistle were determined by 1,1-diphenyl-2-picrylhydrazyl (DPPH) assay (Prescha et al. 2014). Lipophilic fractions of the oils containing lipophilic antioxidants such as tocopherols were found to exhibit higher than the hydrophilic portions with hydrophilic antioxidants like phenolic compounds, and those oils rich in polyunsaturated fatty acid (PUFA) contents were exhibiting higher antiradical activity as well. Seed oils from fruits such as pomegranate also have the ability to reduce levels of total cholesterol and high-density lipoprotein (HDL), suppress colon carcinogenesis, and exhibit chemopreventive efficacy against experimental colon cancers (Khoddami and Roberts 2015). Peony seed oils obtained from *Paeonia suffruticosa* Andr. contain high amount of α-linolenic acid and γ-tocopherol and are reported to be even more potent than extra-virgin olive oil in their free radical scavenging power demonstrating their nutraceutical potential (Yang et al. 2017). Even in traditional Chinese medicine (TCM), seed oils such as camellia oil are considered to be superior nutritional and dietary supplement that improves digestive and immune systems and reduces blood cholesterol while regulating nervous system (Yang et al. 2016). These data and various other studies thus suggest that seed oils are potential dietary sources of beneficial phytochemicals and natural antioxidants (Parry et al. 2005; Mohd-Esa et al. 2010; Patel 2013; Jorge et al. 2016).

8.4.2 Anticancer Activity

Seed oils and seed extracts obtained from various sources have been investigated as potential anticancer agent. *Eruca sativa* seed oil and its bio-active principles have been demonstrated to exhibit antimelanoma and antimutagenic activities against B16F10 melanoma cells induced in C57BL/6 mice (Bansal et al. 2015). The isothiocyanates rich *E. sativa* seed oils were also reported to inhibit melanoma growth and angiogenesis in experimental mice without any major toxicity to the animals (Khoobchandani et al. 2011). Bitter gourd (*Momordica charantia* L.) seed extract was found to possess antiproliferative activity against adult T-cell leukemia cell lines (Su9T01, HUT-102, and Jurkat), and α-eleostearic acid was suggested to be the compound responsible for the activity (Kai et al. 2011; Kai et al. 2014). The antiproliferative activity of seed flours from various fruits containing α-linolenic acid against HT-29 colon cancer cells was also reported (Parry et al. 2006). Jacaranda seed oil was found to contain bio-active fatty acids *cis*-8, *trans-10*, *cis-12*, and octadecatrienoic acid which induces apoptosis in human leukemia HL-60 cells through induction of oxidative stress (Yamasaki et al. 2013).

Brucea javanica seed oil (BJO) is a traditional medicine of China which has been used in lung cancer in combination with chemotherapy or radiotherapy. Study has revealed that BJO inhibited proliferation of A549 and H446 cells and induced G0/G1 arrest partly via regulating p53 and cyclin D1 establishing the rationale behind their use in the treatment of lung cancer (Wang et al. 2016a). Another seed oil from *Prunus dulcis* (almond oil) which was also used in traditional therapy for various health benefits was investigated for in vitro anticancer activity and was found to be active against Colo-320 and Colo-741 cells (Mericli et al. 2017).

8.4.3 Antimicrobial Activity

Antibiotic resistance is a major health concern which compelled scientist to look for an alternative solution from plant resources (Gulfraz et al. 2011). Several plant seed oils have been investigated as antimicrobial agent with promising results. *E. sativa* seed oil has been reported to possess excellent antimicrobial activity, and a stable cream containing *E. sativa* oil was formulated that exhibits antimicrobial activity against *Staphylococcus aureus*, *Escherichia coli*, *S. aureus* (MRSA), and *Malassezia furfur* with no skin irritation (Sanad and Mabrouk 2016). Cold-pressed seed oils of argan, pomegranate, date, flax, safflower, golden berry, walnut, and grape seeds also exhibit strong antibacterial activity against pathogenic bacteria and aflatoxigenic molds such as *S. enteridis*, *L. monocytogenes*, *Aspergillus parasiticus*, and *A. parasiticus* (Farsi and Lee 2011; Gecgel et al. 2016).

Seed oils and extracts obtained from different plant sources have been investigated to exert antiviral and antifungal activities against various pathogenic strains. Bio-active components of neem (*Azadirachta indica*) seed oil such azadiradione, nimbin, salannin, and epoxy-azadiradione were not active individually, but when mixed or combined, they exhibit synergistic antifungal effects (Govindachari et al. 1998). Oil and extracts obtained from carrot (*Daucus carota* L. var. Perfekcja) seeds, *Centratherum anthelminticum*, *Ocimum sanctum*, and *Momordica charantia* L. were also shown to be active against various pathogenic fungi, and these sources represent sustainable alternative to the use of synthetic fungicides (Jasicka-Masiak et al. 2004; Gopalkrishna et al. 2016; Wang et al. 2016b). Foodborne viral diseases are mostly caused by human noroviruses which also contribute to non-bacterial gastroenteritis, and seed oil fatty acid mixtures from *Zanthoxylum schinifolium* were found to exhibit inhibitory effect to control these foodborne viral diseases (Oh and Chung 2014). The fatty acid mixtures of the seed oils from camellia, neem tree, black raspberry, and pumpkin and their polyphenolic and triterpene fractions are also shown to be responsible for their antiviral property (Akihisa et al. 2004; Badr et al. 2011; Koriem 2013; Lee et al. 2016).

8.4.4 Antidiabetic Activity

Some seed oils and seed extracts are also potent hypoglycemic agents. Oral administration of chloroform extract of *Terminalia chebula* seeds produced dose-dependent reduction of blood glucose in diabetic rats that was comparable to an established standard antidiabetic drug, glibenclamide (Rao and Nammi 2006). Cactus pear (*Opuntia ficus-indica* (L.) Mill.) seed oils also exhibit antihyperglycemic activity in diabetic rats by inhibiting intestinal absorption of glucose (Berraaouan et al. 2014). Intraperitoneal administration of *Citrus sinensis* seed oil to alloxan-induced hyperglycemic rats is also able to reduce blood glucose significantly (Chilaka et al. 2015).

Oral administration of seed oil obtained from peony (*Paeonia lactiflora* Pall.) was evaluated in streptozocin-induced diabetic model and found that the seed oil significantly reduced blood glucose comparable to the standard glibenclamide (Su et al. 2015). *Sanbai* melon seed oil, which is part of TCM used for treating diabetes mellitus, alleviates oxidative stress in streptozocin-induced diabetic mice and attenuates the liver and renal injury in these mice. The mechanism by which the islet cells are protected from apoptotic damage was also elucidated confirming their beneficial effects on diabetic mellitus (Wang et al. 2018). *Nigella sativa* and olive seed oils were also diabetic complications at an enzymatic level through a mechanism attributed to their antioxidant activity (Samarji and Balbaa 2014).

8.4.5 Antihypertensive and Cardioprotective Activity

Seed oils are rich dietary source of several bio-active phytoconstituents that provide positive health benefits on blood pressure (Khalesi et al. 2015). A controlled clinical trial on the effect of flaxseed supplements on blood pressure (BP) showed a significant reduction in systolic BP and diastolic BP following supplementation with various flaxseed products (Ursoniu et al. 2016). Oil was extracted from pumpkin seeds which was used in traditional medicine and was evaluated for antihypertensive and cardioprotective activities (El-Mosallamy et al. 2012). Antihypertensive and cardioprotective effects of the pumpkin seed oil were observed which may be demonstrated through a mechanism that involves generation of NO. Sesame seed oil reduces the detrimental effects of diabetes in experimentally induced diabetic rats, and its consumption was suggested to improve the glucose control, hepatic stress, and renal and cardiac health (Aslam et al. 2017). Due to the antioxidant activity of tocopherols in sunflower seed oil, consumption of the oil is suggested to reduce the risk of cardiovascular diseases and certain types of cancer (Guo et al. 2017). Omega-3 polyunsaturated fatty acids are important dietary component for improving health as they reduce cardiovascular diseases and heart disease risk, and since seed oils are rich in omega-3 fatty acid contents, they are key element of functional foods (Ruxton et al. 2004; Imran et al. 2016).

8.4.6 Anti-inflammatory and Wound Healing Activity

Many seed oils contain essential omega-3 fatty acids and other bio-actives that are proven to exhibit anti-inflammatory and antithrombic activities (Nyam et al. 2015). The anti-inflammatory and antioxidant activities of the seed oils promote wound healing and repairing of the skin barrier (Lin et al. 2018). The synergistic effect of antibacterial, antioxidant, and anti-inflammatory property of seed oils has also been attributed to their wound healing property (Rekik et al. 2016). Seed oils are also reported to possess efficient wound healing property against second-degree burn models in rats (Bardaa et al. 2016b).

In traditional medicine, seed oils such as those from *Amburana cearensis* have been used in the treatment of respiratory diseases. This seed oil has been found to contain coumarin (1,2-benzopyrone) which exhibits anti-inflammatory property due to its ability to inhibit vascular permeability and the resultant migration of inflammatory cells (Pereira et al. 2017). Cold-pressed seed oils such as coriander seed oil or black cumin oil are promising anti-inflammatory agent without producing any ulcerogenic effect which makes them highly beneficial in the treatment of inflammation as many anti-inflammatory agents exhibit ulcerogenic effect apart from their anti-inflammatory activity (Ibrahim et al. 2017). The phenolic, β-carotene, unsaturated fatty acids, and tocopherol contents of the pumpkin seed oil have attributed their anti-inflammatory activity as they reduce the expression of inflammatory biomarkers in rat arthritis model (Al-Okbi et al. 2017b).

8.5 Cosmetics Applications of Seed Oils

Fats and oils have always been an important component of cosmetic formulations acting as solvent and vehicles for other ingredients while providing emoliency, moisturizing and grooming or skin conditioning value to the product (Berdick 1972). Cosmetic industry is a multibillion-dollar business, and there is a growing interest toward natural product-based ingredients driven by the customer awareness to the chemicals contained in some commercial products (Vermaak et al. 2011). In addition, most of the seed oils are also proven to be the rich sources of polyunsaturated fatty acids, tocopherols, and flavonoids which impart antioxidant, antiaging, and nutritional benefits to the cosmetic products. Therefore, seed oils such as castor oil continue to be one of the most important unmodified oils in cosmetic industry.

Oxidative damage is considered to be the major cause of various human diseases especially skin problems such as inflammation and aging (Chaikul et al. 2017). The dermocosmetology benefits of seed oil such as sunflower seed oil are well known (Eichenfield et al. 2009; Del Rosso 2011). Natural sunflower (*Helianthus annuus*) seed oil contains essential fatty acids such as linoleic acid which activates peroxisome proliferator-activated receptor-α (PPAR-α) that stimulates keratinocyte differentiation, improves barrier function, and enhances lipid metabolism in the skin (Eichenfield et al. 2009). Such properties make sunflower oil an important cosmetic skin care and therapeutic formulation ingredient. Studies also confirmed that the principal fatty acid contents of mango (*Mangifera indica*) seed oil are stearic acid and oleic acid which showed good deodorizing effect against two malodorous compounds 2-nonenal and isovaleric acid (Wu et al. 2015). It also possesses low iodine number, suggesting its stability against deterioration and its potential in cosmetic application. Due to their favorable physicochemical properties, along with their beneficial fatty acids and phenolic contents, two melon (*Acanthosicyos horridus* and *C. lanatus*) seed oils from Namibia have also been implied to hold good potential to replace major commercial vegetable oils as cosmetic formulation ingredients (Cheikhyoussef et al. 2017). As cosmetics are applied on the skin, evaluation of the safety of their ingredients such as seed oils against allergic and other toxic reactions is highly important. Results from various studies indicate that most of the seed oils used in lipsticks, creams, hairs, body lotions, and various other cosmetic preparations are safe and are not dermal irritant (Burnett et al. 2017). The applications of different seed oils in cosmetic industry are listed in Table 8.2.

8.6 Conclusions and Future Prospects

Seeds are promising renewable and cheaper sources of edible and industrial oils. Many traditional systems utilize seed oils for their multiple health benefits and bioactivities (Yang et al., 2016; Pereira et al. 2017). The importance of consuming long-chain PUFAs in daily recommended amount toward cardiovascular health and thereby reducing heart diseases is well recognized by dietary guidelines of different countries (Ruxton et al. 2004). Seed oils are rich and renewable sources of healthy PUFAs and MUFAs such as omega-3-PUFA (Belayneh et al. 2015; Sande et al. 2018). Depending on the extraction and refining methods, high-quality oils that retain bio-active components can be obtained from various plant seeds. The fatty acid contents, tocopherols, phenolics, and other bio-active components in seed oils rendered exhibit several beneficial activities including antioxidant, anti-inflammatory, antidiabetic, anticancer, antimicrobial, wound healing, antihypertensive, and dermo-cosmetic along with various other valuable applications. The prospect of bio-active seed oils as antimicrobial agent is highly promising as antibiotic resistance is a growing menace and worldwide health threat. A complete understanding of the fatty acid biosynthetic pathways and application of plant biotechnology approaches are required to tailor properties and produce high-end seed oils that meet the end-use requirements (Baud 2018). Moreover, increased consumer awareness toward natural remedy and the emergence of natural-based cosmetics along with plethora of other merits make these seed oils promising industrial and health-promoting oil of the future.

Table 8.2 Some seed oils and their applications in cosmetics industry

Seed oil sources	Cosmetic application	Active constituents	References
Adansonia digitata (baobab oil)	Bath oil preparations, moisturizer, emollient and massage oil, and hot oil soaks are used for hair and nail conditioning. Helpful in eczema and psoriasis	Rich in vitamins A, D, E, and F. Natural source of vitamin D3	Vermaak et al. (2011)
Argania spinosa (argan oil)	Emollient, wound healer, antiaging, antioxidant, hair oils (nourishing for hair)	Vitamin E, phenolics, major fatty acids (oleic, palmitic, stearic, linoleic, and linolenic acids)	Barve and Dighe (2016)
Brassica napus (rapeseed oil)	Skin antiaging formulation	Proteins, phenolics, lipids, and vitamins	Rivera et al. (2015)
Camellia sp. (tea seed oil)	Antioxidant, emollient, hair products, antiaging, astringent	Polyphenols	Wang et al. (2017)
Cannabis sativa (hemp oil)	Skin regenerative, for treating eczema and acne, antiaging	Rich in omega-6 and omega-3 fatty acid	Leizer et al. (2000)
Citrullus lanatus (Kalahari melon oil)	Light skin moisturizer, skin regeneration, emollient, foaming agent; used to treat skin tanning and acne vulgaris	Carrier oil, rich source of vitamin E	Vermaak et al. (2011)
Cucurbita maxima (pumpkin seed oil)	Antioxidant, antiaging, emollient	Rich in essential fatty acid, vitamins A and E, minerals; powerhouse of nutrition, rich in amino acids, omega-6 fatty acids	Bardaa et al. (2016a, b)
Hibiscus cannabinus (kenaf seed oil)	Hard soap preparations, can also be used in cosmetic products such as lipsticks and milky lotion. The oil functions to enhance penetration, control moisture evaporation, and hydrate skin	PUFA (palmitic, oleic, and linoleic acids), vitamin E, phenolic acids	Cheng et al. (2016) and Dhar et al. (2015)
Hibiscus sabdariffa (roselle seed oil)	Antioxidant	Phenols, flavonoids, gamma tocopherol	Al-Okbi et al. (2017a), Dhar et al. (2015)

(continued)

Table 8.2 (continued)

Seed oil sources	Cosmetic application	Active constituents	References
Momordica charantia (bitter melon seed oil)	Antifungal, useful in psoriasis, itching, hair loss, dry and itchy scalp, anti-dandruff, antiaging	Naturally rich source of bio-active compounds for nutraceutical purposes. Contains high amounts of total lipids, mainly α-eleostearic acid, and considerable levels of phytosterols	Braca et al. (2008), Grover and Yadav (2004), Xu et al. (2016) and Yoshime et al. (2016)
Nigella sativa (black cumin seed oil)	Good for acne, hair loss, toothache, and headache	Rich in linoleic acid (omega-6 fatty acid), phytosterol, thymoquinone	Ali and Blunden (2003)
Phoenix dactylifera (date seed oil)	In body creams, shaving soap, and shampoos; protect skin against UVA and UVB lights (sunscreen)	PUFAs, phenolics, carotenoid	Farsi and Lee (2011) and Tafti et al. (2017)
Prunus armeniaca (apricot seed oil)	Antioxidant, antiaging, emollient; for making soap, under eye cream, face oil, hair oil	Unsaturated fats, helps in lowering blood cholesterol level, rich in omega-9 fatty acid, vitamin E	Barve and Dighe (2016)
Ricinus communis (castor oil)	Hair oil	Ricinoleic, isoricinoleic, stearic, and dihydroxy-stearic acid	Barve and Dighe (2016)
Sesamum indicum (sesame oil)	Emollient, antioxidant, hair loss and scalp psoriasis	High in MUFA and PUFA; health benefits of sesame oil are attributed to sesamin and sesamolin, not to omega-6 fatty acid	Barve and Dighe (2016)
Simmondsia chinensis (jojoba oil)	Moisturizer for hairs	Wax esters of fatty acids	Barve and Dighe (2016)
Vaccinium oxycoccos (cranberry seed oil)	Highly moisturizing, hand and body creams, shampoo	Omega-3, omega-6, and omega-9 fatty acids (high content of polyunsaturated fatty acids (PUFAs))	Van Hoed et al. (2009)
Vitis vinifera (grape-seed oil)	It is a light oil and moisturizes skin; for acne problems; as antiaging and skin lightening; promotes hair growth	Polyunsaturated fatty acids (PUFAs), vitamin E, linoleic acid, polyphenolic proanthocyanidins	Aburjai and Natsheh (2003); Shinagawa et al. (2015) and Takahashi et al. (1998)
Yucca aloifolia	Moisture retentive, anti-inflammatory, antioxidant and anti-acne	Linoleic acid, oleic acid, palmitic acid, vitamin E	Mokbli et al. 2017

Conflict of Interests There is no conflict of interest to declare in writing this chapter.

References

Aburjai T, Natsheh FM (2003) Plants used in cosmetics. Phytother Res 17:987–1000
Adams GG, Imran S, Wang S et al (2012) Extraction, isolation and characterisation of oil bodies from pumpkin seeds for therapeutic use. Food Chem 134:1919–1925
Akihisa T, Tokuda H, Ukiya M et al (2004) 3-Epicabraleahydroxylactone and other triterpenoids from Camellia oil and their inhibitory effects on Epstein-Barr Virus activation. Chem Pharm Bull 52:153–156
Ali BH, Blunden G (2003) Pharmacological and toxicological properties of *Nigella sativa*. Phytother Res 17:299–305
Al-Okbi SY, Abdel-Razek AG, Mohammed SE et al (2017a) Roselle seed as a potential new source of healthy edible oil. J Biol Sci 17:267–277
Al-Okbi SY, Mohamed DA, Kandil E et al (2017b) Anti-inflammatory activity of two varieties of pumpkin seed oil in an adjuvant arthritis model in rats. Grasas Aceites 68:e180
Amri IN (2011) The lauric (Coconut and Palm Kernel) Oils. In: Gunstone FD (ed) Vegetable oils in food technology. Composition, properties and uses, 2nd edn. Wiley-Blackwell, West Sussex, pp 169–194
Arya SS, Ramanujam S, Vijayaraghavan PK (1969) Refractive index as an objective method for evaluation of rancidity in edible oils and fats. J Am Oil Chem Soc 46:28–30
Aslam F, Iqbal S, Nasir M et al (2017) Evaluation of White Sesame Seed Oil on Glucose control and biomarkers of hepatic, cardiac, and renal functions in Male Sprague-Dawley Rats with chemically induced diabetes. J Med Food 20:448–457
Assiri AMA, Elbanna K, Abulreesh HH et al (2016) Bioactive compounds of cold-pressed thyme (*Thymus vulgaris*) oil with antioxidant and antimicrobial properties. J Oleo Sci 65:629–640
Badr SEA, Shaaban M, Elkholy YM et al (2011) Chemical composition and biological activity of ripe pumpkin fruits (*Cucurbita pepo* L.) cultivated in Egyptian habitats. Nat Prod Res 25:1524–1539
Baiao NC, Lara LJC (2005) Oil and fat in broiler nutrition. Rev Bras Cienc Avic 7:129–141
Bansal P, Medhe S, Ganesh N et al (2015) Antimelanoma potential of *Eruca sativa* seed oil and its bioactive principles. Ind J Pharm Sci 77:208–217
Bardaa S, Halima NB, Aloui F et al (2016a) Oil from pumpkin (*Cucurbita pepo* L.) seeds: evaluation of its functional properties on wound healing in rats. Lipids Health Dis 15:1–12
Bardaa S, Moalla D, Khedir BS et al (2016b) The evaluation of the healing proprieties of pumpkin and linseed oils on deep second-degree burns in rats. Pharm Biol 54:581–587
Barve K, Dighe A (2016) The Chemistry and Applications of Sustainable Natural Hair Products. Springer, Cham
Baud S (2018) Seeds as oil factories. Plant Reprod 31:213. https://doi.org/10.1007/s00497-018-0325-6
Belayneh HD, Wehling RL, Cahoon E et al (2015) Extraction of omega-3-rich oil from *Camelina sativa* seed using supercritical carbon dioxide. J Supercrit Fluids 104:153–159
Berdick M (1972) The role of fats and oils in cosmetics. J Am Oil Chem Soc 49:406–408
Berraaouan A, Ziyyat A, Mekhfi H et al (2014) Evaluation of antidiabetic properties of cactus pear seed oil in rats. Pharm Biol 52:1286–1290
Bhuiya MMK, Rasul MG, Khan MMK et al (2015) Optimisation of oil extraction process from Australian native Beauty leaf seed (*Calophyllum inophyllum*). Energy Procedia 75:56–61
Braca A, Siciliano T, Arrigo DM et al (2008) Chemical composition and antimicrobial activity of *Momordica charantia* seed essential oil. Fitoterapia 79:123–125
Bromke MA, Hochmuth A, Tohge T et al (2015) Liquid chromatography high-resolution mass spectrometry for fatty acid profiling. Plant J 81:529–536

Burnett CL, Fiume MM, Bergfeld WF et al (2017) Safety assessment of plant-derived fatty acid oils. Int J Toxicol 36(Supplement 3):51S–129S

Cavdar HK, Yanik DK, Gok U et al (2017) Optimisation of microwave-assisted extraction of pomegranate (*Punica granatum* L.) seed oil and evaluation of its physicochemical and bioactive properties. Food Technol Biotechnol 55:86–94

Chaikul P, Lourith N, Kanlayavattanakul M (2017) Antimelanogenesis and cellular antioxidant activities of rubber (*Hevea brasiliensis*) seed oil for cosmetics. Ind Crop Prod 108:56–62

Chaiyasit W, Elias RJ, McClements DJ et al (2007) Role of physical structures in bulk oils on lipid oxidation. Crit Rev Food Sci Nutr 47:299–317

Chaliha B, Lahkar L, Doley A et al (2017) Screening of some lesser known tree-borne oilseed plants from North-East India for their oil content and major fatty acid components. Prostaglandins Leukot Essent Fatty Acids 126:9–19

Cheikhyoussef N, Kandawa-Schulz M, Bock R et al (2017) Characterization of *Acanthosicyos horridus* and *Citrullus lanatus* seed oils: two melon seed oils from Namibia used in food and cosmetics applications. 3 Biotech 7:297

Cheng WY, Akanda JMH, Nyam KL (2016) Kenaf seed oil: a potential new source of edible oil. Trend Food Sci Technol 52:57–65

Chilaka KC, Ifediba EC, Ogamba JO (2015) Evaluation of the effects of *Citrus sinensis* seed oil on blood glucose, lipid profile and liver enzymes in rats injected with alloxan monohydrate. J Acute Dis 4:129–134

Choudhary BR, Haldhar SM, Maheshwari SK et al (2015) Phytochemicals and antioxidants in watermelon (*Citrullus lanatus*) genotypes under hot arid region. Ind J Agric Sci 85:414–417

Cmolik J, Pokorny J (2000) Physical refining of edible oils. Eur J Lipid Sci Technol 102:472–486

Czaplicki S, Ogrodowska D, Derewiaka D et al (2011) Bioactive compounds in unsaponifiable fraction of oils from unconventional sources. Eur J Lipid Sci Technol 113:1456–1464

Dabas D, Shegog RM, Zeigler GR et al (2013) Avocado (*Persea americana*) seed as a source of bioactive phytochemicals. Curr Pharm Des 19:6133–6140

Dandawate PR, Subramaniam D, Padhye SB et al (2016) Bitter melon: a panacea for inflammation and cancer. Chinese J Nat Med 14:81–100

Rosso D (2011) Repair and maintenance of the epidermal barrier in patients diagnosed with atopic dermatitis. An evaluation of the components of a body wash-moisturizer skin care regimen directed at management of atopic skin. J Clin Aesthet Dermatol 4:45–55

Dhar P, Kar CS, Ojha D et al (2015) Chemistry, phytotechnology, pharmacology and nutraceutical functions of kenaf (*Hibiscus cannabinus* L.) and roselle (*Hibiscus sabdariffa* L.) seed oil: an overview. Ind Crop Prod 77:323–332

Dolatowski ZJ, Stadnik J, Stasiak D (2007) Applications of ultrasound in food technology. Acta Sci Pol Technol Aliment 6:89–99

Durdevic S, Savikin K, Zivkovic J et al (2018) Antioxidant and cytotoxic activity of fatty oil isolated by supercritical fluid extraction from microwave pretreated seeds of wild growing *Punica granatum* L. J Supercrit Fluids 133:225–232

Easmin MS, Sarker MZI, Ferdosh S et al (2015) Bioactive compounds and advanced processing technology: *Phaleria macrocarpa* (sheff.) Boerl, a review. J Chem Technol Biotechnol 90:981–991

Eichenfield LF, McCollum A, Msika P (2009) The benefits of sunflower oleodistillate (SOD) in pediatric dermatology. Pediatr Dermatol 26:669–675

El-Mosallamy AEMK, Sleem AA, Abdel-Salam OME et al (2012) Antihypertensive and cardioprotective effects of pumpkin seed oil. J Med Food 15:180–189

Evans WC (1996) Trease and Evans' Pharmacognosy, 14th edn. WB Saunders Company Ltd, London, p 180

Farsi MA, Lee CY (2011) Usage of Date (*Phoenix dactylifera* L.) seeds in human health and animal feed. In: Preedy VR, Watson RR, Patel VB (eds) Nuts and seeds in health and disease prevention. Academic Press, London, pp 447–452

Faugno S, Del Piano L, Crimaldi M et al (2016) Mechanical oil extraction of *Nicotiana tabacum* L seeds: analysis of main extraction parameters on oil yield. J Agric Eng 47:142–147

Freitas IR, Cattelan MG, Rodrigues ML et al (2017) Effect of grape seed extract (*Vitis labrusca* L.) on soybean oil under thermal oxidation. Nutr Food Sci 47:610–622

Gecgel U, Demirci AS, Dulger GC et al (2016) Some physicochemical properties, fatty acid composition and antimicrobial characteristics of different cold-pressed oils. Riv Ital Sostanze Gr 92:187–200

Ghazani SM, Llatas GG, Marangoni AG (2013) Minor constituents in canola oil processed by traditional and minimal refining methods. J Am Oil Chem Soc 90:743–756

Gil-Chavez GJ, Villa JA, Ayala-Zavala JF et al (2013) Technologies for extraction and production of bioactive compounds to be used as nutraceuticals and food ingredients: an overview. Compr Rev Food Sci Food Saf 12:5–22

Gopalkrishna AH, Seshagiri M, Muddaiah S et al (2016) *In vitro* antifungal activity of different components of *Centratherum anthelminticum* and *Ocimum sanctum* seed oils and their synergism against oral pathogenic fungi. J Dent Res Dent Clin Dent Prospect 10:92–98

Govindachari TR, Suresh G, Gopalakrishna G et al (1998) Identification of antifungal compounds from the seed oil of *Azadirachta indica*. Phytoparasitica 26:109–116

Grover JK, Yadav SP (2004) Pharmacological actions and potential uses of *Momordica charantia*: a review. J Ethnopharmacol 93:123–132

Gulfraz M, Sadiq A, Tari H et al (2011) Phytochemical analysis and antibacterial activity of *Eruca sativa* seed. Pak J Bot 43:1351–1359

Guo S, Ge Y, Jom KN (2017) A review of phytochemistry, metabolite changes, and medicinal uses of the common sunflower seed and sprouts (*Helianthus annuus* L.). Chem Cent J 11:95

Hrabovski N, Sinadinovic-Fiser S, Nikolovski B et al (2012) Phytosterols in pumpkin seed oil extracted by organic solvents and supercritical CO_2. Eur J Lipid Sci Technol 114:1204–1211

Ibrahim FM, Attia HN, Maklad YAA et al (2017) Biochemical characterization, anti-inflammatory properties and ulcerogenic traits of some cold-pressed oils in experimental animals. Pharm Biol 55:740–748

Imran M, Nadeem M, Manzoor MF et al (2016) Fatty acids characterization, oxidative perspectives and consumer acceptability of oil extracted from pre-treated chia (*Salvia hispanica* L.) seeds. Lipids Health Dis 15:162

Indian Pharmacopoeia (2007) Saponification value, vol I. Indian Pharmacopoeial Commission, Ghaziabad, p 89

Jasicka-Masiak I, Lipok J, Nowakowska EM et al (2004) Antifungal activity of the carrot seed oil and its major sesquiterpene compounds. Z Naturforsch C 59:791–796

Jorge N, Da Silva AC, Aranha CPM (2016) Antioxidant activity of oils extracted from orange (*Citrus sinensis*) seeds. An Acad Bras Cienc 88:951–958

Kai H, Akamatsu E, Torii E et al (2011) Inhibition of proliferation by agricultural plant extracts in seven human adult T-cell leukaemia (ATL)-related cell lines. J Nat Med 65:651–655

Kai H, Akamatsu E, Torii E et al (2014) Identification of a bioactive compound against adult T-cell leukemia from Bitter Gourd seeds. Plan Theory 3:18–26

Kaki SS, Jabeen T, Reddy JRC et al (2016) Isolation and physico-chemical characterization of *Butea parviflora* seed oil. Int J Fats Oils 67:1–6

Kardash E, Tur'yan Y (2005) Acid value determination in vegetable oils by Indirect titration in aqueous-alcohol media. Croat Chem Acta 78:99–103

Khalesi S, Irwin C, Schubert M (2015) Flaxseed consumption may reduce blood pressure: a systematic review and meta-analysis of controlled trials. J Nutr 145:758–765

Khan LM, Hanna MA (1983) Expression of oil from oil seeds – a review. J Agric Engng Res 28:495–503

Khoddami A, Roberts TH (2015) Pomegranate oil as a valuable pharmaceutical and nutraceutical. Lipid Technol 27:40–42

Khoobchandani M, Ganesh N, Gabbanini S et al (2011) Phytochemical potential of *Eruca sativa* for inhibition of melanoma tumor growth. Fitoterapia 82:647–653

Khoobchandani M, Ojeswi BK, Ganesh N et al (2010) Antimicrobial properties and analytical profile of traditional *Eruca sativa* seed oil: Comparison with various aerial and root plant extracts. Food Chem 120:217–224

Kim J, Kim DN, Lee SH et al (2010) Correlation of fatty acid composition of vegetable oils with rheological behaviour and oil uptake. Food Chem 118:398–402
Koriem KMM (2013) Review on pharmacological and toxicologyical effects of oleum azadirachti oil. Asian Pac J Trop Biomed 3:834–840
Kothari V (2013) Bioactive compounds from plant seeds/seed extracts. In: Martirosyan D (ed) Introduction to functional food science, 1st edn. Food Science Publisher, Texas, pp 86–93
Lee JH, Bae SY, Oh M et al (2016) Antiviral effects of black raspberry (*Rubus coreanus*) seed extract and its polyphenolic compounds on norovirus surrogates. Biosci Biotechnol Biochem 80:1196–1204
Leizer C, Ribnicky D, Poulev A et al (2000) The composition of Hemp seed oil and its potential as an important source of nutrition. J Nutraceuticals Funct Med Foods 2:35–53
Lin TK, Zhong L, Santiago JL (2018) Anti-inflammatory and skin barrier repair effects of topical application of some plant oils. Int J Mol Sci 19:70
Markovic VV, Bastic LV (1976) Characteristics of pumpkin seed oil. J Am Oil Chem Soc 53:42–44
Matthaus B (2012) The database Seed Oil Fatty Acids (SOFA) is back on the Internet! Eur J Lipid Sci Technol 114:701–702
McClements DJ (1995) Advances in the application of ultrasound in food analysis and processing. Trend Food Sci Technol 6:293–299
Mericli F, Becer E, Kabadayi H et al (2017) Fatty acid composition and anticancer activity in colon carcinoma cell lines of *Prunus dulcis* seed oil. Pharm Biol 55:1239–1248
Mohd-Esa N, Hern FS, Ismail A et al (2010) Antioxidant activity in different parts of roselle (*Hibiscus sabdariffa* L.) extracts and potential exploitation of the seeds. Food Chem 122:1055–1060
Mokbli S, Nehdi IA, Sbihi H (2017) *Yucca aloifolia* seed oil: A new source of bioactive compounds. Waste Biomass Valor 9:1087. https://doi.org/10.1007/s12649-017-9892-2
Morgan DA (1942) Smoke, fire, and flash points of cottonseed, peanut, and other vegetable oils. J Am Oil Chem Soc 19:193–199
Navaratne SB, Subasinghe DSJ (2014) Determination of fatty acid profile and physicochemical properties of Watermelon and Soursop seed oils. Eur Int J App Sci Technol 1:26–32
Nyam KL, Sin LN, Kamaria L (2015) Phytochemical analysis and anti-inflammatory effect of Kenaf and Roselle seeds. Mal J Nutr 22:245–254
Nyam KL, Tan CP, Lai OM et al (2009a) Physicochemical properties and bioactive compounds of selected seed oils. LWT Food Sci Technol 42:1396–1403
Nyam KL, Tan CP, Lai OM et al (2011) Optimization of supercritical CO_2 extraction of phytosterol-enriched oil from Kalahari melon seeds. Food Bioprocess Technol 2:1432–1441
Nyam KL, Tan CP, Man YBC et al (2009b) Physiochemical properties of Kalahari melon seed oil following extractions using solvent and aqueous enzymatic methods. Int J Food Sci Technol 44:694–701
Nyam KL, Teh YN, Tan CP et al (2012) *In vitro* antioxidant activities of extract and oil from Roselle (*Hibiscus sabdariffa* L.) seed against Sunflower oil autoxidation. Mal J Nutr 18:265–274
Nyam KL, Wong MM, Long K et al (2013) Oxidative stability of sunflower oils supplemented with kenaf seeds extract, roselle seeds extract and roselle extract, respectively under accelerated storage. Int Food Res J 20:645–651
Ogrodowska D, Czaplicki S, Zadernowski R et al (2012) Phenolic acids in seeds and products obtained from *Amaranthus cruentus*. J Food Nutr Res 51:96–101
Ogrodowska D, Zadernowski R, Czaplicki S et al (2014) Amaranth seeds and products – The source of bioactive compounds. Pol J Food Nutr Sci 64:165–170
Ogunniyi DS (2006) Castor oil: A vital industrial raw material. Bioresour Technol 97:1086–1091
Oh M, Chung MS (2014) Effects of oils and essential oils from seeds of *Zanthoxylum schinifolium* against foodborne viral surrogates. Evid Based Complement Alternat Med 135797:1–6
Padmashree A, Sharma GK, Semwal AD (2011) Studies on the antioxygenic activity of bitter gourd (*Momordica charantia*) and its fractions using various *in vitro* models. J Sci Food Agric 91:776–782

Parry J, Su L, Luther M et al (2005) Fatty acid composition and antioxidant properties of cold-pressed marionberry, boysenberry, red raspberry, and blueberry seed oils. J Agric Food Chem 53:566–573

Parry J, Su L, Moore J et al (2006) Chemical compositions, antioxidant capacities, and antiproliferative activities of selected fruit and seed flours. J Agric Food Chem 54:3773–3778

Patel S (2013) Pumpkin (*Cucurbita* sp.) seeds as nutraceutic: a review on status quo and scopes. Mediterr J Nutr Metab 6:183–189

Paz SM, Aguilar FM, Gimenez MDG et al (2014) Hemp (*Cannabis sativa* L.) seed oil: Analytical and phytochemical characterization of unsaponifiable fraction. J Agric Food Chem 62:1105–1110

Pereira MP, De Melo-Junior AF, Royo VA et al (2017) Histochemical and phytochemical study of *Amburana cearensis* (Fabaceae) seed oil. Nat Prod Commun 12:907–910

Perez C, Del Castillo LR, Gil C et al (2015) Supercritical fluid extraction of grape seeds: extract chemical composition, antioxidant activity and inhibition of nitrite production in LPS -stimulated Raw 264.7 cells. Food Funct 6:2607–2613

Pieszka M, Tombarkiewicz B, Roman A et al (2013) Effect of bioactive substances found in rapeseed, raspberry and strawberry seed oils on blood lipid profile and selected parameters of oxidative status in rats. Environ Toxicol Pharmacol 36:1055–1062

Porto CD, Natolino A, Decorti D (2015) Effect of ultrasound pre-treatment of hemp (*Cannabis sativa* L.) seed on supercritical CO_2 extraction of oil. J Food Sci Technol 52:1748–1753

Prado JM, Dalmolin I, Carareto NDD et al (2012) Supercritical fluid extraction of grape seed: Process scale-up, extract chemical composition and economic evaluation. J Food Eng 109:249–257

Prescha A, Grajzer M, Dedyk M et al (2014) The antioxidant activity and oxidative stability of Cold-Pressed oils. J Am Oil Chem Soc 91:1291–1301

Rao NK, Nammi S (2006) Antidiabetic and renoprotective effects of the chloroform extract of *Terminalia chebula* Retz. seeds in streptozotocin-induced diabetic rats. BMC Compliment Altern Med 6:17

Rekik DM, Khedir SB, Moalla KK et al (2016) Evaluation of wound healing properties of grape seed, sesame, and fenugreek oils. Evid Based Complement Alternat Med 2016:7965689

Rivera D, Rommi K, Fernandes MM et al (2015) Biocompounds from rapeseed oil industry co-stream as active ingredients for skin care applications. Int J Cosmet Sci 37:496–505

Rodriguez G, Lama A, Rodriguez R (2008) Olive stone an attractive source of bioactive and valuable compounds. Bioresour Technol 99:5261–5269

Romero LAF, Piantino CR, Grimaldi R et al (2009) Supercritical CO_2 extraction of omega-3 rich oil from Sachainchi (*Plukenetia volubilis* L.) seeds. J Supercrit Fluids 49:323–329

Ruxton CHS, Reed SC, Simpson MJA et al (2004) The health benefits of omega-3 polyunsaturated fatty acids: a review of the evidence. J Hum Nutr Dietet 17:449–459

Saavedra MJ, Aires A, Dias C et al (2013) Evaluation of the potential of squash pumpkin by-products (seeds and shell) as sources of antioxidant and bioactive compounds. J Food Sci Technol 52:1008–1015

Sadoudi R, Ahmed AD (2017) Studies on physico-chemical characteristics and fatty acid composition of commercially available Algerian frying edible oils. Int Food Res J 24:60–67

Saini RK, Assefa AD, Keum YS (2017) Fatty acid and carotenoid composition of bitter melon (*Momordica charantia* L.) seed arils: a potentially valuable source of lycopene. J Food Meas Charact 11:1266–1273

Salanta LC, Tofana M, Socaci S (2015) Evaluation of polyphenol content and antioxidant properties of some fruit seeds. Bulletin UASVM Food Sci Technol 72:267–268

Samarji R, Balbaa M (2014) Anti-diabetic activity of different oils through their effect on arylsulfatases. J Diabetes Metab Disord 13:116

Sanad RAB, Mabrouk MI (2016) Development and assessment of stable formulations containing two herbal antimicrobials: *Allium sativum* L. and *Eruca sativa* miller seed oils. Drug Dev Ind Pharm 42:958–968

Sande D, Colen G, Santos D et al (2018) Production of omega 3, 6, and 9 fatty acids from hydrolysis of vegetable oils and animal fat with *Colletotrichum gloeosporioides* lipase. Food Sci Biotechnol 27:537–545
Savoire R, Lanoiselle JL, Vorobiev E (2013) Mechanical continuous oil expression from oilseeds: a review. Food Bioprocess Tech 6:1–16
Shah BN, Seth AK (2010) Textbook of Pharmacognosy and Phytochemistry. Elsevier, New Delhi, p 343
Shim YY, Gui B, Wang Y et al (2015) Flaxseed (*Linum usitatissimum* L.) oil processing and selected products. Trend Food Sci Technol 43:162–177
Shinagawa FB, Santana FC, Torres LRO et al (2015) Grapeseed oil: a potential functional food? Food Sci Technol Campinas 35:399–406
Silva ACD, Jorge N (2016) Bioactive compounds of oils extracted from fruits seeds obtained from agroindustrial waste. Eur J Lipid Sci Technol 118:1–5
Simurdiak M, Olukoga O, Hedberg K (2016) Obtaining the Iodine value of various oils via Bromination with Pyridinium tribromide. J Chem Educ 93:322–325
Singh J, Bargale PC (2000) Development of a small capacity double stage compression screw press for oil expression. J Food Eng 43:75–82
Sora AC, Villamiel M (2010) Effect of ultrasound on the technological properties and bioactivity of food: a review. Trend Food Sci Technol 21:323–331
Stanisavljevic IT, Velickovic DT, Todorovic ZB et al (2009) Comparison of techniques for the extraction of tobacco seed oil. Eur J Lipid Sci Technol 111:513–518
Stoia M, Oancea S (2013) Health reasons for improving the oxidative stability of sunflower oil. Rev Oxid Commun 36:636–668
Su J, Wang H, Ma C et al (2015) Anti-diabetic activity of peony seed oil, a new resource food in STZ-induced diabetic mice. Food Funct 6:2930–2938
Sun X, Zhang L, Li P et al (2015) Fatty acid profiles based adulteration detection for flaxseed oil by gas chromatography mass spectrometry. LWT Food Sci Technol 63:430–436
Swisher MC, Fiero GW (1932) The solvent extraction of castor oil seeds. J Pharm Sci 21:579–582
Tafti GA, Dahdivan SN, Ardakani YSA (2017) Physicochemical properties and applications of date seed and its oil. Int Food Res J 24:1399–1406
Takahashi T, Kamiya T, Yokoo Y (1998) Proanthocyanidins from grape seeds promote proliferation of mouse hair follicle cells *in vitro* and convert hair cycle *in vivo*. Acta Derm Venereol 78:428–432
Teixeira A, Baenas N, Perles RD et al (2014) Natural bioactive compounds from winery by-products as health promoters: a review. Int J Mol Sci 15:15638–15678
United States Pharmacopoeia (2005) <841> Specific gravity. United States Pharmacopeial Convention, Maryland
Ursoniu S, Sahebkar A, Andrica F et al (2016) Effects of flaxseed supplements on blood pressure: A systematic review and meta-analysis of controlled clinical trial. Clin Nutr 35:615–625
Vaisali C, Charanyaa S, Belur PD et al (2014) Refining of edible oils: a critical appraisal of current and potential technologies. Int J Food Sci Technol 50:12–23
Van Hoed V, De Clercq N, Echim C et al (2009) Berry seeds: A source of specialty oils with high content of bioactives and nutritional value. J Food Lipid 16:33–49
Vermaak I, Kamatou GPP, Komane-Mofokeng B et al (2011) African seed oils of commercial importance – Cosmetic applications. S Afr J Bot 77:920–933
Veronezi CM, Jorge N (2012) Bioactive compounds in lipid fractions of pumpkin (*Cucurbita* sp) seeds for use in food. J Food Sci 77:653–657
Wang D, Qu X, Zhuang X et l (2016a) Seed oil of *Brucea javanica* induces cell cycle arrest and apoptosis via Reactive Oxygen Species-mediated mitochondrial dysfunction in human lung cancer cells. Nutr Cancer 68:1394–1403
Wang F, Li H, Zhao H et al (2018) Antidiabetic activity and chemical composition of *Sanbai* melon seed oil. Evid Based Complement Alternat Med 5434156:1–14
Wang S, Zheng Y, Xiang F et al (2016b) Antifungal activity of *Momordica charantia* seed extracts toward the pathogenic fungus *Fusarium solani* L. J Food Drug Anal 24:881–887

Wang T (2011) Soybean oil. In: Gunstone FD (ed) Vegetable oils in food technology. Composition, properties and uses, 2nd edn. Wiley-Blackwell, West Sussex, pp 59–98
Wang X, Zeng Q, Contreras MDM et al (2017) Profiling and quantification of phenolic compounds in *Camellia* seed oils: Natural tea polyphenols in vegetable oil. Food Res Int 102:184–194
Watanabe S, Hayano S, Akiya T et al (1973) Determination of fatty acid composition by gas chromatography: I. Analysis with use of thermal conductivity detector. J Am Oil Chem Soc 50:357–359
Wu S, Tokuda M, Kashiwagi A et al (2015) Evaluation of the fatty acid composition of the seeds of *Mangifera indica* L. and their application. J Oleo Sci 64:479–484
Xu Y, Xu L, Chen XT et al (2016) Bitter melon seed oil may reduce the adiposity through the hypothalamus mTOR signaling in mice fed a high fat diet. J Nutr Intermed Metab 6:16–21
Yamasaki M, Motonaga C, Yokoyama M et al (2013) Induction of apoptotic cell death in HL-60 cells by Jacaranda seed oil derived fatty acids. J Oleo Sci 62:925–932
Yang C, Liu XM, Chen Z et al (2016) Comparison of oil content and fatty acid profile of ten new *Camellia oleifera* cultivars. J Lipids 3982486:1–6
Yang X, Zhang D, Song L et al (2017) Chemical profile and antioxidant activity of the oil from Peony seeds (*Paeonia suffruticosa* Andr.). Oxidative Med Cell Longev 9164905:1–11
Yoshime LT, de Melo ILP, Sattler JAG et al (2016) Bitter gourd (*Momordica charantia* L.) seed oil as a naturally rich source of bioactive compounds for nutraceutical purposes. Forum Nutr 41:1–7
Yu LL, Zhou KK, Parry J (2005) Antioxidant properties of cold-pressed black caraway, carrot, cranberry, and hemp seed oils. Food Chem 91:723–729
Yusuf KA, Akhigbe KE, Izuagie FI et al (2017) Performance evaluation of a screw press for extraction of Groundnut (*Arachease hypogeal*) seeds and Cashew (*Anarcardium Occidentale*) Kernel. Int J Eng Inf Syst 1:1–8
Zahir E, Saeed R, Hameed MA et al (2017) Study of physicochemical properties of edible oil and evaluation of frying oil quality by Fourier Transform-Infrared (FT-IR) Spectroscopy. Arab J Chem 10:S3870–S3876
Zhang HF, Yang XH, Wang Y (2011) Microwave assisted extraction of secondary metabolites from plants: Current status and future directions. Trend Food Sci Technol 22:672–688
Zhang Z, Liu Y, Che L (2018) Optimization of supercritical carbon dioxide extraction of *Eucommia ulmoides* seed oil and quality evaluation of the oil. J Oleo Sci 67:255–263

Essential Oils Extracted from Medicinal Plants and Their Applications

9

Desam Nagarjuna Reddy

Contents

Abstract

Essential oils (EOs) are extracted from flowers, leaves, barks, roots, and fruits of the medicinal plants using hydrodistilation or steam distillation and continuous solvent extraction. EOs are mixture of chemical constituents which have less molecular weight substances, such as alcohols, polyphenols, terpenoids, carbonyl compounds, and aliphatic compounds which provide smell and possess

D. N. Reddy (✉)
Natural Products and Analytical Chemistry, Anantha Lakshmi Institute of Technology and Sciences, Anantapur, Andhra Pradesh, India

Centre for Environmental Research and Studies, Jazan University, Jizan, Saudi Arabia
e-mail: dndnrchem@gmail.com

M. S. Akhtar et al. (eds.), *Natural Bio-active Compounds*,
https://doi.org/10.1007/978-981-13-7154-7_9

biological properties. EOs have been used as folk medicine throughout the history. Nowadays, EOs are widely used as an alternative medicine in varied industries such as pharmaceutical, agricultural, sanitary, and food industries due to their antibacterial, antifungal, antiviral, antiparasitical, antidiabetic, anticancer (cytotoxic), insect repellent, food industry (flavoring), aromatherapy, antioxidant, perfume, and cosmetic properties. EOs have a great demand and interest as cosmetic and pharmaceutical substances. The isolation, identification, and characterization of major components of EOs have a premier significance. Individual compounds present in EOs mixture such as thymol, camphor, limonene, α-pinene, terpinolene, menthol, menthone, etc. exhibit wide-ranging biological properties. Commercially, still synthetic chemicals are widely used as biological activities than the EOs from the plants. However, EOs from natural sources are more effective and safe for human health and the environment compared to the synthetic chemicals. The aim of the present chapter is to discuss the specific chemical compounds occurring in EOs, their medical applications, and economic importance.

Keywords
Essential oils · Natural products · Antibacterial · Antifungal · Anticancer

9.1 Introduction

Essential oils (EOs) are obtained from aromatic and medicinal plants as a volatile mixture of chemical compounds with strong odor. EOs are extracted from the aromatic and medicinal plants using steam or hydrodistillation or Soxhlet extraction (solvent extraction or continuous extraction) methods developed in the middle ages by Arabs (Bakkali et al. 2008; Raut and Karuppayil 2014). EOs are considered as one of the most predominant plant products in agriculture, as they exhibit antifungal, antibacterial, antioxidant, anticancer, antidiabetic, antiviral, insect repellent, and anti-inflammatory properties (Buchbauer 2010; Teixeira et al. 2013; Raut and Karuppayil 2014; Said et al. 2016; Swamy et al. 2016).

Research on artificial pharmaceutical substances reveals the significance of EOs extracted from medicinal and aromatic plants, as their therapeutic properties have numerous applications. Consequently, researchers and farmers have been motivated to expand the cultivation and market these substances (Swamy and Sinniah 2015, 2016). Presently, about 100 herbs are known for their EOs, while more than 2000 herbs scattered across 60 families, such as Umbelliferae, Lamiaceae, Lauraceae, Myrtaceae, etc., could produce medicinally valued EOs. In global markets, only 300 among 3000 known types of EOs are deemed to be of commercial importance. EOs have found application in agricultural sectors and can be potentially used in other industries, such as pharmaceuticals, drugs, food, perfumes, makeup products, sanitary products, dentistry, food preservatives, additives, cosmetics, and natural remedies (Swamy et al. 2016; Mahmoudi 2017). EOs, in perfumes, creams, soaps, in flavor and fragrance for foods, sanitary products and industrial solvents phytocompounds, such as limonene, patchoulol, geranyl acetate, etc., derived from have been widely used. Moreover, essential oil blends are used in bath products and in aromatherapy. Further, many EOs are particularly valued for their medicinal properties (Swamy and Sinniah 2015, 2016; Arumugam et al. 2016). For example, menthol EOs are used as natural bug repellent, as well as for treating joint pain, respiratory allergies, muscle pain, headache, hair growth, and fever relief, as well as in cancer treatment (menthol protects against cell death and DNA damage).

EOs or natural products are widely used as fragrances. However, their application in human health, agricultural industry, and environmental protection requires better understanding of their biological properties. Some of the EOs and their chemical constituents are viable as alternatives to the synthetic compounds, presently widely used in the chemical industry. This is because EOs are not associated with harmful side effects (Carson and Riley 2003). In nature, EOs play an important role in providing plant protection against pathogenic bacteria, viruses, and fungi and preventing the attack by insect pests. In addition, EOs can attract or repel insects when present in pollen and seeds. To protect chemical compounds' ecological equilibrium, the use of EOs in pharmaceutical, food, bactericidal, and fungicidal is becoming more prevalent in recent times. EOs yielding medicinal and aromatic plants are normally native to warm countries, where they represent an important traditional pharmacopeia (Arumugam et al. 2016). EOs are less dense than water. They are volatile and mostly colorless, as well as soluble in organic solvents. All plant parts, such as buds, leaves, fruits, bark, root, stems, twigs, and flowers, can contain EOs.

Different methods can be applied for essential oil extraction, such as hydrodistillation, steam distillation, and solvent extraction (including liquid carbon dioxide or microwave extraction). For example, hydrodistillation or steam distillation is typically used for Citrus and Lamiaceae family members. Various factors, such as the extraction method, geographical conditions, type of soil, plant material, and harvesting stage, are being reported to influence on the occurrence of number of chemical constituents in EOs and variations in EO quality and yield (Masotti et al. 2003; Angioni et al. 2006; Swamy and Sinniah 2015; Swamy et al. 2016). In order to ensure a constant chemical composition, quality, and quantity, EOs should be extracted under the same conditions, such as using same plant organs, extraction method, harvesting period or season, and growing plants in the same soil types. Many of the EOs are commercialized and chemotyped by gas chromatography mass spectrometry (GC-MS), and the results have been published in international organizations like the ISO, WHO, EP (European pharmacopoeia), and Council of Europe (Smith et al. 2005) to protect good grade and amount of EOs.

Apiaceae, Lamiaceae, Myrtaceae, Poaceae, and Rutaceae families are of particular importance for medicinal applications. For example, some of the EOs, like anise, caraway, black caraway, clove, oregano, cumin, coriander, sage, basil, dill, lemon balm, peppermint, thyme, and tea oils, already have widespread medicinal applications. Some of the essential oil containing plant families, like Liliaceae, Fabaceae, Pinaceae, Piperaceae, Cupressaceae, and Hypericaceae, also exhibit a considerable medicinal potential (Hammer and Carson 2011). The aim of the present chapter is to discuss the specific chemical compounds occurring in EOs, their medical applications, and economic importance.

9.2 Chemical Composition of Essential Oils

EOs are volatile liquids that are rarely colored. They are complex mixtures comprising of different concentrations, quantities, and compositions of 20–60 chemical components (Bakkali et al. 2008). Among these, two to three major chemical compounds are known to occur in prominent concentrations (20–70%), while other components are present in less concentration. For example, menthone (39.55%) and isopulegone (30.49%) are the major components of *Mentha longifolia* essential oil (Nagarjuna et al. 2017), while cinnamyl acetate (41.98%) is extracted from *Cinnamomum zeylanicum* Blume (Jayaprakash et al. 2000). Similarly, eugenol (86.02%) is obtained from *Cinnamomum verum* (Patel et al. 2013), whereas linalool (46.97%) and 1,8-cineole (14.97%) are the major components of *Ocimum basilicum* (Santoro et al. 2007a, b). Likewise, *Pogostemon cablin* essential oil possesses mainly the patchouli alcohol, also called as patchoulol (32–37%), a tricyclic sesquiterpene (Swamy and Sinniah 2015). While, the leaf essential oil of *Plectranthus amboinicus* is rich in carvacrol (43%), thymol (7%) (a phenolic monoterpenes) (Arumugam et al. 2016). Mainly, higher concentrations of chemical constituents govern the biological properties of the EOs. Most of the EOs also constitute low molecular weight chemical components, such as terpenes and terpenoids (Croteau et al. 2000; Betts 2001; Bowels 2003; Pichersky et al. 2006; Swamy and Sinniah 2015; Arumugam et al. 2016). Terpenes and terpenoids, along with other of aliphatic and aromatic chemical constituents, are shown in Fig. 9.1.

Terpenes are biosynthetically derived isoprene (2-methyl 1,3-butadiene) units. The molecular formula of isoprene unit is C_5H_8. Thus, the basic molecular formula of terpenes comprises of multiples of isoprene units, such as $(C_5H_8)n$, where n denotes the number of isoprene units. This is known as biogenetic rule or C_5 rule. The isopentenyl diphosphate (IPP) molecule has a major role in the terpenes biosynthesis. As chains of IPP units accumulate (acyclic or cyclic), the resulting terpenes are classified based on the size into hemiterpenes (C_5), monoterpenes (C_{10}), sesquiterpenes (C_{15}), diterpenes (C_{20}), triterpenes (C_{30}), and tetraterpenes (C_{40}). Terpene that is having oxygen is called oxygenated terpenoid. EOs consist of 90% monoterpene (a combination of two isoprene units) molecules, thus allowing for a variety of structures and functions. Sesquiterpenes are also present in EOs, but they are not like monoterpenes as main. Sesquiterpenes can also assume a variety of structures and functions, as shown in Table 9.1. When a chemical constituent is optically active, the two optical isomers are frequently obtained in various plants. For example, optical isomers of (+)-α-pinene and (−)-β-pinene can be obtained from *P. palustris* and *P. caribaea,* respectively, while optical isomers of linalool obtained (−)linalol is sourced from *C. sativum* and (+)linalool from a few *C. camphora* plants. Sometimes, a racemic mixture is also encountered, whereby (±)-citronellol is very common. In particular, (+) citronellol from *Eucalyptus citriodora* and the rose and geranium EOs (−) citronellol form is common.

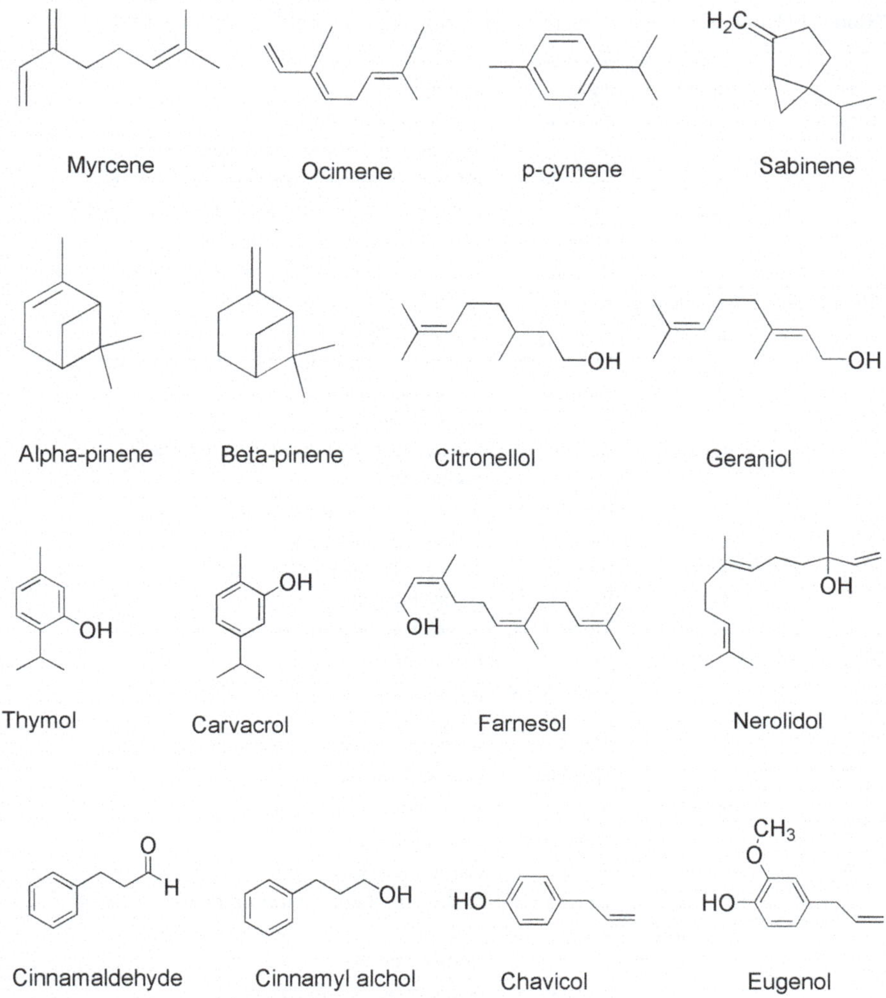

Fig. 9.1 Chemical structures of selected components of essential oils

The EOs terpenes are major chemical constituents than aromatic hydrocarbons. In plants, the biosynthetic pathways of aromatic hydrocarbons (phenyl propane) and terpene derivatives are completely different. For example, cinnamaldehyde is a major compound in cinnamon and clove oil, while eugenol is a minor constituent. Aromatic hydrocarbons generally occur in plants, namely, *C. sativum*, *S. aromaticum*, *P. anisum*, *F. vulgare*, *M. fragrans*, *P. crispum*, *S. albidum*, and *L. verum*, and some plant families, such as Myrtaceae, Rutaceae, and Lamiaceae. In addition, EOs constitute aldehydes (cinnamaldehyde, cuminic aldehyde, perillaldehyde, etc.), alcohols (cinnamic alcohol, terpinenol, menthol, etc.), phenols (eugenol, carvacrol, etc.), and methoxy derivatives (anethole, estragole, etc.); compounds occur on aromatic hydrocarbons.

Table 9.1 Essential oils major chemical class with few examples of phytocompounds

Class	Functional group	Structure type	Examples
Terpenes	Carbures	Acyclic	Myrcene, ocimene, etc.
		Monocyclic	Terpenes, *p*-cymene, phellandrenes, etc.
		Bicyclic	Pinenes, 3-carene, camphene, sabinene, etc.
	Alcohols	Acyclic	Geraniol, linalol, citronellol, lavandulol, nerol, etc.
		Monocyclic	Menthol, α-terpineol, carveol, etc.
		Bicyclic	Borneol, fenchol, chrysanthenol, thuyan-3-ol, etc.
Monoterpenes	Aldehydes	Acyclic	Geranial, neral, citronellal, etc.
	Ketone	Acyclic	Tegetone, etc.
		Monocyclic	Menthones, carvone, pulegone, piperitone, etc.
		Bicyclic	Camphor, fenchone, thuyone, ombellulone, pinocamphone, pinocarvone, etc.
	Esters	Acyclic	Linalyl acetate or propionate, citronellyl acetate, etc.
		Monocyclic	Menthyl or α-terpinyl acetate, etc.
		Bicyclic	Isobornyl acetate, etc.
	Ethers	Bicyclic	1,8-cineole, menthofuran, etc.
	Peroxides	Bicyclic	Ascaridole, etc.
	Phenols	Monocyclic	Thymol, carvacrol, etc.
Sesquiterpenes	Carbures	Acyclic	Farnesenes, etc.
		Monocyclic	β-bisabolene, curcumenes, elemenes, zingiberene, etc.
		Bicyclic	Azulene, cadinenes, b-caryophyllene, etc.
		Tricyclic	Longifolene, etc.
	Alcohols	Acyclic	β-nerolidol, farnesol, etc.
		Monocyclic	Bisabolol, etc.
		Bicyclic	Carotol, β-santalol, etc.
		Tricyclic	Cedrol, patchoulol, viridiflorol, etc.
	Ketones	Monocyclic	Germacrone, cis-longipinan-2,7-dione, turmerones, etc.
		Bicyclic	Nootkatone, β-vetinone, etc.
	Epoxides	Bicyclic	Humulene epoxides, etc.
		Tricyclic	Caryophyllene oxide, etc.

9.3 Biological Effects of Essential Oils

At present, around 60 plant families are known to produce EOs, which are valued in medicinal, pharmaceutical, flavor and fragrance, and agricultural industries. Several plant species belonging to the Apiaceae, Alliaceae, Asteraceae, Lamiaceae, Myrtaceae, Poaceae, and Rutaceae family produce EOs with medicinal and industrial values (Vigan 2010; Hammer and Carson 2011). Details of EOs produced from medicinal and aromatic plants and their medicinal importance are mentioned in

Table 9.2. EOs are rich in terpenes, while phenylpropanoids more frequently occur in Apiaceae, Alliaceae, Lamiaceae, Myrtaceae, and Rutaceae plant families (Chami et al. 2004). These family plants are used for the commercial level manufacture of EOs. For example, patchoulol, coriander, anise, dill, and fennel EOs are extracted from *P. cablin, C. sativum, P. anisum, A. graveolens* and *F. vulgare,* respectively. These EOs are well known for their antimicrobial and anticancer activities. The plants belonging to the Lamiaceae and Apiaceae family are popular for antimicrobial, anticancer, antibacterial, antimutagenic, anti-inflammatory, and antioxidant activities (Swamy and Sinniah 2015; Swamy et al. 2016). Some of the plants from Lamiaceae family produce EOs (Burt 2004; Hammer et al. 2006; Hussain et al. 2008), such as *M. piperita, R. officinalis, O. basilicum, S. officinalis, M. officinalis, S. hortensis, T. vulgaris, L. angustifolia,* and *O. vulgore* (Swamy and Sinniah 2015; Swamy et al. 2016). Likewise, EOs from Lauraceae and Myrtaceae families also exhibit antimicrobial, antitumor, anticancer, antibacterial, and antiviral activities (Burt 2004; Hammer et al. 2006). *Cinnamomum verum* (Lauraceae) and *Syzygium aromaticum* (Myrtaceae) EOs are particularly rich in eugenol. Many EOs have been screened for their pharmacological potential, and in the following sections, studies showing different pharmacological activities of EOs are discussed.

9.3.1 Essential Oils as Antibacterial Agents

Many essential oils have been investigated for their antibacterial and antifungal activities, as well as their potential against Gram-positive and Gram-negative bacteria (Swamy et al. 2016). EOs show good antibacterial properties against *Salmonella, Staphylococcus,* and other bacterial pathogens. Thus, it is essential to study their effects as very good alternatives to antibiotics (Fujita et al. 2015; Karbach et al. 2015; Sienkiewicz et al. 2015). *O. basilicum* essential oil exhibits good antibacterial properties against Gram-positive bacteria (Al Abbasy et al. 2015; Avetisyan et al. 2017). In the investigations of antibacterial effects, manuka oil has been shown to exhibit good antibacterial activity. Similarly, eucalyptus, rosmarinus, *Lavandula* oil, and tree oil were found effective against *Streptococcus mutans, S. sobrinus, Fusobacterium nucleatum,* and *Porphyromonas gingivalis* (Takarada et al. 2004). Tea tree (*Melaleuca alternifolia*) oil is demonstrated to be sensitive to 15 genera of oral bacteria, indicating its potential applications in oral hygiene (Hammer et al. 2003). *Pittosporum undulatum* and *Hedychium gardnerianum* EOs show the highest antibacterial activities against *Staphylococcus epidermis* and *S. aureus.*

Despite the discovery of new antibiotics, bacterial infectious/diseases still pose a serious threat to human health, predominantly due to the appearance of antibiotic-resistant strains. In addition, as the global people continues to expand, this will result in a greater prevalence of bacterial diseases, low immunity, and increased drug resistance. Therefore, bacterial infections will be more likely to be fatal (Ahmad and Beg 2001; Hall-Stoodley et al. 2004; Swamy et al. 2016; Rudramurthy et al. 2016). To decrease the risk of infectious diseases, high concentrations of

Table 9.2 Essential oils from different plant families and their major medicinal importance

Essential oil sources	Plant family	Medicinal importance	References
Origanum vulgare (oregano); *Melissa officinalis* (lemon balm); *Salvia officinalis* (sage); *Mentha* sp.; *Mentha longifolia* (wild mint); *M. piperita* (peppermint); *M. spicata* (spearmint); *Ocimum basilicum* (sweet basil); *O. sanctum*; *Rosmarinus officinalis* (rosemary); *Lavandula officinalis* (lavender); *Lavandula* sp.; *Salvia sclarea* (sage Clary)	Lamiaceae/ Labiatae	Antibacterial; antifungal; anticancer; antiviral; antidiabetic; antimutagenic; antiprotozoal; anti-inflammatory; antioxidant	Bakkali et al. (2008), Raut and Karuppayil (2014), Swamy et al. (2016) and Nagarjuna et al. (2017)
Cinnamomum sp. (cinnamon)	Lauraceae	Antimicrobial; anti-inflammatory; antimutagenic	Raut and Karuppayil (2014) and Toscano-Garibay et al. (2017)
Allium sativum (garlic); *Allium cepa* (onion)	Liliaceae	Antifungal; antiviral; antiprotozoal	Raut and Karuppayil (2014) and Swamy et al. (2016)
Syzygium aromaticum (clove); *Thymus vulgaris* (thyme);*Thymus* sp.; *Melaleuca alternifolia* (tea tree); *Eucalyptus globulus* (blue gum); *Myristica fragrans* (nutmeg)	Myrtaceae	Antibacterial; antifungal; anticancer; antiviral; antimutagenic; anti-inflammatory; antiprotozoal	Bakkali et al. (2008), Raut and Karuppayil (2014) and Swamy et al. (2016)
Foeniculum vulgare (fennel); *Carum nigrum* (black caraway); *Anethum graveolens* (dill); *Cuminum cyminum* (cumin); *Pimpinella anisum* (anise); *Apium graveolens* (celery); *Coriandrum sativum* (coriander)	Apiaceae	Antidiabetic; anticancer; antibacterial; antifungal; antiviral	Bakkali et al. (2008), Raut and Karuppayil (2014) and Swamy et al. (2016)
Artemisia judaica; *A. annua*; *A. absinthium* (wormwood); *A. dracunculus* (tarragon)	Asteraceae	Antifungal; anticancer; antiviral	Bakkali et al. (2008), Raut and Karuppayil (2014) and Swamy et al. (2016)
Pelargonium graveolens (rose geranium)	Geraniaceae	Antibacterial	Raut and Karuppayil (2014) and Swamy et al. (2016)
Jasminum sp.; *Olea europaea* (olive)	Oleaceae	Antibacterial, anticancer	Raut and Karuppayil (2014)

(continued)

Table 9.2 (continued)

Essential oil sources	Plant family	Medicinal importance	References
Piper nigrum (black pepper)	Piperaceae	Antibacterial; antifungal; anticancer; antiprotozoal	Bakkali et al. (2008) and Raut and Karuppayil (2014)
Cedrus libani (cedarwood oil)	Pinaceae	Antifungal	Bakkali et al. (2008), Raut and Karuppayil (2014) and Swamy et al. 2016
Cymbopogon martini (palmarosa); *Cymbopogon citrates*(lemongrass); *Cymbopogon nardus* (citronella grass)	Poaceae	Antifungal; anticancer	Bakkali et al. (2008), Raut and Karuppayil (2014) and Swamy et al. 2016
Citrus sp. (lemon); *C. paradisi* (grape fruit)	Rutaceae	Antibacterial; antifungal; anticancer	Bakkali et al. (2008), Raut and Karuppayil (2014) and Swamy et al. 2016
Rosa sp.	Rosaceae	Antifungal	Bakkali et al. (2008), Raut and Karuppayil (2014) and Swamy et al. 2016
Santalum sp.; *Santalum album* (sandalwood)	Santalaceae	Antiviral	Bakkali et al. (2008) and Raut and Karuppayil (2014)
Zingiber officinale (ginger); *Zingiber montanum*; *Curcuma longa* (turmeric); *Elettaria cardamomum* (cardamom)	Zingiberaceae	Antifungal; anticancer; antioxidant; antimutagenic	Bakkali et al. (2008), Raut and Karuppayil (2014) and Swamy et al. 2016

antibacterial drugs are usually employed, resulting in toxicity and adverse side effects. Hence, there is a need to explore alternative approaches and develop new molecules against human pathogenic bacteria (Galvao et al. 2012; Rudramurthy et al. 2016). In this context, plant EOs exhibit a good potential due to their proven activity against both Gram-positive and Gram-negative bacteria as shown in Table 9.3 (Edris 2007; Lang and Buchbauer 2012; Hassanshahian et al. 2014; Teixeira et al. 2013). Some EOs show potential antibacterial activity against Gram-positive bacteria only, such as *Santalum album*, *Leptospermum scoparium,* and *Chrysopogon zizanioides* (Hammer and Carson 2011). According to the available

Table 9.3 Essential oils as antibacterial agents

Essential oil sources	Bacteria	References
Origanum vulgare (oregano); *Pelargonium graveolens* (rose geranium); *Piper nigrum* (black pepper); *Syzygium aromaticum* (clove); *Thymus vulgaris* (thyme); *Thymus* sp.	*Aeromonas hydrophila* *Alcaligenes faecalis*	Dorman and Deans (2000), Tepe et al. (2004), Lopez et al. (2005, 2007), Bozin et al. (2006) and Rosato et al. (2007)
Carum nigrum (black caraway); *Santolina rosmarinifolia* (cotton lavender)	*Bacillus cereus*	Singh et al. (2006) and Ioannou et al. (2007)
Juglans regia (common walnut); *Melissa officinalis* (lemon balm); *Myristica fragrans* (nutmeg); *Origanum vulgare* (oregano); *Pelargonium graveolens* (rose geranium); *Piper nigrum* (black pepper); *Rosa* sp.; *Syzygium aromaticum* (clove); *Ziziphora clinopodioides* (blue mint); *Thymus vulgaris* (thyme); *Thymus* sp.	*Bacillus subtilis*	Dorman and Deans (2000), Mimica-Dukic et al. (2004), Tepe et al. (2004), Lopez et al. (2005,2007), Bozin et al. (2006), Sonboli et al. (2006), Rosato et al. (2007), Hirulkar and Agrawal (2010) and Rather et al. (2012)
Anethum graveolens (dill); *Apium graveolens* (celery); *Eucalyptus robusta* (swamp mahogany); *E. saligna*; *E. globulus* (blue gum); *Juglans regia* (common walnut); *Melaleuca alternifolia* (tea tree); *Melissa officinalis* (lemon balm); *Mentha longifolia* (wild mint); *M. piperita* (peppermint); *M. spicata* (spearmint); *Pimpinella anisum* (aniseed); *Myristica fragrans* (nutmeg); *Origanum vulgare*(oregano); *Pelargonium graveolens* (rose geranium); *Pinus densiflora* (Japanese red pine); *Pinus koraiensis* (Korean pine); *Piper nigrum* (black pepper); *Rosa* spp.; *Salvia sclarea* (sage clary); *S. officinalis* (sage); *S. lavandulifolia*; *S. rosifolia*; *Santolina rosmarinifolia* (cotton lavender); *Syzygium aromaticum* (clove); *Tamarix boveana* (salt cedar); *Ziziphora clinopodioides* (blue mint); *Thymus vulgaris* (thyme); *Thymus* sp.	*Escherichia coli*	Dorman and Deans (2000), Delaquis et al. (2002), Singh et al. (2002), Dryden et al. (2004), Hong et al. (2004), Mimica-Dukic et al. (2004), Rota et al. (2004), Tepe et al. (2004), Bozin et al. (2006), Carson et al. (2006), Sonboli et al. (2006), Fabio et al. (2007), Lopez et al. (2005,2007), Ioannou et al. (2007), Rafii and Shahverdi (2007), Rosato et al. (2007), Sartorelli et al. (2007), Saidana et al. (2008), Roller et al. (2009), Hirulkar and Agrawal (2010), Baananou et al. (2013), Djenane et al. (2012), Galvao et al. (2012) and Rather et al. (2012)
Mentha longifolia (wild mint); *M. piperita* (peppermint); *M. spicata* (spearmint); *Origanum vulgare* (oregano); *Pelargonium graveolens* (rose geranium); *Piper nigrum* (black pepper); *Rosa* spp.; *Syzygium aromaticum* (clove)	*Enterobacter aerogenes; E. cloacae*	Dorman and Deans (2000), Singh et al. (2002), Tepe et al. (2004), Lopez et al. (2005), Bozin et al. 2006), Fabio et al. (2007), Rafii and Shahverdi (2007), Rosato et al. (2007), Hirulkar and Agrawal (2010) and Djenane et al. (2012)

(continued)

Table 9.3 (continued)

Essential oil sources	Bacteria	References
Melaleuca alternifolia (tea tree); *Origanum vulgare* (oregano); *Pelargonium graveolens* (rose geranium); *Piper nigrum* (black pepper); *Syzygium aromaticum* (clove); *Ziziphora clinopodioides* (blue mint); *Thymus vulgaris* (thyme); *Thymus* sp.	*Enterococcus faecalis*	Dorman and Deans (2000), Singh et al. (2002), Dryden et al. (2004), Tepe et al. (2004), Lopez et al. (2005, 2007), Bozin et al. (2006), Carson et al. (2006), Sonboli et al. (2006), Fabio et al. (2007), Rosato et al. (2007) and Shan et al. (2007)
Eucalyptus robusta (swamp mahogany); *E. saligna; E. globulus* (blue gum); *Eugenia caryophyllus* (clove); *Melaleuca alternifolia* (tea tree); *Mentha longifolia* (wild mint); *M. piperita* (peppermint); *M. spicata* (spearmint); *Salvia sclarea* (sage clary); *S. officinalis* (Sage); *S. lavandulifolia; S. rosifolia*	*Haemophilus influenzae*	Rota et al. (2004), Carson et al. (2006), Fabio et al. (2007), Sartorelli et al. (2007), Shan et al. (2007), Rafii and Shahverdi (2007), Roller et al. (2009), Djenane et al. (2012) and Galvao et al. (2012)
Anethum graveolens (dill); *Eucalyptus robusta* (swamp mahogany); *E. saligna; E. globulus* (blue gum); *Eugenia caryophyllus* (clove); *Juglans regia* (common walnut); *Mentha longifolia* (wild mint); *M. piperita* (peppermint); *M. spicata* (spearmint); *Myristica fragrans* (nutmeg); *Origanum vulgare* (oregano); *Pelargonium graveolens* (rose geranium); *Pinus densiflora* (Japanese red pine); *Pinus koraiensis* (Korean pine); *Piper nigrum* (black pepper); *Rosa* spp.; *Salvia sclarea* (sage clary); *S. officinalis* (sage); *S. lavandulifolia; S. rosifolia; Syzygium aromaticum* (clove); *Ziziphora clinopodioides* (blue mint); *Thymus vulgaris* (thyme); *Thymus* sp.	*Klebsiella pneumoniae*	Dorman and Deans (2000), Delaquis et al. (2002), Hong et al. (2004), Rota et al. (2004), Tepe et al. (2004), Bozin et al. (2006), Carson et al. (2006), Sonboli et al. (2006), Fabio et al.(2007), Lopez et al. (2005,2007), Rafii and Shahverdi (2007), Rosato et al. (2007), Shan et al. (2007), Roller et al. (2009), Hirulkar and Agrawal (2010), Djenane et al. (2012), Galvao et al. (2012) and Rather et al. (2012)
Melaleuca alternifolia (tea tree)	*Mycobacterium avium*	Dryden et al. (2004) and Carson et al. (2006)
Lantana fucata; L. trifolia	*Mycobacterium tuberculosis*	Juliao et al. (2009)
Juglans regia (common walnut); *Myristica fragrans* (nutmeg); *Pelargonium graveolens* (rose geranium); *Rosa* sp.; *Syzygium aromaticum* (clove); *Thymus vulgaris* (thyme); *Thymus* sp.	*Proteus vulgaris*	Dorman and Deans (2000), Hirulkar and Agrawal (2010) and Rather et al. (2012)

(continued)

Table 9.3 (continued)

Essential oil sources	Bacteria	References
Apium graveolens (celery); *Carum nigrum* (black caraway); *Juglans regia* (common walnut); *Melaleuca alternifolia* (tea tree); *Origanum vulgare* (oregano); *Pelargonium graveolens* (rose geranium); *Piper nigrum* (black pepper); *Rosa* spp.; *Syzygium aromaticum* (clove); *Tamarix boveana* (salt cedar); *Ziziphora clinopodioides* (blue mint); *Thymus vulgaris* (thyme); *Thymus* sp.	*Pseudomonas aeruginosa*; drug-resistant *P. aeruginosa*	Dorman and Deans (2000), Singh et al. (2006), Dryden et al. (2004), Tepe et al. (2004), Bozin et al. (2006), Carson et al. (2006), Sonboli et al. (2006), Lopez et al. (2005,2007), Rosato et al. (2007), Saidana et al. (2008), Hirulkar and Agrawal (2010), Baananou et al. (2013) and Rather et al. (2012)
Apium graveolens (celery); *Croton cajucara; Eucalyptus robusta* (swamp mahogany); *E. saligna; E. globulus* (blue gum); *Eugenia caryophyllus* (clove); *Juglans regia* (common walnut); *Lavandula angustifolia* (common lavender); *L. latifolia; L. luisieri; Melaleuca alternifolia* (tea tree); *Melissa officinalis* (lemon balm); *Mentha longifolia* (wild mint); *M. piperita* (peppermint); *M. spicata* (spearmint); *Myristica fragrans* (nutmeg); *Origanum vulgare* (oregano); *Pelargonium graveolens* (rose geranium); *Pinus densiflora* (Japanese red pine); *Pinus koraiensis* (Korean pine); *Piper nigrum* (black pepper); *Rosa* spp.; *Rosmarinus officinalis* (rosemary); *Salvia sclarea* (sage clary); *S. officinalis* (sage); *S. lavandulifolia; S. rosifolia; Santolina rosmarinifolia* (cotton lavender); *Skimmia laureola; Syzygium aromaticum* (clove); *Tamarix boveana* (salt cedar); *Ziziphora clinopodioides* (blue mint); *Thymus vulgaris* (thyme); *Thymus* sp.	*Staphylococcus aureus; methicillin-resistant Staphylococcus aureus* (MRSA)	Dorman and Deans (2000), Delaquis et al. (2002), Dryden et al. (2004), Mimica-Dukic et al. (2004), Rota et al. (2004), Tepe et al. (2004), Alviano et al. (2005), Bozin et al. (2006), Carson et al. (2006), Sonboli et al. (2006), Fabio et al. (2007), Lopez et al. (2005,2007), Ioannou et al. (2007), Rafii and Shahverdi (2007), Rosato et al. (2007), Sartorelli et al. (2007), Shan et al. (2007), Saidana et al. (2008), Roller et al. (2009), Hirulkar and Agrawal (2010), Tohidpour et al. (2010), Baananou et al. (2013), Djenane et al. (2012), Galvao et al.(2012), Rather et al. (2012) and Shah et al. (2013)
Juglans regia (common walnut); *Skimmia laureola; Tamarix boveana* (salt cedar); *Ziziphora clinopodioides* (blue mint)	*S. epidermidis*	Sonboli et al. (2006), Saidana et al. (2008), Rather et al. (2012) and Shah et al. (2013)
Eucalyptus robusta (swamp mahogany); *E. saligna; E. globulus* (blue gum); *Eugenia caryophyllus* (clove); *Melaleuca alternifolia* (tea tree); *Mentha longifolia* (wild mint); *M. piperita* (peppermint); *M. spicata* (spearmint); *Rosa* spp.; *Salvia sclarea* (sage clary); *S. officinalis* (sage); *S. lavandulifolia; S. rosifolia; Thymus vulgaris* (thyme); *Thymus* sp.; *Coriandrum sativum* (coriander)	*Streptococcus pneumonia; S. pyogenes; S. agalactiae; S. haemolyticus*	Delaquis et al. (2002), Singh et al. (2002), Dryden et al. (2004), Lo Cantore et al. (2004), Rota et al. (2004), Carson et al. (2006), Fabio et al. (2007), Rafii and Shahverdi (2007), Sartorelli et al. (2007), Shan et al. (2007), Roller et al. (2009), Hirulkar and Agrawal (2010), Djenane et al. (2012), Galvao et al. (2012), Rather et al. (2012) and Shah et al. (2013)

(continued)

Table 9.3 (continued)

Essential oil sources	Bacteria	References
Coriandrum sativum (coriander); *Juglans regia* (common walnut); *Melissa officinalis* (lemon balm); *Pinus densiflora* (Japanese red pine); *Pinus koraiensis* (Korean pine); *Rosa* spp.; *Salvia sclarea* (sage clary); *S. officinalis* (sage); *S. lavandulifolia; S. rosifolia; Tamarix boveana* (salt cedar)	*Salmonella typhimurium*	Delaquis et al. (2002), Singh et al.(2002), Hong et al.(2004), Lo Cantore et al.(2004), Mimica-Dukic et al. (2004), Rota et al. (2004), Fabio et al. (2007), Roller et al. (2009), Hirulkar and Agrawal (2010) and Saidana et al. (2008)
Myristica fragrans (nutmeg); *Origanum vulgare* (oregano); *Pelargonium graveolens* (rose geranium); *Piper nigrum* (black pepper); *Syzygium aromaticum* (clove); *Thymus vulgaris* (thyme); *Thymus* sp.	*Serratia marcescens*	Dorman and Deans (2000), Tepe et al. (2004), Lopez et al. (2005,2007), Bozin et al. (2006) and Rosato et al. (2007)
Juglans regia (common walnut); *Ocimum basilicum* (sweet basil); *O. gratissimum* (African basil)	*Shigella dysenteriae*	Iwalokun et al. (2003), Bozin et al. (2006) and Rather et al. (2012)
Coriandrum sativum (coriander); *Pinus densiflora* (Japanese red pine); *Pinus koraiensis* (Korean pine)	*Listeria monocytogenes*	Delaquis et al. (2002), Singh et al. (2002), Hong et al. (2004) and Lo Cantore et al.(2004)
Myristica fragrans (nutmeg); *Pelargonium graveolens* (rose geranium); *Piper nigrum* (black pepper); *Syzygium aromaticum* (clove); *Tamarix boveana* (salt cedar); *Thymus vulgaris* (thyme); *Thymus* sp.	*Micrococcus luteus*	Dorman and Deans (2000) and Saidana et al.(2008)
Myristica fragrans (nutmeg); *Origanum vulgare* (oregano); *Pelargonium graveolens* (rose geranium); *Piper nigrum* (black pepper); *Syzygium aromaticum* (clove); *Thymus vulgaris* (thyme); *Thymus* sp.	*Moraxella* sp.	Dorman and Deans (2000), Tepe et al. (2004), Lopez et al. (2005,2007), Bozin et al. (2006) and Rosato et al. (2007)
Myristica fragrans (nutmeg); *Pelargonium graveolens* (rose geranium); *Piper nigrum* (black pepper); *Syzygium aromaticum* (clove); *Thymus vulgaris* (thyme); *Thymus* sp.	*Yersinia enterocolitica*	Dorman and Deans (2000)

evidence, cinnamon, lemongrass, thyme, clove, rosewood, orange, rosemary, peppermint, bay, basil, and eucalyptus EOs exhibit the most effective antimicrobial activity. EOs are very active at $^{<}1\%$ minimum inhibition concentrations (MICs). *Escherichia coli* exhibits zone of inhibition at 0.02, 0.04, and 0.06% concentrations against clove, grass, oregano, bay, and thyme EOs, respectively (Hammer and Carson 2011). Some EOs show less activity, but their major constituent molecules are observed to possess higher activity. For example, eugenol, carvacrol, and 4-terpinenol display greater antibacterial activity than their corresponding EOs. In extant literature, phenols and aldehydes are reported potential antimicrobial activity (Lambert et al. 2001; Ultee et al. 2002; Carson et al. 2006). A large number of the EOs have been shown to be successful against drug-resistant strains, antibiotics, and biofilms (May et al. 2000; Bozin et al. 2006; Galvao et al. 2012). EOs of *A. fragrantissima, A. ligustica, A. absinthium, A. biennis, A. cana, A. dracunculus,*

A. longifolia, *A. frigida*, *C. officinalis*, *C. sativum*, *C. cyminum*, *C. longus*, *D. littoralis*, *E. erythropapps*, *E. rostkoviana*, *F. margarita*, *L. nobilis*, *L. angustifolia*, *L. longifolia*, *J. excelsa*, *M. suaveolens*, *N. sativa*, *O. vulgare*, *T. vulgaris*, *O. basilicum*, *P. cablin*, *T. kotschyanus*, *S. cumini*, *T. ammi*, and *S. sparganophora* show potential antibacterial activity against *S. aureus*, *S. epidermidis*, *E. coli*, *S. mutans*, *B. thermosphacta*, *L. innocua*, *L. monocytogenes*, *P. putida*, *B. cereus*, *B. subtilis*, *N. gonorrhoeae*, *K. pneumoniae*, *C. botulinum*, *C. perfringens*, *S. sonnei*, *S. lutea*, *P. putida*, *M. flavus*, *L. innocua*, *E. faecalis*, and *S. putrefaciens* (Lopes-Lutz and Alviano 2008; Maggi et al. 2009; Matasyoh et al. 2009; Begnami et al. 2010; Runyoro et al. 2010; Ait-Ouazzou et al. 2012; Bejaoui et al. 2013; Teixeira et al. 2013; Yang et al. 2013; Amatiste et al. 2014; Andrade et al. 2014; Bilcu et al. 2014; Bisht et al. 2014; Flores et al. 2014; Kasim et al. 2014; Khoury et al. 2014; Petretto et al. 2014; Pullagummi et al. 2014; Santurio et al. 2014; Singh et al. 2014;; yousefbeyk et al. 2014; Zeedan et al. 2014; Ahmadi et al. 2015; Beatovia et al. 2015; Santos et al. 2015; Ibrahim et al. 2015a, b; Novy et al. 2015).

9.3.2 Essential Oils as Antioxidant Agents

Modern era has brought about different health problems, such as noncommunicable diseases (e.g., cancer, diabetes, and Alzheimer's, Parkinson's, and heart diseases) which are attributed to oxidative stresses. EOs exhibit a significant antioxidant activity due to their phtocompounds, such as flavonoids, terpenoids, and phenolic compounds (McCord 2000; Tomaino et al. 2005; Edris 2007; Ferguson and Philpott 2008; Ruan et al. 2008; Miguel 2010; Cavar et al. 2012; Andrade et al. 2013; Sanchez-Vioque et al. 2013; Aleksic and Knezevic 2014; Bouzabata et al. 2015). Among many EOs, *O. majorana*, *T. filifolia*, *B. monnieri*, *C. longa*, *S. cryptantha*, *A. millefolium*, *S. multicaulis*, *M. officinalis*, *M. alternifolia*, *Ocimum*, and *Mentha* sp. have been reported to possess significant antioxidant activity (Mau et al. 2003; Tepe et al. 2004; Kim et al. 2004; Maheshwari et al. 2006; Maestri et al. 2006; Gulluce et al. 2007; Tripathi et al. 2007; Politeo et al. 2007; Hussain et al. 2008; Aqil et al. 2012; Mohamed et al. 2013; Toscano-Garibay et al. 2017;). Thymol and carvacrol containing EOs in particular show strong antioxidant properties (Tepe et al. 2004; Miguel 2010). Likewise, EOs of *Cuminum cyminum*, *Petroselinum sativum*, *S. cumini*, and *Coriandrum sativum* also exhibit efficient antioxidant (Romeilah et al. 2010; Eshwarappa et al. 2014). In addition, clove oil shows a much stronger antioxidant and radical scavenging activity compared to cinnamon, basil, oregano, nutmeg, and thyme EOs (Tomaino et al. 2005).

9.3.3 Essential Oils as Anticancer Agents

As cancer is a growing problem globally, many curing and preventive therapies have been developed over the years. In the human body, cancer is characterized by uncontrolled proliferation of abnormal cells. The malignant cells have the potential to be metastatic, requiring urgent treatment, such as radiotherapy, and chemotherapy. Among these, chemotherapy treatment is most challenging and can be difficult for patients to tolerate due to extreme side effects. Therefore, many alternative treatments and therapies have been explored. In both developed and developing countries, herbal medicines have been historically used for traditional medicinal treatments. For thousands of years, African and Asian populations have used medicinal plants in folk medicine. Even developed nations are starting to recognize the health benefits medicinal plants, according to the WHO. Plants identified for their anticancer properties have been chemically characterized to reveal the occurrence of many bio-active compounds, such as polyphenols, taxols, brassinosteroids, etc.

Flavonoids, tannins, curcumin resveratrol, and gallocatechins are some of the plant-derived polyphenolic compounds possessing anticancer properties. A regular intake of healthy diet can improve the human health as they are rich in natural antioxidants and can thus reduce the risk of developing cancer. For example, gallocatechins found in green tea and resveratrol found in peanuts, grapes, and red wine are effective in preventing cancer (Azmi et al. 2006; Apostolou et al. 2013). Polyphenols have been shown to regulate cancer cell growth through modifications of acetylation, methylation, or phosphorylation processes involved in the regulation of chromatin function. For example, *C. longa* EOs has been treated various cancer cell lines shown to suppression the tumor necrosis factor (TNF) impression along interaction with various stimuli (Gupta et al. 2014). Flavonoids, another class of plant secondary metabolites, possess therapeutic efficacy and scientifically prove to impart health benefits to humans. In traditional Chinese medicine, litchi leaf (*Litchi chinensis*) is used in cancer treatment (Cao et al. 2013; Wen et al. 2014). Litchi leaf is rich in flavonoids, such as flavones, flavonols, and chalcones (Wen et al. 2014). The essential oil of *Dryopteris erythrosora* showed potential anticancer activity against human lung cancer cells (A456 cell line) (Kloog and Cox 2004; Cao et al. 2013).

Plant-derived compounds also show potential activity against cancer cell lines. These compounds occur naturally and are easily available and nontoxic to the healthy human cells. Thus, they could be administrated to patients orally (Cornblatt et al. 2007; Amin et al. 2009). Still, there are a few exceptions, such as glycosides, lectins, saponins, lignans, lectins, and taxanes (Unnati et al. 2013). BR compounds, such as sulforaphane, isothiocyanates, isoflavones, and pomiferin, are considered histone deacetylase (HDAC) inhibitors. For example, sulforaphane has been used against breast cancer proliferation (Pledgie-Tracy et al. 2007; Seidel et al. 2012). In the studies on inhibition of cancer cell proliferation, taxols (plant molecules) were shown effective against different types of malignancies, like colon cancer, gastric cancer, breast cancer, leukemia, and human liver and pulmonary tumors (Edris 2007; Kaefer and Milner 2008; Hamid et al. 2011). In Table 9.4, details of different medicinal and aromatic plant EOs possessing anticancer properties are cited. For example, *Cymbopogon martini* EOs are rich in geraniol. Geraniol is used against ion homeostasis which interferes with membrane function as well as cancer cell line signaling. *Atractylodes lancea* oils are used for the treatment of malignant tumors (Tsuneki et al. 2005), whereas *Myristica fragrans* (*M. fragrans*) oils contain myristicin and are used for their hepatoprotective activities (Morita et al. 2003).

Table 9.4 Anticancer and antitumor activities of essential oils

Essential oil sources	Antitumor/anticancer	References
Alpinia officinarum (galangal/China root); *Citrus hystrix* (Thai lime); *C. paradise* (grape fruit tree); *Curcuma longa* (turmeric); *Cymbopogon nardus* (citronella grass); *Cymbopogon martini* (palmarosa); *Lavandula angustifolia; Mentha spicata; Ocimum basilicum; O. americanum; O. sanctum; Piper nigrum; P. betle* (beetle leaf); *Zingiber montanum; Vetiveria zizanioides* (Khus)	Inhibition of proliferation of murine leukemia and human mouth epidermal carcinoma cell lines	Hata et al. (2003), Carnesecchi et al. (2004), Koo et al. (2004), Manosroi et al. (2006)
Artemisia annua	Induction of apoptosis in cultured hepatocarcinoma cells	Li et al. (2004)
Atractylodes lancea	Anti-angiogenesis properties	Tsuneki et al. (2005)
Curcuma longa (turmeric)	Inhibition of primary liver cancer	Koo et al. (2004) and Manosroi et al.(2006)
Elettaria cardamomum (cardamom); *Eucalyptus globulus* (eucalyptus)	Induction of apoptosis in human leukemia cells	Moteki et al. (2002)
Allium sativum; Elaeis guineensis (palm oil)	Chemoprevention of various cancers	Milner (2001) and Luk et al. (2011)
Eugenia caryophyllata (i.e., *Syzygium aromaticum*)	Inhibition of proliferation of cancerous cells	Yoo et al. (2005)
Foeniculum vulgare; Myristica fragrans	Hepatoprotective activity	Ozbek et al. (2003), Morita et al. (2003) and Lee et al. (2005)
Foeniculum vulgare	Inhibition of growth of different human cancer cell lines like breast cancer and liver cancer	Ozbek et al. (2003)
Matricaria chamomilla	Induction of apoptosis in highly malignant glioma cell	Cavalieri et al. (2004)
Melaleuca alternifolia	Induction of caspase dependent apoptosis inhuman melanoma	Calcabrini et al. (2004)
Myrica gale (myrtle/bayberry)	Activity against lung and colon cancer cell lines	Sylvestre et al. (2005,2006)
Melissa officinalis	Activity against a series of human cancer cell lines and a mouse cell line	De Sousa et al. (2004)
Myristica fragrans	Induction of apoptosis in human neuroblastoma	Morita et al. (2003) and Lee et al. (2005)

(continued)

Table 9.4 (continued)

Essential oil sources	Antitumor/anticancer	References
Nigella sativa	Inhibition of cancer proliferation in rats	Salim and Fukushima (2003); Mansour et al. (2001)
Olea europaea (olive oil)	Protection against colorectal cancer	Gill et al. (2005)

Lemongrass oil mainly consists of citral, which is used for in vivo studies on the initial phases of hepatocarcinogenesis (Puatanachokchai et al. 2002). EOs extracted from *E. ciliata* shows potential anticancer activity against human glioblastoma (U87), pancreatic cancer (Panc-1), and triple negative breast cancer (MIDA-MB231) (Pudziuvelyte et al. 2017). *A. fragrantissima* EOs show potential anticancer activity against human breast cancer cell line (MCF-7) and colon cancer cell line (HCT116) and the IC_{50} (μg/ml) MCF7 for 0.51 and HCT116 for 0.62 μg/ml. Compared to solvent extracts, EOs have shown better anticancer activity (Choucry. 2017), and also *A. aucheri, M. communis,* and *O. vulgare* EOs show efficient anticancer activity against human promyelocytic leukemia cell lines (HL-60, NB4), lymphocytes, tumor HeLa cells, and Ehrlich ascites carcinoma cells (EACC) (Taherkhani 2015; Romeilah 2016). Orange peel EOs have been investigated by 3-(4,5-dimethylthiazol-2-yl)-2,5-diphenyltetrazolium bromide (MTT) assay against lung cancer cell line (A549) and prostate cancer cell line (22RV-1); it showed good inhibition of the proliferation of a lung and prostate cancer cell lines (Yang et al. 2017).

Individual chemical constituents in EOs show potential anticancer and antitumor activity such as D-limonene (in vivo), geraniol (in vitro and in vivo), thymol and carvacrol (in vitro), thymoquinone (in vitro *and* in vivo), farnesol (in vivo and in vitro), (−)-α-bisabolol (in vitro and in vivo), (−)-β-elemene (in vitro and in vivo), (−)-β-caryophyllene (in vitro), α-humulene (in vitro), nerolidol (in vitro), germacrone (in vitro), and eugenol (in vitro and in vivo) studies which show efficient anticancer activity against stomach (mice), lung (mice), breast (rats, MCF-7), prostate (PC3, mice), skin (B16F10, rats, SCC VII, A431), colon (Caco-2, mice, rats, V79), pancreas (MIAPaca-2, hamsters), kidney (rats), mouth (hamsters), bones (MG63), brain (DBTRG-05MG), liver (HepG2, Caco-2), and blood (HL60) cancer types (Malíková et al. 2008; Lesgards et al. 2014; Bayala et al. 2014; Gautam et al. 2014) respectively.

9.3.4 Essential Oils as Antifungal Agents

Many EOs have been investigated for their antifungal activities. Fungi are very difficult to target because of cellular and molecular levels, human pathogenic fungi, and eukaryotes which are very similar with their host. However, eukaryotes and human pathogenic fungi and their hosts have similarities at molecular and cellular levels (Routh et al. 2011). Some of the human fungal pathogens, such as *Aspergillus* spp., *Cryptococcus* sp., and *Candida* spp., are very problematic for immunocompromised patients. Hence, limited numbers of antifungal drugs are available against fungi (Kathiravan et al. 2012). Currently, the prescribed drugs are resistant to fungal strains and may lead to cause biofilm infections and adverse side effects. Consequently, fungal infections are associated with high morbidity and mortality rates (Sardi et al. 2013; Swamy et al. 2016). Plant EOs that are effective against human pathogenic fungi, plant fungi, and yeast are mentioned in Table 9.5. Based on the EO efficiency, the zone of inhibition to different the targeted organisms varies. For instance, EOs of plants, such as coriander, anise, and fennel, though belonging to the same family, i.e., Apiaceae, show differences in their antifungal activity against *Candida albicans* with MICs (minimum inhibitory concentrations) of 0.25%, 0.5%, and 1%, respectively. Among the EOs, Japanese mint, ginger grass, cinnamon, lemongrass, clove, anise, and geranium oils are particularly encouraging against *C. albicans* and the essential concentration range between 0.01% and 0.15% (Devkatte et al. 2005; Hammer and Carson 2011). EOs can rapidly inhibit growth of dermatophytes and their spores. This is an attribute to the occurrence of high levels of phytocompounds, i.e., α-bisabolol (an alcohol) and eugenol (a phenylpropanoid), in their EOs (Bajpai et al. 2009; Maxia et al. 2009; Pragadheesh et al. 2013). *C. citratus* EOs show a potential activity against many filamentous fungi at the concentration range of 0.006–0.03%. Also, it inhibits the growth of *Aspergillus niger*, *A. flavus*, *P. chrysogenum*, and *P. verrucosum* below 1% concentration (Viuda-Martos et al. 2008). Eucalyptus oil rich in citral, geraniol, geranyl acetate, and citronelol components was found to inhibit the growth of *C. albicans* by blocking the S phase of its life cycle (Zore et al. 2011a). All chemical constituents of tea tree (*Melaleuca alternifolia*) oil, except β-myrcene, exhibit in vitro antifungal activity. Tea tree oils show potential antifungal activity against dermatophytes and filamentous fungi (Hammer et al. 2003). Likewise, the growth of *A. niger* was significantly by the EOs of *Melaleuca ericifolia* fresh leaves. EOs from various plants that are generally used for the flavor and fragrance including *Mentha piperita*, *Brassica niger*, *Angelica archangelica*, and *Cymbopogon citratus* have been tested for their antifungal activity and found that they exhibit very strong antifungal activity. EOs extracted from *A. marmelos*, *C. sativum*, *D. foetidum*, *E. erythropappus*, *E. rostkoviana*, *F. vulgare*, *G. spathulata*, *M. alternifolia*, *M. pulegium*, *M. communis*, *N. sativa*, *O. vulgare*, *P. graveolens*, *P. cablin*, *R. officinalis*, *S. sclarea*, and *S. aromaticum* show efficient antifungal activity of *C. albicans*, *A. niger*, *F. oxysporum*, *C. gattii*, *C. neoformans*, *S. cerevisiae*, *A. alternate*, *F. oxysporum*, *A. flavus*, *T. rubrum*, *E. floccosum*, *C. neoformans*, *M. furfur*, *M. canis*, *M. sympodialis*, *M. gypseum*, *R. rubra*, *T. rubrum*, *T. tonsurans*, *C. zemplinina*, *K. apiculata*, *T. phaffii*, *F. moniliforme*, *F. graminearum*, *P. viridicatum*, *T. violaceum*, *B. cinerea*, *P. oryzae*, *C. tropicalis*, *C. krusei*, and *C. glabrata* (Hammer et al. 2002; Matasyoh et al. 2009; Begnami et al. 2010; Hammer and Carson 2011; Berka-zougali et al. 2012; Wang et al. 2012;

Table 9.5 Essential oils as antifungal agents

Essential oil sources	Fungi	References
Cedrus libani (cedarwood oil); *Cymbopogon martini* (ginger grass); *C. citrates* (lemongrass); *Tamarix boveana; Rosmarinus officinalis* (rosemary); *Foeniculum vulgare* (fennel)	*Alternaria alternata*	Mimica-Dukic et al. (2004), Rota et al. (2004), Ozcan and Chalchat (2008), Rosato et al. (2007), Rasooli et al. (2008), Saidana et al. (2008) and Peighami-Ashnaei et al. (2008)
Allium sativum (garlic); *Artemisia judaica* (wormwood); *A. absinthium; A. biennis; Carum nigrum* (black caraway); *Cedrus libani* (cedarwood oil); *Chenopodium ambrosioides; Cymbopogon martini* (ginger grass); *C. citrates* (lemongrass); *Eugenia caryophyllus* (clove); *Foeniculum vulgare* (fennel); *Juniperi aetheroleum* (juniper); *Matricaria chamomilla* (chamomile); *Zingiber officinale* (ginger); *Tamarix boveana*	*Aspergillus niger*	Saikia et al. (2001), Benkeblia (2004), Mimica-Dukic et al. (2004), Kordali et al. (2005), Pepeljnjak et al. (2005), Kumar et al. (2007), Agarwal et al. (2008), Bansod and Rai (2008), Lopes-Lutz and Alviano (2008), Saidana et al.(2008), Singh et al. (2008), Cetin et al. (2009), Irkin and Korukluoglu (2009), Peighami-Ashnaei et al. (2008) and Tolouee et al. (2010)
Satureja hortensis (summer savory); *Rosmarinus officinalis* (rosemary)	*Aspergillus parasiticus*	Rota et al. (2004), Rosato et al. (2007), Ozcan and Chalchat (2008), Rasooli et al. (2008) and Razzaghi-Abyaneh et al. (2008)
Carum nigrum (black caraway); *Cedrus libani* (cedarwood oil); *Cuminum cyminum* (cumin); *Nigella sativa* (black cumin); *Zingiber officinale* (ginger); *Satureja hortensis* (summer savory)	*Aspergillus flavus*	Singh et al. (2006), Singh et al. (2010), Razzaghi-Abyaneh et al. (2008) and Khosravi et al. (2011)
Cedrus libani (cedarwood oil); *Chenopodium ambrosioides; Cuminum cyminum* (cumin); *Eugenia caryophyllus* (clove); *Nigella sativa* (black cumin)	*Aspergillus fumigatus*	Kumar et al. (2007), Bansod and Rai (2008) and Khosravi et al. (2011)
Rosmarinus officinalis (rosemary); *Foeniculum vulgare* (fennel); *Artemisia judaica* (wormwood); *A. absinthium*; *A. biennis*; *other Artemisia* sp.	*Botrytis cinerea; Botrytis fabae*	Rosato et al. (2007), Ozcan and Chalchat (2008), Lopes-Lutz and Alviano (2008), Rasooli et al. (2008), Cetin et al. (2009), Irkin and Korukluoglu (2009) and Peighami-Ashnaei et al. (2008)

(continued)

Table 9.5 (continued)

Essential oil sources	Fungi	References
Cinnamomum sp.; *Croton cajucara*; *Cymbopogon martini* (ginger grass); *C. citrates* (lemongrass); *Eucalyptus saligna* (saligna); *Eugenia caryophyllus* (clove); *Juniperi aetheroleum* (juniper); *Lavandula* sp.; *Melaleuca alternifolia; Melissa officinalis; Mentha piperita; M. longifolia; M. viridis; Ocimum* sp.; *Ocimum sanctum* (holy basil/tulsi); *Pimpinella anisum; Piper nigrum* (black pepper); *Ziziphora clinopodioides; Santolina rosmarinifolia*	*Candida albicans; C. glabrata; Candida* sp.	Saikia et al.(2001), Singh et al. (2002), Dryden et al. (2004), Mimica-Dukic et al.(2004), Alviano et al. (2005), Devkatte et al. (2005), Pepeljnjak et al. (2005), Carson et al. (2006), Ioannou et al. (2007), Sartorelli et al. (2007), Agarwal et al. (2008), Bansod and Rai (2008), Irkin and Korukluoglu (2009), Mkaddem et al. (2009), Khosravi et al. (2011), Zore et al. (2011b), Zuzarte et al. (2011,2012) and Rabadia et al. (2012)
Cedrus libani (cedarwood oil); *Artemisia judaica* (wormwood); *A. absinthium; A. biennis; Artemisia* sp.	*Cladosporium cladosporioides; C. herbarum*	Kordali et al. (2005), Lopes-Lutz and Alviano (2008), Cetin et al. (2009) and Irkin and Korukluoglu (2009)
Lavandula sp.; *Ziziphora clinopodioides*	*Cryptococcus neoformans*	Khosravi et al. (2011) and; Zuzarte et al. (2011,2012)
Allium sativum (garlic); *Artemisia judaica* (wormwood); *A. absinthium; A. biennis;* other *Artemisia* sp.; *Tamarix boveana; Carum nigrum* (black caraway); *Cymbopogon martini* (ginger grass); *C. citrates* (lemongrass)	*Penicillium cyclopium; P. purpurogenum; P. madriti; P. viridicatum; P. roqueforti; Penicillium* sp.	Saikia et al. (2001), Benkeblia (2004), Kordali et al. (2005), Singh et al. (2006), Agarwal et al. (2008), Lopes-Lutz and Alviano (2008); Saidana et al. (2008), Cetin et al. (2009) and Irkin and Korukluoglu (2009)
Allium sativum (garlic); *Artemisia judaica* (wormwood); *A. absinthium; A. biennis; other Artemisia* sp.; *Chenopodium ambrosioides; Cymbopogon martini* (ginger grass); *C. citrates* (lemongrass); *Tamarix boveana; Rosmarinus officinalis* (rosemary); *Zingiber officinale* (ginger); *Salvia fruticosa; S. officinalis; S. rosifolia*	*Fusarium oxysporum; F. moniliforme; F. solani; F. proliferatum*	Saikia et al. (2001), Benkeblia (2004), Rota et al. (2004), Kordali et al. (2005), Fabio et al. (2007), Kumar et al. (2007), Rosato et al. (2007), Agarwal et al. (2008), Lopes-Lutz and Alviano (2008), Ozcan and Chalchat (2008), Rasooli et al. (2008), Saidana et al. (2008), Singh et al. (2008), Cetin et al. (2009),; Irkin and Korukluoglu (2009) and Ozek et al. (2010)
Artemisia judaica (wormwood); *A. absinthium; A. biennis; and* other *Artemisia* sp.	*Fonsecaea pedrosoi*	Kordali et al. (2005), Lopes-Lutz and Alviano (2008), Cetin et al. (2009) and Irkin and Korukluoglu (2009)
Artemisia judaica (wormwood); *A. absinthium; A. biennis*; other *Artemisia* sp.	*Geotrichum candidum*	Kordali et al. (2005), Lopes-Lutz and Alviano (2008); Cetin et al. (2009) and Irkin and Korukluoglu (2009)

(continued)

Table 9.5 (continued)

Essential oil sources	Fungi	References
Artemisia judaica (wormwood); *A. absinthium; A. biennis*; other *Artemisia* sp.	*Rhizoctonia solani*	Kordali et al. (2005), Lopes-Lutz and Alviano (2008), Cetin et al. (2009) and Irkin and Korukluoglu (2009)
Chenopodiumam brosioides	*Macrophomina phaseolina*	Kumar et al.(2007)
Artemisia judaica (wormwood); *A. absinthium; A. biennis; other Artemisia* sp.; *Cinnamomum* sp.; *Croton argyrophylloides; C. zehntneri; C. cajucara; Syzygium aromaticum; Daucus carota* (wild carrot)	*Microsporum canis; Microsporum gypseum*	Dorman and Deans (2000), Alviano et al. (2005), Kordali et al. (2005), Fontenelle et al. (2008), Lopes-Lutz and Alviano (2008), Tavares et al. (2008), Cetin et al. (2009), Irkin and Korukluoglu (2009) and Pinto et al. (2009)
Mentha piperita; M. longifolia; M. viridis	*Mucor ramannianus*	Agarwal et al. (2008) and Mkaddem et al. (2009)
Artemisia judaica (wormwood); *A. absinthium; A. biennis;* other *Artemisia* sp.	*Pythium debaryanum*	Kordali et al. (2005), Lopes-Lutz and Alviano (2008); Cetin et al. (2009) and Irkin and Korukluoglu (2009)
Artemisia judaica (wormwood); *A. absinthium; A. biennis; Artemisia* sp.	*Trichophyton rubrum T. mentagrophytes; T. roseum*	Dorman and Deans (2000), Kordali et al. (2005), Lopes-Lutz and Alviano (2008), Cetin et al. (2009), Irkin and Korukluoglu(2009) and Pinto et al. (2009)

Santos et al. 2015; Hammer et al. 2012; Hristova et al. 2013; Kocevski et al. 2013; Petretto et al. 2014; Pullagummi et al. 2014; Singh et al. 2014; Ibrahim et al. 2015a, b; Latifah-Munirah et al. 2015; Novy et al. 2015; Papajani et al. 2015;; Venturi et al. 2015; Souza et al. 2016). Most research on the antifungal activity is in the initial phases of clinical trials. Thus, EOs are functioning as an alternative for the existing antifungal drugs (Samber et al. 2015).

9.3.5 Essential Oils and Their Antiviral Activity

In addition to antimicrobial and anticancer activity, EOs also exhibit antiviral activity (see Table 9.6). EOs show potential inhibition against viral replication, as they consist of monoterpenes, sesquiterpenes, and phenylpropanoid chemical constituents (Astani et al. 2011). Eucalyptus, thyme, and *M. alternifolia* (tea tree oil) EOs show potential antiviral activity against herpes virus. Their activity has also been established against viral envelope structures (Carson et al. 2001; Reichling et al. 2005; Schnitzler et al. 2007, 2011). For example, oregano oils exhibit potential antiviral activity against herpes simplex virus (HSV) and yellow fever virus (Meneses et al. 2009). Monoterpenes of EOs, such as isoborneol, have been shown

Table 9.6 Essential oil exhibiting antiviral activities

Essential oil	Antiviral effect	References
Artemisia arborescens; A. vulgaris; Lippia origanoides (wild marjoram); *Origanum vulgare*	Inactivation of yellow fever virus	Sinico et al. (2005), Meneses et al.(2009)
Artemisia arborescens; A. vulgaris; Allium cepa (onion); *A. sativum* (garlic); *Coriandrum sativum* (cilantro/dhania); *Cuminum cyminum; Ocimum basilicum; O. americanum; O. sanctum*	Activity against herpes simplex virus type 1 (HSV-1)	Sinico et al.(2005), Meneses et al. (2009) and Romeilah et al. (2010)
Eugenia caryophyllata (i.e. *Syzygium aromaticum); Eucalyptus globulus* (eucalyptus oil); *Leptospermum scoparium* (manuka oil); *Melaleuca alternifolia; M. armillaris; Origanum vulgare; Santalum* sp. (sandal wood)	Activity against HSV-1 and HSV-2	Benencia and Courreges (2000), Schnitzler et al. (2001), Reichling et al. (2005), Cermelli et al. (2008), Garozzo et al.(2009) and Meneses et al. (2009)
Eucalyptus globulus (eucalyptus oil)	Activity against respiratory viruses	Schnitzler et al.(2001) and Cermelli et al. (2008)
Houttuynia cordata (fishwort/chameleon plant); *Melaleuca alternifolia*	Virucidal effect on influenza virus and HSV-1	Garozzo et al. (2009,2011)
Cymbopogon citrate and other species	Inhibition of HSV-1 replication	Minami et al. (2003)
Mentha piperita	Virucidal activity against HSV-1 and HSV-2	Schuhmacher et al. (2003)
Melissa officinalis L.	Prevention of Replication of HSV-2	Allahverdiyev et al. (2004)
Santolina insularis	Inactivation of viral particles of HSV-1 and HSV-2	De Logu et al. (2000)
Thymus sp.	Inhibition of replication of Epstein-Barr virus (EBV)	Hamid et al. (2011)

effective against HSV-1 virus (Armaka et al. 1999). HSV-2 virus is more delicate than HSV-1 virus to the pine, tea tree, manuka, lemon balm, and santolia EOs in small concentrations, i.e., in the range between 0.0001 and 0.0009% of the IC_{50} value (Garcia et al. 2003; Saddi et al. 2007; Koch et al. 2008; Schnitzler et al. 2011). In a report published by Benencia and Courreges (2000), clove oil eugenol was used against HSV-induced keratitis. Antiviral activity through plaque reduction assay against African green monkey, EOs such as *Melaleuca armillaris* (*M. armillaris*), *Melaleuca ericifolia* (*M. ericifolia*), and *Melaleuca styphelioides* (*M. styphelioides*) showed 99%, 91.5% and 92% effectiveness, respectively (Deans and Ritchie 1987). *A. fragrantissima*, *A. arborescens*, *F. margarita*, *G. marifolia*, *H. mutabilis*, *L. salviifolia*, *M. officinalis*, *M. mollis*, *O. campechianum*, *P. cablin*, and *T. ammi* EOs show potential antiviral activity against ORF virus (a parapox virus), HSV-I, avian influenza A virus (H5N1), HSV-1, HSV-2, avian influenza virus (AIV), subtype (H9N2), influenza A (H2N2) virus, and Japanese encephalitis virus (JEV) (Allahverdiyev et al. 2004; Sinico et al. 2005; Wu et al. 2011, 2013; Kiyohara et al. 2012; Zeedan et al. 2014; Ibrahim et al. 2015a, b; Venturi et al. 2015 Roy et al. 2015; Brand et al. 2016).

9.3.6 Essential Oils as Antidiabetic Agents

Diabetes mellitus (DM), generally known as diabetes, is a metabolic disorder that is becoming increasing prevalent in modern society due to unhealthy lifestyle. Insulin-dependent diabetes is called Type-I diabetes, its causes do not produce insulin, and it damages the pancreas. Type-II diabetes is non-insulin diabetes, it causes insulin resistance in the liver, peripheral tissues, and reduced β-cell mass (Srinivasan and Ramarao 2007; Matthaei et al. 2000), respectively. Diabetes causes changes in metabolism of carbohydrates, fats, and proteins, which results in hyperglycemia, glycosuria, and hyperlipidemia (Baradaran et al. 2013; Behradmanesh et al. 2013; Mirhoseini et al. 2013). Diabetes can be successfully managed with a proper diet and consuming drugs. Also, the use of traditional medicines can effectively control the risk of diabetes with reduced side effects. The essential oil of *Vaccinium arctostaphylos* containing high levels of anthocyanoside myrtillin is used in the traditional medicine for diabetes control (Murray 1997). In mouse models, *Securigera securidaca* essential oil had significantly found effective in reducing the blood glucose level (Hisseinzadeh et al. 2002). Likewise, *Gymnema sylvestre* essential oil has been reported to be effective against both Type 1and Type 2 diabetes. It influenced the absorption of glucose in the digestive track and regenerated and proliferated β-cells (Lirussi et al. 2002; Amini et al. 2012; Madihi et al. 2013; Nasri et al. 2013; Nasri and Shirzad 2013; Rafieian-Kopaei et al. 2013). Similarly, *Atriplex halimus* has been used for Type 2 diabetes treatment in animal models, as it contains fibers, proteins, and other trace elements, such as chromium. Consumption of *A. halimus*, 3 grams dried leaves can reduce blood sugars in Type 2 diabetes patients (Bahmani et al. 2014). *Camellia sinensis* seed oil containing flavonoids like catechin, epicatechin, epigallocatechin, and gallocatechin can increase insulin levels, and their polyphenolic compounds act as antioxidants (Asadi et al. 2013; Parsaei et al. 2013). EOs or natural products have had major impact on diabetes, whereby flavonoids, metformin, anthocyanin, catechin, quercetin, flavone, phenylpropanoids, lipoic acids, and coumarin metabolites are particularly effective (Arabbi et al. 2004; Rafieian-Kopaei et al. 2013; Singab et al. 2014).

Foeniculum vulgare (*F. vulgare*) EOs show potential antidiabetic activity in rates corrected the hyperglycemia from 162.5 ± 3.19 mg/dl to 81.97 ± 1.97 mg/dl and also high activity of serum glutathione peroxide from 59.72 ± 2.78 U/g Hb to

99.60 ± 6.38 U/g Hb (El-Soud et al. 2011). Yen et al. (2015) from Taiwan described that different families of commercial EOs purchased from local market such as (Lamiaceae, Rutaceae, Myrtaceae, Cupressaceae, Piperaceae, Burseraceae, Zingiberaceae, Geraniaceae, Apiaceae, Asteraceae, Pinaceae, and Lauraceae) show potential antidiabetic activity after 24 hrs in culture medium of 3 T3-L1 adipocytes. Alpha amylase inhibition assay shows efficient antidiabetic activity with *S. aromaticum* and *C. cyminum* EOs (Tahir et al. 2016).

9.3.7 Essential Oils as Insect Repellents

Insect repellent is a substance applied to the surface of skin or on clothing to prevent insect bites (Blackwell et al. 2003; Choochote et al. 2007; Nerio et al. 2010). Generally, repellents work as vapor barriers, preventing the insects from coming into contact with the surface. Currently, many synthetic chemicals are used to control insects and arthropods; however, they are causing concerns regarding human health and environmental pollution. Plant molecules or plant EOs are an alternative to this and are used as insect and arthropod repellants. Because of their natural origin, they are relatively safe for human health and environmental friendly. These insect repellent plant molecules have been isolated from a large number of plants, mainly from their essential oils. Some of them have been commercialized in certain formulations as insect repellents (Chaubey 2007). Synthetic insect repellents are widely used to prevent infestation of stored grains, fruits, and other cellulosic materials by different pests, mostly arthropods. Similar circumstances occur for animals and human health. To control insects different insecticides have been used; these insecticides transmit to human pathogens. These days many of these insects are resistant to the chemicals, and it should be applied to larger amounts, due to the temperature changes, global warming, etc.; actually global warming has moved the mosquitoes to transmit malaria, dengue, and yellow fever into high altitude, some temperatures affecting these diseases. EOs are volatile and complex mixtures of hydrocarbons (monoterpenes and sesquiterpenes) with different functional groups (ethers, alcohols, aldehydes, esters, ketones, phenols, and phenol ethers). However, these chemicals can act as insect repellents, in particular if combined with other natural products. For example, to increase protection time, vanillin could be used with *C. winterianus* EOs. Among plant families, *Ocimium* spp., *Eucalyptus* spp., and *Cymbopogon* spp. are widely used as insect repellents. Similarly, some major compounds, such as citronellol, camphor, thymol, α-pinene, and limonene, have shown good insect repellent activity. Among the plant families that contain about 3000 EOs, approximately 10% of these EOs have commercial importance in pharmaceutical, food, and cosmetics industries. The United States Food and Drug Administration (FDA) considers EOs as insect repellents that are safe for human health and environmentally friendly (Trongtokit et al. 2005; Nerio et al. 2010). EOs from a large number of plant families showing the potential insect repellent activity are shown in Table 9.7.

Table 9.7 Essential oils exhibiting insect repellent activities

Family	Insect scientific name	Essential oil sources	References
Diptera	*Anopheles annularis; Anopheles culicifacies; C. quinquefasciatus*	*Mentha piperita*	Ansari et al. (2000)
Diptera	*A. aegypti*	*Z. piperitum*	Choochote et al. (2007)
Diptera	*Culex pipiens*	*Pimpinella anisum; O. basilicum; Eucalyptus camaldulensis*	Erler et al. (2006)
Diptera	*A. aegypti*	*Baccharis spartioides, Aloysia citriodora*	Gillij et al. (2008)
Diptera	*Mansonia*	*Eucalyptus maculata citriodora*	Hadis et al. (2003)
Diptera	*A. gambiae*	*Croton pseudopulchellus, Mkilua fragrans, Endostemon tereticaulis, Ocimum forskolei, Ocimum fischeri, Plectranthus longipes*	Odalo et al. (2005)
Diptera	*A. gambiae*	*Conyza newii, Tarchonanthus camphoratus, Tetradenia riparia, Lippia javanica, Lippia ukambensis, Plectranthus marrubioides*	Omolo et al. (2004)
Diptera	*A. aegypti*	*C. citratus*	Oyedele et al. (2002)
Diptera	*A. braziliensis*	*O. selloi*	Padilha de Paula et al. (2003)
Diptera	*Anopheles stephensi, A. aegypti, C. quinquefasciatus,*	*O. basilicum*	Prajapati et al. (2005)
Diptera	*Anopheles stephensi, A. aegypti, C. quinquefasciatus,*	*Rosmarinus officinalis*	Prajapati et al. (2005)
Diptera	*Anopheles stephensi, A. aegypti, C. quinquefasciatus*	*Cinnamomum zeylanicum*	Prajapati et al. (2005)
Diptera	*C. quinquefasciatus*	*C. citratus*	Pushpanathan et al. (2006)
Diptera	*C. quinquefasciatus*	*Zingiber officinalis*	Pushpanathan et al. (2008)
Diptera	*C. quinquefasciatus*	*Moschosma polystachyum*	Rajkumar and Jebanesan (2005)
Diptera	*C. quinquefasciatus*	*Solanum xanthocarpum*	Rajkumar and Jebanesan (2005)
Diptera	*A. dirus, C. quinquefasciatus*	*Curcuma longa L, C. winterianus, O. americanum*	Tawatsin et al. (2001)

(continued)

Table 9.7 (continued)

Family	Insect scientific name	Essential oil sources	References
Diptera	*A. dirus, C. quinquefasciatus*	*Z. limonella*	Trongtokit et al. (2005)
Diptera	*A. aegypti, C. quinquefasciatus, A. dirus*	*Pogostemon cablin*	Trongtokit et al. (2005)
Diptera	*A. aegypti, C. quinquefasciatus, A. dirus*	*Syzygium aromaticum*	Trongtokit et al. (2005)
Diptera	*A. aegypti*	*Z. limonella, C. nardus*	Trongtokit et al. (2005)
Diptera	*A. albopictus*	*E. globulus*	Yang and Ma (2005)
Diptera	*A. aegypti*	*D. caryophyllum*	Tunón et al. (2006)
Coleoptera	*T. castaneum*	*Nigella sativa, Trachyspermum ammi, Anethum graveolens,*	Chaubey (2007)
Coleoptera	*T. castaneum*	*B. salicifolia*	García et al. (2005)
Coleoptera	*T. castaneum*	*Artemisia annua*	Goel et al. (2007)
Coleoptera	*L. serricorne*	*Perilla frutescens, Thymus vulgaris, Satureia hortensis, Mentha piperita, Cinnamomum cassia, Litsea cubeba, Perilla frutescens*	Hori (2003)
Coleoptera	*Acanthoscelides obtectus*	*Laurus nobilis, Rosmarinus officinalis, E. globulus, Juniperus oxycedrus, Lavandula hybrid, Mentha microphylla, Mentha viridis, Apium graveolens*	Papachristos and Stamopoulos (2002)
Coleoptera	*Callosobruchus maculatus*	*O. basilicum*	Pascual and Ballesta (2003)
Coleoptera	*T. castaneum*	*Artemisia vulgaris*	Wang et al. (2006)
Phthiraptera	*P. humanus capitis*	*Mentha pulegium*	Toloza et al. (2006)
Isoptera	*Coptotermes formosanus*	*Calocedrus macrolepis, Cryptomeria japonica, Chamaecyparis obtusa*	Cheng et al. (2007)
Thysanoptera	*Thrips tabaci*	*Rosmarinus officinalis*	Koschier and Sedy (2003)

Worldwide, *Cymbopogon* spp. produce the most widely used natural insect repellent. In tropical or forest regions, these families are used as mosquito repellents (Trongtokit et al. 2005; Moore et al. 2007). EOs extracted from these plant families have been tested against the arthropod, *Cymbopogon excavates*. The results showed that EOs were 100% efficient for insect repellent activity up to 2 hrs. Likewise, when applied against *Anopheles arabiensis*, the repellent efficacy of these oils was decreased by 59.3% after 4 h (Govere et al. 2000). Vanillin (5%) mixed with *C. winterianus* EOs shows 100% efficacy up to 6 hrs against *A. aegypti*, *C. quinquefasciatus*, and *A. dirus* (Tawatsin et al. 2001). However, *C. nardus* and *C. flexuosus* oils were inactive against *Cydia pomonella* (Lepidoptera) and *Lasioderma serricorne* (cigarette beetle) (Landolt et al. 1999). Eucalyptus oils show high repellent activity against Mansonia mosquitoes, *Pediculus humanus capitis*, *Ixodes ricinus*, and *Aedes albopictus* (Hadis et al. 2003; Yang and Ma 2005; Jaenson et al. 2006; Toloza et al. 2008), as well as moderate activity against *A. aegypti* and *C. pomonella* (Trongtokit et al. 2005; Gillij et al. 2008), but no activity against *L. serricorne* (Hori 2003). *Ocimum* spp. EOs are also found as an efficient insect repellent (Padilha de Paula et al. 2003). The insect pests, such as *A. aegypti*, *A. dirus*, and *C. quinquefasciatus*, were potentially repelled by the *O. americanum*) essential oil (Tawatsin et al. 2001). Similarly, *O. selloi*, *O. basilicum*, and *O. gratissimum* EOs were potentially effective repellents against *A. braziliensis* (Padilha de Paula et al. 2003) while exhibiting no repellent activity against *L. serricorne* and *C. pomonella* (Landolt et al. 1999; Hori 2003). Fresh leaves of EOs extracted from *O. sanctum*, *M. piperita*, *E. globulus*, and *P. amboinicus* oils show potential insect repellent against *Aedes aegypti* (Lalthazuali and Mathew 2017). Some of the EOs such as *J. procera*, *C. citrates*, *C. zeylanicum*, *R. officinalis*, *Z. officinale*, *A. marmelos*, *L. acidissima*, *S. indicus*, *S. amaranthoides*, *C. odorata*, *D. elata*, *C. longa*, *P. heyneanus*, and *Z. limonella* show against *Anopheles arabiensis*, *Culex tritaeniorhynchus*, *Anopheles subpictus*, *Culex quinquefasciatus* (mosquitoes), *Aedes aegypti* (mosquitoes), *Anopheles stephensi* (malaria), and *Aedes albopictus* (mosquitos) strong insect repellent activity in laboratory level, respectively (Govindarajan 2011; Govindarajan et al. 2015; Karunamoorthi et al. 2014; Reegan et al. 2015; Das et al. 2015).

9.3.8 Antimutagenic Properties of Essential Oils

Antimutagenic properties arise due to the inhibition of diffusion of the mutagens into the cells, antioxidant and direct radical scavenging activity, inactivation of mutagens and produced by a mutagen, antioxidant activity of enzyme cell activation, and inhibition of metabolic conversation of P450 of promutagens into mutagens, for instance, by plant extracts (Sharma et al. 2001; Ipek et al. 2005). Plant extract constituents, such as superoxide dismutase (enzyme), glutathione, N-acetylcysteine, retinoids, carotenoids, flavonoids, and other polyphenols, are known to function as reactive oxygen species (ROS) scavengers that can prevent mutagenesis (Racchi 2013; Toscano-Garibay et al. 2017). EOs or their individual components, such as α-bisabolol, aflatoxin B1, 2-aminoanthracene, benzo-a-pyrene, and 2-aminofluorene, potentially inhibit induced mutagenesis and moderate N-oxide, 4-nitroquinoline, and 2-nitrofluorene induced mutagenesis while having less or no induced mutagenesis for sodium azide and nitro-o-phenylenediamine

(Gomes-Carneiro et al. 2005). The antimutagenic effect of α-bisabolol is due to the interaction of it with promutagen biotransformation enzymes. *Salvia officinalis* EOs show the potential inhibition of UVC-induced mutagenesis in *Salmonella typhimurium*, *E. coli*, and *Saccharomyces cerevisiae* (Dudai et al. 2005; Vukovic-Gacic et al. 2006). In an experiment, Idaomar et al. (2002) treated *Drosophila melanogaster,* with EOs of *Ledum groenlandicum*, *Ravensara aromatica,* and *Helichrysum italicum* significantly reduced the induced mutation frequency. Likewise, *Origanum compactum* EOs showed a potential antimutagenic effect against the mutagen, urethane (Mezzoug et al. 2007). *O. majorana*, *C. sinensis*, *C. latifolia*, *A. aucheri*, *A. ciniformis*, and *J. leptoloba* EOs show potential antimutagenic activity against *S. typhimurium* strains TA97a, TA98, TA100, TA100, and TA1535 (Fernandesa et al. 2015; Taherkhani. 2015, 2016; Dantas et al. 2016; Toscano-Garibay et al. 2017).

9.3.9 Phototoxicity

Plant molecules, such as furanocoumarins and coumarins present in grapefruit peel oil and citrus plants oils, are photoactive in nature. For example, citrus EOs contain psoralen (a furocoumarin) that binds with DNA under ultraviolet A (UVA) light, causing it to become highly cytotoxic and mutagenic due to the formation of mono and diadducts in DNA (Lang and Buchbauer 2012; Raut and Karuppayil 2014). However, in dark conditions, cytotoxic and mutagenic activity is not detected (Dijoux et al. 2006; Bakkali et al. 2008). Cytotoxicity and phototoxicity depend on the EOs which contain the chemical constituents that produce free radicals according to the sunlight exposure. Wood oil (*Fusanus spicatus*) is cytotoxic but not phototoxic (Dijoux et al. 2006; Bakkali et al. 2008; Raut and Karuppayil 2014). EOs producing reactive oxygen species (ROS) destruct the cellular and organelle membranes, prooxidants on proteins, and DNA. Under sunlight, oxygen singlets occur, due to reactive oxygen species producing energy on excitation. This may be due to the destruction of the polysaccharides, nucleic acids proteins, and enzymes and, in some times, causes the formation of adducts with DNA and lipid membranes. With or without light, free radical generation depends on the chemical constituents present in the EOs. *Citrus aurantium* and *Cymbopogon citratus* EOs show potential cytotoxic and phototoxic activity (Dijoux et al. 2006). Thus, these photoactive compounds find their application in biomedicine fields including photochemotherapy.

9.3.10 Carcinogenicity of Essential Oils

EOs are potentially cytotoxic without being mutagenic. Thus, carcinogenicity of EOs and their constituents are considered carcinogenic as they are involved in the metabolic activation of secondary carcinogens (Guba 2001). EOs like *S. sclarea* and *M. quinquenervia* produce estrogen secretions, which could induce estrogen-dependent cancer. Major chemical compounds, such as flavins, porphyrins, hydrocarbures, and cyanins, are photosensitive molecules that could cause skin cancer. Under ultraviolet A radiation, EOs containing psoralen are also photosensitive to light and therefore could induce cancer and DNA adducts (Lesgards et al. 2014; Romeilah. 2016; Choucry. 2017). EOs of *Mentha* species containing pulegone as one of the major constituents is known to induce carcinogenicity through metabolism generating glutathione (Zhou et al. 2004). *Sassafras albidum* and *Ocotea pretiosa* EOs contain safrole, and *Laurus nobilis* and *Melaleuca leucadendron* EOs contain methyl eugenol as the major constituent. Safrole and methyl eugenol (phenylpropenes) could induce carcinogenic metabolites in rodents (Burkey et al. 2000; Liu et al. 2000; Gautam et al. 2014). D-limonene from citrus and estragole from *O. basilicum* EOs could induce carcinogenic mutations in male rats and mouse models (Anthony et al. 1987; Miller et al. 1983; Bayala et al. 2014; Chen et al. 2017).

9.3.11 Essential Oils as Antiprotozoal Agents

Different protozoan diseases are very important to public health, such as malaria, trichomoniasis, giardiasis, and leishmaniasis caused by *Plasmodium* sp., *Entamoeba histolytica*, and *Trypanosoma cruzi* species, respectively. Availability of antiprotozoal drugs is limited, and their prolonged use causes side effects (Sauter et al. 2012). Hence, plant extracts and EOs could be a safer treatment alternative for protozoal diseases (Sauter et al. 2012). For example, *Thymus vulgaris* essential oil, thymol, is the major component that inhibits trypanosomal parasite through damage of plasma membrane (Santoro et al. 2007a, b; Saeidnia and Gohari 2012). Compared to *T. vulgaris*, *C. citrates* and *O. gratissimum* EOs show a better antitrypanosomal activity. Terpenoids like thymol, carvacrol, and linalool are known to inhibit *Entamoeba histolytica*. EOs from *M. alternifolia*, *C. copticum,* and *L. angustifolia* show potential protozoal effects (Carson et al. 2006; Mansoor et al. 2011). *C. citrates*, *Origanum* spp., *L. multiflora*, *O. gratissimum*, and *S. thymbra* EOs exhibit a potential antimalarial activity (Tchoumbougnang et al. 2005; El babili et al. 2011). Specifically, *C. cajucara*, *C. citrates*, *O. gratissimum*, *A. millefolium*, *A. abrotanum*, *C. ambrosioides*, *P. caribaea*, and *Piper* spp. EOs show antileishmanial activity (Santin et al. 2009; Santos et al. 2010; Ahmed et al. 2011; Tariku et al. 2011).

9.4 Economic Importance of Essential Oils

In the global markets, EOs and their derived molecules are widely used, such as in perfumes and cosmetics, as well as in food, pharmaceutical, and agricultural industries. Throughout Europe, Africa, and Asia, as well as in the USA, EOs have been used in cosmetics (skin creams, body lotions, soaps, perfumes, shampoos, etc.), medicinal industry (pharmaceutical and bulk drug industry, aromatherapy products, and medicinal supplements), and food industry (herbs, spices, and additives) (Nakatsu et al. 2000; Hussain et al. 2008; Teixeria et al. 2013; Swamy and Sinniah 2015, 2016). Essential oil production has exceeded 70,000 tons per annum, and the main producers are the USA, Brazil, China, India, Australia, Indonesia, Malaysia, Thailand, Sri Lanka, South Africa, Italy, Russia, Nepal, Bangladesh, Germany, and Pakistan. For example, clove, celery, basil, and lemongrass EOs are mainly produced in India; rosemary and lavender EOs are usually grown in Spain and France; geranium and rose geranium EOs are endemic to Africa; and tea oils are grown mostly in Australia and South Wales (Bedi and Tanuja 2010). Among 3000 EOs, only 10% have been commercially exploited (Djilani and Dicko 2012). EOs such as basil, orange oil, corn mint, peppermint, eucalyptus, citronella, lemon, clove, camphor, and cumin oils are medicinally important worldwide (Hussain et al. 2008; Bedi and Tanuja 2010). Based on purity, composition, and material sources, their market value can vary considerably. Globally, more than 80% of the people are depending on the plant-based traditional medicine (Akhtar et al. 2014; Arumugam et al. 2016; Swamy and Sinniah 2015; Swamy et al. 2012). Generally, anise and coriander oil cost $20 to $30 per pound, while thyme, dill, and calendula might cost >$100. Moreover, sweet basil, fennel, clary sage, lavender, and caraway EOs could cost $50–$80 per pound. World wide EOs in global markets estimate more than 62 billion USD per year, and it is imagined by the year 2050 to grow up to 5 trillion USD per year. World wide, plant-based (Natural Products) molecules have high demand in the food industry, perfumes, and cosmetic and pharmaceutical substances. Internationally, more than 250 EOs trade at the value of 1.2 billion USD per year (Akhtar et al. 2014; Arumugam et al. 2016; Bhattacharya et al. 2014; Swamy et al. 2016). Still there are wide differences between low- and high-income peoples in the worldwide annual economic demand of pharmaceutical substances. An average spending by each person on pharmaceutical substances per year in low-income countries, such as India, Nigeria and Sri Lanka is about 0.75 US$, 1.2 US$, 0.58 US$, respectively. While, the same in high-income countries like Japan, the USA, and Germany is 38.5 US$, 35.10 US$, and 53.4 US$, respectively. This indicates clearly that there is a significant difference in the expenditure for pharmaceutical products between low- and high-income nations. This is because people in low-income nations largely depend on herb-based medicines (Lu et al. 2011; Bukar et al. 2016).

9.5 Conclusions and Future Prospects

The EOs could act as antimicrobial agents in personal hygiene, air purification, internal use, insecticides and preservations of crops, and food products because of its non-genotoxic risks. However, some EOs showed antimutagenic, anticarcinogenic, photosensitive, and antidiabetic activities. Recent studies on EOs and their chemical constituents, such as polyphenols, flavonoids, and alcohols, showed their potential in reducing tumor cell proliferation, and murine leukemia. They have also been used in cosmetic, food, and pharmaceutical industries, chemotherapy, and prevention of drug resistance against infectious and noninfectious diseases. Although lot of works have been done on the EOs, but still future researches are needed to optimize their doses in combination with existing drugs for the safer use as a medicine for patients.

References

Agarwal V, Lal P, Pruthi V (2008) Prevention of *Candida albicans* biofilm by plant oils. Mycopathologia 165:13–19

Ahmad I, Beg AZ (2001) Antimicrobial and phytochemical studies on 45 Indian medicinal plants against multi-drug resistant human pathogens. J Ethnopharmacol 74:113–123

Ahmadi R, Alizadeh A, Ketabchi S (2015) Antimicrobial activity of the essential oil of *Thymus kotschyanus* grown wild in Iran. Int J Biosci 6:239–248

Ahmed SBH, Sghaier RM, Guesmi F, Kaabi B, Mejri M, Attia H, Laouini D, Smaali I (2011) Evaluation of antileishmanial, cytotoxic and antioxidant activities of essential oils extracted from plants issued from the leishmaniasis- endemic region of Sned (Tunisia). Nat Prod Res 12:1195–1201

Ait-Ouazzou A, Lor'an S, Arakrak A, Laglaoui A, Rota C, Herrera A, Pagán R, Conchello P (2012) Evaluation of the chemical composition and antimicrobial activity of *Mentha pulegium*, *Juniperus phoenicea*, and *Cyperus longus* essential oils from Morocco. Food Res Int 45:313–319

Akhtar MS, Degaga B, Azam T (2014) Antimicrobial activity of essential oils extracted from medicinal plants against the pathogenic microorganisms: a review. Biol Sci Pharm Res 2:1–7

Al Abbasy WD, Pathare N, Al-Sabahi NJ, Khan AS (2015) Chemical composition and antibacterial activity of essential oil isolated from Omani basil (*Ocimum basilicum* Linn). Asian Pac J Trop Dis 5:645–649

Aleksic V, Knezevic P (2014) Antimicrobial and antioxidative activity of extracts and essential oils of *Myrtus communis* L. Microbiol Res 169:240–254

Allahverdiyev A, Duran N, Ozguven M, Koltas S (2004) Antiviral activity of the volatile oils of *Melissa officinalis* L. against Herpes simplex virus type-2. Phytomed 11:657–661

Alviano WS, Mendonca-Filho RR, Alviano DS, Bizzo HR, Souto-Padron T, Rodrigues ML, Bolognese AM, Alviano CS, Souza MMG (2005) Antimicrobial activity of *Croton cajucara* Benth linalool rich essential oil on artificial biofilms and planktonic microorganisms. Oral Microbiol Immunol 20:101–105

Amatiste S, Sagrafoli D, Giacinti G, Rosa G, Carfora V, Marri N, Tammaro A, Bovi E, Rosati R (2014) Antimicrobial activity of essential oils against *Staphylococcus aureus* in fresh sheep cheese. Italian J Food Saf 3:1696

Amin A, Gali-Muhtasib H, Ocker M, Schneider-Stock R (2009) Overview of major classes of plant-derived anticancer drugs. Int J Biomed Sci 5:1–11

Amini GF, Rafieian-Kopaei M, Nematbakhsh M, Baradaran A, Nasri H (2012) Ameliorative effects of metformin on renal histologic and biochemical alterations of gentamicin-induced renal toxicity in Wistar rats. J Res Med Sci 17:621–625

Andrade MA, Cardoso MG, de Andrade J, Silva LF, Teixeira ML, Resende JMV, Figueiredo ACS, Barroso JG (2013) Chemical composition and antioxidant activity of essential oils from *Cinnamodendron dinisii* Schwacke and *Siparuna guianensis* Aublet. Antioxidants 2:384–397

Andrade BFMT, Barbosa LN, Probst IS, Junior AF (2014) Antimicrobial activity of essential oils. J Essent Oil Res 26:34–40

Angioni A, Barra A, Coroneo V, Dessi S, Cabras P (2006) Chemical composition, seasonal variability, and antifungal activity of *Lavandula stoechas* L. ssp. stoechas essential oils from stem/leaves and flowers. J Agric Food Chem 54:4364–4370

Ansari MA, Vasudevan P, Tandon M, Razdan RK (2000) Larvicidal and mosquito repellent action of peppermint (*Mentha piperita*) oil. Bioresour Technol 71:267–271

Anthony A, Caldwell G, Hutt AG, Smith RL (1987) Metabolism of estragole in rat and mouse and influence of dose size on excretion of the proximate carcinogen 10-hydroxyestragole. Food Chem Toxicol 25:799–806

Apostolou A, Stagos D, Galitsiou E, Spyrou A, Haroutounian S, Portesis N, Trizoglou I, Hayes AW, Tsatsakis AM, Kouretas D (2013) Assessment of polyphenolic content, antioxidant activity, protection against ROS-induced DNA damage and anticancer activity of *Viti vinifera* stem extracts. Food Chem Toxicol 61:60–68

Aqil F, Gupta A, Munagala R, Jeyabalan J, Kausar H, Sharma RJ, Singh IP, Gupta RC (2012) Antioxidant and antiproliferative activities of anthocyanin/ellagitannin-enriched extracts from *Syzygium cumini* L. (Jamun, the Indian Blackberry). Nutr Cancer 64:428–438

Arabbi PR, Genoves MI, Lajolo FM (2004) Flavonoids in vegetable foods commonly consumed in Brazil and estimated ingestion by the Brazilian population. J Agric Food Chem 52:1124–1131

Armaka M, Papanikolaou E, Sivropoulou A, Arsenakis M (1999) Antiviral properties of isoborneol, a potent inhibitor of herpes simplex virus type 1. Antivir Res 43:79–92

Arumugam G, Swamy MK, Sinniah UR (2016) *Plectranthus amboinicus* (Lour.) Spreng: Botanical, phytochemical, pharmacological and nutritional significance. Molecules 21:369. https://doi.org/10.3390/molecules21040369

Asadi Y, Karimi M, Parsaei P, Ezzati S, Khadivi R, Zamiri A, Rafieian-Kopaei M (2013) Effects of *Camellia sinensis* ethanolic extract on histometric and histopathological healing process of burn wound in rat. Middle-East J Sci Res 13:14–19

Astani A, Reichling J, Schnitzler P (2011) Screening for antiviral activities of isolated compounds from essential oils. Evidence-Based Compl Altern Med 2011:253643. https://doi.org/10.1093/ecam/nep187

Avetisyan A, Markosian A, Petrosyan M, Sahakyan N, Babayan A, Aloyan S, Trchounian A (2017) Chemical composition and some biological activities of the essential oils from basil *Ocimum* different cultivars. BMC Compl Altern Med 17:60. https://doi.org/10.1186/s12906-017-1587-5

Azmi AS, Bhat SH, Hanif S, Hadi SM (2006) Plant polyphenols mobilize endogenous copper in human peripheral lymphocytes leading to oxidative DNA breakage: A putative mechanism for anticancer Properties. FEBS Lett 580:533–538

Baananou S, Bouftira I, Mahmoud A, Boukef K, Marongiu B, Boughattas NA (2013) Antiulcerogenic and antibacterial activities of *Apium graveolens* essential oil and extract. Nat Prod Res 27:1075–1083

Bahmani M, Golshahi H, Saki K, Rafieian-Kopaei M, Delfan B, Mohammadi T (2014) Medicinal plants and secondary metabolites for diabetes mellitus control. Asian Pac J Trop Dis 4:S687–S692

Bajpai VK, Yoon JI, Kang SC (2009) Antifungal potential of essential oil and various organic extracts of *Nandina domestica* Thunb., against skin infectious fungal pathogens. Appl Microbiol Biotechnol 83:1127–1133

Bakkali F, Averbeck S, Averbeck D, Idaomar M (2008) Biological effects of essential oils-a review. Food Chem Toxicol 46:446–475

Bansod S, Rai M (2008) Antifungal activity of essential oils from Indian medicinal plants against human pathogenic *Aspergillus fumigatus* and *A. niger*. World J Med Sci 3:81–88

Baradaran A, Madihi Y, Merrikhi A, Rafieian-Kopaei M, Nasri H (2013) Serum lipoprotein (a) in diabetic patients with various renal function not yet on dialysis. Pak J Med Sci 29:354–357

Bayala B, Bassole IHN, Scifo R, Gnoula C, Morel L, Lobaccaro JMA, Simpore J (2014) Anticancer activity of essential oils and their chemical components-a review. Am J Cancer Res 4:591–607

Beatovic D, Krstic-Milosevic D, Trifunovic S, Šiljegovic J, Glamoclija J, Ristic M, Jelacic S (2015) Chemical composition, antioxidant and antimicrobial activities of the essential oils of twelve *Ocimum basilicum* L. cultivars grown in Serbia. Rec Nat Prod 9:62–75

Bedi S, Tanuja VSP (2010) A handbook of aromatic and essential oil plants: Cultivation, chemistry, processing and uses. Agrobios, Jodhpur

Begnami AF, Duarte MCT, Furletti V, Rehder VLG (2010) Antimicrobial potential of *Coriandrum sativum* L. against different *Candida* species *in vitro*. Food Chem 118:74–77

Behradmanesh S, Horestani MK, Baradaran A, Nasri H (2013) Association of serum uric acid with proteinuria in type 2 diabetic patients. J Res Med Sci 18:44–46

Bejaoui A, Chaabane H, Jemli M, Boulila A, Boussaid M (2013) Essential oil composition and antibacterial activity of *Origanum vulgare* subsp. *glandulosum* Desf. at different phonological stages. J Med Food 16:1115–1120

Benencia F, Courreges MC (2000) *In vitro* and *in vivo* activity of eugenol on human herpes virus. Phytother Res 14:495–500

Benkeblia N (2004) Antimicrobial activity of essential oil extracts of various onions (*Allium cepa*) and garlic (*Allium sativum*). LWT-Food Sci Technol 37:263–268

Berka-Zougali B, Ferhat MA, Hassani A, Chemat F, Allaf KS (2012) Comparative study of essential oils extracted from Algerian *Myrtus communis* L. leaves using microwaves and hydrodistillation. Int J Mol Sci 13:4673–4695

Betts TJ (2001) Chemical characterization of the different types of volatile oil constituents by various solute retention ratios with the use of conventional and novel commercial gas chromatographic stationary phases. J Chromatogr A 936:33–46

Bhattacharya R, Reddy KRC, Mishra AK (2014) Export strategy of Ayurvedic products from India. Int J Ayurved Med 5:125–128

Bilcu M, Grumezescu AM, Oprea AE, Popescu RC, Mogoanu GD, Hristu R, Stanciu GA, Mihailescu DF, Lazar V, Bezirtzoglou E, Chifiriuc MC (2014) Efficiency of vanilla, patchouli and ylang essential oils stabilized by iron oxide@C14 nanostructures against bacterial adherence and biofilms formed by *Staphylococcus aureus* and *Klebsiella pneumoniae* clinical strains. Molecules 19:17943–17956

Bisht DS, Menon KRK, Singhal MK (2014) Comparative antimicrobial activity of essential oils of *Cuminum cyminum* L. and *Foeniculum vulgare* Mill. seeds against *Salmonella typhimurium* and *Escherichia coli*. J Essent Oil-Bearing Plants 17:617–622

Blackwell A, Stuart AE, Estambale BA (2003) The repellant and antifeedant activity of *Myrica gale* oil against *Aedes aegypti* mosquitoes and its enhancement by the addition of salicyluric acid. Proc R Colloid Phys 33:209–214

Bouzabata A, Cabral C, Goncalves MJ, Cruz MT, Bighelli A, Cavaleiro C, Casanova J, Tomi F, Salgueiro L (2015) *Myrtus communis* L. as source of a bioactive and safe essential oil. Food Chem Toxicol 75:166–172

Bowles EJ (2003) Chemistry of Aromatherapeutic Oils. Allen & Unwin, Crows Nest

Bozin B, Mimica-Dukic N, Simin N, Anackov G (2006) Characterization of the volatile composition of essential oils of some Lamiaceae spices and the antimicrobial and antioxidant activities of the entire oils. J Agric Food Chem 54:1822–1828

Brand YM, Roa-Linares VC, Betancur-Galvis LA, Duran-Garcıa DC, Stashenko E (2016) Antiviral activity of Colombian Labiatae and Verbenaceae family essential oils and monoterpenes on human herpes viruses. J Essent Oil Res 28:130–137

Buchbauer G (2010) Biological activities of essential oils. In: Baser KHC, Buchbauer G (eds) Handbook of essential oils: Science, technology, and applications. CRC Press, Boca Raton, pp 235–280

Bukar BB, Dayom DW, Uguru MO (2016) The growing economic importance of medicinal plants and the need for developing countries to harness from it: A mini review. IOSR J Pharma 6:42–52

Burkey JL, Sauer JM, McQueen CA, Sipes IG (2000) Cytotoxicity and genotoxicity of methyleugenol and related congeners – a mechanism of activation for methyleugenol. Mutat Res 453:25–33

Burt S (2004) Essential oils: their antibacterial properties and potential applications in foods–a review. Int J Food Microbiol 94:223–253

Calcabrini A, Stringaro A, Toccacieli L, Meschini S, Marra M, Colone M, Salvatore G, Mondello F, Arancia G, Molinari A (2004) Terpinen-4-ol,themain component of *Melaleuca alternifolia* (tea tree)oil inhibits the *in vitro* growth of human melanoma cells. J Invest Dermatol 122:349–360

Cao J, Xia X, Chen X, Xiao J, Wang Q (2013) Characterization of flavonoids from *Dryopteris erythrosora* and evaluation of their antioxidant, anticancer and acetylcholinesterase inhibition activities. Food Chem Toxicol 51:242–250

Carnesecchi S, Bras-Goncalves R, Bradaia A, Zeisel M, Gosse F, Poupon MF, Raul F (2004) Geraniol, a component of plant essential oils, modulates DNA synthesis and potentiates 5-fluorouracil efficacy on human colon tumor xenografts. Cancer Lett 215:53–59

Carson CF, Riley TV (2003) Non-antibiotic therapies for infectious diseases. Commun Dis Intell 27:S143–S146

Carson CF, Ashton L, Dry L, Smith DW, Riley TV (2001) *Melaleuca alternifolia* (tea tree) oil gel (6%) for the treatment of recurrent herpes labialis. J Antimicrob Chemother 48:450–451

Carson CF, Hammer KA, Riley TV (2006) *Melaleuca alternifolia* (Tea Tree) oil: a review of antimicrobial and other medicinal properties. Clin Microbiol Rev 19:50–62

Cavalieri E, Mariotto S, Fabrizi C, De Prati AC, Gottardo R, Leone S, Berra LV, Lauro GM, Ciampa AR, Suzuki H (2004) α-Bisabolol, a nontoxic natural compound, strongly induces apoptosis in gliomacells. Biochem Biophys Res Commun 315:589–594

Cavar S, Maksimovic M, Vidic D, Paric A (2012) Chemical composition and antioxidant and antimicrobial activity of essential oil of *Artemisia annua* L. from Bosnia. Ind Crop Prod 37:479–485

Cermelli C, Fabio A, Fabio G, Quaglio P (2008) Effect of eucalyptus essential oil on respiratory bacteria and viruses. Curr Microbiol 56:89–92

Cetin B, Ozer H, Cakir A, Li D (2009) Chemical composition of hydrodistilled essential oil of *Artemisia incana* (L.) Druce and antimicrobial activity against food borne microorganisms. Chem Biodivers 6:2302–2310

Chami F, Chami N, Bennis S, Trouillas J, Remmal A (2004) Evaluation of carvacrol and eugenol as prophylaxis and treatment of vaginal candidiasis in an immunosuppressed rat model. J Antimicrob Chemother 54:909–914

Chaubey MK (2007) Insecticidal activity of *Trachyspermum ammi* (Umbelliferae), *Anethum graveolens* (Umbelliferae) and *Nigella sativa* (Ranunculaceae) essential oils against stored product beetle *Tribolium castaneum* Herbst (Coleoptera: Tenebrionidae). Afr J Agric Res 2:596–600

Chen L, Liu P, Evans TC Jr, Ettwiller LM (2017) DNA damage is a pervasive cause of sequencing errors, directly confounding variant identification. Science 355:752–756

Cheng S, Chang H, Wu C, Chang S (2007) Anti-termitic activities of essential oils from coniferous trees against *Coptotermes formosanus*. Bioresour Technol 98:456–459

Choochote W, Chaithong U, Kamsuk K, Jitpakdi A, Tippawangkosol P, Tuetun B, Champakaew D, Pitasawat B (2007) Repellent activity of selected essential oils against *Aedes aegypti*. Fitoterapia 78:359–364

Choucry MA (2017) Chemical composition and anticancer activity of *Achillea fragrantissima* (Forssk.) Sch. Bip. (Asteraceae) essential oil from Egypt. J Pharmacogn Phytother 9:1–5

Cornblatt BS, Ye L, Dinkova-Kostova AT, Erb M, Fahey JW, Singh K, Chen MA, Stierer T, Garrett-Mayer E, Argani P, Davidson NE, Talalay P, Kensler TW, Visvanathan K (2007) Preclinical and clinical evaluation of sulforaphane for chemoprevention in the breast. Carcinogenesis 28:1485–1490

Croteau R, Kutchan TM, Lewis NG (2000) Natural products (secondary metabolites). In: Buchanan B, Gruissem W, Jones R (eds) Biochemistry and Molecular Biology of Plants. Wiley, Blackwell, Chichester

Dantas AD, Klein-Júnior LC, Machado MS, Guecheva TN, Dos Santos LD, Zanette RA, de Mello FB, Pêgas Henriques JA, de Mello JR (2016) *Origanum majorana* essential oil lacks mutagenic activity in the *Salmonella*/Microsome and Micronucleus Assays. Sci World J 2016:3694901

Das NG, Dhiman S, Talukdar PK, Rabha B, Goswami D, Veer V (2015) Synergistic mosquito-repellent activity of *Curcuma longa*, *Pogostemon heyneanus* and *Zanthoxylum limonella* essential oils. J Infect Public Health 8:323–328

De Logu A, Loy G, Pellerano ML, Bonsignore L, Schivo ML (2000) Inactivation of HSV-1 and HSV-2 and prevention of cell-to-cell virus spread by *Santolina insularies* essential oil. Antivir Res 48:177–185

De Sousa AC, Alviano DS, Blank AF, Alves PB, Alviano CS, Gattass CR (2004) *Melissa officinalis* L. essential oil: antitumoral and antioxidant activities. J Pharm Pharmacol 56:677–681

Deans SG, Ritchie G (1987) Antibacterial properties of plant essential oils. Int J Food Microbiol 5:165–180

Delaquis PJ, Stanich K, Girard B, Mazza G (2002) Antimicrobial activity of individual and mixed fractions of dill, cilantro, coriander and eucalyptus essential oils. Int J Food Microbiol 74:101–109

Desam NR, Al-Rajab AJ, Sharma M, Mylabathula MM, Gowkanapalli RR, Mohammed A (2017) Chemical composition, antibacterial and antifungal activities of Saudi Arabian *Mentha longifolia* L. essential oil. J Coast Life Med 5:441–446

Devkatte A, Zore GB, Karuppayil SM (2005) Potential of plant oils as inhibitors of *Candida albicans* growth. FEMS Yeast Res 5:867–873

Dijoux N, Guingand Y, Bourgeois C, Durand S, Fromageot C, Combe C, Ferret PJ (2006) Assessment of the phototoxic hazard of some essential oils using modified 3T3 neutral red uptake assay. Toxicol In Vitro 20:480–489

Djenane D, Aider M, Yanguela J, Idir L, Gomez D, Roncales P (2012) Antioxidant and antibacterial effects of Lavandula and Mentha essential oils in minced beef inoculated with *E. coli* O157: H7and *S. aureus* during storage at abuse refrigeration temperature. Meat Sci 92:667–674

Djilani A, Dicko A (2012) The therapeutic benefits of essential oils. In: Bouayed J, Bohn T (eds) Nutrition, Well-being and Health. InTech, Rijeka, pp 155–178

Dorman HJD, Deans SG (2000) Antimicrobial agents from plants: antibacterial activity of plant volatile oils. J Appl Microbiol 88:308–316

Dryden MS, Dailly S, Crouch M (2004) A randomized, controlled trial of tea tree topical preparations versus a standard topical regimen for the clearance of MRSA colonization. J Hosp Infect 58:86–87

Dudai N, Weinstein Y, Krup M, Rabinski T, Ofir R (2005) Citral is a new inducer of caspase-3 in tumor cell lines. Planta Med 71:484–488

Edris AE (2007) Pharmaceutical and the rapeutic potentials of essential oils and their individual volatile constituents: A review. Phytother Res 21:308–323

El Babili F, Bouajila J, Souchard JP, Bertrand C, Bellvert F, Fouraste I, Valentin A (2011) Oregano: chemical analysis and evaluation of its antimalarial, antiox idant, and cytotoxic activities. J Food Sci 76:C512–C518

El-Soud NA, El-Laithy N, El-Saeed G, Wahby MS, Khalil M, Morsy F, Shaffie N (2011) Antidiabetic activities of *Foeniculum Vulgare* Mill. essential oil in Streptozotocin-induced diabetic rats. Maced J Med Sci 4:139–146

Erler F, Ulug I, Yalcinkaya B (2006) Repellent activity of five essential oils against *Culex pipiens*. Fitoterapia 77:491–494

Eshwarappa RSB, Iyer RS, Subbaramaiah SR, Richard SA, Dhananjaya BL (2014) Antioxidant activity of *Syzygium cumini* leaf gall extracts. Bio Impacts 4:101–107

Fabio A, Cermelli C, Fabio G, Nicoletti P, Quaglio P (2007) Screening of the antibacterial effects of a variety of essential oils on microorganisms responsible for respiratory infections. Phytother Res 21:374–377

Ferguson LR, Philpott M (2008) Nutrition and mutagenesis. Annu Rev Nutr 28:313–329

Fernandesa FH, Guterresb ZDR, Violantec IMP, Lopesb TFS, Garceza WS, Garceza FR (2015) Evaluation of mutagenic and antimicrobial properties of brown propolis essential oil from the Brazilian Cerrado biome. Toxicol Rep 2:1482–1488

Flores CR, Pennec A, Nugier-Chauvin C, Daniellou R, Herrera-Estrella L, Chauvin AL (2014) Chemical composition and antibacterial activity of essential oils extracted from plants cultivated in Mexico. J Mex Chem Soc 58:452–455

Fontenelle ROS, Morais SM, Brito EHS, Brilhante RSN, Cordeiro RA, Nascimento NRF, Rocha MFG, Sidrim JJC (2008) Antifungal activity of essential oils of *Croton* species from the Brazilian *Catinga* biome. J Appl Microbiol 104:1383–1390

Fujita KI, Chavasiri W, Kubo I (2015) Anti-salmonella activity of volatile compounds of Vietnam coriander. Phytother Res 29:1081–1087

Galvao LCDC, Furletti VF, Bersan SMF, Da Cunha MG, Ruiz ALTG, Carvalho JED, Sartoratto A, Rehder VLG, Figueira GM, Teixeira Duarte MC, Ikegaki M, DeAlencar SM, Rosalen PL (2012) Antimicrobial activity of essential oils against *Streptococcus mutans* and their anti-proliferative effects. Evidence-Based Compl Altern Med 2012:751435. https://doi.org/10.1155/2012/751435

Garcia CC, Talarico L, Almeida N, Colombres S, Duschatzky C, Damonte EB (2003) Virucidal activity of essential oils from aromatic plants of San Luis, Argentina. Phytother Res 17:1073–1075

García M, Donadel OJ, Ardanaz CE, Tonn CE, Sosa ME (2005) Toxic and repellent effects of *Baccharis salicifolia* essential oil on *Tribolium castaneum*. Pest Manag Sci 61:612–618

Garozzo A, Timpanaro R, Bisignano B, Furneri PM, Bisignano G, Castro A (2009) *In vitro* antiviral activity of *Melaleuca alternifolia* essential oil. Lett Appl Microbiol 49:806–808

Garozzo A, Timpanaro R, Stivala A, Bisignano G, Castro A (2011) Activity of *Melaleuca alternifolia* (tea tree)oil on Influenza virus A/PR/8:study on the mechanism of action. Antivir Res 89:83–88

Gautam N, Mantha AK, Mittal S (2014) Essential oils and their constituents as anticancer agents: A mechanistic view. BioMed Res Int 2014:154106

Gill CI, Boyd A, Mc Dermott E, Mc Cann M, Servili M, Selvaggini R, Taticchi A, Rowland I (2005) Potential anticancer effects of virgin olive oil phenols on colorectal carcinogenesis models *in vitro*. Int J Cancer 117:1–7

Gillij YG, Gleiser RM, Zygadlo JA (2008) Mosquito repellent activity of essential oils of aromatic plants growing in Argentina. Bioresour Technol 99:2507–2515

Goel D, Goel R, Singh V, Ali M, Mallavarapu G, Kumar S (2007) Composition of the essential oil from the root of *Artemisia annua*. J Nat Med 61:458–461

Gomes-Carneiro MR, Dias DM, De-Oliveira ACAX, Paumgartten FJ (2005) Evaluation of mutagenic and antimutagenic activities of α-bisabolol in the Salmonella microsome assay. Mut Res 585:105–112

Govere J, Durrheim DN, Du TN, Hunt RH, Coetzee M (2000) Local plants as repellents against *Anopheles arabiensis*, in Mpumalanga Province, South Africa. Afr J Med 46:213–216

Govindarajan M (2011) Larvicidal and repellent properties of some essential oils against *Culex tritaeniorhynchus* Giles and *Anopheles subpictus* Grassi (Diptera: Culicidae). Asian Pac J Trop Med 4:106–111

Govindarajan M, Rajeswary M, Sivakumar R (2015) Repellent properties of *Delonix elata* (L.) Gamble (Family: Fabaceae) against malaria vector *Anopheles stephensi* (Liston) (Diptera: Culicidae). J Saudi Soc Agric Sci 14:128–133

Guba R (2001) Toxicity myths–essential oils and their carcinogenic potential. Int J Aromather 11:76–83

Gulluce M, Sahin F, Sokmen M, Ozer H, Daferera D, Sokmen A, Polissiou M, Adiguzel A, Ozkan H (2007) Antimicrobial and antioxidant properties of the essential oils and methanol extract from *Mentha longifolia* L. sp. *longifolia*. Food Chem 103:1449–1456

Gupta SC, Tyagi AK, Deshmukh-Taskar P, Hinojosa M, Prasad S, Aggarwal BB (2014) Down regulation of tumor necrosis factor and other proinflammatory biomarkers by polyphenols. Arch Biochem Biophys 559:91–99

Hadis M, Lulu M, Mekonnen Y, Asfaw T (2003) Field trials on the repellent activity of four plant products against mainly *Mansonia* population in Western Ethiopia. Phytother Res 17:202–205

Hall-Stoodley L, Costerton JW, Stoodley P (2004) Bacterial biofilms: from the natural environment to infectious diseases. Nat Rev Microbiol 2:95–108

Hamid AA, Aiyelaagbe OO, Usman LA (2011) Essential oils: its medicinal and pharmacological uses. Int J Curr Res 33:86–98

Hammer KA, Carson CF (2011) Antibacterial and antifungal activities of essential oils. In: Thormar H (ed) Lipids and Essential Oils as Antimicrobial Agents. Wiley, Chichester, pp 255–306

Hammer KA, Carson CF, Riley TV (2002) *In vitro* activity of *Melaleuca alternifolia* (tea tree) oil against dermatophytes and other filamentous fungi. J Antimicrob Chemother 50:195–199

Hammer KA, Dry L, Johnson M, Michalak EM, Carson CF, Riley TV (2003) Susceptibility of oral bacteria to *Melaleuca alternifolia* (tea tree) oil *in vitro*. Oral Microbiol Immunol 18:389–392

Hammer KA, Carson CF, Riley TV, Nielsen JB (2006) A review of the toxicity of *Melaleuca alternifolia* (tea tree) oil. Food Chem Toxicol 44:616–625

Hammer KA, Carson CF, Rileya TV (2012) Effects of *Melaleuca alternifolia* (tea tree) essential oil and the major monoterpene component terpinen-4-ol on the development of single and multistep antibiotic resistance and antimicrobial susceptibility. Antimicrob Agents Chemother 56:909–915

Hassanshahian M, Bayat Z, Saeidi S, Shiri Y (2014) Antimicrobial activity of *Trachyspermum ammi* essential oil against human bacterial. Int J Adv Biol Biom Res 2:18–24

Hata T, Sakaguchi I, Mori M, Ikeda N, Kato Y, Minamino M, Watabe K (2003) Induction of apoptosis by *Citrus paradisi* essential oil in human leukemic(HL-60) cells. In Vivo 17:553–559

Hirulkar NB, Agrawal M (2010) Antimicrobial activity of rose petals extract against some pathogenic bacteria. Int J Pharm Biol Arch 1:478–484

Hisseinzadeh H, Ramezani M, Danaei AR (2002) Antihyperglycaemic effect and acute toxicity of *Securigera secuidaca* L. seed extracts in mice. Phytother Res 16:745–747

Hong EJ, Na KJ, Choi IG, Choi KC, Jeung EB (2004) Antibacterial and antifungal effects of essential oils from *coniferous* trees. Biol Pharm Bull 27:863–866

Hori M (2003) Repellency of essential oils against the cigarette beetle, *Lasioderma serricorne* (*Fabricius*) (Coleoptera: Anobiidae). Appl Entomol Zool 38:467–473

Hristova Y, Gochev V, Wanner J, Jirovetz L, Schmidt E, Girova T, Kuzmanov A (2013) Chemical composition and antifungal activity of essential oil of *Salvia sclarea* L. from Bulgaria against clinical isolates of Candida species. J Biosci Biotech 2:39–44

Hussain AI, Anwar F, Hussain Sherazi ST, Przybylski R (2008) Chemical composition, antioxidant and antimicrobial activities of basil (*Ocimum basilicum*) essential oils depends on seasonal variations. Food Chem 108:986–995

Ibrahim NA, El-Hawary SS, Mohammed MMD, Farid MA, Abdel-Washed NAM, Ali MA, El-Abd EAW (2015a) Chemical composition, antiviral against avian influenza (H5N1) virus and antimicrobial activities of the essential oils of the leaves and fruits of *Fortunella margarita*, lour. swingle, growing in Egypt. J Appl Pharma Sci 5:6–12

Ibrahim NA, El-Sakhawy FS, Mohammed MMD, Farid MA, Abdel-Wahed NAM, Deabes DAH (2015b) Chemical composition, antimicrobial and antifungal activities of essential oils of the leaves of *Aegle marmelos* (L.) Correa growing in Egypt. J Appl Pharma Sci 5:1–5

Idaomar M, El Hamss R, Bakkali F, Mezzoug N, Zhiri A, Baudoux D, Munoz-Serrano A, Liemans V, Alonso-Moraga A (2002) Genotoxicity and antigenotoxicity of some essential oils evaluated by wing spot test of *Drosophila melanogaster*. Mutat Res 513:61–68

Ioannou E, Poiata A, Hancianu M, Tzakou O (2007) Chemical composition and *in vitro* antimicrobial activity of the essential oils of flower heads and leaves of *Santolina rosmarinifolia* L. from Romania. Nat Prod Res 21:18–23

Ipek E, Zeytinoglu H, Okay S, Tuylu BA, Kurkcuoglu M, Husnu Can Baser K (2005) Genotoxicity and antigenotoxicity of Origanum oil and carvacrol evaluated by Ames *Salmonella*/microsomal test. Food Chem 93:551–556

Irkin R, Korukluoglu M (2009) Effectiveness of *Cymbopogon citratus* L. essential oil to inhibit the growth of some filamentous fungi and yeasts. J Med Food 12:193–197

Iwalokun BA, Gbenle GO, Adewole TA, Smith SI, Akinsinde KA, Omonigbehin EO (2003) Effects of *Ocimum gratissimum* L. essential oil at sub-inhibitory concentrations on virulent and multidrug-resistant *Shigella* strains from Lagos, Nigeria. APMIS 111:477–482

Jaenson TG, Palsson K, Borg-Karlson AK (2006) Evaluation of extracts and oils of mosquito (*Diptera*: *Culicidae*) repellent plants from Sweden and Guinea-Bissau. J Med Entomol 43:113–119

Jayaprakasha GK, Jagan Mohan Rao L, Sakariah KK (2000) Chemical composition of the flower oil of *Cinnamomum zeylanicum* blume. J Agric Food Chem 48:4294–4295

Juliao LS, Bizzo HR, Souza AM, Lourenco MC, Silva PE, Tavares ES, Rastrelli L, Leitao SG (2009) Essential oils from two Lantana species with antimycobacterial activity. Nat Prod Comm 4:1733–1736

Kaefer CM, Milner JA (2008) The role of herbs and spices in cancer prevention. J Nutr Biochem 19:347–361

Karbach J, Ebenezer S, Warnke PH, Behrens E, Al-Nawas B (2015) Antimicrobial effect of Australian antibacterial essential oils as alternative to common antiseptic solutions against clinically relevant oral pathogens. Clin Lab 61:61–68

Karunamoorthi K, Girmay A, Hayleeyesus SF (2014) Mosquito repellent activity of essential oil of Ethiopian ethnomedicinal plant against Afro-tropical malarial vector *Anopheles arabiensis*. J King Saud Univer Sci 26:305–310

Kasim LS, Olaleye KO, Fagbohun AB, Ibitoye SF, Adejumo OE (2014) Chemical composition and antibacterial activity of essential oils from *Struchium sparganophora* Linn. Ktze asteraceae. Adv Biol Chem 4:246–252

Kathiravan MK, Salake AB, Chothe AS, Dudhe PB, Watode RP, Mukta MS, Gadhwe S (2012) The biology and chemistry of antifungal agents: A review. Bioorg Med Chem 20:5678–5698

Khosravi AR, Minooeianhaghighi MH, Shokri H (2011) The potential inhibitory effect of *Cuminum cyminum*, *Ziziphora clinopodioides* and *Nigella sativa* essential oils on the growth of *Aspergillus fumigatus* and *Aspergillus flavus*. Braz J Microbiol 42:216–224

Khoury M, El Beyrouthy M, Ouaini N, Iriti M, Eparvier V, Stien D (2014) Chemical composition and antimicrobial activity of the essential oil of *Juniperus excelsa* M. Bieb. growing wild in Lebanon. Chem Biodivers 11:825–830

Kim EY, Baik IH, Kim JH, Kim SR, Rhyu MR (2004) Screening of the antioxidant activity of some medicinal plants. Kor J Food Sci Technol 36:333–338

Kiyohara H, Ichino C, Kawamura Y, Nagai T, Sato N, Yamada H (2012) Patchouli alcohol: *in vitro* direct anti-influenza virus sesquiterpene in *Pogostemon cablin Benth*. J Nat Med 66:55–61

Kloog Y, Cox AD (2004) Prenyl-binding domains: potential targets for Ras inhibitors and anticancer drugs. Semin Cancer Biol 14:253–261

Kocevski D, Du M, Kan J, Jing C, Lacanin I, Pavlovic H (2013) Antifungal effect of *Allium tuberosum*, *Cinnamomum cassia*, and *Pogostemon cablin* essential oils and their components against population of *Aspergillus* species. J Food Sci 78:M731–M737

Koch C, Reichling J, Schneele J, Schnitzler P (2008) Inhibitory effect of essential oils against herpes simplex virus type 2. Phytomedicine 15:71–78

Koo JY, Kim HJ, Jung KO, Park KY (2004) Curcumin inhibits the growth of AGS human gastric carcinoma cells in vitro and shows synergism with5-fluorouracil. J Med Food 7:117–121

Kordali S, Kotan R, Mavi A, Cakir A, Ala A, Yildirim A (2005) Determination of the chemical composition and antioxidant activity of the essential oil of *Artemisia dracunculus* and of the antifungal and antibacterial activities of Turkish *Artemisia absinthium*, *A. dracunculus*, *Artemisia santonicum*, and *Artemisia spicigera* essential oils. J Agric Food Chem 53:9452–9458

Koschier E, Sedy K (2003) Labiate essential oils affecting host selection and acceptance of *Thrips tabaci* lindeman. Crop Prot 22:929–934

Kumar R, Mishra AK, Dubey NK, Tripathi YB (2007) Evaluation of *Chenopodium ambrosioides* oil as a potential source of antifungal, antiaflatoxigenic and antioxidant activity. Int J Food Microbiol 115:159–164

Lalthazuali, Mathew N (2017) Mosquito repellent activity of volatile oils from selected aromatic plants. Parasitol Res 116:821–825

Lambert RJW, Skandamis PN, Coote PJ, Nychas GJE (2001) A study of the minimum inhibitory concentration and mode of action of oregano essential oil, thymol and carvacrol. J Appl Microbiol 91:453–462

Landolt PJ, Hofstetter RW, Biddick LL (1999) Plant essential oils as arrestants and repellents for neonate larvae of the codling moth (*Lepidoptera*: Tortricidae). Environ Entomol 28:954–960

Lang G, Buchbauer G (2012) A review on recent research results (2008–2010) on essential oils as antimicrobials and antifungals: areview. Flav Frag J 27:13–39

Latifah-Munirah B, Himratul-Aznita WH, Zain NM (2015) Eugenol, an essential oil of clove, causes disruption to the cell wall of *Candida albicans* (ATCC 14053). Front Life Sci 8:231–240

Lee BK, Kim JH, Jung JW, Choi JW, Han ES, Lee SH, Kwang HK, Ryu JH (2005) Myristicin-induced neuro toxicity in human neuroblastoma SK-N-SHcells. Toxicol Lett 157:49–56

Lesgards JF, Baldovini N, Vidal N, Pietri S (2014) Anticancer activities of essential oils constituents and synergy with conventional therapies: A review. Phytother Res 28:1423–1446

Li Y, Li MY, Wang L, Jiang ZH, Li WY, Li H (2004) Induction of apoptosis of cultured hepatocarcinoma cell by essential oil of *Artemisia Annul* L. J Sichuan Univ Med Sci Ed 35:337–339

Lirussi F, Beccarello A, Zanette G, De Monte A, Donadon V, Velussi M, Crepaldi G (2002) Silybin-beta-cyclodextrin in the treatment of patients with diabetes mellitus and alcoholic liver disease. Efficacy study of a new preparation of an anti-oxidant agent. Diabetes Nutr Metab 15:222–231

Liu CJ, Chen CL, Chang KW, Chu CH, Liu TY (2000) Safrole in betel quid may be a risk factor for hepatocellular carcinoma: case report. Can Med Ass J 162:359–360

Lo Cantore P, Iacobellis NS, De Marco A, Capasso F, Senatore F (2004) Antibacterial activity of *Coriandrum sativum* L. and *Foeniculum vulgare* Miller Var. *vulgare* (Miller) essential oils. J Agric Food Chem 52:7862–7866

Lopes-Lutz D, Alviano DS, Alviano CS. Kolodziejczyk PP (2008) Screening of chemical composition, antimicrobial and antioxidant activities of Artemisia essential oils. Phytochemistry 69:1732–1738

Lopez P, Sanchez C, Batlle R, Nerin C (2005) Solid and vapor phase antimicrobial activities of six essential oils: susceptibility of selected food borne bacterial and fungal strains. J Agric Food Chem 53:6939–6946

Lopez P, Sanchez C, Batlle R, Nerin C (2007) Vapor-phase activities of cinnamon, thyme, and oregano essential oils and key constituents against food borne microorganisms. J Agric Food Chem 55:4348–4356

Lu Y, Hernandez P, Abegunde D, Edejer T (2011) Medicine Expenditure. In: World medicine situation. Geneva, pp 43–76

Luk SU, Lee TKW, Liu J, Lee DTW, Chiu YT, Ma S, Ng IOL, Wong YC, Chan FL, Ling MT (2011) Chemopreventive effect of PSP through targeting of prostate cancer stem cell-like population. PLoS One 6:e19804

Madihi Y, Merrikhi A, Baradaran A, Rafieian-Kopaei M, Shahinfard N, Ansari R, Baradaran J (2013) Impact of sumac on postprandial high-fat oxidative stress. Pak J Med Sci 29:340–345

Maestri DM, Nepote V, Lamarque AL, Zygadlo JA (2006) Natural products as antioxidants. In: Filippo I (ed) Phytochemistry: Advances in research. Signpost, Trivandrum, pp 105–135

Maggi F, Bramucci M, Cecchini C, Coman MM, Cresci A, Cristalli G, Lupidi G, Papa F, Quassinti L, Sagratini G, Vittori S (2009) Composition and biological activity of essential oil of *Achillea ligustica* All. (Asteraceae) naturalized in central Italy: ideal candidate for anti-cariogenic formulations. Fitoterapia 80:313–319

Maheshwari RK, Singh AK, Gaddipati J, Srimal RC (2006) Multiple biological activities of *curcum*: a short review. Life Sci 78:2081–2087

Mahmoudi R (2017) Application of medicinal plants: From past to present. MOJ Biol Med 1:16

Malíková J, Swaczynová J, Kolář Z, Strnad M (2008) Anticancer and antiproliferative activity of natural brasinosteroids. Phtyochemistry 69:418–426

Manosroi J, Dhumtanom P, Manosroi A (2006) Anti-proliferative activity of essential oil extracted from Thai medicinal plants on KB and P388 cell lines. Cancer Lett 235:114–120

Mansoor A, Ibrahim MA, Zaidi MA, Ahmed M (2011) Antiprotozoal activities of *Vincetoxicum stocksii* and *Carum copticum*. Bangladesh J Pharmacol 6:51–54

Mansour MA, Ginawi OT, El-Hadiyah T, El-Khatib AS, Al-Shabanah OA, Al-Sawaf HA (2001) Effects of the volatile oil constituents of *Nigella sativa* on carbon tetrachloride induced hepatotoxicity in mice: evidence for antioxidant effects of thymoquinone. Res Commun Mol Pathol Pharmacol 110:239–251

Masotti V, Juteau F, Bessie're JM, Viano J (2003) Seasonal and phenological variations of the essential oil from the narrow endemic species *Artemisia molinieri* and its biological activities. J Agric Food Chem 51:7115–7121

Matasyoh JC, Maiyo ZC, Ngure RM, Chepkorir R (2009) Chemical composition and antimicrobial activity of the essential oil of *Coriandrum sativum*. Food Chem 113:526–529

Matthaei S, Stumvoll M, Kellerer M, Häring HU (2000) Pathophysiology and pharmacological treatment of insulin resistance. Endocr Rev 2:585–618

Mau JL, Lai EY, Wang NP, Chen CC, Chang CH, Chyau CC (2003) Composition and antioxidant activity of the essential oil from *Curcuma zedoaria*. Food Chem 82:583–591

Maxia A, Marongiu B, Piras A, Porcedda S, Tuveri E, Goncalves MJ, Cavaleiro C, Salgueiro L (2009) Chemical characterization and biological activity of essential oils from *Daucus carota* L. subsp. *Carota* growing wild on the Mediterranean coast and on the Atlantic coast. Fitoterapia 80:57–61

May J, Chan C, King A, Williams L, French GL (2000) Time kill studies of tea tree oils on clinical isolates. J Antimicrob Chemother 45:639–643

McCord JM (2000) The evolution of free radicals and oxidative stress. Am J Med 108:652–659

Meneses R, Ocazionez RE, Martinez JR, Stashenko EE (2009) Inhibitory effect of essential oils obtained from plants grown in Colombia on yellow fever virus replication *in vitro*. Ann Clin Microbiol Antimicrob 8:8. https://doi.org/10.1186/1476-0711-8-8

Mezzoug N, Elhadri A, Dallouh A, Amkiss S, Skali NS, Abrini J, Zhiri A, Baudoux D, Diallo B, El Jaziri M, Idaomar M (2007) Investigation of the mutagenic and antimutagenic effects of *Origanum compactum* essential oil and some of its constituents. Mutat Res 629:100–110

Miguel MG (2010) Antioxidant and anti-inflammatory activities of essential oils: A short review. Molecules 15:9252–9287

Miller EC, Swanson AB, Phillips DH, Fletcher TL, Liem A, Miller JA (1983) Structure activity studies of the carcinogenicities in the mouse and rat of some naturally occurring and synthetic alkenylbenzene derivatives related to safrole and estragole. Cancer Res 43:1124–1134

Milner JA (2001) A historical perspective on garlic and cancer. J Nutr 131:S1027–S1031

Mimica-Dukic N, Bozin B, Sokovic M, Simin N (2004) Antimicrobial and antioxidant activities of *Melissa officinalis* L. (Lamiaceae) essential oil. J Agric Food Chem 52:2485–2489

Minami M, Kita M, Nakaya T, Yamamoto T, Kuriyama H, Imanishi J (2003) The inhibitory effect of essential oils on herpes simplex virus type-1 replication *in vitro*. Microbiol Immunol 47:681–684

Mirhoseini M, Baradaran A, Rafieian-Kopaei M (2013) Medicinal plants, diabetes mellitus and urgent needs. J Herb Med Pharmacol 2:53–54

Mkaddem M, Bouajila J, Ennajar M, Lebrihi A, Mathieu F, Romdhane M (2009) Chemical composition and antimicrobial and antioxidant activities of *Mentha (longifolia L.* and *viridis*) essential oils. J Food Sci 74:M358–M363

Mohamed A, Ali SI, El-Baz FK (2013) Antioxidant and antibacterial activities of crude extracts and essential oils of *Syzygium cumini* Leaves. PLoS One 8(4):e60269

Moore SJ, Hill N, Ruiz C, Cameron MM (2007) Field evaluation of traditionally used plant-based insect repellents and fumigants against the malaria vector *Anopheles darlingi* in Riberalta, Bolivian Amazon. J Med Entomol 44:624–630

Morita T, Jinno K, Kawagishi H, Arimoto Y, Suganuma H, Inakuma T, Sugiyama K (2003) Hepatoprotective effect of myristicin from nutmeg (*Myristica fra grans*) on lipopolysaccharide/d-galactosamine-induced liver injury. J Agric Food Chem 51:1560–1565

Moteki H, Hibasami H, Yamada Y, Katsuzaki H, Imai K, Komiya T (2002) Specific induction of apoptosis by 1,8-cineole in two human leukemia cell lines, but not a in human stomach cancer cell line. Oncol Rep 9:757–760

Murray M (1997) Bilberry (*Vaccinium myrtillus*). Am J Nat Med 4:18–22

Nakatsu T, Lupo AT, Chinn JW, Kang RKL (2000) Biological activity of essential oils and their constituents. Stud Nat Prod Chem 21:571–631

Nasri H, Shirzad H (2013) Toxicity and safety of medicinal plants. J Herb Med Plarmacol 2:21–22

Nasri H, Baradaran A, Ardalan MR, Mardani S, Momeni A, Rafieian-Kopaei M (2013) Bright reno protective properties of metformin: beyond blood glucose regulatory effects. Iran J Kidney Dis 7:423–428

Nerio SL, Olivero-Verbel J, Stashenko E (2010) Repellent activity of essential oils: A review. Bioresour Technol 101:372–378

Novy P, Davidova H, Serrano-Rojero CS, Rondevaldova J, Pulkrabek J, Kokoska L (2015) Composition and antimicrobial activity of *Euphrasia rostkoviana* hayne essential oil. Evidence-Based Compl Altern Med 2015:734101. https://doi.org/10.1155/2015/734101

Odalo JO, Omolo MO, Malebo H, Angira J, Njeru PM, Ndiege IO, Hassanali A (2005) Repellency of essential oils of some plants from the Kenyan coast against *Anopheles gambiae*. Acta Trop 95:210–218

Omolo MO, Okinyo D, Ndiege IO, Lwande W, Hassanali A (2004) Repellency of essential oils of some Kenyan plants against *Anopheles gambiae*. Phytochemistry 65:2797–2802

Oyedele AO, Gbolade AA, Sosan MB, Adewoyin FB, Soyelu OL, Orafidiya OO (2002) Formulation of an effective mosquito-repellent topical product from Lemongrass oil. Phytomedicine 9:259–262

Ozbek H, Ugras S, Dulger H, Bayram I, Tuncer I, Ozturk G, Ozturk A (2003) Hepatoprotective effect of *Foeniculum vulgare* essential oil. Fitoterapia 74:317–319

Ozcan MM, Chalchat JC (2008) Chemical composition and antifungal activity of rosemary (*Rosmarinus officinalis* L.) oil from Turkey. Int J Food Sci Nutr 59:691–698

Ozek G, Demirci F, Ozek T, Tabanca N, Wedge DE, Khan SI, Baser KHC, Duran A, Hamzaoglu E (2010) Gas chromatographic–mass spectrometric analysis of volatiles obtained by four different techniques from *Salvia rosifolia* Sm. and evaluation for biological activity. J Chromatogr A 1217:741–748

Padilha de Paula J, Gomes-Carneiro MR, Paumgartten FJR (2003) Chemical composition, toxicity and mosquito repellency of *Ocimum selloi* oil. J Ethnopharmacol 88:253–260

Papachristos DP, Stamopoulos DC (2002) Repellent, toxic and reproduction inhibitory effects of essential oil vapours on *Acanthoscelides obtectus* (Say) (Coleoptera: Bruchidae). J Stored Prod Res 38:117–128

Papajani V, Haloci E, Goci E, Shkreli R, Manfredini S (2015) Evaluation of antifungal activity of *Origanum vulgare* and *Rosmarinus officinalis* essential oil before and after inclusion in β-cyclodextrine. Int J Pharm Pharm Sci 7:270–273

Parsaei P, Karimi M, Asadi SY, Rafieian-Kopaei M (2013) Bioactive components and preventive effect of green tea (*Camellia sinensis*) extract on post laparotomy intra-abdominal adhesion in rats. Int J Surg 11:811–815

Pascual M, Ballesta M (2003) Chemical variation in an *Ocimum basilicum* germplasm collection and activity of the essential oils on *Callosobruchus maculatus*. Biochem Syst Ecol 31:673–679

Patel K, Ali S, Sotheeswaran S, Dufour JP (2013) Composition of the leaf essential oil of *Cinnamomum verum* (Lauraceae) from Fiji Islands. J Essent Oil Bearing Plants 10:374–377

Peighami-Ashnaei S, Farzaneh M, Sharifi Tehrani A, Behboudi K (2008) Effect of essential oils in control of plant diseases. Commun Agric Appl Biol Sci 74:843–847

Pepeljnjak S, Kosalec I, Kalodera Z, Blazevic N (2005) Antimicrobial activity of juniper berry essential oil (*Juniperus communis* L., Cupressaceae). Acta Pharma 55:417–422

Petretto GL, Fancello F, Zara S, Foddai M, Mangia NP, Sanna ML, Omer EA, Menghini L, Chessa M, Pintore G (2014) Antimicrobial activity against beneficial microorganisms and chemical composition of essential oil of *Mentha suaveolens ssp. insularis* grown in Sardinia. J Food Sci 79:M369–M377

Pichersky E, Noel JP, Dudareva N (2006) Biosynthesis of plant volatiles: nature's diversity and ingenuity. Science 311:808–811

Pinto E, Vale-Silva L, Cavaleiro C, Salgueiro L (2009) Antifungal activity of the clove essential oil from *Syzygium aromaticum* on *Candida*, *Aspergillus* and dermatophyte species. J Med Microbiol 58:1454–1462

Pledgie-Tracy A, Sobolewski MD, Davidson NE (2007) Sulforaphane induces cell type-specific apoptosis in human breast cancer cell lines. Mol Cancer Ther 6:1013–1021

Politeo O, Jukic M, Milos M (2007) Chemical composition and antioxidant capacity of free volatile a glycones from basil (*Ocimum basilicum* L.) compared with its essential oil. Food Chem 101:379–385

Pragadheesh VS, Saroj A, Yadav A, Chanotiya CS, Alam M, Samad A (2013) Chemical characterization and antifungal activity of *Cinnamomum camphora* essential oil. Ind Crop Prod 49:628–633

Prajapati V, Tripathi AK, Aggarwal KK, Khanuja SPS (2005) Insecticidal, repellent and oviposition-deterrent activity of selected essential oils against *Anopheles stephensi*, *Aedes aegypti* and *Culex quinquefasciatus*. Bioresour Technol 96:1749–1757

Puatanachokchai R, Kishida H, Denda A, Murata N, Konishi Y, Vinitketkumnuen U, Nakae D (2002) Inhibitory effects of lemongrass *Cymbopogon citrates* extract on the early phase of hepatocarcinogenesis after initiation with diethyl-nitrosamine in male Fischer 344 rats. Cancer Lett 183:9–15

Pudziuvelyte L, Stankevicius M, Maruska A, Petrikaite V, Ragazinskiene O, Draksiene G, Bernatoniene J (2017) Chemical composition and anticancer activity of *Elsholtzia ciliata* essential oils and extracts prepared by different methods. Ind Crop Prod 107:90–96

Pullagummi C, Rao NB, Singh BCS, Bheemagani AJ, Kumar P, Venkatesh K, Rani AR (2014) Comparitive studies on antibacterial activity of Patchouli [*Pogostemon cablin* (Blanco) Benth] and Geranium (*Pelargonium graveolens*) aromatic medicinal plants. Afr J Biotechnol 13:2379–2384

Pushpanathan T, Jebanesan A, Govindarajan M (2006) Larvicidal, ovicidal and repellent activities of *Cymbopogan citratus* Stapf. (Graminae) essential oil against the filarial mosquito *Culex quinquefasciatus* (Say) (Diptera: *Culicidae*). Trop Biomed 23:208–212

Pushpanathan T, Jebanesan A, Govindarajan M (2008) The essential oil of *Zingiber officinalis* Linn. (Zingiberaceae) as a mosquito larvicidal and repellent agent against the filarial vector *Culex quinquefasciatus* Say (Diptera: *Culicidae*). Parasitol Res 102:1289–1291

Rabadia AG, Kamat S, Kamat D (2012) Antifungal activity of essential oils against fluconazole resistant fungi. Int J Phytomed 3:506–510

Racchi ML (2013) Antioxidant Defenses in Plants with Attention to *Prunus* and *Citrus* spp. Antioxidants 2:340–369

Rafieian-Kopaei M, Baradaran A, Merrikhi A, Nematbakhsh M, Madihi Y, Nasri H (2013) Efficacy of co-administration of garlic extract and metformin for prevention of gentamicin-renal toxicity in wistar rats: A biochemical study. Int J Prev Med 4:258–264

Rafii F, Shahverdi AR (2007) Comparison of essential oils from three plants for enhancement of antimicrobial activity of nitrofurantoin against *entero bacteria*. Chemotherapy 53:21–25

Rajkumar S, Jebanesan A (2005) Repellency of volatile oils from *Moschosma polystachyum* and *Solanum xanthocarpum* against filarial vector *Culex quinquefasciatus* Say. Trop Biomed 22:139–142

Rasooli I, Fakoor MH, Yadegarinia D, Gachkar L, Allameh A, Rezaei MB (2008) Antimycotoxigenic characteristics of *Rosmarinus officinalis* and *Trachyspermum copticum* L. essential oils. Int J Food Microbiol 122:135–139

Rather MA, Dar BA, Dar MY, Wani BA, Shah WA, Bhat BA, Bashir AG, Bhat KA, Anand R, Qurishi MA (2012) Chemical composition, antioxidant and antibacterial activities of the leaf essential oil of *Juglans regia* L. andits constituents. Phytomed 19:1185–1190

Raut JS, Karuppayil SM (2014) A status review on the medicinal properties of essential oils. Ind Crop Prod 62:250–264

Razzaghi-Abyaneh M, Shams-Ghahfarokhi M, Yoshinari T, Rezaee MB, Jaimand K, Nagasawa H, Sakuda S (2008) Inhibitory effects of *Satureja hortensis* L. essential oil on growth and aflatoxin production by *Aspergillus parasiticus*. Int J Food Microbiol 123:228–233

Reegan AD, Gandhi MR, Paulraj MG, Ignacimuthu S (2015) Ovicidal and oviposition deterrent activities of medicinal plant extracts against *Aedes aegypti* L. and *Culex quinquefasciatus* Say mosquitoes (Diptera: Culicidae). Osong Public Health Res Perspect 6:64–69

Reichling J, Koch C, Stahl-Biskup E, Sojka C, Schnitzler P (2005) Virucidal activity of a beta-triketone-richess entail oil of *Leptospermum scoparium* (manuka oil) against HSV-1and HSV-2 in cell culture. PlantaMed 71:1123–1127

Roller S, Ernest N, Buckle J (2009) The antimicrobial activity of high-necrodane and other lavender oils onmethicillin-sensitive and-resistant *Staphylococcus aureus* (MSSA and MRSA). J Altern Compl Med 15:275–279

Romeilah RM (2016) Chemical compositions, antioxidant, anticancer activities and biological effects of *Myrtus communis* L. and *Origanum vulgare* essential oils. Asian J Biochem 11:104–107

Romeilah RM, Fayed SA, Mahmoud GI (2010) Chemical compositions, antiviral and antioxidant activities of seven essential oils. J Appl Sci Res 6:50–62

Rosato A, Vitali C, De Laurentis N, Armenise D, Antonietta MM (2007) Antibacterial effect of some essential oils administered alone or in combination with norfloxacin. Phytomedicine 14:727–732

Rota C, Carraminana JJ, Burillo J, Herrera A (2004) *In vitro* antimicrobial activity of essential oils from aromatic plants against selected food borne pathogens. J Food Prot 67:1252–1256

Routh MM, Raut JS, Karuppayil SM (2011) Dual properties of anticancer agents: an exploratory study on the *in vitro* anti-Candida properties of thirty drugs. Chemotherapy 57:372–380

Roy S, Chaurvedi P, Chowdhary A (2015) Evaluation of antiviral activity of essential oil of *Trachyspermum Ammi* against Japanese encephalitis virus. Pharm Res 7:263–267

Ruan ZP, Zhang LL, Lin YM (2008) Evaluation of the antioxidant activity of *Syzygium cumini* leaves. Molecules 13:2545–2556

Rudramurthy GR, Swamy MK, Sinniah UR, Ghasemzadeh A (2016) Nanoparticles: alternatives against drug-resistant pathogenic microbes. Molecules 21:836

Runyoro D, Ngassapaa O, Vagionas K, Aligiannis N, Graikou K, Chinou I (2010) Chemical composition and antimicrobial activity of the essential oils of four Ocimum species growing in Tanzania. Food Chem 119:311–316

Saddi M, Sanna A, Cottiglia F, Chisu L, Casu L, Bonsignore L, DeLogu A (2007) Antiherpevirus activity of *Artemisia arborescens* essential oil and inhibition of lateral diffusion in vero cells. Ann Clin Microbiol Antimicrob 6:10. https://doi.org/10.1186/1476-0711-6-10

Saeidnia S, Gohari AR (2012) Trypanocidal monoterpenes: lead compounds to design future trypanocidal drugs. Stud Nat Prod Chem 37:173–189

Said ZB, Haddadi-Guemghar H, Boulekbache-Makhlouf L, Rigou P, Remini H, Adjaoud A, Khoudja NK, Madani K (2016) Essential oils composition, antibacterial and antioxidant activities of hydro distillated extract of *Eucalyptus globulus* fruits. Ind Crop Prod 89:167–175

Saidana D, Mahjoub MA, Boussaada O, Chriaa J, Chéraif I, Daami M, Mighri Z, Helal AN (2008) Chemical composition and antimicrobial activity of volatile compounds of *Tamarix boveanai* (Tamaricaceae). Microbiol Res 163:445–455

Saikia D, Khanuja SPS, Kahol AP, Gupta SC, Kumar S (2001) Comparative antifungal activity of essential oils and constituents from three distinct genotypes of *Cymbopogon* spp. Curr Sci 80:1264–1265

Salim EI, Fukushima S (2003) Chemopreventive potential of volatile oil from black cumin (*Nigella sativa* L.) seeds against rat colon carcinogenesis. Nutr Cancer 45:95–202

Samber N, Khan A, Varma A, Manzoor N (2015) Synergistic anticandidal activity and mode of action of *Mentha piperita* essential oil and its major components. Pharm Biol 53:1496–1504

Sanchez-Vioque R, Polissiou M, Astraka K, Mozos-Pascual M, Tarantilis P, Herraiz-Penalver D, Santana-Meridas O (2013) Polyphenol composition and antioxidant and metal chelating activities of the solid residues from the essential oil industry. Ind Crop Prod 49:150–159

Santin RM, Oliveira SA, Nakamura VC, Dias Filho BP, Ferreira ICP, Ueda Nakamura T (2009) *In vitro* activity of the essential oil of *Cymbopogon citrates*. Parasitol Res 105:1489–1496

Santoro GF, Cardoso MG, Guimarães LG, Mendonça LZ, Soares MJ (2007a) *Trypanosoma cruzi*: Activity of essential oils from *Achillea millefolium* L., *Syzygium aromaticum* L. and *Ocimum basilicum* L. on epimastigotes and trypomastigotes. Exp Parasitol 116:283–290

Santoro GF, Cardoso MG, Guimares LGL, Mendonca LZ, Soares MJ (2007b) Effect of oregano(*Origanum vulgare* L.) and thyme (*Thymus vulgare* L.) essential oils on *Trypanosoma cruzi* (Protozoa: Kinetoplastida) growth and ultrastructure. Parasitol Res 100:783–790

Santos AO, Santin AC, Yamaguchi MU, Cortez LER, Ueda Nakamura T, Dias Filho BP, Nakamura CV (2010) Antileishmanial activity of an essential oil from the leaves and flowers of *Achillea millefolium*. Ann Trop Med Parasitol 104:475–483

Santos NO, Mariane B, Lago JH, Sartorelli P, Rosa W, Soares MG, Da Silva AM, Lorenzi H, Vallim MA, Pascon RC (2015) Assessing the chemical composition and antimicrobial activity of essential oils from Brazilian plants-*Eremanthus erythropappus* (Asteraceae), *Plectrantuns barbatus*, and *P. amboinicus* (Lamiaceae). Molecules 20:8440–8452

Santurio DF, De Jesus FPK, Zanette RA, Schlemmer KB, Fraton A, Fries LLM (2014) Antimicrobial activity of the essential oil of thyme and of thymol against *Escherichia coli* strains. Acta Sci Vet 42:1–4

Sardi JCO, Scorzoni L, Bernardi T, Fusco Almeida AM, Giannini MM (2013) *Candida* species: current epidemiology, pathogenicity, biofilm formation, natural antifungal products and new therapeutic options. J Med Microbiol 62:10–24

Sartorelli P, Marquioreto AD, Amaral-Baroli A, Lima MEL, Moreno PRH (2007) Chemical composition and antimicrobial activity of the essential oils from two species of *Eucalyptus*. Phytother Res 21:231–233

Sauter IP, Rossa GE, Lucas AM, Cibulski SP, Roehe PM, Da Silva LAA, Rott MB, Vargas RMF, Cassel E, Von Poser GL (2012) Chemical composition and amoebicidal activity of *Piper hispidinervum* (Piperaceae) essential oil. Ind Crop Prod 40:292–295

Schnitzler P, Schon K, Reichling J (2001) Antiviral activity of Australian tea tree oil and eucalyptus oil against herpes simplex virus in cell culture. Pharmazie 56:343–347

Schnitzler P, Koch C, Reichling J (2007) Susceptibility of drug-resistant clinical herpes simplex virus type 1 strains to essential oils of ginger, thyme, hyssop, and sandalwood. Antimicrob Agents Chemother 51:1859–1862

Schnitzler P, Astani A, Reichling J (2011) Antiviral effects of plant-derived essential oils and pure oil components. In: Thormar H (ed) Lipids and essential oils as antimicrobial agents. Wiley, Chichester, pp 239–254

Schuhmacher A, Reichling J, Schnitzler P (2003) Virucidal effect of peppermint oil on the enveloped viruses, herpes simplex virus type1and type 2 *in vitro*. Phytomedicine 10:504–510

Seidel C, Florean C, Schnekenburger M, Dicato M, Diederich M (2012) Chromatin-modifying agents in anti-cancer therapy. Biochimie 94:2264–2279

Shah WA, Dar MY, Zagar MI, Agnihotri VK, Qurishi MA, Singh B (2013) Chemical composition and antimicrobial activity of the leaf essential oil of *Skimmia laureola* growing wild in Jammu and Kashmir, India. Nat Prod Res 27:1023–1027

Shan B, Cai YZ, Brooks JD, Corke H (2007) The *in vitro* antibacterial activity of dietary spice and medicinal herb extracts. Int J Food Microbiol 117:112–119

Sharma N, Trikha P, Athar M, Raisuddin S (2001) Inhibition of benzo(a)pyrene- and cyclophosphamide-induced mutagenicity by *Cinnamomum cassia*. Mutat Res 480:179–188

Sienkiewicz M, Głowacka A, Poznanska-Kurowska K, Kaszuba A, Urbaniak A, Kowalczyk E (2015) The effect of clary sage oil on *Staphylococci* responsible for wound infections. Postepy Dermatol Alergol 32:21–26

Singab AN, Youssef FS, Ashour ML (2014) Medicinal plants with potential antidiabetic activity and their assessment. Med Aromat Plants 3:151. https://doi.org/10.4172/2167-0412.1000151

Singh G, Kapoor IPS, Pandey SK, Singh UK, Singh RK (2002) Studies on essential oils: part 10; antibacterial activity of volatile oils of some spices. Phytother Res 16:680–682

Singh G, Marimuthu P, De Heluani CS, Catalan CA (2006) Antioxidant and biocidal activities of *Carum nigrum* (seed) essential oil, oleoresin, and their selected components. J Agric Food Chem 54:174–181

Singh G, Kapoor IPS, Singh P, De Heluani CS, De Lampasona MP, Catalan CA (2008) Chemistry, antioxidant and antimicrobial investigations on essential oil and oleoresins of *Zingiber officinale*. Food Chem Toxicol 46:3295–3302

Singh P, Shukla R, Kumar A, Prakash B, Singh S, Dubey NK (2010) Effect of *Citrus reticulata* and *Cymbopogon citratus* essential oils on *Aspergillus flavus* growth and aflatoxin production on *Asparagus racemosus*. Mycopathologia 170:195–202

Singh S, Das SS, Singh G, Schuff C, De Lampasona MP, Catalan CAN (2014) Composition, *in vitro* antioxidant and antimicrobial activities of essential oil and oleoresins obtained from black cumin seeds (*Nigella sativa* L.). Bio Med Res Int 2014:918209

Sinico C, De Logu A, Lai F, Valenti D, Manconi M, Loy G, Fadda AM (2005) Liposomal incorporation of *Artemisia arborescens* L. essential oil and *in vitro* antiviral activity. Eur J Pharm Biopharm 59:161–168

Smith RL, Cohen SM, Doull J, Feron VJ, Goodman JI, Marnett LJ, Portoghese PS, Waddell WJ, Wagner BM, Hall RL, Higley NA, Lucas-Gavin C, Adams TB (2005) A procedure for the safety evaluation of natural flavor complexes used as ingredients in food: essential oils. Food Chem Toxicol 43:345–363

Sonboli A, Babakhani B, Mehrabian AR (2006) Antimicrobial activity of six constituents of essential oil from *Salvia*. Z Naturforsch 61:160–164

Souza CM, Pereira Junior SA, Moraes Tda S, Damasceno JL, Amorim Mendes S, Dias HJ, Stefani R, Tavares DC, Martins CH, Crotti AE, Mendes-Giannini MJ, Pires RH (2016) Antifungal activity of plant-derived essential oils on Candida tropicalis planktonic and biofilms cells. Med Mycol 54:515–523

Srinivasan K, Ramarao P (2007) Animal models in type 2 diabetes research: an overview. Indian J Med Res 125:451–472

Swamy MK, Sinniah UR (2015) A comprehensive review on the phytochemical constituents and pharmacological activities of *Pogostemon cablin* Benth.: an aromatic medicinal plant of industrial importance. Molecules 20:8521–8547

Swamy MK, Sinniah UR (2016) Patchouli (*Pogostemon cablin* Benth.): Botany, agrotechnology and biotechnological aspects. Ind Crop Prod 87:161–176

Swamy MK, Sudipta KM, Lokesh P, Neeki MA, Rashmi W, Bhaumik SH, Darshil SH, Vijay R, Kashyap NSS (2012) Phytochemical screening and in vitro antimicrobial activity of *Bougainvillea spectabilis* flower extracts. Int J Phytomed 4:375–379

Swamy MK, Sinniah UR, Akhtar MS (2016) Antimicrobial properties of plant essential oils against human pathogens and their mode of action: An updated review. Evidence-Based Compl Altern Med 2016:3012462. https://doi.org/10.1155/2016/3012462

Sylvestre M, Legault J, Dufour D, Pichette A (2005) Chemical composition and anticancer activity of leaf essential oil of *Myrica gale* L. Phytomedicine 12:299–304

Sylvestre M, Pichette A, Longtin A, Nagau F, Legault J (2006) Essential oil analysis and anticancer activity of leaf essential oil of *Croton flavens* L. from Guadeloupe. J Ethnopharmacol 103:99–102

Taherkhani M (2015) Anticancer, cytotoxic activity, mutagenic and anti-mutagenic activities of *Artemisia aucheri* essential oil. TEOP 18:1329–1337

Taherkhani (2016) Chemical constituents, antimicrobial, cytotoxicity, mutagenic and antimutagenic effects of *Artemisia ciniformis*. Iranian J Pharma Res 15:471–481

Tahir HU, Sarfraz RA, Ashraf A, Adil S (2016) Chemical composition and antidiabetic activity of essential oils obtained from two spices (*Syzygium aromaticum* and *Cuminum cyminum*). Int J Food Propert 19:2156–2164

Takarada K, Kimizuka R, Takahashi N, Honma K, Okuda K, Kato T (2004) A comparison of the antibacterial efficacies of essential oils against oral pathogens. Oral Microbiol Immunol 19:61–64

Tariku Y, Hymete A, Hailu A, Rohloff J (2011) *In vitro* evaluation of antileishmanial activity and toxicity of essential oils of *Artemisia absinthium* and *Echinops kebericho*. Chem Biodivers 8:614–623

Tavares AC, Goncalves MJ, Cavaleiro C, Cruz MT, Lopes MC, Canhoto J, Salgueiro LR (2008) Essential oil of *Daucus carota* subsp. *Halophilus*: composition, antifungal activity and cytotoxicity. J Ethnopharmacol 119:129–134

Tawatsin A, Wratten SD, Scott RR, Thavara U, Techandamrongsin Y (2001) Repellency of volatile oils from plants against three mosquito vectors. J Vector Ecol 26:76–82

Tchoumbougnang F, Zollo PH, Dagne E, Mekonnen Y (2005) *In vivo* antimalarial activity of essential oils from *Cymbopogon citratus* and *Ocimum gratissimum* on mice infected with *Plasmodium berghei*. Planta Med 71:20–23

Teixeira B, Marques A, Ramos C, Neng NR, Nogueira JMF, Saraiva JA, Nunes ML (2013) Chemical composition and antibacterial and antioxidant properties of commercial essential oils. Ind Crop Prod 43:587–595

Tepe B, Donmez E, Unlu M, Candan F, Daferera D, Vardar-Unlu G, Polissiou M, Sokmen A (2004) Antimicrobial and antioxidative activities of the essential oils and methanol extracts of *Salvia cryptantha* (Montbret et Aucherex Benth.) and *Salvia multicaulis* (Vahl). Food Chem 84:519–525

Tohidpour A, Sattari M, Omidbaigi R, Yadegar A, Nazemi J (2010) Antibacterial effect of essential oils from two medicinal plants against methicillin-resistant *Staphylococcus aureus* (MRSA). Phytomed 17:142–145

Tolouee M, Alinezhad S, Saberi RR, Eslamifar A, Zad SJ, Jaimand K, Taeb J, Razzaghi-Abyaneh M (2010) Effect of *Matricaria chamomilla* L. flower essential oil on the growth and ultrastructure of *Aspergillus niger* van Tieghem. Int J Food Microbiol 139:127–133

Toloza AC, Zygadlo J, Mougabure Cueto G, Biurrun F, Zerba E, Picollo MI (2006) Fumigant and repellent properties of essential oils and component compounds against permethrin-resistant *Pediculus humanus capitis* (Anoplura: Pediculidae) from Argentina. J Med Entomol 43:889–895

Toloza AC, Lucia A, Zerba E, Masuh H, Picollo MI (2008) Interspecific hybridization of eucalyptus as a potential tool to improve the bioactivity of essential oils against permethrin-resistant head lice from Argentina. Bioresour Technol 99:7341–7347

Tomaino A, Cimino F, Zimbalatti V, Venuti V, Sulfaro V, De Pasquale A, Saija A (2005) Influence of heating on antioxidant activity and the chemical composition of some spice essential oils. Food Chem 89:549–554

Toscano-Garibay JD, Arriaga-Alba M, Sánchez-Navarrete J, Mendoza-García M, Flores-Estrada JJ, Moreno-Eutimio MA, Espinosa-Aguirre JJ, González-Ávila M, Ruiz-Pérez NJ (2017) Antimutagenic and antioxidant activity of the essential oils of *Citrus sinensis* and *Citrus latifolia*. Sci Rep 7:11479

Tripathi R, Mohan H, Kamat JP (2007) Modulation of oxidative damage by natural products. Food Chem 100:81–90

Trongtokit Y, Rongsriyam Y, Komalamisra N, Apiwathnasorn C (2005) Comparative repellency of 38 essential oils against mosquito bites. Phytother Res 19:303–309

Tsuneki H, Ma EL, Kobayashi S, Sekizaki N, Maekawa K, Sasaoka T, Wang MW, Kimura I (2005) Antiangiogenic activity of β-eudesmol *in vitro* and *in vivo*. Eur J Pharmacol 512:105–115

Tunón H, Thorsell W, Mikiver A, Malander I (2006) Arthropod repellency, especially tick (*Ixodes ricinus*), exerted by extract from *Artemisia abrotanum* and essential oil from flowers of *Dianthus caryophyllum*. Fitoterapia 77:257–261

Ultee A, Bennik MHJ, Moezelaar R (2002) The phenolic hydroxyl group of carvacrol is essential for action against the food-borne pathogen *Bacillus cereus*. Appl Environ Microbiol 68:1561–1568

Unnati S, Ripal S, Sanjeev A, Niyati A (2013) Novel anticancer agents from plant sources. Chin J Nat Med 11:16–23

Venturi CR, Danielli LJ, Klein F, Apel MA, Montanha JA, Bordignon SA, Roehe PM, Fuentefria AM, Henriques AT (2015) Chemical analysis and *in vitro* antiviral and antifungal activities of essential oils from *Glechon spathulata* and *Glechon marifolia*. Pharm Biol 53:682–688

Vigan M (2010) Essential oils: renewal of interest and toxicity. Eur J Dermatol 20:685–692

Viuda-Martos M, Ruiz-Navajas Y, Fernandez-Lopez J, Perez-Alvarez J (2008) Antifungal activity of lemon (*Citrus lemon* L.), mandarin (*Citrus reticulata* L.), grapefruit (*Citrus paradisi* L.) and orange (*Citrus sinensis* L.) essential oils. Food Cont 19:1130–1138

Vukovic-Gacic B, Nikcevic S, Beric-Bjedov T, Knezevic-Vukcevic J, Simic D (2006) Antimutagenic effect of essential oil of sage (*Salvia officinalis* L.) and its monoterpenes against UV-induced mutations in *Escherichia coli* and *Saccharomyces cerevisiae*. Food Chem Toxicol 44:1730–1738

Wang J, Zhua F, Zhoua XM, Niua CY, Lei CL (2006) Repellent and fumigant activity of essential oil from *Artemisia vulgaris* to *Tribolium castaneum* (Herbst) (Coleoptera: Tenebrionidae). J Stored Prod Res 42:339–347

Wang GS, Deng JH, Ma YH, Shi M, Li B (2012) Mechanisms, clinically curative effects, and antifungal activities of cinnamon oil and pogostemon oil complex against three species of *Candida*. J Tradit Chin Med 32(1):19–24

Wen L, Wu D, Jiang Y, Prasad KN, Lin S, Jiang G, He J, Zhao M, Luo W, Yang B (2014) Identification of flavonoids in litchi (*Litchi chinensis* Soon.) leaf and evaluation of anticancer activities. J Funct Foods 6:555–563

Wu H, Li B, Wang X, Jin M, Wang G (2011) Inhibitory effect and possible mechanism of action of patchouli alcohol against influenza a (H2N2) virus. Molecules 16:6489–6501

Wu XL, Ju DH, Chen J, Yu B, Liu KL, He JX, Dai CQ, Wu S, Chang Z, Wang YP, Chen XY (2013) Immunologic mechanism of patchouli alcohol anti-H1N1 influenza virus may through regulation of the RLHsignal pathway *in vitro*. Curr Microbiol 67:431–436

Yang P, Ma Y (2005) Repellent effect of plant essential oils against *Aedes albopictus*. J Vector Ecol 30:231–234

Yang X, Zhang X, Yang SP, Liu WQ (2013) Evaluation of the antibacterial activity of patchouli oil. Iran J Pharm Res 12:307–316

Yang C, Chen H, Chen H, Zhong B, Luo X, Chun J (2017) Antioxidant and anticancer activities of essential oil from Gannan navel orange peel. Molecules 22:1391

Yen HF, Hsieh CT, Hsieh TJ, Chang FR, Wang CK (2015) *In vitro* anti-diabetic effect and chemical component analysis of 29 essential oils products. J Food Drug Anal 23:124–129

Yoo CB, Han KT, Cho KS, Ha J, Park HJ, Nam JH, Kil UH, Lee KT (2005) Eugenol isolated from the essential oil of *Eugenia caryophyllata* induces a reactive oxygen species mediated apoptosis in HL-60 human promyelocytic leukemia cells. Cancer Lett 225:41–52

Yousefbeyk F, Gohari AR, Sourmaghi MHS, Amini M, Jamalifar H, Amin M, Golfakhrabadi F, Ramezani N, Amin G (2014) Chemical composition and antimicrobial activity of essential oils from different parts of *Daucus littoralis* Smith subsp. *Hyrcanicus Rech*. f. J Essent Oil-Bearing Plants 17:570–576

Zeedan GSG, Abdalhamed AM, Ottai ME, Abdelshafy S, Abdeen E (2014) Antimicrobial, antiviral activity and GC-MS analysis of essential oil extracted from *Achillea fragrantissima* plant growing in Sinai Peninsula, Egypt. J Microbiol Biochem Technol S8:006. https://doi.org/10.4172/1948-5948.S8-006

Zhou S, Koh HL, Gao Y, Gong ZY, Lee EJ (2004) Herbal bioactivation: the good, the bad and the ugly. Life Sci 74:935–968

Zore GB, Thakre AD, Jadhav S, Karuppayil SM (2011a) Terpinoids inhibit *Candida albicans* growth by affecting membrane integrity and arrest of cell cycle. Phytomed 18:1181–1190

Zore GB, Thakre AD, Rathod V, Karuppayil SM (2011b) Evaluation of anti-*Candida* potential of geranium oil constituents against clinical isolates of *Candida albicans* differentially sensitive to fluconazole: inhibition of growth, dimorphism and sensitization. Mycoses 54:e99–e109

Zuzarte M, Maria JG, Carlos C, Canhoto J, Vale-Silva L, Silva MJ, Pinto E, Salgueiro L (2011) Chemical composition and antifungal activity of the essential oils of *Lavandula viridis* L'Her. J Med Microbiol 60:612–618

Zuzarte M, Goncalves MJ, Cruz MT (2012) *Lavandula luisieri* essential oil as a source of antifungal drugs. Food Chem 135:1505–1510

Cellulose-Based Hydrogels: Present and Future

10

Dilipkumar Pal, Amit Kumar Nayak, and Supriyo Saha

Contents

D. Pal (✉)
Department of Pharmaceutical Sciences, Guru Ghasidas Vishwavidyalaya (A Central University), Bilaspur, Chhattisgarh, India
e-mail: drdilip2003@yahoo.co.in

A. K. Nayak
Department of Pharmaceutics, Seemanta Institute of Pharmaceutical Sciences, Mayurbhanj, Odisha, India

S. Saha
School of Pharm Science & Technology, Sardar Bhagwan Singh University, Dehradun, Uttarakhand, India

M. S. Akhtar et al. (eds.), *Natural Bio-active Compounds*,
https://doi.org/10.1007/978-981-13-7154-7_10

Abstract

Cellulose, the most abundant natural polymer is extracted from various renewable resources, such as plant fibers (cotton, jute, hemp, linen, wood fibers, etc.) and is also generated by some bacteria, fungi, and animals. Currently, cellulose is extracted from tea, bamboo, linseed, and cellulose lignin as a natural polymer in various industrial applications, such as paper, textiles, adsorbents, drug delivery, cosmetics, tissue engineering, etc., because of its easy availability from natural sources, low price, easy extraction and processing, renewability, non-toxicity, biodegradability, ecological friendliness, and favorable physico-mechanical properties. Recently, hydrogels composed of cellulose have been used in several industrial applications. These hydrogels are mainly prepared through crosslinking, polymer-blending, formation of interpenetrating polymer networks (IPNs), and graft-copolymerization. This chapter provides an overview on the present scenario and future aspects of cellulose hydrogels and their applications.

Keywords

Cellulose · Chemical modifications · Crosslinking, Hydrogel · Polymer

10.1 Introduction

Hydrogels are three dimensional polymeric networks that contain a large amount of water trapped in their structure. Hydrogels are procured from biopolymers, synthetic polymers, and/or polyelectrolytes. Hydrogels are divided into chemical gels and physical gels based on the way of crosslinking through ionic or hydrogen bonds and through covalent bonds, respectively. Nowadays, owing to the biodegradable nature of hydrogels, they have wide applications in the fields of drug delivery, tissue engineering, biosensors, contact lenses, absorbable materials, purification tool, etc. Synthetic polymer-based hydrogels are obtained via a cross-linkage reaction of polyethylene glycol, polyvinyl alcohol, polyamide amine, poly-N-isopropylacrylamide, polyacrylamide, and polyacrylic acid along with their copolymers. Photo polymerization, adjustable mechanical properties, controllable scaffold architecture, and chemical compositions are the prima focus properties of synthetic polyethylene glycol (PEG)-based hydrogels. Hydrogels fabricated with hyaluronic acid, alginate, starch, gelatin, cellulose, and chitosan derivatives show potential applicability owing to their biocompatibility and biodegradability. Various cellulose derivatives, such as methylcellulose, carboxymethylcellulose, and hydroxymethylcellulose, are extensively used in pharmaceutical, biotechnological, and agricultural fields due to their abundance, low cost, and nontoxic nature (Pal and Nayak 2017a, b). This chapter provides an overview on the present scenario and future aspects of cellulose hydrogels and their applications.

10.2 Classifications of Cellulose Derivatives

Cotton, softwoods, linen, and jute are cellulose-containing natural biopolymers, which are procured from some bacteria, fungi, and animals. Natural fibers with variable chain length and molecular weight are the principal cellulose obtained from plant sources. The molecular weight of cellulose is a maximum of 1500 Da; at which 40–50 glucose units are linked in longitudinal formations known as crystallites that are parallel to the longitudinal length of 600–650 nanometer. The glucose monomers in pyranose sugar are connected with unbranched chains via β-1,4-glucosidic linkages in a zigzag manner to comply with the nature of the linear structure of a polysaccharide polymer (Fig. 10.1). High crystalline structure and structural

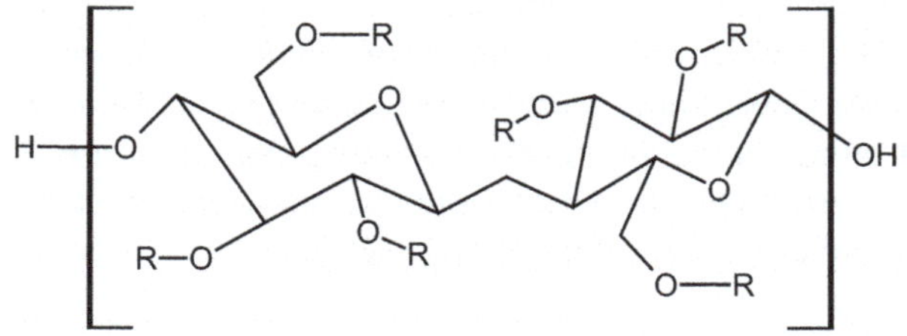

Fig. 10.1 Cellulose derivatives. (*R: H, CH_3* = Methylcellulose*; R: H, CH_2CH_3* = Ethylcellulose*; R: H, $[CH_2CH_2O]_nH$* = Hydroxyethylcellulose*; R: H, CH_2COONa* = Carboxymethylcellulose*; R: H, $[CH_2CH(CH_3)O]_nH$*)

rigidity are the primary features of cellulose, which makes the molecule insoluble in water and organic solvents. The esterifcation reaction is used to modify cellulose to flourish its applicability. The maximum value of degree of substitution in cellulose structure is 3. Etherification, esterification, electrolytic dissociation, and water solubility are used to modify the parent cellulose structure (Chang and Zhang 2011).

10.3 Hydrogels Obtained from Unprocessed Cellulose

Cellulose hydrogels are prepared from a cellulose solution via physical crosslinking. Highly complicated hydrogen bonds in the cellulose structure make it insoluble in any solvent. Nowadays, developed solvents, such as N-methylmorpholine-N-oxide (NMMO), ionic liquids (ILs), and alkali/urea (or thiourea) aqueous systems, are used to prepare cellulose hydrogels.

10.3.1 LithiumChloride/Dimethylacetamide (DMAc) System

When a lithiumchloride-dimethylacetamide system is used for cellulose hydrogel development, it is prepared as a hydrogel bead by adding a drop wise composite solution of cellulose-lithiumchloride-dimethylacetamide into methanol or isopropanol as an azeotropic non-solvent system. Larger molecular weight cellulose is dissolved in the LiCl/DMAc solution. In the LiCl/DMAc system, the cellulose concentration is 7 wt% and the transparency, strength, and water content of hydrogels depend upon the ratio of the coagulation/regeneration bath. In non-aqueous organic solution, a maximum of 97% transparency (97%) is obtained and water has a very strong influence on the physical properties of cellulose.

10.3.2 N-Methylmorpholine-N-Oxide System

The N-methylmorpholine-N-oxide system is very useful for the production of fibers, films, food casings, membranes, sponges, and beads without noxious byproducts. At high temperature, it is used to dissolve high molecular weight cellulose to obtain a transparent solution. At 100 °C, after reaction with N-methylmorpholine-N-oxide, cellulose loses its crystalline nature. Regenerated cellulose products are obtained by addition of a slight excess of water into the cellulose-N-methylmorpholine-N-oxide-water solution.

10.3.3 Ionic Liquids System

1-Butyl-3-methylimidazolium chloride (BMIMCl) and 1-allyl-3-methylimidazolium chloride as hydrophilic liquids are used to dissolve cellulose. After the addition of water, ethanol, and acetone, cellulose is regenerated with the same degree of polymerization as initial cellulose, and it can be used as films, beads, and gels. Ionic liquids such as 1-allyl-3-methylimidazolium chloride with deionized water as coagulant are used to obtain cellulose hydrogels. Flexible gels are procured from submerging cellulose in 1-butyl-3-methylimidazolium chloride, and they keep for seven days at room temperature.

10.3.4 Aqueous Systems of Alkali-Urea/Thiourea

A mixture of sodium hydroxide (7%) and urea (12%) in water is used to develop the aqueous system of alkali-urea/thiourea, which is pre-cooled to (−) 12 °C, and cellulose with less than 120,000 Da molecular weight quickly dissolves. At low temperature, cellulose is dissolved in sodium hydroxide, urea, and water composite. In cellulose hydrogels, the stringent behavior of the cellulose solution is agitated with the abrupt increase in temperature up to 50 °C and sudden cooling down to (−) 20 °C. The tendency of the cellulose solution to transfer into gel is reported by the molecular weight, concentration, and temperature of cellulose. The transformation of the cellulose solution into gel is obtained by decreasing the temperature from 60.3 °C to 30.5 °C with variation in concentrations from (3–5) wt%. Furthermore, the temperature of the cellulose solution (5%) is decreased from 59.4 °C–30.5 °C with stepping molecular weight from 4.5×10^4 to 11.4×10^4 Da.

10.3.5 Bacterial Cellulose Hydrogels

Nonpathogenic microbial strains such as *A. xylinum* are used to produce hydrogels with an ultra-fine networking structure, which have greater tension with greater H_2O captivating capacity, crystalline structure, and biological compatibility. On the basis of the high purity and unusual physicochemical properties, hydrogels procured from bacterial cellulose have various implications in the field of tissue scaffolding and tissue and dental implants. Synthesis of tubular bacterial cellulose gel is exerted by a bacterial culture of cellulose in an aerobic silicone tube with less than 8 mm inner diameter. The longitudinal orientation of fibrils in the silicone tube is independently related to gravitational force, availability of oxygen, and the topology of the inner surface of the silicone tube with dependency on the curvature of the silicone tube.

10.4 Cellulose Derived Hydrogel

Biocompatible cellulose derivatives with water solubility characteristics behave with a thickening, binding, emulsifying, film forming, suspension aiding, lubricating, and stabilizing nature; especially in food, pharmaceutical, and cosmetic industries. Certain cellulose derived materials, such as methyl cellulose, hydroxypropyl cellulose, hydroxypropylmethyl cellulose, and carboxymethyl cellulose are used to develop cellulose hydrogels via physical and chemical cross-linked behavior and also by interaction with a hydrogen bonding, polymer–polymer reaction. Hydrogels can be developed by participating in a crosslinking reaction between two or more kinds of polymer chains with a crosslinking agent or under UV light irradiation.

10.4.1 Physical Crosslinking

Replacing methyl groups or hydroxylpropyl groups with hydroxyl groups using hydrogen bonds leads to water soluble hydrogels. The water soluble methylcellulose solution exerts a reversible physical gel property, if it is heated above a specified temperature via hydrophobic interactions. A higher gelation property is reported for hydroxypropyl methylcellulose than methylcellulose, which can deliver a solid gel with a proportional ratio in substitution and molecular weight. The gel forming nature of cellulose derivatives exerts a heavily methoxylated portion of polymer due to exclusion of water from polymer. The second cycle of heating has a greater gel formation time than the first cycle of heating. In an experiment for the hatching of human embryonic stem cells, a methylcellulose coated polystyrene dish is used for embryo development in a culture of liquid suspension. The heat sensitive hydroxypropyl cellulose microgels are developed via a chemical crosslinking reaction of hydroxypropyl cellulose chains with a dodecyltrimethylammonium bromide solution.

10.4.2 Chemical Crosslinking Reaction

10.4.2.1 Reagents Responsible for Chemical Crosslinking

The swelling behavior of cellulose hydrogels requires a chemical crosslinking reaction. In cellulose, molecules with two functional groups are used as crosslinking agents to form a covalent linkage with different polymer molecules via hydrophilic three dimensional networks. Superabsorbent hydrogels of cellulose are prepared via the crosslinking reaction of carboxymethyl cellulose and hydroxyethyl cellulose with divinylsulfone, which illustrates as a highly sensitive absorptive material to treat edemas in the body, and its activity fluctuates with variations in pH and strength of the external solution. Single-phase HPC hydrogels remain nonporous in nature compared to biphasic cross-linked micro-porus HPC hydrogels. A thermally sensitive hydrogel is developed by a reaction between hydropropyl cellulose and poly(ethylene glycol) diglycidyl ether, which swells at 20 °C and constricts at 60 °C. Hydroxypropyl cellulose hydrogels are prepared via a crosslinking reaction between hydroxypropyl cellulose and epichlorohydrin; ammonia showed prominent activity as an anionic dye absorbent.

10.4.2.2 Radical Crosslinking Reaction

Chemical hydrogels are also prepared via crosslinking by irradiation of solid polymer. Radical cross-linkers have limited applications in the market of FMCG and medicines considering toxicity problems. The concentration of CMC, HPC, and MC in aqueous solutions undergoes crosslinking reactions by ionization irradiation to develop cellulose hydrogels. Highly substituted and high concentration carboxymethyl cellulose undergoes a chemical crosslinking reaction to develop hydrogels by imposing ionic radiation. Gels are developed in a fast gelation technique at the primary stage of a lower dose of γ rays radiation. The electron beam technique and anaerobic γ rays are used to irradiate medium to high concentration carboxymethyl cellulose, which forms 90–95% of the gel fraction after ionizing irradiation. The aerobic irradiation process abolished the greater concentration of gel fraction and lower ratio of crosslinking reaction for high concentration solutions radiated by an electron beam, which also exhibited 90% of the gel fraction, while 65% γ irradiation is observed for preparation of hydroxypropyl cellulose hydrogels.

10.5 Hydrogels Composed of Cellulose Polymer

10.5.1 Blending Composites

10.5.1.1 Natural Polymers as Blending Agent

Natural polymers, such as alginates and hyaluronic acid, are used to develop a novel compound to remove heavy metals from solution, to remove starch from potatoes and rice, and to amalgamate with sodium alginate for tissue scaffolding. To produce phase separating composite hydrogels within the same solvent system, a combination of chitin and cellulose is developed. Beads composed of chitin and cellulose are developed by coagulation mixing of 4 wt% of cellulose and 2 wt% of chitin solution in a 6 wt% of sodium hydroxide/5 wt% water soluble thiourea solution. The range of qe adsorption value of heavy metal ions on the beads is: Pb^{2+}, Cd^{2+}, Cu^{2+}. The adsorptivity of the Pb ion on formulation is greater than chitin because of the broad surface area and greater hydrophilicity of formulation. The hydrophilic background and microsized porous formulation of cellulose improves the adsorptivity of Pb ion over chitin. Cellulose fused with chitosan formulations are developed via reaction of cellulose into chitosan solution. Formulations are chemically reacted with ethylene glycol diglycidyl ether solvent, with greater adsorptivity for the Cu ion. Chitosan are mixed with a concentrated carboxymethyl cellulose solution to form physical formulations with an irradiated crosslinking reaction. The surface area of the low concentration formulation is smooth, and a higher amount of chitosan is brittle in nature. An injectable matrix develops for chitosan and methylcellulose under mild conditions in the absence of organic solvent, high temperature or abrupt pH. Chitosan fused with methylcellulose and sodium phosphate formulations reveal good cell vulnerability and proliferation. Hydrogels made up of hyaluronan–methyl cellulose, effectively applied as injection with normal swell, with greater loading of residual particle, slow in vitro degradability, and good diffusion up to 150 kg/mol, can be used as polymeric carriers.

10.5.1.2 Polyvinyl Alcohol as Blending Agent

Polyvinyl alcohol cross-linked hydrogels are prepared by reagents, electron imposing, and falling of gamma radiation by the heat cycle method. The physical crosslinking reaction has a greater residual toxic chemical crosslinking agent and greater withstand strength than polyvinyl alcohol. Crosslinking methods directly affect the chemical structure of cellulose fused with polyvinyl alcohol formulation. Chemical crosslinking reactions between cellulose and polyvinyl alcohol are observed with high swelling behavior and less resistance. Formulations are developed by crosslinking of cellulose and polyvinyl alcohol with repeated freezing and thawing.

10.5.2 Polyelectrolyte Complexes

A water soluble polyelectrolyte complex can be prepared by the electrostatic reaction between different poles of the electrolyte in water. Polymeric complex materials are developed as matrix and hydrogels. Carboxymethyl cellulose is a negatively charged polymer, so these formulations are developed by a positively charged polyelectrolyte in the presence of carboxymethyl cellulose. A formulation membrane is prepared by blending chitosan and carboxymethyl cellulose solutions followed by reaction with glutaraldehyde. Depending on the pH of the solution, a hydrogel can bend toward either anode or cathode, and its electromechanical behavior is influenced by the ionic and electric field strength. The ionic strength of 0.2 M (pH 6) in the Britton-Robinson buffer solution shows a greater angle for formulation bending. The permeation flux of membrane is based on the temperature and water content in the aqueous ethanol medium. The formulation is observed with high permeation.

10.5.3 Interpenetrating Polymer Networks (IPNs)

The interpenetrating polymer matrix illustrates as a close combination of more than one polymer within the same formulation. IPNs are mainly two types: sequential and semi-IPN. For the sequential type, cellulose is taken as polymer source, whereas semi-IPN is developed by polymeric reaction within different networks of cellulose. A linear, branched cellulose crosslinked network is known as a semi-interpenetrating network. This formulation of hydrogel is developed by solubilizing bacterial cellulose within a water soluble gelatin solution and reacting it with N-(3-dimethylaminopropyl)-N′-ethylcarbodiimide hydrochloride to produce a sequential interpenetrating formulation. Stress–strain curves of BC/gelatin double network formulation show typical compression and elongation. The double network observed with 3.7 mega Pascal strength is perpendicular to the stratified structure, and the tonicity is 3.0 mega Pascal. Formulation with the double network exhibits low frictional strength (Hebeish et al. 2015).

10.6 Various Cellulose-Based Hydrogels

10.6.1 Cellulose Hydrogels from Tea Residue

Using a 1-allyl-3-methylimidazolium chloride solvent, tea cellulose is used to prepare composite hydrogels, and the prepared hydrogels are characterized by application of polymer κ-carrageenan, chitosan, and guar gum. Salicylic acid sodium salt as a model drug is applied to check the swelling behavior, drug loading capacity, and chemical kinetics of release for the prepared hydrogels. Cell cytotoxicity and biocompatibility is evaluated by the thiazolyl blue tetrazolium bromide assay. The thermostability and mechanical characteristics of the composite hydrogels is improved by chitosan and guar gum, whereas starch (soluble) is used to improve the equilibrium swelling ratio, drug loading, and releasing properties. Guar gum and chitosan are used to increase permeable resistance and are good for release of hydrogels (Fig. 10.2, Table 10.1). Therefore, it is concluded that addition of chitosan, κ-carrageenan, guar gum, and soluble starch improve cell compatibility and non-cytotoxic behavior (Zhijun and Huihua 2016).

10.6.2 Graphene Embeded Tea Residue Cellulose Hydrogel

Hydrogels can be prepared by the homogeneous mixture of tea cellulose and graphene oxide in the ionic liquid 1-allyl-3-methylimidazolium chloride. X-ray diffraction, FTIR, thermogravimetry analysis, and morphology of the formulation are used to characterize the formulation. Methylene blue is used as a reagent to check the adsorption parameter. The effects of temperature on formulation are also investigated. The TGA curve shows a peak around 280 °C, shifted to 320 °C, and

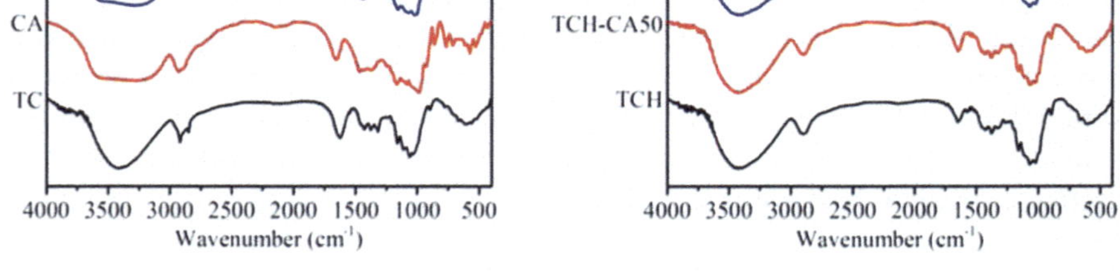

Fig. 10.2 FTIR characteristics of additives; (**a**) tea cellulose powder, and (**b**) corresponding hydrogels prepared from tea cellulose and additives. (Source: Zhijun et al. copyright © 2016 with permission from Elsevier B.V)

Table 10.1 Formula of composite hydrogels prepared from tea cellulose and additives

Marked sample	Additives	Additive amount (% of tea cellulose)
TCH	–	0
TCH-CA25	*K*-carrageenan	25
TCH-CA50		50
TCH-CA100		100
TCH-SS25	Soluble starch	25
TCH-SS50		50
TCH-SS100		100
TCH-GG25	Guar gum	25
TCH-GG50		50
TCH-GG100		100
TCH-CH25	Chitosan	25
TCH-CH50		50
TCH-CH100		100

Source: Zhijun et al. copyright © 2016 with permission from Elsevier B.V.

the hardness, ability to withstand, and gum like nature are observed 12.7, 4.1, and 17.8 times, respectively (Figs. 10.3, and 10.4). This formulation is observed with 46.35 mg/g of adsorption capacity and methylene blue shows 92.7% adsorbed by hydrogel, which confirms the pseudo-second order kinetic model (Zhijun et al. 2017a, b).

10.6.3 Nanofibrillar Cellulose Hydrogels for Controlled Drug Release

An anionic nanofibrillar cellulose (ANFC) hydrogel matrix with (3–6.5) % concentration is used for drug delivery and is achieved by lyophilization of excipients embedded in porous aerogel, which is easily administered in hydrogel formulations with constant rheological parameters. Release of medicine from the designed formulation is not affected by the freeze drying technique. Higher ANFC fiber content is correlated with smaller diffusion coefficients in the case of large molecules, which also indicates that concentration of anionic nanofibrillar cellulose is observed with control release behavior (Fig. 10.5, Tables 10.2 and 10.3). Therefore, the formulation is successfully lyophilized and used prominently (Paukkonen et al. 2017).

10.6.4 Photoluminescent Carboxymethyl Cellulose Hydrogels

Carboxymethyl cellulose hydrogels are developed to demonstrate the activity profile of self-cure, creating luminescence of light using an organic approach. Formulations are observed with bluish-green emission under UV light, and 95% healing efficiency is observed under good environment. A healed hydrogel is two and half times

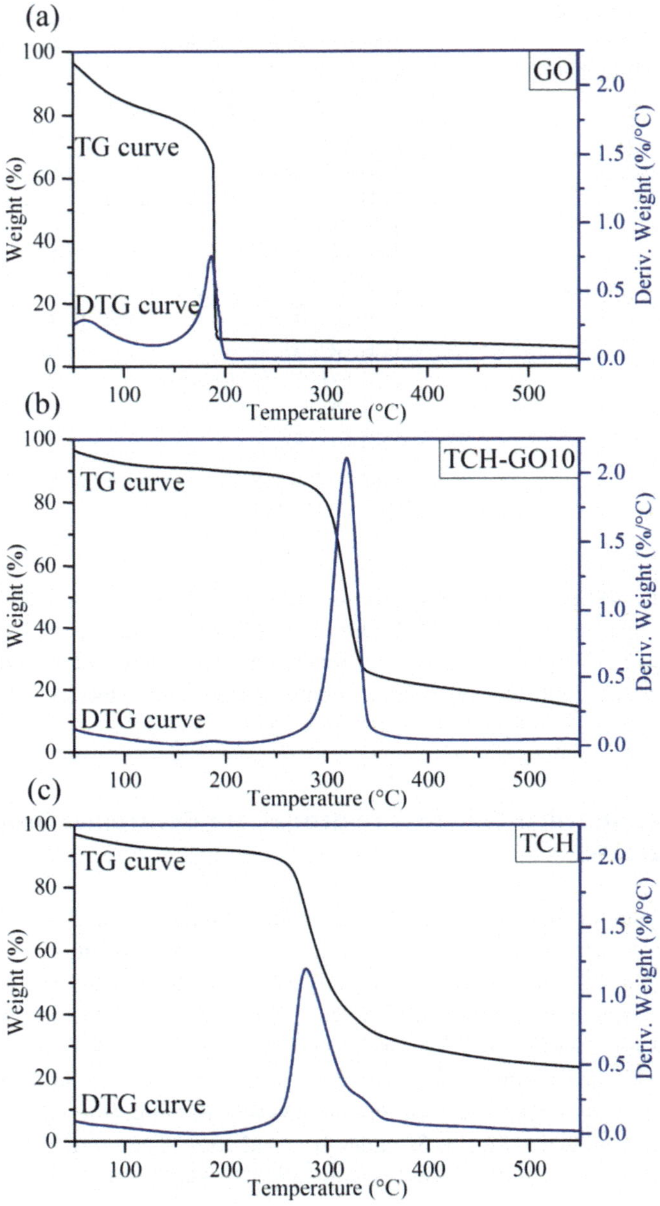

Fig. 10.3 TG and DTG curves of graphene oxide (**a**), graphene oxide/tea cellulose hydrogels (**b**), and tea cellulose hydrogels (**c**). (Zhijun et al. copyright © 2017 with permission from Elsevier B.V)

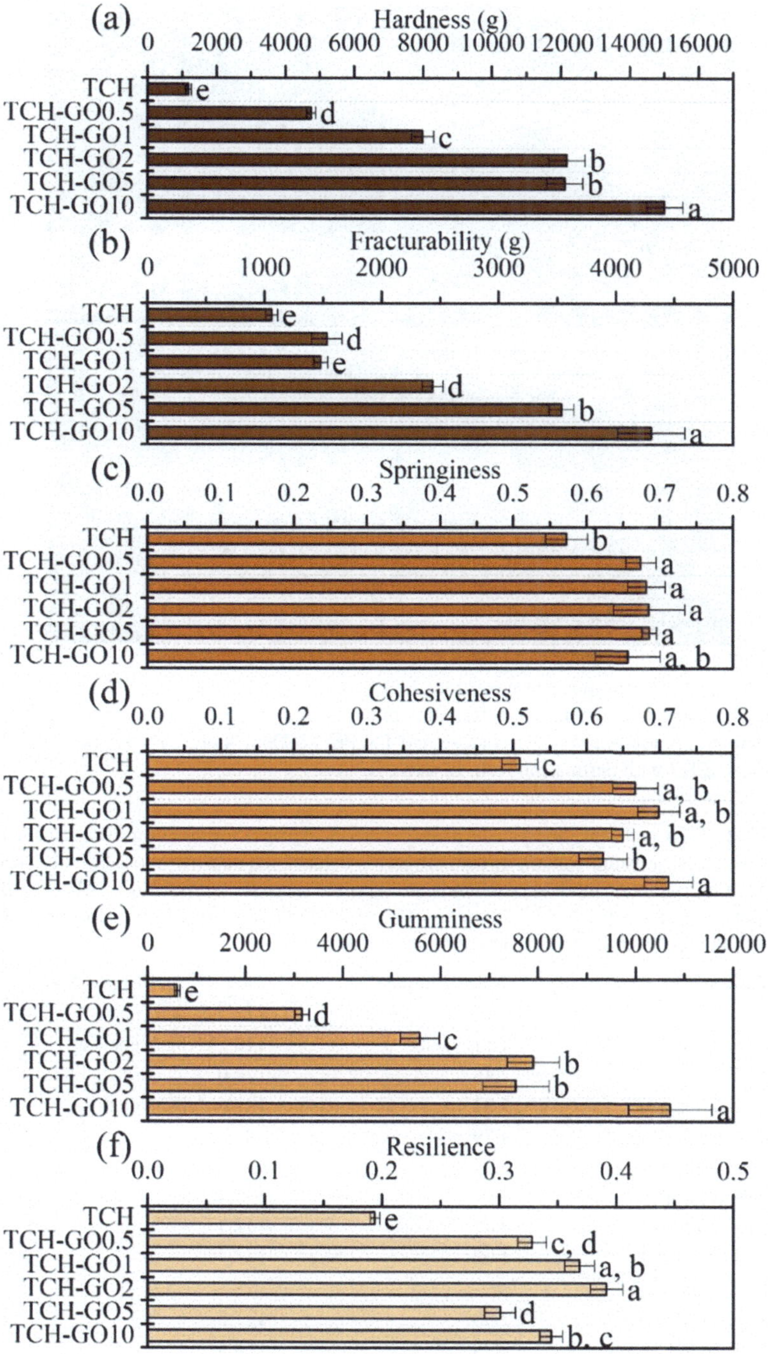

Fig. 10.4 The TPA of the prepared hydrogels (**a**) for hardness, (**b**) for fracturability, (**c**) for springiness, (**d**) for cohesiveness, (**e**) for gumminess and (**f**) for resilience). Values are expressed as means ± SD. Bars marked by different letters mean significant differences between the values ($p < 0.05$). (Source: Zhijun et al. copyright © 2017 with permission from Elsevier B.V.)

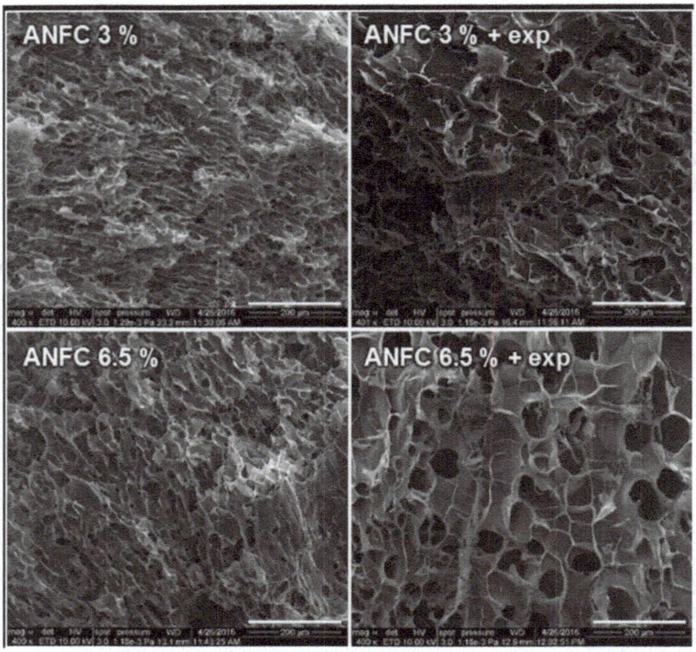

Fig. 10.5 SEM micrographs of freeze-dried highly porous ANFC aerogels. 3% and 6.5% ANFC hydrogels are freeze-dried without (left) and with the excipients (right). (Scale bar is 200 μm) Abbreviation exp. refers to 0.3% trehalose and 1% PEG 6000. (Source: Paukkonen et al. (2017) copyright © 2017 with permission from Elsevier B.V.)

Table 10.2 TGA and DSC analysis of different ANFC aerogel formulations after freeze-drying (n = 1). Residual moisture content of the aerogels is determined as a mass loss (%) of evaporated water. Melting points of model compounds are reported as onset temperatures. All formulations except pure 3% and 6.5% ANFC contained 1% PEG 6000/0.3% trehalose in the final aerogel. Abbreviation exp. refers to 1% PEG 6000/0.3% trehalose

Formulation	Water content (%)	Tm (°C) for drug or ANFC
ANFC 3%	6.77	171.5
ANFC 3% + exp	4.11	170 (ANFC/trehalose)
ANFC 6.5%	7.48	174.0
ANFC 6.5% + exp	5.81	187 (ANFC/trehalose)
BSA 1%/ANFC 3%	4.79	nd
BSA 1%/ANFC 6.5%	6.05	nd
MZ 100% reference	nd	159.4
MZ 2%/ANFC 3%	2.42	148.7
MZ 2%/ANFC 6.5%	4.98	143.5
NAD 100% reference	nd	128.8
NAD 1.7%/ANFC 3%	2.90	127.3
NAD 1.7%/ANFC 6.5%	4.79	128.3
KETO 100% reference	nd	94.6
KETO 3.4%/ANFC 3%	1.96	81
KETO 3.4%/ANFC 6.5%	3.84	84

Source: Paukkonen et al. (2017) copyright © 2017 with permission from Elsevier B.V.

Table 10.3 Diffusion coefficients for model compounds in different ANFC hydrogels. Values are presented for 3% and 6.5% ANFC hydrogels with and without excipients. Freeze-dried formulations are re-gelled prior to diffusion studies. The net charge of compounds at pH 7 is in parenthesis. The abbreviation exp. refers to PEG 6000 and trehalose

	Diffusion coefficients for model compounds in different ANFC hydrogels (10^{-8} cm^2/s)					
Compound	3% ANFC	3% ANFC/exp	FD 3% ANFC/exp	6.5% ANFC	6.5% ANFC/exp	FD 6.5% ANFC/exp
Ketoprofen (−)	62.9	60.9	54.5	58.0	52.8	47.4
NAD (+)	365.6	393.4	383.4	301.1	322.3	349.7
Metronidazole (ø)	745.7	823.0	841.5	733.3	761.8	779.6
BSA (−)	7,7	23.4	22.0	7.4	15.2	9.5
Lysozyme (+)	4.0	nd	nd	nd	nd	nd
4 kDa FITCdextran (ø)	58.9	nd	nd	35.6	nd	nd

Source: Paukkonen et al. (2017) copyright © 2017 with permission from Elsevier B.V.

more stretched than normal. The self-cure formulations adhere to hard materials, like glass and plastic, along with soft tissues, and it is also used as a sealing agent and for mucosal surface adhesion in bioengineering fields (Yong et al. 2017)

10.6.5 Papain-Fused Magnetic Cellulose Hydrogels

Papain is frozen within the hydrogels, obtained from tea by solubilizing in 1-allyl-3-methylimidazolium chloride as solvent, and coated with ferric oxide by reaction with a mixture of ferric and ferrous chloride in an ammonium hydroxide solution. The formulation is characterized via vibration magnetometer, SEM, X-ray powder diffraction, FT-IR, TGA, and DSC analytical processes. The characteristics of papain-free and papain-fused formulations are compared by thermal exposure, pH, and temperature of formulation along with rate of reaction. The magnetic behavior has paramagnetic characteristics, greater stability toward heat, and lower affinity toward substrate. A pH of 8.0 at 90 °C is required for formulation development (Zhijun et al. 2017a, b).

10.6.6 Potassium Copper Hexacyanoferrate-Fused Cellulose Hydrogel

Potassium copper hexacyanoferrate-cellulose hydrogels are prepared from immobilized potassium copper hexacyanoferrate (KCuHCF) in a polymer mixture of carboxymethyl cellulose and hydroxyethyl cellulose to adsorb cesium ions. Formulation freezing is accelerated by nanosized potassium copper hexacyanoferrate, which shows good surface activity. The formulation shows greater cesium adsorption

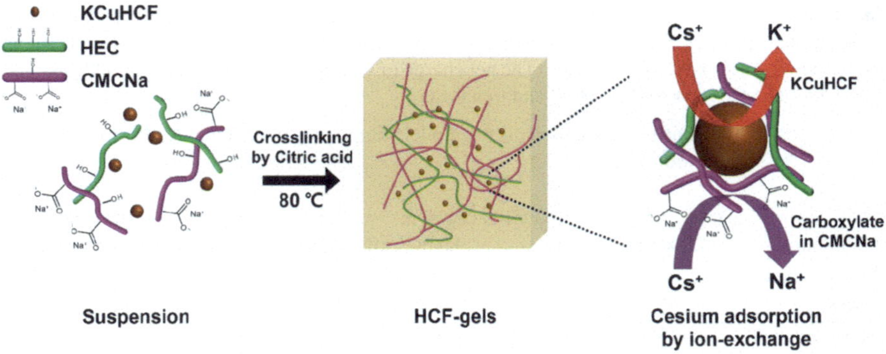

Fig. 10.6 Representation of synthesis and cesium adsorption of HCF-gels. (Source: Yonghwan et al. (2017) copyright © 2016 with permission from Elsevier B.V.)

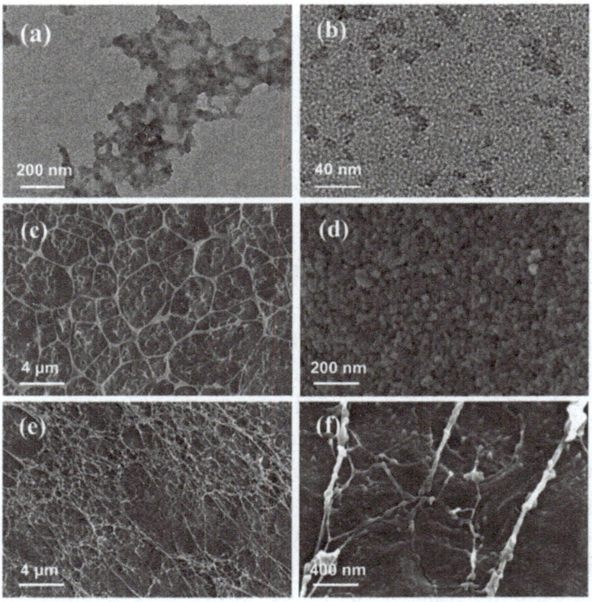

Fig. 10.7 TEM images of (**a**) HCF-gel-1 (**b**) HCF-gel-1 (higher magnification), and SEM images of (**c**) cellulose hydrogel (**d**) bulk KCuHCF (**e**) HCF-gel-1 (**d**) HCF-gel-1 (highermagnification). (Source: Yonghwan et al. (2017) copyright © 2016 with permission from Elsevier B.V.)

capacities correlated with the number of ion-exchangeable sites (Figs. 10.6, 10.7 and Table 10.4), and it demonstrates good removal of cesium, greater than 90% from sea water containing 0.11 mmol/l of cesium (Yonghwan et al. 2017).

10.6.7 Superabsorbent Crosslinked Carboxymethyl Cellulose-PEG Hydrogels

Carboxymethyl cellulose-PEG hydrogels are prepared by carboxymethyl cellulose (CMC), polyethylene glycol, and citric acid (CA) for wound dressing and skin repair substitutes. Also, carboxymethyl cellulose hydrogels obtain 10 wt% of

Table 10.4 Cs^+ adsorption capacities from other studies related to immobilization of metal hexacyanoferrates

Co-metal	Support	Adsorption capacity (mmol Cs/g capacity)
Cu	HEC + CMC	2.06–2.32
Cu	Silica	0.13
Cu	PAN	0.19
Cu	Chitin	0.89
Fe	Graphene oxide/Fe_3O_4	0.42
Fe	Polyacrylic acid	0.55
Ni	Silica	1.69
Ni	Biomass	1.07

Source: Yonghwan et al. (2017) copyright © 2016 with permission from Elsevier B.V.

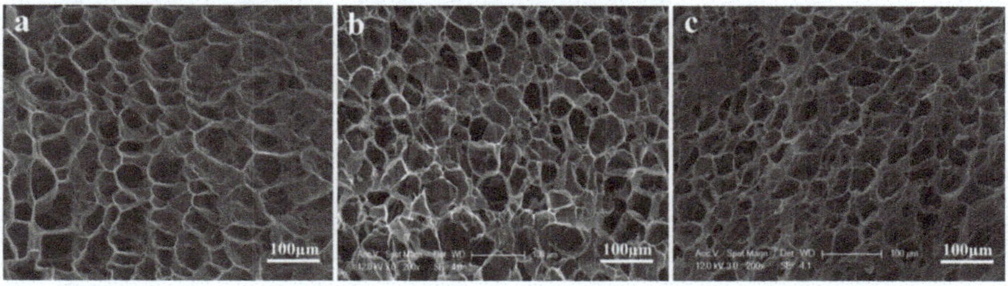

Fig. 10.8 SEM images of cellulose hydrogel (**a**), CT hydrogel (**b**) and CT hydrogel after desorption (**c**). (Source: Ying et al. (2017) copyright © 2017 with permission from Elsevier B.V.)

polyethylene glycol. Outcomes show greater absorbing criteria with good swelling behavior depends upon citric acid and polyethylene glycol. The characterization data reveals that the presence of hydroxyl groups plays a prominent role in the formation of the hybrid network. Morphological features of formulation depend upon the amount of crosslinking and elastic nature. These formulations have shown over 95% responses against human embryonic kidney cell line (Capanema et al. 2018).

10.6.8 Tannin-Immobilized Cellulose Hydrogel

Tannin-fused cellulose hydrogels are prepared by a reaction involving equilibrium immobilization and crosslinking. SEM and mechanical test are used to characterize the structures and properties of hydrogels. A cationic dye model such as methylene blue (MB) is used to evaluate the ability to adsorb. Immobilized tannin along with methylen blue is used to improve the adsorption facility as evaluated by the isotherm theory of Langmuir. Six continuous repetitions of the adsorption–desorption cycle for formulation show no significant loss of adsorption capacity (Fig. 10.8 and 10.9; Table 10.5). A facile homogeneous reaction is used to prepare both tannin fusion and development of formulation, which provides a path to form tannin-fused materials for waste water management (Ying et al. 2017).

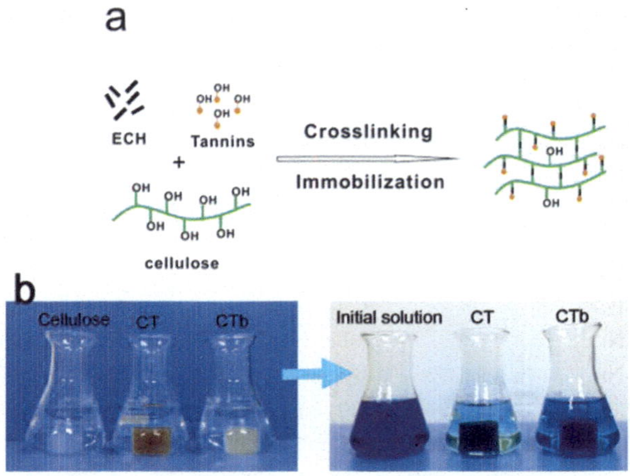

Fig. 10.9 The mechanism for the formation of CT hydrogel (**a**); digital images of cellulose hydrogel, CT and CTb hydrogel in distilled water and the adsorption effect of CT and CTb hydrogel (**b**). (Source: Ying et al. (2017) copyright © 2017 with permission from Elsevier B.V.)

Table 10.5 The compressive strain and compressive strength of cellulose hydrogel and CT hydrogel

Sample	The compress strain (%)	The compress strength (kPa)
Cellulose	82 ± 10	204 ± 10
CT0.5/50	82 ± 10	195 ± 5
CT1/50	79 ± 5	190 ± 6
CT1.5/50	77 ± 10	186 ± 5

Source: Ying et al. (2017) copyright © 2017 with permission from Elsevier B.V.

10.6.9 Gelatin-Based Bacterial Cellulose Hydrogel

Hydrogel composite material is developed using bacterial cellulose and gelatin, which is obtained by the polymer reaction between gelatin bacteria derived cellulose. The uniform size and shape of bacterial cellulose chains are characterized by scanning electron microscopy (SEM) images using a glutaraldehyde cross-linker. Amine groups and hydroxyl groups of gelation bacterial cellulose are used to form hydrogen bonds. The hydrogel composite presents with thermal stability from heat, resistance from chemical reaction, and mechanical changes. In water, 400–600% swelling is observed, which correlates with the development of a candidate with a better drug delivery system (Treesuppharat et al. 2017).

10.6.10 Rice and Oat Husks Cellulose Fibres

Rice and oat husks are used to procure cellulose fibers, characterized by composition, morphology, functional group, crystal structure of structure, and heat studies. Extracted fibers are used to prepare formulations with polyvinyl alcohol. The structural, crystal, and thermal behavior of these fibers depends on the source of the cellulose. The hydrogel extracted from rice cellulose fibers has a porous sponge-like formulation similar to the formulation extracted from oat husk. The absorption capacity of water, observed from hydrogel formulation obtained from rice and oat husks, is 141.6% and 392.1%, respectively. At 25 °C, the rice hydrogel displays greater water absorbing and maximum stress, which is promising for biomaterials and the agricultural field (Jean et al. 2017).

10.6.11 Injectable Carboxymethyl Cellulose-Pullulan Hydrogels

An injectable adhesive hydrogel is developed from carboxymethyl cellulose (CMC) and pullulan, which is used as a postoperative barrier against adhesion. Tyramine is used as a carboxymethyl cellulose modifier to introduce a crosslinking reaction. This formulation is developed by an enzyme linked reaction of tyramine frozen carboxymethyl cellulose with horse radish peroxidase and hydrogen peroxide. Pullulan improvised formulation is used to evaluate adhesiveness of the hydrogel solution with respect to wound area and to accelerate biodegradation. An attenuated total reflection (ATR)-Fourier-transform infrared spectroscopy (FTIR) (ATR-FTIR) technique is used to confirm CMC modification. The amounts of horse radish peroxide and hydrogen peroxide in hydrogels are measured by time to gel formation, storage modulus, and loss of weight. The formulation shows much less cell proliferation and cell toxicity. Therefore, carboxymethyl cellulose fused with pullulan formulation is used as an injectable in situ barrier against adhesion (Bang et al. 2017).

10.6.12 Linseed Gum Embeded in Cellulose Hydrogels

An amalgamation of cellulose and linseed gum is solubilized within an aqueous solution of sodium hydroxide/urea with the reaction of epichlorohydrin to obtain a composite formulation. SEM, FT-IR, X-ray diffraction, and thermogravimetric analysis are used to characterize structural topology of formulation. The behavior of swelling and water absorption capabilities are also investigated. Outcomes suggested that linseed gum uses water adsorptivity. This porous superabsorbent formulation is used for waste water management and for water conservation in agriculture (Hao et al. 2017)

10.6.13 Bamboo Cellulose Hydrogels

Bamboo pulp fibers are used to prepare dimensionally stable hydrogels by purification within a short time from ultrasonic treatment. Ultra-fine fibers are more dispersed within the formulation than in cellulose nanofibrils. During the pulp forming process, using nitric acid and potassium chlorate, hydroxyl groups are transferred into carboxylic acid by oxidation. Sodium carboxylate forms in the presence of a sodium hydroxide solution. The negatively charged COO^- groups regulate the formation of formulation. Equilibrium formulations form at pH 7, 9, and 11 (Fig. 10.10 and 10.11) (Xiaofang et al. 2014).

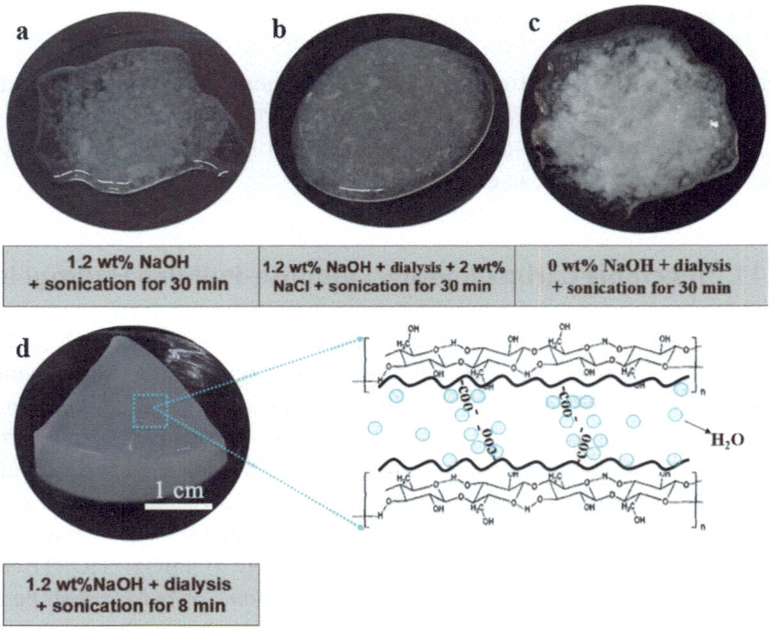

Fig. 10.10 Modulating the macroscopic gelation properties of cellulose hydrogels at different conditions ((**a**)–(**d**)) and a scheme for the possible gelation mechanism (**d**). (Source: Xiaofang et al. (2014) copyright © 2014 with permission from Elsevier B.V.)

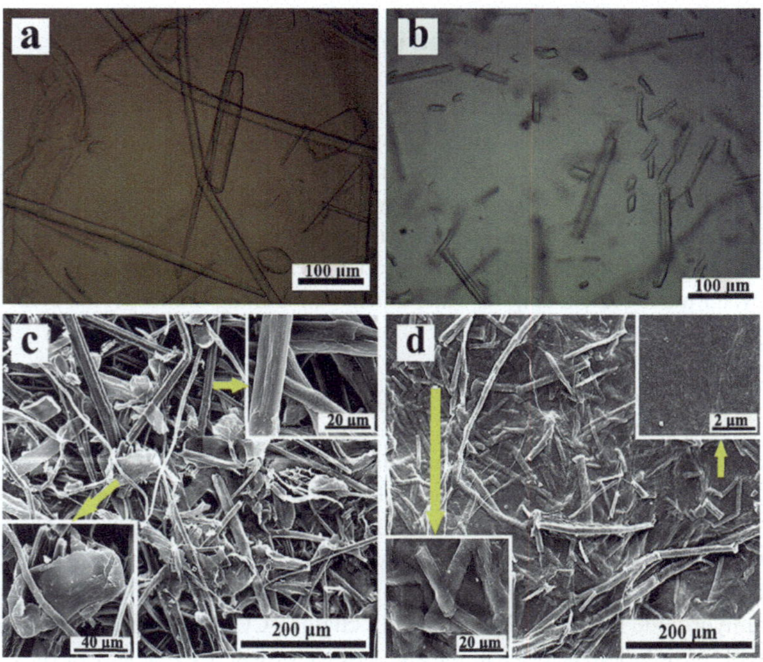

Fig. 10.11 Optical microscope images for (**a**) the purified cellulose fibers and (**b**) the obtained hydrogel; SEM images of (**c**) purified cellulose fibers and (**d**) freeze-dried hydrogel. (Source: Xiaofang et al. (2014) copyright © 2014 with permission from Elsevier B.V.)

10.6.14 Cellulose Hydrogels Consisting of Primary Rare-Earth Compounds

Cellulose hydrogels embedded with lanthanide nanoparticles are developed from a low temperature reaction between aqueous solutions of alkali hydroxide and urea. Concentrations of the lanthanide luminescence correlate with the generation of red, green, and blue colors developed at different ratios. This fluorescence emission from the formulation is used for bio imaging and fluoroimmunoassay (Qiyang et al. 2017)

10.6.15 Biocompatible Cellulose-Based Superabsorbent Hydrogels

A new cellulose hydrogel is developed using the reaction between quaternized cellulose and pure cellulose in the aqueous solution of sodium hydroxide and urea. The formulation observed with greater absorbance, strength to resist, biological compatibility, and a proper bactericidal effect against *S. cerevisiae*. The presence of quaternized cellulose shows good bactericidal activity by the interaction between the anionic membrane and internalized pores of polycationic formulation (Fig. 10.12 and Table 10.6). These hydrogels are used in hygienic disposable diapers (Na et al. 2017).

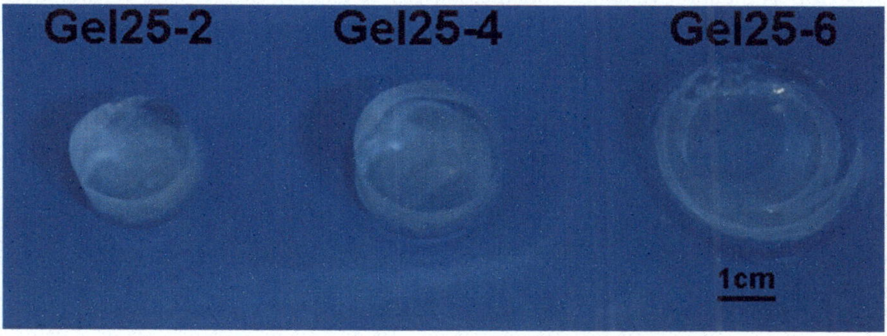

Fig. 10.12 Photographs of QC/cellulose hydrogels after swelling equilibrium in distilled water. (Source: Na et al. (2017) copyright © 2017 with permission from Elsevier B.V.)

Table 10.6 Conditions for the preparation of QC/cellulose hydrogels

Code	QC		Cellulose Mw ($\times10^{-4}$)	Weight ratio QC/cellulose	Swelling cellulose g/g
	Mw ($\times10^{-4}$)	DS			
Gel9–2	9.4	0.23	9.4	9:1	206.6
Gel9–4	9.4	0.42	9.4	9:1	337.3
Gel9–6	9.4	0.61	9.4	9:1	433.9
Gel25–2	25.6	0.28	9.4	9:1	247.1
Gel25–4	25.6	0.44	9.4	9:1	607.5
Gel25–6	25.6	0.69	9.4	9:1	983.9

Source: Na et al. (2017) copyright © 2017 with permission from Elsevier B.V.

10.6.16 Microwave Assisted Cellulose Hydrogel

Microwave irradiation is used to create acidic hydrolysis of a cellulose derived hydrogel, which is developed by dissolute cellulose within an aqueous solution of sodium hydroxide and urea. This method is useful to obtain glucose from cellulose; micron-sized crystalline cellulose and absorbable cotton. At 160 °C, hydrogel hydrolysis for 10 min produces an acidic concentration of glucose accelerated from (0.42–44.6) %. After the treatment with ozone, formulation significantly increases the yield. Furthermore, the formulation easily accesses the β-glycosidic bonds (Binzhe et al. 2015).

10.6.17 Bacterial Cellulose Hydrogel from G. Xylinum

Bacterial cellulose hydrogels are biosynthesized from *Gluconacetobacter xylinum* (ATCC53582), cultured in a Hestrin and Schramm (HS) medium, and incubated for 7 days at 30 °C. Time-dependent behavior of formulation is computed through compression–relaxation tests. Micro-morphological analysis illustrates structure–function relationships with the reflection of two different processes as movement of water and aggregation of different layers. Outcomes correlate with the fraction exponential analytical model, which shows sensitive and flexible time dependent behavior. The data reveals the thickness stress–relaxation behavior of formulation (Xing et al. 2016).

10.6.18 Supermolecular Hydroxyethyl Cellulose Hydrogel

Supermolecular hydroxyethyl cellulose hydrogels are developed by the reaction between the lauryl group embedded on hydroxyethyl cellulose and poly-β-cyclodextrin. The fixation concentrations of hydroxyethyl cellulose and poly-β-cyclodextrin are 30 mg mL^{-1}, with good viscosity, rheological, and swelling behavior. The synthesized samples are characterized by FT-IR, ^{1}H- ^{1}H NMR, SEM, and gel permeation chromatography studies. Encapsulation capacity of formulation is 21.89 wt%, as determined by the phenolphthalein probe method. Loading and in vitro release of Eugenol (EG) are investigated as a model. Bacteriostasis against *E. coli* is quantified by the agar diffusion method. The formulation possesses good bacteriostatic activity for biomedical applications (Nan et al. 2017).

10.6.19 Pineapple Peel Cellulose Hydrogels

Sepiaink fused with pineapple peel cellulose are used to prepare novel composite hydrogels in an ionic (1-butyl-3-methylimidazolium chloride) liquid. FTIR, field emission SEM, XRD, TG analysis, and DSC are used to evaluate the chemistry of formulations. The effectivity is evaluated by time and the temperature of acetylation reaction, presence of acetic anhydride within anhydroglucose unit, and adsorptivity of sepia ink on methylene blue, maintaining the pseudo-second order kinetic model, whereas the adsorptivity is increased from 53.72 to 138.25 mg/g (Hongjie and Huihua 2017).

10.6.20 Cellulose Acetate and EDTA Hydrogel

Biodegradable hydrogels as a substrate of sodium, potassium, and phosphorous in soil are developed from the acetate derivative of cellulose with a greater degree of substitution (2.5), and ethylene diaminetetraacetic dianhydride is catalyzed by triethylamine. FT-IR spectroscopy and thermogravimetric analysis are used to evaluate the ester forming reaction between cellulose acetate and ethylenediaminetetraacetic dianhydride. The biodegradation is examined in soil composed of 23% sand, 23% cattle manure, 23% soil, and 31% water. The reducing effect of the formulation on leaching of fertilizers is used to grow eucalyptus young plant sporophytes with lower mortality. The formulation acts as a biodegradable, nontoxic fertilizer (Andre and Vagner 2017).

10.6.21 Cellulose Hydrogel Films from Ionic Liquid

Cellulose-based hydrogel film is developed by dissolving a micron-sized crystal of cellulose in acetate and a chloride derivative of 1-butyl-3-methylimidazolium acetate, and the optical, resist of change, and adsorptivity are evaluated. In mild conditions, both formulations are clear, whereas the acetate derivative of formulation shows greater tension and elongates at a breakpoint of 9.5 and 2.5 times, respectively, compared with the chloride derivative of formulation. The spongy porous surface of acetate formulation is observed with greater absorptivity, whereas chloride formulation is the best substance for freezing pyronin B, acting as an optical indicator to detect the biologically active one (Svetlana et al. 2016).

10.6.22 Carboxymethyl Cellulose Hydrogel Beads from Anionic Surfactant

Hydrogel beads are developed by carboxymethyl cellulose using an aluminum ion directed gelation process with or without sodium dodecyl sulfate and sodium chloride. The formation of hydrogels is characterized by Fourier-transform-infrared spectroscopy and energy dispersive X-ray spectroscopy. The porous structure of the formulation is evaluated by SEM. The addition of sodium dodecyl sulphate maximizes the swelling nature. The adsorption of methylene blue within the structure is associated with the porosity of the formulation. The formulations show a pseudo-second order kinetic model with 82 mg g^{-1} adsorptivity. Formulations are displayed with a sudden increase in removal, with a maximum value of 350 mg g^{-1}

correlated with the entrapped dye. Therefore, the optimized hydrogels promise effective adsorption and regeneration as low cost dye adsorbents (Benhalima et al. 2017).

10.6.23 Maleimide-Grafted Cellulose Nanocrystals

Bionanocomposite hydrogels are prepared using the Diels-Alder "click" reaction and developed from gelatin (furan modified) using maleimide-fused cellulose nanocrystals. The X-ray photoelectron spectroscopic technique is used to evaluate the formulation. Swelling and rheology are the main parameters used to evaluate the hydrogel. The Diels-Alder "click" reaction helps form a strict network with less swelling behavior due to additional cross-linkage points. This formulation is authorized for formation of renewable nanocomposite hydrogels (Garcia et al. 2016).

10.6.24 Cellulose Composite Hydrogels with Graphene Oxide

Cellulose fused with graphene composite hydrogels are obtained from wood pulp and reduced graphene oxide using vitamin C as reducing agent. Mechanical and thermal withstand is enhanced in the composite hydrogel and the mechanical behavior of formulation directly corresponds to reduced graphene oxide. Young's modulus has over four times higher compressive strength in cellulose (Fig. 10.13 and Scheme 10.1). This method shows wide applicability in the biology and energy related fields (Manman et al. 2015).

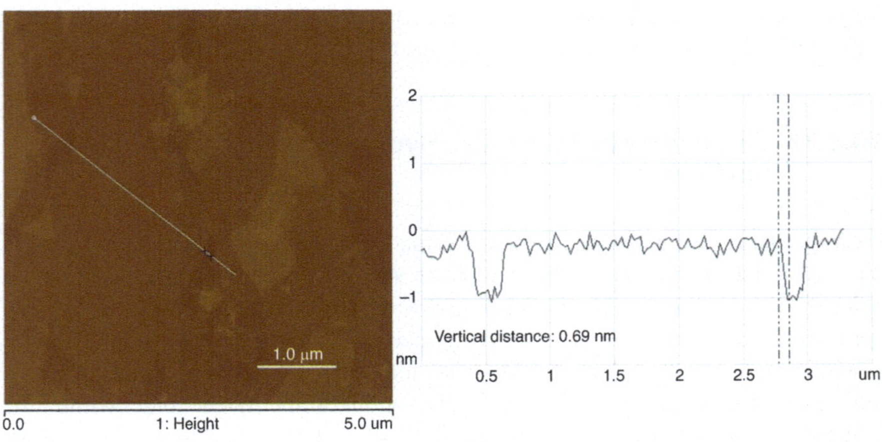

Fig. 10.13 Tapping mode AFM topographic images and height profiles of a single layer GO. (Source: Manman et al. (2015) copyright © 2015 with permission from Elsevier B.V.)

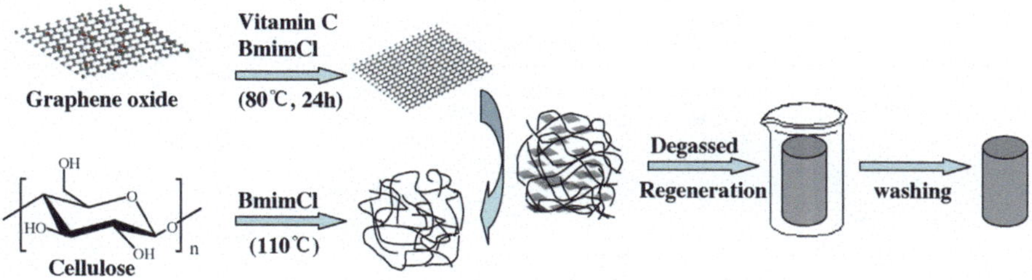

Scheme 10.1 Schematic diagram of the green synthesis of CGH. (Source: Manman et al. (2015) copyright © 2015 with permission from Elsevier B.V.)

10.6.25 Ionic Liquid-Based Cellulose Hydrogel

Conductive hydrogels are developed with microcrystalline cellulose and polypyrrole in ionic liquid and are characterized by Fourier transform-IR, SEM, X-ray diffraction, and thermogravimetric analysis. When amalgamated with toluenesulphonyloxy sodium, the formulations exhibit proper electrical conduction. Cellulose content decreases the swelling behavior with increasing methylcellulose concentration of the hydrogels. Furthermore, the composite formulations are observed with significant mechanical strength (Xiangtao et al. 2015).

10.6.26 Lignosulfonate Embeded Polyacrylic Acid-Hydroxyethyl Cellulose Hydrogels

Semi-interpenetrating hydrogels are prepared by hydrogen bonding using lignosulfonate (waste product) embedded in the chain of polyacrylic acid and a hydroxyethyl cellulose polymer. Shear stress, compression, and elasticity are evaluated. Hydrogels return to their original shape after removal of compression stress. The porous nature of the hydrogel shows good swelling behavior, which also accelerates absorption and removal of dyes (Jiaojiao et al. 2017).

10.6.27 Polyethylen Glycol Cellulose Hydrogels

Cellulose hydrogels with cations are obtained from the reaction between quaternized cellulose and polyethylene glycol diglycidyl ether. Hydrogels show a good swelling profile in aqueous medium, independent of the solution pH and temperature. Adsorption of anionic dyes increases with gradual concentration of degree of substitution with the pseudo-second order Langmuir adsorption model (Fig. 10.14). Also, they are used as adsorbents for cleaning of waste water (Hiroyuki et al. 2016).

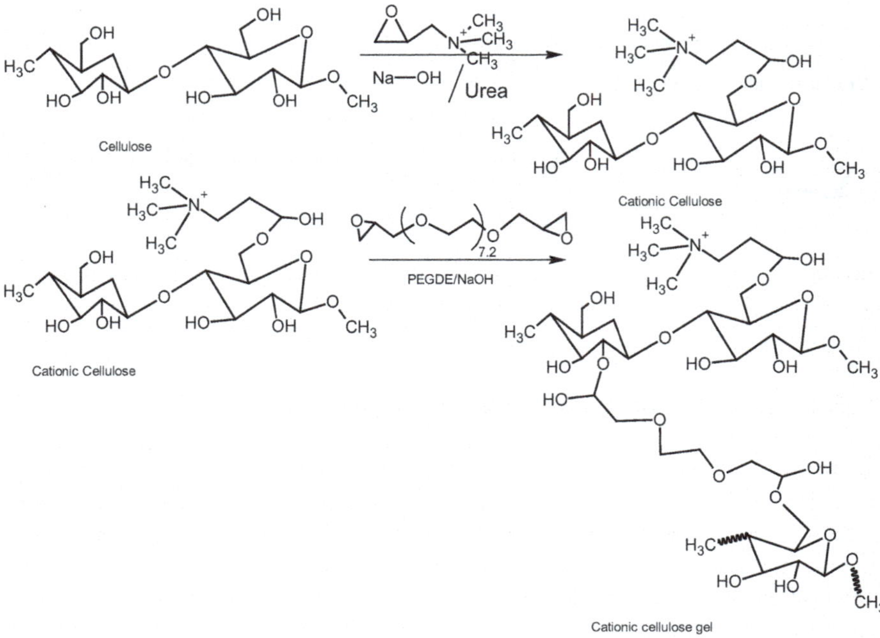

Fig. 10.14 Top: Preparation of CCGs from cellulose by a two-step reaction. (Source: Hiroyuki et al. (2016) copyright © 2016 with permission from Elsevier B.V.)

10.6.28 Deuterated Cellulose Composite Hydrogels

Deuterated cellulose composite hydrogels are developed from the reaction between arabinoxylan, xyloglucan, and mixed linkage glucans and evaluated by small angle X-ray scattering, small angle neutron scattering, X-ray powder diffraction, and NMR spectroscopy. The results indicate the ribbon surface of cellulose-arabinoxylan interactions occur via the adsorption process, whereas xyloglucan and mixed linked glucan interact with cellulose by (i) interfibrillar domain interactions between interfibril and cellulose microfibrils and (ii) surface area. Xyloglucan is crystallized with cellulose by formation of interleukin beta rich microfibrils. Furthermore, MLG interacts with cellulose to form a paracrystalline structure. XG and MLG directly play an important role in the cell growth (Marta et al. 2017).

10.6.29 Cellulose–Lignin Hydrogel

Cellulose embedded with lignin hydrogel beads are developed by solubilizing cellulose and lignin in 1-ethyl-3-methylimidazolium acetate (in water). The lipase enzyme procured from *Candida rugosa* is frozen on cellulose–lignin beads, which reveals that lipase frozen beads show more activity and stability than the enzyme frozen on cellulose beads. Increasing the concentration of lignin in formulation directly correlates with an increase in action and stabilization of frozen lipase. The activity of hydrogel, loading of protein, and site specific activity of frozen lipase on maximum cellulose–lignin formulations are 2.6, 2.2, and 1.2, respectively, times higher than lipase frozen cellulose beads. The impact of lignin is regulated by the property of cellulose–lignin formulation. After incubation for 12 h at 40 °C, the activity of cellulose–lignin formulation is 3.2 and 1.9 times higher than free lipase and lipase frozen on cellulose beads, respectively. At pH 3.0, $t_{1/2}$ of lipase frozen on cellulose–lignin beads is 24, which is three times higher than cellulose beads. Outcomes show that cellulose–lignin formulation could have great applications for mankind (Saerom et al. 2015).

10.6.30 Carboxymethyl Cellulose Embedded Graphene Oxide Hydrogel

The biodegradable carboxymethyl cellulose embedded with graphene oxide nanocrystalline composition of hydrogels are developed using a crosslinking reaction with ferric chloride for the controlled release behavior of doxorubicin. The π-π interactions between doxorubicin and graphene oxide result in greater drug loading and proper controlled release behavior of doxorubicin loaded from carboxymethyl cellulose and graphene oxide hydrogels. At pH 6.8 and 7.4, the release characteristic of doxorubicin from pH dependent formulations and interactions between graphene oxide and doxorubicin with hydrogen bonding are unstable under acidic conditions with greater release (pH 6.8). The graphene oxide embedded nanocomposite in formulations is characterized by X-ray powder diffraction and morphology is confirmed by FT-IR, SEM, and TEM. Furthermore, the swelling characteristic of the hydrogels is identified in a phosphate buffer (Monireh and Hassan 2017).

10.6.31 Amoxicillin Embeded Carboxymethyl Cellulose Hydrogel

The carboxymethyl cellulose hydrogel combined with polymers of lactic acid and itaconic acid is prepared by microwave radiation grafting between N,N^1-methylene-bis-acrylamide and potassium persulphate as the crosslinking agent and starter, respectively. Variable parameters are used to optimize yield of reaction. FT-IR, field emission SEM, X-ray powder diffraction, thermogravimetric analysis, and TEM are used to characterize the formation of hydrogel. Amoxycilin as a model drug is embedded in hydrogels, and its antimicrobial activities are evaluated against *Staphylococcus aureus* and *Escherichia coli*; outcomes report about 95% bactericidal activity is observed after the first 24 hours. An amoxicillin controlled drug delivery system from formulation is tested as a function of the media pH and time for total release, with the highest release of 98% at pH 2.2 after seven hours with non-Fickian diffusion characteristics (Sood et al. 2017).

10.6.32 Pineapple Peel Carboxymethyl Cellulose Hydrogel

New hydrogel formulations are obtained from pineapple (peel), carboxymethyl cellulose, polyvinyl alcohol (PVA), and mesoporous silica (SBA-15) with sudden freezing and sudden warming to prepare the papain frozen carrier system. The concentration of enzyme, system pH, concentration of crosslinking agent, and time of crosslinking are maximized, which shows optimum activity at 40 °C and pH responsive activity within a narrow range from pH 7.0 to 7.5. Frozen papain has a higher pH value and more thermo-responsive storage than free papain. At 80 °C, incubation for 2 h of frozen papain is contained with 56% of the primary activity state; apparently, free papain is contained with 16%. After ten days of immobilization, freeze-thawed papain is observed with 79% and free papain retains 27% of its initial activity (Hongjie et al. 2017).

10.6.33 Hyaluronic Acid Cellulose Hydrogels for Transdermal Delivery System

Novel hydrogel systems are prepared by hyaluronic acid–hydroxyethyl cellulose to investigate a transdermal delivery system of isoliquiritigenin. SEM and surface area analyzer (SAA) are used to characterize reaction and networking features of hydrogels. The concentration of cross-linker and swelling medium regulates the morphology and swelling behavior of hyaluronic acid–hydroxyethyl cellulose hydrogels. Hyaluronic acid-hydroxyethyl cellulose hydrogels exist in the anhydrous form. Rheological studies are used to characterize both viscous and elastic properties and stabilization of hydrogels, and they also follow the Fickian mechanism of diffusion. In vitro skin permeation suggests that formulation is observed with proper delivery of isoliquiritigenin into the skin. Outcomes indicate that the formulation composed of hyaluronic acid and hydroxyethyl cellulose behaves as a potential transdermal delivery system (Bong et al. 2016).

10.6.34 Bentonite Carboxymethyl Cellulose Hydrogel

Composite hydrogels are obtained from the reaction between carboxymethyl cellulose and citric acid in the presence of bentonite to potentiate the release of thiamethoxam by an ex situ encapsulation process. NMR, FT-IR, X-ray powder diffraction, and SEM-energy dispersive X-ray spectroscopy techniques are used to characterize the formulations. HPLC is used to evaluate the release profile of thiamethoxam from formulations within pH 7 to 11, and chemical kinetics are obtained by the Gallagher-Corrigan equation, which reveals that maximum release is within alkaline to neutral pH. These triggered release formulations of thiamethoxam are very important as pesticides when alkaline pH is present in the gut (Sarkar and Singh 2017).

10.6.35 Cellulose-Based Double-Network Hydrogels

Novel cellulose-based double-network hydrogels are prepared by the ultraviolet ray mediated copolymerization of polyacrylic acid intercalized quaternized cellulose (QCE) and polyvinyl alcohol as a bactericidal double networking hydrogel system with greater withstand strength and self-cure behavior. QCE is used as an antibacterial agent to protect from microbial growth in nature. The tension and shear stress of the hydrogel at break point are 465.37% and 1.13 MPa, respectively, with the polyvinyl alcohol concentration and QCE contents of 8% and 1.5%. The formulation contains prominent self-cure properties due to formation of an ionic bond between the Fe^{3+} ion and –COOH group and H-bonds form between the polyvinyl alcohol molecules. The formulation shows proper response against change in pH value and water retaining capacity. These hydrogels are applied in various biomedical fields (Yixi et al. 2017).

10.6.36 Hydrogel Obtained from Bamboo Shoot and Cyclodextrin

Carboxymethyl cellulose hydrogels are obtained from cellulose (procured from bamboo shoot) cross-linked with cyclodextrin using epichlorohydrin as the crossing agent. FT-IR spectroscopy, thermogravimetric analysis, and SEM are used to characterize the proper structure of hydrogel. The pH of the solution, temperature, and ionic strength of the media are correlated with the hydrogel sensitivity. With a minimal temperature and maximum pH value, the prepared formulation displays high swelling behavior of 23,338% at 15 °C and 6937% at pH 8.0, respectively. In a 0.1 M sodium chloride solution, the formulation has the maximum water retention rate of 48.73%. The behavior of hydrogel adsorption is studied by sodium salicylate as the model drug and release occurred at the intestinal pH of 7.4 and gastric pH of 1.8. The formulation shows greater drug release of 63.09% after 380 min in intestinal fluid and 22.09% after 400 min in gastric juice (Figs. 10.15 and 10.16). Therefore, the formulation exhibits pH responsiveness, mainly as a control release carrier of drugs (Shumin et al. 2016).

10.6.37 Hydrogel Derived from Polymethacrylic Acid, Carboxymethyl Cellulose with Gold Nanoparticles

The hydrogel is derived from in situ intercalation of gold nanoparticles amalgamated with carboxymethyl cellulose chemically reacted with polymethacrylic acid to potentiate skin permeability and enhance the drug loading efficacy for transdermal drugs. Variations in reaction factors and grades of cross-linked hydrogels are formed and optimized on the basis of lower swelling characteristic and higher crosslinking density. FT-IR and NMR spectroscopy, filed emission SEM, X-ray powder diffraction, thermogravimetric analysis, ultraviolet–visible diffuse reflectance study, atomic force microscopy, and TEM analysis are used to characterize the composite hydrogel.

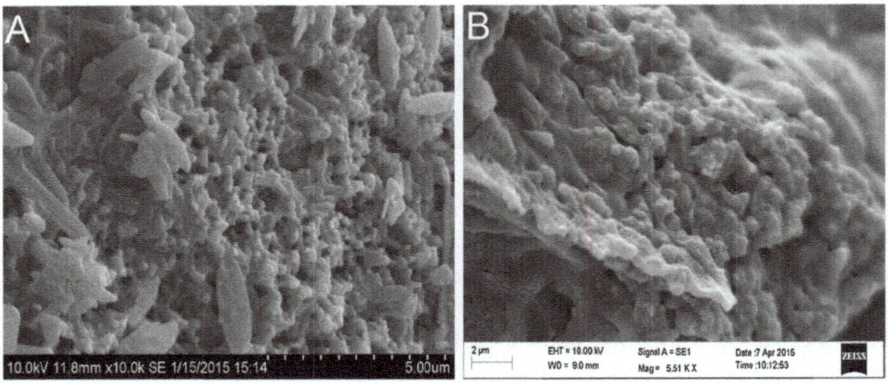

Fig. 10.15 SEM images of composite hydrogel prepared at 7:3 ratio of modified bamboo shoot cellulose/β-CD. (**a**) Prior to saturation of sodium salicylate. (**b**) After saturation of sodium salicylate. (Source: Sumin et al. copyright © 2016 with permission from Elsevier B.V.)

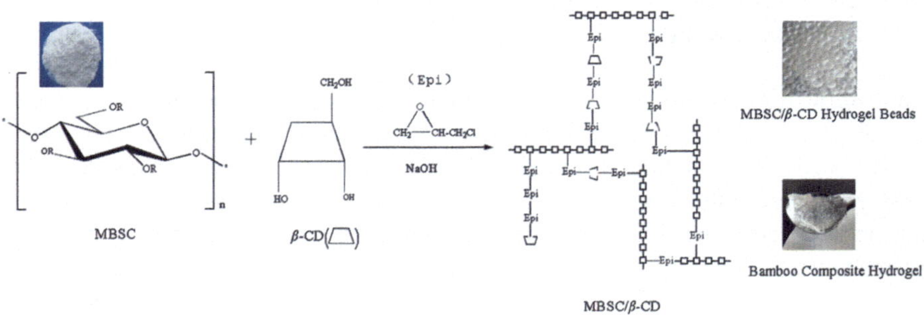

Fig. 10.16 Synthesis of the bamboo composite hydrogels. [Source: Sumin et al. copyright © 2016 with permission from Elsevier B.V.]

A rheological study is performed to examine the gel strength. Human mesenchymal stem cells are used to evaluate the cytotoxic behavior of the hydrogel, which confirms the non-cytotoxic nature. Diltiazem hydrochloride and diclofenac sodium are used as therapeutic agents to perform in vitro release of drugs and in situ intercalation of gold nanoparticles on cross-linked carboxymethyl cellulose, which is observed with proper release of drugs in a controlled way, such as 85% for diltiazem and 79% for diclofenac sodium are released in a span of three days (Mandal et al. 2017).

10.6.38 Carboxymethyl Cellulose Hydrogels with Polyethylene Glycol

Novel carboxymethyl cellulose hydrogels are developed from the chemical reaction between sodium carboxymethyl cellulose and polyethylene glycol diglycidyl ether. FT-IR and solid state NMR are used to characterize the structures of the formulation. Crosslinking degree is increased with increasing feed ratio of sodium carboxymethyl cellulose and polyethylene glycol diglycidyl ether, which is increased strength of hydrogel. The formulation shows enzyme degradability, but after three days of incubation with the cellulose enzyme, the carboxymethyl cellulose content is decreased from 62 to 28 wt%. Furthermore, the formulation is observed with optimum adsorption of protein adsorption with release ability and release pattern depends on the size of protein and degree of crosslinking in the hydrogels (Fig. 10.17). Therefore, carboxymethyl cellulose-based hydrogels perform as a delivery system for protein-based drug molecules (Hiroyuki 2014).

10.6.39 Carboxymethyl Cellulose/ZnO Nanocomposite Hydrogels

Carboxymethyl cellulose-zinc oxide nanocomposite hydrogels are formed by zinc oxide nanoparticles embedded within a swollen carboxymethyl cellulose matrix. X-ray powder diffraction, ultraviolet–visible spectroscopy, and SEM studies are

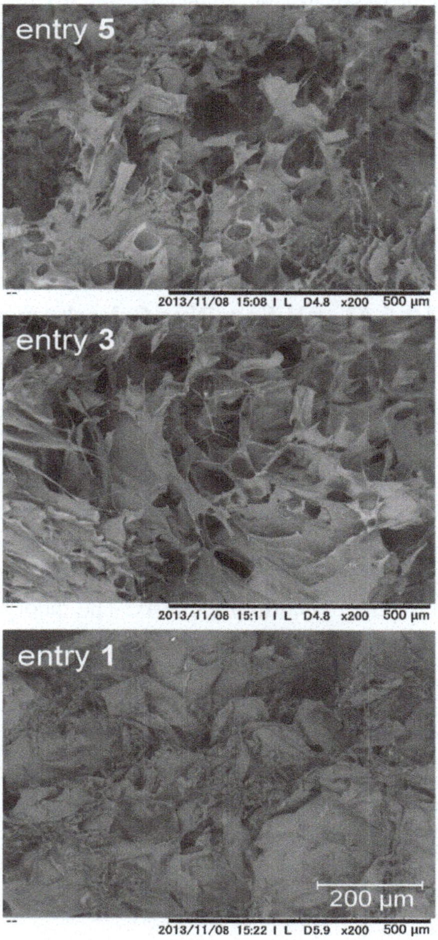

Fig. 10.17 SEM images of a cross-section of CMC hydrogels (entries 1, 3, and 5). (Source: Hiroyuki et al. (2016) copyright © 2014 with permission from Elsevier B.V.)

vused to characterize the formation of ZnO nanoparticles in the hydrogels. ZnO nanoparticles are formed within the size range of 10 nm to 20 nm in the matrix as revealed by SEM micrographs. The swelling characteristic of hydrogels is sensitive to pH and the amount of salt in the medium. The ZnO nanocomposite hydrogels possess greater swelling behavior in different aqueous mediums. The nanocomposite matrixes are observed with good bactericidal effects against *E. coli* and *S. aureus* (Mehdi et al. 2015).

10.6.40 Superabsorbent Cellulosic Hydrogels

Superabsorbent cellulosic hydrogels are prepared by crosslinking of softwood kraft fibers with polyvinylmethylether-co-maleic acid and polyethylene glycol. Superabsorbent cellulosic hydrogels are characterized by fiber length, time during crosslinking, and the amount of polyvinylmethylether-co-maleic acid on water absorption and retention value. The results show that decreased softwood fiber length from 2.41 to 0.50 mm along with weight ratio of fiber to polymers is equivalent to 1.00:1.28, subsequently increasing the water absorption and retention value from 86.50 to 189.20 g/g. The cross-linked fibers are analyzed by a scanning electron and light microscope, which reveals that the polymers and fibers form a fibrous hydrogel. The esteric linkage between PVMEMA and PEG/SW kraft pulp fibers is characterized by FT-IR spectroscopy. The outcomes show ester cross-linked pulps with good water absorption, and it is concluded that milled bleached softwood kraft fibers are utilized as refiner dust or pulp fines for novel water absorbent applicability (Shaobo and Ragauskas 2012).

10.6.41 pH-Sensitive Caborboxymethyl Cellulose–Zinc Oxide Nanoparticle

Biologically feasible nanocomposite hydrogels are developed on the basis of carboxymethyl cellulose and zinc oxide nanoparticles for the application of controlled release drug delivery. The Fe (III) ion is used as a cross-linker to develop a biologically formed nanocomposite hydrogel matrix. Propranolol hydrochloride is used as a therapeutic agent. X-ray powder diffraction, FTIR, thermogravimetric assay, and SEM are used to characterize intercalation of zinc oxide nanoparticles into a carboxymethyl cellulose matrix. The incorporation efficiency of propranolol is evaluated by UV–visible spectroscopy. Furthermore, the prepared nanocomposite beads are observed with a pH sensitive swelling behavior and optimum water absorbing property at pH 7.4. The swelling ratio of zinc oxide–carboxymethyl cellulose hydrogels in different aqueous solutions is higher than the pure hydrogel. In vitro drug release data show sustained and controlled release profiles for zinc oxide nanoparticles embedded with sodium carboxymethyl cellulose bead, with increased concentration of zinc oxide nanoparticles (Zhila et al. 2016).

10.6.42 Sugar Cane Bagasse-Based Composite Hydrogels

Sugarcane bagasse cellulose and gelatin incorporated into the polyacrylic network (in situ process) are used to develop composite hydrogels by the free radical polymerization technique. FT-IR, X-ray powder diffraction, thermogravimetric assay, SEM, and measurement of pH at null point are used to characterize the hydrogels. The absorption of Cu^{2+} ions from water by hydrogels is recorded by the response surface methodology with the Box-Behnken design (Maity and Ray 2017).

10.6.43 Cellulose Derived Nanocrystals from Rice Husks

Hydrogel sensitivity toward slight pH change is developed by reinforcing gelatin with cellulose nanocrystals, where glutaraldehyde is used as a crosslinking agent due to the presence of the amino group on gelatin. The effects of cellulose nanocrystals on the frequent withstand properties and swelling behavior of the hydrogel matrix are quantified by cellulose nanocrystal ratios of 0%, 5%, 10%, 15%, 20%, and 25%. The FT-IR spectrum shows stretching near 1630 $cm^{-}1$ due to the reaction between gelatins within the formulation. The crystalline character and mechanical properties of hydrogel networks are increased with the higher cellulose nanocrystal concentration. The total crystalline nature improves the storage modulus of the cellulose nanocrystal–gelatin hydrogel from 122 Pa to 468 Pa with the addition of 25% cellulose nanocrystal. At pH 3, the maximum swelling ratio of cellulose nanocrystal gelatin hydrogels is observed with excellent sensitivity toward pH change. Cellulose nanocrystal–gelatin hydrogels are highly responsive toward pH changes, and drug release behavior is evaluated by theophylline. The outcomes suggest that gelatin hydrogels composed of 15% cellulose nanocrystal are the suitable formulation for controlled drug delivery systems (Shok et al. 2016).

10.6.44 Technetium Labeled Cellulose Hydrogel

Technetium-99 m nanofiber-fused cellulose is prepared as an injectable hydrogel. After subcutaneous injection in mice, drug release and distribution is studied with a single photon emission computed tomography/computed tomography imaging device. WinNonlin, 1-compartmental model is used to simulate drug release profiles. The nanofibrillar cellulose hydrogel remains static at the injection site during study as compared with saline solutions. The nanofibrillar cellulose hydrogel decreases the rate of elimination with technetium-99 m-labeled human serum albumin by two times, but the release rate of a small compound 123I-b-CIT (a cocaine analogue) remains constant. Exploring the interactions between nanofibrillar cellulose and compounds are possible with further investigation (Patrick et al. 2014).

10.6.45 Carboxymethyl Cellulose–Zinc Oxide Nanocomposite Hydrogel

Carboxymethyl cellulose–zinc oxide embeded MCM-41 nanocomposite hydrogels are developed via amalgamation of zinc oxide within mesoporous silica as a nanocarrier with carboxymethyl cellulose hydrogel, where the cross-linker citric acid is used to negate cytotoxicity occurred owing to conventional cross-linkers. X-ray powder diffraction, SEM, TEM, zeta potential, and UV–vis spectroscopy are used to evaluate nanocomposite hydrogels. Data from swelling and erosion tests reveal that carboxymethyl cellulose–zinc oxide nanocomposite hydrogels are particulated within the first hours of the test, where MCM-41 is used as a substrate for zinc oxide nanoparticles to overcome this problem, and it also shows great improvement in tensile strength by 12%, swelling by 100%, erosion by 53%, and gas permeability by 500 percent. Tetracycline is used to evaluate site directed delivery and antibacterial properties of the sample. Cell compatibility is observed by adipose tissue derived stem cells, and cell compatibility of nanocomposite hydrogel films are analyzed and the outcomes reveal higher compatibility. Therefore, outcomes reveal that the prepared hydrogel behaves as a promising wound healing material (Rasul and Hassan 2017).

10.6.46 Wheat Straw-Based Cellulose Hydrogel

A novel wheat straw-based semi-interpenetrating polymer network hydrogel is prepared by polymerization technique with the notion of the slow release characteristic of nitrogen and phosphorus from fertilizer. The swelling behavior and release of fertilizer are characterized by influences of particle size, salt solution ionic strength, and changes in the pH values. The sodium ion has a greater swelling and fertilizer releasing nature than potassium and calcium ions. Products can hold more water within pH range 6 to 9. Schott's second order swelling kinetics model is used to evaluate the variability of ionic strength and swelling behavior of the product. The diffusion characteristics are directly correlated with diffusion coefficients in the kinetic equation (Li et al. 2016).

10.6.47 Polyvinyl Alcohol Cellulose Hydrogels

Polyvinyl alcohol–cellulose nanowhisker (CNW) nanocomposite hydrogels are prepared by the freeze thawing technique as wound dressing materials. CNW is developed by acidic hydrolysis of commercially available crystalline microcellulose, and it is characterized by the size, shape, morphological, structural, and thermal properties of the network. Polyvinyl alchohol–cellulose nanowhisker nanocomposites are developed, and their morphology, thermo responsiveness, and chemical and physical properties are characterized. The addition of cellulose nanowhisker to the hydrogel allows control of the pore morphology of the samples. The samples are observed with greater transparency, high thermal stability, good withstanding properties, and an optimum water vapor transmission rate for wound dressing after cellulose nanowhisker intercalation within the polyvinyl alcohol network. Microbial invasion data reveals that the developed hydrogels act as a wall against microorganisms and can be used as wound dressing materials (Jimena et al. 2014).

10.6.48 Cellulose Hydrogels for Ocular Bandage

Cellulose-based hydrogels are developed by procuring cellulose from wooden pulp, cotton, and bacterial waste, which is then dissolved in LiCl/N,N-dimethylacetamide with excellent wound healing properties. Owing to its transparency, these materials are a suitable option for ocular applications. The amount of water present, optical transparency, and tensile strength are the primary characters to evaluate. The time of activation, dissolution, relative humidity (RH), and concentration of cellulose also have an impact on the properties of hydrogels. Time of overnight activation improves the optical transparency of the hydrogels from 77% to 97% at 550 nm wavelength, while the control cellulose concentration improves their tear strength by 200%. Avicel PH 101, Sigma-Aldrich microcrystalline cellulose 435,236, and bacterial cellulose types are selected for the biological compatibility tests (Patchan et al. 2013).

10.6.49 Hydrogel Used for Slow Release NPK Fertilizer and Water Retention Capacity

A hydrogel of cellulose fused with its acetate form is developed by reaction between the acetate derivative of cellulose and ethylenediamine tetraacetic dianhydride (catalyst triethylamine) in dimethylformamide (DMF) as solvent. Subsequent conversion (Q2) of unreacted carboxyl groups into sodium carboxylates occurs with the addition of aqueous sodium bicarbonate solution to increase the water loving property of hydrogels. The hydrogel is submerged in the aqueous solution containing potassium chloride (2.5% K^+) and ammonium dihydrogen phosphate (NH_4^+ 2.5% and 13.5% $H_2PO_4^-$) at normal temperature for a time span of 2 h. The outcomes reveal that the product contains 120.5 mg of potassium ion/g, 104.3 mg of phosphate ion/g, and 84.0 mg of ammonium ion/g. The prepared sample shows a slow release characteristic but is also able to absorb a huge amount of H_20 and preserve the soil moisture at the same time. The outcomes reveal that the hydrogel network should be used in farming crops, flowers, and vegetables, mainly in drought hit zones (Andre et al. 2015).

10.6.50 Methyl Cellulose-Based Injectable Hydrogels

Thermosensitive injectable hydrogels of methylcellulose are developed by incorporating polyethylene glycol, carboxymethyl cellulose, and chitosan sulfate in a methylcellulose solution. The viscosity of the methylcellulose solutions depend on the amount of carboxymethyl cellulose in soil. Potassium ion concentrations regulate the gelation temperature to gain a methylcellulose-based solution that transformed into a gel-like structure at body temperature. The composition of the solution affects the gel strength. Addition of polyethylene glycol lowers the repulsive nature between the carboxymethyl cellulose and chitosan sulfate, which increases the gel strength. The activity of the methylcellulose-based injectable hydrogels is evaluated by a rat cecal abrasion model to reduce postsurgical adhesions and reduce adhesiolytic difficulties (Yongli et al. 2014).

10.6.51 Collagen/Cellulose Hydrogel Beads Reconstituted from Ionic Liquid Solution

Collagen embedded cellulose hydrogel beads are developed from a 1-butyl, 3-methylimidazolium chloride solution and adsorption of Cu (II) from aqueous solutions are considered. They have 3D porous macrocrystalline structure, in which amino groups act as binding sites for Cu^{2+} ions. The adsorption equilibrium capacity (qe) of collagen embedded cellulose hydrogel beads is highly influenced by the mass ratio of collagen and cellulose, and gradually increases until the ratio is above 2:1. The optimum adsorption is observed at pH 6. The (qe) value of Cu^{2+} ions increases with higher initial concentration of the solution. The Langmuir theory on isotherms suggests that the maximum adsorption capacity (qm) of collagen embedded cellulose hydrogel beads is 1.06 mmol/g, and it also maintained a good adsorption nature after the completion of the fourth adsorption–desorption cycle (Jilei et al. 2013).

10.6.52 Biodegradable Cellulose-Based Hydrogel Membranes with Antibacterial Property

Biodegradable hydrogels are developed using chloramphenicol and 2, 3-dialdehyde cellulose followed by characterization and evaluation of bacteriostatic activity. Bacterial cellulose is isolated from *Acetobacter xylinum* and modified into 2, 3-dialdehyde cellulose using the sodium metaperiodate oxidation process. The interaction between chloramphenicol–bacterial cellulose and chloramphenocol-2, 3-dialdehyde cellulose is evaluated by the attenuated total reflectance-FTIR technique. The water holding capacity of bacterial cellulose and 2, 3-dialdehyde cellulose membranes are 65.6% and 5.3%, respectively. HPLC analysis is used to analyze chloramphenicol release profiles. The drug-loading capacities of bacterial cellulose and 2, 3-dialdehyde cellulose membranes are 5 mg/cm^2 and 0.1 mg/cm^2, respectively. Membranes show a 99 to 99.5% rate of release of chloramphenicol within the first 24 h without causing any burst release effects. In vitro bacteriostatic evaluation of bacterial cellulose and 2, 3-dialdehyde cellulose, both chloramphenicol loaded matrixes, reveals their ability to inhibit bacterial growth over a prolonged duration. Activity against bacteria is still prominent after 3 days of incubation with disk diffusion tests. MTT assay reveals that fibroblast adhesion and proliferation of fibroblast on chloramphenicol loaded 2, 3-dialdehyde cellulose membranes are higher than on chloramphenicol loaded bacterial cellulose and 2, 3-dialdehyde cellulose membrane. 2, 3-dialdehyde cellulose membranes are highly effective as wound healers (Lacin 2014).

10.6.53 Fe^{3+} Crosslinked Alginate and Carboxymethyl Cellulose Hydrogel Beads

Sodium alginate embedded with sodium carboxymethyl cellulose hydrogel is prepared in a ferric chloride solution to exhibit pH sensitive deliver of metformin hydrochloride. The $FeCl_3$ solution regulates the encapsulation power, swelling behavior, and in vitro release kinetics of the beads. At 37 °C and pH 7.4, the beads showed greater swelling property than at pH 1.2. A scanning electron microscope is used to quantify external morphology, which suggests that the beads are spherical with a smooth outer line and bead size is significantly decreased with higher concentration of crosslinking agent. The crosslinking between sodium alginate and sodium carboxymethyl cellulose with ferric chloride is confirmed by FTIR. X-ray powder diffraction indicates that metformin is thoroughly contained in the polymer network. In vitro release data reveals that metformin loaded beads are observed with prominent release profiles at pH 7.4 compared to pH 1.2. The polymeric matrix has little deviation with the Fickian diffusion mechanism (Swamy and Yun 2015).

10.6.54 Carboxymethyl Cellulose-Based Hydrogel with Nanosilver Composite

Hydrogel composites with silver nanoparticles (AgNPs) are prepared via carboxymethyl cellulose, polyvinyl alcohol, and a crosslinking agent, such as ethylene glycol diglycidyl ether (EGDE), with the incorporation of hydrogel composite silver nanoparticles by the microwave radiation technique. The resulting hydrogels and hydrogel composite silver nanoparticles are characterized by spectroscopy, thermal studies, microscopic analysis, and X-ray powder diffraction analysis. Scanning electron and transmission electron microscopy outcomes reveal that synthesized hydrogel composite silver nanoparticles are spherical with 8 to 14 nm internal diameter. X-ray powder diffraction shows the presence of nanocrystals of silver with face-centered cubic structure. Energy dispersive spectroscopy confirms the presence of silver without impurities. The Kirby-Bauer method is used to evaluate the antibacterial activity of hydrogel composite silver nanoparticles against urinary tract infection pathogens. Rheological experiments reveal that the storage modulus has higher values than the loss modulus. The hydrogel composite silver nanoparticles show greater antibacterial activity against *E. coli*, *K. pneumoniae*, *P. aeruginosa*, *P. vulgaris*, *S. aureus*, and *P. mirabilis* compared to pure hydrogel (Alshehri et al. 2016).

10.6.55 Modified Bacterial Cellulose Hydrogel with Dextran

A novel dextran/bacterial cellulose (BC) hydrogel is prepared as a perfect wound healing material to effectively improve the healing process. Hydrogel morphology, thermostable nature, withstand properties, cell toxicity, cell proliferation, and healing of wounds are assessed for hydrogels. The results reveal that the dextran affects the networking of bacterial cellulose resulting in decreased temperature of decomposition from 339 °C to 261 °C, decreased water content from 98.7% to 89.2%, and decreased tensile strength from 23 MPa to 0.61 MPa. Rate of elongation is approximately 33% to 28% in bacterial cellulose, and 10% to 20% in dextran modified groups. Cell assisted experiments confirm that dextran modified bacterial cell hydrogels have increased cell proliferation without cell toxicity compared to natural bacterial cellulose. The in vivo wound repairing test illustrates that dextran–modified bacterial cellulose hydrogel can increase the wound repairing process and accelerate skin maturation, which suggests that dextran-bacterial cellulose hydrogels are promising dressing materials for medical application (Lin et al. 2017).

10.6.56 Cyclodextrin Embeded Gallic Acid Cellulose Hydrogel

A hydrogel composed of cyclodextrin is developed for the controlled release system of biomolecules as wound dressing materials. The hydrogel properties show a high amount of swelling and good encapsulation capacity of cyclodextrin. β-cyclodextrin or hydroxypropyl-β cyclodextrin is chemically crosslinked with hydroxypropyl methylcellulose under normal conditions. The swelling characteristics, FTIR, and differential scanning colorimetry are used to evaluate the hydrogels. Antibacterial activity and cytotoxicity, gallic acid loading, and release from the polymeric networks are also analyzed. The obtained hydrogels are strong and clear with proper swelling characteristic. The gel hydroxypropyl-β cyclodextrin had more hydrophilic surface. Both hydrogel networks are able to intercalate gallic acid. Their sustained release behavior is observed for 48 h, with efficient bacteriostatic activity after its adsorption within the polymeric matrix, and where gallic acid directly affects fibroblast growth. Therefore, gels amalgamated with β-cyclodextrin or hydroxypropyl-β cyclodextrin with the conjugation of gallic acid are confirmed as good antibacterial wound dressing materials (Pinho et al. 2014).

10.6.57 Carboxymethyl Cellulose with Double Hydroxide Layer as Hydrogel Beads

Novel nanobeads with a pH sensitive property are developed via amalgamation of layered double hydroxides and carboxymethyl cellulose by using ibuprofen as a model drug, which is interlinked between layered double hydroxides through the co-precipitation method. FTIR, XRD, and SEM are used to characterize the synthesized layered double hydroxides–ibuprofen nanohybrids and nanobeads. In vitro drug delivery tests in simulating conditions are used to check the potential of the nanocomposite beads as a controlled drug delivery system. The drug release profile is confirmed an efficient protection against gastric pH with controlled liberation of drug in the intestinal tract (Soroush et al. 2014).

10.6.58 A New Galangin Delicery System Prepared by Cyclosophoraose-Cellulose Hydrogels

A novel cyclophorase/cellulose hydrogel is prepared via a crosslinking reaction between cyclophorase and cellulose for efficient galangin delivery to facilitate the bactericidal activity, where cyclosophoraose (Cys) is a highly water-soluble complex forming agent, which increases the solubility of drug molecules that had minimum water solubility with encapsulation of molecule via host–guest complexation. Galangin is used directly against β-lactamase activity and to overcome the bacterial strain. The retardation of β-lactamase activity by galangin is used to potentiate the activity of the methicillin-resistant *S. aureus* strain. In the aqueous solution of galangin, an efficient delivery is observed with enhanced antibacterial activity due to its very poor solubility in water and poor bioavailability. Rheological parameters, FTIR, solid state nuclear magnetic resonance spectroscopy, X-ray powder diffraction pattern, and FE-SEM are used to evaluate the structure and morphology of hydrogels. The drug loading concentration of galangin is increased with higher cyclophorase concentration, which is released within 48 h. The loading and cumulative amounts of galangin release from the cyclophorase–cellulose hydrogel are 1.62 and 1.64 times greater than galangin release from the pure cellulose hydrogel, respectively. Furthermore, the galangin-loaded cyclophorase-cellulose hydrogel possesses bacteriostatic activity against *S. aureus* with maintained growth inhibition up to 72 h but did not present any cytotoxic behavior toward human dermal fibroblasts. Therefore, the cyclophorase–cellulose hydrogel is observed as a potential antibacterial dressing material for the delivery of galangin to site directed targets for wound healing in a sustained manner over a long period without frequent change in dressing material (Daham et al. 2016).

10.6.59 Carboxymethyl Cellulose Hydrogel Matrix

A carboxymethyl cellulose hydrogel matrix is obtained by ionic reaction between different carboxymethyl cellulose chains using iron, calcium salt as a root targeted delivery system, solubilized in water within liquid fertilizer. Root targeted delivery vehicle (RTDV) is used to develop a low cost controlled release device to deliver fertilizers to plant roots, increase efficient use of fertilizer, and evaluate fertilizer use on wheat growing in nutrient depleted soil media. RTDV permitted a 78% reduction in the amount of fertilizer used to achieve growing wheat in nutrient depleted media. Quantifying the losses associated with the RTDV synthesis shows an optimized manufacturing as high as 94%. Furthermore, the system is used for delivering formulation in soil for the plant's growth. These results illustrate that fertilizer absorption is facilitated and the efficiency of fertilizers in soil with a significant reduction of costs and environmental damage. Therefore, RTDV is used as an effective tool to increase fertilizer use efficiency in agriculture (Davidson et al. 2013).

10.6.60 Cellulose Nanofibers–Polyvinyl Alcohol Hydrogels with pH Sensitive and Controlled Release Pattern

Cellulose and polyvinyl alcohol are used to develop nanofiber hydrogels for the evaluation of the in vitro release characteristic of cisplatin with pH sensitivity and a controlled release system based on cellulose-containing nanofibers and polyvinyl alcohol. This novel delivery system is evaluated by Fourier-transform infrared spectroscopy X-ray powder diffraction, and SEM analytical techniques are used to characterize it. The content of the cellulose-containing nanofibers and the pH of the media affect the release characteristic of the drug. At pH 7.4, a hydrogel with 1 wt% cellulose-containing nanofibers in the release media sustained the drug release. The experimental data is best fitted with the Korsmeyer-Peppas kinetic model, which suggests that the controlled release follows the diffusion mechanism. The hydrogel network is used as the delivery system of cisplatin in the small intestinal region in a controlled manner (Azhar et al. 2017).

10.6.61 Nanotube Hydrogels Network Composed of Cellulose and Halloysite

A nanotube composite hydrogel made of cellulose and halloysite is prepared via introduction of halloysite nanotubes in a cellulose sodium hydroxide/urea solution to develop a hydrogel network with epichlorhydrine cross-linked at increased temperature. Cellulose–halloysite nanotube composite hydrogels are evaluated by shear stress viscosity, withstand properties, microcrystalline structure, swelling behavior, cell compatibility, and delivering system of drug molecules. Addition of halloysite nanotubes into the hydrogel increases the viscosity of the network. The strength of compression of the hydrogel network with 66.7 percent and halloysite nanotubes is 128 KPa; the cellulose only hydrogel is 29.8 KPa compressive power. The ability to withstand deformation is improved for this network as suggested by rheological measurement. X-ray diffract pattern and FTIR show unchanged crystalline nature and structure of halloysite nanotubes in the composite network. H-bond formation occurs between halloysite nanotubes and cellulose in the network. Scanning electron microscopy confirms the porosity of 200–400 μm in composite hydrogels. Swelling ratios in the sodium chloride solution and pure water are decreased by the addition of halloysite nanotubes for the composite hydrogels. Cellulose-fused halloysite nanotube hydrogels display a good biological feasibility with osteoblast precursor cell line (MC3T3-E1) and breast cancer cell line (MCF-7). Curcumin-fused formulation is developed by physical adsorption technique. The curcumin embedded hydrogels show an efficient inhibition on cancer cell lines. All the outcomes illustrate that the cellulose–halloysite nanotube composite hydrogels have good applications, such as delivery of anticancer drugs, anti-inflammatory agents, and effectiveness in wound dressings (Huang et al. 2017).

10.6.62 Thermosensitive Hydrogel of Hydrophobically-Modified Methylcellulose

A thermosensitive hydrogel of methylcellulose modified by stearic acid (MCS) is prepared by esterification of methylcellulose (MC) with stearic acid in dry DMSO using DCC and DMAP and characterized by ^{1}H NMR spectra. According to the ^{1}H NMR spectra, the prepared MCS contained 4.6 stearyl groups per chain, i.e., the stearyl groups occupy 2.1% (w/w) of the MCS. MCS is developed for intra-vaginal drug delivery to prevent sexual transmission of human immunodeficiency virus and other vaginal infections. Here, the formulation of stearic acid-fused methylcellulose is evaluated by sodium chloride and phosphates, with transition of solution to gel formation at 37 °C. In vitro cytotoxic assessment and in vivo irritation of mucosa layer are evaluated and the results show that the MCS hydrogel possesses good biocompatibility similar to the hydroxyethyl cellulose (HEC) gel. The release studies confirm that MCS hydrogel controlled tenofovir sustained release for 10 h without burst release, longer than that from the HEC gel or poloxamer 407 hydrogel. Therefore, this thermo-responsive formulation is used as a carrier for intra-vaginal delivery of antiviral drugs with controlled release behavior (Ning et al. 2012).

10.6.63 Hydroxypropylmethyl Cellulose Embeded Polyacrylamide Hydrogel

In the presence of potassium persulfate, a novel hydrogel is prepared via grafting polyacrylamide chains onto hydroxypropyl methylcellulose using the solution polymerization technique in homogeneous aqueous medium. The percentage and efficiency of grafting are used to characterize the effect of reaction parameters. The developed hydrogels are evaluated by intrinsic measurement of viscosity by a Ubbelohde type viscometer with 0.58 mm capillary diameter at 25 °C and justified by CHN, FTIR, ^{13}C NMR, and thermal studies. Efficient and equilibrium swelling behaviors of hydrogels are illustrated by pH and time in diversified buffer solutions similar to gastric juice and intestinal fluid. Outcomes reveal that increasing the grafting percent and percent grafting efficiency value decreased the rate of swelling. The hydrogels are utilized as matrices for controlled/sustained/targeted drug delivery (Das et al. 2012).

10.6.64 Hydrogel Microparticles Composed of Microwave Assisted Bacterial Cellulose

Novel hydrogel microparticles are prepared from bacteria derived cellulose and a polyacrylamide polymer as initiator to cover thick burns. Fourier-transform infrared spectroscopy, X-ray light diffraction, elemental analysis, and SEM are used to evaluate microparticulate carrier structure and surface morphology. L929 cells are used to evaluate the cytocompatibility of microparticles. The non skin irritant hydrogel has a prominent effect on contraction of wound compared to the control group. However, in vivo histological examination of the burn healing process reveals that samples show increased formation of epithelial tissue and enhanced fibroblast growth with wound healing, and new prominent skin is achieved after hydrogel treatment. The outcomes from in vitro and in vivo experiments show the significant biological compatibility and effectiveness of this network as a dressing material for burn wounds. (Pandey et al. 2016)

10.6.65 Carboxymethyl Cellulose Embeded Carboxymethyl Polyvinyl Alcohol Hydrogel

A novel smart hydrogel is prepared from carboxymethyl cellulose and a carboxymethyl polyvinyl alcohol polymer with adipic dihydrazide cross-linker. The smart hydrogel composed of 3% carboxymethyl cellulose and 1% carboxymethyl polyvinyl alcohol exerts a response to different conditions of media and exerts a high swelling capacity. At pH 1, 7, and 11, swelling percentages are 360, 1440, and 2277%, respectively, and the hydrogel can resist its shape. This smart copolymer hydrogel is pH dependent and utilized in water purification management, agricultural processes, and in drug delivering protocol. The network is biologically feasible with living cells, presenting a greater rate of survival by stepping the concentration of polymer, which is applicable in biomedical sciences, such as site directed drug delivery and formation of scaffolds in the tissue engineering business (Dahlan et al. 2017).

10.6.66 Cellulose-Based Hydrogel Network

This study focused on the synthesis of a cheap biologically degraded cellulose hydrogel to evaluate as a controlled release of fertilizer. To achieve the goal, modification of cellulose is done by reaction with urea. The formed network undergoes loading with fertilizer to evaluate the controlled release characteristic. Samples are analytically characterized by Fourier-transform infrared spectroscopy, analysis of elements, analysis of thermogravimetric effect, and by the scanning electron microscope technique. A justified swelling behavior of hydrogel is observed in distilled water, tap water, and a 0.9% sodium chloride solution. Furthermore, the water holding capacity and water retaining characteristic of the hydrogel are evaluated. Therefore, the release characteristic of fertilizer from loaded hydrogel is evaluated, and it exerts an excellent controlled release behavior. As per the outcomes, this network is a suitable option as a moisture retaining additive for soil improvement purposes (Khoo et al. 2016).

10.6.67 Cellulose Hydrogel Film Regenerated from Sugarcane Bagasse Waste

A cellulose hydrogel film is prepared from sugarcane bagasse waste, procured from Okinawa, Japan, and an in vivo experiment is conducted in mice. A cellulose-containing fiber is extracted from the dry pulp of sugarcane and solubilized in lithium chloride/N,N-dimethyl acetamide solvent mixture, which is embedded in the hydrogel film with a phase inverted process under ethanolic vapor. Hydrogel films are implanted in the intraperitoneal region of mice with a span of 4 weeks to evaluate biocompatibility; outcomes reveal that implanted hydrogel films have little impact on mice growth. As per the postmortem evaluation, no sign of inflammation is observed around the intraperitoneal region. In a span of 3 to 4 weeks, molecular weight decreased from 5.7×10^5 Dalton to 3.9×10^5 Dalton, which confirms that cellulose fragments are biologically degradable in nature. Viscoelasticity data show that hydrogel films assure their mechanical withstanding in the living body. Therefore, the biocompatible cellulose hydrogel films extracted from sugarcane bagasse waste show greater withstand properties in the body. (Nakasone et al. 2016)

10.6.68 Scaffold Hydrogels Prepared from *Agave Tequilana* Weber Bagasse with Fibroblast Compatibility

A cellulose hydrogel film with greater transparency and higher flexibility is developed using *Agave tequilana* Weber var. azul bagasse fibers as the tissue scaffolding agent and evaluated by NIH 3 T3 fibroblast cells assay. The phase inversion method is employed to obtain hydrogel film without using a crosslinking agent in a dimethylacetamide/lithium chloride system with (4–12) wt% of variable concentration.

Water contents are observed within the range of (239 to 323) % with film mechanics of (50–66) N/mm^2. Better cytocompatibility is observed in development of these films than the control polystyrene dish. Images obtained from the atomic force microscope illustrate that the hydrogel films possess lower lithium chloride, but clearly it contained an ordered and aggregated fiber alignment and influenced fibroblast cells spreading (Carrillo et al. 2013).

10.6.69 Interpenetrated Biohydrogel Composed of Hydroxyethyl Cellulose and Wooden Pulp

An interpenetrated hydroxyethyl cellulose biohydrogel film is prepared from wooden pulp, cellulose, and hydroxyethyl cellulose. A higher concentration of hydroxyethyl cellulose directly affects the withstanding properties of the hydrogel from 10 wt% to 50 wt% as per the content of cellulose. In vitro biocompatibility attributes, such as adsorption of protein, fibrin forming time, and platelet aggregation, are used to affect the increment of HEC. Better cytocompatibility is observed in the obtained hydrogels than the reference in a polystyrene tissue culture dish. Therefore, decreased content of hydroxyethyl cellulose is observed in the case of fibroblast cells and HEC content, which showed different pulp aggregation. Therefore, the cell concentration of hydrogel significantly depends upon the interpenetrating capacity of HEC (Carrillo et al. 2014).

10.6.70 Superabsorbent Nanocomposite Hydrogels of Cellulose and Clay

A nanocomposition of cellulose and clay is used to develop hydrogels via a chemical reaction between cellulose, carboxymethyl cellulose, and interjected clay in an aqueous solution of sodium hydroxide/urea with greater dye absorbing, withstand performance, and removal efficiency. As per the pseudo-second order and Langmuir model, hydrogels that show a greater absorption tendency for methylene blue (MB) solution undergo a voluntary physical absorption process. Hydrogels have 96.6% and 98.0% efficiencies at MB solution concentrations of 10 mg L^{-1} and 100 mg L^{-1}, respectively. Therefore, a nanocomposite hydrogel composed of cellulose and clay is an innovative technique platform for decontamination of dye (Peng et al. 2016).

10.7 Conclusions and Future Prospects

The detailed information in the area of cellulose hydrogels obtained from natural and bacterial source, derivatized natural hydrogels, and nanocomposites can be used for controlled release drug delivery targeting specific sites. Therefore, biomedical scientists are focusing on the naturally derived biodegradable hydrogels to develop the formulation of medicines for site directed and controlled release delivery systems as well as tissue and molecular engineering. Naturally derived polymers are widely used in the world due to their biodegradable and biocompatible nature as well as their abundances in the arena of a controlled drug delivering system. Full and semi-IPNs of cellulose hydrogels have a wide range of applicability, and when formed with clay, gallangin, gallic acid, silver, and chloramphenicol they possess various properties, such as antibacterial, antifungal, release characteristic of NPK fertilizer and retaining water in soil, transdermal delivery of isoliquiritigenin, and many more. In the future, more research is needed to explore the natural biodegradable polymers with their own activity profile, which can synergize the activity of active pharmaceutical ingredients.

References

Alshehri SM, Aldalbahi A, Abdullah BA, Chaudhary AA, Panhuis M, Alhokbany N, Ahamad M (2016) Development of carboxymethyl cellulose-based hydrogel and nanosilver composite as antimicrobial agents for UTI pathogens. Carbohydr Polym 138:229–236

Andre MS, Janaina BC, Silva JMS, Botaro VR (2015) Synthesis, characterization and application of hydrogel derived from cellulose acetate as a substrate for slow-release NPK fertilizer and water retention in soil. J Environ Chem Eng 3:996–1002

Andre MS, Vagner RB (2017) Biodegradable hydrogel derived from cellulose acetate and EDTA as a reduction substrate of leaching NPK compound fertilizer and water retention in soil. J Control Release 260:194–201

Azhar FF, Shahbazpour E, Olad A (2017) pH sensitive and controlled release system based on cellulose nanofibers-polyvinyl alcohol hydrogels for cisplatin delivery. Fiber Polym 18:416–423

Bang S, Ko YG, Kim WI, Cho D, Park WH, Kwon OH (2017) Preventing postoperative tissue adhesion using injectable carboxymethyl cellulose-pullulan hydrogels. Int J Biol Macromol 105:886–893

Benhalima T, Ferfera-Harrar H, Lerari D (2017) Optimization of carboxymethyl cellulose hydrogels beads generated by an anionic surfactant micelle templating for cationic dye uptake: swelling, sorption and reusability studies. Int J Biol Macromol 105:1025–1042

Binzhe S, Lian D, Gege P, Xiaoxia L, Aihua X (2015) Efficient production of glucose by microwave-assisted acid hydrolysis of cellulose hydrogel. Bioresour Technol 192:253–256

Bong J, Kong A, Kim S, Nam P (2016) Properties and *in vitro* drug release of hyaluronic acid-hydroxyethyl cellulose hydrogels for transdermal delivery of isoliquiritigenin. Carbohydr Polym 147:473–481

Capanema NSV, Mansur AAP, de Jesus AC, Carvalho SM, de Oliveira LC, Mansur HS (2018) Superabsorbent crosslinked carboxymethyl cellulose-PEG hydrogels for potential wound dressing applications. Int J Biol Macromol 106:1218–1234

Carrillo KLT, Sueyoshi SS, Tagaya M, Kobayashi T (2013) Fibroblast compatibility on scaffold hydrogels prepared from *Agave tequilana* weber bagasse for tissue regeneration. Ind Eng Chem Res 52:11607–11613

Carrillo KLT, Sueyoshi SS, Tagaya M, Kobayashi T (2014) Biohydrogels interpenetrated with hydroxyethyl cellulose and wooden pulp for biocompatible materials. Ind Eng Chem Res 53:4650–4659

Chang C, Zhang L (2011) Cellulose-based hydrogels: present status and application prospects. Carbohydr Polym 84:40–53

Daham J, Hyun K, Jeong J, Dindulkar SD, Cho E, Yung-Hun Yang YH, Seunho J (2016) Cyclosophoraose/cellulose hydrogels as an efficient delivery system for galangin, a hydrophobic antibacterial drug. Cellulose 23:2609–2625

Dahlan NA, Pushpamalar J, Veeramachineni AK, Muniyandy S (2017) Smart hydrogel of carboxymethyl cellulose grafted carboxymethyl polyvinyl alcohol and properties studied for future material applications. J Polym Environ 26:2061–2071

Das R, Panda AB, Pal S (2012) Synthesis and characterization of a novel polymeric hydrogel based on hydroxypropyl methyl cellulose grafted with polyacrylamide. Cellulose 19:933–945

Davidson DW, Verma MS, Gu FX (2013) Controlled root targeted delivery of fertilizer using an ionically crosslinked carboxymethyl cellulose hydrogel matrix. Springerplus 2:318

Garcia AC, Gonzalez TK, Gurrea OG, Algar AI, Eceiza NG (2016) Maleimide-grafted cellulose nanocrystals as cross-linkers for bio nanocomposite hydrogels. Carbohydr Polym 149:94–101

Hao Z, Qian L, Qingde H, Hu T, Fenghong H, Wenlin L, Chuyun W, Changsheng L, Jiqu X, Pingmei G, Qi Z (2017) A facile and efficient strategy for the fabrication of porous linseed gum/cellulose superabsorbent hydrogels for water conservation. Carbohydr Polym 157:1830–1836

Hebeish A, Farag S, Sharaf FS, Shaheen TI (2015) Radically new cellulose nanocomposite hydrogels: temperature and pH responsive characters. Int J Biol Macromol 81:356–361

Hiroyuki K (2014) Characterization and properties of carboxymethyl cellulose hydrogels crosslinked by polyethylene glycol. Carbohydr Polym 106:84–93

Hiroyuki K, Kota O, Ryo K, Kazuhiro O, Hisaho H, Yuuichi S (2016) Cationic cellulose hydrogels cross-linked by poly(ethylene glycol): preparation, molecular dynamics, and adsorption of anionic dyes. Carbohydr Polym 152:170–180

Hongjie D, Shiyi O, Zhijun L, Huihua H (2017) Pineapple peel carboxymethyl cellulose/polyvinyl alcohol/mesoporous silica SBA-15 hydrogel composites for papain immobilization. Carbohydr Polym 169:504–514

Hongjie D, Huihua H (2017) Modified pineapple peel cellulose hydrogels embedded with sepia ink for effective removal of methylene blue. Carbohydr Polym 148:1–10

Huang B, Mingxian L, Changren Z (2017) Cellulose–halloysite nanotube composite hydrogels for curcumin delivery. Cellulose 24:2861–2875

Jean PO, Graziella PB, Karina OL, Shanise LMEH, Gabriela SR, Alvaro RGD, Elessandra RZ (2017) Cellulose fibers extracted from rice and oat husks and their application in hydrogel. Food Chem 221:153–160

Jiaojiao Z, Kun Z, Jingya N, Chao T, Ying C, Yong H (2017) Synthesis and characterization of lignosulfonate-graft-poly (acrylic acid)/ hydroxyethyl cellulose semi-interpenetrating hydrogels. React Funct Polym 115:28–35

Jilei W, Ligang W, Yingchong M, Kunlan L, Minghui L, Yachen Y, Lei W, Hui Q (2013) Collagen/cellulose hydrogel beads reconstituted from ionic liquid solution for cu (II) adsorption. Carbohydr Polym 98:736–743

Jimena SG, Leandro NL, Alejandra P, Vera AA (2014) Poly(vinyl alcohol)/cellulose nanowhiskers nanocomposite hydrogels for potential wound dressings. Mater Sci Eng C 34:54–61

Khoo SM, Moghadam PN, Fareghi AR, Movagharnezhad N (2016) Synthesis of a cellulose-based hydrogel network: characterization and study of urea fertilizer slow release. J Appl Polym Sci 133:42935. https://doi.org/10.1002/app.42935

Lacin NT (2014) Development of biodegradable antibacterial cellulose based hydrogelmembranes for wound healing. Int J Biol Macromol 67:22–27

Li X, Li Q, Xu X, Su Y, Yue Q, Gao B (2016) Characterization, swelling and slow release properties of a new controlled release fertilizer based on wheat straw cellulose hydrogel. J Taiwan Inst Chem Eng 60:564–572

Lin SP, Kung HN, Tsai YS, Tseng TN, Hsu KD, Cheng KC (2017) Novel dextran modified bacterial cellulose hydrogel accelerating cutaneous wound healing. Cellulose 24:4927–4937

Maity J, Ray SK (2017) Removal of cu (II) ion from water using sugarcane bagasse cellulose and gelatin based composite hydrogels. Int J Biol Macromol 97:238–248

Mandal B, Rameshbabu AP, Dhara S, Pal S (2017) Nanocomposite hydrogel derived from poly (methacrylic acid)/carboxymethyl cellulose/AuNPs:a potential transdermal drugs carrier. Polymer 120:9–19

Manman X, Quanbo H, Xiaohui W, Runcang S (2015) Highly tough cellulose/graphene composite hydrogels prepared fromionic liquids. Ind Crop Prod 70:56–63

Marta MS, Deirdre M, Bernadine MF, Michael JG, Elliot PG (2017) Multi-scale characterisation of deuterated cellulose composite hydrogels reveals evidence for different interaction mechanisms with arabinoxylan, mixed-linkage glucan and xyloglucan. Polymer 124:1–11

Mehdi Y, Iman G, Hassan N, Mohammad A (2015) Synthesis and characterization of antibacterial carboxymethyl cellulose/ZnO nanocomposite hydrogels. Int J Biol Macromol 74:136–141

Monireh R, Hassan N (2017) Carboxymethyl cellulose/graphene oxide bio-nanocomposite hydrogel beads as anticancer drug carrier agent. Carbohydr Polym 168:320–326

Na P, Yanfeng W, Qifa Y, Lei L, Yuxing A, Qiwei L, Chunyu C (2017) Biocompatible cellulose-based superabsorbent hydrogels with antimicrobial activity. Carbohydr Polym 137:59–64

Nakasone K, Ikematsu S, Takaomi Kobayashi T (2016) Biocompatibility evaluation of cellulose hydrogel film regenerated from sugarcane bagasse waste and its *in vivo* behavior in mice. Ind Eng Chem Res 55:30–37

Nan S, Ting W, Xiufeng Y (2017) Self-assembled supermolecular hydrogel based on hydroxyethyl cellulose: formation, *in vitro* release and bacteriostasis application. Carbohydr Polym 172:49–59

Ning L, Meihua Y, Liandong D, Yang J, Anjie D (2012) Thermosensitive hydrogel of hydrophobically-modified methylcellulose for intravaginal drug delivery. J Mater Sci Mater Med 23:1913–1919

Pal D, Nayak A (2017a) Plant polysaccharides-blended ionotroLpically-gelled alginate multiple-unit systems for sustained drug release. In: Thakur VJ, Thakur MK, Kessler MR (eds) Handbook of composites from renewable materials, vol 6, 1st edn. Wiley-Scrivener, Beverly, pp 399–440

Pal D, Nayak A (2017b) Natural starch blended ionotropically gelled microparticles/beads for sustained drug release. In: Thakur VJ, Thakur MK, Kessler MR (eds) Handbook of composites from renewable materials, vol 6, 1st edn. Wiley-Scrivener, USA, pp 527–560

Pandey M, Mohamad N, Low WL, Martin C, Amin MMC (2016) Microwaved bacterial cellulose-based hydrogel microparticles for the healing of partial thickness burn wounds. Drug Deliv Transl Res 7:89–99

Patchan M, Graham JL, Xia Z, Maranchi JP, McCally R, Schein O, Elisseeff JH, Trexler MM (2013) Synthesis and properties of regenerated cellulose-based hydrogels with high strength and transparency for potential use as an ocular bandage. Mater Sci Eng C 33:3069–3076

Patrick L, Yan RL, Mari R, Arto U, Kim B, Marjo Y (2014) Technetium-99m-labeled nanofibrillar cellulose hydrogel for in vivo drug release. Eur J Pharm Sci 65:79–88

Paukkonen H, Kunnari M, Lauren P, Hakkarainen T, Auvinen VV, Oksanen T, Koivuniemi R, Yliperttula M, Laaksonen T (2017) Nanofibrillar cellulose hydrogels and reconstructed hydrogels as matrices for controlled drug release. Int J Pharm 532:269–280

Peng N, Hu D, Zeng J, Li Y, Liang L, Chang C (2016) Superabsorbent cellulose-clay nanocomposite hydrogels for high efficient removal of dye in water. ACS Sustain Chem Eng 4:7217–7224

Pinho E, Mariana H, Graca S (2014) Cyclodextrin/cellulose hydrogel with gallic acid to prevent wound infection. Cellulose 21:4519–4530

Qiyang W, Jinhua G, Yixiang W, Lingyun C, Jie C, Lina Z (2017) Creation of the tunable color light emission of cellulose hydrogelsconsisting of primary rare-earth compounds. Carbohydr Polym 161:235–243

Rasul R, Hassan N (2017) A potential bioactive wound dressing based on carboxymethyl cellulose/ZnO impregnated MCM-41 nanocomposite hydrogel. Mater Sci Eng C 73:456–464

Saerom P, Sung HK, Ji HK, Hyejeong Y, Hyung JK, Yung HY, Hyungsup K, Yong HK, Sung HH, Sang HL (2015) Application of cellulose/lignin hydrogel beads as novel supports for immobilizing lipase. J Mol Catal B Enzym 119:33–39

Sarkar DJ, Singh A (2017) Base triggered release of insecticide from bentonite reinforced citric acid crosslinked carboxymethyl cellulose hydrogel composites. Carbohydr Polym 156:303–311

Shaobo P, Ragauskas AJ (2012) Preparation of superabsorbent cellulosic hydrogels. Carbohydr Polym 87:1410–1418

Shok YO, Ishak A, Iqbal MC, Amin MC (2016) Cellulose nanocrystals extracted from rice husks as a reinforcing material in gelatin hydrogels for use in controlled drug delivery systems. Ind Crop Prod 93:227–234

Shumin L, Wenchao L, Huihua H (2016) Characterization and behavior of composite hydrogel prepared from bamboo shoot cellulose and cyclodextrin. Int J Biol Macromol 89:527–534

Sood S, Gupta VK, Agarwal S, Dev K, Pathania D (2017) Controlled release of antibiotic amoxicillin drug using carboxymethyl cellulose-cl- poly(itaconic acid-co-lactic acid) hydrogel. Int J Biol Macromol 101:612–620

Soroush B, Mehdi Y, Hassan N (2014) pH sensitive nanocomposite hydrogel beads based on carboxymethyl cellulose/layered double hydroxide as drug delivery systems. J Polym Res 21:454

Svetlana VM, Dina AM, Sergei GK, Tatyana NS (2016) Evaluation of novel applications of cellulose hydrogel films reconstituted from acetate and chloride of 1-butyl-3-methylimidazolium by comparing their optical, mechanical and adsorption properties. Mater Tod Commun 8:108–117

Swamy BY, Yun YS (2015) *In vitro* release of metformin from iron (III) cross-linked alginate–carboxymethyl cellulose hydrogel beads. Int J Biol Macromol 77:114–119

Treesuppharat W, Rojanapanthu P, Siangsanoh C, Manuspiya H, Ummartyotin S (2017) Synthesis and characterization of bacterial cellulose and gelatin-based hydrogel composites for drug delivery systems. Biotechnol Rep 15:84–91

Xiangtao L, Bing Q, Junrong L, Huining X, Beihai H, Liying Q (2015) Preparation of cellulose-based conductive hydrogels with ionic liquid. React Funct Polym 86:1–6

Xiaofang Z, Yaru W, Canhui L, Wei Z (2014) Cellulose hydrogels prepared from micron-sized bamboo cellulose fibers. Carbohydr Polym 114:166–169

Xing G, Piotr K, Zhijun S, Changqing L, Guang Y, Igor S, Vadim VS (2016) Through-thickness stress relaxation in bacterial cellulose hydrogel. J Mech Behav Biomed Mater 59:90–98

Ying P, Shan C, Yue C, Zhidong L, Jin Z, Shuqi L, Xingjun W, Jie L, Xuejing Z, Keyong T (2017) Tannin-immobilized cellulose hydrogel fabricated by a homogeneous reaction as a potential adsorbent for removing cationic organic dye from aqueous solution. Int J Biol Macromol 103:254–260

Yixi W, Zhicun W, Keliang W, Jianning W, Guihua M, Zhiyong L, Xuhong G (2017) Synthesis of cellulose-based double-network hydrogels demonstrating high strength, self-healing, and antibacterial properties. Carbohydr Polym 168:112–120

Yong MC, Lei S, Shao AY, Lei S, Wen JZ, Zhao W, Chen H (2017) Self-healing and photoluminescent carboxymethyl cellulose-based hydrogels. Eur Polym J 94:501–510

Yonghwan K, Yun KK, Sungjun K, David H, Jae WL (2017) Nanostructured potassium copper hexacyanoferrate-cellulose hydrogel for selective and rapid cesium adsorption. Chem Eng J 313:1042–1050

Yongli Z, Chunjuan G, Xiulan L, Chen X, Yang Z, Zhiming S, Yu L, Jianping G (2014) Thermosensitive methyl cellulose-based injectable hydrogelsfor post-operation anti-adhesion. Carbohydr Polym 101:171–178

Zhijun L, Dianxin L, Hongjie D, Huihua H (2017a) Enhanced properties of tea residue cellulose hydrogels by addition of graphene oxide. J Mol Liq 244:110–116

Zhijun L, Dianxin L, Hongjie D, Huihua H (2017b) Preparation and characterization of papain embedded in magnetic cellulose hydrogels prepared from tea residue. J Mol Liq 232:449–456

Zhijun L, Huihua H (2016) Preparation and characterization of cellulose composite hydrogels from tea residue and carbohydrate additives. Carbohydr Polym 147:226–233

Zhila Z, Akbari H, Farhadnejad B, Furughi N, Saeedeh AM, Mohammad Y, Khorsand G (2016) pH-sensitive bionanocomposite hydrogel beads based on carboxymethyl cellulose/ZnO nanoparticle as drug carrier. Int J Biol Macromol 93:1317–1327

Influence of Elicitors and Eustressors on the Production of Plant Secondary Metabolites

11

Aurora Mariana Alvarado, Humberto Aguirre-Becerra, Ma. Cristina Vázquez-Hernández, Ernesto Magaña-Lopez, Ixchel Parola-Contreras, Laura Helena Caicedo-Lopez, Luis Miguel Contreras-Medina, Juan Fernando Garcia-Trejo, Ramon G. Guevara-Gonzalez, and Ana A. Feregrino-Perez

Contents

Abstract

Plants are the main source of secondary metabolites, which can be used in different sectors such as pharmaceutics, food, cosmetics, agriculture, etc. turning them into an attractive source of income. Primary metabolites (carbohydrates, lipids, and proteins) have been linked to vital processes such as growth, development, and fruiting, while the secondary metabolites (phenols, flavonoids, and carotenoids, to name a few) are the result of adaptation and evolution of the plants with respect to changes in the environment. Therefore, it can be said that the

Ramon G. Guevara-Gonzalez and Ana A. Feregrino-Perez have equally contributed for this chapter.

A. M. Alvarado · H. Aguirre-Becerra · M. C. Vázquez-Hernández · E. Magaña-Lopez
I. Parola-Contreras · L. H. Caicedo-Lopez · L. M. Contreras-Medina · J. F. Garcia-Trejo
R. G. Guevara-Gonzalez (✉) · A. A. Feregrino-Perez
School of Engineering, Autonomous University of Querétaro, El Marques, Querétaro, México
e-mail: ramonggg66@gmail.com; feregrino.angge@hotmail.com

M. S. Akhtar et al. (eds.), *Natural Bio-active Compounds*,
https://doi.org/10.1007/978-981-13-7154-7_11

secondary metabolites are secreted when the plant is under biotic and abiotic stresses. These secondary metabolites possess several bio-active functions and, hence, are well recognized for industrial applications. Due to difficulties for extracting these natural plant bio-active compounds, they are usually produced alternatively through plant cell/tissue culture methods. Further, several approaches, such as the use of plant and tissue cell cultures, the application of metabolism-inducing factors or elicitation, control of biological factors, such as intensity of light, induction by sound waves, and application of nanoparticles are used to alter/enhance in vitro production of these bio-active metabolites. Overall, these approaches interact with the biochemical routes of the plant either in positive or negative ways to produce secondary metabolites in high quantities. Although there is information on this aspect, the effect of these strategies cannot be generalized, since it has been observed that the metabolism of the plant response depends on the study model, the concentration and time of use of the stimulus, as well as its nature. Considering the above facts, this chapter discusses on the most current strategies for the production of secondary metabolites in plants in a continuous and reliable manner.

Keywords
Biological factors · Elicitors · Eustressors · Nanoparticles · Secondary metabolites

11.1 Introduction

The plants are sessile organisms which evolutionarily have developed a specific defense mechanism against the attack of potential pathogens such as viruses, bacteria, fungi, oomycetes, nematodes, and insects (Jamwal et al. 2018). Nature poses more than 250,000 diverse plant species that are repository of probably hundreds of thousands of low molecular compounds called secondary metabolites (Srivastava 2017). The commercial importance of plant secondary metabolites as a natural source of pharmaceuticals; food additives – colorants, flavorings, and odorants; and other industrial materials has focused the attention of the scientific community in finding new techniques that trigger secondary metabolism with techniques that modify the environment of the plant (Tiwari and Rana 2015). These compounds can be directly extracted from plants and/or chemically synthesized for industrial use (Ramirez-Estrada et al. 2016; Acevedo et al. 2018; Joo et al. 2018). Nowadays, the knowledge of secondary metabolites is ever expanding, and they find diverse applications in various domains (Narayani and Srivastava 2017).

The traditional classification of plant chemicals is divided into primary and secondary/specialized metabolites. Although the boundary is not clearly defined, some metabolites of interest are subject to discussion on the role they carry out in the plant, and it is discussed if they are fundamental to be considered primary or secondary. Fundamentally, secondary metabolites are not required in plant developmental processes, such as respiration and photosynthesis, and have a specific function as defense against pathogens or energy accumulation (Ho et al. 2018; Jamwal et al. 2018; Singh and Dwivedi 2018). These molecules are usually classified according to their biosynthetic pathway in three large molecule families: terpenoids, phenolics, and alkaloids (Kabera et al. 2014).

The plants are overlooked for a long time, but in the past few decades, it gained popularity due to presence of more than over 100,000 known secondary metabolites with diverse functions (Narayani and Srivastava 2017). Secondary metabolites are fundamental for the ecological adaptation. These include free radical scavengers, antiproliferative agents, UV protectants, and defense against microbial infection or herbivores attack and even crosstalk with other plants. These molecules are synthesized as a result of a cascade of biochemical reactions that are triggered when they are exposed to different types of stress (Narayani and Srivastava 2017; Wiesel et al. 2014).

In plant ecophysiology, the term "stress" could be defined as any factor that modifies plant functioning, growth, or reproduction (Cheynier et al. 2013; Wiesel et al. 2014; Narayani and Srivastava 2017). Abiotic factors include chemical and physical stimulators, while biotic stresses include vertebrate and invertebrate herbivores, grazers, and, also, diseases caused by microbes, such as fungi, bacteria, and viruses, and competition among plants (Cheynier et al. 2013; Ramirez-Estrada et al. 2016). Very often plants are exposed to two or more abiotic and/or biotic stress factors, and these generally trigger a series of morphological, physiological, biochemical, and molecular changes in plants affecting its growth (Pandey et al. 2017; Kyriacou and Rouphael 2018). Hence, to survive under such conditions and to cope with such multitude of challenges, plants have evolved complex mechanisms to perceive external signals that consequently cause modifications in signaling components, gene transcription, noncoding RNAs, proteins, and secondary metabolites production (Wiesel et al. 2014; Long 2011; Narayani and Srivastava 2017). Phytohormones such as salicylic acid, jasmonic acid, ethylene, and abscisic acid are endogenous, low-molecular-weight molecules that primarily regulate the protective responses of plants against both biotic and abiotic stresses via synergistic and antagonistic actions, which are referred to as signaling crosstalk (du Jardin 2015; Singh and Dwivedi 2018). In addition to the aforementioned molecules, polyphenolic compounds, terpenes, and nitrogen derivatives that act as response signals to environmental stress are also considered of great importance (García-Mier et al. 2015; Papadopoulou et al. 2018; Talebi et al. 2018). When a plant is subjected to a degree of stress either biotic or abiotic, various biochemical mechanisms are induced in response to these stimuli. Plants throughout their evolution have developed defense systems that perceive and respond rapidly to changes sensed, modulating the expression of genes at the transcriptional and post-transcriptional levels (Walley and Dehesh 2010; Ramirez-Estrada et al. 2016).

Nowadays, there has been an increase in the interest on the production of specific secondary metabolites in different plant species commercially due to their beneficial effects. For instance, they induce pre-existing and induced systemic resistance in plants to generate resistance to pathogens (Garcia-Mier et al. 2015; Naik and Al-Khayri 2016). Secondary metabolites are activated only during a particular stage of growth and development or during periods of plant stress (García-Mier et al. 2015).

Elicitation is a method that involves exogenous application (via foliar, irrigation, directly soil or growth medium), which simulates an environmental stress that consequently triggers a series of biochemical and molecular responses that enhance the production of secondary metabolites (Narayani and Srivastava 2017; Cardenas-Manríquez et al. 2016; Effect of blue light treatment on fruit et al. 2014).

Elicitation has proved one of the most effective strategies in plant crop; an advantage is that it constitutes the only sustainable and eco-friendly system to obtain complex chemical structures biosynthesized by rare or endangered plant species (Ramirez-Estrada et al. 2016). Elicitation effect on the production of metabolites depends on the application protocol which includes method (foliar sprinkled, directly on substrate, or by exposure to several physical stimuli), application time, concentration, temperature, etc. This chapter describes the definitions, characteristics, applications, and effects of the eustressor factors in several vegetal crops for the production of metabolites of interest in diverse areas, putting in context the reader and offering a perspective of use of the elicitors to enhance the production of these molecules

11.2 Elicitors as Eustress Factors

Naturally, when a plant is subjected to some kind of stress, it immediately ignites its defense system (both constitutive and induced). Through a signaling cascade, plant recognizes a wide range of chemical factors called as "elicitors" (Naik and Al-Khayri 2016). Recently, the researchers have been oriented to use biotic and abiotic types of elicitors to enhance the production of value-added secondary metabolite in plants. Studies have been conducted to characterize these elicitors and scale up their use at the industrial level. Originally, the term, elicitor was used for molecules capable of inducing the production of phytoalexins, but it is now commonly used for compounds stimulating any types of plant defense systems. The first biotic elicitors were described in the early 1970 (Naik and Al-Khayri 2016; Garcia-Mier et al. 2014). Concept of elicitors and their applications in plants have changed from their description and uses. They are sometimes mentioned in various publications as inducers, metabolic modifying factors, or factors inducing metabolism (Garcia-Mier et al. 2014; 2015). Currently it is about homogenizing these concepts, so that the different areas of study have a firm research base. In this paper we suggest the term "eustress factors" to refer generally to molecules or stimuli capable of inducing any type of stress (Fig. 11.1). Besides enhancing productivity, elicitation involves manipulation of biochemical and metabolic pathways via stress induction. Eustress factors are chemicals, biofactors, or biostimulators from various sources that can induce physiological changes of the target living organism. In a broader sense, elicitors for a plant refer to chemicals from various sources that can trigger physiological and morphological responses and phytoalexin accumulation. It may include eustressors, such as metal ions and inorganic compounds, and biotic elicitors from fungi, bacteria, viruses, or herbivores, plant cell wall components, as well as chemicals that are released at the attack site by plants upon pathogen or herbivore attack (Table 11.1).

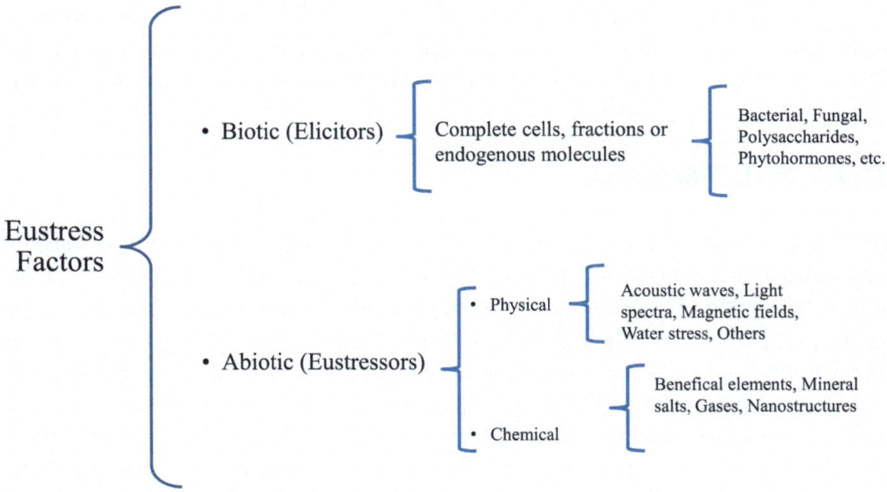

Fig. 11.1 Classification of "eustress factors" based on their nature of molecules or stimuli capable of inducing any type of stress

Table 11.1 Elicitation and production of metabolites on several vegetables species

Plants	Elicitors	Metabolites (primary, secondary, or specialized)	References
Periwinkle	PGPR (B1 and B2) isolated from a healthy tomato plant rhizosphere soil and healthy cucumber rhizosphere soil	Vincristine and vinblastine	Al-Zahrani et al. (2018)
Grapes	Methyl jasmonate	Flavonoid and terpene compounds	Gutiérrez-Luna et al. (2018)
Nicotiana	Herbivore regurgitant *Manduca sexta* and *Tupiocoris notatus*	Terpenoids and fatty acid derivatives	Joo et al. (2018)
Soybean	*B. simplex* strain Sneb545	4-Vinylphenol, L-methionine, piperine, and palmitic acid	Kang et al. (2018)
Tomato plants	*Meloidogyne incognita*	Furostanol glycosides	Chialva et al. (2018)
Wheat	*Azospirillum brasilense*	Plant growth regulators, siderophores, volatile organic compounds, and protective enzymes as chitinase and glucanase	Chávez-Herrera et al. (2018)
Ocimum	*Trichoderma harzianum*	Enhance plant growth, antioxidants, and oil content	Gupta et al. (2018)
Datura	*Agrobacterium rhizogenes*	Tropane alkaloids	Moussous et al. (2018)

It is well-known that treatment of plants with elicitors or attack by incompatible pathogens causes an array of defense reactions leading to secrete secondary metabolites (du Jardin 2015; Narayani and Srivastava 2017; Vázquez-Hernández et al. 2019).

11.2.1 Biotic Elicitation

Biotic elicitors are of biological origin and can be complete cells, cell fractions, or molecules which are naturally produced by plants, or they are naturally found in the soil of cultivation (e.g., phytoalexins, salicylic acid, jasmonic acid, gibberellic acid, hydrogen peroxide, and plant growth-promoting rhizobacteria among many others). The composition of biotic elicitors is very complex, and its origin is considered endogenous. Also, a large number of distinct compounds, such as oligosaccharides and lipo- and glycoproteins, are considered as biotic. Elicitors of exogenus origin include microorganism (pathogenic or beneficial) or substances produce by them (Chialva et al. 2018; Joo et al. 2018; Le et al. 2018). Phytohormones, present in trace amounts in plant tissues, not only regulate plant developmental processes but also play important roles in plant responses to biotic stresses. Gibberellins produced by fungi and bacteria are secondary metabolites that elicit signals to establish symbiotic interactions with host plants (Waqas et al. 2015). Phytoalexins are induced by plant-pathogen interactions and very rarely by plant-herbivore interactions (Baetz and Martinoia 2014; Schmelz et al. 2014). The signaling of plant defense hormones, such as salicylic acid and jasmonic acid, fluctuated with the application of sole or combined endophytes in the diseased plants; they play an essential role in the ignition of the signaling cascades that induce the production of secondary metabolites to serve as a defense to plants (Table 11.1) (Talebi et al. 2018). In the following section, plant hormones of interest in research and their applications as well as the microorganisms that are currently being used as elicitors are discussed in detail.

11.2.1.1 Salicylic Acid

Salicylic acid (SA, 2-hydroxy benzoic acid) is one of a diverse group of phenolic compounds synthesized by plants (endogenous). It regulates different aspects of plant growth and development. It is a well-known inducer of plant systematic-acquired resistance (SAR) in plant-pathogen interaction, but it is not a universal inducer for production of plant defensive metabolites. SA quickly accumulates at the site of infection during pathogen attack and plant hypersensitive reaction, and it spreads to other parts of the plant to induce a wide range of defense responses (Figueroa-Pérez et al. 2018; Ho et al. 2018; Papadopoulou et al. 2018).

SA mediates the phenylpropanoid pathways and is effective in triggering the stress response through elicitation and plays an important role in biochemical processes. When applied exogenously, SA causes accumulation of hydrogen peroxide. Diabetic nephropathy is ameliorated with peppermint (*Mentha piperita*) infusions prepared from salicylic acid-elicited plants (Figueroa-Pérez et al. 2018; Ho et al. 2018).

Exogenously applied SA influences plant functions in a dose-dependent manner, for example, in *Matricaria chamomilla*, 50 and 250 μM SA concentrations were reported to, respectively, promote and inhibit growth. At concentrations of 0.1 and 0.5 mM, SA promoted photosynthesis and growth of *Vigna radiata*. Applied SA was evidenced to produce a positive effect on *Punica granatum* increasing total phenolic compounds (Khan et al. 2015).

11.2.1.2 Nitric Oxide

NO (nitric oxide) is a gaseous ROS; it facilitates plant secondary metabolite production. NO plays significant roles in elicitor-induced secondary metabolite production in tissue and cell cultures of medicinal plants (Jamwal et al. 2018). NO is involved in embryogenesis, stomatal closure, seed germination and development, and flowering time. In the vegetative growth phase, low concentrations of NO increase the biomass, with the excess of NO being adverse. The intracellular level of NO is regulated by several biosynthetic and elimination routes, with the enzyme nitrate reductase being the natural source. The NO and related molecules like reactive nitrogen species are signals of the development of the root system and are active components of stress responses induced by heavy metals (Kolbert 2016).

Compounds such as SNP (sodium nitroprusside) and 2-4-carboxyphenyl-4455-tetramethylimidazoline-1-oxyl-3-oxide (cPTIO) cause pleiotropic responses. SNP and NO donors are used to activate NO bioregulation functions (Fatma et al. 2016). SNP as a signaling compound has protective and toxic effects, which depend on its concentration and exposure time. In turn, SNP regulates the production of the endogenous proline and polyamine metabolites (Domingos et al. 2015) and altered the nonenzymatic antioxidants (GSH and AsA) and reduced GSH contents in *Jatropha curcas* (Gadelha et al. 2017).

The cross-protection role of nitric oxide (SNP, 1 mM) in conferring metal tolerance in mung bean (*Vigna radiata* L. cv. BARI Mung-2) seedlings had been studied. SNP reduced the Cd metal uptake, achieved the levels of phytochelatin (PC) content, and reduced oxidative damage-enhancing nonenzymatic antioxidants (AsA and GSH) and activities of enzymes like ascorbate peroxidase (APX), glutathione peroxidase (GPX), glutathione S-transferase (GST), and glutathione reductase (GR) through improved physiology and growth and glyoxalase system in detoxifying MG (Nahar et al. 2016). NO protects against the detrimental effects of heavy metals; some investigations conclude that heavy metal triggers accumulation of NO. The exogenous application of 100 μM SNP in "mustard" (*Brassica juncea* L.) plants reversed the effects of application of 50 μM Cd through its response of glutathione reductase, ascorbate peroxidase, and superoxide dismutase (Per et al. 2017).

The biochemical mechanisms of NO increased paraquat tolerance in *Brassica napus* seedlings at 500 μM SNP for 24 h. Also, applying paraquat, the synthesis of water-soluble antioxidant groups such as ascorbate-AsA and reduced glutathione-GSH was stopped; also paraquat decreased the activities of glutathione reductase (GR) in a dose-dependent manner (Hasanuzzaman et al. 2018).

11.2.1.3 Chitosan

Chitosan is a nontoxic, biodegradable biopolymer showing antimicrobial and plant immunity-eliciting properties. Thus, its function in the defense system may be related to various pathogenesis-related proteins, defense-related enzymes, and secondary metabolite accumulation (Xing et al. 2014). This molecule affects several physiological responses, such as immunity, defense mechanisms that involve several enzymes such as phenylalanine ammonium lyase, polyphenol oxidase, catalase, and hydrogen peroxide. Recently, it has been discovered that this molecule induces the mechanisms in plants against various types of biotic stress since it helps in the formation of barriers and increases the productivity of plants (Katiyar et al. 2015). Plants are the primary sources of medicinally important compounds. The hydrolysable tannins are the main therapeutically active constituent of *Phyllanthus amarus*. Pharmacological screening revealed hydrolysable tannins as antioxidant, anticancer, antimicrobial, and anti-inflammatory compounds. The chitosan elicitor was able to induce more than threefold increase in the accumulation of hydrolysable tannin when compared to the untreated control cultures (Malayaman et al. 2017). Chitosan showed to be effective during table grape in postharvest storage, by replacing the SO_2 treatments, resulting in the control of fungal diseases. Chitosan spray treatment at cluster level was able to enhance phenylpropanoid and mevalonate pathways. Chitosan has been demonstrated to induce defense mechanism in tomato, cucumber, and rose shrubs so induces the accumulation of phytoalexins resulting in antifungal responses and enhanced protection from further infections (Katiyar et al. 2015)

11.2.1.4 Herbivore-Associated Elicitors (HAEs)

Several species of plants can distinguish the mechanical damage from the damage caused by insects or herbivorous predators through the perception of chemical agents (elicitors associated with herbivores, HAEs, or herbivore-associated molecular patterns, HAMPs). This perception is made through specific receptors that identify saliva proteins. The perception of HAEs is the result of the activation of specific responses of the plant for defense against the attack of herbivores. This mechanism of action generates changes in the metabolism which causes changes in the pattern of growth and development of the plant. Pioneers in these studies have reported that volicitin binds to maize (*Zea mays*) plasma membranes with properties that resemble those of a ligand-receptor interaction and responses are triggered by means of ion flux through the plasma membrane, activation of proteins, and generation of reactive oxygen species (Xu et al. 2015). Interestingly, bioassays with pea leaf miner *Liriomyza huidobrensis* initiated rhythmic HIPV emission in *Phaseolus lunatus* (lima bean) under diurnal conditions and that a greater concentration of volatiles increased parasitoid (*Opius dissitus*) locomotion and oviposition under light, resulting in coordinated rhythmic parasitoid behavior synchronized by light–dark cycles (Joo et al. 2018).

11.2.1.5 Microorganisms

The infection of vegetal tissues provoked by any microorganism, both pathogenic and nonpathogenic, initiates a series of complex processes in the physiological interactions, which originate characteristic responses at the cellular, tissue, and plant organ level. This induces the production of secondary metabolites that trigger defense process of plant in response to perceived stimulus (Gómez and Reis 2011). Physiological developments in plants are influenced by interactions that are established between bacterial populations present in the soil of cultivation, substrate, and nutritive solutions (hydroponics) and those that are dispersed in the environment and that come into direct contact. Cultivars sense perceived stimuli and respond by inducing pathways of defense against pathogens to prevent parasitic diseases and stimulate the development of roots when microorganisms that are recognized are associated with a beneficial effect on plant (Chamam et al. 2015). There is a great diversity of microorganisms which form very complex communities with both beneficial and pathogenic interactions. Most microorganisms are inactive in their metabolism while they are dispersed in soil and are activated when they come into direct contact with roots of host plant (Chialva et al. 2018). Among the beneficial microorganisms found in the soil, plant growth-promoting rhizobacteria (PGPR) are able to colonize plant root systems and enhance plant growth and nutrition. They protect plant against root and/or foliar pathogens through several mechanisms that included induction of systemic resistance. In the literature, there are many examples of PGPR uses in agriculture. Nowadays, these microorganisms can be found on the commercial preparations marketed as "biostimulants." This term is refereed to solutions or formulations with varying ingredients including PGPR, humic substances (HS), hormone-containing products (HCP), and amino acid-containing products (AACP) (du Jardin 2015). The role of biostimulants is enhanced growth, modulation of development and of quality traits, and increased tolerance to environmental stress (du Jardin 2012, 2015). In addition to PGPR, there are microorganisms that are used to induce production of several metabolites of interest. Sesame plants are infected by numerous pathogenic fungi, including *Fusarium*. These fungi can disrupt normal metabolism of amino acids and fatty acid in plants. Fatty acid signaling plays a very important role in plant defense. Sesamin and sesamolina are unique secondary metabolites in sesame that prevent oxidative damage induced by pathogens. *Fusarium*-infected sesame plants have significantly higher contents of amino acids, compared with negative control. Co-inoculation with *Penicillium* mitigated the *Fusarium*-induced changes in fatty acids and chlorophyll (Radhakrishnan et al. 2013). In addition to PGPR, several soil-borne microbes such as mycorrhizal fungi enhanced production of secondary metabolites, promoted growth, and induced resistance. These interactions in outdoor crops and protected agriculture are currently increasing their attention because of the potential for application in the agronomic industry. The application of PGPR, endophytic fungi, and plant growth-promoting fungi (PGPF) is very important because their use reduces the need of chemical fertilizers and thus reduces the pollution of soil and water. *Trichoderma* sp. is a PGPF widely used in agriculture. It promotes growth and induces resistance against pathogens as well as acts as a control agent, so

Trichoderma strains induce root branching and increase shoot biomass by the presence of fungal auxin-like compounds (Contreras-Cornejo et al. 2016). The amount of microorganisms (bacteria, fungi, actinomycetes, protozoa, and algae) in soil is influenced by the conditions of temperature, humidity, salts, and other compounds as well as by plants that are dispersed in it. Additionally PGPR, PGPF can be found in plant growth-promoting bacteria (PGPB) which have a specific symbiotic relationship with plants directly promoting growth, facilitating acquisition of nutrients, modulating hormone levels, and also acting as a biocontrol bacterium against pathogens (Chávez-Herrera et al. 2018; Glick 2012). Endophytic fungi are remarkably multifarious organisms associated with various plant tissues and organs. They can be surface associated, and not all live between or within host cells, and none of their interacting partners are discernibly harmed. The fungal endophyte possesses a synergistic effect of promoting plant growth and improves biotic and abiotic stress resistance as well. They also induce host plant defenses against phytopathogenic organisms through regulating plant physiological responses. Some examples are *Piriformospora indica* in barley, *Penicillium simplicissimum* in *Arabidopsis thaliana*, *Fusarium verticillioides* in maize, and *Trichoderma harzianum* in cucumber. Beside reactive oxygen species, siderophore, and organic acid, the gibberellin-producing ability of *P. citrinum* and *Aspergillus terreus* may offer extra benefits to sunflower and promote plant growth (Waqas et al. 2015). The application of microorganisms improves the growth of plants as well as stimulates the defense system. They also facilitate the development of food safety to obtain high quality of foods through the use of these biostimulants and inducers as they induce increased secondary metabolite accumulation in foods. Some of the important human nutrients such as lycopene in tomato, capsaicin in pepper, stevia glycosides in stevia, and other phytopharmaceutical metabolites are induced by treating plants with microorganisms (Kilam et al. 2015; Mazzei et al. 2016; Garcia-Mier et al. 2015).

11.2.2 Abiotic Elicitors

11.2.2.1 Chemical Eustress

Chemical eustressors are known to alleviate the biotic and abiotic stresses in plants. The application of them in low doses or for a short period of time in plants activates the defense response to resist stresses, and can improve plant's productivity (Demkura and Ballaré 2012). These abiotic factors trigger the formation, accumulation and biosynthesis related to the expression of genes related to secondary metabolites in plants (Pavarini et al. 2012). These chemical stress factors are classified as beneficial elements, mineral salts, gases and nanostructure.

11.2.2.1.1 Nanotechnology

Starting to talk about nanotechnology as an elicitation technique in agriculture is a good way to make the transition between biotic and abiotic eustressors. However, it is convenient to highlight that an appropriate use of this technology might reduce the production costs considerably, thanks to the inherent properties that this involves, as well as the possibility of converting the traditional agricultural in a precision agriculture, being used as a tool to diminish the negative effects of the excesses that the green revolution has left to date and at the same time to produce secondary metabolites (now also known as specialized metabolites) with the same or better efficiency than metabolic engineering does. All these tools are used in order to offer better foods (functional or nutraceutical) that provide benefits beyond basic nutrition and supply the needs of generation of compounds for industries (Colinas and Goossens 2018; Fuentes et al. 2018).

One of the main reasons why nanotechnology is a good transitional issue between biotic and abiotic eustressors is because this subject of multidisciplinary study is so recent that it cannot be classified easily. Coupled with the great diversity that exists both of nanomaterials (usually the vehicles of other substances) and of nanoparticles (usually the substances delivered by nanomaterials), the inclusion of the prefix "nano" is complicated to categorize when it comes to applying it to other fields of study. An example of this is the existence of nanotechnology of abiotic origin, such as those based on metals or carbon. And at the same time, there is nanotechnology of biotic origin, as recent as can be the nanoemulsions of essential oils of plants or as well-known as chitosan. Chitosan, which when synthesized on a nanometric scale, shows in the plants different and more powerful effects to those provoked on a larger scale, activating the defense system of the plants and increasing the production of specialized metabolites in *Vitis vinifera*, *Curcuma longa*, *Phyllanthus debilis*, *Eucalyptus citriodora*, *Mentha piperita*, and *Thymus daenensis* (Asgari-Targhi et al. 2018; Feregrino-Perez et al. 2018).

The characteristics of nanotechnology of biotic origin, such as biocompatibility and its nontoxic effect, are those that are mainly sought for applications in biotechnology and agriculture, where safety for final consumers is of vital importance. Even so, while these investigations are progressing, we have access to nanotechnology of abiotic origin, which, although it is also scarce in its agricultural approach, has interesting eustress effects, some incidental and some directly focused on its effect on plant species. As is the case of an investigation carried out in Iran, where they worked with callus of *Satureja rechingeri*, they knew that plants produced high amounts of phenolic compounds, especially rosmarinic acid. In that case, multiple-walled carbon nanotubes (MWCNT) with a carboxyl functional group (COOH) were tested to increase solubility and activated carbon (AC). The materials showed that they can act as elicitors that improve rosmarinic acid, especially in concentrations of 100 μg/mL of MWCNT-COOH and 250 μg/mL of CA. The plants increased the primary metabolism (such as the amino acids tyrosine and phenylalanine), and, consequently, they were reflected in the production of rosmarinic and caffeic acid. However, it still needs more research at the molecular level. For the moment, the safety of the use of MWCNT is debated, although the production of secondary metabolites in the tissues is an opportunity to produce them even if the nanotubes are not safe (Esmaeili et al. 2015).

An important source of information on the effect caused by nanoparticles (NPs) in plants is the research on nano-toxicity, which evaluates some variables among which is the content of specialized metabolites and the changes they suffer according to characteristics such as concentration, time of exposition, and size and shape of the NPs. And there is varied information that shows that the effects are as diverse as the materials and conditions in which they are applied. For example, it is demonstrated that some nanoparticles enhance the antioxidant enzymatic activities but at the same time modify the genetic expression of metabolic processes in plants (Chen et al. 2018).

Recently, in research where the graphene (carbon nanomaterial) was tested in tomato, the seedling was longer than the control. The researchers speculate that the elongation may be due to the high biosynthesis of gibberellic acid in seedlings treated with graphene, but there is still a need to perform more studies at the molecular level to clarify the mechanisms that are activated (Zhang et al. 2014) (Fig. 11.2). In plants of *Arabidopsis thaliana*, the high production of anthocyanin (flavonoid) may be due to the oxidative stress and act as an antioxidant that protects the cell (Caverzan et al. 2016).

A case in which structural arrangements have different effects on plants, although there is no difference in their chemical behavior, is observed in TiO_2 nanoparticles (anatase, rutile, and brookite), where anatase exhibits the highest catalytic activity and promotes the synthesis of carotene and chlorophyll in cucumber (Siddiqi and Husen 2017). Something similar happened with the application of anatase (TiO_2 NPs) in flaxseed (*Linum usitatissimum* L.), where an increase in the content of chlorophyll and carotenoids of the leaves of plants was generated, both in conditions of water stress and under normal conditions, compared to a control (Aghdam et al. 2016).

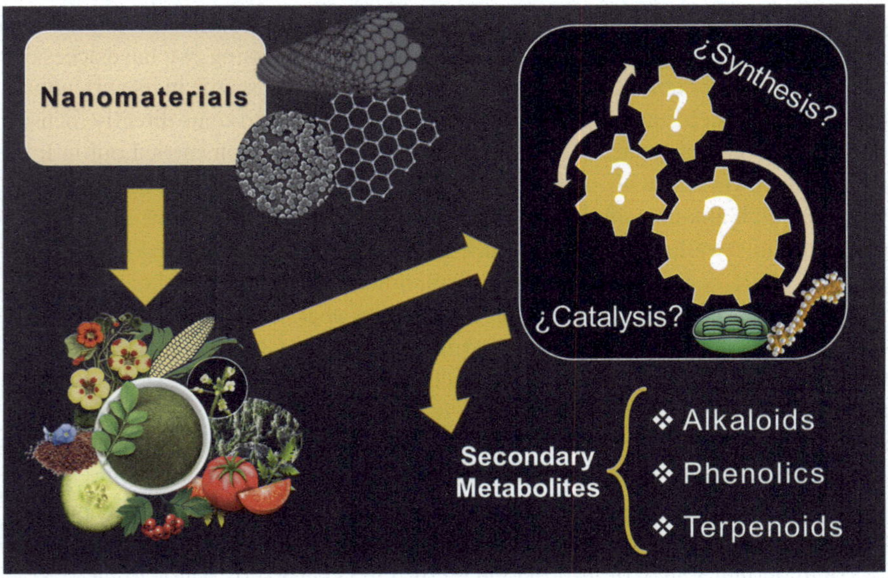

Fig. 11.2 Nanomaterials affect secondary metabolites (although mechanisms of action are not yet clear)

In ornamental plants of geranium (*Pelargonium zonale*), it has been reported that when applying nanoparticulate silver, the contents of leaf chlorophyll and carotenoids resulted higher in comparison with the control, with concentrations of 20–60 mg/L of nanoparticulate silver, that would help to resist the post-storage conditions (Hatami and Ghorbanpour 2013). In other occasions, the response of nanomaterials has not been what is expected, as in plants of hawthorn (*Crataegus* sp.) where there is a problem of adaptation to arid areas and drought conditions at the time of transplanting. An experiment with SiO_2 NPs to increase the resistance of these forest plants, by improving the physiological and biochemical response, showed no significant results in chlorophyll and carotenoids measurement, although more information is needed about the interactions between NPs and the effect of stress due to drought in the soil (Ashkavand et al. 2015).

The moringa (*Moringa peregrine*) has an economic and medical importance. The researchers tried to improve their resistance to stress, but this time to a saline stress, and it was sought to achieve with foliar applications with Hoagland solution added with Zn and Fe nanoparticles, at different concentrations. In the investigation, a total increase of chlorophyll, carotenoids, proline, carbohydrates, and percentage of crude protein was achieved with the Hoagland treatments with or without stress, getting an alternative to relieve salt stress (Soliman et al. 2015)

Currently, in all areas the environmental consequences inspire and demand more and better solutions to the problems of the present. Therefore, it is not enough to offer a large amount of food, since now safe and functional foods must be offered, which generate confidence in the production methods and at the same time provide enough food to feed the 7500 million people in 2018. Now there is a time of change, and nanotechnology is here to stay, but much more research is needed. In the future, nanotechnology that is safe and friendly to the environment will be favored (as chitosan or mineral NPs). For example, analcite, with a natural origin of volcanic mineral, was tested in wheat and corn to see the effect of its application against drought and found a strong accumulation of protective antioxidants (flavonoids, carotenoids) and catalase activation (in corn), which can bring great benefits and applications (Zaimenko et al. 2014). As well as this material, several more will be tested each time more frequently and with more specific objectives, until achieving a wider and more complete vision that ensures its free and safe use.

11.2.2.1.2 Beneficial Elements

The beneficial elements include heavy metals, rare elements, and other elements such as aluminum (Al), cobalt (Co), sodium (Na), selenium (Se), and silica (Si). The researchers have been carrying out with different proportions of these elements to increase plant yields and secondary metabolites (Ávila-Juárez and Miranda-Rodríguez 2018). With conventional agriculture, the use of universal nutrient solutions in crops has been extended. These universal nutrient solutions are composed of macronutrients and micronutrients. Macronutrients are composed by nitrogen (N), phosphorus (P), potassium (K), calcium ions (Ca), magnesium (Mg), and sulfur (S). Micronutrients are used in trace; its deficiency or excess can lead to diseases and decrease yields, like iron (Fe), copper (Cu), zinc (Zn), boron (B), Co, molybdenum (Mo), and manganese (Mn) (Steiner 1984).

Seeing that, macronutrients like N at 10 mM and 20 mM in "mustard" (*Brassica juncea* L.) plants increased the proline content production (Iqbal et al. 2015). Although silicon helps in the synthesis of lignin and stabilizes the cell wall through polyuronides providing turgor to the stem and leaves, Si improves the use of water and protects against pathogens and diseases. The interaction of silicon with Fe, Al, Cd, and Zn can alleviate the toxicity in apoplast and symplast (Haddad and Kamangar 2016).

Consequently, metal excess in the roots causes alterations in their growth and morphology as part of their stress-induced morphogenic responses (SIMR). Likewise, the cell division is blocked, cell elongation is inhibited, cell division of the pericycle is induced, and cell differentiation is altered with the application of nutrient solutions. This also suggests that the emergence of root growth responses and the formation of lateral roots are independent from the property (e.g., redox-active, nonmetal, redox-inactive metal) and the type (essential or nonessential) of the element. The redistribution of auxins alters the metabolism in the root system, while cytokinin and ethylene regulate the root system architecture under heavy metals (Kolbert 2016).

For example, "rice" seed germination decreased at 100 μM Cd; as a result of application of Cd, contents of malondialdehyde (MDA) are reduced; meanwhile addition of 30 μM SNP increased accumulation of proline in roots and shoots at the same concentrations (He et al. 2014). Moreover, Co and bacteria in root nodules of legumes help to synthesize B_{12} and fix nitrogen from air (El-Metwally and El-Saidy 2016). Furthermore, the effects of cysteine in *Ocimum basilicum* L. under cobalt stress were studied. Concentrations of 100 and 500 μM Co improved proline content, MDA, and aldehydes compared with the control (Azarakhsh et al. 2015).

In addition, exogenous applications of selenium (Se) help tolerate oxidative stress by improving cadmium defense and the methylglyoxal detoxification system. Se triggers the regulatory interaction between the metabolism of ethylene, proline, and glutathione S-transferase (GSTs) that can reverse oxidative stress by cadmium and As toxicity (Khan et al. 2015).

Likewise aluminum (Al) triggers gene expression and modifies the protein function; it can alleviate proton toxicity and improves the activity of antioxidant enzymes (Liu et al. 2014). It had been proved in soluble aluminum ions (Al^{3+}) in "rice" (*Oryza sativa* L.). In this crop were identified 700 proteins of which 106 were expressed upon Al^{3+} toxicity. In *Potamogeton crispus* Al was used to prove the phytotoxic effects. The root activity and chlorophyll content are indirectly proportional at Al increases at 1.5 mg/L concentration (Lin et al. 2017). In "tomato" genotypes, *Solanum lycopersicum* var. *cerasiforme* (CNPH 0082) and *Solanum lycopersicum* var. *esculentum* (Calabash Rouge) in clay soils, concentrations of 2, 5.2, and 0.3 mmolc kg^{-1} Al were applied. As a result of these applications, glutathione reductase (GR), ascorbate peroxidase (APX), and guaiacol peroxidase (GPOX) enzymes increased in shoots and roots (bands I, II, and III) when the levels of available Al decreased (Nogueirol et al. 2015).

Chromium (Cr) is a nonessential element for plants; it is taken up via sulfate transporters and causes biochemical imbalances in plant. In high doses, Cr interrupts respiration and photosynthesis processes, causing oxidative damage, and the modification of water relations, increases reactive oxygen species (ROS), inhibits enzymatic activities, and changes the mineral nutrient uptake by the root and death plant (Martínez-Trujillo and Carreón-Abud 2015; Islam et al. 2016). In maize (*Zea mays* L.), leaf flavonoid content and total phenolic content (TPC) decreased when maize plants were exposed to Cr treatment. However, selected plant growth-promoting bacteria and salicylic acid (SA) application help to increment 1.07 times in leaf flavonoid content than Cr treatment (Islam et al. 2016). The effect of stress was investigated by Cr in two "maize" genotypes: Wandan 13 and Runnong 35. The doses at 150 μmol L^{-1} Cr enhanced soluble sugars and lipid peroxidation as MDA, proline, and TPC and decreased soluble protein contents. Wandan 13 showed hyperactivity of CAT, SOD, POD, APX, glutathione peroxidase (GPX), and GR (Anjum et al. 2017).

11.2.2.1.3 Heavy Metals

There are nonessential elements, their concentration is an important factor in the growth, and their excess may lead to reduce growth and metabolic process in plants. According to their physicochemical properties, heavy metals can be divided into two groups, redox-active and non-redox-active, depending on the redox synthesis (Emamverdian et al. 2015). The excess of heavy metals induces cellular oxidation, disturbance of the metabolism, and enzymatic inhibition, causing the delay of the growth deriving in the death of the plants. The exposure of plants to heavy metal stress increased the production of secondary metabolites and stimulate the immune response (Lajayer et al. 2017).

Arsenic (As) in *Artemisia annua* on leaves in response to 100 μM arsenate (Na_2HAsO4 7 H_2O) for 3 days increased the total of sesquiterpenes to 50.13%. The total of monoterpene content was reduced, but β-farnesene, camphene, β-caryophyllene, α- and β-pinene, γ-selinene, and pinocarvone showed small increases (Kumari et al. 2017). "Soybean" was irrigated with solutions at 25 μM sodium arsenate ($AsHNa_2O_4$ $7H_2O$) and 25 μM sodium arsenite ($NaAsO_2$) concentrations. Soluble sugar and proline content increased in the leaves of soybean plants, although there was no increment in glycine betaine content (Vezza et al. 2018).

"Mustard" (*Brassica juncea* L.) plants were induced at 50 μM Cd and 100 μM Na nitroprusside (a donor of NO) (SNP). Its compounds, as SOD, APX, and GR, were stimulated and reduced GSH. SNP increased the antioxidant enzyme activity in the presence of Cd by about 80%, APX by 140%, and GR by 125% (Per et al. 2017).

Ag+ ions can stimulate the accumulation of phenolic acids. $AgNO_3$ at 2.5, 5, and 10 ppm concentrations were used in mature wild seeds of *Salvia virgata*. The highest contents of phenolic compounds were obtained in the treatment with 2.5 ppm of Ag^+ ions on the 5th day after elicitation. Maximum caffeic acid accumulation was obtained with the addition of 2.5 ppm Ag^+ ions (Dowom et al. 2017).

11.2.2.1.4 Rare Elements

Xue et al. (2012) cited: "Rare earth elements (REEs) are a glump of 17 chemical elements, including scandium (Sc), yttrium (Y), and the 15 lanthanides." These elements are being reported as hyperaccumulation as cerium (Ce) and lanthanum (La) (van der Ent et al. 2013). REEs can influence positive or negative physiological effects depending on the conditions and doses in the plants. The application of REEs in appropriate concentrations promotes germination of seeds, root growth, foliar area, and improvement of the quality of the fruits. High concentrations of $nCeO_2$ accumulated Cu, Zn, Mn, and P but reduced Ca and Na (Peralta-Videa et al. 2014). The rooting medium of cultures of "loquat" (*Eriobotrya japonica*) in vitro at 3.0 mM of Eu $(NO_3)3$ and La $(NO_3)3$ increased the activities of peroxidase (POD) and nitrate reductase. Effects on germination of "tomato" (*Solanum lycopersicum*) were examined at 0.5 mg kg^{-1} Ce. This application had a decrease on SG at high pH values (Thomas et al. 2014).

11.2.2.1.5 Mineral Salts

Salinity is controversial in the production of vegetables because it decreases quality and growing. The most critical requirement carried out by plants is the ability to detoxify radicals under salinity conditions. The accumulation of methylated metabolites is crucial for the radical role of scavengers and osmoprotectants in some species tolerant to salinity. The effects of high salt concentration cause physiological changes in the plant growth, such as the alteration of the secondary metabolism, osmotic stress, nutritional deficiencies, and ionic toxicity (Mustafa et al. 2017). In this subclassification are NaCl, sulfates, phosphates, and phosphites.

11.2.2.1.5.1 NaCl

Osmotic potential decreases when salinity increases. NaCl reduces chlorophyll b and carotenoid content. Three categories have been studied to manage stress tolerance by plants. The first mechanism is ion exclusion, where Na^+ transporters reduce the agglomeration of toxic NA^+ in leaves and roots before arriving to the photosynthetic tissue in the bud. The second one is tissue tolerance; the high concentrations of Na in the leaves are distributed in the cell and outside it, to reduce the deterioration of NA in the cytosol and carrying water to the cells. At last, the third mechanism is osmotic stress tolerance; through long distance signals, it keeps compatible solutes in large quantities to have water availability (Almeida et al. 2017).

"Swiss chard" (*Beta vulgaris* L. var. cicla L.) was evaluated for 14 days at 4 °C at 100 kg N/ha; this dose increases the level of nitrate and ascorbic acid content (Miceli and Miceli 2014). This plant under moderate salinity (100 mmol L^{-1} NaCl) on nitrate (NO^{3-}) had increased its growth and the NR activity (Kaburagi et al. 2015). In treatment "Swiss chard" at 50 mM of NaCl increased the proline content on leaves; meanwhile at 50 mM of KCl improved the proline concentration in the roots (Kaburagi et al. 2014).

Two genotypes (W and M) of "soybeans" (*Glycine max* L.) were treated with neutral salt stress (NaCl and Na_2SO_4, at 1:1 molar ratio, 45 mmol·L^{-1} Na^+) and alkali salt stress (Na_2CO_3 and $NaHCO_3$, at a 1:1 molar ratio, 45 mmol·L^{-1} Na^+).

Organic acids as citric acid, proline, citraconic acid, mucic acid, galactonic acid, fumaric acid, dehydroascorbic acid, 4-aminobutyric acid, and ferulic acid were measured in leaves and were highest in genotype W (Almeida et al. 2017). In "barley" at 100 mmol L^{-1} NaCl concentration did not increase NO^{3-} (Kaburagi et al. 2015).

The secondary metabolite accumulation was studied in "peppermint" (*Mentha piperita* L.) plants at two concentrations: 100 mM of NaCl and 0.5 mg l^{-1} of brassinosteroid and the second 2.5 mg l^{-1} of brassinosteroid for 150 mM NaCl. The reduction of essential oil content was by increase in salinity, while the increase of NaCl raised antioxidant enzyme activities, proline, lipid peroxidation, and total phenolic contents (TPC) (Çoban and Göktürk Baydar 2016). "Eggplant" genotypes (Saadia-tolerant and Black Beauty-sensitive) were treated with NaCl. Black Beauty genotype increased in of MDA at 12 and 15 dS m^{-1}. The antioxidants (SOD, CAT, and POD) in both genotypes had their maximum enzymatic activity at 15 dS m^{-1} of NaCl (Mustafa et al. 2017).

11.2.2.1.5.2 Sulfates

The sulfates (S) are absorbed by the plants as inorganic sulfate, reducing it and adding it to the methionine (MET) and cysteine (Cys) in the primary assimilation of the sulfate (Gigolashvili and Kopriva 2014). Sulfated secondary metabolites, such as glucosinolates as a glutathione (GSH), a class of phytochelatins, S-amino acids, hydrogen sulfide, and S-rich proteins, are synthesized when the activity of APS kinase is reduced; these products help the plants against detoxification of heavy metals, interactions with pathogens, and herbivore and oxidative stress (Capaldi et al. 2015).

The amount of S available in the secondary metabolic pathway would affect flavonoids and GSH. These metabolites when present in the root assist functions such as the gravitropic response, response to phosphate deprivation, lateral development, nodulation, and the inhibition of auxin transport. The S in high concentrations come to cause hyperosmotic and hyperionic stresses causing a disorganized metabolism, inhibition of essential ions, and decrease in turgor (Capaldi et al. 2015). Whereby, the influence of S at 100 mg S kg^{-1} soil and 200 mg S kg^{-1} soil was investigated on "mustard" (*Brassica juncea* L.) plants with or without 100 mM NaCl. Excess 200 mg S kg^{-1} soil alleviated the effects of salt stress and increased photosynthesis and production of GSH. However, the increased ATP sulfurylase (ATPS) activity and Cys content were at 100 mg S kg^{-1} soil. Salt treatment increased GSH accumulation (Fatma et al. 2014).

11.2.2.1.5.3 Phosphates

The accumulation of amino acids, alanine, proline, and GABA, and the phosphoesters, glucose-6-phosphate and glycerol-3-phosphate, was observed as well as changes in the levels of minor sugars and various organic acids (Ding et al. 2017). Inorganic pyrophosphate is involved in the movement of auxins to the leaves, in homeostasis in the phloem and the roots, and in seed germination and development. The overexpression of membrane-bound H+-pumping pyrophosphatases in some plants improved their tolerance to salt stress and drought and crop production and seed development. These stress conditions induce changes in the secondary metabolism like shikimate and the phenylpropanoid biosynthetic pathways (Gutiérrez-Luna et al. 2018).

Camellia sinensis (L.) cv. Bixiangzao, Nokangzao, and Ruixue were treated with phosphorus (KH_2PO_4) at 50 μM and pH 5.5. The metabolites as flavonoids and glucosides, sugars, alcohols, organic acids, and amino acids had increased. The metabolism of sugar and organic acid was enriched (Ding et al. 2017). Phosphorus in "garden sage" (*Salvia officinalis*) increases the biomass of the leaf, total TPC, and rosmarinic acid concentrations but does not have any effect on quality and quantity of essential oils (Verma and Shukla 2015).

11.2.2.1.6 Gases

Some gases may regulate plant growth and serve as an alarm function. The detection of these gas emissions by the plants enhances the resistance for defending from pathogens. Previous studies concluded that a foliar spray of 10% methanol affects the expression of hundreds of genes, activating multiple detoxification and signaling pathways (Dorokhov et al. 2012).

11.2.2.1.6.1 Nitric Oxide

NO is a growth regulator and developer in that it has a protective role for the effects of biotic and abiotic stress. The application of NO can decrease the toxic effects of salt and NA^+ in plants. In maize (*Zea mays* L.) cultivars, salt treatment (100 mM) was applied with SNP at 3 and 6 mg L^{-1}. The exogenous applications of NO increased biomass production, total chlorophyll content, and maximum fluorescence yield (Fv/Fm), and it decreased the Na^+ accumulation. With the application of NO, maize decrease in antioxidant activities was obtained (Kaya et al. 2015).

In spinach (*Spinacia oleracea* L.) was added 200 nl l^{-1} and maintained between 162 and 173 nl l^{-1} after 3 h of NO injection under salinity conditions at 200 mmol L^{-1} (NaCl). NO application decreased the toxic effect of salinity and restore the chlorophyll synthesis, in spite of the H_2O_2 and MDA levels in leaves decreased. The application of NaCl and NO increased the antioxidant capacity like POD, SOD, and CAT. The proline level had increased threefolds, while the NO application had increased onefold, and the DPPH free radical improved activity by 12% with NO application (Du et al. 2015).

In sunflower plants was applied 0, 1, 10, or 100 μM of the NO donor sodium nitroprusside (SNP) under drought. The best dose was 10 μM SNP; this dose increased the proline content and the activity of PG-POD, yet it decreased the level of MDA (Cechin et al. 2015). *Arabidopsis thaliana* (L.), Heynh. improved lipid catabolism and accumulation of phospholipids, the content of sugars, levels of polyamines, and chlorophyll breakdown after 6 h of NO exposure (Leon et al. 2016).

11.2.2.2 Eustressors: Physical Factors

11.2.2.2.1 Acoustic Emissions (AE)

There is an extensive type of abiotic stress (UV irradiation, high light, high and low temperature, drought, alkalinity, salinity, and nutrient deficiencies) that is known to possibly damage plants. These abiotic factors have been also used as elicitors, now defined as "eustressors" to develop tolerance strategies against them. Generally, new eliciting methods used remain as a challenge to the low-level production of bio-active compounds (less than 1% dry weight).

Elicitation mainly depends on the physiological state of a plant. Moreover, the type of inductor, dose, and application technique represent limitations for its uses, as well do their costs. Given the fact that plants have an evolutionary mechanism to self-protect, interact, and communicate through the environment, some studies have established their ability to sense signals. Plants have the ability to respond toward key physical stimulus like light, temperature, and mechanical stimuli (Heggie and Halliday 2005; Telewski 2006). The sensitivity of reaction in plants has evolved complexity levels and still-unknown mechanism. One of the recently physical stimuli discussed is sound, which is generated by consequence of its physiological processes and perceived from external factors.

11.2.2.2.1.1 Plant Signals and Acoustic Emissions (AE)

Signaling events in plants are due to combinations of several stress factors. Considering that being sessile organisms, plants are susceptible to be exposed to various kinds of environmental stimuli, including unfavorable spurs (Prasch and Sonnewald 2015). It is demonstrated that acoustic emissions (AE) from ecological conditions or applied as novel technology in agriculture can initiate diverse signals, which themselves activate signal transduction cascades, similar to other abiotic stress factors.

Different signaling events result in appropriate plant responses leading to adaption processes, including defense, growth, and reproduction. Examples of signals involved in these responses are ABA (abscisic acid), ROS (reactive oxygen species), JA (jasmonic acid), SA (salicylic acid), Ca^{2+} (calcium), MAPK (mitogen-activated protein kinase), CPK (Ca^{2+}-dependent protein kinases), SnRK (sucrose non-fermenting-1-related protein kinase), TOR-1 (target of rapamycin) (Prasch and Sonnewald 2015), and volatile organic compounds (Widhalm et al. 2015). Favorably, AE is a signal with a lower metabolic cost of generation in plants and has been currently analyzed (Gagliano 2013, 2015).

The interest for applying AE as a "eustressor" is linked to their varied ecological roles that are potentially measurable in terms of acoustic waves by using adequate sensors. In fact, the application of secondary metabolites should, in the future, upregulate and homogenize the levels of desired active components by introducing a variety of elicitors and "eustressors" in carefully controlled environments (Kennedy and Wightman 2011).

11.2.2.2.1.2 Acoustic Emissions (AE) as New Eustressor

Indeed, sound is an emerging physical trigger in plants beyond chemical triggers, such as plant hormones and other immune activators which have been used to improve plant health (Jung et al. 2018). Currently, it is reported that sound vibrations (SV), acoustic waves (AW), or acoustic emissions (AE) have been proposed to stimulate defense responses at physiological, biochemical, and transcriptional levels (Mishra et al. 2016; Fernandez-Jaramillo et al. 2018).

First, science has related this sensory and communicative issue from a bioacoustics perspective. Similar to flies buzzing, the sound effect results from specific frequencies of bee buzzing. It is also known that buzzing sound facilitates the pollination of flowers (releasing pollen from plant anthers) (De Luca and Vallejo-Marín 2013). As well, Cocroft (2014) established that AE produced by insect's chewing serves as an alarm signal to plants, after demonstrating that applying recorded insect chewing sounds caused an increase of phytochemical production (glucosinolates and flavonoids) (Martínez-Ballesta et al. 2013; Cocroft 2014). Glucosinolates are related with both plant defense reactions and human health benefits after its consumption such as the reduction of the risk of certain cancers and diseases owed to oxidative stress (Dinkova-Kostova and Kostov 2012).

In the same way, other studies reported that amplifying at 100 decibels (intensity), a wide range of frequencies (oscillatory pattern) between 0 and 1.5 kHz could improve natural protection responses in rice plants, specifically against drought (Jeong et al. 2014). Also, Hassanien et al. (2014) found a higher disease resistance in pepper, cucumber, and tomato after AE treatments (Hassanien et al. 2014).

Likewise, previous reports showed increasing amounts of IAA (indole-3-acetic acid) showing an increase of phytohormone signaling in plants, which is comparable with other abiotic eustressor factors. However, the phenotypic and biochemical results have not been the only ones explored. Previous research supports that 1 h of AE could improve the activity of cell wall structure and fluidity (Bochu et al. 2004a). Furthermore, some researches have reported a significant yield quality improvement in rice and strawberry crops (Qi et al. 2010; Yu et al. 2013). Besides, some studies are focused on self-tolerance and resistance of plants against disease (Choi et al. 2017). The stated discoveries have proposed the AE as a new agricultural technology that could decrease requirements for chemical fertilizers and biocides (Zhang 2012). Respecting this, AE can be used as "green" eustressor.

For instance, applying frequencies of 500 Hz resulted in significant amounts of growth-related hormones in *Arabidopsis*, showing indole-3-acetic acid (IAA), gibberellin (GA), as well as defense-related hormones such as salicylic acid (SA) and jasmonic acid (JA) (Ghosh et al. 2017a). These results indicate that mainly effects of AE are related to plant hormone signaling (Kim et al. 2010; Hassanien et al. 2014).

Another type of AE application is related to ripening control. The sound may affect fruit development slowing up tomato and wheat ripening rates signifying a new potential as growth regulator (Hassanien et al. 2014; Jeong et al. 2014). Other studies showed that playing different types of music has the significant impact on all studied traits. As a result of playing classical music compared to control treatment (nonmusic), variables such as chlorophyll (47%), gibberellin hormone (81%), nitrogen (44%), and calcium (21%) increased (Alavijeh et al. 2016). But, in this study are not well-defined the characteristics of the applied AE. This has been a main difficulty about the precise experimental conditions that were not specified by different researchers. Anyhow, the proposal of AE signaling in plants as a new adaptation factor is still under discussion (Ghosh et al. 2017c; Jung et al. 2018).

It is remarkable that AE has an impact over phytohormones and these compounds are responsible for biomass production as regulators of downstream signaling cascades throughout a plant's life cycle in growth, flowering, ripening, senescence, and defense responses (Hou et al. 2009). Also, it has been reported important bioactivities of these plant hormones in human health (Chanclud and Lacombe 2017). Plus, this leads us to a big picture about AE potential, including the "eustressor" action, modifying plants behavior and bringing adaptation changes mediated by accumulation of interest compounds with health benefits to humans, in the same manner phytohormones do. Some outstanding reports have established AE parameters and subsequently increase of specific compounds, as we show in Table 11.2. In addition, this table presents the linkage suggested between phytohormones (that can be induced by AE) and the impact on human well-being, once ingested through the diet. Plant hormone effects are principally studied in microbiota modulation, cancer, and anti-inflammatory diseases, among brain functions and hormonal regulation (Blaser et al. 2016; Chanclud and Lacombe 2017). These are the most important topics in health science and nutrition today.

Moreover, AE emitted by plants could be differentiated in particular conditions that have not been clearly interpreted yet. Above and beyond, playback studies are needed to confirm ecological applications of AE in order to encourage further uses as elicitors, remaining the huge sustainable potential of this "environmentally friendly" agricultural technology. Until now, is barely known the optimal sound treatment. The impact of AE treatments could differ depending on several factors such as plant model, amplitude and frequencies of AE, time and duration of treatment, and application distance among others that are not explained in most abovementioned studies.

11.2.2.2.1.3 Emission and Perception of AE Signals: Mechanisms Proposed

Growing evidence about the highly sensitive ability of plants of generating and reacting to sound signals from environment have being discovered (De Roo et al. 2016; Jung et al. 2018). But, the discussion is how sound affect plants. The mechanism of sensing AE in plants, like ones emitted by insects (500–2000 Hz) or wind, is still in discussion.

Perception: So far, it is being proposed as a mechano-stimuli perception of waves. However, a reliable explanation of sound-specific structure for recognition by plants has not been completely elucidated. Nonetheless, these emission-perception signals could vary from species to species, like another factor does, and also by stage and ecological conditions.

The molecular mechanism proposed for AE based on the second messenger includes calcium ion (Ca^{2+}) signals. The channels that mediate Ca^{2+} flux (MSLs and

Table 11.2 Elicitor potential of acoustic emissions (AE)

Sample	Emission/ frequency (Hz) intensity (dB)	Compounds	Health benefits	References
Cucumber and cabbage	180 min daily and 20 kHz/75 dBA	The level of polyamines (PAs), vitamin C, and uptake of oxygen O_2	Antioxidant activity, immune system stimulation	Qin et al. (2003) and Shigeoka et al. (2002)
Chrysanthemum sp.	1.4 kHz 1 h at 95 dB	ABA	ABA dampens inflammation caused by obesity, intestinal bowel disease (IBD), and influenza infection and impacts inflammatory response	Bochu et al. (2004), Guri et al. (2010), and Hontecillas et al. (2013)
Arabidopsis thaliana	1.4 kHz, 10 days at 0.095 kdB	Increase indole-3-acetic acid (IAA)-auxin, gibberellin acid (GA)	Anticarcinogenic properties	Ester et al. (2009), Cocroft (2014), and Reihill et al. (2016)
	500 Hz at 80 dB		Anti-inflammatory effects	
Arabidopsis thaliana	500, 1000, and 3000 Hz during 3 h for 10 days 100-dB	Increase salicylic acid (SA)	Glucose metabolism regulation. Limits insulin resistance and potentially ameliorate the unbalanced microbiota of diabetic patients	Goldfine et al. (2013), Utzschneider et al. (2016), and Choi et al. (2017)
Arabidopsis thaliana	(100–1000 Hz) Chewing vibrations	The increase of glucosinolates	Anticancer activity	Cocroft (2014) and Sánchez-Pujante et al. (2017)
Arabidopsis thaliana	(100–1000 Hz) Chewing vibrations	Anthocyanins	Ameliorate cardiovascular disease risk and oxidative stress biomarkers	Alvarez-Suarez et al. (2014) and Cocroft (2014)

MCA membrane exchange ion channels) are possibly located in the plasmatic membrane, facilitating the ions efflux/influx (Ca^{2+}). The generation of Ca^{2+} transport is thought as a critical step for sound signaling. Ca^{2+} is sensed possibly through various Ca^{2+} sensors and/or CDPKs (Calcium-dependent protein kinase), which pass the message throughphosphorylation/dephosphorylation to different signaling proteins or to transcription factors, eventually resulting into gene expression (Mishra et al. 2016).

Emission: Plants can release spontaneous frequencies that could be linked to its physiological conditions, for example, frequencies of 50–120 Hz are generated probably from self-cavitation (Ulrich et al. 2002; Laschimke et al. 2006). The cavitation (gas bubbles) emissions from plants are supported in the energy liberated during xylem tension (27–36 dB). The origin of this tension is the water evaporation in the stomatal region. When the stomata in the leaves are open, water is transpired due to the difference in water vapor pressure between the atmosphere and the substomatal cavity. When water evaporates, spaces at the small capillaries (nanometer scale) located adjacently to cell walls, are retracted due to capillary forces (strong adhesion), causing forces that generates energy changes, causing sound emissions (De Roo et al. 2016).

After the aforementioned explanations, it is strongly suggested that AE can influence an adaptive behavior and consequently increase the synthesis of secondary metabolites, similar to stress-induced ones: abscisic acid (ABA) and protective enzymes (SOD, POT, CAT, APX) (Bochu et al. 2004).

11.2.2.2.2 Light Spectra

Harsh environmental factors cause adverse effects in plant growth and influence the production of secondary metabolites. The most studied include temperature, humidity, water supply, minerals, CO_2, and light conditions. The latter is one of the most important and obvious requirements for plant correct development. The energy of sunlight is used for plants to grow and develop via photosynthesis. Plants sense light through specific molecules called photoreceptors that trigger specific signals for photomorphogenesis or other defense systems. Photoreceptors are specially designed proteins that perceive light and produce signals for certain biological effects in plants. A small cofactor or chromophore molecule is what allows photoreceptors to sense specific light wavelengths of over a continuous spectral range (Burgie et al. 2014). Five photosensory systems have been identified: phytochromes, cryptochromes, phototropins, members of the Zeitlupe family, and UV Resistance Locus 8 (UVR8) (Bantis et al. 2018). The general information of these photoreceptors is summarized in Table 11.3.

Depending on the dose rate and exposure time, either insufficient or excess levels, light can become a type of eustress, producing several effects, from damage to cellular components to triggering of defense systems for secondary metabolite production (Ramakrishna and Ravishankar 2011; Müller-Xing et al. 2014). Ultraviolet (UV) radiation has focused the interest of the scientific community in finding the effects it might have in organisms due to the depletion of the stratospheric ozone layer (Häder et al. 2015). UV radiation is about the seven percent of the light that comes from the sun, where up to 98.7% of UV that reaches the Earth surface corresponds to UV-A. UV-B can cause serious damage to DNA, membrane, and proteins, whereas UV-A induces DNA damage less efficiently because of the activation of photoreactions forming reactive oxygen species (ROS) (Häder et al. 2015; Hideg and Strid 2016). Organisms generate additional UV-absorbing pigments, where higher plants produce secondary metabolites such as flavonoids and anthocyanins to cope with this issue (Jiang et al. 2010).

Table 11.3 Brief description of plant photoreceptors

Photoreceptor	Light wavelength response	Function in plants	Reference
Phytochromes	Inactive Pr – red (660–700 nm) Active Pr – far red (700–750 nm)	Seed germination, seedling de-etiolation, dormancy, circadian rhythm, phototropism, flowering, stomata development, flowering, transition, senescence, shade avoidance, neighbor detection, and elongation inhibition	Possart et al. (2014), Li et al. (2015a), and Casal (2013)
Cryptochromes	Blue (495–400 nm) UV-A (400–315 nm)	Circadian rhythm, hypocotyl elongation and anthocyanin biosynthesis, stomatal opening, photoperiodic flowering, seed dormancy and germination, adaptation to environments enriched in green light	Barrero et al. (2014), Devlin and Kay (2000), and Zhang et al. (2011)
Phototropins	Blue (495–400 nm) UV-A (400–315 nm)	Phototropism, inhibition of hypocotyl elongation, growth regulation toward a directional light orientation, chloroplast accumulation, and stomatal opening	Sztatelman et al. (2016) and Li et al. (2015b)
Members of the Zeitlupe family	Blue (495–400 nm) UV-A (400–315 nm)	Circadian rhythm and photoperiodic flowering	Zoltowski and Imaizumi (2014)
UVR8	UV-B (315–280 nm)	UV protection, shade avoidance inhibition, protection from oxidative stress, flavonoid induction	Hayes et al. (2014); Coffey et al. (2017)

11.2.2.2.2.1 Light Technology in Crop Production

Light is defined as that within the visible spectrum; nevertheless that definition depends upon the sense of sight involving the response of individuals; moreover, ultraviolet (UV) and infrared (IR) parts are not visible to humans (Stevens 2013). In this chapter, visible light will essentially be considered as radiant energy covering a segment of the electromagnetic spectrum encompassing the range from 380 (violet) to 620–760 nm (red) (Khanna 2014). Ultraviolet (UV) and infrared (IR) parts, although not visible, will be also included in the definition of light.

Supplemental lighting is gaining acceptance among producers in protected agriculture where LEDs and high intensity discharge (HID) lamps are the most common. LEDs have allowed a sustainable and highly efficient use of energy (Singh et al. 2015) and have offered new perspectives for the scientific community, since the emitted spectra depend on the properties of the semiconductor material they are made of (Olle and Viršilė 2013), allowing the true spectral composition of blue, green, red, and far-red wavelengths that matches with plant-specific photoreceptors (Singh et al. 2015). Other light technologies are not new but are still used in greenhouse and plant experimentation. High sodium pressure (HSP) and other HID lamps are widely implemented since they emit a wide light spectrum, in the visible (400–700 nm) and invisible (700–850 nm) ranges, with a peak at yellow (around 589 nm) (Giedrė et al. 2013). However, these lamps include disadvantages related with heat generation and high-energy consumption; hence, LEDs are replacing HPS lamps in most applications as a result of high-energy costs, heat generation, and suboptimal spectrum for photosynthesis (Jiang et al. 2017).

The main goal of conventional agriculture is to increase yield, paying no attention to the phytochemical content which has beneficial effect on human health. Increasing yield has been an option for the need of food of a growing population in the world, which is estimated to be of 8.3 billion in 2030 and 9.3 billion in 2050 (Bruinsma 2017). However, a new trend has been emerging in the last years where both yield and phytochemicals converge into a new vision of agricultural management in a framework of integrated agricultural practices (Garcia-Mier et al. 2015). In this way light has become a viable solution for increasing plant secondary metabolism in intensive culture systems since it is a physical noninvasive method that requires technology that nowadays is becoming more available. Advances in lighting systems have made possible new research of plant responses due to specific wavelengths, most of them related with growth and development; however, we will focus on the production of secondary metabolites in plants that are mostly produced in greenhouses, where lighting systems could be applied for generating hermetic conditions, making possible to achieve the goal of having an increased yield with acceptable nutraceutical value.

11.2.2.2.2.2 Light Effect on Phenolic Compound Content

Phenolic compounds are a large class of secondary metabolites that are widely studied in the chemical, biological, agricultural, and medical fields (Kabera et al. 2014). These molecules protect the plant from oxidative stress, but they are also of considerable interest due to their antioxidant properties and beneficial health potential in humans as they have been associated with reduced risk of cancer and cardiovascular and neurodegenerative diseases (Shahidi and Ambigaipalan 2015; Lorenzo and Munekata 2016). Phenolic compounds have in common the hydroxylated aromatic rings, but they include a vast group of structures that are divided into four main groups: phenolic group with one aromatic ring, with two aromatic rings, quinones and polymers (Kabera et al. 2014).

An attack on plant tissue or stressful environments triggers the defense mechanisms that end in the production of phenolic compounds (Dias et al. 2016). Biosynthesis of phenolic compounds is sensitive to light conditions, serving as a photoprotection in plants (Zoratti et al. 2014). A family of cryptochromes (flavoproteins) cause the production of other phenolic compounds (Bravo et al. 2012), finding that they are active UV filters in light stressful conditions (Dias et al. 2016). Moreover, different light spectra interfere the acetate-malonate and shikimate pathways to stimulate or inhibit the production of phenolic compounds (Leal-Costa et al. 2015).

Flavonoids are water-soluble pigments found in vacuoles of plant cells and mainly classified into six major subgroups: chalcones, flavones, flavonols, flavandiols, anthocyanins, and proanthocyanidins or condensed tannins. In higher plants, one of the most important functions they are involved is in UV filtration (Kabera et al. 2014); therefore light quality and quantity are two of the most important environmental factors that regulate flavonoid synthesis and accumulation (Zoratti et al. 2014). Some specific flavonoids have been found to have an important role in light stress induction. Dihydroxy-B-ring-substituted flavonoids have great potential to inhibit the generation and reduce levels of reactive oxygen species (ROS) caused by light stress, performing antioxidant functions; moreover, these molecules can be found in the chloroplast, suggesting a role as scavengers of singlet oxygen and stabilizers of the chloroplast outer envelope membrane (Agati et al. 2012).

Fini et al. (2011) proposed that flavonoids do not only have a primary function as UV-B screening photoprotection but they work under severe excess light conditions, irrespective of the relative proportions of the solar wave bands, enhancing the biosynthesis of dihydroxy-B-ring-substituted flavonoid glycosides; furthermore, H_2O_2 may diffuse out of the chloroplast and be transported to the vacuole by tonoplast intrinsic proteins. In correct concentrations, H_2O_2 is a signaling molecule responsible for increasing tolerance to stress, and flavonoids may have a role in keeping the concentration of H_2O_2 at a sublethal level (Fini et al. 2011). Chalcone synthase, a key enzyme involved in flavonoid biosynthesis, is also regulated by several environmental and endogenous stimuli, including light (A-H-Mackerness et al. 1999). UV-A radiation and blue light each act synergistically with UV-B to stimulate CHS transcript accumulation in *A. thaliana* (Ebisawa et al. 2008). Quercetin, which synthesis is also stimulated by UV light exposure, is another one of the most prominent flavonoids with the antioxidant ability of counteracting oxidative stress generated as a result of reactive oxygen species (ROS) and binding transition metal ions (Alrawaiq and Abdullah 2014).

Anthocyanins are water-soluble antioxidant pigments giving the red, purple, and blue colors of many flowers and fruit, serving as attract pollinators and seed dispersers. Anthocyanins are also produced under light stress conditions counteracting the negative effects of nitrogen and ROS, protecting photosynthetic tissues against oxidative damage by absorbing light in the visible and UV-A (Guo and Wang 2010; Zhang et al. 2014). In human health, they are associated with protection against certain cancers, cardiovascular diseases, and other chronic disorders. Since anthocyanin bioavailability is low, there are therefore excellent reasons for engineering anthocyanin biosynthesis, understanding chemical and physical factors that influence their functional properties in plants for a diversity of applications (Zhang et al. 2014).

Many experiments on phenolic content where the light was the stress factor have been performed. Evaluation of the light effect on secondary metabolites of several crops has been proved in all productive stages, from seedling to postharvesting. Table 11.4 shows examples of diverse light condition experiments on tomato,

Table 11.4 Effect on phenolic compound content of experiments where light was the stress factor on plants with commercial interest

Cultivar	Light condition (treatments)	Result	References
Red leaf lettuce "Sunmang"	M1: 23R/77B M2: 47R/53B M3: 73R/27B 100R, 100B 171 ± 7 μmol·m^{-2}·s^{-1} 12-h light photoperiod 300 μmol·m^{-2}·s^{-1}	(↑) TPC. Best result with R followed by M3 of 9 mg GAE/shoot both. R and M3 got the highest content and concentration of chicoric acid and caffeic acid	Son et al. (2017)
Lettuce "Grizzly"	W, R, B, 70R/30B Control: normal greenhouse 300 μmol·m^{-2}·s^{-1} 12-h light photoperiod	(↑) TP. Best result with W of 1.18 mgg^{-1} Better results with R, B, and 70R:30B than control, but not significant	Amoozgar et al. (2017)
Lamb's lettuce "Nordhollandse"	100R, 90R/10B, 70R/30B, 50R/50B, cool W (3500K), warm W (4700K+2700K), HSP.200 μmol·m^{-2}·s^{-1} 16-h light photoperiod Autumn and winter season	(↑) Total phenolic acids. Best results in winter with 90R/10B and 50R/50B of 92.81 and 85.33 mg 100 g^{-1} FW, respectively (↑) Total flavonoid glycosides. Best results in winter with HPS, warm W and 50R/50B of 24.44, 23.09 and 22.50 mg 100 g^{-1} FW, respectively (↑) Total free flavonoids. Best results in autumn with 70R/30B, 50R/50B, and HPS of 3.07, 2.43, and 2.37 mg 100 g^{-1} FW, respectively	Długosz-Grochowska et al. (2016)
Red leaf lettuce "Sunmang"	B: R LEDs 171 ± 7 μmol·m^{-2}·s^{-1} 12-h light photoperiod	(↑) TPC. Best result in 47B: 53R, 1.4 and 2.4 times higher compared with the control (FL + HSP) and 0 blue, respectively (↑) TFC. Highest value under 47B: 53R of 0.6 (+) -catechin mgg^{-1}FW	Son and Oh (2013)
Green leaf lettuce "Grand Rapid TBR"	B: R LEDs 171 ± 7 μmol·m^{-2}·s^{-1} 12-h light photoperiod	(↑) TPC. Best result in 26B: 74R and 59B:51, 2.2–2.7 times higher compared with the control (FL + HSP) (↑) TFC. Highest value under 59B: 41R of 0.4 (+) -catechin mgg^{-1} FW	Son and Oh (2013)

(continued)

Table 11.4 (continued)

Cultivar	Light condition (treatments)	Result	References
Red leaf lettuce "Banchu Red Fire"	B: 100 μmol·m^{-2}·s^{-1} B/R: 50/50 μmol·m^{-2}·s^{-1} R: 100 μmol·m^{-2}·s^{-1} FL: 100 μmol·m^{-2}·s^{-1} All 14-h light photoperiod	(↑) TP. Best result with B/R of 161.6 nmolmg^{-1} DW (↓) TP. 47 nmolmg^{-1} DW with R. Lower than FL (control) (↑) Chlorogenic acid. Best result with B of 15 nmolmg^{-1} DW (↓) Chlorogenic acid. 1.9 nmolmg^{-1} DW with R. Lower than FL (control) (↑) Anthocyanin. Best result with B/R of 0.27 OD530 mg^{-1} DW (↓) Anthocyanin. 0.06 OD530 mg^{-1} DW with R. Lower than FL	Johkan et al. (2010)
Red leaf lettuce "Hongyeom"	UV-A: continuous UV-A irradiation for 7 days UV-B1: 4 h irradiation of UV-B per day for 6 days UV-B2: gradual increased irradiation of UV-B from 1 to 7 h over 6 days UV-C: 2 h irradiation of UV-C per day for 3 days	(↑) TPC. All better than control at day 2 after transplant. Best result with UV-C at 4.5 mg GAEg^{-1} FW (↑) Total anthocyanin concentration. Best result at day 3 after transplant 9.0 mgg^{-1} FW	Lee et al. (2014)
Romaine green baby leaf lettuce "Thumper"	Basal: combination of B (455 nm – 8 μmolm^{-2}s^{-1}), R (638 nm – 150 μmolm^{-2}s^{-1}), R (670 nm – 12 μmolm^{-2}s^{-1}) and FR (735 nm – 4 μmolm^{-2}s^{-1}) Basal + UV (380 nm – 4 μmolm^{-2}s^{-1}) Basal + G (520 nm– 12 μmolm^{-2}s^{-1}) Basal + Y (595 nm– 10 μmolm^{-2}s^{-1}) Basal + O (622 nm – 28 μmolm^{-2}s^{-1}) 16 h photoperiod	(↑) TPC. Best results with basal + UV and basal + O of 2.32 and 2.64 mgg^{-1}, respectively (↑) Anthocyanins. Best result with basal + G of 104.15 mgg^{-1} Note: growth chamber experiment	Samuolienė et al. (2013)

(continued)

Table 11.4 (continued)

Cultivar	Light condition (treatments)	Result	References
Romaine green baby leaf lettuce "Thumper"	HSP: 90 μmolm^{-2}s^{-1} HSP + B (455 nm – 210 μmolm^{-2}s^{-1}) HSP + B (470 nm – 210 μmolm^{-2}s^{-1}) HSP + G (505 nm – 210 μmolm^{-2}s^{-1}) HSP + G (530 nm – 210 μmolm^{-2}s^{-1})	(↓) TPC. All treatments with lower value than HSP (1.00 mgg^{-1}) (↑) Anthocyanins. Best result with HSP + G (530 nm) and HSP + B (470 nm) of 80.09 and 62.94 mgg^{-1} Note: greenhouse experiment	Samuolienė et al. (2013)
Tomato "Toy-mini tomato"	W, B, R, and G LEDs All 150 μmol·m^{-2}·s^{-1} All 18 h light photoperiod	(↑) TPC in leaves and stem. Best result with B of 90 mgml^{-1}, 1.3-fold than W in leaves; and 19 mgml^{-1}, 1.2-fold than W in stem (↓) TPC in stem with R and G, 49% and 37% lower than W, respectively	Kim et al. (2013)
Tomato "Micro-Tom"	Visible light: 50 Wm^{-2} UV-A: 7 Wm^{-2} 24-h exposure	(↑) Anthocyanin. Best result with UV-A and combination of UV-A with visible light, 0.4 mgmg^{-1} FW both (↓) Anthocyanin. Worst result with visible light, 0.1 mgmg^{-1} FW	Guo and Wang (2010)
Tomato "San Marzano"	PL1: 1 Jcm^{-2}side^{-1} PL2: 2 Jcm^{-2}side^{-1} PL4: 4 Jcm^{-2}side^{-1} PL8: 8 Jcm^{-2}side^{-1} UV2: UV-C 2 Jcm^{-2}side^{-1} UV4: UV-C 4 Jcm^{-2}side^{-1} Control: non-irradiated PL: polychromatic light (200–1100 nm) Storage time: 21 days	(↑) Polyphenol content. Best results with UV2 and UV4, 408.6 and 393.9 mg GAEkg^{-1} FW, respectively Note: postharvest experiment	Pataro et al. (2015)
Cherry tomato "Cuty"	Five treatments: R, B, G, W, FL. 205 μmol·m^{-2}·s^{-1} each, 12-h photoperiod during 27 days (Seedlings)	(↑) TPC. Best results with B and W, 1.2 and 1.1 GAE mgg^{-1} FW, respectively (↑) TFC. Best results with B and W, 0.7 and 0.6 (+) -catechin mgg^{-1}FW	Kim et al. (2014)

(continued)

Table 11.4 (continued)

Cultivar	Light condition (treatments)	Result	References
Tomato "Moneymaker"	Outdoor: garden grown UV-A+B: four side surrounded with a combination of UV-A and UV-B, 17.42 kJm^{-2}day^{-1}, 12-h photoperiod	(↑) TPC. Best result with UV-A+B of 7.39 mg GAEg^{-1} DW (↑) Rutin concentration. Best result with UV-A+B of 119.2 μgg^{-1}DW	Dzakovich et al. (2016)
Tomato "Daniela"	UV-C-1: 1 h – 1 kJm^{-2} UV-C-3: 3 h – 3 kJm^{-2} UV-C-12: 12 h – 12.2 kJm^{-2} Control 1: day/night cycle Control 2: total darkness	(↑) TPC. Best result with UV-C-12 of 366.35 mg GAEkg^{-1} FW (↑) Chlorogenic acid. Best results with UV-C-1 and Control 2, of 10.35 and 10.24 mgkg^{-1} FW, respectively Note: postharvest experiment	Bravo et al. (2012)
Tomato "Moneymaker"	UV-B: 1.69 Wm^{-2}, 1 h, 6.08 kJm^{-2} day Control: UV-B lamps screened with benzophenone-treated polyethylene film, known to block UV-B radiation Stages: mature green (MG), turning tomatoes (TU), and red ripe (RR) stage	(↑) TPC in peel. Best result with UV-B in MG of 240 mg100 g^{-1}FW. No statistical difference in TU (280 mg100 g^{-1} FW) (↑) Flavonoids in flesh. Best result with UV-B in MG of 3 mg100 g^{-1} FW. No statistical difference in TU (4 mg100 g^{-1} FW) (↑) Flavonols in flesh. Best result with UV-B in MG of 3 mg100 g^{-1} FW. No statistical difference in TU (3 mg100 g^{-1} FW) Note: postharvest experiment Note: irradiation was carried out daily until RR stage which was achieved after 10 and 18 days for TU and MG	Castagna et al. (2013)
Strawberry "Fengguang"	B: 40 μmol·m^{-2}·s^{-1} Control: darkness 12 days at 5 °C	(↑) TPC: higher level with B light after 2 days of storage, 0.5 mg GAEg^{-1} FW Note: postharvest experiment	Xu et al. (2014)

(continued)

Table 11.4 (continued)

Cultivar	Light condition (treatments)	Result	References
Strawberry "Sulhyang"	Control: non-irradiation Four treatments: 385, 470, 525, and 630 nm Five days with corresponding wavelength	(↑) TPC. Best value of 172.75 mg/100 g with 470 nm, 13% higher than control. 470 nm range appeared to be most effective for the increase in TPC (↑) Anthocyanin. Best value of 12 mg/100 g with 470, 525, and 630nm ranges, 21% higher than control Note: postharvest experiment	Kim and Juvik (2011)
Strawberry "Daewang"	B: 200 μmol·m^{-2}·s^{-1} R: 200 μmol·m^{-2}·s^{-1} B:R (3:7): 200 μmol·m^{-2}·s^{-1} 16-h photoperiod Duration: 94 days	(↑) TPC. Best result with R of 170 mg100 g^{-1} No statistical difference was found in flavonoids and anthocyanins content, with values of 18.0 and 15 mg100 g^{-1}, respectively Note: growth chamber experiment	Choi et al. (2015)
Strawberry "Daewang"	B: 75 μmol·m^{-2}·s^{-1} R: 75 μmol·m^{-2}·s^{-1} B:R (3:7): 75 μmol·m^{-2}·s^{-1}, 8 h, 17:00–23:00 Control: 8 μmol·m^{-2}·s^{-1}. weak incandescent light 20 μmol·m^{-2}·s^{-1} (8:00–17:00) Duration: 151 days	(↑) TCP. Slightly better with R of 190 mg100 g^{-1} but no statistical difference among treatments (↑) Flavonoids. Slightly better with control of 75 mg100 g^{-1} but no statistical difference among treatments No statistical difference was found in anthocyanin content, with value of 25 mg100 g^{-1} Note: greenhouse experiment Note: no statistical difference was found in secondary metabolite content, but author emphasizes a higher amount compared with growth chamber (previous row)	Choi et al. (2015)

Used abbreviations. *B* blue, *R* red, *G* green, *W* white, *FR* far red, *FL* fluorescent light, *HSP* high sodium pressure, *TPC* total phenolic compounds, *TFC* total flavonoid concentration, *TP* total phenols, *QC* quercetin content, *DW* dry weight, *FW* fresh weight, (↑) increment of, (↓) decrement of

lettuce, and strawberry, plants that are produced under intensive culture systems and therefore are of commercial interest. Tomato (*Solanum lycopersicum* L.) is the second most important vegetable crop next to potato and an important source of phenolic compounds, particularly flavonoids (Desneux et al. 2011; Coyago-Cruz et al. 2017). Lettuce constitutes an important part of greenhouse production since its worldwide demand is increasing because of its crisp texture, pleasant aroma and flavor, fresh appearance, and richness in phytochemicals, such as phenolic compound (Llorach et al. 2008; Fu et al. 2012); therefore, many studies on lettuce have been conducted in order to increase its nutraceutical value. The strawberry (*Fragaria* × *ananassa*) is the most consumed berry with a huge commercial and economic impact throughout the world; moreover, it can be considered unequivocally the most studied berry since it contains a remarkable nutritional composition with a vast variety of phenolic constituents (Giampieri et al. 2014).

11.2.2.2.2.3 Light Effect on Terpenoid Content

Most carotenoids share a common C40 backbone structure of isoprene units (terpenoids), with a chain of conjugated double bonds, which creates a chromophore that absorbs light in the blue range of the spectrum (Neuman et al. 2014; Gong and Bassi 2016). Some primary carotenoids serve as accessory pigments to transfer absorbed energy to chlorophylls, and secondary carotenoids function as protective molecules because of their antioxidant properties, maintaining suitable levels of reactive radicals, preventing lipid peroxidation, and promoting the stability and functionality of the photosynthetic apparatus(Gong and Bassi 2016).

Many experiments on terpenoid content where the light was the stress factor have been performed. Table 11.5 summarizes relevant results on plants of commercial interest; among the most important crops are lettuce and tomato, which importance was previously described. Tomato is an important source of carotenoids, particularly lycopene and β-carotene (Frusciante et al. 2007); in the postharvest period, carotenoids accumulate during ripening due to the degradation of the chlorophyll and the transformation of the chloroplasts into chromoplasts during the lag phase preceding the maturation (Pataro et al. 2015).

11.2.2.2.2.4 Light Effect on Antioxidant Capacity

Antioxidant defense systems in plants include various molecules that serve as antioxidants, such as carotenoids, tocopherol, flavonoids, ascorbate, and phenolic compounds, which play important roles in protection from photooxidative damage (Ashry and Mohamed 2011; Kim et al. 2013). The antioxidant content of fruits depends not only on the cultivar and farming methods but mostly on the postharvest handling practices (Jagadeesh et al. 2011). The development of effective methods aimed at prolonging the fresh status or even increasing the content and activity of antioxidant compounds of plant products has been important in the market as well as for the improvement of the positive effects of the consumption of fruits and vegetables on human health (Pataro et al. 2015). In that way, UV light irradiation has risen as a promising and environmentally friendly technology for the preservation of fresh horticultural crops using UV hormesis, inducing beneficial stress responses

Table 11.5 Effect on terpenoid content of experiments where light was the stress factor on plants with commercial interest

Cultivar	Light condition (treatment)	Result	References
Lettuce "Grizzly"	W, R, B, 70R/30B Control: normal greenhouse 300 μmol·m^{-2}·s^{-1} 12-h light photoperiod	(↑) CC. Best result with 70R:30B of 0.59 mgg^{-1}. Better results with W and R than control, but not significant	Amoozgar et al. (2017)
Red leaf lettuce "Banchu Red Fire"	B: 100 μmol·m^{-2}·s^{-1} B/R: 50/50 μmol·m^{-2}·s^{-1} R: 100 μmol·m^{-2}·s^{-1} FL: 100 μmol·m^{-2}·s^{-1} All 14-h photoperiod	(↑) CC. Best result with B/R of 304 μmolmg^{-1} DW (↓) CC. Worst result with R (231 μmolmg^{-1} DW) lower than FL (control)	Johkan et al. (2010)
Romaine green baby leaf lettuce "Thumper"	Basal: combination of B (455 nm – 8 μmolm^{-2}s^{-1}), R (638 nm – 150 μmolm^{-2}s^{-1}), R (670 nm – 12 μmolm^{-2}s^{-1}) and FR (735 nm – 4 μmolm^{-2}s^{-1}) Basal + UV (380 nm – 4 μmolm^{-2}s^{-1}) Basal + G (520 nm – 12 μmolm^{-2}s^{-1}) Basal + Y (595 nm – 10 μmolm^{-2}s^{-1}) Basal + O (622 nm – 28 μmolm^{-2}s^{-1}) 16-h photoperiod	(↑) β-Car: best value with basal + Y of 2.78 μgg^{-1} (higher than basal + UV, basal + G, basal + O), but no significant difference with basal (↑) α-Car: best value with basal + UV of 1.35 μgg^{-1} Note: growth chamber experiment	Samuolienė et al. (2013)
Romaine green baby leaf lettuce "Thumper"	HSP: 90 μmolm^{-2}s^{-1} HSP + B (455 nm – 210 μmolm^{-2}s^{-1}) HSP + B (470 nm – 210 μmolm^{-2}s^{-1}) HSP + G (505 nm – 210 μmolm^{-2}s^{-1}) HSP + G (530 nm – 210 μmolm^{-2}s^{-1})	(↓) β-Car: all treatments with lower value than HSP (0.28 mgg^{-1}) (↑) α-Car: best value with HSP + 505 nm of 2.88 μgg^{-1} Note: greenhouse experiment	Samuolienė et al. (2013)
Tomato "San Marzano"	PL1: 1 Jcm^{-2}side^{-1} PL2: 2 Jcm^{-2}side^{-1} PL4: 4 Jcm^{-2}side^{-1} PL8: 8 Jcm^{-2}side^{-1} UV2: UV-C 2 Jcm^{-2}side^{-1} UV4: UV-C 4 Jcm^{-2}side^{-1} Control: non-irradiated PL: polychromatic light (200–1100 nm) Storage time: 21 days	(↑) Lyc. Best result with PL8 and UV2, 74.15 and 67.65 mgkg^{-1} FW, respectively (↑) CC. Best result with PL4 and PL2, 156.78 and 105.97 mgkg^{-1} FW, respectively Note: postharvest experiment	Pataro et al. (2015)

(continued)

Table 11.5 (continued)

Cultivar	Light condition (treatment)	Result	References
Tomato "Moneymaker"	UV-B: 1.69 Wm^{-2}, 1 h, 6.08 kJm^{-2} day Control: UV-B lamps screened with benzophenone-treated polyethylene film, known to block UV-B radiation Stages: mature green (MG), turning tomatoes (TU), and red ripe (RR) stage	(↑) Lut in peel. Best result with UV-B in MG and TU of 2.9 and 2.5 mg100 g^{-1}, respectively (↑) Lyc in flesh. Best result with UV-B in MG and TU of 9 and 13 mg100 g^{-1}, respectively (↑) β-Car in peel: Best result with UV-B in MG and TU of 40 mg100 g^{-1} both Note: postharvest experiment Note: irradiation was carried out daily until RR stage which was achieved after 10 and 18 days for TU and MG	Castagna et al. (2013)
Tomato "Capello"	Control: No UV-C UV-C-3: 3.7 kJm^{-2} (hormic dose) UV-C-24: 24.4 kJm^{-2} (hyper dose) 38 days	(↓) Lyc content. Best value with control at day 36 of 42 μgg^{-1} FW. Lower values with UV-C-3 and UV-C-24 (28 and 25 μgg^{-1} FW, respectively) (↑) CC. Best value with UV-C-3 and UV-C-24 (9 μgg^{-1} FW both) Note: postharvest experiment	Maharaj et al. (2010)
Tomato "Moneymaker"	Outdoor: garden grown UV-A+B: four side surrounded with a combination of UV-A and UV-B, 17.42 $kJm^{-2}day^{-1}$, 12-h photoperiod	(↑) Lyc concentration. Best result with UV-A+B of 0.91 mgg^{-1} DW (↑) β-Car concentration. Best result with UV-A+B of 0.25 mgg^{-1} DW. No statistical difference with control treatment	Dzakovich et al. (2016)
Tomato "Daniela"	UV-C-1: 1 h – 1 kJm^{-2} UV-C-3: 3 h – 3 kJm^{-2} UV-C-12: 12 h – 12.2 kJm^{-2} Control 1: day/night cycle Control 2: total darkness	(↓) β-Car. All UV-C treatments were lower than controls. Control 1 and control 2 were 1.40 and 1.23 $mgkg^{-1}$ FW, respectively (↑) Total Lyc. Best results with UV-C-1 and UV-C-3, of 59.91 and 63.17 $mgkg^{-1}$ FW, respectively Note: postharvest experiment	Bravo et al. (2012)

(continued)

Table 11.5 (continued)

Cultivar	Light condition (treatment)	Result	References
Satsuma mandarin	B: 50 $\mu mol \cdot m^{-2} \cdot s^{-1}$ R:50 $\mu mol \cdot m^{-2} \cdot s^{-1}$ Control: 24 h darkness Experiment: 6 days	(↑) β-Car: best result of 4.5 $\mu ggFW^{-1}$ with B (↓) β-Car: worst result of 2.0 $\mu ggFW^{-1}$ with R light. Worse than control (↑) β-Cry: best result of 15.0 $\mu ggFW^{-1}$ with R (↑) T-vio: 13.0 $\mu ggFW^{-1}$ with B light. Slightly better than control (↓) C-vio: worst result of 30.0 $\mu ggFW^{-1}$ with B light. Worse than control (↑) α-Car: best result of 1.2 $\mu ggFW^{-1}$ with B. (↓) α-Car: worst result of 0.4 $\mu ggFW^{-1}$ with R light. Worse than control (↑) Lut: 16.0 $\mu ggFW^{-1}$ with B light. Slightly better than control	Ma et al. (2012)

B blue, *R* red, *G* green, *O* orange, *Y* yellow, *FL* fluorescent light, *HSP* high sodium pressure, *CC* carotenoid content, *β-Car* β-carotene, *FW* fresh weigh, *β-Cry* β-cryptoxanthin, *T-vio* all-trans-violaxanthin, *C-vio* 9-cis-violaxanthin, *Lyc* lycopene, *α-Car* α-carotene, *Lut* lutein, *DW* dry weight, (↑) increment of, (↓) decrement of

(Jagadeesh et al. 2011; Shama 2007). Table 11.6 shows examples of diverse light condition experiments on tomato, lettuce, and strawberry where antioxidant capacity was measured.

11.3 Plant Cell and Plant Tissue Culture for Secondary Metabolite Production

The plants face diverse challenges, which make necessary adaptation and defense responses to factors in their environment. To do this, plants produce chemical compounds through complex metabolic pathways. These compounds, called secondary metabolites, are important for man as well, due to their bioactivity, which can be exploited in pharmaceuticals, flavorings, fragrances, food, and agrochemicals (Paek et al. 2014). Therefore, large amounts of these compounds are demanded for industries. Secondary metabolites are obtained through the processing of large amounts of vegetable raw material. This is because they are found in lower concentration (< 1% dry weight of the plant) than the primary metabolites (Wilson and Roberts 2012). The plant material is obtained through the collection of wild populations and cultivation. The crop can be open field, in areas of natural distribution of the species and/or protected agricultural systems (Lubbe and Verpoorte 2011). However, these three forms have disadvantages, for example,

Table 11.6 Effect on antioxidant capacity of experiments where light was the stress factor on plants with commercial interest

Cultivar	Light condition (treatment)	Result	References
Red leaf lettuce "Banchu Red Fire"	B: 100 μmol·m^{-2}·s^{-1} B/R: 50/50 μmol·m^{-2}·s^{-1} R: 100 μmol·m^{-2}·s^{-1} FL: 100 μmol·m^{-2}·s^{-1} All 14-h photoperiod	(↑) AC. Best result with R followed by M3 of mM TEAC shoot^{-1}	Son et al. (2017)
Lamb's lettuce "Nordhollandse"	100R, 90R/10B, 70R/30B, 50R/50B, cool W (3500K), warm W (4700K+2700K), HSP,200 μmol·m^{-2}·s^{-1} 16-h light photoperiod Autumn and winter season	(↑) AC. In autumn, best result with 50R/50B and HPS of 254.05 and 240.79 mmol TEg^{-1} FW, respectively. In winter, best results with 90R/10B, 70R/30B, and 50R/50B of 307.62, 337.35, and 365.11, respectively	Długosz-Grochowska et al. (2016)
Read leaf lettuce "Sunmang"	B: R LEDs 171 ± 7 μmol·m^{-2}·s^{-1} 12-h photoperiod	(↑) AC. Highest value under 47B: 53R of 4 mM TEAC g^{-1} FW	Son and Oh (2013)
Green leaf lettuce "Grand Rapid TBR"	B: R LEDs 171 ± 7 μmol·m^{-2}·s^{-1} 12-h photoperiod	(↑) AC. High but no significant difference between treatments with B LEDs above 26% of 1.8 mM TEAC g^{-1} FW	Son and Oh (2013)
Red leaf lettuce "Hongyeom"	UV-A: continuous UV-A irradiation for 7 days UV-B1: 4-h irradiation of UV-B per day for 6 days UV-B2: gradual increased irradiation of UV-B from 1 to 7 h over 6 days UV-C: 2-h irradiation of UV-C per day for 3 days	(↑) AC. All better than control at day 2 after transplant. Best result with UV-C and UV-B1 with 7.5 and 7.0 mmol TEACg^{-1} FW	Lee et al. (2014)
Romaine green baby leaf lettuce "Thumper"	Basal: combination of B (455 nm – 8 μmolm^{-2}s^{-1}), R (638 nm – 150 μmolm^{-2}s^{-1}), R (670 nm – 12 μmolm^{-2}s^{-1}) and FR (735 nm – 4 μmolm^{-2}s^{-1}) Basal + UV (380 nm – 4 μmolm^{-2}s^{-1}) Basal + G (520 nm – 12 μmolm^{-2}s^{-1}) Basal + Y (595 nm – 10 μmolm^{-2}s^{-1}) Basal + O (622 nm – 28 μmolm^{-2}s^{-1}) 16-h photoperiod	(↓) AC. All treatments slightly lower than basal (10.02 DPPH μmol g^{-1}), but not significantly different Note: growth chamber experiment	Samuolienė et al. (2013)

(continued)

Table 11.6 (continued)

Cultivar	Light condition (treatment)	Result	References
Romaine green baby leaf lettuce "Thumper"	HSP: 90 μmolm^{-2}s^{-1} HSP + B (455 nm – 210 μmolm^{-2}s^{-1}) HSP + B (470 nm – 210 μmolm^{-2}s^{-1}) HSP + G (505 nm – 210 μmolm^{-2}s^{-1}) HSP + G (530 nm – 210 μmolm^{-2}s^{-1})	(↑) AC. Best value with HSP + G (530 nm) of 10.91 DPPH μmol g^{-1} Note: greenhouse experiment	Samuolienė et al. (2013)
Tomato "Toy-mini tomato"	Four treatments: W, B, R, and G LEDs All 150 μmol·m^{-2}·s^{-1} All 18-h light photoperiod	(↑) AC. Activity of CAT, APX, and GR increased with B light (15%, 7%, and 1.4-fold, respectively, compared to W) (↓) AC. Activity of CAT, APX, and GR decreased with G and R light compared to W	Kim et al. (2013)
Tomato "San Marzano"	PL1: 1 Jcm^{-2}side^{-1} PL2: 2 Jcm^{-2}side^{-1} PL4: 4 Jcm^{-2}side^{-1} PL8: 8 Jcm^{-2}side^{-1} UV2: UV-C 2 Jcm^{-2}side^{-1} UV4: UV-C 4 Jcm^{-2}side^{-1} Control: non-irradiated PL: polychromatic light (200–1100 nm) Storage time: 21 days	(↑) AC. Best results with UV4 and PL8, 56.67 and 55.09% DPPH inhibition, respectively Note: postharvest experiment	Pataro et al. (2015)
Cherry tomato "Cuty"	Five treatments: R, B, G, W, FL 205 μmol·m^{-2}·s^{-1} each, 12 photoperiod during 27 days (seedlings)	(↑) AC. Best results with B and W, 3.5 and 3.0 GAE mM TEACg^{-1} FW (↓) AC. Worst result with G. 1.5 GAE mM TEACg^{-1} FW. Worse than control	Kim et al. (2014)
Tomato "Daniela"	UV-C-1: 1 h – 1 kJm^{-2} UV-C-3: 3 h – 3 kJm^{-2} UV-C-12: 12 h – 12.2 kJm^{-2} Control 1: day/night cycle Control 2: total darkness	(↑) AC. Best result with UV-C-12, of 1.7 mmol Fe (II) equivalents kg^{-1} FW and 1.7 mmolTroloxkg^{-1} FW Note: postharvest experiment	Bravo et al. (2012)

(continued)

Table 11.6 (continued)

Cultivar	Light condition (treatment)	Result	References
Strawberry "Fengguang"	B: 40 μmol·m^{-2}·s^{-1} Control: darkness 12 days at 5 °C	Superoxide dismutase activity in control declined gradually, but was markedly higher with B light during the whole storage Cat activity. Significantly higher activity was found with B light after 2 days of storage APX activity. B light treatment maintained significantly higher activity after 4 days of storage	Xu et al. (2014)

B blue, *R* red, *G* green, *FL* fluorescent light, *HSP* high sodium pressure, *AC* antioxidant capacity, *DW* dry matter weight, (↑) increment of, (↓) decrement of, *CAT* catalase, *APX* ascorbate peroxidase, *GR* glutathione reductase

genetic erosion and risk of wild populations, seasonal variability in the content of metabolites, difficulties in propagating and managing the species, high costs of domestication, decrease in agricultural land, and others. Therefore, it was necessary to develop strategies that would allow the production of secondary metabolites, minimizing the disadvantages.

Bioengineering has provided strategies such as cell and plant tissue culture and synthetic biology to produce secondary metabolites (Pérez-Alonso and Jiménez 2011). Cell and tissue culture allows mass production, manipulation of the microenvironment, and use of space independent from climate and soil conditions (Gerolino et al. 2015; Dias et al. 2016; Grąbkowska et al. 2016). For example, it is considered a fast way to produce active compounds from traditional Chinese medicine (Liu 2017). It is also considered an alternative to produce phenolic compounds due to the fact that it allows the application of elicitation, requires less space, and allows a greater control of environmental conditions (Dias et al. 2016). The following is a summary of the production of secondary metabolites through the cultivation of plant cells and tissues for secondary metabolite production and perspectives.

11.3.1 In Vitro Production Versus Field Production

In vitro culture of plant cell and plant tissue can be defined as the manipulation of cells and organs in an aseptic culture medium under controlled conditions of light, humidity, and temperature (Smetanska 2008; Dias et al. 2016). Biological foundation of this technique is the cellular totipotentiality, cellular property referred to the ability of some cells to cause the growth and development of a new individual. Cultured cells can be differentiated or dedifferentiated depending on the desired end product (cells, tissues, organs, or whole plants). Cells have the genetic machinery to trigger the biosynthesis of compounds; therefore the cells can be used for the production of secondary metabolites (Shahzad et al. 2017). In addition, cell culture is capable of accumulating equal or greater amounts of metabolites than plants in open field (Grąbkowska et al. 2016).

In vitro techniques for production of secondary metabolites present advantages on field production. First, field production is affected by climatic conditions, so yield may vary according to the time of year or be seasonal. Some species are geographically restricted. On the other hand, large quantities of vegetable raw material are required; this implies a greater area sown, which is a disadvantage considering that since 1991 the percentage of agricultural land has decreased (Altieri 2018). In vitro techniques to produce secondary metabolites arise as an alternative to field production of plants. This is due to the advantages it presents in in vitro production compared to the conventional production of plants.

According to Hussain et al. (2012), the main advantages of producing secondary metabolites through the in vitro culture of cells, tissues, and organs are:

- Control of environmental conditions which allows independency of geographical, climatic, and soil factors.
- The culture is carried out aseptically, which contributes to the sanitary safety of the product.
- Isolation of compounds can be done more quickly and efficiently.
- Yields are similar or superior to those obtained from the whole plant.

The in vitro culture of cells, tissues, and organs affords to have highly controlled microenvironments. For example, cell culture system establishment for metabolite production requires to determine aspects such as high-producing cell lines or clones (natural or transgenic genotypes), optimization both culture media (carbohydrate and nitrogen sources; phosphate and growth regulator levels and salt) and environment conditions (temperature, light, pH, agitation, and aeration) for arise growth (Murthy et al. 2014). Control in the accumulation phase focuses on the use of one or more elicitors that detonate the metabolic responses necessary for the biosynthesis of the desired compound.

11.3.1.1 Cell Suspension Culture

Plant cell culture is a technique based on totipotency property of cells. Therefore, cell possesses genetic machinery necessary for secondary metabolite biosynthesis. This can be used to obtain chemical factories that biosynthesize secondary metabolites. It is possible to use dedifferentiated cells that can reproduce and activate the biosynthesis pathways (Yue et al. 2016). This technique has been used for secondary metabolite production of pharmaceutical importance such as those present in *Waltheria americana* (Mundo et al. 2017).

11.3.1.2 Adventitious Root Culture

Some secondary metabolites are biosynthesized by specialized cells. Therefore, in vitro production needs to be done by culturing whole tissues or organs

(Hernández-Ramírez. 2016). Adventitious root culture presents higher rates of growth rate and secondary metabolite production than those produced by cell culture (Baque et al. 2012). In adventitious root culture, eustress factors are used to stimulate the production of secondary metabolites.

For example, Sivanandhan et al. (2012) studied elicitors and eustressor effect on withanolide production in cultivation of adventitious roots of *Withania somnifera* (L) Dunal. They used aluminum chlorine (10 mgl^{-1}) as eustressor and chitosan (100 mgl^{-1}) as elicitor. Both eustress factors increased the amount of withanolides in adventitious root culture of *W. somnifera*. This is useful for larger-scale production of withanolides (Sivanandhan et al. 2012). Likewise, elicitation effect on secondary metabolite production of *Fagonia indica* Burm.f. was tested. Results have shown a positive elicitation effect with methyl jasmonate in all stages of the culture (Khan et al. 2017).

11.3.1.3 Hairy Root Culture

Hairy root (HR) culture is a biotechnological technique of in vitro culture, which is characterized by the cultivation of roots transformed by the infection with the soil bacterium *Agrobacterium rhizogenes*; the transformed root cultures are a kind of tissue culture since the cells cultured are differentiated (Pistelli et al. 2010). This technique is used for the production of secondary metabolites (Rady et al. 2018). Table 11.7 shows some metabolites produced through hairy root culture. The hairy root cultures are preferred over other in vitro culture techniques of metabolite production, because these tissue cultures are a good biocatalyst because it has genetic stability, hormone autotrophy, and multienzyme biosynthetic potential mimicking that of the parent plants (Grzegorczyk-Karolak et al. 2018).

Table 11.7 Production of secondary metabolites through hairy roots culture

Compound	Secondary metabolites	Species	References
Alkaloids	Hyoscyamine	*Datura stramonium*	Sun et al. (2013)
	Scopolamine		
	Vincamine	*Hyoscyamus muticus*	Zolala et al. (2010)
		Catharanthus roseus	Verma et al. (2013)
Flavonoids	Flavonoid	*Ginkgo biloba*	Hao et al. (2010)
		Isatis tinctoria	Jiao et al. (2018)
		Raphanus sativus	Balasubramanian et al. (2018)
Phenolic acids	Rosmarinic acid	*Eryngium planum*	Szopa and Ekiert (2014)
		Salvia miltiorrhiza	Sheng and Chen (2013)
	p-Hydroxybenzoic acid	*Daucus carota*	Sircar et al. (2007)
Quinones	Anthraquinones	*Morinda officinalis*	Zheng et al. (2014)
		Rubia tinctorum	Perassolo et al. (2017)
Terpenoids	Glycyrrhizic acid	*Glycyrrhiza uralensis*	Yang et al. (2014)
	Valerenic acid	*Valeriana officinalis*	Torkamani et al. (2014)
	Tanshinones	*Salvia miltiorrhiza*	Hao et al. (2015)

11.3.2 Eustress Factors in Plant In Vitro Culture

Eustress factors (elicitors and eustressors) have been used for two main purposes in in vitro culture. These are research and increase in the yields of metabolites produced (Patel and Krishnamurthy 2013). Research is focused on elucidating the mechanism of action of the main eustress factors used in in vitro culture (Ramirez-Estrada et al. 2016). Likewise, in the field of research, it is necessary to know the mechanisms that provide such varied metabolic responses (Ramirez-Estrada et al. 2016). For this the use of eustressor factors allows to investigate regulation and enzymology of secondary metabolism (Patel and Krishnamurthy 2013; Grąbkowska et al. 2016).

On the other hand, eustress factor use in in vitro culture to increase the yield of the target compound is common. The cell has genetic machinery for production of both primary and secondary metabolites. However, in a first phase culture, an increase in biomass is required; therefore the cell biosynthesizes primary metabolites. After that phase in many cases, the intervention of an elicitor is required (Ramirez-Estrada et al. 2016). The elicitor induces the genetic expression so that the cells initiate and increase the production of secondary metabolites (Ghosh et al. 2017b).

Eustress not only improve yield but also induces the synthesis of other metabolites, for example, in *Lavandula officinalis* cell cultures that grew with anoxic stress (eustressors) and jasmonic acid (elicitor). Rosmarinic acid production was increased, and also the production of caffeic acid was induced (Gonçalves and Romano 2013). This increases the potential of the production of secondary metabolites through in vitro techniques.

11.3.2.1 Elicitors In Vitro

11.3.2.1.1 Microorganisms

Under natural conditions the roots of the plants are in contact with a set of microorganisms. The microorganisms produce substances associated with pathogenicity to a host. These substances induce a response in the plant's defense system. Therefore, the microorganisms act as elicitors. Awad et al. (2014) conducted a study in which they applied soil microorganisms (bacteria and fungi) as elicitor in root culture of *Taverniera cuneifolia* for glycyrrhizic acid (GA) production. Results showed that microorganism culture use improves GA yield. *Mucor hiemalis* (fungus) and *Rhizobium leguminosarum* (bacteria) induced higher GA yields (Awad et al. 2014).

Elicitation can also be done with extracts of microorganisms. For example, yeast extract (YE) effect on cryptotanshinone and tanshinone IIA production in adventitious root culture of *Perovskia abrotanoides* was studied. Tanshinone production was achieved with 200 mg L^{-1} YE (Zaker et al. 2015). However, the application of microorganism in in vitro culture is complex, mainly because it is difficult to control the growth of the microorganism (Jiao et al. 2018). Microorganism immobilization is an alternative to establish co-culture systems in in vitro culture. Immobilization restricts the microorganism movement but allows its metabolic products to diffuse toward aqueous phase. Flavonoid production improved in *Isatis tinctoria* hairy root cultures using a co-cultivation system with immobilized *Aspergillus niger* (Jiao et al. 2018).

11.3.2.1.2 Chitosan

Chitosan has been used to stimulate secondary metabolite production in in vitro culture. This elicitor has been used in dedifferentiated cultures and differentiated cell cultures. Chitosan increased threefold hydrolysable tannin accumulation of *Phyllanthus debilis*. Chitosan application (150 mg) was carried out during the stationary phase of the plant cell suspension (Malayaman et al. 2017). Other researchers studied chitosan elicitation effect on plumbagin production in *Plumbago indica* root culture. Results showed that chitosan treatment (150 mg L^{-1}) applied at 14-day-old culture increased plumbagin production up to 6.6-fold compared to untreated root culture (Jaisi and Panichayupakaranant 2017).

11.3.2.1.3 Hormones

Hormonal elicitors have been used to stimulate secondary metabolite production of cell cultures. It was found that the application of methyl jasmonate and growth in the dark stimulated the production of volatile compounds of *Ajuga bracteosa* (Ali et al. 2018). Likewise, methyl jasmonate (50 mM) application of methyl jasmonate (50 μM) on shoot culture of *Prunus salicina* × *Prunus persica* significantly increased anthocyanin production (Lucioli et al. 2017).

11.3.2.2 Eustressor

11.3.2.2.1 Light Spectra

UV radiation effect of UV on the secondary metabolite production in in vitro culture has been examined. Callus cultures of *Vitis vinifera* L Öküzgözü exposed to different UV-C treatments showed an increase in the production of phenolic compounds, ferulic acid and trans-resveratrol (Cetin 2014). On the other hand, study on the effect of the light spectrum on secondary metabolite production in *Stevia rebaudiana* (Bert) callus culture showed that cultures exposed to blue light increased the content of phenols and total flavonoids (Ahmad et al. 2016).

Light wavelength effect on secondary metabolite biosynthesis of *Salvia miltiorrhiza* was studied. Results show that blue light significantly reduces tanshinone IIA content of *S. miltiorrhiza*. This is through gene regulation (Chen et al. 2018).

11.3.2.2.2 Irradiation

Gamma irradiation effect on biomass production and metabolite accumulation was tested in *Hypericum triquetrifolium* Turra callus cultures. Results revealed stimulating effect of gamma irradiation on phenolic compounds and naphthoadiantrone production (Azeez et al. 2017).

11.3.2.2.3 Ozone

Ozone (O_3) is a highly oxidizing molecule. The plant tissue exposure to O_3 causes oxidative stress which induces secondary metabolite biosynthesis such as phenolic compounds. Ozone application effect on in vitro shoot of *Melissa officinal* plants was tested. The ozone treatment (200 ppb for 3 h) increased the accumulation of phenolic compounds. Specifically, rosmarinic acid increased threefold (Tonelli et al. 2015).

11.3.2.2.4 Drought

Drought is a eustressor that affects growth and productivity of plants. Therefore, drought effect on essential oil production in in vitro culture of *Carum copticum* was evaluated. Drought stress can be used to stimulate metabolites production in callus culture of *C. copticum* (Razavizadeh and Adabavazeh 2017).

11.4 Conclusions and Future Prospects

New tendencies aim toward the production of food with high nutritional and nutraceutical value with benefits in human health. New production technique that allows a feasible application in the crop industry has emerged. In this way eustressors and elicitors have become a viable solution for increasing plant metabolism in intensive culture systems, allowing the techniques such as sound and light in order to satisfy the demand for natural compounds since it is a noninvasive physical method that requires technology that nowadays is becoming more available. Both methods are physical abiotic factors and show a positive tendency to improve or provoke the growth process and the responses of the immune system in the plant to be used as a strategy in production systems; however more studies are needed to consider the selective restrictions and other relevant environmental components. The use of chemical factors such as gases, mineral salts, beneficial elements, and nanostructures has generated higher production of secondary metabolites, although there is controversy in different works where it is specified that the application of these compounds is harmful to the plant or crop. Likewise, it has been suggested that the application of this type of eustressors and their effect on the crop will depend on the dose, the type of crop, the type of application, as well as the period of application. The same tendency has been observed when applying elicitors, since studies indicate that the presence of this type of biotic factors increases the growth and defense of the plant before different factors; however the response is not generalized; thus more studies are necessary in this respect, with different experimental models, given that the requirements and responses of each one are different depending on the factor with which they interact.

In addition to the above, continuous biotechnological advances in in vitro culture have made it possible to convert cells into "factories" of natural compounds. However, the current production of secondary metabolites through cell and tissue culture has a series of challenges. To deal with these challenges requires the elucidation of genetic and physiological aspects such as gene expression and regulation of biosynthesis pathways. This will allow obtaining high production cell lines and elicitor treatments for specific compounds and plants. Likewise, the scaling-up of cultivation techniques at industrial level will be more feasible due to the optimization of culture media, control, and management of bioreactors.

Therefore, the various strategies such as controlled elicitation and the in vitro tissue culture have been and must be promoted for their uses for the massive production of plants and for the production of secondary metabolites.

References

Acevedo FE, Peiffer M, Ray S, Meagher R, Luthe DS, Felton GW (2018) Intraspecific differences in plant defense induction by fall armyworm strains. New Phytol 218:310–321

Agati G, Azzarello E, Pollastri S, Tattini M (2012) Flavonoids as antioxidants in plants: location and functional significance. Plant Sci 196:67–76

Aghdam MTB, Mohammadi H, Ghorbanpour M (2016) Effects of nanoparticulate anatase titanium dioxide on physiological and biochemical performance of *Linum usitatissimum* (Linaceae) under well-watered and drought stress conditions. Rev Bras Bot 39:139–146

Ahmad N, Rab A, Ahmad N (2016) Light-induced biochemical variations in secondary metabolite production and antioxidant activity in callus cultures of *Stevia rebaudiana* (Bert). J Photochem Photobiol B Biol 154:51–56

Alavijeh RZ, Sadeghipour O, Riahi H, Dinparvar SV (2016) The effect of sound and music on some physiological and biochemical traits, leaf nutrient concentration and grain yield of cowpea. Iioabj 7:447–458

Ali H, Khan MA, Ullah N, Khan RS (2018) Impacts of hormonal elicitors and photoperiod regimes on elicitation of bioactive secondary volatiles in cell cultures of *Ajuga bracteosa*. J Photochem Photobiol B Biol 183:242–250

Almeida DM, Oliveira MM, Saibo NJM (2017) Regulation of Na+ and K+ homeostasis in plants: towards improved salt stress tolerance in crop plants. Genet Mol Biol 40(1 suppl 1):326–345

Alrawaiq NS, Abdullah A (2014) A review of flavonoid quercetin: metabolism, bioactivity and antioxidant properties. Int J Pharm Tech Res 6:933–941

Altieri MA (2018) Agroecology: the science of sustainable agriculture. CRC Press

Alvarez-Suarez JM, Giampieri F, Tulipani S, Casoli T, Di Stefano G, González-Paramás AM, Santos-Buelga C, Busco F, Quiles JL, Cordero MD, Bompadre S, Mezzetti B, Battino M (2014) One-month strawberry-rich anthocyanin supplementation ameliorates cardiovascular risk, oxidative stress markers and platelet activation in humans. J Nutr Biochem 25:289–294

Al-Zahrani HSM, Elbeshehy EKF, Aldhebiani AY, Elbeaino T (2018) Effect of Cucumber mosaic virus (CMV) infection on antineoplastic alkaloids from periwinkle (*Catharanthus roseus* L.) cultured in the Mecca region and resistance induction by plant-growth-promoting rhizobacteria (PGPR). Biotechnol Biotechnol Equip 32:49–57

Amoozgar A, Mohammadi A, Sabzalian MR (2017) Impact of light-emitting diode irradiation on photosynthesis, phytochemical composition and mineral element content of lettuce cv. Grizzly. Photosynthetica 55:85–95

Anjum SA, Ashraf U, Khan I, Tanveer M, Shahid M, Shakoor A, Wang L (2017) Phyto-toxicity of chromium in maize: oxidative damage, Osmolyte accumulation, anti-oxidative defense and chromium uptake. Pedosphere 27(2):262–273

Asgari-Targhi G, Iranbakhsh A, Ardebili ZO (2018) Potential benefits and phytotoxicity of bulk and nano-chitosan on the growth, morphogenesis, physiology, and micropropagation of *Capsicum annuum*. Plant Physiol Biochem 127:393–402

Ashkavand P, Tabari M, Zarafshar M, Tomášková I, Struve D (2015) Effect of SiO2 nanoparticles on drought resistance in hawthorn seedlings. For Res Pap 76:350–359

Ashry NA, Mohamed HI (2011) Impact of secondary metabolites and related enzymes in flax resistance and or susceptibility to powdery mildew. World J Agric Sci 7(1):78–85

Ávila-Juárez L, Miranda-Rodríguez H (2018) Variations in bioactive content in different tomato trusses due to elicitor effects. J Chem 2018:1–9

Awad V, Kuvalekar A, Harsulkar A (2014) Microbial elicitation in root cultures of *Taverniera cuneifolia* (Roth) Arn. for elevated glycyrrhizic acid production. Ind Crops Prod 54:13–16

Azarakhsh MR, Asrar Z, Mansouri H (2015) Effects of seed and vegetative stage cysteine treatments on oxidative stress response molecules and enzymes in *Ocimum basilicum* L. under cobalt stress. J Soil Sci plant Nutr 15:651–662

Azeez H, Ibrahim K, Pop R, Pamfil D, Hârţa M, Bobi O (2017) Changes induced by gamma ray irradiation on biomass production and secondary metabolites accumulation in *Hypericum triquetrifolium* Turra callus cultures. Indust Crops Prod 108:183–189

Baetz U, Martinoia E (2014) Root exudates: the hidden part of plant defense. Trends Plant Sci 19(2):90–98

Balasubramanian M, Anbumegala M, Surendran R, Arun M (2018) Elite hairy roots of *Raphanus sativus* (L.) as a source of antioxidants and flavonoids. 3Biotech 8:128. https://doi.org/10.1007/s13205-018-1153-y

Bantis F, Smirnakou S, Ouzounis T, Koukounaras A, Ntagkas N, Radoglou K (2018) Current status and recent achievements in the field of horticulture with the use of light-emitting diodes (LEDs). Sci Hortic 235:437–451

Baque MA, Moh SH, Lee EJ, Zhong JJ, Paek KY (2012) Production of biomass and useful compounds from adventitious roots of high-value added medicinal plants using bioreactor. Biotechnol Adv 30:1255–1267

Barrero JM, Downie AB, Xu Q, Gubler F (2014) A role for barley CRYPTOCHROME1 in light regulation of grain dormancy and germination. Plant Cell 26:1094–1104

Blaser MJ, Cardon ZG, Cho MK, Dangl JL, Donohue TJ, Green JL, Knight R, Maxon ME, Northen TR, Pollard KS, Brodie EL (2016) Toward a predictive understanding of Earth's microbiomes to address 21st century challenges. Am Soc Microbiol 7:1–16

Bochu W, Jiping S, Biao L, Jie L, Chuanren D (2004) Soundwave stimulation triggers the content change of the endogenous hormone of the *Chrysanthemum* mature callus. Colloids Surfaces B 37:107–112

Bravo S, García-Alonso J, Martín-Pozuelo G, Gómez V, Santaella M, Navarro-González I, Periago MJ (2012) The influence of post-harvest UV-C hormesis on lycopene, β-carotene, and phenolic content and antioxidant activity of breaker tomatoes. Food Res Int 49:296–302

Bruinsma J (2017) World agriculture: towards 2015/2030: an FAO study. Routledge, London

Burgie ES, Bussell AN, Walker JM, Dubiel K, Vierstra RD (2014) Crystal structure of the photosensing module from a red/far-red light-absorbing plant phytochrome. Proc Nat'l Acad Sci USA 111:10179–10184

Capaldi FR, Gratão PL, Reis AR, Lima LW, Azevedo RA (2015) Sulfur metabolism and stress defense responses in plants. Trop Plant Biol 8:60–73

Cardenas-Manríquez G, Vega-Muñoz I, Villagomez-Aranda AL, Leon-Galvan MF, Cruz-Hernandez A, Torres Pacheco I, Rangel-Cano RM, Rivera-Bustamante RF, Guevara-Gonzalez RG (2016) Proeomic and metabolomic profiles in transgenic tobacco (N. tabacum santhi NC) to CchGLP from Capsicum chinense BG-3821 resistant to biotic and abiotic stresses. Environ Exp Bot 130:33–41

Casal JJ (2013) Photoreceptor signaling networks in plant responses to shade. Annu Rev Plant Biol 64:403–427

Castagna A, Chiavaro E, Dall'Asta C, Rinaldi M, Galaverna G, Ranieri A (2013) Effect of postharvest UV-B irradiation on nutraceutical quality and physical properties of tomato fruits. Food Chem 137(1):151–158

Caverzan A, Casassola A, Brammer SP (2016) Reactive oxygen species and antioxidant enzymes involved in plant tolerance to stress. In: Shanker AK, Shanker C (eds) Abiotic and biotic stress in plants-recent advances and future perspectives. InTech. https://doi.org/10.5772/61368. https://www.intechopen.com/books/abiotic-and-biotic-stress-in-plants-recent-advances-and-future-perspectives/reactive-oxygen-species-and-antioxidant-enzymes-involved-in-plant-tolerance-to-stress. Accessed 21 Sept 2018

Cechin I, Cardoso GS, de Fumis T, Corniani N (2015) Nitric oxide reduces oxidative damage induced by water stress in sunflower plants. Bragantia 74:200–206

Cetin ES (2014) Induction of secondary metabolite production by UV-C radiation in *Vitis vinifera* L. Öküzgözü callus cultures. Biol Res 47:37. https://doi.org/10.1186/0717-6287-47-37

Chamam A, Wisniewski-Dyé F, Comte G, Bertrand C, Prigent-Combaret C (2015) Differential responses of *Oryza sativa* secondary metabolism to biotic interactions with cooperative, commensal and phytopathogenic bacteria. Planta 242(6):1439–1452

Chanclud E, Lacombe B (2017) Plant hormones: key players in gut microbiota and human diseases. Trends Plant Sci 22:754–758

Chávez-Herrera E, Hernández-Esquivel AA, Castro-Mercado E, García-Pineda E (2018) Effect of *Azospirillum brasilense* sp245 lipopolysaccharides on wheat plant development. J Plant Growth Regul 37:859–866

Chen X, O'Halloran J, Jansen MAK (2018) Time matters: the toxicity of zinc oxide nanoparticles to *Lemna minor* L. increases with exposure time. Water Air Soil Pollut 229:99. https://doi.org/10.1007/s11270-018-3759-4

Cheynier V, Comte G, Davies KM, Lattanzio V, Martens S (2013) Plant phenolics: recent advances on their biosynthesis, genetics, and Â ecophysiology. Plant Physiol Biochem 72:1–20

Chialva M, Salvioli di Fossalunga A, Daghino S, Ghignone S, Bagnaresi P, Chiapello M, Novero M, Spadaro D, Perotto S, Bonfante P (2018) Native soils with their microbiotas elicit a state of alert in tomato plants. New Phytol (Online). https://doi.org/10.1111/nph.15014

Choi HG, Moon BY, Kang NJ (2015) Effects of LED light on the production of strawberry during cultivation in a plastic greenhouse and in a growth chamber. Sci Hortic 189:22–31

Choi B, Ghosh R, Gururani MA, Shanmugam G, Jeon J, Kim J, Park SC, Jeong MJ, Han KH, Bae DW, Bae H (2017) Positive regulatory role of sound vibration treatment in *Arabidopsis thaliana* against *Botrytis cinerea* infection. Sci Rep 7:1–14

Çoban Ö, Baydar NG (2016) Brassinosteroid effects on some physical and biochemical properties and secondary metabolite accumulation in peppermint (Mentha piperita L.) under salt stress. Ind Crop Prod 86:251–258

Cocroft HMARB (2014) Plants respond to leaf vibrations caused by insect herbivore chewing. Oecologia 175:257–1266

Coffey A, Prinsen E, Jansen MAK, Conway J (2017) The UVB photoreceptor UVR8 mediates accumulation of UV-absorbing pigments, but not changes in plant morphology, under outdoor conditions. Plant Cell Environ 40:2250–2260

Colinas M, Goossens A (2018) Combinatorial transcriptional control of plant specialized metabolism. Trends Plant Sci 23:324–336

Contreras-Cornejo HA, Macías-Rodríguez L, del-Val E, Larsen J (2016) Ecological functions of Trichoderma spp. and their secondary metabolites in the rhizosphere: interactions with plants. FEMS Microbiol Ecol 92:1–17

Coyago-Cruz E, Corell M, Stinco CM, Hernanz D, Moriana A, Meléndez-Martínez AJ (2017) Effect of regulated deficit irrigation on quality parameters, carotenoids and phenolics of diverse tomato varieties (*Solanum lycopersicum* L.). Food Res Int 96:72–83

De Luca PA, Vallejo-Marín M (2013) What's the "buzz" about? The ecology and evolutionary significance of buzz-pollination. Curr Opin Plant Biol 16:429–435

De Roo L, Vergeynst LL, De Baerdemaeker NJF, Steppe K (2016) Acoustic emissions to measure drought-induced cavitation in plants. Appli Sci 6:71. https://doi.org/10.3390/app6030071

Desneux N, Luna MG, Guillemaud T, Urbaneja A (2011) The invasive South American tomato pinworm, Tuta absoluta, continues to spread in Afro-Eurasia and beyond: the new threat to tomato world production. J Pest Sci 84:403–408

Demkura PV, Ballaré CL (2012) UVR8 mediates UV-B-induced Arabidopsis defense responses against Botrytis cinerea by controlling Sinapate accumulation. Mol Plant 5(3):642–652

Devlin PF, Kay SA (2000) Cryptochromes are required for phytochrome signaling to the circadian clock but not for rhythmicity. Plant Cell 12:2499–2509

Dias MI, Sousa MJ, Alves RC, Ferreira ICFR (2016) Exploring plant tissue culture to improve the production of phenolic compounds: a review. Indust Crops Prod 82:9–22

Ding Z, Jia S, Wang Y, Xiao J, Zhang Y (2017) Phosphate stresses affect ionome and metabolome in tea plants. Plant Physiol Biochem 120:30–39

Dinkova-Kostova AT, Kostov RV (2012) Glucosinolates and isothiocyanates in health and disease. Trends Mol Med 18:337–347

Długosz-Grochowska O, Kołton A, Wojciechowska R (2016) Modifying folate and polyphenol concentrations in Lamb's lettuce by the use of LED supplemental lighting during cultivation in greenhouses. J Funct Foods 26:228–237

Domingos P, Prado Ana M, Wong A, Gehring C, Feijo Jose A (2015) Nitric oxide: a multitasked signaling gas in Plants. Mol Plant 8(4):506–520. https://doi.org/10.1016/j.plaphy.2017.09.007

Dorokhov YL, Komarova TV, Petrunia IV, Frolova OY, Pozdyshev DV, Gleba YY, Vance V (2012) Airborne signals from a wounded leaf facilitate viral spreading and induce antibacterial resistance in neighboring plants. PLoS Pathog 8(4):e1002640

Dowom SA, Abrishamchi P, Radjabian T, Salami SA (2017) Enhanced phenolic acids production in regenerated shoot cultures of *Salvia virgata* Jacq. After elicitation with ag+ ions, methyl jasmonate and yeast extract. Ind Crop Prod 103:81–88

du Jardin, P. (2012). The Science of Plant Biostimulants–A bibliographic analysis, Ad hoc study report. European Commission.

du Jardin P (2015) Plant biostimulants: definition, concept, main categories and regulation. Sci Hortic 196:3–14

Du ST, Liu Y, Zhang P, Liu HJ, Zhang XQ, Zhang RR (2015) Atmospheric application of trace amounts of nitric oxide enhances tolerance to salt stress and improves nutritional quality in spinach (*Spinacia oleracea* L.). Food Chem 173:905–911

Dzakovich MP, Ferruzzi MG, Mitchell CA (2016) Manipulating sensory and phytochemical profiles of greenhouse tomatoes using environmentally relevant doses of ultraviolet radiation. J Agric Food Chem 64:6801–6808

Ebisawa M, Shoji K, Kato M, Shimomura K, Goto F, Yoshihara T (2008) Supplementary ultraviolet radiation B together with blue light at night increased quercetin content and flavonol synthase gene expression in leaf lettuce (*Lactuca sativa* L.). Environ Control Biol 46:1–11

El-Metwally IM, El-Saidy AEA (2016) Physiological and chemical response of faba bean to water regime and cobalt supplement in sandy soils. Int J PharmTech Res 9:158–169

Emamverdian A, Ding Y, Mokhberdoran F, Xie F (2015) Heavy metal stress and some mechanisms of plant defense response. Sci World J 2015:1–18. https://doi.org/10.1155/2015/756120

Esmaeili H, Hadian J, Mirjalili MH, Rezadoost H (2015) The effect of activated charcoal and multi-walled carbon nanotubes on the growth, rosmarinic and caffeic acid content, total phenol, total flavonoid and antioxidant activity of *Satureja rechingeri* calluses. J Bio Env Sci 6:297–304

Ester K, Ćurković-Perica M, Kralj M (2009) The phytohormone auxin induces G1 cell-cycle arrest of human tumor cells. Planta Med 75:1423–1426

Fatma M, Asgher M, Masood A, Khan NA (2014) Excess sulfur supplementation improves photosynthesis and growth in mustard under salt stress through increased production of glutathione. Environ Exp Bot 107:55–63

Fatma M, Masood A, Per TS, Rasheed F, Khan NA (2016) Interplay between nitric oxide and sulfur assimilation in salt tolerance in plants. Crop J 4(3):153–161. https://doi.org/10.1016/j.cj.2016.01.009

Feregrino-Perez AA, Magaña-López E, Guzmán C, Esquivel K (2018) A general overview of the benefits and possible negative effects of the nanotechnology in horticulture. Sci Hortic 238:126–137

Fernandez-Jaramillo AA, Duarte-Galvan C, Garcia-Mier L, Jimenez-Garcia SN, Contreras-Medina LM (2018) Effects of acoustic waves on plants: an agricultural, ecological, molecular and biochemical perspective. Sci Hortic 235:340–348

Figueroa-Pérez MG, Pérez-Ramírez IF, Enciso-Moreno JA, Gallegos-Corona MA, Salgado LM, Reynoso-Camacho R (2018) Diabetic nephropathy is ameliorated with peppermint (*Mentha piperita*) infusions prepared from salicylic acid-elicited plants. J Funct Foods 43:55–61

Fini A, Brunetti C, Di Ferdinando M, Ferrini F, Tattini M (2011) Stress-induced flavonoid biosynthesis and the antioxidant machinery of plants. Plant Signal Behav 6:709–711

Frusciante L, Carli P, Ercolano MR, Pernice R, Di Matteo A, Fogliano V, Pellegrini N (2007) Antioxidant nutritional quality of tomato. Mol Nutr Food Res 51:609–617

Fu W, Li P, Wu Y (2012) Effects of different light intensities on chlorophyll fluorescence characteristics and yield in lettuce. Sci Hortic 135:45–51

Fuentes P, Armarego-Marriott T, Bock R (2018) Plastid transformation and its application in metabolic engineering. Curr Opin Biotechnol 49:10–15

Gadelha CG, Miranda RS, Alencar NLM, Costa JH, Prisco JT, Gomes-Filho E (2017) Exogenous nitric oxide improves salt tolerance during establishment of *Jatropha curcas* seedlings by ameliorating oxidative damage and toxic ion accumulation. J Plant Physiol 212:69–79

Gagliano M (2013) Green symphonies: a call for studies on acoustic communication in plants. Behav Ecol 24:789–796

Gagliano M (2015) The flowering of plant bioacoustics: how and why. Behav Ecol 24:800–801

Garcia-Mier L, Jimenez-Garcia SN, Chapa-Oliver AM, Mejia-Teniente L, Ocampo-Velazquez RV, Rico-García E, Feregrino-Pérez AA, Guevara-Gonzalez R, Torres-Pacheco I (2014) Strategies for sustainable plant food production: facing the current agricultural challenges—agriculture for today and tomorrow. In: Biosystems engineering: biofactories for food production in the century XXI. Springer, pp 1–50

Garcia-Mier L, Jimenez-garcia SN, Guevara-gonzález RG, Feregrino-perez AA, Contreras-medina LM, Torres-pacheco I (2015) Compounds, antioxidant activity, and quality parameters in Sweet bell pepper. J Chem 2015:1–8

Gerolino EF, Chierrito TPC, Santana Filho A, Souto ER, Gonçalves RAC, AJBd O (2015) Evaluation of limonoid production in suspension cell culture of *Citrus sinensis*. Rev Bras 25(5):455–461

Ghosh D, Lin Q, Xu J, Hellmann HA (2017a) Editorial: how plants Deal with stress: exploration through proteome investigation. Front Plant Sci 8:1176

Ghosh D, Lin Q, Xu J, Hellmann HA (2017b) How plants deal with stress: exploration through proteome investigation. Front Plant Sci 8:1–2

Ghosh R, Gururani MA, Ponpandian LN, Mishra RC (2017c) Expression analysis of sound vibration-regulated genes by touch treatment in *Arabidopsis*. Front Plant Sci 8:100

Giampieri F, Alvarez-Suarez JM, Battino M (2014) Strawberry and human health: effects beyond antioxidant activity. J Agric Food Chem 62:3867–3876

Giedrė S, Aušra B, Ramūnas S, Akvilė V, Jurga S, Sandra S, Pavelas D (2013) LED illumination affects bioactive compounds in romaine baby leaf lettuce. J Sci Food Agric 93:3286–3291

Gigolashvili T, Kopriva S (2014) Transporters in plant sulfur metabolism. Front Plant Sci 5(442). https://doi.org/10.3389/fpls.00442

Glick BR (2012) Plant growth-promoting Bacteria: mechanisms and applications. Scientifica 2012:1–15. https://doi.org/10.6064/2012/96340

Goldfine AB, Fonseca V, Jablonski KA, Chen YDI, Tipton L, Staten MA, Shoelson SE (2013) Salicylate (Salsalate) in patients with type 2 diabetes: a randomized trial. Ann Intern Med 159(1):1–12

Gómez DE, Reis EM (2011) Inductores abióticos de resistencia contra fitopatógenos abiotic inductor resistance. Quim Viva 1:6–17

Gonçalves S, Romano A (2013) *In vitro* culture of lavenders (*Lavandula* spp.) and the production of secondary metabolites. Biotechnol Adv 31:166–174

Gong M, Bassi A (2016) Carotenoids from microalgae: a review of recent developments. Biotechnol Adv 34:1396–1412

Grąbkowska R, Matkowski A, Grzegorczyk-Karolak I, Wysokińska H (2016) Callus cultures of *Harpagophytum procumbens* (Burch.) DC. ex Meisn: production of secondary metabolites and antioxidant activity. South Afr J Bot 103:41–48

Grzegorczyk-Karolak I, Kuźma L, Skała E, Kiss AK (2018) Hairy root cultures of *Salvia viridis* L. for production of polyphenolic compounds. Ind Crop Prod 117:235–244. https://doi.org/10.1016/j.indcrop.2018.03.014

Guo J, Wang M-H (2010) Ultraviolet A-specific induction of anthocyanin biosynthesis and PAL expression in tomato (*Solanum lycopersicum* L.). Plant Growth Regul 62:1–8

Gupta R, Singh A, Kanaujia R, Kushwaha S, Pandey R (2018) Trichodermaharzianum ThU and its metabolites underscore alteration in essential oils of *Ocimum basilicum* and *Ocimum sanctum*. Proc Natl Acad Sci India Sect B Biol Sci 88:219–227

Guri AJ, Hontecillas R, Bassaganya-Riera J (2010) Abscisic acid ameliorates experimental IBD by downregulating cellular adhesion molecule expression and suppressing immune cell infiltration. Clin Nutr 29:824–831

Gutiérrez-Luna FM, Hernández-Domínguez EE, Valencia-Turcotte LG, Rodríguez-Sotres R (2018) Pyrophosphate and pyrophosphatases in plants, their involvement in stress responses and their possible relationship to secondary metabolism. Plant Sci 267:11–19

Haddad R, Kamangar A (2016) The ameliorative effect of silicon and potassium on drought stressed grape (*Vitis vinifera* L.) leaves. Iran J Genet Plant Breed 4:48–58

Häder D-P, Williamson CE, Wängberg S-Å, Rautio M, Rose KC, Gao K, Helbling EW, Sinha RP, Worrest R (2015) Effects of UV radiation on aquatic ecosystems and interactions with other environmental factors. Photochem Photobiol Sci 14:108–126

Hao G, Du X, Zhao F, Ji H (2010) Fungal endophytes-induced abscisic acid is required for flavonoid accumulation in suspension cells of *Ginkgo biloba*. Biotechnol Lett 32:305–314

Hao X, Shi M, Cui L, Xu C, Zhang Y, Kai G (2015) Effects of methyl jasmonate and salicylic acid on tanshinone production and biosynthetic gene expression in transgenic Salvia miltiorrhiza hairy roots. Biotechnol Appl Biochem 62:24–31

Hasanuzzaman M, Nahar K, Rahman A, Al Mahmud J, Alharby HF, Fujita M (2018) Exogenous glutathione attenuates lead-induced oxidative stress in wheat by improving antioxidant defense and physiological mechanisms. J Plant Interact 13:203–212

Hassanien RHE, Hou T-z, Li Y-f, Li B-m (2014) Advances in effects of sound waves on plants. J Integr Agric 13:335–348

Hatami M, Ghorbanpour M (2013) Effect of nanosilver on physiological performance of Pelargonium plants exposed to dark storage. J Hortic Res 21:15–20

Hayes S, Velanis CN, Jenkins GI, Franklin KA (2014) UV-B detected by the UVR8 photoreceptor antagonizes auxin signaling and plant shade avoidance. Proc Natl Acad Sci U S A 111:11894–11899

He J, Ren Y, Chen X, Chen H (2014) Protective roles of nitric oxide on seed germination and seedling growth of rice (*Oryza sativa* L.) under cadmium stress. Ecotoxicol Environ Saf 108:114–119

Heggie L, Halliday KJ (2005) The highs and lows of plant life: temperature and light interactions in development. Int J Dev Biol 49:675–687

Hideg É, Strid Å (2016) The effects of UV-B on the biochemistry and metabolism of plants. In: Jordan B (ed) UV-B radiation and plant life molecular biology to ecology. CAB, Bosto, pp 90–110

Ho TT, Lee JD, Jeong CS, Paek KY, Park SY (2018) Improvement of biosynthesis and accumulation of bioactive compounds by elicitation in adventitious root cultures of Polygonum multiflorum. Appl Microbiol Biotechnol 102:1–11

Hontecillas R, Roberts PC, Carbo A, Vives C, Horne WT, Genis S, Velayudhan B, Bassaganya-Riera J (2013) Dietary abscisic acid ameliorates influenza-virus-associated disease and pulmonary immunopathology through a PPAR γ-dependent mechanism. J Nutr Biochem 24:1019–1027

Hou T, Baoming L, Guanghui T, Qing Z, Yingping X, Lirong Q (2009) Application of acoustic frequency technology to protected vegetable production. Trans Chinese Soc Agric Eng 25:156–160

Hussain MS, Fareed S, Saba Ansari M, Rahman A, Ahmad IZ, Saeed M (2012) Current approaches toward production of secondary plant metabolites. J Pharm Bioallied Sci 4(1):10

Islam F, Yasmeen T, Arif MS, Riaz M, Shahzad SM, Imran Q, Ali I (2016) Combined ability of chromium (Cr) tolerant plant growth promoting bacteria (PGPB) and salicylic acid (SA) in attenuation of chromium stress in maize plants. Plant Physiol Biochem 108:456–467

Iqbal N, Umar S, Khan NA (2015) Nitrogen availability regulates proline and ethylene production and alleviates salinity stress in mustard (*Brassica juncea*). J Plant Physiol 178:84–91

Jagadeesh SL, Charles MT, Gariepy Y, Goyette B, Raghavan GSV, Vigneault C (2011) Influence of postharvest UV-C Hormesis on the bioactive components of tomato during post-treatment handling. Food Bioprocess Technol 4(8):1463–1472

Jaisi A, Panichayupakaranant P (2017) Chitosan elicitation and sequential Diaion® HP-20 addition a powerful approach for enhanced plumbagin production in *Plumbago indica* root cultures. Process Biochem 53:210–215

Jamwal K, Bhattacharya S, Puri S (2018) Plant growth regulator mediated consequences of secondary metabolites in medicinal plants. J Appl Res Med Aromat Plants 9:26–38

Jeong M-J, Cho J-I, Park S-H, Kim K-H, Lee SK, Kwon T-R, Park S-C, Siddiqui ZS (2014) Sound frequencies induce drought tolerance in rice plant. Pak J Bot 46:2015–2020

Jiang Y, Wang X, Bao S, Guo R, Johnson DG, Shen X, Li L (2010) INO80 chromatin remodeling complex promotes the removal of UV lesions by the nucleotide excision repair pathway. Proc Natl Acad Sci U S A 107:17274–17279

Jiang N, Grundy S, Bian Z, & Lu C (2017) Investigation of LED light effects on plant growth in improved protected horticulture system

Jiao J, Gai QY, Wang W, Zang YP, Niu LL, Fu YJ, Wang X (2018) Remarkable enhancement of flavonoid production in a co-cultivation system of *Isatis tinctoria* L. hairy root cultures and immobilized *Aspergillus niger*. Indust Crops Prod 112:252–261

Johkan M, Shoji K, Goto F, Hashida S, Yoshihara T (2010) Blue light-emitting diode light irradiation of seedlings improves seedling quality and growth after transplanting in red leaf lettuce. HortScience 45:1809–1814

Joo Y, Schuman MC, Goldberg JK, Kim S, Yon F, Brütting C, Baldwin IT (2018) Herbivore-induced volatile blends with both "fast" and "slow" components provide robust indirect defense in nature. Funct Ecol 32:136–149

Jung J, Kim SK, Kim JY, Jeong MJ, Ryu CM (2018) Beyond chemical triggers: evidence for sound-evoked physiological reactions in plants. Front Plant Sci 9:1–7

Kabera JN, Semana E, Mussa AR, He X (2014) Plant secondary metabolites: biosynthesis, classification, function and pharmacological properties. J Pharm Pharmacol 2:377–392

Kaburagi E, Morikawa Y, Yamada M, Fujiyama H (2014) Sodium enhances nitrate uptake in Swiss chard (*Beta vulgaris var. cicla* L.). Soil Sci Plant Nutr 60(5):651–658

Kaburagi E, Yamada M, Fujiyama H (2015) Sodium, but not potassium, enhances root to leaf nitrate translocation in Swiss chard (*Beta vulgaris var. cicla* L.). Environ Exp Bot 112:27–32

Kang W, Zhu X, Wang Y, Chen L, Duan Y (2018) Transcriptomic and metabolomic analyses reveal that bacteria promote plant defense during infection of soybean cyst nematode in soybean. BMC Plant Biol 18:86

Katiyar D, Hemantaranjan A, Singh B (2015) Chitosan as a promising natural compound to enhance potential physiological responses in plant: a review. Indian J Plant Physiol 20(1):1–9

Kaya C, ASHRAF M, SÖNMEZ O, Tuna AL, Aydemir S (2015) Exogenously applied nitric oxide confers tolerance to salinity-induced oxidativestress in two maize (*Zea mays* L.) cultivars differing in salinity tolerance. Turk J Agric For 39:909–919

Kennedy DO, Wightman EEL (2011) Herbal extracts and phytochemicals: plant secondary metabolites and the enhancement of human brain function. Adv Nutr 2:32–50

Khan MIR, Fatma M, Per TS, Anjum NA, Khan NA (2015) Salicylic acid-induced abiotic stress tolerance and underlying mechanisms in plants. Front Plant Sci 6:462

Khan T, Abbasi BH, Khan MA, Azeem M (2017) Production of biomass and useful compounds through elicitation in adventitious root cultures of *Fagonia indica*. Ind Crops Prod 108:451–457

Khanna VK (2014) Fundamentals of solid-state lighting: LEDs, OLEDs, and their applications in illumination and displays. CRC press

Kilam D, Saifi M, Abdin MZ, Agnihotri A, Varma A (2015) Combined effects of *Piriformospora indica* and *Azotobacter chroococcum* enhance plant growth, antioxidant potential and steviol glycoside content in *Stevia rebaudiana*. Symbiosis 66(3):149–156

Kim HS, Juvik JA (2011) Effect of selenium fertilization and methyl jasmonate treatment on glucosinolate accumulation in broccoli florets. J Am Soc Hortic Sci 136(4):239–246

Kim T-H, Böhmer M, Hu H, Nishimura N, Schroeder JI (2010) Guard cell signal transduction network: advances in understanding abscisic acid, CO_2, and Ca^{2+} signaling. Annu Rev Plant Biol 61:561–591

Kim K, Kook H, Jang Y, Lee W, Kamala-kannan S, Chae J, Lee K (2013) The Effect of Blue-light-emitting diodes on antioxidant properties and resistance to *Botrytis cinerea* in tomato. Plant Pathol Microbiol 4:203. https://doi.org/10.4172/2157-7471.1000203

Kim EY, Park SA, Park BJ, Lee Y, Oh MM (2014) Growth and antioxidant phenolic compounds in cherry tomato seedlings grown under monochromatic light-emitting diodes. Hortic Environ Biotechnol 55:506–513

Kolbert Z (2016) Implication of nitric oxide in excess element-induced morphogenic responses of the root system. Plant Physiol Biochem 101:149–161

Kumari A, Pandey N, Rai S (2017) Protection of *Artemisia annua* roots and leaves against oxidative stress induced by arsenic. Biol Plant 61:367–377

Kyriacou MC, Rouphael Y (2018) Towards a new definition of quality for fresh fruits and vegetables. Sci Hortic 234:463–469

Lajayer BA, Ghorbanpour M, Nikabadi S (2017) Heavy metals in contaminated environment: destiny of secondary metabolite biosynthesis, oxidative status and phytoextraction in medicinal plants. Ecotoxicol Environ Saf 145:377–390

Laschimke R, Burger M, Vallen H (2006) Acoustic emission analysis and experiments with physical model systems reveal a peculiar nature of the xylem tension. J Plant Physiol 163:996–1007

Le KC, Im WT, Paek KY, Park SY (2018) Biotic elicitation of ginsenoside metabolism of mutant adventitious root culture in *Panax ginseng*. Appl Microbiol Biotechnol 102:1687–1697

Leal-Costa MV, Tavares ES, Kuster RM, Lage CLS (2015) Light spectra affect the morphoanatomical and chemical features of clonal *Phyllanthus tenellus* Roxb. grown *in vitro*. J Med Plants Res 9:111–121

Lee M, Son JE, Oh M (2014) Growth and phenolic compounds of *Lactuca sativa* L. grown in a closed-type plant production system with UV-A,-B, or -C lamp. J Sci Food Agric 94:197–204

Leon J, Costa A, Castillo MC (2016) Nitric oxide triggers a transient metabolic reprogramming in *Arabidopsis*. Sci Rep 6:1–14

Li FW, Rothfels CJ, Melkonian M, Villarreal JC, Stevenson DW, Graham SW, Wong GK-S, Mathews S, Pryer KM (2015a) The origin and evolution of phototropins. Front Plant Sci 6:637

Li F, Melkonian M, Rothfels CJ, Villarreal JC, Stevenson DW, Graham SW, Wong GK, Pryer KM, Mathews S (2015b) origin of canonical plant phytochromes. Nat Commun 6:1–12

Lin Q, Liu B, Min F, Zhang Y, Ma J, He F et al (2017) Effects of aluminate flocculant on turion germination and seedling growth of *Potamogeton crispus*. Aquat Toxicol 193:236–244

Liu C-x (2017) Plant Tissue Culture and Biosynthesis Provide a Fast Way to Produce Active Constituents of Traditional Chinese Medicines. Chinese Herbal Medicines 2:002

Liu J, Piñeros MA, Kochian LV (2014) The role of aluminium sensing and signalling in plant aluminium resistance. J Integr Plant Biol 56(3):221–230

Llorach R, Martínez-Sánchez A, Tomás-Barberán FA, Gil MI, Ferreres F (2008) Characterisation of polyphenols and antioxidant properties of five lettuce varieties and escarole. Food Chem 108:1028–1038

Long TA (2011) Many needles in a haystack: cell-type specific abiotic stress responses. Curr Opin Plant Biol 14(3):325–331

Lorenzo JM, Munekata PES (2016) Phenolic compounds of green tea: health benefits and technological application in food. Asian Pac J Trop Biomed 6:709–719

Lubbe A, Verpoorte R (2011) Cultivation of medicinal and aromatic plants for specialty industrial materials. Ind Crops Prod 34:785–801

Lucioli S, Di Bari C, Nota P, Frattarelli A, Forni C, Caboni E (2017) Methyl jasmonate promotes anthocyanins' production in *Prunus salicina* × *Prunus persica in vitro* shoot cultures. Plant Biosyst 151:788–791

Ma G, Zhang L, Kato M, Yamawaki K, Kiriiwa Y, Yahata M, Ikoma Y, Matsumoto H (2012) Effect of blue and red LED light irradiation on β-cryptoxanthin accumulation in the flavedo of citrus fruits. J Agric Food Chem 60:197–201

Mackerness SAH, Surplus SL, Blake P, John CF, Buchanan-Wollaston V, Jordan BR, Thomas B (1999) Ultraviolet-B induced stress and changes in gene expression in *Arabidopsis thaliana*: role of signalling pathways controlled by jasmonic acid, ethylene and reactive oxygen species. Plant Cell Environ 22:1413–1423

Maharaj R, Arul J, Nadeau P (2010) UV-C irradiation of tomato and its effects on color and pigments. Adv Environ Biol 4:308–315

Malayaman V, Sisubalan N, Senthilkumar RP, Ranjithkumar R (2017) Chitosan mediated enhancement of hydrolysable tannin in *Phyllanthus debilis* Klein ex Willd via plant cell suspension culture. Int J Biol Macromol 104:1656–1663

Martínez-Ballesta MC, Moreno DA, Carvajal M (2013) The physiological importance of glucosinolates on plant response to abiotic stress in Brassica. Int J Mol Sci 14:11607–11625

Martínez-Trujillo M, Carreón-Abud Y (2015) Effect of mineral nutrients on the uptake of Cr(VI) by maize plants. New Biotechnol 32(3):396–402

Mazzei P, Vinale F, Woo SL, Pascale A, Lorito M, Piccolo A (2016) Metabolomics by H-HRMAS-NMR of tomato plants treated with two secondary metabolites isolated from Trichoderma. J Agric Food Chem 64(18):3538–3545

Miceli A, Miceli C (2014) Effect of nitrogen fertilization on the quality of Swiss chard at harvest and during storage as minimally processed produce. J Food Qual 37(2):125–134

Mishra RC, Ghosh R, Bae H (2016) Plant acoustics: in the search of a sound mechanism for sound signaling in plants. J Exp Bot 67:4483–4494

Moussous A, Paris C, Khelifi-Slaoui M, Bekhouche M, Zaoui D, Rosloski SM, Makhzoum A, Desobry S, Khelifi L (2018) *Pseudomonas* spp. increases root biomass and tropane alkaloid yields in transgenic hairy roots of *Datura* spp. In Vitro Cell Dev Biol 54:117–126

Müller-Xing R, Xing Q, Goodrich J (2014) Footprints of the sun: memory of UV and light stress in plants. Front Plant Sci 5:474

Mundo J, Villeda-Hernández J, Herrera-Ruiz M, del Carmen GM, Arellano-García J, León-Rivera I, Perea-Arango I (2017) Neuropharmacological and neuroprotective activities of some metabolites produced by cell suspension culture of *Waltheria americana* Linn. Biomed Pharmacother 94:129–139

Murthy HN, Lee E-J, Paek K-Y (2014) Production of secondary metabolites from cell and organ cultures: strategies and approaches for biomass improvement and metabolite accumulation. Plant Cell, Tissue and Organ Culture (PCTOC) 118(1):1–16

Mustafa Z, Ayyub C, Amjad M, Ahmad R (2017) Assesment of biochemical and attibutes against salt stress in eggplant (*Solanum melongena* L.) genotypes japs. J Anim Plant Sci 27(2):503–509

Nahar K, Hasanuzzaman M, Rahman A, Alam M, Mahmud J-A, Suzuki T, Fujita M (2016) Polyamines confer salt tolerance in mung bean (*Vigna radiata* L.) by reducing sodium uptake, improving nutrient homeostasis, antioxidant defense, and methylglyoxal detoxification systems. Front Plant Sci 7:1104

Naik PM, Al-Khayri JM (2016) Abiotic and biotic elicitors–role in secondary metabolites production through in vitro culture of medicinal plants. In: Abiotic and biotic stress in plants-recent advances and future perspectives. InTech

Narayani M, Srivastava S (2017) Elicitation: a stimulation of stress in in vitro plant cell/tissue cultures for enhancement of secondary metabolite production. Phytochem Rev 16(6):1227–1252

Neuman H, Galpaz N, Cunningham FX, Zamir D, Hirschberg J (2014) The tomato mutation nxd1 reveals a gene necessary for neoxanthin biosynthesis and demonstrates that violaxanthin is a sufficient precursor for abscisic acid biosynthesis. Plant J 78:80–93

Nogueirol RC, Monteiro FA, Gratão PL, Borgo L, Azevedo RA (2015) Tropical soils with high aluminum concentrations cause oxidative stress in two tomato genotypes. Environ Monit Assess 187:73

Olle M, Viršilė A (2013) The effects of light-emitting diode lighting on greenhouse plant growth and quality. Agric Food Sci 22:223–234

Paek KY, Murthy HN, Zhong JJ (2014) Production of biomass and bioactive compounds using bioreactor technology. Production of Biomass and Bioactive Compounds Using Bioreactor Technology. https://doi.org/10.1007/978-94-017-9223-3

Pandey P, Irulappan V, Bagavathiannan MV, Senthil-Kumar M (2017) Impact of combined abiotic and biotic stresses on plant growth and avenues for crop improvement by exploiting physio-morphological traits. Front Plant Sci 8:1–15
Papadopoulou GV, Maedicke A, Grosser K, van Dam NM, Martínez-Medina A (2018) Defence signalling marker gene responses to hormonal elicitation differ between roots and shoots. AoB Plants 10:ply031. https://doi.org/10.1093/aobpla/ply031
Pataro G, Sinik M, Capitoli MM, Donsì G, Ferrari G (2015) The influence of post-harvest UV-C and pulsed light treatments on quality and antioxidant properties of tomato fruits during storage. Innov Food Sci Emerg Technol 30:103–111
Patel H, Krishnamurthy R (2013) Journal of pharmacognosy and phytochemistry elicitors in plant tissue culture. J Pharmacogn Phyochem 2:60–65
Pavarini DP, Pavarini SP, Niehues M, Lopes NP (2012) Exogenous influences on plant secondary metabolite levels. Anim Feed Sci Technol 176(1–4):5–16. https://doi.org/10.1016/j.anifeedsci.2012.07.002
Per TS, Masood A, Khan NA (2017) Nitric oxide improves S-assimilation and GSH production to prevent inhibitory effects of cadmium stress on photosynthesis in mustard (*Brassica juncea* L.). Nitric Oxide 68:111–124
Peralta-Videa JR, Hernandez-Viezcas JA, Zhao L, Diaz BC, Ge Y, Priester JH, Holden PA, Gardea-Torresdey JL (2014) Cerium dioxide and zinc oxide nanoparticles alter the nutritional value of soil cultivated soybean plants. Plant Physiol Biochem 80:128–135. https://doi.org/10.1016/j.plaphy.2014.03.028
Perassolo M, Cardillo AB, Mugas ML, Núñez Montoya SC, Giulietti AM, Rodríguez Talou J (2017) Enhancement of anthraquinone production and release by combination of culture medium selection and methyl jasmonate elicitation in hairy root cultures of Rubia tinctorum. Ind Crops Prod 105:124–132
Pérez-Alonso N, Jiménez E (2011) Producción de metabolitos secundarios de plantas mediante el cultivo in vitro. Biotecnología vegetal 11(4)
Pistelli L, Giovannini A, Ruffoni B, Bertoli A, Pistelli L (2010) Hairy root cultures for secondary metabolites production. In: Giardi MT, Rea G, Berra B (eds) Bio-farms for nutraceuticals: functional food and safety control by biosensors. Springer US, Boston, MA, pp 167–184
Possart A, Fleck C, Hiltbrunner A (2014) Shedding (far-red) light on phytochrome mechanisms and responses in land plants. Plant Sci 217–218:36–46
Prasch CM, Sonnewald U (2015) Signaling events in plants: stress factors in combination change the picture. Environ Exp Bot 114:4–14
Qi L, Teng G, Hou T, Zhu B, Liu X (2010) Influence of sound wave stimulation on the growth of strawberry in sunlight greenhouse. In: Li D, Zhao C (eds) Computer and computing technologies in agriculture III. CCTA 2009, IFIP advances in information and communication technology, vol 317. Springer, Berlin/Heidelberg, pp 449–454
Qin Y, Lee W, Choi Y, Kim T (2003) Biochemical and physiological changes in plants as a result of different sonic exposures. Ultrasonics 41(5):407–411
Radhakrishnan R, Pae S-B, Shim K-B, Baek I-Y (2013) Penicillium sp. mitigates fusarium-induced biotic stress in sesame plants. Biotechnol Lett 35(7):1073–1078
Rady MR, Saker MM, Matter MA (2018) In vitro culture, transformation and genetic fidelity of Milk thistle. J Genet Eng Biotechnol. https://doi.org/10.1016/j.jeb.2018.02.007
Ramakrishna A, Ravishankar GA (2011) Influence of abiotic stress signals on secondary metabolites in plants. Plant Signal Behav 6:1720–1731
Ramirez-Estrada K, Vidal-Limon H, Hidalgo D, Moyano E, Golenioswki M, Cusidó RM, Palazon J (2016) Elicitation, an effective strategy for the biotechnological production of bioactive high-added value compounds in plant cell factories. Molecules 21:182
Razavizadeh R, Adabavazeh F (2017) Effects of sorbitol on essential oil of Carum copticum L. under *in vitro* culture. Rom Biotechnol Lett 22:12281–12289
Reihill JA, Malcomson B, Bertelsen A, Cheung S, Czerwiec A, Barsden R, Elborn JS, Dürkop H, Hirsch B, Ennis M, Kelly C, Schock BC (2016) Induction of the inflammatory regulator A20 by gibberellic acid in airway epithelial cells. Br J Pharmacol 173:778–789

Samuoliene G, Brazaityte A, Sirtautas R, Viršile A, Sakalauskaite J, Sakalauskiene S, Duchovskis P (2013) LED illumination affects bioactive compounds in romaine baby leaf lettuce. J Sci Food Agric 93:3286–3291

Sánchez-Pujante PJ, Borja-Martínez M, Pedreño MÁ, Almagro L (2017) Biosynthesis and bioactivity of glucosinolates and their production in plant *in vitro* cultures. Planta 246:19–32

Schmelz EA, Huffaker A, Sims JW, Christensen SA, Lu X, Okada K, Peters RJ (2014) Biosynthesis, elicitation and roles of monocot terpenoid phytoalexins. Plant J 79(4):659–678

Shahidi F, Ambigaipalan P (2015) Phenolics and polyphenolics in foods, beverages and spices: antioxidant activity and health effects-a review. J Funct Foods 18:820–897

Shama G (2007) Process challenges in applying low doses of ultraviolet light to fresh produce for eliciting beneficial hormetic responses. Postharvest Biol Technol 44(1):1–8

Sheng DF, Chen L (2013) Effects of PEG-6000 stress on tanshinones accumulation in hairy roots of Salvia miltiorrhiza. Chin Tradit Herb Drugs 44:1181–1185

Shigeoka S, Ishikawa T, Tamoi M, Miyagawa Y, Takeda T, Yabuta Y, Yoshimura K (2002) Regulation and function of ascorbate peroxidase isoenzymes. J Exp Bot 53:1305–1319

Shahzad A, Sharma S, Parveen S, Saeed T, Shaheen A, Akhtar R, Yadav V, Upadhyay A, Ahmad Z (2017) Historical perspective and basic principles of plant tissue culture. In: Plant biotechnology: principles and applications. Springer, pp 1–36

Siddiqi KS, Husen A (2017) Plant response to engineered metal oxide nanoparticles. Nanoscale Res Lett 12(1):92

Singh A, Dwivedi P (2018) Methyl-jasmonate and salicylic acid as potent elicitors for secondary metabolite production in medicinal plants: a review. J Pharmacogn Phytochem 7:750–757

Singh D, Basu C, Meinhardt-Wollweber M, Roth B (2015) LEDs for energy efficient greenhouse lighting. Renew Sustain Energy Rev 49:139–147

Sircar D, Roychowdhury A, Mitra A (2007) Accumulation of p-hydroxybenzoic acid in hairy roots of *Daucus carota*. J Plant Physiol 164:1358–1366

Sivanandhan G, Arun M, Mayavan S, Rajesh M, Mariashibu TS, Manickavasagam M, Selvaraj N, Ganapathi A (2012) Chitosan enhances withanolides production in adventitious root cultures of *Withania somnifera* (L.) Dunal. Indust Crops Prod 37:124–129

Smetanska I (2008) Production of secondary metabolites using plant cell cultures. In: Food biotechnology. Springer, Berlin, Heidelberg, pp 187–228

Soliman AS, El-feky SA, Darwish E (2015) Alleviation of salt stress on *Moringa peregrina* using foliar application of nanofertilizers. J Hortic For 7:36–47

Son KH, Oh MM (2013) Leaf shape, growth, and antioxidant phenolic compounds of two lettuce cultivars grown under various combinations of blue and red light-emitting diodes. HortSci 48:988–995

Son KH, Lee JH, Oh Y, Kim D, Oh MM, In BC (2017) Growth and bioactive compound synthesis in cultivated lettuce subject to light-quality changes. Hortscience 52(4):584–591

Srivastava MNS (2017) Elicitation: a stimulation of stress in *in vitro* plant cell / tissue cultures for enhancement of secondary metabolite production. Phytochem Rev 16:1227–1252

Sun JW, Zhang H, Wang FY, Sun YM, Sun M (2013) Effects of methyl jasmonate on accumulation and release of main tropane alkaloids in liquid cultures of *Datura stramonium* hairy root. China J Chinese Mater medica 38:1712–1718

Steiner AA (1984) The universal nutrient solution. In: 6. International Congress on Soilless Culture, Lunteren (Netherlands), 29 Apr-5 May 1984. ISOSC

Stevens WR (2013) Building physics: lighting: seeing in the artificial environment. Elsevier

Szopa A, Ekiert H (2014) Production of biologically active phenolic acids in Aronia melanocarpa (Michx.) Elliott *in vitro* cultures cultivated on different variants of the Murashige and Skoog medium. Plant Growth Regul 72:51–58

Sztatelman O, Łabuz J, Hermanowicz P, Banaś AK, Bazant A, Zgłobicki P, Aggarwal C, Nadzieja M, Krzeszowiec W, Strzałka W, Gabrys H (2016) Fine tuning chloroplast movements through physical interactions between phototropins. J Exp Bot 67:4963–4978

Talebi M, Moghaddam M, Pirbalouti AG (2018) Methyl jasmonate effects on volatile oil compounds and antioxidant activity of leaf extract of two basil cultivars under salinity stress. Acta Physiol Plant 40(2):34

Telewski FW (2006) A unified hypothesis of mechanoperception in plants. Am J Bot 93:1466–1476
Thomas PJ, Carpenter D, Boutin C, Allison JE (2014) Rare earth elements (REEs): effects on germination and growth of selected crop and native plant species. Chemosphere 96:57–66
Tiwari R, Rana C (2015) Plant secondary metabolites: a review. Int J Eng Res Gen Sci 3(5):661–670
Tonelli M, Pellegrini E, D'Angiolillo F, Petersen M, Nali C, Pistelli L, Lorenzini G (2015) Ozone-elicited secondary metabolites in shoot cultures of *Melissa officinalis* L. Plant Cell Tissue Org Cult 120:617–629
Torkamani M, Jafari M, Abbaspour N, Heidary R, Safaie N (2014) Enhanced production of valerenic acid in hairy root culture of *Valeriana officinalis* by elicitation. Open Life Sci 9:853–863
Ulrich Z, Heike SH, Wegner L, Hans Jürgen W, Michael S, Axel H, Friedrich Wilhelm B (2002) What are the driving forces for water lifting in the xylem conduit. Physiol Plant 114:327–335
Utzschneider KM, Kratz M, Damman CJ, Hullarg M (2016) Mechanisms linking the gut microbiome and glucose metabolism. J Clin Endocrinol Metab 101:1445–1454
van der Ent A, Baker AJ, Reeves RD, Pollard AJ, Schat H (2013) Hyperaccumulators of metal and metalloid trace elements: facts and fiction. Plant Soil 362(1–2):319–334
Vázquez-Hernández MC, Parola-Contreras I, Montoya-Gómez LM, Torres-Pacheco I, Schwarz D, Guevara-González RG (2019) Eustressors: chemical and physical stress factors used to enhance vegetables production. Sci Hortic 250:223–229
Verma N, Shukla S (2015) Impact of various factors responsible for fluctuation in plant secondary metabolites. J Appl Res Med Aromat Plants 2(4):105–113
Verma P, Mathur AK, Masood N, Luqman S, Shanker K (2013) Tryptophan over-producing cell suspensions of *Catharanthus roseus* (L) G. Don and their up-scaling in stirred tank bioreactor: detection of a phenolic compound with antioxidant potential. Protoplasma 250:371–380
Vezza ME, Llanes A, Travaglia C, Agostini E, Talano MA (2018) Arsenic stress effects on root water absorption in soybean plants: physiological and morphological aspects. Plant Physiol Biochem 123:8–17
Walley JW, Dehesh K (2010) Molecular mechanisms regulating rapid stress signaling networks in Arabidopsis. J Integr Plant Biol 52(4):354–359
Waqas M, Khan AL, Hamayun M, Shahzad R, Kim Y-H, Choi K-S, Lee I-J (2015) Endophytic infection alleviates biotic stress in sunflower through regulation of defence hormones, antioxidants and functional amino acids. Eur J Plant Pathol 141(4):803–824
Widhalm JR, Jaini R, Morgan JA, Dudareva N (2015) Rethinking how volatiles are released from plant cells. Trends Plant Sci 20(9):545–550
Wiesel L, Newton AC, Elliott I, Booty D, Gilroy EM, Birch PRJ, Hein I (2014) Molecular effects of resistance elicitors from biological origin and their potential for crop protection. Front Plant Sci 5(November):1–13
Wilson SA, Roberts SC (2012) Recent advances towards development and commercialization of plant cell culture processes for the synthesis of biomolecules. Plant Biotechnol J 10(3):249–268
Xing K, Zhu X, Peng X, Qin S (2014) Chitosan antimicrobial and eliciting properties for pest control in agriculture: a review. Agron Sustain Dev 35(2):569–588
Xu F, Shi L, Chen W, Cao S, Su X, Yang Z (2014) Effect of blue light treatment on fruit quality, antioxidant enzymes and radical-scavenging activity in strawberry fruit. Sci Hortic 175:181–186
Xu S, Zhou W, Pottinger S, Baldwin IT (2015) Herbivore associated elicitor-induced defences are highly specific among closely related Nicotiana species. BMC Plant Biol 15(1):2
Xue W, Yousheng, L, Dongwu L, Hengjian X, Tao L, Fengyun Z (2012) Cerium toxicity, uptake and translocation in Arabidopsis thaliana seedlings. J Rare Earth 30(6):579–585
Yang R, Wang LQ, Liu Y (2014) Research progress on tissue culture of *Glycyrrhizae radix* et *rhizoma*. Chin Tradit Herb Drugs 45:1796–1802
Yu S, Jiang S, Zhu L, Zhang J, Jin Q (2013) Effects of acoustic frequency technology on rice growth, yield and quality. Trans Chinese Soc Agric Eng 29:141–147
Yue W, Ming Q, Lin B, Rahman K, Zheng CJ, Han T, Qin L (2016) Medicinal plant cell suspension cultures: pharmaceutical applications and high-yielding strategies for the desired secondary metabolites. Crit Rev Biotechnol 36:215–232

Zaimenko NV, Didyk NP, Dzyuba OI, Zakrasov OV, Rositska NV, Viter AV (2014) Enhancement of drought resistance in wheat and corn by nanoparticles of natural mineral analcite. Ecol Balk 6:1–10

Zaker A, Sykora C, Gössnitzer F, Abrishamchi P, Asili J, Mousavi SH, Wawrosch C (2015) Effects of some elicitors on tanshinone production in adventitious root cultures of *Perovskia abrotanoides* Karel. Indust Crops Prod 67:97–102

Zhang (2012) Application progress of plant audio control technology in modern agriculture. Ningxia J Agric For Sci Technol 53:80–81

Zhang T, Maruhnich SA, Folta KM (2011) Green light induces shade avoidance symptoms. Plant Physiol 157:1528–1536

Zhang Y, Butelli E, Martin C (2014) Engineering anthocyanin biosynthesis in plants. Curr Opin Plant Biol 19:81–90

Zheng CJ, Wu XY, Wang ZJ (2014) Optimization of inducement conditions for hairy roots of *Morinda officinalis*. Mod Agric Sci Tech 10:77–78

Zolala J, Farsi M, Gordan HR, Mahmoodnia M (2010) Producing a high scopolamine hairy root clone in *Hyoscyamus muticus* through transformation by *Agrobacterium rhizogenes*. J Agric Sci Technol 9:327–339

Zoltowski BD, Imaizumi T (2014) Structure and function of the ZTL/FKF1/LKP2 group proteins in Arabidopsis. Enzymes. Elsevier 35:213–239

Zoratti L, Karppinen K, Luengo Escobar A, Häggman H, Jaakola L (2014) Light-controlled flavonoid biosynthesis in fruits. Front Plant Sci 5:534

KRAS as Potential Target in Colorectal Cancer Therapy

12

Shu-Kee Eng, Teng Hern Tan Loh, Bey-Hing Goh, and Wai-Leng Lee

Contents

Abstract

Colorectal cancer (CRC) is among the most commonly diagnosed cancers affecting both genders in the world. It is characterized by genetic instability, which drives tumor formation via the activation of oncogenes, such as *KRAS* and B-*RAF*. Approximately 40% of CRC patients harbor the mutated *KRAS* oncogene as the predominant form of *RAS* mutation which plays a key role in the early development of cancer adenoma. The constitutively mutated K-Ras protein is

S.-K. Eng · W.-L. Lee (✉)
School of Science, Monash University Malaysia, Subang Jaya, Selangor, Malaysia
e-mail: lee.wai.leng@monash.edu

T. H. T. Loh
Biomedical Research Laboratory, Jeffrey Cheah School of Medicine and Health Sciences, Monash University Malaysia, Subang Jaya, Selangor, Malaysia

Biofunctional Molecule Exploratory Research Group (BMEX), School of Pharmacy, Monash University Malaysia, Subang Jaya, Selangor, Malaysia

Novel Bacteria and Drug Discovery Research Group (NBDD), School of Pharmacy, Monash University Malaysia, Subang Jaya, Selangor, Malaysia

M. S. Akhtar et al. (eds.), *Natural Bio-active Compounds*,
https://doi.org/10.1007/978-981-13-7154-7_12

able to bypass signals from its upstream effector, epidermal growth factor receptor (EGFR), thus rendering the existing anti-EGFR treatments (cetuximab and panitumumab) ineffective. However, there is no direct anti-KRAS treatment showing clinical benefits despite several decades of comprehensive efforts. To date, many efforts have been done to target the aberrant *KRAS* signaling at different levels including (1) inhibit RAS membrane association and (2) inhibit downstream KRAS effectors. The present chapter highlights the existing approaches in the management of *KRAS*-driven cancers by targeting RAS, and also discusses the potential drug candidates on this horizon.

Keywords
Cancer · EGFR effector · Oncogene · RAS biology · Tumor formation

12.1 Introduction

Colorectal cancer (CRC) is one of the most commonly diagnosed cancers in both male and female worldwide (Siegel et al. 2017). It is commonly classified as colon or bowel cancer which involves the development of tumors in the tissues lining the lower part of the large intestine. Genetic instability is the one of the major characteristics of CRC, whereby a series of genetic and epigenetic alterations can take place, driving the formation of tumors by activating oncogenes such as *KRAS* and B-*RAF* and by inactivating tumor suppressor genes, *TP53* (Pino and Chung 2010, Sieber et al. 2003). The accumulation of these altered genetic materials in CRC often associated with the development of invasive and metastatic tumor from an adenoma (Fearon and Vogelstein 1990). Particularly, the mutation in RAS, including the *KRAS*, *NRAS*, and *HRAS* as the three main RAS family, has been found to drive CRC formation and progression. Being the most frequently mutated RAS isoform, *KRAS* is detected in 40–45% of CRC (Bokemeyer et al. 2009; Karapetis et al. 2008; Van Cutsem et al. 2011). Accounting for 95% of all mutation

B.-H. Goh
Biomedical Research Laboratory, Jeffrey Cheah School of Medicine and Health Sciences, Monash University Malaysia, Subang Jaya, Selangor, Malaysia

Biofunctional Molecule Exploratory Research Group (BMEX), School of Pharmacy, Monash University Malaysia, Subang Jaya, Selangor, Malaysia

Novel Bacteria and Drug Discovery Research Group (NBDD), School of Pharmacy, Monash University Malaysia, Subang Jaya, Selangor, Malaysia

Asian Centre for Evidence Synthesis in Population, Implementation and Clinical Outcomes, Health and Well-Being Cluster, Global Asia in the 21st Century Platform, Monash University Malaysia, Subang Jaya, Selangor, Malaysia

Center of Health Outcomes Research and Therapeutic Safety, School of Pharmaceutical Sciences, University of Phayao, Phayao, Thailand

types in CRC, codon 12 and 13 are two hotspots (occurring 80% and 20%, respectively) of *KRAS* mutations with G12D and G12V being the most frequent point mutation patterns in codon 12 and G13D in codon 13 (Neumann et al. 2009). On the other hand, a low mutation rate of *NRAS* was found in CRC (approximately 3%) and no reported *HRAS* mutations in CRC so far (Irahara et al. 2010; Vaughn et al. 2011; Cox et al. 2014). The detection of mutated *KRAS* in the early stages of CRC suggested the involvement of *KRAS* mutation in the early events of tumor development and is described as a key event for the progression of adenoma to CRC (McLellan et al. 1993; Nash et al. 2010). Thus, *KRAS* could be an attractive therapeutic target for CRC harboring mutated *KRAS*.

For the past decades, the development of targeted therapies such as the two monoclonal antibodies – cetuximab and panitumumab – which target the epidermal growth factor receptor (EGFR) have improved the survival of patients with CRC. Ligand-induced EGFR activation plays a significant role in tumor proliferation, invasion, migration, and neovascularization mediated by downstream events including the RAS-RAF-MAPK and PI3K-AKT-mTOR pathways. However, the mutations in the *KRAS* oncogene have rendered nonresponsiveness to cetuximab and panitumumab among the CRC patients with mutated *KRAS*. CRC patients with tumors bearing high expression level of EGFR and wild-type *KRAS* showed an overall response rate of 35% to treatment with cetuximab and panitumumab, whereas a very low response rate (3%) was observed in patients carrying *KRAS* mutations (Amado et al. 2008; Lievre et al. 2008; Raponi et al. 2008; Saif et al. 2009). Moreover, progression-free survival and overall survival in response to anti-EGFR monoclonal antibody (cetuximab and panitumumab) therapy showed improvement only in patients carrying wild-type *KRAS* compared to those bearing mutant *KRAS* (Deschoolmeester et al. 2010). Clearly, EGFR inhibitors only block the receptor signaling and its downstream events, including those mediated by *KRAS*; the mutations in *KRAS* gene lead to constitutively activated GTP-bound RAS proteins which in turn bypass the upstream EGFR signals and render the blocking of EGFR ineffective. Several retrospective studies demonstrated that *KRAS* mutations in exon 2 (codons 12 and 13) are associated with high resistance to cetuximab or panitumumab in CRC treatment (Moroni et al. 2005).

To date, targeting KRAS signaling in human cancer remains difficult and challenging. In fact, there is still no approved direct inhibitor on oncogenic *KRAS*. As K-Ras is the key regulator of many effector proteins such as phosphoinositide 3-kinase (PI3K) and kinase rapidly accelerated fibrosarcoma (Raf) which are involved in cancer cell proliferation and survival, the development of new therapeutic strategies for treating tumor bearing mutated KRAS is therefore a pressing need (Prenen et al. 2010). Furthermore, early attempts to block oncogenic *RAS* by interfering farnesylation process (an essential step in posttranslational modification for Ras activity) were hindered by the presence of an alternative prenylation system (geranylgeranyltransferase) that can recover the activity of K-Ras and N-Ras after the inhibition of farnesyltranferase. A number of attempts have also been made to target the downstream effector proteins in the *KRAS* signaling pathway. Targeting downstream pathways of *RAS* is complicated due to its complex networks involving a series of downstream effectors and the presence of feedback loops. Although early efforts yielded limited clinical outcome, the search of different approaches to inhibit Ras signaling is still ongoing with the hope of finding a successful treatment. The present chapter highlights the existing approaches in the management of *KRAS*-driven cancers by targeting RAS and also discusses the potential drug candidates on the horizon.

12.2 RAS Biology

In the 1960s, *RAS* genes were originally identified as viral oncogenes transduced from the genome of Harvey and Kirsten rat sarcoma viruses (Fig. 12.1). The study of these viral genes and their encoded proteins has gained momentum in the 1980s with the identification of constitutively activated and potently transformed RAS genes in human cancer cell lines (Karnoub and Weinberg 2008). *HRAS* and *KRAS* were the first two human oncogenes detected in later years after the discovery of these oncogenic genetic materials in Harvey and Kirsten murine sarcoma viruses in 1964 and 1967, respectively (Harvey 1964; Kirsten et al. 1970). Subsequently, further research identified *NRAS* as the third human *RAS* gene which was first found in human neuroblastoma cells. In summary, the human genome comprised of three *RAS* gene which encodes four different protein isoforms: H-Ras, N-Ras, K-Ras4A, and K-Ras4B. As shown in Fig. 12.2, each Ras protein is made up of two main domains, membrane-targeting domain and G domain. Almost all Ras proteins are highly conserved and closely related to each isoforms (85% amino sequence similarity at the G domain) with major differences detected at the hypervariable region of their C-terminal membrane-targeting domains (Cox et al. 2015). The difference in the two K-*Ras* isoforms were due to alternative exon 4 splicing in the *KRAS* locus with K-Ras4B being the major isoform expressed in most tissues (Tsai et al. 2015).

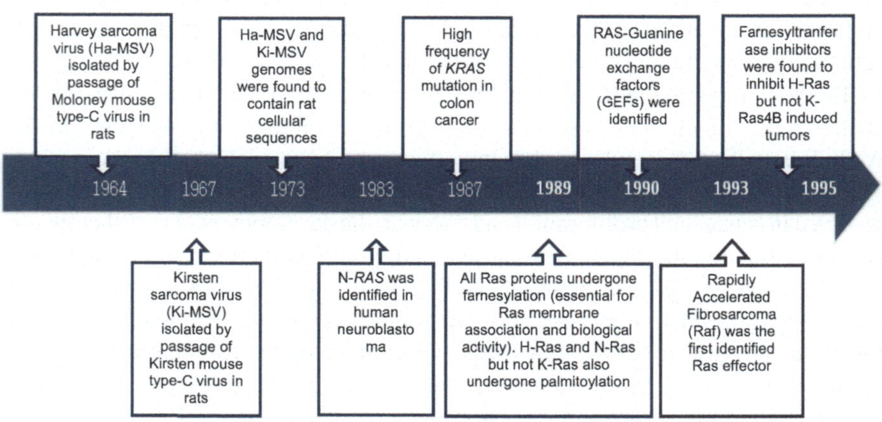

Fig. 12.1 Timeline of key discoveries in Ras research. (Adopted from Malumbres and Barbacid 2003; Cox and Der 2010)

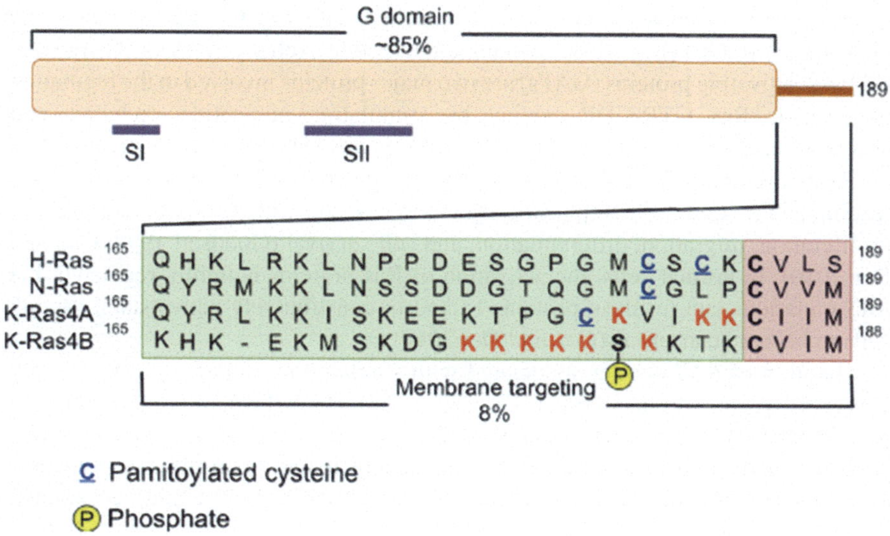

Fig. 12.2 All human Ras proteins consist of two major domains, the G domain and membrane-targeting domain. The G domain (amino acid 1–164) includes protein regions responsible for the nucleotide binding and GTP hydrolysis particularly the switch regions (switch I (SI), amino acid 30–38; and switch II (SII), amino acid 60–76) which encounter conformational changes during GDP-GTP exchange. The membrane-targeting domain is made up of the remaining 24–25 C-terminal residues whereby the first 20–21 amino acids (in green box) refer to the hypervariable (HV) domain and the latter (in pink box) consist of a CAAX tetrapeptide motif (C, cysteine; A, any aliphatic residue; and X, terminal amino acid). The HV domain is the region where the four Ras protein isoforms exhibit the greatest differences in terms of amino acid sequences. It consists of important membrane-targeting sequencing element that includes palmitoylatable cysteines and stretches of positively charged (polybasic) amino acid residues (PBR). Unlike the other three Ras protein isoforms, K-Ras4B does not own a palmitoylatable cysteines but a serine (amino acid 181) that can be phosphorylated and regulates the protein association to the plasma membrane. The CAAX motif is involved in posttranslational lipid modification of Ras protein whereby the cysteine residue is covalently attached with a C15 farnesyl group by the enzyme farnesyltransferases. (Adapted from Zeitouni et al. 2016)

Ras proteins are founding members of the superfamily of small G proteins with intrinsic GTPases activity that governs a series of cellular signal transduction pathways. A common characteristic of Ras proteins is that they govern a series of cellular signal transduction across membrane, particularly in transducing signals induced by growth factors. In particular, Ras is an important downstream effector in the EGFR signaling pathway. It is known to play key roles in signal transduction cascade, especially in activation of PI3K, the mitogen-activated extracellular signal-regulated kinases (ERK1 and ERK2), and Raf which is associated with cell proliferation (Schubbert et al. 2007). Nascent Ras traffics to plasma membrane from cytosol where it stays and connects signals from cell surface receptors to intracellular effector pathways including phosphoinositide 3-kinase (PI3K)-Akt, Son of Sevenless (SOS), and Raf-MEK-ERK mitogen-activated protein kinase (MAPK) (Hanahan and Weinberg 2011). Individual Ras proteins are activated when bound to guanosine triphosphate (GTP) and are inactive with the binding of guanosine diphosphate (GDP) (Fig. 12.3a). Guanine nucleotide exchange factors (GEFs) and GTPase-activating proteins (GAPs) are two major proteins involved in the regulation of intrinsic Ras GTP-GDP cycling by stimulating nucleotide exchange and accelerating intrinsic GTP hydrolysis activity, respectively. The activated Ras GTPases preferentially interacts with a wide spectrum of downstream effectors, resulting in a series of cytoplasmic signaling cascade which initiate cell growth, proliferation, migration, differentiation, and cell survival (Goodsell 1999; Cox and Der 2010). In normal cells, the activation of Ras proteins is tightly regulated, but *RAS* mutations can result in proteins locked in constitutively active state, driving cells to become cancerous (Cox and Der 2010).

Mutation of *RAS* is often associated with single point mutation at their highly conserved region around codon 12, 13, or 61 which favors the formation of permanent GTP-bound (active state) of Ras proteins, leading to unintended and persistent activation of a series of Ras-dependent downstream effector pathways even without the presence of incoming signals (Fig. 12.3b) and, eventually,

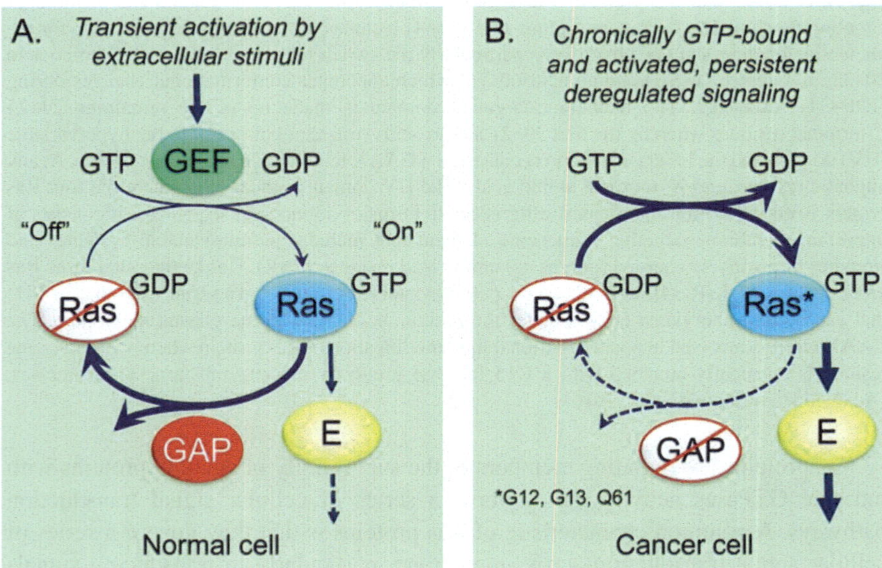

Fig. 12.3 Ras GDP-GTP cycle in normal and cancer cells. (**a**) *RAS* in normal cells. Wild-type Ras protein cycles between active GTP-bound and inactive GDP-bound states, and the proteins exist mainly in inactive form in nondividing cells. Ras is activated transiently by growth factors via the activation of guanine nucleotide exchange factors (GEFs) which facilitate GTP binding to Ras. Ras-GTP then binds preferentially to a series of downstream effectors (labeled E) such as Son of Sevenless (Sos) and phosphoinositide 3-kinase (PI3K). The signaling cascade of Ras is attenuated by GTPase-activating proteins (GAPs) which promote intrinsic hydrolysis of GTP, returning Ras protein to its inactive state (Ras-GDP). (**b**) *RAS* mutations in cancer cells. Missense mutations at residues G12, G13, or G61 impair intrinsic GTP hydrolysis activity of GAPs, thus resulting in Ras protein locked in constitutively active GTP-bound state (Adapted from Baines et al. 2011)

contributing to the formation of cancer (Schramm et al. 2000; Vigil et al. 2010). Belonging to the family of *RAS*, *KRAS*, *NRAS*, and *HRAS* are three most common oncogenes found in approximately 30% of all human cancers with *KRAS* being the most commonly mutated isoforms (22%) followed by *NRAS* (8%) and *HRAS* (3%). These *RAS* genes are involved in almost every steps of tumorigenesis with specific mutated isoforms varies among cancer cell types (Prior et al. 2012; Cox et al. 2014). The mutation of *KRAS* genes is found predominantly in cancers of the colon, lung, and pancreas, while *HRAS* gene mutation is highly associated with cancers of the head, neck, and skin, and the mutations of *NRAS* often lead to hematopoietic malignancies (Rajalingam et al. 2007).

12.3 *KRAS* Mutation in Colorectal Cancer

Vogelstein et al. (1988) described point mutations in *KRAS* as an early event in pathogenesis of CRC. In fact, *KRAS* mutations have been identified in approximately 30–50% of CRC cases. Undoubtedly, *KRAS* mutation represents an important biomarker that predicts the suitability for treatment using EGFR inhibitors in metastatic CRC. However, the prognostic value of *KRAS* mutations in CRC remains controversial due to the discrepancies demonstrated by several reports in defining the prognosis. Thereby, some studies revealed the association of poorer prognosis with *KRAS* mutations, while some others reported negative association. Majority of the findings supported the poor prognostic value of *KRAS* mutations in CRC as there is no significant difference in the survival rate of patients carrying either wild-type *KRAS* or mutated *KRAS*, while several clinical trials found that the presence of *KRAS* mutation on codon 13 lowered the survival rate of stages I and II CRC patients as compared to those with wild-type *KRAS* (Russo et al. 2005; Dinu et al. 2014). The analysis of the *KRAS* mutational statuses has proven to be very strong negative predictive biomarker for selection of therapeutic approaches in CRC.

Epidermal growth factor receptor (EGFR, also known as human EGF receptor, HER or c-erbB1) is a tyrosine kinase receptor which has been the primary targets to be exploited for treatment in CRC with monoclonal antibodies (Cohen 2003). Acting as a membrane signal transducer, EGFR is commonly overexpressed in approximately 80% of tumors in CRC with poor prognosis (Prenen et al. 2010). Cetuximab (Erbitux) and panitumumab (Vectibix) are the first two anti-epidermal growth factor receptor (EGFR)-targeted monoclonal antibodies approved by the US Food and Drug Administration (FDA) in metastatic CRC treatment. They function to inhibit the ligand-dependent activation of EGFR by binding to its extracellular receptor site (Goldstein et al. 1995; Li et al. 2005). Cetuximab (approved by FDA in 2004), a human mouse IgG1 monoclonal antibody, is used as a second-line treatment in treating EGFR-positive metastatic CRC patients who are intolerant with irinotecan-based therapy or in combination drug therapy with 5-FU, irinotecan, and oxaliplatin, while panitumumab (approved in 2007 by FDA) is used as a third-line treatment in metastatic CRC. Patient response to clinical drug is one of the key factors in a successful treatment; however, EGFR-positive metastatic CRC patients with tumors bearing *KRAS* mutation were shown to be resistant to anti-EGFR treatment (Lievre et al. 2006; Amado et al. 2008; De Roock et al. 2008).

12.4 Treatment of *KRAS*-Driven Cancer

KRAS is one of the major oncogenes associated with invasive cancers such as colon, pancreatic, and non-small-cell lung cancers which makes it an ideal target for cancer therapeutics. After three decades of characterization, *KRAS* mutations still poses a great therapeutic challenge as the oncogenic proteins are the key regulators of many downstream pathways involved in cell proliferation and survival, and their surfaces lack of major sites for the binding of small molecules (Cooper 1982). This led to the debates on the efficacy of therapeutic approaches targeting Ras in treating *RAS*-driven cancers (Schaeffer et al. 1994). The key role of mutant *KRAS* in promoting tumor growth was supported by one of the very first studies involving the disruption of mutated *KRAS* genes in HCT116 and DLD-1 CRC cell lines via homologous recombination. The removal of the mutant *KRAS* allele (without affecting the wild-type allele) significantly altered the cell morphology and impaired anchorage-independent growth both in vitro and in nude mice (Shirasawa et al. 1993). Another study utilized short hairpin RNA (shRNA) to silent mutant *KRAS* expression in both CAPAN-1 human pancreatic ductal adenocarcinoma and SW480 CRC cell lines, resulting in the impaired tumor xenograft growth in mice (Lim et al. 2005). To date, *KRAS* has remained notoriously "undruggable" whereby drug developed to inhibit the oncogenic mutation has not been able to stop the progression of cancer in a clinical setting (Stephen et al. 2014).

12.4.1 Targeting KRAS Membrane Association

Ras protein is initially synthesized in the cytosol and activated via several posttranslational modification processes (Fig. 12.4). The first step is the farnesylation process which involves the addition of isoprenoid farnesyl (15-carbon farnesyl) to the cysteine resides in the CAAX box at the C-terminal end (C, cysteine; A, aliphatic amino acid; X, terminal amino acid typically methionine or serine) (Shirasawa et al. 1993). The AAX amino acids are then cleaved off by Ras-converting CAAX endopeptidase 1 (Rce1) followed by carboxymethylation of the farnesylated cysteine by isoprenylcysteine carboxyl methyltransferase (Icmt) (Prior and Hancock 2001). The protein is eventually palmitoylated and anchored to the plasma membrane in the presence of palmitoyltransferase (for K-Ras4A, H-Ras, and N-Ras) or polybasic domain (K-Ras4B). The protein is then activated at the plasma membrane via the binding of GTP (Hancock et al. 1990). Farnesylation of Ras at the C-terminal is an essential process in facilitating the localization of Ras protein to the plasma membrane which then binds to various downstream effectors of signaling cascades. The insertion of the isoprenoid farnesyl moiety into the plasma membrane was found to be a targeted event, whereby the prenylated Ras protein binds specifically

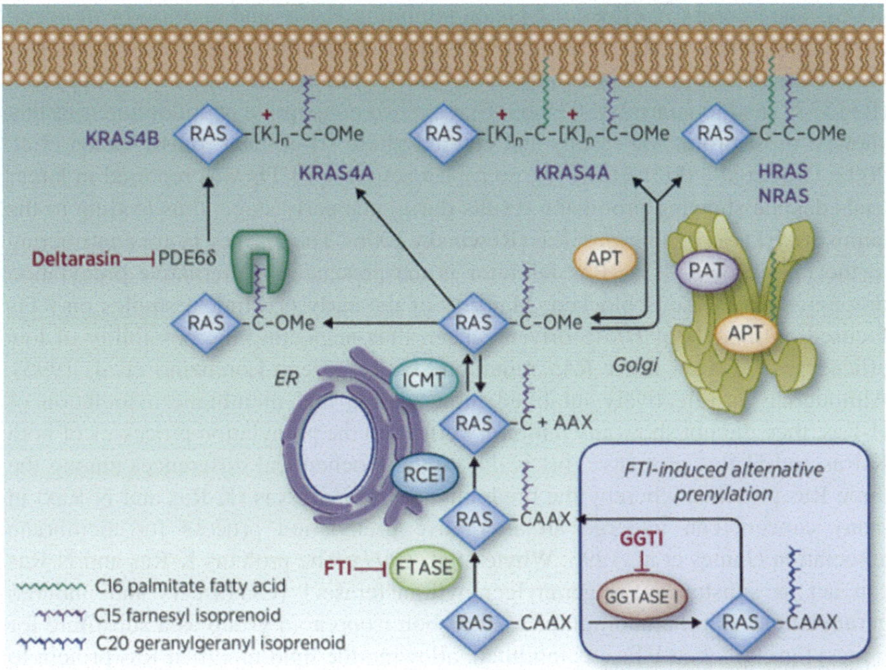

Fig. 12.4 Trafficking pathway of Ras proteins. Nascent Ras proteins synthesized in cytosol are activated via three posttranslational modifications. The first modification of Ras involves the addition of a C-15 farnesyl isoprenoid lipid to the cysteine residues in the CAAX motif followed by Ras-converting CAAX endopeptidase 1 (Rce1)-mediated proteolytic cleavage of AXX residues and finally carboxymethylation of the farnesylated cysteine catalyzed by isoprenylcysteine carboxyl methyltransferase (Icmt). The presence of FTase inhibitors (FTIs) prevents the first and subsequent steps of posttranslational modifications but induces the alternative prenylation of K-Ras and N-Ras via geranylgeranyltransferase I (GGTase I). GGTase I catalyzes the attachment of C-20 isoprenoid group to Ras which allow the same subsequent posttranslational modifications of Ras. To be trafficked to the plasma membrane, N-Ras and H-Ras with one and two palmitoylatable cysteine, respectively, undergo reversible acylation catalyzed by Golgi-resident protein acyltransferase (PAT). K-Ras4B with no palmitoylatable cysteines but stretches of six adjoining lysines (polybasic region, PBR) is trafficked and anchored directly to the plasma membrane via electrostatic charges without going through acylation at Golgi apparatus. K-Ras4A which consists of a combination of palmitoylatable cysteine and two branches of PBR undergoes both trafficking pathways. (Adapted from Cox et al. 2015)

to the prenyl receptors on the plasma membrane (Marshall 1993). The critical role of the enzyme farnesyltransferase (FTase) in determining the proper functioning of Ras has led to joint effort in identifying potential inhibitors of this enzyme with the aim to inhibit Ras binding to the plasma membrane, thus interfering its signaling pathway (Samatar and Poulikakos 2014).

For the past two decades, a series of chemically distinct farnesyltransferase inhibitors (FTIs) were identified which include farnesyl diphosphate analogs, bisubstrate inhibitors, no peptide peptidomimetics, and CAAX peptidomimetics with several of these advancing to clinical trials for cancer treatment (Basso et al.

2006). The impressive effect of FTIs on inhibiting H-Ras and tumor growth in cell culture and mouse models bearing *HRAS*-driven cancer has led to the progression of these FTIs into clinical trials in the early 1920s. Among the tested FTIs, tipifarnib (R115777) and lonafarnib (SCH66336) were two nonpeptide peptidomimetrics that showed most promising results and entered phase III clinical trials (Awada et al. 2002; Crul et al. 2002). However, no robust activity of FTIs was reported in larger trials despite showing promising results during the early stage, thus leading to the demise of FTIs as inhibitors of Ras (Rowinsky 2006). The first key factor contributing to the failure of FTIs as Ras inhibitor is the presence of alternative prenylation process when FTase is blocked. Majority of the early preclinical studies on FTIs focused on models of *HRAS*-driven cancer, thus neglecting the possibility of low efficacy of FTIs on other *RAS* mutation statuses (Sepp-Lorenzino et al. 1995). Although FTIs effectively inhibited farnesylation and membrane association of H-Ras, they did not show any inhibitory effect on the prenylation processes of both K-Ras and H-Ras proteins. This is due to the biochemical differences among the three Ras proteins, whereby the predominant forms of Ras (K-Ras and N-Ras) in many cancers can undergo an alternative prenylation process for membrane association (James et al. 1996; Whyte et al. 1997). The proteins K-Ras and N-Ras can act as substrates for geranylgeranyltransferase-1 (GGTase-1) and undergo geranylation with the addition of a 20-carbon isoprenoid group as a substitute for farnesyl group when FTase is inhibited, allowing the lipid modified Ras protein to be anchored to the plasma membrane for proper functioning, thus rendering FTIs ineffective (Cox et al. 1992). Another factor contributing to the demise of FTIs is the unspecific binding of the inhibitors. FTIs are generally assumed as "anti-Ras" inhibitors, but they can bind not only to FTase but also to other farnesylated proteins such as ARHI/NOEY2 and Di-Ras1/Rig which are important tumor suppressor proteins, thus complicating the antitumor effect of FTIs (Ellis et al. 2002; Luo et al. 2003). The inactivation of the gene encoding for GGTase-1 was shown to reduce tumor growth and enhance survival rate in mice *KRAS*-driven lung cancer (Sjogren et al. 2007). Moreover, the combinational inhibition of FTase and GGTase-1 greatly reduced *KRAS*-induced tumor growth and enhanced the survival of mice in comparison with either FTI or geranylgeranyltransferase-1 inhibitor (GGTI) alone (Zverina et al. 2012). A knockout study conducted by Liu and his team also supported the effect of dual inhibitions of FTase and GGTase-1 on diminishing tumor size and extending life spans of mice with *KRAS*-induced lung cancer (Liu et al. 2010). By far, dual prenylation inhibitors (DPIs) were developed and showed significant effect on blocking the prenylation of KRas but at high concentration that showed toxicity in mice (Lobell et al. 2001; deSolms et al. 2003). Therefore, developing a drug that can inhibit the prenylation enzymes (FTase and GGTase-1) with negligible cytotoxicity may be the next strategy in blocking *KRAS* signaling.

On top of that, the enzymes Rce1 and Icmt involved in the cleavage and methylation of farnesylated protein moiety during posttranslational modification of Ras have also been considered as candidates for anti-Ras inhibitors (Winter-Vann and Casey 2005). They are less explored targets mainly because the inhibition of both enzymes only reduced half of the transforming activity and membrane association of Ras as compared to FTIs (Kato et al. 1992). Although the inhibition of these enzymes displayed low inhibitory effect on Ras, there are studies supporting the potential use of Rce1 and Icmt inhibitors for blocking Ras-induced tumorigenesis. Deficiency of Rce1 in mouse embryo fibroblast was found to inhibit the membrane association of Ras and reduce Ras-mediated cell transformation, but the inhibitory effect on Ras was found to be low in cancer cells (Bergo et al. 2000; Kim et al. 1999). Peptidyl(acyloxy)methyl ketones is one of the very few identified substrate analogs that showed significant inhibitory effect of Rce1 activity (Kato et al. 1992). However, study found that depletion of Rce1 in mice accelerates mutant *KRAS*-driven myeloproliferative disease, and this may be due to the dysfunction of Rce1 substrate, Ras-proximate 1 (Rap1), in inhibiting *RAS* signaling, suggesting that Rce1 may not be a good anti-Ras target for drug development (Kometani et al. 2004). Similar to Rce1, depletion of Icmt inhibited membrane association of Ras and resulted in reduced K-Ras-mediated tumor growth in mice. Another animal study also provided evidence on the reduction of tumor growth in mice with *KRAS*-driven lung cancer in response to Icmt deficiency (Winter-Vann and Casey 2005). Through the screening of the chemical library, a compound named cysmethynil (2-[5-(3-methylphenyl)-1-octyl-1H-indol-3-yl] acetamide) was identified as molecular inhibitor of Icmt which showed inhibitory effect on the Icmt-dependent growth and interrupted the localization of Ras in cancer cells (Kato et al. 1992). Moreover, cystmethynil treatment showed remarkable effect on reducing growth of the xenograft tumors derived from human PC3 prostate cancer (Kim et al. 1999). However, the mechanism of action of the inhibitors of both Rce1 and Icmt is one of the major issues that require further investigations. Proteins other than Ras also undergo post-modifications such as proteolysis and methylation, and the presence of Icmt inhibitor can inhibit the function of these proteins. Bergo and colleagues found that the inactivation of Icmt greatly reduced cancer cell growth and transformation induced not only by *KRAS* but also by B-*RAF*. The same study also showed that the production of RhoA proteins was significantly reduced as a consequence of increased protein turnover, suggesting that Icmt inhibition may not be a good therapeutic strategy as studies supported the beneficial role of activated RhoA proteins in suppressing *KRAS*-induced cancer (Marshall 1993; Bergo et al. 2000). The findings on Rce1 and Icmt suggested the importance of better understanding the signaling events of these enzymes to further validate their potentials as drug candidates and to determine the target populations of cancer in which their inhibitors may be best applied.

The failure of the post-modification enzyme inhibitors to block oncogenic *KRAS* signaling has prompted the discovery of alternative approaches to prevent K-Ras protein from binding to the plasma membrane. One such inhibitor, Salirasib (also known as S-trans, trans-farnesylthiosalicylic acid, FTS), is an S-farnesyl cysteine analog that interrupts the anchorage of active K-Ras to the plasma membrane by competing with the oncogenic protein for its receptor, galectin-1, located on the inner surface of the plasma membrane (Weisz et al. 1999). This results in the degradation of active GTP-bound K-Ras in the cytoplasm which impairs K-Ras-dependent downstream signaling pathways and potentially alters various cellular activities such as migration, proliferation, and cell survival (Marom et al. 1995; Reif et al. 1999; Shalom-Feuerstein et al. 2004). Importantly, FTS has shown potential as anti-*KRAS* inhibitors in clinical trials of K-Ras-induced pancreatic cancer (Laheru et al. 2012). However, Schmukler and colleagues have demonstrated that FTS alone promotes cell growth and induces autophagy in HCT116 (human CRC) and Panc-1 (human pancreatic cancer) carrying *KRAS* mutations. These observations may limit the use of FTS as anti-*RAS* inhibitor in cancer therapy as the drug partially protects the cancer cells from death (Schmukler et al. 2013). Interestingly, the same research team found that the use of chloroquine (an autophagy inhibitor) synergizes the effect of FTS in inhibiting cell growth and survival of *KRAS*-bearing cancer cells. The combination of FTS and chloroquine significantly increased caspase-induced apoptotic cell death while inhibiting tumor growth, suggesting the potential of the use of combined drug treatment in cancer therapy (Schmukler et al. 2014). CRC is commonly characterized by genetic defects, whereby majority of CRCs (70–95%) hyperactivate the Wnt pathway, resulting in overexpression of β-catenin which promotes tumor growth (Frattini et al. 2004; Markowitz and Bertagnolli 2009). Combination therapeutic treatment involving FTS was proposed to treat CRC cells carrying both *Wnt* and *KRAS* mutation. Moreover, a significant fraction of patients with activated *Wnt* pathway carry activated *KRAS* mutation (Frattini et al. 2004; Wood et al. 2007). The combined use of β-catenin inhibitor (PKF115-584) and FTS showed remarkable effect on inhibiting anchorage-independent growth and promoting cell death and cell cycle arrest as compared to the use either FTS or β-catenin inhibitor alone (Mologni et al. 2012).

Another inhibitor, deltarasin is a small pyrazolopyridazinone molecule that interferes the localization of farnesylated K-Ras protein to the plasma membrane by binding to the farnesyl-binding sites of guanine nucleotide dissociation inhibitor (GDI)-like solubilizing factor, PDEδ (Zimmermann et al. 2013). *KRAS*-mediated signaling pathways are strongly dependent on the enrichment of K-Ras proteins at the plasma membrane (Schmick et al. 2015). PDEδ plays important role in facilitating the localization of K-Ras to the plasma membrane by binding to the farnesyl moiety at the cytosol as a protective step against the binding of farnesylated-K-Ras to the extensive endomembrane surfaces (Schmick et al. 2014). However, it was found that deltarasin is less target-specific as it can bind to other proteins such as transport channels and G-protein-coupled receptors and causes cell cytotoxicity at high dosage. A new potential inhibitor of PDEδ named Deltazinone 1 was then developed to overcome the weakness of deltarasin. The new drug is highly selective to its target and displays low unspecific cytotoxic effect as compared deltarasin. The inhibitor induced specific cell death in K-Ras-dependent pancreatic cancer cell lines with minimal cytotoxic effects. However, it was found that Deltazinone 1 is unsuitable for in vivo study as it is metabolized rapidly in mice (Papke et al. 2016). The development of these small inhibitors shed light on the direct inhibition of *KRAS* signaling, but further work will be required to optimize and improve their biochemical properties and selectivity for *KRAS* while reducing their cytotoxic potential.

12.4.2 Targeting Downstream Pathways of KRAS

In addition to the inhibitors developed to block membrane association of K-Ras, several other approaches of inhibiting K-Ras activity by targeting downstream pathways of *KRAS* were pursued. The most adopted approach is targeting the Raf-MEK-ERK and phosphoinositide 3-kinase (PI3K)-AKT-mammalian target of rapamycin (mTOR) pathways which are actively involved in driving oncogenesis in cancers expressing high level of K-Ras (Roberts and Der 2007; Wong et al. 2010).

12.4.2.1 Inhibiting Raf-MEK-ERK Pathway

Raf-MEK-ERK cascade (one of mitogen-activated protein kinase (MAPK) pathways) is the most intensively studied and best characterized effector pathway of K-Ras (Roberts and Der 2007; Samatar and Poulikakos 2014). The arrival of extracellular signal at the plasma membrane activates the transphosphorylated cell surface receptor tyrosine kinases (RTKs) which in turn recruit guanine exchange factors (Son of Sevenless 1 and 2 (SOS1 and SOS2)) and adaptor protein Grb2 to promote the activation of K-Ras. The activated K-Ras (K-Ras-GTP) promotes the phosphorylation of Raf serine/threonine protein kinases (A-Raf, B-Raf, and C-Raf-1) which then activate MEK1 and MEK2 protein kinases. MEK1/MEK2 phosphorylate and propagate the signal down the cascade by phosphorylating ERK1 and ERK2 protein kinase, thus regulating the activities of a wide spectrum of substrates particularly involved in cell proliferation and survival (Robinson and Cobb 1997; Yoon and Seger 2006). Due to its essential role in cellular growth and survival, the Raf-MEK-ERK signaling cascade is regulated by complex control and feedback systems involving scaffolding proteins, phosphatases, and protein kinases. The abnormal upregulation of the Raf-MEK-ERK pathway is therefore highly prevalent in oncogenesis, contributing to almost one third of human cancers (Wan et al. 2004). The importance of this pathway leads to intensive efforts in discovering drugs or inhibitors targeting the downstream effectors, Raf, MEK, and ERK, and multiple inhibitors are currently undergoing preclinical and clinical evaluations.

B-*RAF* has been the focus of research on drug development as it is the most common oncogenic mutations found in human cancers as compared to another two isoforms of RAF (A-*RAF* and C-*RAF*) (Davies et al. 2002). It is highly prevalent in cancers such as non-small-cell lung carcinoma (NSCLC; ~5%), CRC (~10%), melanoma (~60%), and hairy cell leukemia (100%) (Samatar and Poulikakos 2014). The major mutational pattern of B-*RAF* (approximately 90% of tumors bearing B-*RAF* mutations) is the substitution of a valine with glutamic acid (V600E) within the protein kinase activation segment (Davies et al. 2002; Wan et al. 2004). The effects of Raf inhibitors are highly dependent on the B-*RAF* mutational status, and the Ras-GTP levels as the key activation steps (homodimerization and heterodimerization) of Raf are promoted by upstream effector K-Ras (Weber et al. 2001). In cells bearing wild-type B-*RAF*, the binding of Raf inhibitors stabilize the conformation of the active Raf dimmers (Lavoie et al. 2013). At low saturation of Raf inhibitors, the drug binds to one protomer which in turn transactivates the other drug-free protomer. When the drug concentrations are highly saturated in cells, both protomers of Raf are occupied, thus resulting in the inhibition of the activity and downstream signaling of Raf (Poulikakos et al. 2010). Unlike wild-type B-*RAF*, dimerization is not required for the activation of mutant B-Raf (V600E). The mutated isoforms exist predominantly as hyperactive monomers even at low level of K-Ras-GTP, allowing effective binding of Raf inhibitors to its monomeric site, thus inhibiting its activity and downstream pathways (Poulikakos et al. 2010; Freeman et al. 2013).

12.4.2.1.1 Raf Inhibitors

Sorafenib (BAY43-9006) is one of the very first drugs developed as Raf-1 inhibitor which targets both wild type and mutant B-Raf (Lyons et al. 2001). The crystallographic analysis of sorafenib revealed the inhibitory action of the drug by binding to the ATP-binding site of Raf, thus inhibiting Raf activation and phosphorylation (Wan et al. 2004). However, no anticancer activity of sorafenib was reported in a randomized phase III clinical trial on patients with metastatic melanoma bearing mutated B-Raf (V600E) (Eisen et al. 2006). A later study found that sorafenib is able to inhibit other kinases involved in tumor angiogenesis such as FGFR-1, VEGFR2, and VEGFR3 (Wilhelm et al. 2004). The drug was approved for the treatment of advanced renal cell carcinomas (RCC) and unresectable hepatocellular carcinoma (HCC) in year 2005 and 2007, respectively. Due to the low occurrence rate of *RAS* and B-*RAF* (V600E) mutations in these cancers, the inhibitory action of sorafenib on Raf is unclear, but the anti-angiogenesis activity is most likely its main anticancer effect (Tannapfel et al. 2003; Karnoub and Weinberg 2008; Gattenlohner et al. 2009). On top of that, sorafenib has displayed promising anticancer effect on preclinical models of CRC cell lines including those bearing *KRAS* mutations (Kometani et al. 2004). The compound is able to potently inhibit the activation of Ras-dependent Raf-MEK-ERK pathway even in the presence of *KRAS* mutations but is unable to completely block Ras signaling (Wilhelm et al. 2004). The promising anticancer activity of sorafenib has led to combination drug study involving both sorafenib and irinotecan (existing chemotherapy for metastatic CRC). The combined regimen may help to overcome drug resistance to existing drug therapy and act as second or third-line treatment of metastatic CRC and tumors bearing *KRAS* mutations. However, the promising synergic effect of the combined regimen in a phase III clinical trials was restricted by the toxicity of irinotecan enhanced by sorafenib. Further work involving pharmacokinetic studies will be required to better understand the synergy of sorafenib and irinotecan.

Vemurafenib (previously known as PLX4032) was the first B-Raf (V600E)-selective inhibitor to undergo clinical evaluation followed by dabrafenib (Bollag et al. 2010; Falchook et al. 2012; Hauschild et al. 2012). In vivo analysis revealed the high selectivity of both drugs against mutated B-Raf, suppressing cellular activity and downstream signaling only in B-Raf (V600E)-mutated cells and tumors. However, these Raf inhibitors (at non-saturated concentrations) activated Raf and its downstream pathways in cells bearing wild-type B-Raf instead of suppressing Raf activity (Joseph et al. 2010; Hatzivassiliou et al. 2010). The clinical evaluations of both vemurafenib and dabrafenib found that both inhibitors significantly enhance progression-free survival of patients with mutated B-Raf (V600E) melanoma (Tsai et al. 2008; Flaherty et al. 2010; Chapman et al. 2011). However, a portion of patients developed benign skin tumors after the drug treatments seemingly due to the activation of Raf-MEK-ERK signaling in normal cells. Apart from that, the efficacy of the drugs varied among patients owing to the development of resistance toward the Raf inhibitors (Su et al. 2012). To date, the main resistance mechanisms to Raf inhibitors resulted from abrasion that increases dimerization of Raf kinases, thus restricting the effect of these drugs in suppressing the activity of Raf and its downstream pathways (Poulikakos and Rosen 2011).

Following the promising results of vemurafenib and dabrafenib (first generation of Raf inhibitors), several potential new candidates of Raf inhibitors have entered preclinical and clinical evaluations. LGX818 is the second generation of B-Raf (V600E)-selective inhibitor that is currently in phase I clinical trials with extremely slow off-rate compared to other Raf inhibitors. In preclinical studies, LGX818 showed sustained and strong (above 60%) inhibition of phosphorylated MEK in human melanoma xenograft models (Mamounas et al. 2010). Another Raf inhibitors that are undergoing clinical trials are MLN2480 and TAK-632 with similar features. Unlike the B-Raf (V600E)-selective inhibitors mentioned above, these two compounds are potent inhibitor of the Raf-MEK-ERK signaling in cells carrying both *RAS* mutation and wild-type B-*RAF*, seemingly due to low saturation of the Raf dimer and slow off-rate of the compound (Nakamura et al. 2013). This type of Raf inhibitors which are more potent to Raf dimer may be potential drug candidates in treating tumors bearing *KRAS* mutations or tumors resistant to the treatment of dabrafenib and vemurafenib.

12.4.2.1.2 MEK Inhibitors

MEK1 and MEK2 are closely related dual-specificity kinases that mediate signal transmission from activated Raf to their only know substrates, ERK1 and ERK2 (Roskoski 2012). The prevalence of activating MEK mutations is very low in human cancers in comparison with B-*RAF* (Marks et al. 2008; Murugan et al. 2009). As downstream effector of K-Ras and Raf, MEK has become a potential candidates in drug development with the assumption that MEK inhibition could be equivalent in blocking Raf-MEK-ERK signaling by K-Ras. A number of non-ATP-competitive, selective, and potent MEK inhibitors are currently under clinical evaluations. Early clinical evaluations with MEK inhibitors showed promising results especially on patients with mutated B-Raf melanoma. The very first few inhibitors entering clinical trials are CI-1040, PD0325901, and selumetinib (AZD6244). However, cellular toxicity was the major concerns restricting the effectiveness of these inhibitors, resulting in common side effects such as visual disturbances and skin rash.

Trametinib (allosteric inhibitor of both MEK1 and MEK2) is the first MEK inhibitor being approved by the US Food and Drug Administration (FDA) for the treatment of melanoma with B-RAF (V600E/K) mutations. It is a potent inhibitor which preferentially binds to unphosphorylated MEK to prevent activation and phosphorylation of MEK. Although the drug response rate (~20%) was found to be lower than Raf inhibitors (dabrafenib, 52% and vemurafenib, 48%), trametinib significantly improved disease-free progression and overall survival rate of patients compared to standard chemotherapy (Carlino et al. 2014). Since the mechanism of actions of MEK and Raf inhibitors are different, it was predicted that the combination of both inhibitors can result in a greater inhibitory effect on Raf-MEK-ERK signaling in cancer cells, while their respective toxicities could be reduced by their opposing effect on normal cells. In a phase III clinical trials, the combined treatment of trametinib (MEK inhibitor) and dabrafenib (Raf inhibitor) displayed higher efficacy in melanoma patients with B-*RAF* (V600E) mutations than either trametinib or dabrafenib alone while significantly reducing the side effects caused by Raf-MEK-ERK signaling in normal cells induced by Raf inhibitors (Flaherty et al. 2012). The combination of trametinib and dabrafenib was then approved by FDA in 2014 for treatment of melanoma patients with B-*RAF* (V600E/K) mutations.

Although MEK inhibitors show therapeutic benefits on B-*RAF* (V600E)-mutated melanomas, they have minimal effect on cancers driven by *KRAS* mutations. One possible explanation is that, in tumors bearing *KRAS* mutations, the reactivation of the feedback mechanism of Raf-ERK-MEK signaling is consistently stronger compared to tumors with B-*RAF* (V600E) mutations. Moreover, the feedback mechanism of Raf-ERK-MEK signaling pathway remains intact in mutant *KRAS*-bearing cells, allowing the integration of signals from RTKs, while the mechanism is impaired in B-Raf-mutated cells (Dougherty et al. 2005; Ritt et al. 2010). K-Ras-mediated Raf-MEK-ERK pathway is able to initiate potent negative feedback loops that attenuate signaling from EGFR and other RTKs (Fig. 12.5). The inhibition of these feedback loops results in overstimulation of the upstream effectors such as Raf and MEK that potentially oppose the antitumor effect of the pathway inhibitors, leading to drug resistance (Rozengurt et al. 2014). Recent studies revealed the potential of MEK inhibitors with different inhibitory mechanisms in overcoming feedback reactivation of Raf-MEK-ERK signaling which in turn improved the inhibitors efficacy in *KRAS*-driven cancer patients. It was found that CH5126766 (an allosteric MEK inhibitor) preferentially bind to unphosphorylated MEK to prevent MEK phosphorylation by Raf, blocking the feedback reactivation of Raf-MEK-ERK signaling and, thus, enhancing the efficacy of the inhibitors in tumors bearing mutated *KRAS* (Ishii et al. 2013). Overall, MEK inhibitors that can inhibit both MEK kinase activity and phosphorylation (induced by Raf) showed enhanced improved efficacy in mutant *KRAS*-bearing tumors (Hatzivassiliou et al. 2013; Lito et al. 2014).

12.4.2.1.3 ERK Inhibitors

ERK is known as downstream effector of MEK. It has resulted in limited development of ERK1-/ERK2-selective inhibitors with the assumption that ERK inhibitors would not bring better clinical benefit as compared to MEK inhibitors. However, the development of ERK inhibitors has recently gained interest with the aim to increase the response rate and potency of Raf-MEK-ERK pathway inhibition. This is mainly due to the fact that the common resistance mechanism of patients to Raf and MEK inhibitors involve the feedback reactivation of Raf-MEK-ERK signaling (Samatar and Poulikakos 2014). Several ERK inhibitors have been reported, and SCH772984

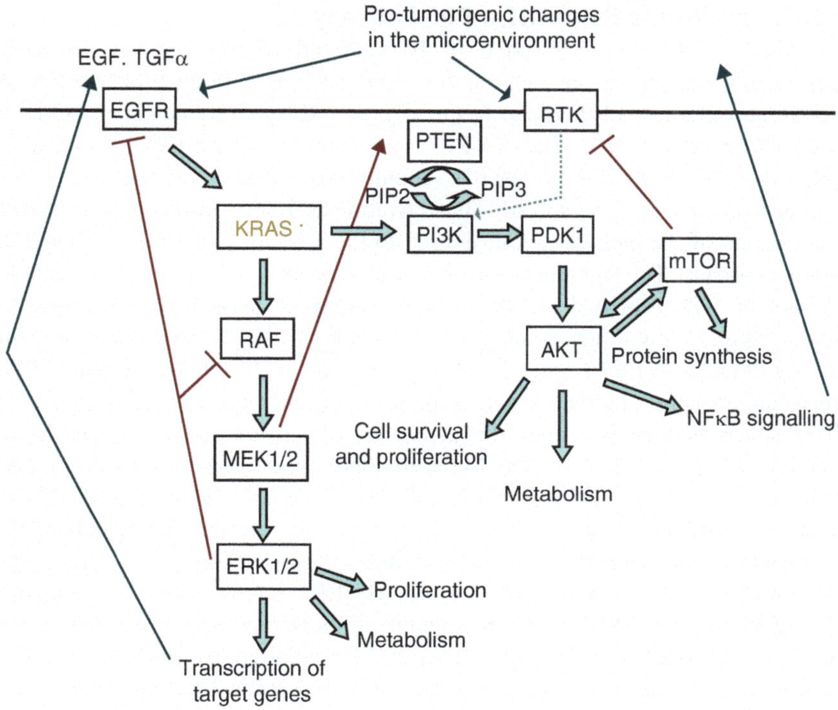

Fig. 12.5 An overview of signaling pathway of oncogenic *KRAS*. The constitutively mutated *KRAS* activates the Raf-MEK-ERK and PI3K-Akt-mTOR pathways via Raf phosphorylation and PI3K activation, respectively, whereby both pathways are actively involved in driving cancer initiation and progression. The activity of KRAS is promoted by the activation of its upstream effector, epidermal growth factor receptor (EGFR) upon arrival of growth factors. The signaling of both Ras-MEK-ERK and PI3K-Akt-mTOR pathways is regulated by negative feedback loops. The existence of feedback loops which arise from effectors of the same or different pathways help to restrict the activity of upstream effectors which in turn adjust the outcome of the signaling cascades. These negative feedback loops are indicated by solid red lines headed by a vertical line, while the stimulation of pro-tumorigenic signaling is represented by arrows in green. The arrows in red depict the stimulation of negative feedback loop of PTEN. The * represents *KRAS* in its mutated form

is the well-characterized inhibitor that showed promising results in in vivo studies. It is an ATP-competitive inhibitor that selectively binds to ERK1 and ERK2 and inhibits the intrinsic activity of the ERK kinases, thus preventing ERK phosphorylation induced by MEK (Chaikuad et al. 2014; Deng et al. 2014). Cell cultures and mouse studies revealed the effect of SCH772984 in inhibiting cellular proliferation and promoting apoptosis selectively on tumors and mice xenografts carrying *RAS* or B-*RAF* mutations (Morris et al. 2013). The clinical analogue of SCH772984 and SCH900353 and two other ERK inhibitors (BVD-523 and RG7842, information unpublished) are currently undergoing phase I clinical trials.

12.4.2.2 Inhibiting PI3K-AKT-mTOR Pathway

PI3K-AKT-mTOR is the second best characterized effector pathway of K-Ras which is involved in various cellular functions such as cellular proliferation and survival, and it is one of the most frequently altered signaling cascades in human cancers (Wong et al. 2010). Class IA PI3Ks are heterodimeric kinases composed of regulatory (p84) and catalytic (p110) subunits, and each of them exists in four different isoforms (α, β, γ, and δ). The activation of K-Ras in response to growth factors is one of the pathways contributing to the activation of PI3K-AKT-mTOR signaling cascade. The signaling cascade can also be initiated by G-protein-coupled receptors or RTKs located on the plasma membrane upon arrival of signaling compounds and growth factors such as angiopoietin 1 (Ang1), vascular endothelial cell growth factor (VEGF), and fibroblast growth factor (FGF). The activated PI3K induces the phosphorylation and activation of phosphoinositide (3,4,5) biphosphate (PIP3) which in turn promotes the activation of a wide range of cytoplasmic signaling proteins particularly the Akt serine/threonine kinases and subsequently mTOR which promotes cellular growth and protein synthesis (Courtney et al. 2010). In addition to *KRAS* mutations, the PI3K-AKT-mTOR pathway can be deregulated by a variety of mechanisms. This often includes the loss-of-function of the tumor suppressor phosphatase and tensin homolog (PTEN), mutation of genes encoding RTKs (*ERBB1* and *ERBB2*), PI3K subunits (*PIK3R1*, *PIK3R2*, *PI3K3CA*, and *PIK3CB*), and AKT (*AKT1*). The frequent deregulation of PI3K-AKT-mTOR pathway in cancer has made its components attractive targets for drug development. A number of inhibitors of PI3K, AKT, and mTOR which have undergone clinical evaluations or been tested clinically are discussed in the following sections.

PI3K inhibitors can be classified into two classes, Pan-class I PI3K and Isoform-specific PI3K inhibitors. Pan-class I PI3K inhibitors targets all four isoforms of class I PI3K (α, β, γ, and δ). Copanlisib (BAY80-6946) is a Pan-class I PI3K inhibitor that targets the catalytic subunits of p110α and p11δ. It showed strong anticancer effect on breast cancer cell lines overexpressed *PIK3CA* and/or *HER2* in comparison to those bearing wild-type *PIK3CA* and missing of *HER2*. In vivo analysis revealed the effect of copanlisib in tumor regression in rats with amplified *HER2* (Liu et al. 2013). The inhibitory agent has undergone phase III clinical evaluation for B-cell lymphoma therapy. In addition, buparlisib (BKM120) being the most advanced drug under the category of Pan-class I PI3K inhibitors has showed impressive antiproliferative effect on a wide range of cancer cell lines (more than 400) (Maira et al. 2012). Buparlisib has been the choice of therapeutic agent in treating glioblastoma multiforme (GBM) tumors and brain metastases derived from breast cancer due to its ability to penetrate the blood-brain barrier (Ando et al. 2014). Recently, studies have found that inhibition of the PI3K-Akt-mTOR pathway is able to enhance the anticancer effect of chemotherapeutic agents in various cancer cell lines including colon cancers (Engelman 2009; Pal and Mandal 2012). A combination drug study involving buparlisib and cetuximab (approved anti-EGFR drug in treating metastatic CRC but ineffective in CRC with *KRAS* mutations) displayed significant inhibitory effect on the growth of colon cancer cells bearing mutant KRAS/wild-type PI3KCA but not those with mutant *KRAS*/mutant *PI3KCA*.

The presence of buparlisib is able to enhance the efficacy of cetuximab by blocking the wild-type *PI3KCA* pathway in CRC patients with mutant KRAS/wild-type PI3KCA (Hong et al. 2016).

Interestingly, many isoform-specific PI3K inhibitors have been developed and entered clinical trials despite the significant antiproliferative effect of Pan-class I PI3K inhibitors. This may be the strategies to maximize the clinical success of different types of inhibitors in targeting the superfamily of PI3K. As compared to Pan-class 1 PI3K inhibitors, the development of isoform-specific PI3K inhibitors aims to target specific changes in the pathway while reducing the toxicities accumulated from multiple inhibition of PI3K isoforms (Dienstmann et al. 2014). The approval of idealalisib (PI3Kδ-specific inhibitor) for acute lymphocytic leukemia (ALL) treatment suggested that inhibition of PI3Kδ alone is sufficient and the use of PI3Kδ-specific inhibitor can help to reduce the metabolic side effect due to inhibition of other PI3K isoforms (Lannutti et al. 2011). Among the inhibitors that have entered clinical trials, alpelisib (BYL719) and taselisib (GDC-0032) display high efficacy in mutant PIK3CA-driven cancers. Combining cetuximab (approved anti-EGFR drug in treating metastatic CRC) and alpelisib (a p110α inhibitor) has shown efficacy in patients with recurrent or metastatic head and neck squamous cell carcinoma (HNSCC) (Massacesi et al. 2016). Taselisib was found to have equal inhibitory effect on p110α, p110δ, and p110γ while 30-fold less potent on p110β (Ndubaku et al. 2013).

In addition, mTOR inhibitors can be classified into two classes which include rapalogs (inhibiting mTORC1 only) and catalytic (inhibiting both mTORC1 and mTORC2). Rapalogs are the improved analogs of the very first mTORC1 inhibitor, rapamycin, and these include everolimus and temsirolimus which were approved for the treatment of renal cell carcinoma (RCC) (Banerji et al. 2012). On the other hand, catalytic mTOR inhibitors are the improved version of rapalogs which suppress the feedback loop of mTOR that activates AKT (Dowling et al. 2010). AZD2014 is the first approved mTORC1/mTORC2 inhibitors that showed positive effect on advanced solid tumors, while other catalytic inhibitors displaying clinical benefits are MLN0128 and CC-223 (Mita et al. 2013; Nayak et al. 2016). AKT is one of the therapeutic targets in PI3K-AKT-mTOR pathway. The developed AKT inhibitors can be grouped as adenosine triphosphate (ATP)-competitive or allosteric. Displaying promising effect on phase I clinical trials, ipatasertib and AZD5363 (ATP-competitive) and MK-2206 (allosteric) are currently being tested in a series of solid tumors (Yan et al. 2011; Yap et al. 2011; Banerji et al. 2012).

Belonged to the superfamily of PI3K-related kinases (PIKK), PI3K and mTOR shared similar structural domains and the inhibitors developed to target both kinases are known as dual-specificity PI3K-mTOR inhibitors. The dual inhibitors target the active sites of both protein kinases simultaneously and interrupt the both upstream and downstream pathways of AKT, thus blocking the negative feedback loop of mTORC1-S6K-IRS1 that activates AKT (O'Reilly et al. 2006; Serra et al. 2008). The dual-specificity inhibitors that have entered clinical trials are GDC-0980 and BEZ235 (Dolly et al. 2010; Peyton et al. 2011). Preclinical cell culture studies supported the broader efficacy of dual-specificity PI3K-mTOR inhibitors in various genotypes as compared to agents that target only one component of the pathway (Serra et al. 2008). Although the dual inhibitors showed clinical benefits in phase I trials, drug bioavailability and toxicities still remained the major challenge in the subsequent studies.

The activity of single inhibitors PI3K, mTOR, PI3K-mTOR, and AKT on various cancer cell lines was found to be moderate based on early observation in clinical trials. However, these single inhibitors were found to be extremely active in tumors with genetically altered *PIK3CA* or *PTEN* (Brachmann et al. 2009; Weigelt et al. 2011). The modest activity of the single agents may have resulted from feedback activation of other pathways. Both preclinical and clinical studies have shown that PI3K and AKT inhibitors can induce feedback activation of human epidermal growth factor receptor (HER) family and extracellular signal-regulated kinase (ERK), and this may reduce the anticancer effect of these inhibitors (Chandarlapaty et al. 2011; Serra et al. 2011; Yan et al. 2013). Moreover, the clinical activity of these single agents is low in tumors bearing *KRAS* mutations. Although K-Ras is the upstream effector of both Raf-MEK-ERK and PI3K-AKT-mTOR pathways, responses to single inhibitors of these pathways are variable in cancers with mutant *RAS*. This is mainly due to close interaction between Raf-MEK-ERK and PI3K-AKT-mTOR pathways which provide a potential escape route when either one of the pathways is inhibited (Britten 2013). The limited activity of the single inhibitors has led to the combination of inhibitors targeting both the Raf-MEK-ERK and PI3K-AKT-mTOR pathways (Mendoza et al. 2011). A cell culture study found that lung tumor with *PIK3CA* mutation but not with *KRAS* mutation responded to NVP-BEZ235 (a dual-specificity PI3K-mTOR inhibitor) treatment. However, a combination of both NVP-BEZ235 and selumetinib (MEK1/MEK2 inhibitor) showed inhibitory effect on the growth of mutant *KRAS*-bearing tumors (Engelman et al. 2008). On top of that, a retrospective study of patients receiving combination treatment of Raf-MEK-ERK and PI3K-AKT-mTOR pathways inhibitors (under phase I clinical study) showed that simultaneous inhibitions of both pathways can improve treatment efficacy as compared to single-pathway inhibition (Shimizu et al. 2012). The combination treatments include buparlisib (pan-PI3K inhibitor) and vemurafenib (Raf inhibitor); pictilisib (pan-PI3K inhibitor) and GDC-0973 (MEK inhibitor); and copanlisib (PI3K inhibitor) and refametinib (allosteric MEK inhibitor) (LoRusso et al. 2012; Algazi et al. 2014; Ramanathan et al. 2014). In phase I clinical studies, combinations of buparlisib and trametinib (allosteric MEK1/MEK2 inhibitor) and pictilisib and GDC-0973 showed promising effects on tumors bearing *KRAS* and B-*RAF* mutations (LoRusso et al. 2012; Bedard et al. 2015). However, toxicity is the major concern in the combination drug treatments, and further investigation is required to improve the anticancer property and tolerability of the combination pathways inhibitors.

12.4.3 Potential Use of Natural Compounds in Treatment of KRAS-Driven Cancers

Although genetic instability is the major characteristic of colon cancers, it was found that only 10% of all cancers are caused by genetic alteration, while the remaining 90% are caused by lifestyle factors such as diet (Lannutti et al. 2011; Rozengurt et al. 2014). For that reason, chemoprevention which refers to the intake or use of agents to inhibit or delay cancer progression or reverse malignancy stage has gained high attention from the cancer researchers (Pal and Mandal 2012). Extensive study has demonstrated that a series of natural compounds identified in plants, vegetables, and spices displayed strong potential in chemoprevention and cancer treatment (Wang et al. 2012). Numerous well-studied natural compounds including resveratrol, curcumin, apigenin, and sulforaphane have been documented to target key signaling pathways such as Wnt, Notch, and P13K/Akt in various cancers (Jakubikova et al. 2005; Aggarwal et al. 2007; Rusin et al. 2009; Tutelian and Lashneva 2013; Ferrucci et al. 2016). In view of the fact that studies focused on inhibition of KRAS by natural compound specified in colon cancer is limited, different cancer cell types were discussed in this section (Table 12.1).

12.4.3.1 Natural Compounds Targeting *KRAS* Activity

One of the natural compounds that was found to be effective against *KRAS* mutation by targeting membrane association of K-Ras protein was lycopene. Lycopene is a naturally existing bright red carotenoid pigment found in tomatoes and other red fruits such as watermelon and pink grapefruit (Chan and Hung 2014). The consumption of tomatoes is well known for its beneficial effect on lowering risk of developing chronic diseases and cancer. In particular, epidemiological studies reported that increased consumption of lycopene-rich tomatoes is associated with a 40% reduction in the risk of developing prostate cancer (Yang et al. 2011). Many in vitro studies on lung, colon, and liver cancer cells supported the anticancer activity of lycopene by inhibiting cancer cell proliferation and by inducing apoptosis, but the exact underlying mechanisms of lycopene still remained controversial. Lately a novel mechanism of lycopene involving the alteration of mevalonate pathway and impairment of Ras membrane localization was proposed (Palozza et al. 2010). They found that lycopene was able to downregulate the expression of 3-hydroxy-3-methylglutaryl coenzyme A (HMG-CoA) reductase which is a key enzyme catalyzing the biosynthesis of cholesterol and isoprenoids (Swanson and Hohl 2006). This in turn reduced the production of farnesyl pyrophosphate and geranylgeranyl pyrophosphate which are important signaling molecules involved in farnesylation of GTPases to ensure proper membrane localization and protein activation (Brown and Goldstein 1980). As mentioned earlier, farnesylation is an essential step in posttranslational modifications of Ras (member of GTPases family) to ensure proper activity (Marshall 1993). It was found that lycopene treatment

Table 12.1 Inhibitory effects of natural compounds on *KRAS* activities and *KRAS*-dependent pathways in cancer cells

Natural compounds	Dietary sources (common name/ botanical name)	Activity of active compounds	Model tested	References
Resveratrol	Grape (*Vitis vinifera*)	Inhibits colon tumor formation and growth by downregulating *KRAS* expression	Mouse with *KRAS* activated sporadic CRC	Saud et al. (2014)
Capsaicin	Chili (*Capsicum frutescens*)	Induces cell apoptosis by suppressing the expression of important signaling pathways such as *KRAS*, *BRAF*, *PTPN*11, and *AKT*	Human acute lymphoblastic leukemia CCRF-CEM cells	Bozok-Cetintas et al. (2014)
Honokiol	Magnolia (*M. officinalis*, *M. obovata*, and *M. grandiflora*)	Induces G1 arrest and apoptosis. Inhibits tumor growth by interfering Raf-MEK-ERK and PI3K-AKT-mTOR pathways mediated by oncogenic *KRAS*	Lung cancer cells bearing *KRAS* mutation	Luo et al. (2017)
Lycopene	Tomato (*Lycopersicum esculentum*)	Inhibits tumor growth by interfering mevalonate pathway and Ras protein activation	Ras-activated prostatic carcinoma LNCaP cells	Palozza et al. (2010)
Quercetin	Indian gooseberry (*Emblica officinalis*)	Induces apoptosis by activating p53 which is regulated by *KRAS*	Lung and colon cancer cells with *KRAS* mutations	Lee et al. (2009)

impaired Ras membrane localization in cancer cells, resulting in cytoplasmic Ras accumulation and, thus, inactivating the protein activity. The reduced Ras activity in response to lycopene treatment inhibits cancer cell growth by reducing the Ras-dependent activation of NF-κB. Moreover, the growth-inhibitory effect was accompanied by apoptosis induction and by an arrest of cell cycle progression as evidenced by increased levels of p21 and p53 as well as decreased levels of pAKT and cyclin D1 levels. Interestingly, the same research team found that the growth-inhibitory effect of lycopene was more effective in cancer cells bearing *KRAS* mutation as compared to those normal and non-transformed cells. The ability of lycopene in reducing the expression of HMG-CoA reductase and imparing the activity of Ras protein was observed in *KRAS*-bearing lung, colon, and prostate cancers (Palozza et al. 2010). These data supported the potential use of lycopene in cancer therapy and provide a novel insight into the growth-inhibitory mechanism of lycopene on human cancers with *KRAS* mutation.

Another natural compound that showed great potential in suppressing K-Ras activity was quercetin. Quercetin (3,3′,4′,5,7-pentahydroxyflavone) is one of the most abundant dietary flavonoids and polyphenols found in various vegetables and fruits such as apples and berries (Lee et al. 2009). It has been shown to exert antiproliferative, anti-inflammatory, and antioxidant effect on many cancers (Jakubikova et al. 2005; Rusin et al. 2009; Jenjaroenpun et al. 2013). Quercetin has drawn more and more attention due to its specific pro-apoptotic effect on tumors cells instead of normal, non-transformed cells, but its anticancer mechanism still remained unclear (Ferrucci et al. 2016). Quercetin was found to induce cell cycle arrest at G1 phase and caused an elevation in the expression of proteins involved in apoptosis such as AMPK and p53 in colon cancer. The tumor size of the treated colon cancer was reduced significantly after 6 weeks of quercetin treatment, supporting the pro-apoptotic effect of the compounds via the activation of AMPK and p53 (Kim et al. 2010). In another study, quercetin was found to specifically induce the activation of p53 in colon cancer cells with mutated *KRAS* but not in those with wild-type *KRAS*. Moreover, they found that suppression of K-Ras activity significantly induced the expression of p53, suggesting the potential of quercetin in the treatment of *KRAS*-induced cancer by interfering K-Ras activity (Lee et al. 2009). However, high dose of quercetin was found to inhibit the expression of GSK-3β which in turn promote the activation of β-catenin and colon cancer development (Clevers 2004; Polakis 2012). Therefore more studies are required in fine-tuning the action mode of quercetin as a potential phytoagent targeting K-Ras activation. Last but not least, resveratrol was found to inhibit cell growth in CRC by downregulating expression of *KRAS*. Resveratrol (3,4′,5-trihydroxy-trans-stilbene) is a phytoalexin present in plants including grapes, peanuts, and berries. Many in vitro studies have supported the anticancer effect of resveratrol on CRC by inhibiting cancer initiation and progression (Carter et al. 2014). In the studies focusing on the effect of resveratrol on miRNA expression, miR-96 was found to regulate K-Ras translation in treated colon cancer cells. This finding supported the ability of resveratrol in the inhibition of colon tumor formation and growth by downregulating *KRAS* expression (Saud et al. 2014). Many clinical studies found that large dose of resveratrol could effectively inhibit the growth of colon tumor and well tolerated by cancer patients with minimal side effects (Bozok-Cetintas et al. 2014; Luo et al. 2017).

12.4.3.2 Natural Compounds Targeting *KRAS*-Dependent Downstream Pathways

Many natural compounds were found to inhibit tumor growth via suppression of Raf-MEK-ERK and PI3K-AKT-mTOR. As mentioned in previous sections, these pathways are downstream of KRAS signaling, and their inhibitions were found in *KRAS*-driven cancer cells treated by some of the natural compounds. Capsaicin (trans-8-methyl-N-vanillyl-6-none-namide) is a component of red chili pepper (*Capsicum annuum* L.) which produces pungent smell and gives a spicy taste. Many in vitro studies have demonstrated the anticancer effect on various cancer cell types including leukemia and pancreatic and urothelial cancer, but there are limited preclinical studies of the anticancer effects of capsaicin on CRC (Ito et al. 2004; Amantini et al. 2009; Pramanik et al. 2011). Early studies found that capsaicin treatment increased the degradation of cyclin D1 (key regulator of the transition of cell cycle from G1 to S phase) and induced G1 arrest in human CRC (Musgrove et al. 2011). Capsaicin was then reported to inhibit growth and induce apoptosis in pancreatic cells bearing *KRAS* mutation by downregulating Ras-dependent Raf-MEK-ERK pathway (Zhang et al. 2013). Similar study on acute lymphoblastic leukemia (ALL) T lymphoblastoid (CCRF-CEM) cells reported a significant downregulation of a series of key signaling pathways including *KRAS*, *BRAF*, and *PTPN11* in response to capsaicin treatment (Bozok-Cetintas et al. 2014). The indirect inhibitory effect on KRAS signaling was also observed in cancer cells treated with honokiol [(3′,5-di-(2-propenyl)-1,1′-biphenyl-2,2′-diol], a natural biphenolic compound isolated from magnolia tree bark (Fried and Arbiser 2009). Honokiol was found to exert anticancer, anti-inflammatory, and anti-angiogenic activity in various cancer cell types including lung, prostate, and breast cancers (Yang et al. 2002; Hahm and Singh 2007; Liu et al. 2008). In vitro studies revealed the pro-apoptotic effect of honokiol on mutant *KRAS*-driven non-small-cell lung cancer (NSCLC). Honokiol was found to induce apoptosis and G1 cell cycle arrest in mutated *KRAS* lung cancer cells by interfering both Raf-MEK-ERK and PI3K-AKT-mTOR pathways. The same study also discovered that honokiol treatment in mutant *KRAS*-driven lung cancer significantly upregulated the expression of Sirt3 which plays important role in regulating cell metabolism and acts as tumor suppressor in cancer development (Luo et al. 2017). Together these studies provide better understanding of the mechanisms associated with the anticancer activity of capsaicin and honokiol and strengthened the potentials of these natural compounds as potential agents for the treatment of *KRAS*-driven cancers.

12.5 Conclusions and Future Prospects

Oncogenic *KRAS* plays a key role in the early events of CRC tumor development. The prevalence of this oncogene has prompted the need to search for new therapeutic strategies as *KRAS*-driven CRC is not responsive to the approved chemotherapeutic agents targeting EGFR. One of the most widely adopted approaches to inhibit mutated *KRAS* signaling is to target the association of K-Ras to plasma membrane for proper protein functioning. The development of small inhibitors that obstruct the binding of K-Ras protein to the plasma membrane has shed light on direct inhibition of *KRAS* signaling. *KRAS* selectivity and toxicity are the main issues related to this category of inhibitors which in turn suggests the need to optimize and improve the biochemical properties of these small inhibitors before going for clinical trials. Targeting the downstream pathways (Raf-MEK-ERK and PI3K-AKT-mTOR) is another commonly adopted approach in blocking *KRAS* signaling as these pathways are actively involved in driving oncogenesis in *KRAS*-driven cancers. However, the clinical efficacy of the developed single-pathway inhibitors in tumors bearing *KRAS* mutations is generally disappointing. The failure of these inhibitors may be due to the feedback activation of the upstream signaling nodes or other connecting pathways. Moreover, the cross-connections between Raf-MEK-ERK and PI3K-AKT-mTOR pathways provide a potential escape route when either one of the pathways is blocked. This has led to the combinational drug treatment involving inhibitors of both pathways and promising antiproliferative effect is observed in KRAS-bearing tumors in early clinical trials. Toxicity is the major issue in the combination drug treatments, suggesting the necessity of improvement drug tolerability in the subsequent trials. The efficacy of dual or multi-targeted inhibitors could be potentially increased by new generations of agents and improved drug combinations. Lately, various natural compounds identified in plants, vegetables, and spices were found to show chemopreventive and therapeutic effects, and some of these compounds have been discovered to protect against colon cancer via inhibiting *KRAS* activity or targeting its downstream pathways Raf-MEK-ERK and PI3K-AKT-mTOR. These natural compounds which are relatively safe and target multiple signaling pathways in cancer cells may provide alternative approach for the treatment of colon cancers with *KRAS* mutations.

Acknowledgments We gratefully acknowledge the Fundamental Research Grant Scheme grants (FRGS/1/2014/SKK01/MUSM/03/2) from the Ministry of Higher Education Malaysia for financial support on our research on KRAS mutation as a drug target in cancers.

References

Aggarwal BB, Sundaram C, Malani N, Ichikawa H (2007) Curcumin: the Indian solid gold. Adv Exp Med Biol 595:1–75

Algazi AP, Posch C, Ortiz-Urda S, Cockerill A, Munster PN, Daud A (2014) A phase I trial of BKM120 combined with vemurafenib in BRAFV600E/k mutant advanced melanoma. J Clin Oncol 32:9101. https://doi.org/10.1200/jco.2014.32.15_suppl.9101

Amado RG, Wolf M, Peeters M, Van Cutsem E, Siena S, Freeman DJ, Juan T, Sikorski R, Suggs S, Radinsky R, Patterson SD, Chang DD (2008) Wild-type KRAS is required for panitumumab efficacy in patients with metastatic colorectal cancer. J Clin Oncol 26:1626–1634

Amantini C, Ballarini P, Caprodossi S, Nabissi M, Morelli MB, Lucciarini R, Cardarelli MA, Mammana G, Santoni G (2009) Triggering of transient receptor potential vanilloid type 1 (TRPV1) by capsaicin induces Fas/CD95-mediated apoptosis of urothelial cancer cells in an ATM-dependent manner. Carcinogenesis 30:1320–1329

Ando Y, Inada-Inoue M, Mitsuma A, Yoshino T, Ohtsu A, Suenaga N, Sato M, Kakizume T, Robson M, Quadt C, Doi T (2014) Phase I dose-escalation study of buparlisib (BKM120), an oral pan-class I PI3K inhibitor, in Japanese patients with advanced solid tumors. Cancer Sci 105:347–353

Awada A, Eskens FALM, Piccart M, Cutler DL, van der Gaast A, Bleiberg H, Wanders J, Faber MN, Statkevich P, Fumoleau P, Verweij J (2002) Phase I and pharmacological study of the oral farnesyltransferase inhibitor SCH 66336 given once daily to patients with advanced solid tumours. Eur J Cancer 38:2272–2278

Baines AT, Xu D, Der CJ (2011) Inhibition of Ras for cancer treatment: the search continues. Future Med Chem 3:1787–1808

Banerji U, Dean EJ, Gonzalez M, Greystoke AP, Basu B, Krebs M, Puglisi M, Grinsted L, Oelmann E, Burke W, Harrington E, Green S, Ranson M (2012) First-in-human phase I trial of the dual mTORC1 and mTORC2 inhibitor AZD2014 in solid tumors. J Clin Oncol 30:3004. https://doi.org/10.1200/jco.2012.30.15_suppl.3004

Basso AD, Kirschmeier P, Bishop WR (2006) Lipid posttranslational modifications. Farnesyl transferase inhibitors. J Lipid Res 47:15–31

Bedard PL, Tabernero J, Janku F, Wainberg ZA, Pazares L, Vansteenkiste J, Van Cutsem E, Perez-Garcia J, Stathis A, Britten CD, Le N, Carter K, Demanse D, Csonka D, Peters M, Zubel

A, Nauwelaerts H, Sessa C (2015) A phase Ib dose-escalation study of the oral pan-PI3K inhibitor buparlisib (BKM120) in combination with the oral MEK1/2 inhibitor trametinib (GSK1120212) in patients with selected advanced solid tumors. Clin Cancer Res 21:730–738

Bergo MO, Leung GK, Ambroziak P, Otto JC, Casey PJ, Young SG (2000) Targeted inactivation of the isoprenylcysteine carboxyl methyltransferase gene causes mislocalization of K-Ras in mammalian cells. J Biol Chem 275:17605–17610

Bokemeyer C, Bondarenko I, Makhson A, Hartmann JT, Aparicio J, De Braud F, Donea S, Ludwig H, Schuch G, Stroh C, Loos AH, Zubel A, Koralewski P (2009) Fluorouracil, leucovorin, and oxaliplatin with and without cetuximab in the first-line treatment of metastatic colorectal cancer. J Clin Oncol 27:663–671

Bollag G, Hirth P, Tsai J, Zhang J, Ibrahim PN, Cho H, Spevak W, Zhang C, Zhang Y, Habets G, Burton EA, Wong B, Tsang G, West BL, Powell B, Shellooe R, Marimuthu A, Nguyen H, Zhang KY, Artis DR, Schlessinger J, Su F, Higgins B, Iyer R, Dandrea K, Koehler A, Stumm M, Lin PS, Lee RJ, Grippo J, Puzanov I, Kim KB, Ribas A, McArthur GA, Sosman JA, Chapman PB, Flaherty KT, Xu X, Nathanson KL, Nolop K (2010) Clinical efficacy of a RAF inhibitor needs broad target blockade in BRAF-mutant melanoma. Nature 467:596–599

Bozok-Cetintas V, Tezcanli-Kaymaz B, Aktug H, Oltulu F, Taskiran D (2014) Capsaicin induced apoptosis and gene expression dysregulation of human acute lymphoblastic leukemia CCRF-CEM cells. J BUON 19:183–190

Brachmann SM, Hofmann I, Schnell C, Fritsch C, Wee S, Lane H, Wang S, Garcia-Echeverria C, Maira SM (2009) Specific apoptosis induction by the dual PI3K/mTor inhibitor NVP-BEZ235 in HER2 amplified and PIK3CA mutant breast cancer cells. Proc Natl Acad Sci USA 106:22299–22304

Britten CD (2013) PI3K and MEK inhibitor combinations: examining the evidence in selected tumor types. Cancer Chemother Pharmacol 71:1395–1409

Brown MS, Goldstein JL (1980) Multivalent feedback regulation of HMG CoA reductase, a control mechanism coordinating isoprenoid synthesis and cell growth. J Lipid Res 21:505–517

Carlino MS, Todd JR, Gowrishankar K, Mijatov B, Pupo GM, Fung C, Snoyman S, Hersey P, Long GV, Kefford RF, Rizos H (2014) Differential activity of MEK and ERK inhibitors in BRAF inhibitor resistant melanoma. Mol Oncol 8:544–554

Carter LG, D'orazio JA, Pearson KJ (2014) Resveratrol and cancer: focus on *in vivo* evidence. Endocrine-Relat Cancer 21:R209–R225

Chaikuad A, McTacconi E, Zimmer J, Liang Y, Gray NS, Tarsounas M, Knapp S (2014) A unique inhibitor binding site in ERK1/2 is associated with slow binding kinetics. Nature Chem Biol 10:853

Chan CM, Hung CF (2014) Lycopene and retinal pigment epithelial cells: molecular aspects. In: Reedy AP, Victor R (eds) Handbook of nutrition, diet and the eye. Academic, San Diego

Chandarlapaty S, Sawai A, Scaltriti M, Rodrik-Outmezguine V, Grbovic-Huezo O, Serra V, Majumder PK, Baselga J, Rosen N (2011) AKT inhibition relieves feedback suppression of receptor tyrosine kinase expression and activity. Cancer Cell 19:58–71

Chapman PB, Hauschild A, Robert C, Haanen JB, Ascierto P, Larkin J, Dummer R, Garbe C, Testori A, Maio M, Hogg D, Lorigan P, Lebbe C, Jouary T, Schadendorf D, Ribas A, O'day SJ, Sosman JA, Kirkwood JM, Eggermont AM, Dreno B, Nolop K, Li J, Nelson B, Hou J, Lee RJ, Flaherty KT, McArthur GA (2011) Improved survival with vemurafenib in melanoma with BRAF V600E mutation. New Engl J Med 364:2507–2516

Clevers H (2004) Wnt breakers in colon cancer. Cancer Cell 5:5–6

Cohen RB (2003) Epidermal growth factor receptor as a therapeutic target in colorectal cancer. Clin Colorectal Cancer 2:246–251

Cooper GM (1982) Cellular transforming genes. Science 217:801–806

Courtney KD, Corcoran RB, Engelman JA (2010) The PI3K pathway as drug target in human cancer. J Clin Oncol 28:1075–1083

Cox AD, Der CJ (2010) Ras history: the saga continues. Small GTPases 1:2–27

Cox AD, Hisaka MM, Buss JE, Der CJ (1992) Specific isoprenoid modification is required for function of normal, but not oncogenic, Ras protein. Mol Cell Biol 12:2606–2615

Cox AD, Fesik SW, Kimmelman AC, Luo J, Der CJ (2014) Drugging the undruggable RAS: mission possible. Nature Rev Drug Dis 13:828–851
Cox AD, Der CJ, Philips MR (2015) Targeting RAS membrane association: back to the future for anti-RAS drug discovery. Clin Cancer Res 21:1819
Crul M, Klerk GJD, Swart M, Veer LJVT, Jong DD, Boerrigter L, Palmer PA, Bol CJ, Tan H, Gast GCD, Beijnen JH, Schellens JHM (2002) Phase I clinical and pharmacologic study of chronic oral administration of the farnesyl protein transferase inhibitor R115777 in advanced cancer. J Clin Oncol 20:2726–2735
Davies H, Bignell GR, Cox C, Stephens P, Edkins S, Clegg S, Teague J, Woffendin H, Garnett MJ, Bottomley W, Davis N, Dicks E, Ewing R, Floyd Y, Gray K, Hall S, Hawes R, Hughes J, Kosmidou V, Menzies A, Mould C, Parker A, Stevens C, Watt S, Hooper S, Wilson R, Jayatilake H, Gusterson BA, Cooper C, Shipley J, Hargrave D, Pritchard-Jones K, Maitland N, Chenevix-Trench G, Riggins GJ, Bigner DD, Palmieri G, Cossu A, Flanagan A, Nicholson A, Ho JW, Leung SY, Yuen ST, Weber BL, Seigler HF, Darrow TL, Paterson H, Marais R, Marshall CJ, Wooster R, Stratton MR, Futreal PA (2002) Mutations of the BRAF gene in human cancer. Nature 417:949–954
De Roock W, Piessevaux H, De Schutter J, Janssens M, De Hertogh G, Personeni N, Biesmans B, Van Laethem JL, Peeters M, Humblet Y, Van Cutsem E, Tejpar S (2008) KRAS wild-type state predicts survival and is associated to early radiological response in metastatic colorectal cancer treated with cetuximab. Ann Oncol 19:508–515
Deng Y, Shipps GW, Cooper A, English JM, Annis DA, Carr D, Nan Y, Wang T, Zhu HY, Chuang CC, Dayananth P, Hruza AW, Xiao L, Jin W, Kirschmeier P, Windsor WT, Samatar AA (2014) Discovery of novel, dual mechanism ERK inhibitors by affinity selection screening of an inactive kinase. J Med Chem 57:8817–8826
Deschoolmeester V, Boeckx C, Baay M, Weyler J, Wuyts W, Van Marck E, Peeters M, Lardon F, Vermorken JB (2010) KRAS mutation detection and prognostic potential in sporadic colorectal cancer using high-resolution melting analysis. Br J Cancer 103:1627–1636
Desolms SJ, Ciccarone TM, Mactough SC, Shaw AW, Buser CA, Ellis-Hutchings M, Fernandes C, Hamilton KA, Huber HE, Kohl NE, Lobell RB, Robinson RG, Tsou NN, Walsh ES, Graham SL, Beese LS, Taylor JS (2003) Dual protein farnesyltransferase-geranylgeranyltransferase-I inhibitors as potential cancer chemotherapeutic agents. J Med Chem 46:2973–2984
Dienstmann R, Rodon J, Serra V, Tabernero J (2014) Picking the point of inhibition: a comparative review of PI3K/AKT/mTOR pathway inhibitors. Mol Cancer Ther 13:1021–1031
Dinu D, Dobre M, Panaitescu E, Bîrlă R, Iosif C, Hoara P, Caragui A, Boeriu M, Constantinoiu S, Ardeleanu C (2014) Prognostic significance of KRAS gene mutations in colorectal cancer-preliminary study. J Med Life 7:581–587
Dolly S, Wagner AJ, Bendell JC, Yan Y, Ware JA, Mazina KE, Holden SN, Derynck MK, De Bono JS, Burris HA (2010) A first-in-human, phase 1 study to evaluate the dual PI3K/mTOR inhibitor GDC-0980 administered QD in patients with advanced solid tumors or non-Hodgkin's lymphoma. J Clin Oncol 28:3079–3079
Dougherty MK, Muller J, Ritt DA, Zhou M, Zhou XZ, Copeland TD, Conrads TP, Veenstra TD, Lu KP, Morrison DK (2005) Regulation of Raf-1 by direct feedback phosphorylation. Mol Cell 17:215–224
Dowling RJ, Topisirovic I, Fonseca BD, Sonenberg N (2010) Dissecting the role of mTOR: lessons from mTOR inhibitors. Biochim Biophys Acta 1804:433–439
Eisen T, Ahmad T, Flaherty KT, Gore M, Kaye S, Marais R, Gibbens I, Hackett S, James M, Schuchter LM, Nathanson KL, Xia C, Simantov R, Schwartz B, Poulin-Costello M, O'Dwyer PJ, Ratain MJ (2006) Sorafenib in advanced melanoma: a Phase II randomised discontinuation trial analysis. Br J Cancer 95:581–586
Ellis CA, Vos MD, Howell H, Vallecorsa T, Fults DW, Clark GJ (2002) Rig is a novel Ras-related protein and potential neural tumor suppressor. Proc Natl Acad Sci USA 99:9876–9881
Engelman JA (2009) Targeting PI3K signalling in cancer: opportunities, challenges and limitations. Nat Rev Cancer 9:550–562

Engelman JA, Chen L, Tan X, Crosby K, Guimaraes AR, Upadhyay R, Maira M, McNamara K, Perera SA, Song Y, Chirieac LR, Kaur R, Lightbown A, Simendinger J, Li T, Padera RF, Garcia-Echeverria C, Weissleder R, Mahmood U, Cantley LC, Wong KK (2008) Effective use of PI3K and MEK inhibitors to treat mutant Kras G12D and PIK3CA H1047R murine lung cancers. Nat Med 14:1351–1356

Falchook GS, Long GV, Kurzrock R, Kim KB, Arkenau TH, Brown MP, Hamid O, Infante JR, Millward M, Pavlick AC, O'day SJ, Blackman SC, Curtis CM, Lebowitz P, Ma B, Ouellet D, Kefford RF (2012) Dabrafenib in patients with melanoma, untreated brain metastases, and other solid tumours: a phase 1 dose-escalation trial. Lancet 379:1893–1901

Fearon ER, Vogelstein B (1990) A genetic model for colorectal tumorigenesis. Cell 61:759–767

Ferrucci V, Boffa I, De Masi G, Zollo M (2016) Natural compounds for pediatric cancer treatment. Naunyn Schmiedebergs Arch Pharmacol 389:131–149

Flaherty KT, Puzanov I, Kim KB, Ribas A, McArthur GA, Sosman JA, O'Dwyer PJ, Lee RJ, Grippo JF, Nolop K, Chapman PB (2010) Inhibition of mutated, activated BRAF in metastatic melanoma. New Engl J Med 363:809–819

Flaherty KT, Infante JR, Daud A, Gonzalez R, Kefford RF, Sosman J, Hamid O, Schuchter L, Cebon J, Ibrahim N, Kudchadkar R, Burris HA, Falchook G, Algazi A, Lewis K, Long GV, Puzanov I, Lebowitz P, Singh A, Little S, Sun P, Allred A, Ouellet D, Kim KB, Patel K, Weber J (2012) Combined BRAF and MEK inhibition in melanoma with BRAF V600 mutations. New Engl J Med 367:1694–1703

Frattini M, Balestra D, Suardi S, Oggionni M, Alberici P, Radice P, Costa A, Daidone MG, Leo E, Pilotti S, Bertario L, Pierotti MA (2004) Different genetic features associated with colon and rectal carcinogenesis. Clin Cancer Res 10:4015–4021

Freeman AK, Ritt DA, Morrison DK (2013) Effects of Raf dimerization and its inhibition on normal and disease-associated Raf signaling. Mol Cell 49:751–758

Fried LE, Arbiser JL (2009) Honokiol, a multifunctional antiangiogenic and antitumor agent. Antioxid Redox Signal 11:1139–1148

Gattenlohner S, Etschmann B, Riedmiller H, Muller-Hermelink HK (2009) Lack of KRAS and BRAF mutation in renal cell carcinoma. Eur Urol 55:1490–1491

Goldstein NI, Prewett M, Zuklys K, Rockwell P, Mendelsohn J (1995) Biological efficacy of a chimeric antibody to the epidermal growth factor receptor in a human tumor xenograft model. Clin Cancer Res 1:1311–1318

Goodsell DS (1999) The molecular perspective: the ras oncogene. Stem Cells 17:235–236

Hahm ER, Singh SV (2007) Honokiol causes G0-G1 phase cell cycle arrest in human prostate cancer cells in association with suppression of retinoblastoma protein level/phosphorylation and inhibition of E2F1 transcriptional activity. Mol Cancer Ther 6:2686–2695

Hanahan D, Weinberg RA (2011) Hallmarks of cancer: the next generation. Cell 144:646–674

Hancock JF, Paterson H, Marshall CJ (1990) A polybasic domain or palmitoylation is required in addition to the CAAX motif to localize p21ras to the plasma membrane. Cell 63:133–139

Harvey JJ (1964) An unidentified virus which causes the rapid production of tumours in mice. Nature 204:1104–1105

Hatzivassiliou G, Song K, Yen I, Brandhuber BJ, Anderson DJ, Alvarado R, Ludlam MJ, Stokoe D, Gloor SL, Vigers G, Morales T, Aliagas I, Liu B, Sideris S, Hoeflich KP, Jaiswal BS, Seshagiri S, Koeppen H, Belvin M, Friedman LS, Malek S (2010) RAF inhibitors prime wild-type RAF to activate the MAPK pathway and enhance growth. Nature 464:431–435

Hatzivassiliou G, Haling JR, Chen H, Song K, Price S, Heald R, Hewitt JF, Zak M, Peck A, Orr C, Merchant M, Hoeflich KP, Chan J, Luoh SM, Anderson DJ, Ludlam MJ, Wiesmann C, Ultsch M, Friedman LS, Malek S, Belvin M (2013) Mechanism of MEK inhibition determines efficacy in mutant KRAS- versus BRAF-driven cancers. Nature 501:232–236

Hauschild A, Grob JJ, Demidov LV, Jouary T, Gutzmer R, Millward M, Rutkowski P, Blank CU, Miller WH Jr, Kaempgen E, Martin-Algarra S, Karaszewska B, Mauch C, Chiarion-Sileni V, Martin AM, Swann S, Haney P, Mirakhur B, Guckert ME, Goodman V, Chapman PB (2012) Dabrafenib in BRAF-mutated metastatic melanoma: a multicentre, open-label, phase 3 randomised controlled trial. Lancet 380:358–365

Hong S, Kim S, Kim HY, Kang M, Jang HH, Lee WS (2016) Targeting the PI3K signaling pathway in KRAS mutant colon cancer. Cancer Med 5:248–255

Irahara N, Baba Y, Nosho K, Shima K, Yan L, Dias-Santagata D, Iafrate AJ, Fuchs CS, Haigis KM, Ogino S (2010) NRAS mutations are rare in colorectal cancer. Diagn Mol Pathol 19:157–163

Ishii N, Harada N, Joseph EW, Ohara K, Miura T, Sakamoto H, Matsuda Y, Tomii Y, Tachibana-Kondo Y, Iikura H, Aoki T, Shimma N, Arisawa M, Sowa Y, Poulikakos PI, Rosen N, Aoki Y, Sakai T (2013) Enhanced inhibition of ERK signaling by a novel allosteric MEK inhibitor, CH5126766, that suppresses feedback reactivation of RAF activity. Cancer Res 73:4050–4060

Ito K, Nakazato T, Yamato K, Miyakawa Y, Yamada T, Hozumi N, Segawa K, Ikeda Y, Kizaki M (2004) Induction of apoptosis in leukemic cells by homovanillic acid derivative, capsaicin, through oxidative stress: implication of phosphorylation of p53 at Ser-15 residue by reactive oxygen species. Cancer Res 64:1071–1078

Jakubikova J, Sedlak J, Mithen R, Bao Y (2005) Role of PI3K/Akt and MEK/ERK signaling pathways in sulforaphane- and erucin-induced phase II enzymes and MRP2 transcription, G2/M arrest and cell death in Caco-2 cells. Biochem Pharmacol 69:1543–1552

James G, Goldstein JL, Brown MS (1996) Resistance of K-RasBV12 proteins to farnesyltransferase inhibitors in Rat1 cells. Proc Natl Acad Sci USA 93:4454–4458

Jenjaroenpun P, Kremenska Y, Nair VM, Kremenskoy M, Joseph B, Kurochkin IV (2013) Characterization of RNA in exosomes secreted by human breast cancer cell lines using next-generation sequencing. PeerJ 1:e201

Joseph EW, Pratilas CA, Poulikakos PI, Tadi M, Wang W, Taylor BS, Halilovic E, Persaud Y, Xing F, Viale A, Tsai J, Chapman PB, Bollag G, Solit DB, Rosen N (2010) The RAF inhibitor PLX4032 inhibits ERK signaling and tumor cell proliferation in a V600E BRAF-selective manner. Proc Natl Acad Sci USA 107:14903–14908

Karapetis CS, Khambata-Ford S, Jonker DJ, O'Callaghan CJ, Tu D, Tebbut NC, Simes RJ, Chalchal H, Shapiro JD, Robitaille S, Price TJ, Shepherd L, Au HJ, Langer C, Moore MJ, Zalcberg JR (2008) K-ras mutations and benefit from cetuximab in advanced colorectal cancer. New Engl J Med 359:1757–1765

Karnoub AE, Weinberg RA (2008) Ras oncogenes: split personalities. Nature Rev Mol Cell Biol 9:517–531

Kato K, Der CJ, Buss JE (1992) Prenoids and palmitate: lipids that control the biological activity of Ras proteins. Semin Cancer Biol 3:179–188

Kim E, Ambroziak P, Otto JC, Taylor B, Ashby M, Shannon K, Casey PJ, Young SG (1999) Disruption of the mouse Rce1 gene results in defective Ras processing and mislocalization of Ras within cells. J Biol Chem 274:8383–8390

Kim HJ, Kim SK, Kim BS, Lee SH, Park YS, Park BK, Kim SJ, Kim J, Choi C, Kim JS, Cho SD, Jung JW, Roh KH, Kang KS, Jung JY (2010) Apoptotic effect of quercetin on HT-29 colon cancer cells via the AMPK signaling pathway. J Agric Food Chem 58:8643–8650

Kirsten WH, Schauf V, McCoy J (1970) Properties of a murine sarcoma virus. Bibl Haematol 2:46–49

Kometani K, Ishida D, Hattori M, Minato N (2004) Rap1 and SPA-1 in hematologic malignancy. Trends Mol Med 10:401–408

Laheru D, Shah P, Rajeshkumar NV, McAllister F, Taylor G, Goldsweig H, Le DT, Donehower R, Jimeno A, Linden S, Zhao M, Song D, Rudek MA, Hidalgo M (2012) Integrated preclinical and clinical development of S-trans, trans-Farnesylthiosalicylic Acid (FTS, Salirasib) in pancreatic cancer. Invest New Drugs 30:2391–2399

Lannutti BJ, Meadows SA, Herman SE, Kashishian A, Steiner B, Johnson AJ, Byrd JC, Tyner JW, Loriaux MM, Deininger M, Druker BJ, Puri KD, Ulrich RG, Giese NA (2011) CAL-101, a p110delta selective phosphatidylinositol-3-kinase inhibitor for the treatment of B-cell malignancies, inhibits PI3K signaling and cellular viability. Blood 117:591–594

Lavoie H, Thevakumaran N, Gavory G, Li JJ, Padeganeh A, Guiral S, Duchaine J, Mao DYL, Bouvier M, Sicheri F, Therrien M (2013) Inhibitors that stabilize a closed RAF kinase domain conformation induce dimerization. Nature Chem Biol 9:428

Lee SJ, Jung YS, Lee SH, Chung HY, Park BJ (2009) Isolation of a chemical inhibitor against K-Ras-induced p53 suppression through natural compound screening. Int J Oncol 34:1637–1643

Li S, Schmitz KR, Jeffrey PD, Wiltzius JJ, Kussie P, Ferguson KM (2005) Structural basis for inhibition of the epidermal growth factor receptor by cetuximab. Cancer Cell 7:301–311

Lievre A, Bachet JB, Le Corre D, Boige V, Landi B, Emile JF, Cote JF, Tomasic G, Penna C, Ducreux M, Rougier P, Penault-Llorca F, Laurent-Puig P (2006) KRAS mutation status is predictive of response to cetuximab therapy in colorectal cancer. Cancer Res 66:3992–3995

Lievre A, Bachet JB, Boige V, Cayre A, Le Corre D, Buc E, Ychou M, Bouche O, Landi B, Louvet C, Andre T, Bibeau F, Diebold MD, Rougier P, Ducreux M, Tomasic G, Emile JF, Penault-Llorca F, Laurent-Puig P (2008) KRAS mutations as an independent prognostic factor in patients with advanced colorectal cancer treated with cetuximab. J Clin Oncol 26:374–379

Lim KH, Baines AT, Fiordalisi JJ, Shipitsin M, Feig LA, Cox AD, Der CJ, Counter CM (2005) Activation of RalA is critical for Ras-induced tumorigenesis of human cells. Cancer Cell 7:533–545

Lito P, Saborowski A, Yue J, Solomon M, Joseph E, Gadal S, Saborowski M, Kastenhuber E, Fellmann C, Ohara K, Morikami K, Miura T, Lukacs C, Ishii N, Lowe S, Rosen N (2014) Disruption of CRAF-mediated MEK activation is required for effective MEK inhibition in KRAS mutant tumors. Cancer Cell 25:697–710

Liu H, Zang C, Emde A, Planas-Silva MD, Rosche M, Kuhnl A, Schulz CO, Elstner E, Possinger K, Eucker J (2008) Anti-tumor effect of honokiol alone and in combination with other anti-cancer agents in breast cancer. Eur J Pharmacol 591:43–51

Liu M, Sjogren AKM, Karlsson C, Ibrahim MX, Andersson KME, Olofsson FJ, Wahlstrom AM, Dalin M, Yu H, Chen Z, Yang SH, Young SG, Bergo MO (2010) Targeting the protein prenyltransferases efficiently reduces tumor development in mice with K-RAS-induced lung cancer. Proc Natl Acad Sci USA 107:6471–6476

Liu N, Rowley BR, Bull CO, Schneider C, Haegebarth A, Schatz CA, Fracasso PR, Wilkie DP, Hentemann M, Wilhelm SM, Scott WJ, Mumberg D, Ziegelbauer K (2013) BAY 80-6946 is a highly selective intravenous PI3K inhibitor with potent p110alpha and p110delta activities in tumor cell lines and xenograft models. Mol Cancer Ther 12:2319–2330

Lobell RB, Omer CA, Abrams MT, Bhimnathwala HG, Brucker MJ, Buser CA, Davide JP, Desolms SJ, Dinsmore CJ, Ellis-Hutchings MS, Kral AM, Liu D, Lumma WC, Machotka SV, Rands E, Williams TM, Graham SL, Hartman GD, Oliff AI, Heimbrook DC, Kohl NE (2001) Evaluation of farnesyl:protein transferase and geranylgeranyl:protein transferase inhibitor combinations in preclinical models. Cancer Res 61:8758–8768

Lorusso P, Shapiro G, Pandya SS, Kwak EL, Jones C, Belvin M, Musib LC, De Crespigny A, Mckenzie M, Gates MR, Chan ITC, Bendell JC (2012) A first-in-human phase Ib study to evaluate the MEK inhibitor GDC-0973, combined with the pan-PI3K inhibitor GDC-0941, in patients with advanced solid tumors. J Clin Oncol 30:2566

Luo RZ, Fang X, Marquez R, Liu SY, Mills GB, Liao WS, Yu Y, Bast RC (2003) ARHI is a Ras-related small G-protein with a novel N-terminal extension that inhibits growth of ovarian and breast cancers. Oncogene 22:2897–2909

Luo LX, Li Y, Liu ZQ, Fan XX, Duan FG, Li RZ, Yao XJ, Leung ELH, Liu L (2017) Honokiol induces apoptosis, G1 arrest, and autophagy in KRAS mutant lung cancer cells. Front Pharmacol 8:199

Lyons JF, Wilhelm S, Hibner B, Bollag G (2001) Discovery of a novel Raf kinase inhibitor. Endocr Relat Cancer 8:219–225

Maira SM, Pecchi S, Huang A, Burger M, Knapp M, Sterker D, Schnell C, Guthy D, Nagel T, Wiesmann M, Brachmann S, Fritsch C, Dorsch M, Chene P, Shoemaker K, De Pover A, Menezes D, Martiny-Baron G, Fabbro D, Wilson CJ, Schlegel R, Hofmann F, Garcia-Echeverria C, Sellers WR, Voliva CF (2012) Identification and characterization of NVP-BKM120, an orally available pan-class I PI3-kinase inhibitor. Mol Cancer Ther 11:317–328

Malumbres M, Barbacid M (2003) RAS oncogenes: the first 30 years. Nature Rev Cancer 3:459

Mamounas EP, Tang G, Fisher B, Paik S, Shak S, Costantino JP, Watson D, Geyer CE, Wickerham DL, Wolmark N (2010) Association between the 21-gene recurrence score assay and risk of

locoregional recurrence in node-negative, estrogen receptor-positive breast cancer: Results from NSABP B-14 and NSABP B-20. J Clin Oncol 28:1677–1683
Markowitz SD, Bertagnolli MM (2009) Molecular origins of cancer: molecular basis of colorectal cancer. New Engl J Med 361:2449–2460
Marks JL, Gong Y, Chitale D, Golas B, Mclellan MD, Kasai Y, Ding L, Mardis ER, Wilson RK, Solit D, Levine R, Michel K, Thomas RK, Rusch VW, Ladanyi M, Pao W (2008) Novel MEK1 mutation identified by mutational analysis of epidermal growth factor receptor signaling pathway genes in lung adenocarcinoma. Cancer Res 68:5524–5528
Marom M, Haklai R, Ben-Baruch G, Marciano D, Egozi Y, Kloog Y (1995) Selective inhibition of Ras-dependent cell growth by farnesylthiosalisylic acid. J Biol Chem 270:22263–22270
Marshall CJ (1993) Protein prenylation: a mediator of protein-protein interactions. Science 259:1865–1866
Massacesi C, Di Tomaso E, Urban P, Germa C, Quadt C, Trandafir L, Aimone P, Fretault N, Dharan B, Tavorath R, Hirawat S (2016) PI3K inhibitors as new cancer therapeutics: implications for clinical trial design. OncoTargets Ther 9:203–210
Mclellan EA, Owen RA, Stepniewska KA, Sheffield JP, Lemoine NR (1993) High frequency of K-ras mutations in sporadic colorectal adenomas. Gut 34:392–396
Mendoza MC, Er EE, Blenis J (2011) The Ras-ERK and PI3K-mTOR pathways: cross-talk and compensation. Trends Biochem Sci 36:320–328
Mita MM, Wolin EM, Meyer T, Nemunaitis JJ, Bergsland EK, Mahipal A, Wong L, Wu X, Carmichael J, Chopra R, Hege K, Bendell JC (2013) Phase I expansion trial of an oral TORC1/TORC2 inhibitor (CC-223) in nonpancreatic neuroendocrine tumors (NET). J Clin Oncol 31:e15004
Mologni L, Brussolo S, Ceccon M, Gambacorti-Passerini C (2012) Synergistic effects of combined Wnt/KRAS inhibition in colorectal cancer cells. PLoS One 7:e51449
Moroni M, Veronese S, Benvenuti S, Marrapese G, Sartore-Bianchi A, Di Nicolantonio F, Gambacorta M, Siena S, Bardelli A (2005) Gene copy number for epidermal growth factor receptor (EGFR) and clinical response to antiEGFR treatment in colorectal cancer: a cohort study. Lancet Oncol 6:279–286
Morris EJ, Jha S, Restaino CR, Dayananth P, Zhu H, Cooper A, Carr D, Deng Y, Jin W, Black S, Long B, Liu J, Dinunzio E, Windsor W, Zhang R, Zhao S, Angagaw MH, Pinheiro EM, Desai J, Xiao L, Shipps G, Hruza A, Wang J, Kelly J, Paliwal S, Gao X, Babu BS, Zhu L, Daublain P, Zhang L, Lutterbach BA, Pelletier MR, Philippar U, Siliphaivanh P, Witter D, Kirschmeier P, Bishop WR, Hicklin D, Gilliland DG, Jayaraman L, Zawel L, Fawell S, Samatar AA (2013) Discovery of a novel ERK inhibitor with activity in models of acquired resistance to BRAF and MEK inhibitors. Cancer Discov 3:742–750
Murugan AK, Dong J, Xie J, Xing M (2009) MEK1 mutations, but not ERK2 mutations, occur in melanomas and colon carcinomas, but none in thyroid carcinomas. Cell Cycle 8:2122–2124
Musgrove EA, Caldon CE, Barraclough J, Stone A, Sutherland RL (2011) Cyclin D as a therapeutic target in cancer. Nat Rev Cancer 11:558–572
Nakamura A, Arita T, Tsuchiya S, Donelan J, Chouitar J, Carideo E, Galvin K, Okaniwa M, Ishikawa T, Yoshida S (2013) Antitumor activity of the selective pan-RAF inhibitor TAK-632 in BRAF inhibitor-resistant melanoma. Cancer Res 73:7043–7055
Nash GM, Gimbel M, Cohen AM, Zeng ZS, Ndubuisi MI, Nathanson DR, Ott J, Barany F, Paty PB (2010) KRAS mutation and microsatellite instability: two genetic markers of early tumor development that influence the prognosis of colorectal cancer. Ann Surg Oncol 17:416–424
Nayak L, Hays JL, Muzikansky A, Gaffey SC, Do KT, Puduvalli VK, Lee EQ, Rinne ML, Norden AD, Beroukhim R, Wen PY, Doyle LA, Chen HX, Shapiro G, Reardon DA (2016) A phase I study of MLN0128 and bevacizumab in patients with recurrent glioblastoma and other solid tumors. J Clin Oncol 34:2013
Ndubaku CO, Heffron TP, Staben ST, Baumgardner M, Blaquiere N, Bradley E, Bull R, Do S, Dotson J, Dudley D, Edgar KA, Friedman LS, Goldsmith R, Heald RA, Kolesnikov A, Lee L, Lewis C, Nannini M, Nonomiya J, Pang J, Price S, Prior WW, Salphati L, Sideris S, Wallin JJ, Wang L, Wei B, Sampath D, Olivero AG (2013) Discovery of 2-{3-[2-(1-isopropyl-3-methyl-

1H-1,2-4-triazol-5-yl)-5,6-dihydrobenzo[f]imidazo[1,2-d][1,4]oxazepin-9-yl]-1H-pyrazol-1-yl}-2-methylpropanamide (GDC-0032): a beta-sparing phosphoinositide 3-kinase inhibitor with high unbound exposure and robust *in vivo* antitumor activity. J Med Chem 56:4597–4610
Neumann J, Zeindl-Eberhart E, Kirchner T, Jung A (2009) Frequency and type of KRAS mutations in routine diagnostic analysis of metastatic colorectal cancer. Pathol Res Pract 205:858–862
O'Reilly KE, Rojo F, She QB, Solit D, Mills GB, Smith D, Lane H, Hofmann F, Hicklin DJ, Ludwig DL, Baselga J, Rosen N (2006) mTOR inhibition induces upstream receptor tyrosine kinase signaling and activates Akt. Cancer Res 66:1500–1508
Pal I, Mandal M (2012) PI3K and Akt as molecular targets for cancer therapy: current clinical outcomes. Acta Pharmacol Sin 33:1441–1458
Palozza P, Colangelo M, Simone R, Catalano A, Boninsegna A, Lanza P, Monego G, Ranelletti FO (2010) Lycopene induces cell growth inhibition by altering mevalonate pathway and Ras signaling in cancer cell lines. Carcinogenesis 31:1813–1821
Papke B, Murarka S, Vogel HA, Martín-Gago P, Kovacevic M, Truxius DC, Fansa EK, Ismail S, Zimmermann G, Heinelt K, Schultz-Fademrecht C, Al Saabi A, Baumann M, Nussbaumer P, Wittinghofer A, Waldmann H, Bastiaens PIH (2016) Identification of pyrazolopyridazinones as PDEδ inhibitors. Nature Commun 7:11360
Peyton JD, Rodon Ahnert J, Burris H, Britten C, Chen LC, Tabernero J, Duval V, Rouyrre N, Silva AP, Quadt C, Baselga J (2011) A dose-escalation study with the novel formulation of the oral pan-class I PI3K inhibitor BEZ235, solid dispersion system (SDS) sachet, in patients with advanced solid tumors. J Clin Oncol 29:3066
Pino MS, Chung DC (2010) The chromosomal instability pathway in colon cancer. Gastroenterol 138:2059–2072
Polakis P (2012) Wnt Signaling in Cancer. Cold Spring Harbor Perspect Biol 4:a008052
Poulikakos PI, Rosen N (2011) Mutant BRAF melanomas-dependence and resistance. Cancer Cell 19:11–15
Poulikakos PI, Zhang C, Bollag G, Shokat KM, Rosen N (2010) RAF inhibitors transactivate RAF dimers and ERK signalling in cells with wild-type BRAF. Nature 464:427–430
Pramanik KC, Boreddy SR, Srivastava SK (2011) Role of mitochondrial electron transport chain complexes in capsaicin mediated oxidative stress leading to apoptosis in pancreatic cancer cells. PLoS One 6:e20151
Prenen H, Tejpar S, Van Cutsem E (2010) New strategies for treatment of KRAS mutant metastatic colorectal cancer. Clin Cancer Res 16:2921–2926
Prior IA, Hancock JF (2001) Compartmentalization of Ras proteins. J Cell Sci 114:1603–1608
Prior IA, Lewis PD, Mattos C (2012) A comprehensive survey of Ras mutations in cancer. Cancer Res 72:2457–2467
Rajalingam K, Schreck R, Rapp UR, Albert Š (2007) Ras oncogenes and their downstream targets. Biochim Biophys Acta 1773:1177–1195
Ramanathan RK, Von Hoff DD, Eskens F, Blumenschein GR, Richards DA, Renshaw FG, Rajagopalan P, Kelly A, Pena CE, Mross KB (2014) A phase 1b trial of PI3K inhibitor copanlisib (BAY 80-6946) combined with the allosteric-MEK inhibitor refametinib (BAY 86-9766) in patients with advanced cancer. J Clin Oncol 32:2588–2588
Raponi M, Winkler H, Dracopoli NC (2008) KRAS mutations predict response to EGFR inhibitors. Curr Opin Pharmacol 8:413–418
Reif S, Weis B, Aeed H, Gana-Weis M, Zaidel L, Avni Y, Romanelli RG, Pinzani M, Kloog Y, Bruck R (1999) The Ras antagonist, farnesylthiosalicylic acid (FTS), inhibits experimentally-induced liver cirrhosis in rats. J Hepatol 31:1053–1061
Ritt DA, Monson DM, Specht SI, Morrison DK (2010) Impact of feedback phosphorylation and Raf heterodimerization on normal and mutant B-Raf signaling. Mol Cell Biol 30:806–819
Roberts PJ, Der CJ (2007) Targeting the Raf-MEK-ERK mitogen-activated protein kinase cascade for the treatment of cancer. Oncogene 26:3291–3310
Robinson MJ, Cobb MH (1997) Mitogen-activated protein kinase pathways. Curr Opin Cell Biol 9:180–186

Roskoski R Jr (2012) MEK1/2 dual-specificity protein kinases: structure and regulation. Biochem Biophys Res Commun 417:5–10

Rowinsky EK (2006) Lately, it occurs to me what a long, strange trip it's been for the farnesyltransferase inhibitors. J Clin Oncol 24:2981–2984

Rozengurt E, Soares HP, Sinnet-Smith J (2014) Suppression of feedback loops mediated by PI3K/mTOR induces multiple over-activation of compensatory pathways: an unintended consequence leading to drug resistance. Mol Cancer Ther 13:2477–2488

Rusin M, Zajkowicz A, Butkiewicz D (2009) Resveratrol induces senescence-like growth inhibition of U-2 OS cells associated with the instability of telomeric DNA and upregulation of BRCA1. Mech Ageing Dev 130:528–537

Russo A, Bazan V, Agnese V, Rodolico V, Gebbia N (2005) Prognostic and predictive factors in colorectal cancer: Kirsten Ras in CRC (RASCAL) and TP53CRC collaborative studies. Ann Oncol 16:S44–S49

Saif MW, Oettle H, Vervenne WL, Thomas JP, Spitzer G, Visseren-Grul C, Enas N, Richards DA (2009) Randomized double-blind phase II trial comparing gemcitabine plus LY293111 versus gemcitabine plus placebo in advanced adenocarcinoma of the pancreas. Cancer J 15:339–343

Samatar AA, Poulikakos PI (2014) Targeting RAS-ERK signalling in cancer: promises and challenges. Nat Rev Drug Discov 13:928–942

Saud SM, Li W, Morris NL, Matter MS, Colburn NH, Kim YS, Young MR (2014) Resveratrol prevents tumorigenesis in mouse model of Kras activated sporadic colorectal cancer by suppressing oncogenic KRAS expression. Carcinogenesis 35:2778–2786

Schaeffer BK, Glasner S, Kuhlmann E, Myles JL, Longnecke DS (1994) Mutated c-K-ras in small pancreatic adenocarcinomas. Pancreas 9:161–165

Schmick M, Vartak N, Papke B, Kovacevic M, Truxius DC, Rossmannek L, Bastiaens PIH (2014) KRas localizes to the plasma membrane by spatial cycles of solubilization, trapping and vesicular transport. Cell 157(2):459–471

Schmick M, Kraemer A, Bastiaens PI (2015) Ras moves to stay in place. Trends Cell Biol 25:190–197

Schmukler E, Grinboim E, Schokoroy S, Amir A, Wolfson E, Kloog Y, Pinkas-Kramarski R (2013) Ras inhibition enhances autophagy, which partially protects cells from death. Oncotarget 4:145–155

Schmukler E, Wolfson E, Haklai R, Elad-Sfadia G, Kloog Y, Pinkas-Kramarski R (2014) Chloroquine synergizes with FTS to enhance cell growth inhibition and cell death. Oncotarget 5:173–184

Schramm K, Krause K, Bittroff-Leben A, Goldin-Lang P, Thiel E, Kreuser ED (2000) Activated K-ras is involved in regulation of integrin expression in human colon carcinoma cells. Intl J Cancer 87:155–164

Schubbert S, Shannon K, Bollag G (2007) Hyperactive Ras in developmental disorders and cancer. Nat Rev Cancer 7:295–308

Sepp-Lorenzino L, Ma Z, Rands E, Kohl NE, Gibbs JB, Oliff A, Rosen N (1995) A peptidomimetic inhibitor of farnesyl: protein transferase blocks the anchorage-dependent and independent growth of human tumor cell lines. Cancer Res 55:5302–5309

Serra V, Markman B, Scaltriti M, Eichhorn PJA, Valero V, Guzman M, Botero ML, Llonch E, Atzori F, Di Cosimo S, Maira M, Garcia-Echeverria C, Parra JL, Arribas J, Baselga J (2008) NVP-BEZ235, a dual PI3K/mTOR inhibitor, prevents PI3K signaling and inhibits the growth of cancer cells with activating PI3K mutations. Cancer Res 68:8022–8030

Serra V, Scaltriti M, Prudkin L, Eichhorn PJ, Ibrahim YH, Chandarlapaty S, Markman B, Rodriguez O, Guzman M, Rodriguez S, Gili M, Russillo M, Parra JL, Singh S, Arribas J, Rosen N, Baselga J (2011) PI3K inhibition results in enhanced HER signaling and acquired ERK dependency in HER2-overexpressing breast cancer. Oncogene 30:2547–2557

Shalom-Feuerstein R, Lindenboim L, Stein R, Cox AD, Kloog Y (2004) Restoration of sensitivity to anoikis in Ras-transformed rat intestinal epithelial cells by a Ras inhibitor. Cell Death Differ 11:244–247

Shimizu T, Tolcher AW, Papadopoulos KP, Beeram M, Rasco DW, Smith LS, Gunn S, Smetzer L, Mays TA, Kaiser B, Wick MJ, Alvarez C, Cavazos A, Mangold GL, Patnaik A (2012) The clinical effect of the dual-targeting strategy involving PI3K/AKT/mTOR and RAS/MEK/ERK pathways in patients with advanced cancer. Clin Cancer Res 18:2316–2325

Shirasawa S, Furuse M, Yokoyama N, Sasazuki T (1993) Altered growth of human colon cancer cell lines disrupted at activated Ki-ras. Science 260:85–88

Sieber OM, Heinimann K, Tomlinson IP (2003) Genomic instability – the engine of tumorigenesis. Nat Rev Cancer 3:701–708

Siegel RL, Miller KD, Fedewa SA, Ahnen DJ, Meester RGS, Barzi A, Jemal A (2017) Colorectal cancer statistics, 2017. CA: A Cancer J Clin 67:177–193

Sjogren AK, Andersson KM, Liu M, Cutts BA, Karlsson C, Wahlstrom AM, Dalin M, Weinbaum C, Casey PJ, Tarkowski A, Swolin B, Young SG, Bergo MO (2007) GGTase-I deficiency reduces tumor formation and improves survival in mice with K-RAS-induced lung cancer. J Clin Invest 117:1294–1304

Stephen Andrew G, Esposito D, Bagni Rachel K, McCormick F (2014) Dragging Ras back in the ring. Cancer Cell 25:272–281

Su F, Viros A, Milagre C, Trunzer K, Bollag G, Spleiss O, Reis-Filho JS, Kong X, Koya RC, Flaherty KT, Chapman PB, Kim MJ, Hayward R, Martin M, Yang H, Wang Q, Hilton H, Hang JS, Noe J, Lambros M, Geyer F, Dhomen N, Niculescu-Duvaz I, Zambon A, Niculescu-Duvaz D, Preece N, Robert L, Otte NJ, Mok S, Kee D, Ma Y, Zhang C, Habets G, Burton EA, Wong B, Nguyen H, Kockx M, Andries L, Lestini B, Nolop KB, Lee RJ, Joe AK, Troy JL, Gonzalez R, Hutson TE, Puzanov I, Chmielowski B, Springer CJ, McArthur GA, Sosman JA, Lo RS, Ribas A, Marais R (2012) RAS mutations in cutaneous squamous-cell carcinomas in patients treated with BRAF inhibitors. New Engl J Med 366:207–215

Swanson KM, Hohl RJ (2006) Anti-cancer therapy: targeting the mevalonate pathway. Curr Cancer Drug Targets 6:15–37

Tannapfel A, Sommerer F, Benicke M, Katalinic A, Uhlmann D, Witzigmann H, Hauss J, Wittekind C (2003) Mutations of the BRAF gene in cholangiocarcinoma but not in hepatocellular carcinoma. Gut 52:706–712

Tsai J, Lee JT, Wang W, Zhang J, Cho H, Mamo S, Bremer R, Gillette S, Kong J, Haass NK, Sproesser K, Li L, Smalley KS, Fong D, Zhu YL, Marimuthu A, Nguyen H, Lam B, Liu J, Cheung I, Rice J, Suzuki Y, Luu C, Settachatgul C, Shellooe R, Cantwell J, Kim SH, Schlessinger J, Zhang KY, West BL, Powell B, Habets G, Zhang C, Ibrahim PN, Hirth P, Artis DR, Herlyn M, Bollag G (2008) Discovery of a selective inhibitor of oncogenic B-Raf kinase with potent antimelanoma activity. Proc Natl Acad Sci USA 105:3041–3046

Tsai FD, Lopes MS, Zhou M, Ponce O, Fiordalisi JJ, Gierut JJ, Cox AD, Haigis KM, Philips MR (2015) K-Ras4A splice variant is widely expressed in cancer and uses a hybrid membrane-targeting motif. Proc Natl Acad Sci 112:779–784

Tutelian VA, Lashneva NV (2013) Biologically active substances of plant origin. Flavonols and flavones: prevalence, dietary sourses and consumption. Vopr Pitan 82:4–22

Van Cutsem E, Kohne CH, Lang I, Folprecht G, Nowacki MP, Cascinu S, Shchepotin I, Maurel J, Cunningham D, Tejpar S, Schlichting M, Zubel A, Celik I, Rougier P, Ciardiello F (2011) Cetuximab plus irinotecan, fluorouracil, and leucovorin as first-line treatment for metastatic colorectal cancer: updated analysis of overall survival according to tumor KRAS and BRAF mutation status. J Clin Oncol 29:2011–2019

Vaughn CP, Zobell SD, Furtado LV, Baker CL, Samowitz WS (2011) Frequency of KRAS, BRAF, and NRAS mutations in colorectal cancer. Genes Chromo Cancer 50:307–312

Vigil D, Cherfils J, Rossman KL, Der CJ (2010) Ras superfamily GEFs and GAPs: validated and tractable targets for cancer therapy. Nat Rev Cancer 10:842–857

Vogelstein B, Fearon ER, Hamilton SR, Kern SE, Preisinger AC, Leppert M, Nakamura Y, White R, Smits AM, Bos JL (1988) Genetic alterations during colorectal-tumor development. New Engl J Med 319:525–532

Wan PT, Garnett MJ, Roe SM, Lee S, Niculescu-Duvaz D, Good VM, Jones CM, Marshall CJ, Springer CJ, Barford D, Marais R (2004) Mechanism of activation of the RAF-ERK signaling pathway by oncogenic mutations of B-RAF. Cell 116:855–867

Wang H, Khor TO, Shu L, Su Z, Fuentes F, Lee JH, Kong ANT (2012) Plants against cancer: a review on natural phytochemicals in preventing and treating cancers and their druggability. Anti-cancer Agents Med Chem 12:1281–1305

Weber CK, Slupsky JR, Kalmes HA, Rapp UR (2001) Active Ras induces heterodimerization of cRaf and BRaf. Cancer Res 61:3595–3598

Weigelt B, Warne PH, Downward J (2011) PIK3CA mutation, but not PTEN loss of function, determines the sensitivity of breast cancer cells to mTOR inhibitory drugs. Oncogene 30:3222–3233

Weisz B, Giehl K, Gana-Weisz M, Egozi Y, Ben-Baruch G, Marciano D, Gierschik P, Kloog Y (1999) A new functional Ras antagonist inhibits human pancreatic tumor growth in nude mice. Oncogene 18:2579–2588

Whyte DB, Kirschmeier P, Hockenberry TN, Nunez-Oliva I, James L, Catino JJ, Bishop WR, Pai JK (1997) K- and N-Ras are geranylgeranylated in cells treated with farnesyl protein transferase inhibitors. J Biol Chem 272:14459–14464

Wilhelm SM, Carter C, Tang L, Wilkie D, Mcnabola A, Rong H, Chen C, Zhang X, Vincent P, McHugh M, Cao Y, Shujath J, Gawlak S, Eveleigh D, Rowley B, Liu L, Adnane L, Lynch M, Auclair D, Taylor I, Gedrich R, Voznesensky A, Riedl B, Post LE, Bollag G, Trail PA (2004) BAY 43-9006 exhibits broad spectrum oral antitumor activity and targets the RAF/MEK/ERK pathway and receptor tyrosine kinases involved in tumor progression and angiogenesis. Cancer Res 64:7099–7109

Winter-Vann AM, Casey PJ (2005) Post-prenylation-processing enzymes as new targets in oncogenesis. Nat Rev Cancer 5:405–412

Wong KK, Engelman JA, Cantley LC (2010) Targeting the PI3K signaling pathway in cancer. Curr Opin Genet Devel 20:87

Wood LD, Parsons DW, Jones S, Lin J, Sjoblom T, Leary RJ, Shen D, Boca SM, Barber T, Ptak J, Silliman N, Szabo S, Dezso Z, Ustyanksky V, Nikolskaya T, Nikolsky Y, Karchin R, Wilson PA, Kaminker JS, Zhang Z, Croshaw R, Willis J, Dawson D, Shipitsin M, Willson JK, Sukumar S, Polyak K, Park BH, Pethiyagoda CL, Pant PV, Ballinger DG, Sparks AB, Hartigan J, Smith DR, Suh E, Papadopoulos N, Buckhaults P, Markowitz SD, Parmigiani G, Kinzler KW, Velculescu VE, Vogelstein B (2007) The genomic landscapes of human breast and colorectal cancers. Science 318:1108–1113

Yan Y, Wagle MC, Punnoose E, Musib L, Budha N, Nannini M, Lin K, Liederer BM, Murli S, Ramakrishnan V, Patel P, Cervantes A, Tabernero J (2011) A first-in-human trial of GDC-0068:A novel, oral, ATP-competitive Akt inhibitor, demonstrates robust suppression of the Akt pathway in surrogate and tumor tissues. Mol Cancer Ther 10:B154–B154

Yan Y, Serra V, Prudkin L, Scaltriti M, Murli S, Rodriguez O, Guzman M, Sampath D, Nannini M, Xiao Y, Wagle MC, Wu JQ, Wongchenko M, Hampton G, Ramakrishnan V, Lackner MR, Saura C, Roda D, Cervantes A, Tabernero J, Patel P, Baselga J (2013) Evaluation and clinical analyses of downstream targets of the Akt inhibitor GDC-0068. Clin Cancer Res 19:6976–6986

Yang SE, Hsieh MT, Tsai TH, Hsu SL (2002) Down-modulation of Bcl-XL, release of cytochrome c and sequential activation of caspases during honokiol-induced apoptosis in human squamous lung cancer CH27 cells. Biochem Pharmacol 63:1641–1651

Yang CM, Yen YT, Huang CS, Hu ML (2011) Growth inhibitory efficacy of lycopene and beta-carotene against androgen-independent prostate tumor cells xenografted in nude mice. Mol Nutr Food Res 55:606–612

Yap TA, Yan L, Patnaik A, Fearen I, Olmos D, Papadopoulos K, Baird RD, Delgado L, Taylor A, Lupinacci L, Riisnaes R, Pope LL, Heaton SP, Thomas G, Garrett MD, Sullivan DM, De Bono JS, Tolcher AW (2011) First-in-man clinical trial of the oral pan-AKT inhibitor MK-2206 in patients with advanced solid tumors. J Clin Oncol 29:4688–4695

Yoon S, Seger R (2006) The extracellular signal-regulated kinase: multiple substrates regulate diverse cellular functions. Growth Fact 24:21–44

Zeitouni D, Pylayeva-Gupta Y, Der C, Bryant K (2016) KRAS mutant pancreatic cancer: no lone path to an effective treatment. Cancers 8(4):45

Zhang JH, Lai FJ, Chen H, Luo J, Zhang RY, Bu HQ, Wang ZH, Lin HH, Lin SZ (2013) Involvement of the phosphoinositide 3-kinase/Akt pathway in apoptosis induced by capsaicin in the human pancreatic cancer cell line PANC-1. Oncol Lett 5:43–48

Zimmermann G, Papke B, Ismail S, Vartak N, Chandra A, Hoffmann M, Hahn SA, Triola G, Wittinghofer A, Bastiaens PIH, Waldmann H (2013) Small molecule inhibition of the KRAS-PDE[dgr] interaction impairs oncogenic KRAS signalling. Nature 497:638–642

Zverina EA, Lamphear CL, Wright EN, Fierke CA (2012) Recent advances in protein prenyltransferases: substrate identification, regulation, and disease interventions. Curr Opin Chem Biol 16:544–552

Recent Insights on the Anticancer Properties of Flavonoids: Prospective Candidates for Cancer Chemoprevention and Therapy

13

Irfan A. Ansari and Mohd Sayeed Akhtar

Contents

Irfan A. Ansari and Mohd Sayeed Akhtar have equally contributed for this chapter.

I. A. Ansari (✉)
Department of Biosciences, Integral University, Lucknow, Uttar Pradesh, India
e-mail: iaansari@iul.ac.in

M. S. Akhtar
Department of Botany, Gandhi Faiz-e-Aam College, Shahjahanpur, Uttar Pradesh, India
e-mail: sayeedbot@gmail.com

M. S. Akhtar et al. (eds.), *Natural Bio-active Compounds*,
https://doi.org/10.1007/978-981-13-7154-7_13

Abstract

Flavonoids are polyphenolic compounds of very important class of plant secondary metabolites having a broad spectrum of biological activities. Because of their antioxidative, anti-inflammatory, antimutagenic, and anticarcinogenic properties, flavonoids have become an indispensable component in nutraceuticals, pharmaceuticals, medicinal, and cosmetic applications. The bioavailability, metabolism, and biological activities of many flavonoids have drawn the attention of researchers to use them as an alternative source of therapeutics for the treatment of various diseases; flavonoids have been shown to disrupt the initiation, promotion, and progression of cancer by modulating various signaling pathways and their downstream components associated with cellular proliferation, differentiation, inflammation, apoptosis, metastasis, angiogenesis, and reversal of multidrug resistance. Many natural flavonoids and their synthetic analogs are being investigated for their potential applications in anticancer therapies, because of their multi-targeted mechanism of action. Thus, the aim of the present chapter is to highlight the new insights on the recent progress of flavonoids as effective candidates in cancer therapeutics and prevention.

Keywords

Apoptosis · Cell cycle · Chemoprevention · Flavonoids · Metastasis

13.1 Introduction

Plants have been an integral part of our daily diet due to their nutritional properties (Namdeo 2007). Since decades, several chemical and biological studies have well elucidated the role of their primary metabolites, such as carbohydrates, amino acids, and lipids in performing critical functions such as cell division, growth, respiration, storage, and reproduction (Bourgaud et al. 2001). Besides this, the plants also synthesize a broad range of low molecular weight chemical compounds, which may be distinctive from primary metabolites and vary from species to species which are known as "secondary metabolites." They are responsible for specific tastes, odors, and colors of plant. Secondary metabolites do not have any significant role in maintenance of basic functions of plants, but play an imperative role in the communication of plant with its environment (Dixon 2001; Oksman-Caldentey and Inze 2004).

Plants are the chief sources of secondary metabolites, which have been frequently used in pharmaceutical, agrochemical, flavor, and aroma industries. Secondary metabolites are known as allelochemicals, which function as chemical defense compounds and influence molecular targets in herbivores or microbes. Secondary metabolites can be grouped into the alkaloids, terpenes, phenolics, and flavanoids (Karuppuswamy 2009; Rattan 2010). Alkaloids are the important class of highly diversified group of secondary metabolites containing a ring structure having a nitrogen atom and are often characterized by their bitter taste. Alkaloids are widely distributed in the plant kingdom mainly in higher plants. Moreover, several alkaloids also exhibited significant biological activities like anti-inflammatory, anticancer, antimetastatic, and antiangiogenic properties (Benyhe 1994; Lee 2011; Huang et al. 2007; Chen et al. 2008). Similarly, terpenoids are more numerous and structurally diverse compounds which have been used in the perfumery and cosmetic industries and also possess several biological and pharmacological properties. Terpenoids are classified on the basis of number and carbon skeleton formed by the joining of isoprene units followed by cyclization and rearrangements into monoterpenes (C10), sesquiterpenes (C15), diterpenes (C20), triterpenes (C30), tetraterpenes (C40), and polyterpenes. The terpenoids such as mono-, di-, and tetraterpenoids are synthesized from 2-C-methyl-D-erythritol 4-phosphate pathway while sesqui- and triterpenoids by the mevalonate pathway (Ludwiczuk et al. 2017). It has been demonstrated that terpenoids exhibit various properties such as antimicrobial activity, anti-inflammatory activity, anticancer activity, etc. (Raut and Karuppayil 2014; Lesgards et al. 2014).

Phenolics, being ubiquitously present in plant organs, are often regarded as integral component of our diet. These are the compounds which possess one or more aromatic rings with one or more hydroxy groups. These are the most abundant plant secondary metabolites. They are prominently involved in plant defense against pathogens, parasites, and predators, and also they are responsible for the colors of the plants. They are the prevalent constituents of plant foods like fruits, vegetables, etc. and beverages such as tea, coffee, beer, wine, etc. Plant phenolics chiefly include phenolics, flavonoids, tannins, and the less common stilbenes and lignans. Phenolics are primarily responsible for the bitterness and astringency of fruits and fruit juices (Nijveldt et al. 2001). The biosynthesis of phenolic compounds initiates with the commitment of glucose to the pentose phosphate pathway (PPP) and transforming glucose-6-phosphate irreversibly to ribulose-5-phosphate. On the other hand, PPP also produces erythrose-4-phosphate along with phosphoenolpyruvate from glycolysis, which is then used through the phenylpropanoid pathway to generate phenolic compounds after being channeled to the shikimic acid pathway to produce phenylalanine (Lin et al. 2010; Vattem et al. 2005). Phenolics can inhibit the absorption of amylase in the treatment of carbohydrate absorption, such as diabetes (Sales et al. 2012). There are many fruits and vegetables that contain phenolic compounds, especially, grapes, berries, and tomatoes. Phenolic compounds, such as phenolic acids and flavonoids, could promote health benefits by reducing the risk of metabolic syndrome and the related complications of type 2 diabetes. However, different groups of phenolic compounds have different biological characteristics, and very little is known about the mechanisms by which they could contribute to the prevention of disease; there still is a need for further studies. Reactive oxygen (ROS) and reactive nitrogen species (RNS) are highly reactive oxidized molecules, which are generated constantly by normal cellular conditions, for instance, the activity of the mitochondrial respiratory chain and inflammation, which could lead to damage in other biological molecules, like proteins and DNA (Halliwell 2002; Urso and Clarkson 2003; Lea et al. 2015).

Flavonoids are widely distributed phenolics in the plant kingdom. More than 5000 different flavonoids have been discovered till date, and on the basis of their chemical structure, they are classified into various groups. Among them, flavones, flavonols, flavanols, flavanones, anthocyanins, and isoflavones are particularly important because of their presence in various fruits and vegetables (Harbone 1993).

Reactive oxygen species (ROS) can damage DNA, and division of cells with unrepaired or misrepaired damage leads to mutations. If these changes appear in critical genes, such as oncogenes or tumor suppressor genes, initiation or progression of carcinogenesis may result. Moreover, ROS can interfere directly with cell signaling and growth (Loft and Poulsen 1997; Pryor 1997). The antitumor activity of flavonoids has recently gained attention because of their potential to quench and neutralize the ROS and ROS-mediated damage in the cellular compartment. Moreover, flavonoids have also been shown to modulate various altered cellular signaling pathways associated with cancer progression and development. Thus, the aim of the present chapter is to highlight the new insights on the recent progress of flavonoids as effective candidates in cancer therapeutics and prevention (Table 13.1 and 13.2).

13.2 Flavonoids

Flavonoids are the most ever-present plant-specific secondary metabolites having a benzo-ūFE;-pyrone in their variable phenolic structures and are predominantly found in fruits, vegetables, grains, bark, roots, stems, flowers, tea, wine, etc. A large number of flavonoids have been recognized, many of which are responsible for the attractive colors of flowers, fruit, and leaves (Nijveldt et al. 2001). These compounds are chiefly involved in various essential functions such as reproduction by recruiting pollinators and seed dispersers. It has been suggested that flavonoids are also responsible for the beautiful display of fall color which as a result protect leaf cells from photo-oxidative damage, thereby enhancing the efficiency of nutrient retrieval during senescence (Winkel-Shirley 2002). The chemical nature of flavonoids depends on their structural class, degree of hydroxylation, other substitutions and conjugations, and degree of polymerization (Kumar and Pandey 2013).

Table 13.1 Dietary sources of different flavonoids

Flavonoid classes	Dietary sources	Examples
Flavonols	Tea	Catechin, Epicatechin, epigallocatechin
Flavone	Fruit skins, red wine, buckwheat, red pepper, and tomato skin	Chrysin, Apigenin, Rutin, luteolin, and luteolin glucosides
Flavonol	Kaempferol, quercetin, myricetin, and tamarixetin	Onion, red wine, olive oil, berries, and grapefruit
Flavanone	Naringin, naringenin, taxifolin, and hesperidin	Citrus fruits, grapefruits, lemons, and oranges
Isoflavone	Genistin, daidzin	Soya bean
Anthocyanidin	Apigenidin, cyanidin	Cherry, raspberry, and strawberry

Table 13.2 Anticancer effects of flavonoids on various human cancer cell lines

Cancers	Cancer cell lines	Flavonoids
Human oral cancer	HSC-2, HSG, SCC-25	Flavanones, isoflavans, EGC, chalcones, EGCG, curcumin, genistein, ECG, quercetin, cisplatin
Human breast cancer	MCF-7	Flavanones, daidzein, genistein, quercetin, luteolin
Human thyroid cancer	ARO, NPA, WRO	Genistein, apigenin, kaempferol, chrysin, luteolin, biochanin A
Human lung cancer	SK-LU1, SW900, H441, H661, haGo-K-1, A549	Catechin, epicatechin, quercetin, kaempferol, luteolin, genistein, apigenin, myricetin, silymarin
Human colon cancer	Caco-2, HT-29, IEC-6, HCT-15	Flavone, quercetin, genistein, anthocyanin
Human prostate cancer	LNCaP, PC3, DU145	Catechin, epicatechin, quercetin, kaempferol, luteolin, genistein, apigenin, myricetin, silymarin
Human leukemia cancer	HL-60, K562, Jurkat	Apigenin, quercetin, myricetin, chalcones

13.2.1 Structure and Classification of Flavonoids

Chemically, flavonoids consist of 15 carbons and two phenyl rings (A and B) and a heterocyclic ring C. Flavonoids are chemically diverse group of secondary metabolites which can be divided into subgroups including anthocyanidins, flavonols, flavones, flavanols, flavanones, chalcones, dihydrochalcones, and dihydroflavonols (Treutter 2006). Flavonoids are classified on the basis of degree of oxidation, annularity of ring C, and connection position of ring B. Flavones and flavonols contain the largest number of compounds, representing the narrow sense flavonoids, namely, 2-benzo-γ-pyrone category. Quercetin is the most extensively studied flavonoids which belong to the flavonol class. The classification of flavonoids is shown in the figure (Fig. 13.1). The class flavanones and flavanonols possess saturated C2 = C3 bonds and frequently coexist with relevant flavones and flavonols in plants. Isoflavones, such as daidzein, are 3-phenyl-chromone substances. Chalcones being as the key precursors of flavonoid biosynthesis are ring C-opening isomers of dihydroflavones which are responsible for color appearance of plants. Although aurones are five-membered ring C benzofuran derivatives, they lack the typical structure of flavonoids. Anthocyanidins are primarily responsible for characteristic color of plants because they belong to a group of important chromene pigments and thus exist in the form of ions. Flavanols are reduction products of dihydroflavonols, particularly with flavan-3-ols widely distributed in the plant kingdom, also known as catechins. Nevertheless, there are still other flavonoids without C6—C3—C6 skeleton, for instance, biflavones, furan chromones, and xanthones (Fig. 13.2).

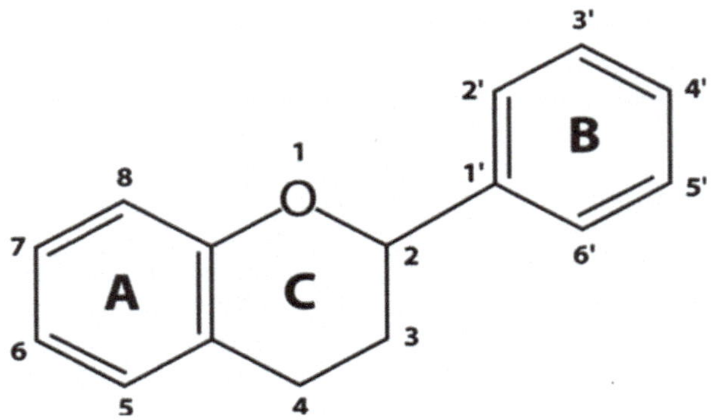

Fig. 13.1 Basic chemical structure of flavonoids

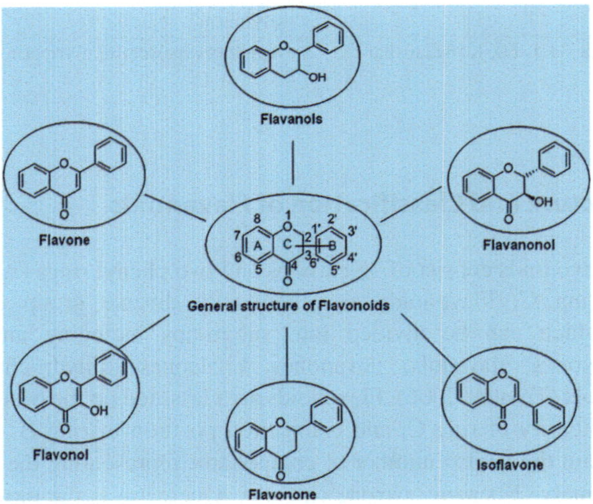

Fig. 13.2 Classification of major classes of flavonoids

13.2.2 Biosynthesis of Flavonoids

Biosynthesis of flavonoids takes place through the phenylpropanoid pathway in which phenylalanine is converted into 4-coumaroyl-CoA, the ultimate precursor of the flavonoid biosynthesis pathway as shown in Fig. 13.3. Chalcone synthase is the first enzyme specific for the flavonoid pathway which produces chalcones. All the flavonoids are derivatives of these chalcones. Although the central pathway for flavonoids biosynthesis is conserved in plants, depending on the species, a group of enzymes, such as isomerases, reductases, hydroxylases, and several Fe^{2+}/2-oxoglutarate-dependent dioxygenases, modify the basic flavonoid skeleton, leading to the different flavonoid subclasses (Martens et al. 2010). Lastly, transferases modify the flavonoid backbone with sugars, methyl groups, and/or acyl moieties,

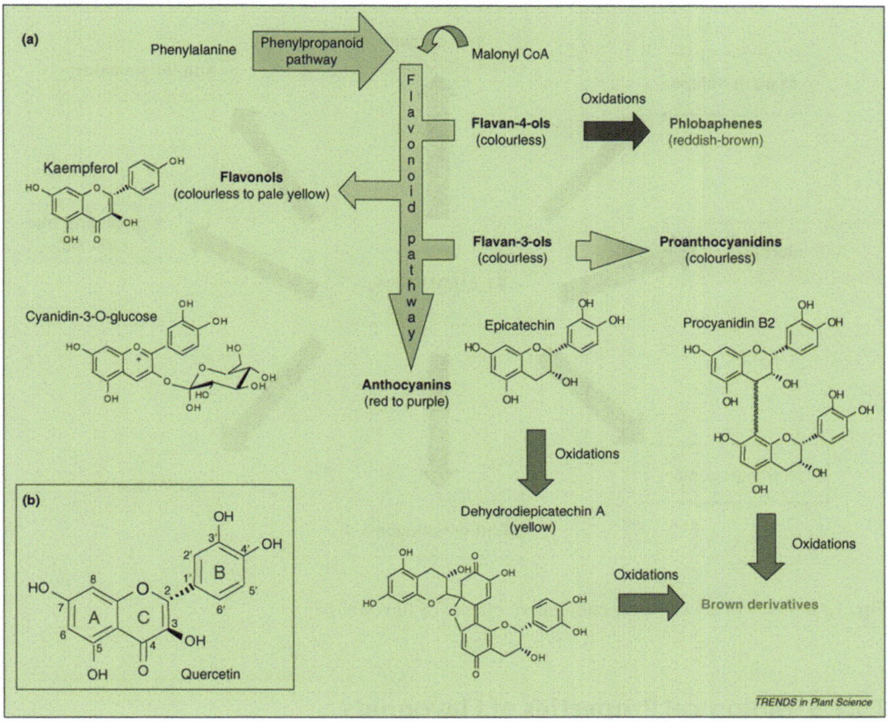

Fig. 13.3 Biosynthesis of major classes of flavonoids through phenylpropanoid pathway

modulating the physiological activity of the resulting flavonoid by altering their solubility, reactivity, and interaction with cellular targets (Bowles et al. 2005; Ferrer et al. 2008).

Biosynthesis of flavonoids begins with the condensation of one molecule of 4-coumaroyl-CoA and three molecules of malonyl-CoA yielding naringenin chalcone, and the reaction is catalyzed by the enzyme chalcone synthase (CHS) (Fig. 13.1). The two immediate precursors of the chalcone, coumaroyl-CoA and malonyl-CoA, originate from two different pathways of primary metabolism. Coumaroyl-CoA is derived from the amino acid phenylalanine by phenylpropanoid pathway and is common to the biosynthesis of a variety of compounds such as lignin, coumarins, stilbenes, and flavonoids. Malonyl-CoA is produced by the carboxylation of acetyl-CoA, a central intermediate in the Krebs cycle. Chalcone is subsequently isomerized by the enzyme chalcone flavanone isomerase (CHI) to yield a flavanone. From these central intermediates, the pathway diverges into several side branches, each yielding a different class of flavonoids (Heller and Forkmann 1988).

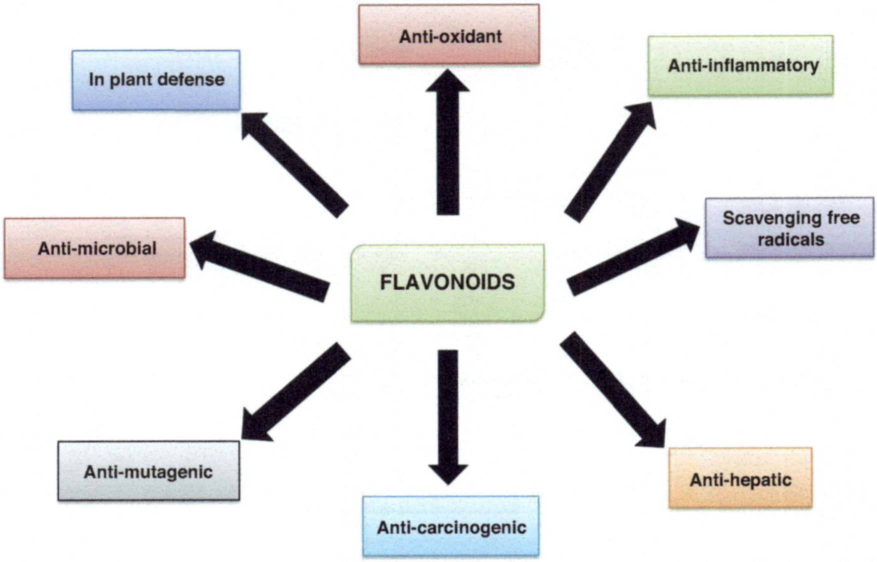

Fig. 13.4 Flowchart showing cumulative effects of flavonoids

13.2.3 Biological Properties of Flavonoids

The biological properties of flavonoids have been attributed to their potential cytotoxic property and their capacity to interact with various enzymes. The cumulative effects of flavonoid are shown in Fig. 13.4. A large number of flavonoids are accountable for providing stress protection via acting as scavengers of free radicals such as reactive oxygen species (ROS), as well as chelating metals that generate ROS through the Fenton reaction (Williams et al. 2004). Flavonoids are also involved in developing resistance to metal toxicity in plants (Kidd et al. 2001). Flavonoids could inhibit polar auxin transport and enhance consequent localized auxin accumulation in plants (Peer and Murphy 2007; Kuhn et al. 2011; Lewis et al. 2011). The various biological properties of flavonoids are as follows:

13.2.3.1 Role of Flavonoids in Root Nodulation

Flavonoids have been shown to be predominantly involved in the initiation of nodulation process. This phenomenon has been observed in transgenic plants having flavonoid-deficient roots produced by knockdown of chalcone synthase enzyme by RNA interference. The flavonoid-deficient roots were not able to initiate formation of nodules (Wasson et al. 2006).

13.2.3.2 Role of Flavonoids in Plant Defense

In several studies, flavonoids have been shown to protect plants from UV rays, and it is attributed to their UV-absorbing property. Moreover, biosynthesis of flavonols has been shown to induce by the UV light in various studies (Ryan et al. 2002; Berli et al. 2010; Stracke et al. 2007; Agati et al. 2011; Kusano et al. 2011). The presence of the hydroxy group in the third position of the flavonoid skeleton is accountable for chelating metal ions such as aluminum, zinc, iron, and copper and, consequently, inhibiting the formation of free radicals as well as to reduce ROS (Verdan et al. 2011). Flavonoids also provide protection to plants against pathogen and herbivores (Cornell and Hawkins 2003; Kliebenstein 2004; Bidart-Bouzat and Imeh-Nathaniel 2008).

13.2.3.3 Role of Flavonoids in Plant Reproduction and Fertility

The role of flavonoids in pollination has been established through the revelation of their role in providing characteristic colors to the pollen grains in different species of plants which can be detected by insects, facilitating successful pollination (Zerback et al. 1989; Van Der Meer et al. 1992). The role of flavonoids in pollen germination and pollen tube formation is elucidated by producing flavonoid-deficient mutants lacking chalcone synthase in maize and petunia (Pollak et al. 1993). Moreover, the silencing of chalcone synthase gene results in parthenocarpy in tomato (Schijlen et al. 2007). The silencing of FLS in tobacco causes production of less-seeded fruits, and silenced lines had lower flavonol and anthocyanidins levels. In addition, the pollen of these silenced lines was unable to produce functional pollen tubes. These experiments revealed that flavonoids have essential roles in pollen germination and consequently in plant fertility (Mahajan et al. 2011).

13.3 Pharmacological Effects of Flavonoids

Various studies have now established the protective role of flavonoids in various human ailments. The polyphenolic structure of flavonoids is attributed to their various pharmacological activities. Flavonoids are considered to be good source of natural antioxidants in human diets, and this property is accredited to the hydroxy (OH-) group present in the flavonoids via scavenging free radicals or by chelating metal ions. Thus, flavonoids could prevent free radical generation that leads to oxidative stress and consequently, many diseases. Flavonoids have been shown to play a protective role against many diseases such as cancer, cardiovascular and respiratory disorders, arthritis, and early aging. They have been accredited to boost the antioxidant defense system via inducing the expression of many antioxidant enzymes. In addition to all these properties, flavonoids also possess diverse biological activities which are important for various health aspects in human, for instance, anti-inflammatory, anti-ulcer, antiviral, anticancer, anti-diabetic, and cytotoxic properties (Nijveldt et al. 2001). The mechanism of action of flavonoids is shown in Fig. 13.5. In the next section, we have explained these biological properties individually in detail.

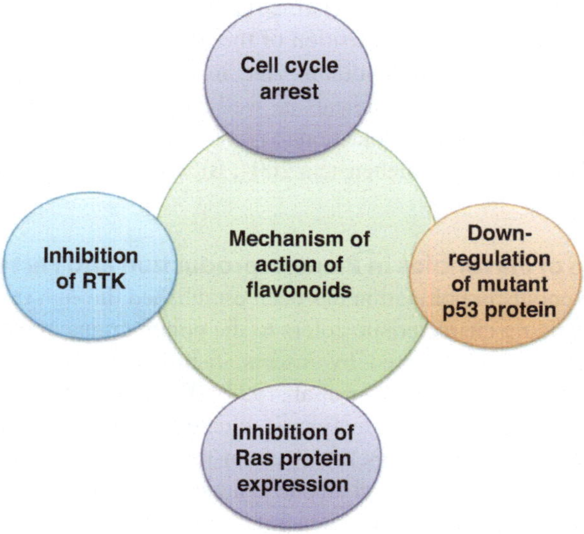

Fig. 13.5 Circular representation showing mechanism of action of flavonoids

13.3.1 Antioxidative Property of Flavonoids

Antioxidant property of almost every group of flavonoids is the most studied property. In this regard, flavones and catechins have been recognized as the most powerful flavonoids having free radical scavenging property against reactive oxygen species. Human cells and tissues are in a continuous threat to be damaged by indigenous free radicals and reactive oxygen species produced during normal metabolism or by exogenous radicals generated due to environmental factors (Groot 1994; Grace 1994). The most important event by which free radicals interfere with cellular functions is lipid peroxidation resulting in cellular membrane damage. This membrane damage causes a shift in the net charge of the cell and osmotic pressure, leading to swelling and eventually cell death. Moreover, free radicals can induce pro-inflammatory mediators and initiate an inflammatory cascade, contributing to a general inflammatory response and tissue damage. To protect themselves from the damage caused by reactive oxygen species, living organisms have evolved an effective antioxidant defense system that includes enzymes such as superoxide dismutase, catalase, and glutathione peroxidase, as well as nonenzymatic counterparts such as glutathione, ascorbic acid, and alpha-tocopherol (Halliwell 1995). The increased production of reactive oxygen species during injury results in consumption and depletion of the endogenous scavenging compounds.

Antioxidants are specific compounds that protect human, animal, and plant cells against the damaging effects of free radicals. Flavonoids are best known phytochemicals that act as antioxidants and may have an additive effect to the endogenous scavenging compounds (Kelly et al. 2002; Kukic et al. 2006). The antioxidant action of flavonoids includes suppression of ROS formation via inhibition of certain enzymes like NADPH oxidases, by scavenging free radicals, and regulation of antioxidant defense system (Mishra et al. 2013). Flavonoids have also been shown to protect the lipid membranes which are damaged due to lipid peroxidation. Thus, the flavonoids act as antioxidants and could play protective role in the onset and prevention of many diseases caused due to oxidative stress (Ramchoun et al. 2009).

13.3.2 Antiviral Property of Flavonoids

Naturally occurring flavonoids have been shown to exhibit significant antiviral property. They have been found to inhibit various enzymes associated with the life cycle of viruses. Flavon-3-ol was found to be more effective than flavones and flavonones in selective inhibition of HIV-1 and HIV-2. Baicalin, another flavonoid, isolated from *Scutellaria baicalensis*, also is known to inhibit immune-deficiency virus infections (Gerdin & Srensso 1983). Anti-dengue virus properties of quercetin, hesperetin, and naringin have also been reported, recently (Zandi et al. 2011).

13.3.3 Free Radical Scavenging Property of Flavonoids

Flavonoids play protective role in preventing injuries caused by free radicals. Flavonoids neutralize the highly reactive oxygen radicals via their hydroxy groups (Korkina and Alfanas'ev 1997). Epicatechin and rutin are powerful radical scavengers (Hanasaki et al. 1994). Moreover, the scavenging property of rutin has been attributed to its inhibitory activity on the enzyme xanthine oxidase. Furthermore, flavonoids can also inhibit LDL oxidation in vitro by neutralizing free radicals. Thus, flavonoids may exert protective effect against atherosclerosis (Kerry and Abbey 1997).

13.3.4 Anti-Inflammatory Effect

Eicosanoids such as prostaglandins, thromboxanes, and leukotrienes are known to be important lipid-derived inflammatory mediators (Moroney et al. 1988). They are produced by the action of cyclooxygenases and lipoxygenases on arachidonic acid released from lipid membrane under the influence of certain inflammatory signals. Flavonoids have been shown to inhibit eicosanoid biosynthesis, thus acting as potent anti-inflammatory agents. Different flavonoids like quercetin have been demonstrated to inhibit both cyclooxygenase and lipoxygenase activities (Kim et al. 1998).

13.3.5 Hepatoprotective Activity

A number of flavonoids such as quercetin, rutin, catechin, apigenin, and naringenin have been previously described for their hepatoprotective properties (Tapas et al. 2008). Zhu et al. demonstrated that anthocyanin cyanidin-3-O-β- glucoside (C3G) increases hepatic glutamate-cysteine ligase (GCLC) expression by increasing cAMP levels to activate protein kinase A (PKA), which in turn upregulates cAMP response element binding protein (CREB) phosphorylation to promote CREB-DNA binding and increase GCLC transcription. Increased GCLC expression results in a decrease in hepatic ROS levels and proapoptotic signaling. Furthermore, C3G treatment lowers hepatic lipid peroxidation, inhibits the release of pro-inflammatory cytokines, and protects against the development of hepatic steatosis (Zhu et al. 2012).

13.3.6 Anticancer Property

The anticancer property of flavonoids has garnered the attention of researchers to use them as anticancer therapeutics. Flavonoids, having antioxidant activity, have been accredited to exert anticarcinogenic activity (Stefani et al. 1999). Some flavonoids such as luteolin, apigenin, and fisetin have been demonstrated to be potent inhibitors of cell proliferation (Fotsis et al. 1997). An inverse association between flavonoid intake and the subsequent incidence of lung cancer has been established in a large clinical study (Knekt et al. 1997). This effect was mainly ascribed to quercetin, which provided >95% of the total flavonoid intake in that particular study. Moreover, quercetin and apigenin inhibited melanoma growth and influenced its invasiveness and metastatic potential in mice (Caltagirone et al. 2000). Furthermore, flavonoids have also been speculated to inhibit angiogenesis (Fotsis et al. 1997).

13.4 Flavonoids as Prospective Anticancer Agents

Flavonoids are a large group of heterogenous polyphenols ubiquitously present in fruits and vegetables having several health benefits. Dietary factors play an imperative role in the prevention of cancers. It has been reported that fruits and vegetables having flavonoids are promising source of cancer chemopreventive agents. The critical relationship of fruit and vegetable intake and cancer prevention has been extensively documented. In one such study, an inverse relationship between consumption of onions and/or apples, two major sources of the flavonol quercetin with the incidence of cancer of the prostate, lung, stomach, and breast, has been observed. Several mechanisms have been proposed for the effect of flavonoids on the initiation and progression of carcinogenesis including regulation of developmental and hormonal actions. Major molecular mechanisms of action of flavonoids are (i) inhibition of mutant p53 protein expression, (ii) abrogation of cell cycle progression, (iii) inhibition of tyrosine kinase signaling, (iv) suppression of heat shock proteins, (v) affinity with estrogen receptor, and (vi) suppression of Ras family protein expression.

Mutations of p53 are among the most common genetic abnormalities in human cancers (Nigro et al. 1989). The inhibition of p53 expression could lead to arrest the cancer cells in the G2/M phase of the cell cycle. Flavonoids have been found to suppress the expression of mutant p53 protein to nearly undetectable levels in human breast cancer cell lines (Avila et al. 1994). Tyrosine kinases are a family of proteins located in or near the cell membrane involved in the transduction of growth factor signals to the nucleus. Their expression is thought to be involved in oncogenesis through an ability to override normal regulatory growth control (Boutin 1994). Drugs inhibiting tyrosine kinase activity are thought to be possible antitumor agents without the cytotoxic side effects seen with conventional chemotherapy. Quercetin was the first tyrosine kinase inhibiting compound tested in a human phase I trial. Heat shock proteins form a complex with mutant p53, which allows tumor cells to bypass normal mechanisms of cell cycle arrest. Heat shock proteins also allow for improved cancer cell survival under different bodily stresses. Flavonoids are known to inhibit production of heat shock proteins in several malignant cell lines, including breast cancer, leukemia, and colon cancer. Recent studies have shown that the flavanol epigallocatechin-3-gallate inhibited fatty acid synthase (FAS) activity and lipogenesis in prostate cancer cells, an effect that is strongly associated with growth arrest and cell death. Contrastingly, expression of FAS is markedly increased as compared to normal tissues in various human cancers. Upregulation of FAS occurs early in tumor development and is further enhanced in more advanced tumors. Quercetin is known to produce cell cycle arrest in proliferating lymphoid cells. It has been reported that in addition to its antineoplastic activity, quercetin exerted growth inhibitory effects on several malignant tumor cell lines in vitro. These included P-388 leukemia cells, gastric cancer cells (HGC-27, NUGC-2, NKN-7, and MKN-28), colon cancer cells (COLON320DM), human breast cancer cells, human squamous and gliosarcoma cells, and ovarian cancer cells (Zhao 2003). Markaverich et al. proposed that tumor cell growth inhibition by quercetin may be due to its interaction with nuclear type II estrogen binding sites (EBS). It has been experimentally proved that increased signal transduction in human breast cancer cells is markedly reduced by quercetin acting as an antiproliferative agent. Barnes has extensively reviewed the anticancer effects of genistein on in vitro and in vivo models. In a study to determine the effects of isoflavones genistein, daidzein, and biochanin A on mammary carcinogenesis, genistein was found to suppress the development of chemically induced mammary cancer without reproductive or endocrinological toxicities. Neonatal administration of genistein (a flavonoid) exhibited a protective effect against the subsequent development of induced mammary cancer in rats (Chung 1995). Hesperidin, a flavanone glycoside, is known to inhibit azoxymethane-induced colon and mammary cancers in rats. The anticancer properties of flavonoids contained in citrus fruits have been reviewed by Carroll et al. (2003). Several flavonols, flavones, flavanones, and the isoflavone biochanin A are reported to have potent antimutagenic activity. A carbonyl function at C-4 of the flavone nucleus was found to be essential for their activity. Flavone-8-acetic acid has also been shown to have antitumor effects. In earlier studies, ellagic acid, robinetin, quercetin, and myricetin have been shown to inhibit the tumorigenicity of BP-7, 8-diol-9, and 10-epoxide-2 on mouse skin (Urso and Clarkson 2003). Higher consumption of phytoestrogens, including isoflavones and other flavonoids, has been shown to provide protection against prostate cancer risk. It is well known that due to oxidative stress, cancer initiation may take place, and thus, potent antioxidants show potential to combat progression of carcinogenesis. Potential of antioxidant as an anticancer agent depends on its competence as an oxygen radical inactivator and inhibitor. Therefore diets rich in radical scavengers would diminish the cancer-promoting action of some radicals (Treutter 2006).

13.5 Anticancer Activities of Some Bio-active Flavonoids

The essential feature of bio-active flavonoids is their free radical scavenging activity. These antioxidant properties are solely responsible for their antitumor effects (Nijveldt et al. 2001). They reportedly prevent cell damage caused by reactive oxygen formed via normal metabolic processes and induced by exogenous factors (e.g., UV radiation, xenobiotics) that can modify transcriptional factor and protein kinase activities and lead to DNA damage that increases mutation probability and mismatch repair. Proto-oncogene activation and changes in suppressor genes can initiate cancer transformation (Nijveldt et al. 2001). These bio-active flavonoids have antiproliferative effects and induce apoptosis in diverse cancer cell lines. As free radical scavengers, flavonoids inhibit invasion and metastasis (Kuntz et al. 1999). Nijveldt et al. (2001) reported that these bio-active flavonoid compounds were cytotoxic for cancer but not for normal cells. Flavones also regulate macrophage function in cancer cell elimination and are potential inhibitors of cell proliferation (e.g., apigenin and luteolin). An inverse relationship exists between bio-active flavonoids in the diet and the occurrence of lung cancer (Nijveldt et al. 2001). The anticancer properties of some bio-active flavonoids have been thoroughly described below.

13.5.1 Quercetin

Quercetin (3, 3′,4′,5,7-pentahydroxyflavone) is a polyphenolic flavonoid widely found in plants. Frequently, quercetin exists as glycosides (sugar derivatives), e.g., rutin, in which the hydrogen of the R-4 hydroxy groups is replaced by a disaccharide. Quercetin is termed as aglycone, or sugarless form of rutin (Cody 1988). Several in vitro and in vivo studies have suggested that quercetin plays an imperative role in protecting cells from oxidative stress induced by reactive oxygen species. Oxidative stress is one of the chief hallmarks of cancer development as shown in Fig. 13.6. It is widely demonstrated that reactive oxygen species (ROS) and reactive nitrogen species (RNS) play key role in cancer development (Wiseman and Halliwell 1996) (Fig. 13.7).

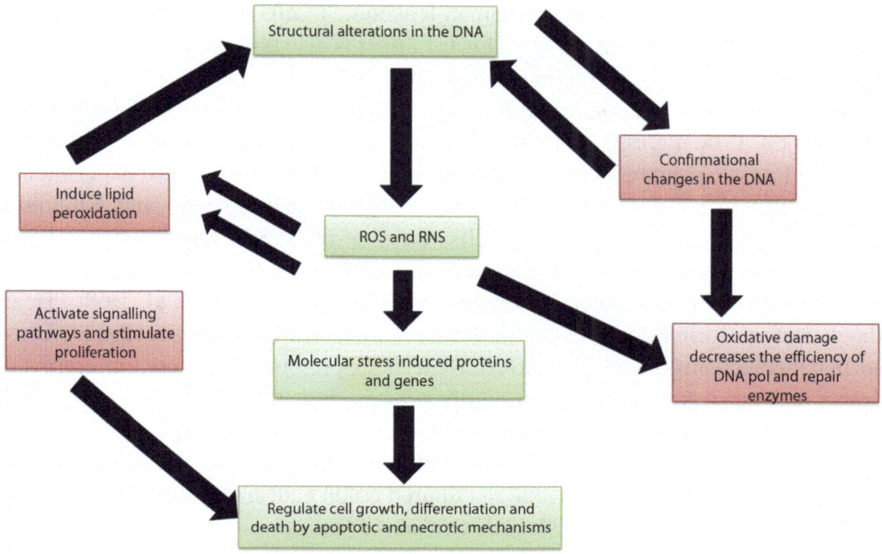

Fig. 13.6 Role of ROS/RNS in cancer development

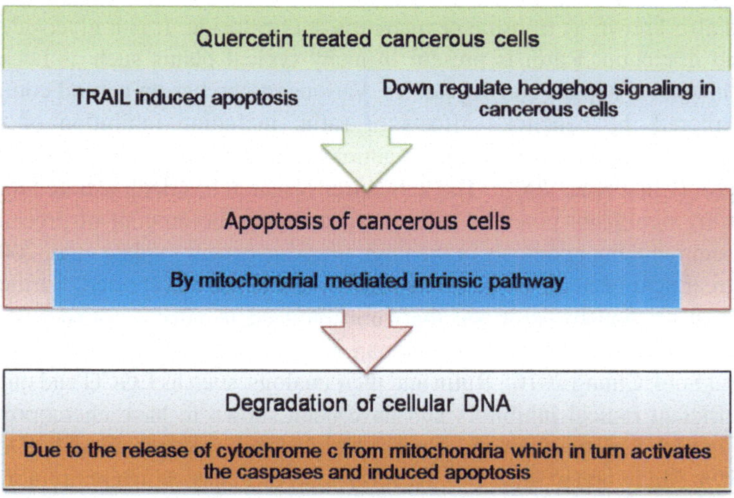

Fig. 13.7 Flowchart model showing apoptotic nature of quercetin on cancer cells

The molecular mechanism of action of quercetin has been reported in the down-regulation of mutant p53 protein expression in various breast cancer cell lines (Avila et al. 1994). The inhibition of expression of p53 has been found to arrest the cells in the G2/M phase of the cell cycle. This downregulation is found to be much less in cells with an intact p53 gene (Avila et al. 1996). The G1 checkpoint controlled by the p53 gene is a major site for the control of cellular proliferation. It has been reported that quercetin arrests the human leukemic T-cells in the late G1 phase of the cell cycle (Yoshida et al. 1992). This G1 arrest is also observed in gastric cancer cells when treated with quercetin (Yoshida et al. 1990).

Various studies have confirmed that the intravenous administration of quercetin (dosages 60–1700 mg/m2) led to the inhibition of lymphocyte tyrosine kinase at 1 h in 9 of 11 cases of human cancer patients (Ferry et al. 1996). In vitro experiments have confirmed these results, both in nonmalignant cells (Yokoo and Kitamura 1997) and in rat mammary tumor cells (Levy et al. 1984). Quercetin has been reported to inhibit the production of heat shock proteins in several malignant cell lines, including breast cancer, leukemia, and colon cancer (Elia et al. 1996). Heat shock proteins form a complex with mutant p53, which allows tumor cells to bypass normal mechanisms of cell cycle arrest. Heat shock proteins also allow for improved cancer cell survival under different bodily stresses (low circulation, fever, etc.) and are associated with shorter disease-free survival and chemotherapy drug resistance in breast cancer (Ciocca et al. 1993). In addition to this, quercetin also inhibits the expression of the p21-ras oncogene in cultured colon cancer cell lines (Oesterreich et al. 1993; Ranelletti et al. 1999; DeVita et al. 1997).

13.5.2 Rutin

Rutin is also known as rutoside or quercetin-3-O-rutoside. It is a glycoside of the flavonoid quercetin. Rutin is present in many typical plants such as buckwheat, apples, black tea, apples, and vegetables. Various researches on natural compounds have explored the beneficial effects of rutin, including inhibition of platelet aggregation, being anti-inflammatory, antioxidant, and reduction of blood fat and cholesterol (Chan et al. 2007). The anticancer research has demonstrated that rutin could exert significant potential effect in decreasing the amount of precancerous lesions and inducing apoptosis in the large intestine cancer (Volate et al. 2005), but for the treatment of neuroblastoma, this effect has not been reported. Furthermore, it has been previously described that rutin induced in vitro cytotoxic effects on various cancer cell lines including human colon cancer cells (Kuntz et al. 1999; Guon and Sook Chung 2016). Rutin and their analogs, such as EGCG and quercetin, act as efficient radical inhibitors and have been shown to have chemopreventive activity in both variety of colonic cancer cell lines and in murine models (Deschner et al. 1993; Mahmoud et al. 2000). Nevertheless, rutin has shown antitumor effects in some in vivo models such as NK/Ly ascites and B16F10 cells (Molnar et al. 1981; Menon et al. 1995). Therefore, it has also been illustrated that rutin exerted significant beneficial effects on decreasing the amount of precancerous lesions and inducing apoptosis in the large intestine cancer and human neuroblastoma LAN-5 cells (Chen et al. 2013).

13.5.3 Hesperitin and Naringenin

Major citrus flavanone naringenin and hesperetin possess antioxidant activities, although lower in comparison to other polyphenols (van Acker et al. 2000; Jeon et al. 2002). Hesperetin has been shown to inhibit chemically induced colon, urinary bladder, and mammary carcinogenesis in in vivo animal models. They have been reported to regulate apolipoprotein B secretion by HepG2 cells, possibly through inhibition of cholesterol ester synthesis, and to inhibit 3-hydroxy-3-methylglutaryl-coenzyme A reductase and acyl coenzyme A:cholesterol O-acyltransferase in rats (Lee et al. 1999; Kurowska et al. 2000a, b). Naringenin has been attributed to possess anti-inflammatory actions and different types of effects on sex hormone metabolism (Ruh et al. 1995; Rosenberg et al. 1998; Dechaud et al. 1999; Yoon et al. 2001). It has been shown to bind to estrogen receptors (Kuiper et al. 1998). Naringenin has been shown to increase the intestinal cytochrome P-450 IIIA (CYP3A4) enzyme expression (Bailey et al. 2000; Dresser et al. 2000).

13.5.4 Apigenin and Scutellarin

Apigenin (4-, 5, 7-trihydroxyflavone) is a member of the flavone class and possesses free radical scavenging, anticarcinogenic, and anti-inflammatory effects (Siddiqui et al. 2008). As a prospective anticancer agent, apigenin is capable of inhibiting cell growth and inducing apoptosis in cancer cells without incurring cytotoxic effects on normal cells (Liu 2004). It has been demonstrated that apigenin possessed growth inhibitory properties in breast cancer by the regulation of the p14ARF-Mdm2-p53 pathway, in colon cancer by increasing the expression of UDP-glucuronosyltransferase, and in pancreatic cancer through the downregulation of NF-κB activity with the suppression of Akt (Lee et al. 2008). In addition, apigenin can also increase the effect of cancer drugs when combined with other therapeutic reagents such as doxorubicin and taxol. Apigenin has also been shown to inhibit in vitro angiogenesis (Jiang and Fang 2006). The flavones derived from *Scutellaria* possess cytostatic and cytotoxic activities against many human cancer cells. They show no toxicity to normal epithelial and peripheral blood and myeloid cells. The combination use of baicalin and scutellarin could exhibit a synergistic effect and significantly improve their antitumor activity (Wu 2007). A new safe natural drug composed by berberine and baicalin has the property of inhibiting carcinogenesis and lowering tyrosine kinase activity (Zhao 2003).

13.5.5 Genistein

Genistein has been shown to exhibit both chemopreventive and chemotherapeutic potentials in multiple tumor types (Dixon and Ferreira 2002; Empie and Gugger 2005). Several in vitro and in vivo studies have established its anticancer activity in colon, prostate, breast, skin, urinary, and bladder cancer (Empie and Gugger 2005). Genistein has shown protective effect against UVR-induced skin sunburns, premature aging, and skin cancer. It has also shown to reduce bone loss in patients with osteoporosis and metastatic bone cancers. Moreover, a pharmaceutical product containing genistein has been shown to prevent epithelial ovarian cancer. The human epidermal growth factor (EGF) exerts its biological effect by binding to a specific 170 kDa cell membrane receptor (EGF-Rc). Conjugates of EGF-genistein have been shown to inhibit the EGF-Rc tyrosine kinase in breast cancer cells and trigger cell apoptosis. In addition, the conjugates had potent antitumor activity against breast cancer xenografts both in SCID mice and monkeys. Furthermore, EGF-genistein conjugates were also used as chemopreventive agent for the development or recurrence of EGF-Rc expressing breast cancer in mammal.

13.5.6 Epigallocatechin Gallate

Green tea mainly contain catechins such as (−)-epicatechin gallate (ECG), (−)-epigallocatechin (EGC), and (−)-epigallocatechin gallate (EGCG). Tea catechins are potent inhibitors of cancer cell proliferation and metastasis against various cancer cell lines like prostate, lung, colon, bladder, and cervical cancer cell lines. Epigallocatechin gallate (EGCG), the most abundantly found catechin in green tea, has been the focus of many investigations. EGCG could inhibit the growth of human lung cancer, particularly for 4-(methylnitrosamino)-1-(3-pyridyl)- 1-butanone (NNK) lung tumorigenesis (Chung 1995). In addition, EGCG is an anti-folate agent, which can inhibit the activity of dihydrofolate reductase (DHFR). EGCG acts by disturbing the folic acid metabolism in cells, causing the inhibition of DNA and RNA synthesis, alteration of DNA methylation, and modulation of cell signaling pathways (Navarro-Perán et al. 2007). Thus, the anti-folate compounds based on EGCG may be useful in the treatment of a range of disorders including cancer (Rodriguez-Lopez et al. 2007). It has been found that EGCG and other tea catechins might exert a synergistic effect in inhibiting tumor cell growth when combined with some active ingredients, such as A3 adenosine receptor agonists or thymidylate synthase inhibitors (Rodriguez-Lopez et al. 2008). Vanillylamine, the head group of capsaicin, combination with the tea catechins, displays unexpected potential utility for the treatment of cancer (Morre and Morre 2007). Ascorbic acid, L-proline, and L-lysine could effectively enhance the activity of tea catechins in blocking cancer cell proliferation and metastasis (Netke et al. 2006; Rath et al. 2006).

13.6 Conclusions and Future Prospects

Flavonoids have fascinated the researchers due to its nontoxic nature and wide range of biological activities. These flavonoids as natural compounds have received great advantage as therapeutic agent because these polyphenolic compounds are consumed daily and their half-life is also long and they are easily absorbed by the intestine. Consequently, the role of dietary flavonoids in cancer prevention is immensely studied. Epidemiological studies have established the association between high dietary intake of flavonoids with low cancer incidence in humans. These studies are supported by large number of in vitro and in vivo studies, which show that flavonoids may inhibit various stages in the carcinogenesis process, namely, tumor initiation, promotion, and progression. Mechanisms behind the anticarcinogenic activity of flavonoids include carcinogen inactivation, antiproliferation, cell cycle arrest, induction of apoptosis and differentiation, inhibition of angiogenesis, antioxidation, and reversal of multidrug resistance or a combination of these mechanisms. Furthermore, the intriguing results from various laboratories have encouraged the development of flavonoids as chemotherapeutic agents in human clinical trials. While these experiences strengthen the notion that flavonoids could be useful anticancer agents, to date, only few clinical studies have demonstrated the anticancer property of flavonoids in vivo. Therefore, more focused clinical studies are required to establish whether the dietary effects of these compounds can be exploited to achieve cancer preventive or therapeutic effects in human. This book chapter can be concluded by considering that many chemotherapeutic agents against tumor cells exhibit cytotoxicity against normal cells which remain a major obstacle in successful chemotherapy. Moreover, development of multidrug resistance further limits chemotherapy in cancer. Thus, the promising results will stimulate the development of flavonoids for cancer chemoprevention and chemotherapy.

References

Agati G, Biricolti S, Guidi L, Ferrini F, Fini A, Tattini M (2011) The biosynthesis of flavonoids is enhanced similarly by UV radiation and root zone salinity in *L. vulgare* leaves. J Plant Physiol 168:204–212

Avila MA, Velasco JA, Cansado J, Notario V (1994) Quercetin mediates the down-regulation of mutant p53 in the human breast cancer cell line MDA-MB468. Cancer Res 54:2424–2428

Avila MA, Velasco JA, Harter KW, Velasco JA, Notario V (1996) Quercetin as a modulator of the cellular neoplastic phenotype. Adv Expl Med Biol 401:101–110

Bailey DG, Dresser GK, Kreeft JH, Munoz C, Freeman DJ, Bend JR (2000) Grapefruit-felodipine interaction: effect of unprocessed fruit and probable active ingredients. Clin Pharmacol Ther 68:468–477

Benyhe S (1994) Morphine: new aspects in the study of an ancient compound. Life Sci 55:969–979

Berli FJ, Moreno D, Piccoli P, Hespanhol-Viana L, Silva MF, Bressan-Smith R, Cavagnaro JB, Bottini R (2010) Abscisic acid is involved in the response of grape (*Vitis vinifera* L.) cv. Malbec leaf tissues to ultraviolet-B radiation by enhancing ultraviolet-absorbing compounds, antioxidant enzymes and membrane sterols. Plant Cell Environ 33:1–10

Bidart-Bouzat MG, Imeh-Nathaniel A (2008) Global change effects on plant chemical defenses against insect herbivores. J Integr Plant Biol 50:1339–1354

Bourgaud F, Gravot A, Milesi S, Gontier E (2001) Production of plant secondary metabolites: a historical perspective. Plant Sci 161:839–851

Boutin JA (1994) Tyrosine protein kinase inhibition and cancer. Int J Biochem 26:1203–1226

Bowles D, Isayenkov J, Lim EK, Poppenberger B (2005) Glycosyltransferases: managers of small molecules. Curr Opin Plant Biol 8:254–263

Caltagirone S, Rossi C, Poggi A, Ranelletti FO, Natali PG, Brunetti M, Aiello FB, Piantelli M (2000) Flavonoids apigenin and quercetin inhibit melanoma growth and metastatic potential. Int J Cancer 87:595–600

Chan HJ, Ji YL, Chul HC, Chang JK (2007) Anti-asthmatic action of quercetin and rutin in conscious Guinea-pigs challenged with aerosolized ovalbumin. Arch Pharma Res 30:1599–1607

Chen J, Zhao H, Wang X, Lee FSC, Yang H, Zheng L (2008) Analysis of major alkaloids in Rhizoma coptidis by capillary electrophoresis-electrospray-time of flight mass spectrometry with different background electrolytes. Electrophoresis 29:2135–2147

Chen H, Miao Q, Geng M, Liu J, Hu Y, Tian L, Pan J, Yang Y (2013) Anti-tumor effect of rutin on human neuroblastoma cell lines through inducing G2/M cell cycle arrest and promoting apoptosis. Sci World J 2013:269165

Chung FL (1995) Inhibition of lung tumorigenesis by administration of a polyphenol. US Patent No. US5391568

Ciocca DR, Clark GM, Tandon AK, Fuqua SA, Welch WJ, McGuire WL (1993) Heat shock protein hsp70 in patients with axillary lymph node-negative breast cancer: prognostic implications. J Natl Cancer Inst 85:570–574

Cody V (1988) Plant flavonoids in biology and medicine part II. Prog Clin Biol Res 280:461

Cornell HV, Hawkins BA (2003) Herbivore responses to plant secondary compounds: a test of phytochemical coevolution theory. Am Nat 161:507–522

de Groot H (1994) Reactive oxygen species in tissue injury. Hepatogastroenterol 41:328–332

Dechaud H, Ravard C, Claustrat F, de la Perriere AB, Pugeat M (1999) Xenoestrogen interaction with human sex hormone–binding globulin (hSHBG). Steroids 64:328–334

Deschner EE, Ruperto JF, Wong GY, Newmark HL (1993) The effect of dietary quercetin and rutin on AOM-induced acute colonic epithelial abnormalities in mice fed a high-fat diet. Nutr Cancer 20:199–204

DeVita NT, Hellman S, Rosenberg SA (1997) Cancer: principles and practice of oncology, 5th edn. Lippencott-Raven, Philadelphia

Dixon RA (2001) Natural products and plant disease resistance. Nature 411:843–847

Dixon RA, Ferreira D (2002) Genistein. Phytochemistry 60:205–211

Dresser GK, Spence JD, Bailey DG (2000) Pharmacokinetic-pharmacodynamic consequences and clinical relevance of cytochrome P450 3A4 inhibition. Clin Pharmacokinet 38:41–57

Elia G, Amici C, Rossi A, Santoro MG (1996) Modulation of prostaglandin A1-induced thermotolerance by quercetin in human leukemic cells: role of heat shock protein. Cancer Res 56:210–217

Empie M, Gugger E (2005) Method of preparing and using compositions extracted from vegetable matter for the treatment of cancer. US Patent No. US6900240

Ferrer J, Austin M, Stewart CJ, Noel J (2008) Structure and function of enzymes involved in the biosynthesis of phenylpropanoids. Plant Physiol Biochem 46:356–370

Ferry DR, Smith A, Malkhandi J, Fyfe DW, deTakats G (1996) Phase I clinical trial of the flavonoid quercetin: pharmacokinetics and evidence for *in vivo* tyrosine kinase inhibition. Clin Cancer Res 2:659–668

Fotsis T, Pepper MS, Aktas E, Breit S, Rasku S, Adlercreutz H, Wähälä K, Montesano R, Schweigerer L (1997) Flavonoids, dietary-derived inhibitors of cell proliferation and *in vitro* angiogenesis. Cancer Res 57:2916–2921

Gerdin B, Srensso E (1983) Inhibitory effect of the flavonoid on increased microvascular permeability induced by various agent in rat skin. Int J Microcir Clin Exp 2:39–46

Grace PA (1994) Ischaemia-reperfusion injury. Br J Surg 81:637–647

Guon TE, Sook Chung HA (2016) Hyperoside and rutin of *Nelumbo nucifera* induce mitochondrial apoptosis through a caspase-dependent mechanism in HT-29 human colon cancer cells. Oncol Lett 11:2463–2470

Halliwell B (1995) How to characterize an antioxidant: an update. Biochem Soc Symp 61:73–101

Halliwell B (2002) Effect of diet on cancer development: is oxidative DNA damage a biomarker. Free Rad Biol Med 32:968–974

Hanasaki Y, Ogawa S, Fukui S (1994) The correlation between active oxygens scavenging and antioxidative effects of flavonoids. Free Rad Biol Med 16:845–850

Harbone JB (1993) The flavonoids. Advance in research since 1986. Chapman and Hall, London

Heller W, Forkmann G (1988) Biosynthesis. In: Harborne JB (ed) The Flavonoids. Chapman and Hall, London, pp 399–425

Huang M, Gao H, Chen Y, Zhu H, Cai Y, Zhang X, Miao Z, Jiang H, Zhang J, Shen H, Lin L, Lu W, Ding J (2007) Chimmitecan, a novel 9-substituted camptothecin, with improved anticancer pharmacologic profiles *in vitro* and *in vivo*. Clin Cancer Res 13:1298–1307

Jeon SM, Bok SH, Jang MK, Kim YH, Nam KT, Jeong TS, Park YB, Choi MS (2002) Comparison of antioxidant effects of naringin and probucol in cholesterol-fed rabbits. Clin Chim Acta 317:181–190

Jiang BH, Fang J (2006) Apigenin for chemoprevention, and chemotherapy combined with therapeutic reagents. US Patent No. US20060189680

Karuppusamy S (2009) A review on trends in production of secondary metabolites from higher plants by *in vitro* tissue, organ and cell cultures. J Med Plants Res 3:1222–1239

Kelly EH, Anthony RT, Dennis JB (2002) Flavonoid antioxidants: chemistry, metabolism and structure-actvity relationships. Nutri Biochem 13:572–584

Kerry NL, Abbey M (1997) Red wine and fractionated phenolic compounds prepared from red wine inhibit low density lipoprotein oxidation *in vitro*. Atherosclerosis 135:93–102

Kidd PS, Llugany M, Poschenrieder C, Gunse B, Barcelo J (2001) The role of root exudates in aluminium resistance and silicon induced amelioration of aluminium toxicity in three varieties of maize (*Zea mays* L.). J Exp Bot 52:1339–1352

Kim HP, Mani I, Iverson L, ZIboh VA (1998) Effects of naturally occurring flavonoids and bio-flavonoids on epidermal cyclooxygenase and lipooxygenase in Guinea pigs. Prostaglandins Leukot. Essent Fatty Acids 58:17–24

Kliebenstein DJ (2004) Secondary metabolites and plant/environment interactions: a view through *Arabidopsis thaliana* tinged glasses. Plant Cell Environ 27:675–684

Knekt P, Jarvinen R, Seppanen R, Hellövaara M, Teppo L, Pukkala E, Aromaa A (1997) Dietary flavonoids and the risk of lung cancer and other malignant neoplasms. Am J Epidemiol 146:223–230

Korkina LG, Afanasev IB (1997) Antioxidant and chelating properties of flavonoids. Adv Pharmacol 38:151–163

Kuhn BM, Geisler M, Bigler L, Ringli C (2011) Flavonols accumulate asymmetrically and affect auxin transport in *Arabidopsis*. Plant Physiol 156:585–595

Kuiper GG, Lemmen JG, Carlsson B, Corton JC, Safe SH, van der Saag PT, van der Burg B, Gustafsson JA (1998) Interaction of estrogenic chemicals and phytoestrogens with estrogen receptor beta. Endocrinol 139:4252–4263

Kukic J, Petrovic C, Niketic M (2006) Antioxidant activity of four endemic *Stachys taxa*. Biol Pharma Bull 29:725–729

Kumar S, Pandey AK (2013) Chemistry and biological activities of flavonoids: an overview. Sci World J 2013:162750

Kuntz S, Wenzel U, Daniel H (1999) Comparative analysis of the effects of flavonoids on proliferation, cytotoxicity, and apoptosis in human colon cancer cell lines. Eur J Nutr 38:133–142

Kurowska EM, Borradaile NM, Spence JD, Carrol CC (2000a) Hypocholesterolemic effects of dietary citrus juices in rabbits. Nutr Res 20:121–129

Kurowska EM, Spence JD, Jordan J, Wetmore S, Freeman DJ, Piche LA, Serratore P (2000b) HDL-cholesterol-raising effect of orange juice in subjects with hypercholesterolemia. Am J Clin Nutr 72:1095–1100

Kusano M, Tohge T, Fukushima A, Kobayashi M, Hayashi N, Otsuki H, Kondou Y, Goto H, Kawashima M, Matsuda F, Niida R, Matsui M, Saito K, Fernie AR (2011) Metabolomics reveals comprehensive reprogramming involving two independent metabolic responses of Arabidopsis to UV-B light. Plant J 67:354–369

Lea AJ, Tung J, Zhou XA (2015) Flexible, efficient binomial mixed model for identifying differential dna methylation in bisulfite sequencing data. PLoS Genet 11:e1005650

Lee MR (2011) The history of *Ephedra* (ma-huang). J R Coll Physians Edinb 41:78–84

Lee SH, Jeong TS, Park YB, Kwon YK, Choi MS, Bok SH (1999) Hypocholesterolemic effect of hesperetin mediated by inhibition of 3-hydroxy-3-methylglutaryl coenzyme A reductase and acyl coenzyme A: cholesterol acyltransferase in rats fed high-cholesterol diet. Nutr Res 19:1245–1258

Lee SH, Ryu JK, Lee KY, Woo SM, Park JK, Yoo JW, Kim YT, Yoon YB (2008) Enhanced antitumor effect of combination therapy with gemcitabine and apigenin in pancreatic cancer. Cancer Lett 259:39–49

Lesgards JF, Baldovini N, Vidal N, Pietri S (2014) Anticancer activities of essential oils constituents and synergy with conventional therapies: a review. Phytother Res 28:1423–1446

Levy J, Teuerstein I, Marbach M, Radian S, Sharoni Y (1984) Tyrosine protein kinase activity in the DMBAinduced rat mammary tumor: inhibition by quercetin. Biochem Biophys Res Commun 123:1227–1233

Lewis DR, Ramirez MV, Miller ND, Vallabhaneni P, Ray WK, Helm RF, Winkel BS, Muday GK (2011) Auxin and ethylene induce flavonol accumulation through distinct transcriptional networks. Plant Physiol 156:144–164

Liu Y (2004) Safe natural pharmaceutical composition for treating cancer. US Patent No. US20040072790A1

Loft S, Poulsen HE (1997) Cancer risk and oxidative DNA damage in man. J Mol Med 74:297–312

Ludwiczuk A, Skalicka-Wozniak K, Georgiev MI (2017) Terpenoids. In: Badal S, Delgoda R (eds) Pharmacognosy: fundamentals, applications and strategied. Elsevier, London, pp 233–266

Mahajan M, Ahuja PS, Yadav SK (2011) Post-transcriptional silencing of flavonol synthase mRNA in tobacco leads to fruits with arrested seed set. PLoS One 6:e28315

Mahmoud NN, Carothers AM, Grunberger D, Bilinski RT, Churchill MR, Martucci C, Newmark HL, Bertagnolli MM (2000) Plant phenolics decrease intestinal tumors in an animal model of familial adenomatous polyposis. Carcinogenesis 21:921–927

Martens S, Preuss A, Matern U (2010) Multifunctional flavonoid dioxygenases: flavonols and anthocyanin biosynthesis in *Arabidopsis thaliana* L. Phytochemistry 71:1040–1049

Menon LG, Kuttan R, Kuttan G (1995) Inhibition of lung metastasis in mice induced by B16F10 melanoma cells by polyphenolic compounds. Cancer Lett 95:221–225

Mishra A, Kumar S, Pandey AK (2013) Scientific validation of the medicinal efficacy of *Tinospora cordifolia*. Sci World J 2013:292934. https://doi.org/10.1155/2013/292934

Molnar J, Beladi I, Domonkos K, Földeák S, Boda K, Veckenstedt A (1981) Antitumor activity of flavonoids on NK/Ly ascites tumor cells. Neoplasma 28:11–18

Moroney MA, Alcaraz MJ, Forder RA, Carry F, Hoult JR (1988) Selectivity of 5-neutrophil oxygenase and lipoxygenase inhibition by an anti-inflammatory flavonoid glycoside and related aglycone flavonoids. J Pharm Pharmacol 40:787–792

Morre DM, Morre DJ (2007) Compositions based on vanilloid catechin synergies for prevention and treatment of cancer. US Patent No. US7192612B2

Namdeo AG (2007) Plant cell elicitation for production of secondary metabolites-a review. Pharmacogn Rev 1:69–79

Navarro-Perán E, Cabezas-Herrera J, del Campo LS (2007) Effects of folate cycle disruption by the green tea polyphenol epigallocatechin-3-gallate. Int J Biochem Cell Biol 39:2215–2225

Netke S, Niedzwiecki A, Rath M, Roomi WM (2006) Nutrient pharmaceutical formulation comprising polyphenols and use in treatment of cancer. US Patent No. US-7041699-B2

Nigro JM, Baker SJ, Preisinger AC, Jessup JM, Hostetter R, Cleary K, Bigner SH, Davidson N, Baylin S, Devilee P, Glover T, Collins FS, Weslon A, Modali R, Harris CC, Vogelstein B (1989) Mutations in the p53 gene occur in diverse human tumour types. Nature 342:705–708

Nijveldt RJ, van Nood E, Danny EC, van Hoorn DE, Boelens PG, van Norren K, van Leeuwen PAM (2001) Flavonoids: a review of probable mechanisms of action and potential applications. Am J Clin Nutr 74:418–425

Oesterreich S, Weng CN, Qui M, Hilsenbeck SG, Osborne CK, Fuqua SAW (1993) The small heat shock protein hsp27 is correlated with growth and drug resistance in human breast cancer cell lines. Cancer Res 53:4443–4448

Oksman-Caldentey KM, Inze D (2004) Plant cell factories in the post-genomic era: new ways to produce designer secondary metabolites. Trends Plant Sci 9:433–440

Peer WA, Murphy AS (2007) Flavonoids and auxin transport: modulators or regulators. Trends Plant Sci 12:556–563

Pollak PE, Vogt T, Mo Y, Taylor LP (1993) Chalcone synthase and flavonol accumulation in stigmas and anthers of *Petunia hybrida*. Plant Physiol 102:925–932

Pryor WA (1997) Cigarette smoke radicals and the role of free radicals in chemical carcinogenicity. Environ Health Perspect 105:S875–S882

Ramchoun M, Harnafi H, Alem C, Benlys M, Elrhaffari L, Amrani S (2009) Study on antioxidant and hypolipidemic effects of polyphenol rich extract from *Thymus vulgaris* and *Lavandula multifida*. Pharm Res 1:106–112

Ranelletti FO, Maggiano N, Serra FG, Ricci R, Larocca LM, Lanza P, Scambia G, Fattorossi A, Capelli A, Piantelli M (1999) Quercetin inhibits p21-ras expression in human colon cancer cell lines and in primary colorectal tumors. Int J Cancer 85:438–445

Rath M, Netke S, Niedzwiecki AA (2006) Nutrient pharmaceutical formulation comprising polyphenols and use in treatment of cancer. US Patent No. US7041699B2

Rattan RR (2010) Mechanism of action of insecticidal secondary metabolites of plant origin. Crop Protect 29:913–920

Raut JS, Karuppayil SM (2014) A status review on the medicinal properties of essential oils. Indust Crops Prod 62:250–264

Rodriguez-Lopez JN, Navarro-Peran EM, Cabezas-Herrera J (2007) Dihydrofolate reductase inhibition by epigallocatechin gallate compounds. US Patent No. US20070249545A1

Rodriguez-Lopez JN, Cabezas-Herrera J, Navarro-Peran EM, Sanchez DCL (2008) Epigallocatechin-3-gallate compositions for cancer therapy and chemoprotection. Patent No. WO2008075201

Rosenberg RS, Grass L, Jenkins DJ, Kendall CW, Diamandis EP (1998) Modulation of androgen and progesterone receptors by phytochemicals in breast cancer cell lines. Biochem Biophys Res Commun 248:935–939

Ruh MF, Zacharewski T, Connor K, Howell J, Chen I, Safe S (1995) Naringenin: a weakly estrogenic bioflavonoid that exhibits antiestrogenic activity. Biochem Pharmacol 50:1485–1493

Ryan KG, Swinny EE, Markham KR, Winefield C (2002) Flavonoid gene expression and UV photoprotection in transgenic and mutant *Petunia* leaves. Phytochemistry 59:23–32

Sales PM, Souza PM, Simeoni LA, Magalhães PO, Silveira D (2012) Amylase Inhibitors: a review of raw material and isolated compounds from plant source. J Pharm Pharm Sci 15:141–183

Schijlen EG, De Vos CH, Martens S, Jonker HH, Rosin FM, Molthoff JW, Tikunov YM, Angenent GC, Van Tunen AJ, Bovy AG (2007) RNA interference silencing of chalcone synthase, the first step in the flavonoid biosynthesis pathway, leads to parthenocarpic tomato fruits. Plant Physiol 144:1520–1530

Siddique YH, Beg T, Afzal M (2008) Antigenotoxic effect of apigenin against anti-cancerous drugs. Toxicol In Vitro 22:625–631

Stefani ED, Boffetta P, Deneo-Pellegrini H, Mendilaharsu M, Carzoglio JC, Ronco A, Olivera L (1999) Dietary antioxidants and lung cancer risk: a case-control study in Uruguay. Nutr Cancer 34:100–110

Stracke R, Ishihara H, Huep G, Barsch A, Mehrtens F, Niehaus K, Weisshaar B (2007) Differential regulation of closely related R2R3-MYB transcription factors controls flavonol accumulation in different parts of the *Arabidopsis thaliana* seedling. Plant J 50:660–677

Tapas AR, Sakarkar DM, Kakde RB (2008) Flavonoids as nutraceuticals: a review. Trop J Pharma Res 7:1089–1099

Treutter D (2006) Significance of flavonoids in plant resistance: a review. Environ Chem Lett 4:147–157

Urso ML, Clarkson PM (2003) Oxidative stress, exercise, and antioxidant supplementation. Toxicology 189:41–54

van Acker FA, Tromp MN, Haenen GR, van der Vijgh WJ, Bast A (2000) Flavonoids can replace a-tocopherol as an antioxidant. FEBS Lett 473:145–148

van Der Meer IM, Stam ME, Van Tunen AJ, Mol JN, Stuitje AR (1992) Antisense inhibition of flavonoid biosynthesis in *Petunia* anthers results in male sterility. Plant Cell 4:253–262

Vattem DA, Randhir R, Shetty K (2005) Cranberry phenolics-mediated antioxidant enzyme response in oxidatively stressed porcine muscle. Process Biochem 40:2225–2238

Verdan AM, Wang HC, García CR, Henry WP, Brumaghim JL (2011) Iron binding of 3-hydroxychromone, 5-hydroxychromone, and sulfonated morin: implications for the antioxidant activity of flavonols with competing metal binding sites. J Inorg Biochem 105:1314–1322

Volate SR, Davenport DM, Muga SJ, Wargovich MJ (2005) Modulation of aberrant crypt foci and apoptosis by dietary herbal supplements (quercetin, curcumin, silymarin, ginseng and rutin). Carcinogenesis 26:1450–1456

Wasson AP, Pellerone FI, Mathesius U (2006) Silencing the flavonoid pathway in *Medicago truncatula* inhibits root nodule formation and prevents auxin transport regulation by *Rhizobia*. Plant Cell 18:1617–1629

Williams R, Spencer J, Rice-Evans C (2004) Flavonoids: antioxidants or signaling molecules. Free Rad Biol Med 36:838–849

Winkel-Shirley B (2002) Biosynthesis of flavonoids and effects of stress. Curr Opin Plant Biol 5:218–223

Wiseman H, Halliwell B (1996) Damage to DNA by reactive oxygen and nitrogen species: role in inflammatory disease and progression to cancer. Biochem J 313:17–29

Wu Y (2007) The synergistically pharmaceutical composition of baicalein and baicalin for inhibiting tumor. Patent No. EP1849467

Yokoo T, Kitamura M (1997) Unexpected protection of glomerular mesangial cells from oxidant triggered apoptosis by bioflavonoid quercetin. Am J Phys 273:206–212

Yoon K, Pallaroni L, Stoner M, Gaido K, Safe S (2001) Differential activation of wild-type and variant forms of estrogen receptor alpha by synthetic and natural estrogenic compounds using a promoter containing three estrogen-responsive elements. J Steroid Biochem Mol Biol 78:25–32

Yoshida M, Sakai T, Hosokawa N, Marui N, Matsumoto K, Fujioka A, Nishino H, Aoike A (1990) The effect of quercetin on cell cycle progression and growth of human gastric cancer cells. FEBS Lett 260:10–13

Yoshida M, Yamamoto M, Nikaido T (1992) Quercetin arrests human leukemic T-cells in late G1 phase of the cell cycle. Cancer Res 52:6676–6681

Zandi K, Teoh BT, Sam SS, Wong PF, Mustafa MR, Abubakar S (2011) Antiviral activity of four types of bioflavonoid against dengue virus type-2. Vir J 8:560. https://doi.org/10.1186/1743-422X-8-560

Zerback R, Dressler K, Hess D (1989) Flavonoid compounds from pollen and stigma of *Petunia hybrida*: inducers of the vir region of the *Agrobacterium tumefaciens* Ti plasmid. Plant Sci 62:83–91

Zhao XX (2003) Plant drug for preventing cancer. US Patent No. US200318039

Zhu W, Jia Q, Wang Y, Zhang Y, Xia M (2012) The anthocyanin cyanidin-3-O-β-glucoside, a flavonoid, increases hepatic glutathione synthesis and protects hepatocytes against reactive oxygen species during hyperglycemia: involvement of a cAMPPKA- dependent signaling pathway. Free Rad Biol Med 2012:314–327

14 Natural Compounds Extracted from *Moringa oleifera* and Their Agricultural Applications

A. Khairulmazmi and A. Tijjani

Contents

Abstract

Natural bio-active compounds synthesized by plants as secondary metabolites are well known and established. Today, their application in various fields such as medicine in the form of drugs and biopesticides in agriculture is well documented.

A. Khairulmazmi and A. Tijjani have equally contributed for this chapter.

A. Khairulmazmi (✉)
Department of Plant Protection, Faculty of Agriculture, Universiti Putra Malaysia, Serdang, Selangor Darul Ehsan, Malaysia

Institute of Plantation Studies, Universiti Putra Malaysia, Serdang, Selangor Darul Ehsan, Malaysia
e-mail: khairulmazmi@upm.edu.my

A. Tijjani
Department of Crop Production, Faculty of Agriculture and Agricultural Technology, Abubakar Tafawa Balewa University, Bauchi, Bauchi State, Nigeria
e-mail: tijjaniahmadu72@gmail.com

M. S. Akhtar et al. (eds.), *Natural Bio-active Compounds*,
https://doi.org/10.1007/978-981-13-7154-7_14

In recent times, the delivery of such compounds is achieved through nanodelivery technology, which is gaining acceptability in both field of drugs and agrochemical industries. The bio-active compounds with chemical diversity are obtained from nature either as homogenous plant crude extracts or as purified compounds. Crude plant extracts exist as a combination of different bio-active compounds with various polarities, and their partition remains a challenge in the process of characterization and identification. Extraction of these compounds from plant species is achieved by different solvents and extraction methods. Analytical methods like HPLC have commonly been utilized with GC-MS and LC-MS/MS chromatography methods to identify the compounds. Crude extracts from different morphological parts of plant species including *Moringa oleifera* are increasingly becoming important in the context of agricultural pest management and human medicine. *M. oleifera* is a medicinal plant that synthesizes such metabolites which include phenolic acids, carotenoids, quinones, antraquinones, flavonoids, flavonols, flavones, tannins, alkaloids, coumarins, terpenoids, amines, cyanogenic glycosides, triterpenoids, non-protein amino acids, glucosinolates, polyacetylenes, polyketides, phenylpropanes, steroids and saponins. They exert biological activities and can potentially be used to retard microbial activities. Other uses of *M. oleifera* are medicinal uses and other purposes such as water purification, fertilizer, biogas and biopesticides. The aim of this chapter is to highlight the uses and profiling of bio-active compounds of *M. oleifera*, their mode of action and prospects in commercial biopesticides for agricultural applications.

Keywords
Bio-active compounds · GC-MS · *M. oleifera* · Mode of action · Nanotechnology

14.1 Introduction

Moringa oleifera Lamarck (syn. *M. pterygosperma* Gaertn) belongs to the monogeneric family *Morigaceae* having about 33 species (Arora et al. 2013; Arora and Onsare 2014). It is locally called Pokok Kelo, drumstick tree, moonga, marango, mulangay, kelor, mlonge, horseradish tree, sajna and saijhan and supposed to be originated from northwest India. It is distributed worldwide, mainly in the tropics and subtropics (Fahey 2005). The most commonly cultivated species are *M. oleifera*, *M. drouhardii*, *M. peregrine*, *M. concanensis*, *M. pygmaea*, *M. borziana*, *M. stenopetala*, *M. rivae*, *M. arborea*, *M. longituba*, *M. ruspoliana*, *M. hildebrandtii* and *M. ovalifolia*. Among the commonly species, *M. oleifera* is well known and grown naturally in Afghanistan, Brazil, Mexico, Pakistan, Paraguay, Peru, Southeast and West Asia, West and East Africa, Sri Lanka, Southern Florida and West Indies (Satish et al. 2014; Chaudhary and Chaurasia 2017). The plant contains natural phytochemicals which could be used as biopesticides and also possesses strong microbial activities (Satish et al. 2007). The crude extract of *M. oleifera* contains phenolic compounds, flavonoids, tannins, saponins and alkaloids (Pereira et al.

2015). The entire plant parts including flowers, roots, bark, leaves, stem, seeds and essential oils have antimicrobial properties and therapeutic properties (Anwar et al. 2007; Dwivedi and Enespa 2012). This plant has the agricultural significance due to (i) strong antimicrobial activity, (ii) non-environmental pollution, (iii) safer for human consumption, (iv) indigenous knowledge of local farmers and (v) good bio-formulating ability.

Plant extracts are the most important source of natural bio-active compounds (Freire et al. 2015). These compounds have been extracted from plants using different solvents that possess antimicrobial, antifungal, and antibacterial activities against various pathogens (Bhattacharjee and Dey 2014) and are alternative to the chemical pesticides (Sogvar et al. 2016). Moreover, they improve growth parameters, crop immunity and postharvest quality (Zhao and Zhang 2013; Mustafa et al. 2014). Extraction of these compounds from plant species is achieved by different solvents and extraction methods. Nearly most or entire bio-active compounds of plant origin are usually saturated organic compounds or aromatic; hence they are frequently collected through initial methanol or ethanol extraction. Researchers have suggested other solvents that include hexane, ethyl acetate, dichloromethane, acetone, chloroform and butanol, and some suggested a combination of the solvents in an appropriate ratio to get the most excellent solvent systems for their extraction (Villas-Boas et al. 2005; Gurjar et al. 2012). One common method for the extraction of these compounds from plants is the filtration method, which involves dissolving of ground wet or dried, fresh plant parts into certain amount of solvent, shaken vigorously and filtrated after sometimes usually after 24 h. Green (2004) and Parekh et al. (2005) highlighted the use of serial exhaustive extraction method that includes consecutive extraction of active compounds with solvents of varying polarity starting from non-polar to higher polar solvents in order to ensure extraction of quantum crucial compounds with ample polarity range. Other researchers make use of Soxhlet extraction using organic solvent for dried plant material (Gurjar et al. 2012). Analytical methods like HPLC have commonly been utilized with GC-MS and LC-MS/MS chromatography methods to identify bio-active compounds (Snyder et al. 2012). GC-MS had high ability in the partitioning and identification of compounds from multiple biological mixtures (Villas-Boas et al. 2005). Additionally, GC-MS technique is able to identify numerous bio-active compounds that have dissimilar functions and are volatile in nature (Wang et al. 2012; Huang et al. 2012). It is also suitable for the analysis of compounds that are volatile in nature with lower molecular weight. LC-MS chromatography is another analytical technique apart from GC-MS in which LC is connected to spectrometry together with atmospheric pressure chemical ionization and electrospray ionization (Villas-Boas et al. 2005). The technique can analyse and identify a number of non-volatile bio-compounds even in minute quantities. Other methods employed in the separation and characterization of active compounds include thin-layer chromatography (TLC) and nuclear magnetic resonance (^{1}H NMR) (Sharma and Paliwal 2013).

In order to confirm the claims of the antimicrobial or biological activity of the natural bio-active and pure phytochemical compounds, several methods have been used to assess their growth inhibition ability (Das et al. 2010). These include poison food technique (Rhouma et al. 2009; Verma et al. 2010), agar disk diffusion (Baris et al. 2006; Gulluce et al. 2006; Das et al. 2010), agar well diffusion (Das et al. 2010), bioautography (Schmourlo et al. 2005; Cos et al. 2002), dilution methods (Gurjar et al. 2012), agar dilution (Cordell and Colvard, 2005; Patwardhan et al. 2005), diffusion (Ergene et al. 2006), broth microdilution assay (Baris et al. 2006) and broth macrodilution assay (Das et al. 2010; Gurjar et al. 2012). The approaches used to determine the response of the fungal pathogens against phytochemical compounds include spore germination measurement because it is the pivotal stage in the life cycle of the fungal pathogens in invading environments to cause infections (Slawecki et al. 2002). Evaluation of the potency of the compounds to inhibit mycelial growth also provides initial information of its mechanism of action. The aim of this chapter is to highlight the uses and profiling of bio-active compounds of *M. oleifera*, their mode of action and prospects in commercial biopesticides for agricultural applications.

14.2 Uses of *Moringa oleifera*

Moringa oleifera is a medicinal plant known for its widely and esteemed uses as food for man, forage for animal, alley cropping, fertilizer, green manure, foliar nutrient, ornamental plantings, biopesticide and various domestic purposes as biogas, blue dye, gum, domestic cleaning agent, water purification, pulp, rope, tannin, machine lubrication (oil), cosmetic products and traditional medicine (Fuglie 1999; Kumar et al. 2010; Luqman et al. 2012; Pinto et al. 2015; Stohs and Hartman 2015). Recent studies on animal, human and in vitro on the potential toxicological effects, genotoxicity, cytotoxicity and acute and sub-acute toxicity on the extracts from different morphological parts of *M. oleifera* revealed that they were extremely safe at the utilized amounts and doses (Ajibade et al. 2013; Araujo et al. 2013; Asiedu-Gyekye et al. 2014; Stohs and Hartman 2015). Similarly, dietary supplementation with *M. oleifera* leaves confirmed its safety effects (Zvinorova et al. 2015; Stohs and Hartman 2015). Because of its biosafety with neither toxicity nor mutagenicity (Luqman et al. 2012; Stohs and Hartman 2015), traditional menders have, for many decades, prescribed various morphological parts of the plant such as leaves, pods, flowers, roots, barks, seeds and oils for treatment of cancer, hypertension, diabetes, skin diseases, respiratory illnesses and ear and dental infections, for water purification and as nutrient supplementary source (Fahey 2005; Anwar et al. 2007; Luqman et al. 2012; Stohs and Hartman 2015). Pinto et al. (2015) and Stohs and Hartman (2015) highlighted on the scientific literature review search for the history of safe use (HOSU) for *M. oleifera* species accessed with different keyword combinations, viz. (a) "*M. oleifera*" and "food safety"; (b) "*M. oleifera*" and "risk assessment"; (c) "*M. oleifera*" and "toxicity"; (d) "*M. oleifera*" and "nutrition"; (e) "*M. oleifera*" and "review"; and (f) "*M. oleifera*" (http://www.ncbi.nlm.nih.gov/pubmed/).

At present, the plant is gaining researchers' interest as it is a good candidate for biological control that will replace the long-standing chemical management of pests and diseases (Satish et al. 2013). The tree is known commonly as "the miracle tree", "horseradish tree" or "benoil tree" (Fuglie 1999), because of the bio-active compounds available from original plant sources which are less expensive, renewable naturally and acceptable because of long-standing historical use and less side effects and better tolerance from patients (Arora and Onsare 2014).

14.3 Phytochemistry/Phytochemical Profiling

The ability of plants to synthesize secondary metabolites is well known and documented (Gurjar et al. 2012). *M. oleifera* is one of such plants that synthesize these metabolites. Chemical analysis of various parts of the plant showed that it is containing a profile of valuable and important vitamins, minerals and protein and a quantum of crucial phytochemicals that have pressing antimicrobial activities and can potentially be used to retard microbial activities (Satish et al. 2013; Arora and Onsare 2014). These metabolites include phenolic acids, phenols, carotenoids, antraquinones, quinones, flavonoids, flavonols, flavones, coumarins, tannins, amines, alkaloids, terpenoids, triterpenoids, glucosinolates, polyacetylenes, polyketides, steroids and saponins (Thompson and Thompson 2010; Gurjar et al. 2012; Arora and Onsare 2014). Other compounds reported present in *M. oleifera* include β-sitosterol, quercetin, zeatin, kaempferol, caffeoylquinic acid, rhamnose, isothiocyanates and glucosinolates (Sharma et al. 2012; Sharma and Paliwal 2013; Iqbal 2015). In addition, the stem barks of the plant are reported to consist of two important alkaloids, known as moringine and moringinine (Sharma and Paliwal 2013). Faizi et al. (1994) reported the successful isolation of vanillin, β-sitostenone, β-sitosterol, hydroxymellin and octacosanoic acid, from the stem bark of *M. oleifera*. Furthermore, reports of Razis et al. (2014) and Teixeira et al. (2014) on the phytochemical constituents of the *M. oleifera* leaves indicated the presence of fatty acids and various antioxidant compounds like ascorbic acid and phenolics (Vongsak et al. 2014; Alhakmani et al. 2013). Different derivatives of gallic acid, salicylic acid, caffeic acid, indole alkaloid *N*,α-L-rhamnopyranosyl vincosamide and four isothiocyanates were isolated and characterized from *M. oleifera* leaves (Panda et al. 2013; Waterman et al. 2014). Mbikay (2012) investigated the phytochemical compounds isolated from different parts and concluded that the major and prominent ones are phenolic acids, myricetin, rutin, chlorogenic acid, glycosides, niaziminin and niazin. Atawodi et al. (2010) also reported the presence of several procyanidin compounds in the root and stem barks of *M. oleifera*. Other phytochemicals non-volatile in nature and found in the leaves of *M. oleifera* include phenylvaleric acid, apigenin 6-C glucoside and quinic acid.

Investigation of the chemical compounds from *M. oleifera* leaf and seed samples collected from Ladang 2, Universiti Putra Malaysia (UPM), with GC-MS led to the identification of different volatile bio-active compounds. From the methanol leaf extracts, 67 different bio-active compounds were detected and identified. The active principal components having higher peak percentages were 6-decanoic acid (Z) (19.87%), 2-dimethyl(trimethylsilylmethyl)silyloxymethyltetrahydrofuran (11.90%), beta-L-rhamnofuranoside (11.07%), malonic acid (6.22%), n-hexadecanoic acid (5.87%), 1,3-propanediol (3.77%), benzeneacetonitrile (3.12%), octadecanoic acid $$, stearic acid (2.95%), 1,2,3,5-cyclohexanetetrol (2.38%), docosanoic acid (2.24%), cyclohexanecarboxylic (2.20%), 2-furancarboxaldehyde (2.14%) and 2,3-dihydro-3,5-dihydroxy-6-methyl-4H-pyran-4-one (2.11%). Their retention time (RT) and molecular formula were presented in Table 14.1. The major bio-active compounds identified in ethanol crude extracts were cis-9-hexadecenal (15.60%); beta-L-rhamnofuranoside, 5-O-acetyl-thio-octyl- (14.14%); phytol (9.76%); 1,3-propanediol, 2-(hydroxymethyl)-2-nitro (8.06%); beta-L-rhamnofuranoside, 5-O-acetyl-thio-octyl- (6.19%); malonic acid (5.78%); n-hexadecanoic acid (4.80%); Vitamin E (2.59); 9,12,15-octadecatrienoic acid, ethyl ester (Z,Z,Z) (2.36); and tetrahydro-3-furanylmethanol (1.42%) (Table 14.1). The major bio-active compounds identified in ethyl acetate crude extracts were cis-9-hexadecenal (39.17%); Vitamin E (13.14%); n-hexadecanoic acid (5.63%); octadecanoic acid (5.41%); cyclohexanone, 2-(1-methyl-2-nitroethyl)- (3.54%); phytol (3.01%); tetracontane (2.77%); oleoyl chloride (2.15%); gamma-sitosterol (2.06%); hexatriacontane (1.90%); 3,7,11,15-tetramethyl-2-hexadecen-1-ol (1.88%); and fucosterol (1.81%). The major bio-active compounds identified in hexane crude extracts were 9,12,15-octadecatrienoic acid, (Z,Z,Z) (18.86%); phytol (13.95%); tetracontane (13.81%); n-hexadecanoic acid (8.32%); alpha-tocopherol-beta-D-mannoside (7.96%); tetratriacontane (5.56%); 9-octadecenamide, (Z) (3.28%); oxirane, heptadecyl- (3.03%); octadecanoic acid (2.23%); palmitamide (2.11%); and tetratetracontane (2.01%). The major volatile bio-active compounds found in the aqueous leaf extracts of the *M. oleifera* were phenol (38.39%); phenylethyl alcohol (7.67%); undecane (7.20%); indole (3.61%); butanoic acid, 2-methyl (2.99%); cyclopentasiloxane, decamethyl- (2.68%); trisiloxane, 1,1,1,5,5,5-hexamethyl-3,3-bis[(trimethylsilyl)oxy]- (2.56%); diethyl phthalate (2.47%); butanoic acid, 3-methyl- (2.25%); and cyclotetrasiloxane, octamethyl- (2.05%) (Table 14.1).

Similarly, the extracts from the Malaysian *M. oleifera* seeds were analysed using GC-MS, and the major bio-active compounds identified were 59 from methanol seed extracts, 47 from ethanol seed extracts, 18 from ethyl acetate seed extracts, 41 from hexane seed extracts and 22 from aqueous seed extracts (Table 14.2) as compared with those stored and available in the National Institute for Standards and Technology (NIST), a mass spectral library built up by using pure substances and the mass spectra from the published literature. The compounds name, percentages, retention time and molecular formula of the materials were established. GC-17A version programming software was utilized to handle the mass spectra and the chromatogram. Benzeneacetonitrile (43.96%), ethyl 1-thio-alpha-l-arabinofuranoside (10.08%), D-allose (9.62%); 1,3-propanediol, 2-(hydroxymethyl)-2-nitro- (8.94%); 2-furancarboxaldehyde, 5-(hydroxymethyl)- (5.68%); 1,6-anhydro-beta-D-glucofuranose (4.22%); and oleic acid (4.05%) were the major components present in the methanol seed extracts. The major ones present in the ethanol seed extracts were 9-octadecenoic acid, (E)- (38.15%); n-hexadecanoic acid (9.40%); 1,3-propanediol, 2-(hydroxymethyl)-2-nitro- (6.60%); ethyl oleate (5.91%); octadecanoic acid (5.12%); 9-octadecenoic acid (Z)-2,3-dihydroxypropyl ester

Table 14.1 Phytochemical compounds identified from *M. oleifera* leaf extracts using GC-MS technique in Malaysia

Extracts	Peak (%)	Retention time	Chemical compounds	Molecular formula
Methanol extract	19.87	13.333	6-Decanoic acid, (Z)-	$C_{18}H_{34}O_2$
	11.90	14.900	2-Dimethyl(trimethylsilylmethyl) silyloxymethyltetrahydrofuran	$C_{11}H_{26}O_2$
	11.07	15.192	β-L-Rhamnofuranoside, 5-O-acetyl-thio-octyl	$C_{16}H_{30}O_5$
	6.22	15.058	Malonic acid	$C_{15}H_{28}O_4$
	5.87	12.467	n-Hexadecanoic acid, methyl ester	$C_{16}H_{32}O_2$
	3.77	9.700	1,3-Propanediol	$C_4H_9NO_5$
	3.13	9.758	Benzeneacetonitrile	C_8H_7NO
	2.95	13.433	Octadecanoic acid	$C_{18}H_{36}O_2$
	2.38	10.725	1,2,3,5-Cyclohexanetetrol	$C_6H_{12}O_4$
	2.24	15.150	Docosanoic acid	$C_{22}H_{44}O_2$
	2.20	14.808	Cyclohexanecarboxylic acid	$C_{18}H_{32}O_2$
	2.14	7.967	2-Furancarboxaldehyde	$C_6H_6O_3$
	2.11	7.317	4H-Pyran-4-one,2,3-dihydro-3,5-dihydroxy-6-methyl-	$C_6H_8O_4$
Ethanol extract	15.60	13.336	cis-9-Hexadecenal	$C_{16}H_{30}O$
	14.14	15.185	beta-L-Rhamnofuranoside,5-O-acetyl-thio-octyl-	$C_{16}H_{30}O_5S$
	9.76	13.211	Phytol	$C_{20}H_{40}O$
	8.06	9.538	1,3-Propanediol, 2-(hydroxymethyl)-2-nitro	$C_4H_9NO_5$
	6.19	15.053	beta-L-Rhamnofuranoside,5-O-acetyl-thio-octyl-	$C_{16}H_{30}O_5S$
	5.78	14.878	Malonic acid	$C_{12}H_{22}O_4$
	4.80	12.461	n-Hexadecanoic acid	$C_{16}H_{32}O_2$
	2.59	17.241	Vitamin E	$C_{29}H_{50}O_2$
	2.36	13.474	9,12,15-Octadecatrienoicacid,ethylester,(Z,Z,Z)-	$C_{20}H_{34}O_2$
	1.42	10.406	Tetrahydro-3-furanylmethanol	$C_5H_{10}O_2$
Ethyl acetate extract	39.17	13.342	cis-9-Hexadecenal	$C_{16}H_{30}O$
	13.14	17.244	Vitamin E	$C_{29}H_{50}O_2$
	5.63	12.465	n-Hexadecanoic acid	$C_{16}H_{32}O_2$
	5.41	13.434	Octadecanoic acid	$C_{18}H_{36}O_2$
	3.54	14.810	Cyclohexanone,2-(1-methyl-2-nitroethyl)-	$C_9H_{15}NO_3$
	3.01	13.214	Phytol	$C_{20}H_{40}O$
	2.77	16.373	Tetracontane	$C_{40}H_{82}$
	2.15	14.606	Oleoyl chloride	$C_{18}H_{33}ClO$
	2.06	18.162	Gamma-sitosterol	$C_{29}H_{50}O$
	1.90	15.657	Hexatriacontane	$C_{36}H_{74}$
	1.88	11.842	3,7,11,15-Tetramethyl-2-hexadecen-1-ol	$C_{20}H_{40}O$
	1.81	18.233	Fucosterol	$C_{29}H_{48}O$
Hexane extract	18.86	13.344	9,12,15-Octadecatrienoic acid, (Z,Z,Z)-	$C_{18}H_{30}O_2$
	13.95	13.211	Phytol	$C_{20}H_{40}O$
	13.81	16.372	Tetracontane	$C_{40}H_{82}$

(continued)

Table 14.1 (continued)

Extracts	Peak (%)	Retention time	Chemical compounds	Molecular formula
	8.32	12.467	n-Hexadecanoic acid	$C_{16}H_{32}O_2$
	7.96	17.241	alpha-Tocopherol-beta-D-mannoside	$C_{35}H_{60}O_7$
	5.56	15.654	Tetratriacontane	$C_{34}H_{70}$
	3.28	14.349	9-Octadecenamide, (Z)-	$C_{18}H_{35}NO$
	3.03	17.643	Oxirane, heptadecyl-	$C_{19}H_{38}O$
	2.23	13.434	Octadecanoic acid	$C_{18}H_{36}O_2$
	2.11	13.540	Palmitamide	$C_{16}H_{33}NO$
	2.01	14.886	Tetratetracontane	$C_{44}H_{90}$
Aqueous extract	38.39	6.931	Phenol	C_6H_6O
	7.67	8.519	Phenylethyl alcohol	$C_8H_{10}O$
	7.20	8.335	Undecane	$C_{11}H_{24}$
	3.61	10.408	Indole	C_8H_7N
	2.99	5.058	Butanoic acid, 2-methyl-	$C_5H_{10}O_2$
	2.68	8.651	Cyclopentasiloxane, decamethyl-	$C_{10}H_{30}O_5Si_5$
	2.56	11.824	Trisiloxane, 1,1,1,5,5,5-hexamethyl-3,3-bis[(trimethylsilyl)oxy]-	$C_{12}H_{36}O_4Si_5$
	2.47	12.943	Diethyl phthalate	$C_{12}H_{14}O_4$
	2.25	4.972	Butanoic acid, 3-methyl-	$C_5H_{10}O_2$
	2.05	6.860	Cyclotetrasiloxane, octamethyl-	$C_8H_{24}O_4Si_4$

(4.79%); malonic acid, 3,3-dimethylbut-2-yl propyl ester (3.92%); 2-thiopheneacetic acid, octyl ester (2.83%); and 2-furancarboxaldehyde, 5-(hydroxymethyl)- (2.11%). Acetic acid (54.23%); oleic acid (13.63%); hexadecanoic acid (8.73%); 1,12-tridecadiene (4.56%); 1,2,3-propanetriol, monoacetate (4.52%); lactonitrile (2.15%); and hemimellitene (2.11%) were the major compounds identified in ethyl acetate seed extracts. The major compounds present in hexane seed extracts were 9-octadecenoic acid, (E)- (47.23%); octadecanoic acid (12.17%); n-hexadecanoic acid (8.00%); cyclohexanecarboxylic acid, undec-10-enyl ester (7.30%); eicosanoic acid (4.37%); oleic anhydride (4.08%); docosanoic acid (2.93%); and cis-13-eicosenoic acid (2.32%). Acetic acid (47.01%), 2,3-butanediol (11.04%), pyrogallol (7.10%), 1,2-benzenediol (5.92%), 6-dodecanone (4.49%), furfural (2.98%) and benzeneacetic acid (2.33%) were the major compounds identified in aqueous seed extracts (Table 14.2).

The quantity and quality of the active phytochemicals in the extracts are determined by the solvent(s) and the extraction method(s) used (Gurjar et al. 2012; Sharma and Paliwal 2012). The solvents diffuse into the plant tissues and solubilize bioactive compounds of similar polarity, and then the active compounds from the plant tissue were separated from the inactive compounds (Gurjar et al. 2012). A variety of solvents, extraction methods and techniques of analysis were used to determine and

Table 14.2 Phytochemical compounds identified from *M. oleifera* seed extracts using GC-MS technique in Malaysia

Extracts	Peak (%)	Retention time	Chemical compounds	Molecular formula
Methanol extract	43.96	9.758	Benzeneacetonitrile	C_8H_7NO
	10.08	14.877	Ethyl 1-thio-. alpha.-l-arabinofuranoside	$C_7H_{14}O_4S$
	9.62	9.871	D-Allose	$C_6H_{12}O_6$
	8.94	9.500	1,3-Propanediol,2-(hydroxymethyl)-2-nitro-	$C_4H_9NO_5$
	5.68	7.961	2-Furancarboxaldehyde,5-(hydroxymethyl)-	$C_6H_6O_3$
	4.22	10.468	1,6-Anhydro-beta-D-glucofuranose	$C_6H_{10}O_5$
	4.05	13.333	Oleic acid	$C_{18}H_{34}O_2$
	1.42	7.047	2-Deoxy-D-galactose	$C_6H_{12}O_5$
	1.12	9.129	1,6-Anhydro-beta-d-talopyranose	$C_6H_{10}O_5$
Ethanol extract	38.15	13.346	9-Octadecenoic acid, (E)-	$C_{18}H_{34}O_2$
	9.40	12.466	n-Hexadecanoic acid	$C_{16}H_{32}O_2$
	6.60	9.594	1,3-Propanediol,2-(hydroxymethyl)-2-nitro-	$C_4H_9NO_5$
	5.91	13.471	Ethyl oleate	$C_{20}H_{38}O_2$
	5.12	13.363	Octadecanoic acid	$C_{18}H_{36}O_2$
	4.79	15.673	9-Octadecenoic acid (Z)-, 2,3-dihydroxypropyl ester	$C_{21}H_{40}O_4$
	3.92	14.878	Malonic acid, 3,3-dimethylbut-2-yl propyl ester	$C_{12}H_{22}O_4$
	2.83	8.982	2-Thiopheneacetic acid, octyl ester	$C_{14}H_{22}O_2S$
	2.11	7.967	2-Furancarboxaldehyde,5-(hydroxylmethyl)-	$C_6H_6O_3$
	1.47	9.757	Benzeneacetonitrile, 4-hydroxy-	C_8H_7NO
Ethyl acetate extract	54.23	5.723	Acetic acid	$C_2H_4O_2$
	13.63	25.803	Oleic acid	$C_{18}H_{34}O_2$
	8.73	24.541	Hexadecanoic acid	$C_{16}H_{32}O_2$
	4.56	25.204	1,12-Tridecadiene	$C_{13}H_{24}$
	4.52	20.279	1,2,3-Propanetriol, monoacetate	$C_5H_{10}O_4$
	3.22	5.438	Butane, 1-propoxy-	$C_7H_{16}O$
	2.15	13.920	Lactonitrile	C_3H_5NO
	2.11	11.798	Hemimellitene	C_9H_{12}
Hexane extract	47.23	13.356	9-Octadecenoic acid, (E)-	$C_{18}H_{34}O_2$
	12.17	13.442	Octadecanoic acid	$C_{18}H_{36}O_2$
	8.00	12.459	n-Hexadecanoic acid	$C_{16}H_{32}O_2$
	7.30	14.807	Cyclohexanecarboxylic acid, undec-10-enyl ester	$C_{18}H_{32}O_2$
	4.37	14.325	Eicosanoic acid	$C_{20}H_{40}O_2$
	4.08	14.601	Oleic anhydride	$C_{36}H_{66}O_3$
	2.93	15.150	Docosanoic acid	$C_{22}H_{44}O_2$
	2.32	14.230	cis-13-Eicosenoic acid	$C_{20}H_{38}O_2$

(continued)

Table 14.2 (continued)

Extracts	Peak (%)	Retention time	Chemical compounds	Molecular formula
Aqueous extract	47.01	5.386	Acetic acid	$C_2H_4O_2$
	11.04	9.342	2,3-Butanediol	$C_4H_{10}O_2$
	7.10	20.849	Pyrogallol	$C_6H_6O_3$
	5.92	16.541	1,2-Benzenediol	$C_6H_6O_2$
	4.49	.442	6-Dodecanone	$C_{12}H_{24}O$
	2.98	9.098	Furfural	$C_5H_4O_2$
	2.33	15.645	Benzeneacetic acid	$C_8H_8O_2$

estimate the existence of such compounds (Satyanarayana 2005; Rodríguez-Perez et al. 2016). The selection of the solvents for the extraction largely depends on the properties of the solvent which include its ease of removal at low temperature or heat, rate of extraction, ability to solubilize antimicrobials from plants, inability of the extracts to form complex with the solvents and toxicity in bioassay (Ncube et al. 2008; Gurjar et al. 2012; Razis et al. 2014). Yassa and Tohamy (2014) showed that traditionally water has been used as a universal solvent for the extraction as aqueous extract, but Das et al. have shown that plant extracts from organic solvents like chloroform have been established to have further consistent antimicrobial effectiveness in comparison to aqueous extract. Methanol and ethanol have been identified by different researchers as the most common and best solvents for the extraction of active compounds from *M. Oleifera* (Ajibade et al. 2013; Vongsak et al. 2014; Rodríguez-Perez et al. 2016). Additionally, they can increase the efficiency of the extraction of active compounds and easily be removed from the final product (Moraes et al. 2013; Rodríguez-Perez et al. 2016). Other solvents commonly used include butanol, hexane, acetone, chloroform, ethyl acetate, petroleum ether and dicholromethane and sometimes a combination of two or more solvents to get the best solvent systems for extraction (Gurjar et al. 2012). The availability of active compounds in the extracts is also determined by the extraction method used; therefore, there is a need to select the extraction methods that are environmentally friendly for the active compounds by using technologies that are safe and effective while giving rise to enhanced yields and minimizing their impact on the quality of the end product. Different extraction methods that range from green extraction technologies such as microwave-assisted extraction, ultrasound-assisted extraction (UAE) and accelerated solvent extraction (Moraes et al. 2013) to conventional methods such as maceration extraction, reflux method (Moghimipour and Handali 2015) solid-liquid extraction (Vongsak et al. 2014; Rodríguez-Perez et al. 2016), Soxhlet extraction (Gurjar et al. 2012; Vongsak et al. 2014), supercritical fluid extraction (Zhao and Zhang 2013) and pressurized hot water extraction (Matshediso et al. 2015) have been recommended and used for bioactive compound extraction from different parts of *M. oleifera*. In addition, analytical tools like thin-layer chromatography (TLC), high-pressure liquid chromatography (HPLC), gas chromatography-mass spectrophotometry (GC-MS) and liquid chromatography-mass spectrophotometry (LC-MS) have been used for the separation and quantification of active compounds in *M. Oleifera* (Chuang et al. 2007; Vongsak et al. 2014).

14.4 Mode of Action

Although plants use different scenarios to conquer such fungal attacks, through the production of proteins and antimicrobial peptide compounds (Zottich et al. 2011; Duan et al. 2013), it will be of enormous importance to treat susceptible host plants with antimicrobial compounds from such valuable and potential plants like *M. oleifera* that possess antimicrobial properties in order to provide them with protection. Broekaert et al. (1992) reported the antifungal effectiveness of some chitin-binding proteins against phytopathogens. They have the ability to deter with the fungal life cycle either by killing the pathogen or by interfering with fungal growth (Morais et al. 2010; Choi et al. 2012). Typical example of such group of chemicals includes that from *Setcreasea purpurea* (lectin-/chitin-binding) that at 1.51 mg/ml inhibits the growth of *Sclerotinia sclerotiorum*, *Penicillium italicum*, *Rhizoctonia solani* and *Helminthosporium maydis* (Yao et al. 2010). Similarly, *Mo*-CBP_3 chitin-binding protein isolated from *M. oleifera* seeds (Batista et al. 2014) represents a group of molecules that inhibit fungal growth because of their affinity to the fungus chitin as confirmed by affinity chromatography (Gifoni et al. 2012), and this enables these molecules to accomplish their function in plant defence mechanisms. Chitin is one major cell wall structural component of the of fungi and also a linear homopolymer of β-(1,4)-linked *N*-acetyl-D-glucosamine residues (Freire et al. 2015). The mechanisms of action by which *Mo-CBP*$_3$ chitin-binding protein performs as an antifungal agent have been demonstrated by Batista et al. (2014) and Gifoni et al. (2012). They show in their studies the ability of the *Mo*-CBP_3 protein to activate the reactive oxygen species (ROS) production and trigger morphological and ultra-structural alterations on the mycelium and spores of *F. solani*, a model filamentous fungus that grips relevance in causing diseases in economically valuable plants. *Mo*-CBP_3 binding to the surface of the fungal chitin as observed after treating cells of *F. solani* with *Mo*-CBP_3 at 0.1 mg/ml concentration is the first step in the mechanism, followed by rapid cell killing at the surface rather than the interior of the cell by the action of antimicrobial molecules (*Mo*-CBP_3) (Jenssen et al. 2006; Batista et al. 2014). The next vital step is the *Mo*-CBP_3 binding to the cell membrane components of *F. solani* via electrostatic interactions (Gifoni et al. 2012), after which it persuaded production of ROS as indicated by pellet (reddish-brown) in the innermost of the *F. solani* spores, as opposed by the controls (negative) (Fig. 14.1).

Further examination on the mechanisms of actions for *Mo*-CBP_3 isolated from *M. oleifera* seeds revealed cell wall shrinkage, vacuolation and cytolysis in contrast to the control cells (Fig. 14.2a–c) (Batista et al. 2014). Vacuoles are organelles that performed the function of either detoxification or for storage of resources (Richards et al. 2010). Additional formation of vacuole might be attributed to the response and defence of *F. solani* to the cytotoxic effectiveness of *Mo*-CBP_3. Other changes noticeable in the cytosol treated and incubated with *Mo*-CBP_3 are the condensation

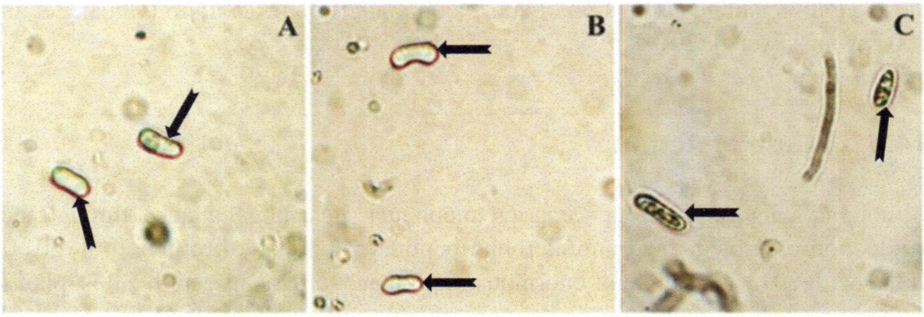

Fig. 14.1 *F. solani* Spores induced with reactive oxygen species (ROS) by treatment with 3,3′-diaminobenzidine (DAB) for ROS detection. Cells were incubated with (**a**) H_2O, (**b**) BSA (0.1 mg mL^{-1}) or (**c**) *Mo*-CBP_3 (0.1 mg mL^{-1}), and uptake of DAB is established by the dark staining (reddish-brown) reaction in conidia, as shown by Fig. 4.1 (A–C). Bars: 2.5 μm. (Adopted from Batista et al. 2014)

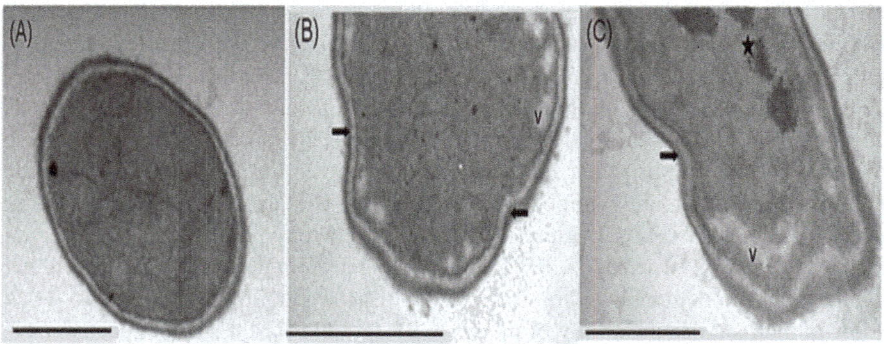

Fig. 14.2 TEM micrograph view of *F. solani* cells cultured either with of *Mo*-CBP_3 (0.05 mg. ml^{-1}) (**a**) or without it (**b** and **c**). Star indicates condensation of the cytosolic content. Vacuole condensation (V) is also shown. Arrows indicate shrinkage of the cell wall. Bars: 0.5 μm (**a**–**c**)

of granular material possibly due to the electrostatic interactions of the cations present in *Mo*-CBP_3 with the negative charges of metabolites in the fungal cell, as a result of the coagulation properties of the protein (Batista et al. 2014) as previously reported (Gifoni et al. 2012) (Fig. 14.2c). The mechanisms of actions of *Mo*-CBP_3 confirmed the antifungal effectiveness of the compounds, as prominent morphological changes have been noticed via interactions with the cell membrane of *F. solani* followed by the induction of ROS and final cell death. Other compounds found in *M. oleifera* that showed antimicrobial activity and served as plant defence mechanisms against phytopathogens are summarized in a tabular form (Table 14.3).

Examination of the changes on the mycelium of *B. cinerea* treated with *M. oleifera* methanol leaf extract at MIC concentration through SEM micrographs revealed that *M. oleifera* has caused some detrimental effects on the morphology of the mycelium compared to the controls (Fig. 14.3). The effects observed include abnormal growth, lysis, shrinkage, disruption, aggregation, reduced hyphal length

Table 14.3 Mode of action of phytochemicals found in *M. oleifera*

Compound	Mechanism
Phenolics	Membrane disruption, substrate deprivation
Phenolic acids	Bind to adhesins, complex with cell wall, inactivate enzymes
Terpenoids	Membrane disruption
Alkaloids	Intercalate into cell wall
Tannins	Bind to proteins, enzyme inhibition, substrate deprivation
Flavonoids	Bind to adhesins, complex with cell wall, inactivate enzymes
Coumarins	Interaction with eucaryotic DNA
Lectins and polypeptides Simple phenols	Form disulphide bridges

Source: Gurjar et al. (2012)

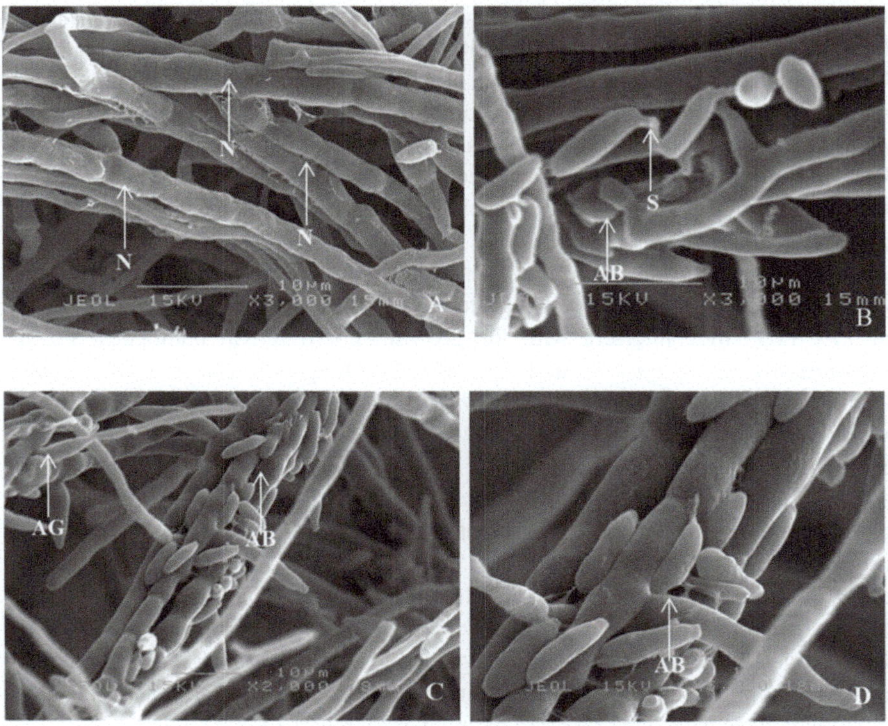

Fig. 14.3 Effect of *M. oleifera* methanol crude extract on *B. cinerea* (BCH07) mycelium: (**a**) control SEM micrographs; (**b**) shrinkage and abnormal growth on treated mycelium; (**c**) abnormal growth and aggregation; (**d**) abnormal growth

and diameters and pore formation on the mycelium of the samples treated with crude methanol leaf extracts. The control micrograph for the mycelium was observed with normal growth and smooth surfaces.

14.5 Application of *M. oleifera* in Agricultural Pest Management

Drawbacks associated with chemical control methods that affect their potency in disease management arose the interest of scientists in developing alternative methods of control, especially those which are eco-friendly, non-toxic to human and animals, biodegradable, within farmers' feasibility and having specific action and broad-spectrum activity (Marino and Bersani 2001; Abhishek et al. 2013). Scientific investigations on different morphological parts of *M. oleifera* indicated the presence of biologically active compounds which possess an array of antimicrobial properties acting on different types of microorganisms. These bio-active compounds were evaluated in vitro and further subjected to validate their efficacy via in vivo method for controlling disease incidence and severity in crops. Potential contributions made by *M. oleifera* in the field of human health and plant disease management should not be overemphasized. Recently, Batista et al. (2014) reported the antifungal inhibitory activity of *Moringa*-chitin-binding protein (*Mo*-CBP_3) purified from the seeds of *M. oleifera* Lam. against mycelial growth and spore germination of *Fusarium solani* at 0.05 mg/ml. Similarly, in another study, Gifoni et al. (2012) also showed the antifungal efficacy of *Mo*-CBP_3 protein purified from *M. oleifera* seed against phyopathogenic fungi *F. oxysporum*, *F. solani*, *Colletotrichum gloeosporioides* and *Pythium oligandrum* at 0.05 mg/ml and 0.1 mg/ml, respectively. The antifungal effect of *M. oleifera* extracts was reported by Dwivedi and Enespa (2012) in which at 75% (v/v) they interfere with the growth of two fungal pathogens, viz. *F. oxysporum* f. sp. *lycopersici* and *F. solani*. The fungicidal effect of *M. oleifera* extracts on *Fusarium*, *Pythium* and *Rhizoctonia*, causing tomato rots, was also reported (Moyo et al. 2012). Again, El-Mohamedy and Abdalla (2014) reported the fungicidal effect of *M. oleifera* against *F. oxysporum*, *F. solani*, *Alternaria alternata*, *A. solani*, *Rhizoctonia solani*, *Sclerotium rolfsii* and *Macrophomina phaseolina* causing rots on fruit and other perishables. Additional evidences on the antifungal properties of *M. oleifera* include that of Tijjani et al. (2014) on the inhibitory effect of *M. oleifera* against mycelial growth of *Aspergillus flavus*, a causal agent of tomato fruit rot. Prapassom et al. (2012) reported the antifungal efficacy of 14 crude leaf extracts including *M. oleifera* against *C. gloeosporioides* (Penz.). Other potential inhibitory attributes of different morphological parts of *M. oleifera* against phytopathogens have been reported by many researchers (Table 14.4).

Other important biological potentials of *M. oleifera* reported by different researchers in the literatures include antimicrobial activities (Arora et al. 2013; Arora and Onsare 2014), antifungal (Kadhim and AL-Shammaa 2014; Prapassom et al. 2012; Chuang et al. 2007), antioxidant (Satish et al. 2013; Verma et al. 2009), antibacterial and antiulcer (Belay and Sisay 2014), anti-inflammatory, diuretic,

Table 14.4 Potential attributes of the morphological parts of *M. oleifera* against some phytopathogens

Part used	Target Pathogen	Biological Activity	References
Fruit	*Alternaria* sp., *Colletotrichum* sp., *Curvularia* sp., *Fusarium* sp.	Antifungal	Mohammed et al. (2012)
Flowers and callus	*Candida albicans*	Antifungal	
Flowers	*Candida albicans*, *Microsporum canis* and *Alternaria* sp.	Antifungal	Rocha et al. (2011)
Seed	*Fusarium solani*	Antifungal	Rahman et al. (2009) and Jabeen et al. (2008)
	Aspergillus niger and *Candida albicans*	Antifungal	Saadabi and Zaid (2011)
	Alternaria alternata, *Aspergillus candidus*, *A. flavus*, *A. niger*, *A. rugulosus*, *A. sydowii*, *A. terreus*, *Chaetomium globosum*, *Fusarium solani*, *Helminthosporium* sp., *Macrophomina phaseolina*, *Nigrospora sphaerica*, *Rhizoctonia solani* and *Rhizopus nigricans*	Antifungal	Sahab and Nawar (2015)
Leaves	*P. aeruginosa*, *Aspergillus niger*, *A. oryzae*, *A .terreus* and *A. nidulans*	Antimicrobial	Prashith et al. (2010)
	Aspergillus tamarii, *Rhizopus solani*, *Mucor mucedo* and *Aspergillus niger*	Antifungal	Peixoto et al. (2011) and Jamil et al. (2007)
	Fusarium solani	Antifungal	Dwivedi and Enespa (2012)
	Sclerotium rolfsii	Antifungal	Adandonon et al. (2006)
	Candida albicans, *Aspergillus niger*, and *Rhizopus stolonifer*	Antifungal	Aisha et al. (2016)
	Rhizopus stolonifer	Antifungal	Tijjani et al. (2013)
	Candida albicans, *Aspergillus niger*, *Aspergillus flavus*, *Aspergillus fumigatus* and *Cryptococcus neoformans*	Antifungal	
	Aspergillus flavus (LBFC299), *A. niger* (LBFC394), *A. niger* aggregate (LBFC396), *A. ochraceus* (LBFC271), *A. versicolor* (LBFC283), *A. ustus* (LBFC402), *A. aculeatus* (LBFC371), *A. fumigatus* (LBFC377), *A. foetidus* (LFC379), *A. nidulans* (LBFC393), *A. terreus* (LBFC402) and *A. carbonarius* (LBFC373)	Antifungal	Orole et al. (2016)
	Pseudomonas aeruginosa	Antibacterial	Dewangan et al. (2010)

(continued)

Table 14.4 (continued)

Part used	Target Pathogen	Biological Activity	References
	Candida albicans	Antifungal	Abdulkadir et al. (2015)
	Rhizoctonia sp., *Pythium* sp. and *Fusarium* sp.	Antifungal	Raj et al. (2014)
Stem	*Aspergillus niger* v	Antifungal	Arowora and Adetunji (2014)
	P. aeruginosa	Antibacterial	Chetia and Gogoi (2011)

antispasmodic (Caceres et al. 1992), antimutagenic and antioxidant (Satish et al. 2013), antistress (Luqman et al. 2012), anticancer (Krishnamurthy et al. 2015), cytotoxic activities (Pinto et al. 2015), antitumour, antipyretic, antiepileptic, antinociceptive and antidiabetic (Vinoth et al. 2012).

14.6 Nanotechnology for Bio-active Compound Delivery

Nanotechnology has been considered as a promising delivery system with respect to eco-friendly pesticide formulation, including those with natural plant products incorporated as active ingredients (Rodríguez-Rojo et al. 2012; Angajala et al. 2014). It is a useful system for delivering chemical compounds across the cuticle, and it has the potential of application in the food industry, disease treatment delivery system, food packaging and bio-active compound delivery to target sites. The widespread commercial application of nanoemulsion technology especially in pharmaceutical and agrochemical industries is due to its small droplet size (10–100 nm) as suggested by some researchers (Polychniatou and Tzia 2014; Rai et al. 2015) that causes a reduction in the Brownian motion and gravitational strength (Tadros et al. 2004; Mishra et al. 2014). Similarly, the small size of droplets spread uniformly on the surface and helps to enhance wetting, spreading and penetration as a result of the low surface tension. Additionally, it can prevent flocculation and coalescence of the droplets and enables the system to remain dispersed without separation. Furthermore, the surfactant film thickness prevents disruption and thinning of liquid film between the droplets. Nanoemulsion can encapsulate active ingredients within their droplets, and this helps in reducing chemical degradation (McClements and Decker 2000) and hence expands cell wall penetration of the fungus, due to their smaller size (Zahid et al. 2012). Other benefits include less amount of energy requirement, increased rate of absorption and elimination of variation on absorption, and increased bioavailability (Mishra et al. 2014). Several reports have been advanced in using nanoemulsion technology to manage plant diseases with active ingredients from different plant sources (Ocsoy et al. 2013), but none was reported on preparation of nanoemulsion using *M. oleifera* as active compound.

14.6.1 Components of Nanoemulsion

Basically, a nanoemulsion consists of an active ingredient, an oil phase, a surfactant and an aqueous phase. The oil phase is usually the carrier(s) which are mineral oils and/or vegetable oils (Wang and Liu 2007; Xu et al. 2011). Today, vegetable oils and their esterified derivatives are gaining interest compared to the mineral oils because of their biodegradability and renewability (Xu et al. 2007). Castor oil, copaiba oil, olive oil, soy oil, Agnique® AMD 10 and Edenol SP 100 are few examples of vegetable oils used in nanoemulsion formulations. The oil phase in nanoemulsion formulations enhances the spread of droplets on the surface of plants and splits open the cuticle to increase both the fluidity of cuticular components and pesticide diffusion rates (Green and Beestman 2007).

14.6.2 Surfactants

Surfactants, also called surface-active agents are organic compounds that are amphiphilic in nature (having both hydrophobic groups and hydrophilic groups) that reduce surface and interfacial tensions by accumulating at the interface of immiscible fluids and increase solubility, bioavailability, mobility and subsequent biodegradation of hydrophobic or insoluble organic compounds (Singh et al. 2007). They can be found and used as emulsifiers, de-emulsifiers, wetting agents, forming agents, detergents in petroleum, petrochemicals, functional food ingredients and environmental management and in foods and beverages, agrochemicals, cosmetic and pharmaceuticals, and the mining and metallurgical industries (Van Hamme et al. 2006; Singh et al. 2007). Surfactant emulsifiers are added to nanoemulsion formulations to ensure spontaneous emulsification with good stability qualities in the spray tank. Surfactants can be classified as ionic and non-ionic. Non-ionic surfactants are preferred in pesticide formulation system (Wang and Liu 2007; Mehmood 2015) than the ionic ones due to their less toxicity, enhanced solubility, spreading, adsorption, translocation, and penetration of active ingredients into the target. Myers (2005) define non-ionic surfactant as a surfactant that carries no electric charge, as its water solubility is derived from the occurrence of polar functionalities capable of significant hydrogen bonding relations with water (e.g. polyglycidols and polyoxyethylenes) (Shafiq et al. 2007). Although stable nanoemulsions are best formulated with surfactants or a combination of surfactants having hydrophilic-lipophilic balance (HLB) values close to that required for the oil phase, it is important to know that there are no specified rules to resolve the ratio of surfactants in the blend surfactants. However, guidance can be obtained from the (HLB) system.

14.6.3 Ternary Phase Diagram

Ternary phase diagram is a graphical representation of the phase behaviour of mixtures containing three components (carrier, surfactant and water) in a triangular diagram (Fig. 14.4). In nanoemulsion formulation, phase diagram describes the

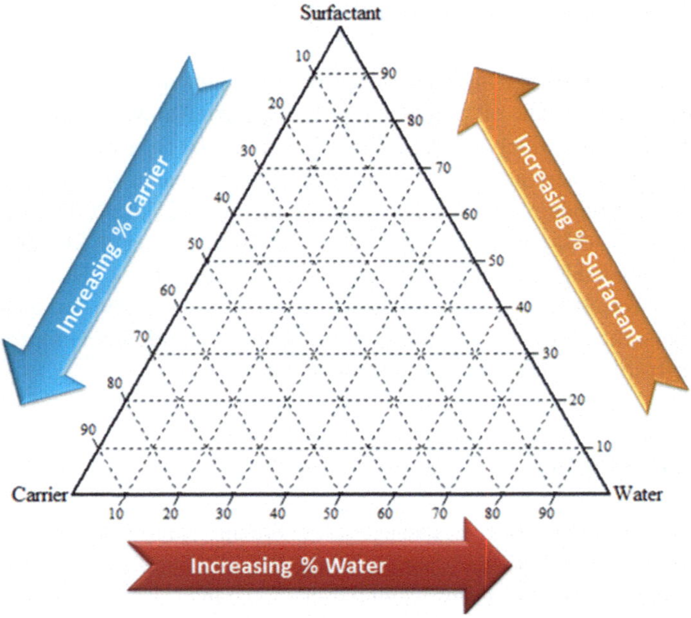

Fig. 14.4 Ternary phase diagram showing the percentage increase in the components of nanoemulsion formulation. (Source: Tijjani 2018)

phase equilibrium and the correlation among the phase performance of a combination and how the compositions are obtained. Primarily, it is used to determine the isotropic region. The isotropic region is the region where the components used form a homogeneous mixture that could be used to develop biopesticide formulation which consists of water, surfactant, carrier and active ingredients. To develop nanoemulsions, a stable area is graphically represented in the triangular phase diagrams in which each corner in the triangle designates a certain component. Ternary phase diagram is needed to compare the three components at once. The diagram is divided into three axes, representing the carrier, surfactant and water. It describes the concentration of the three ingredients of a mixture inside a triangle. At the corners, the concentration of the pure components is 100%. The percentage of a specific compound decreases linearly with increasing distance from this corner (Fig. 14.4). From a given point, the fraction of each of the three components in the composition can be determined. Depending on where one is inside the triangle, all the three components of the mixture are present together in varying concentrations or proportions.

14.7 Utility of *M. oleifera* in Commercial Biopesticides

The farmers and consumers are willing to get organic agricultural produce from organic farming especially through increased utilization of biopesticides. The global attention is attracted by biopesticides as compatible and safer approach to control pest populations like plant pathogens and insects without causing negative impacts to human health and the environment. The biopesticides are formulations that are natural in origin and functionally similar in efficacy to naturally occurring ingredients which are used to mitigate pests primarily through a non-toxic mechanism and in an eco-friendly manner.

The destructive activities of plant pathogens (fungi, bacteria, nematodes, etc.), insects and weeds have plagued agriculture from time immemorial, and this leads to a drastic decrease in yields (Mazid et al. 2011). Management of these pests to increase food security in order to meet the needs of increasing population is imperative, and this should be done in such a way that no damage is done to human health, public goods and the environment (Bastiaans et al. 2008). Over past half a decade, crop protection against pests depends solemnly on chemical pesticides and new legislations on chemical usage, and the evolution of resistance in pest populations has resulted in their declining availability. Besides that, the use of synthetic pesticides is significantly becoming more difficult due to their hazardous and toxic effects on the nontarget biodiversity and the environment (Meng et al. 2010; Marei et al. 2012; Yadav et al. 2013). Strategies used nowadays to overcome this problem include conventional plant breeding, which is practically achieved by crossing plants with desired qualities, and genetic engineering, which aims at obtaining transgenic plants expressing defined features (Qaim 2010; Zhang et al. 2010; Wang et al. 2014).

In the light of recent scientific developments throughout the world, scientists are investigating plants with antimicrobial properties, because of their low toxicity, potent medicinal activities and economic viability, in comparison with the synthetic ones (Janmeda et al. 2011). Regarding this issue, many compounds have been tested and their effectiveness confirmed against phytopathogens. Rotenone, nicotine, pyrethrum and saponins are compounds of plant origin used in the preparation of plant-based biopesticides. Typical example of plant-based biopesticide is that of Milsana® and Regalia® formulated from anthraquinone-containing extracts of giant knotweed (*Reynoutria sachalinensis*). Both biopesticides are commercially developed and marketed by Marrone Bio Innovations Inc. sold as Milsana® and Regalia®. Both have antifungal and antibacterial properties against various fungal and bacterial pathogens. In addition, they also serve as plant defence inducers and act in the accumulation of fungistatic phenolic compounds in the plants. Furthermore, research activities were geared towards understanding of proteins with antifungal properties to be used in the production of variety of crop resistant to pathogens through genetic engineering (Batista et al. 2014; Pinto et al. 2015). Today, several researches have shown the effectiveness of some proteins with antifungal activity on various susceptible under different growing conditions (Lacerda et al. 2014).

Moringa oleifera Lamarck (*Syn Moringa pterygosperma* Gaertn), is among such plants with several applications especially in the field of medicine and industry (Salem and Makkar 2009; Kumar et al. 2010). *Mo*-CBP_3 is a protein with binding property to the fungal cell wall chitin isolated from *M. oleifera* seeds that have as low as 18.0 kDa molecular weight and potent broad-spectrum antifungal efficacy against important fungal pathogens (Gifoni et al. 2012). It is a protein that performs antifungal activity at different pH and temperature (Batista et al. 2014). It is a promising bio-active compound that can be explored to confer resistance against phytopathogenic fungi to nutritionally and economically important crops (Pinto et al. 2015).

To develop appropriate commercial production and formulation protocols (Spadaro et al. 2010) as well as to minimize the potential use of antimicrobial biocontrol agents, a better understanding of the mechanisms of action is necessary. It is essential to understand how the pathogens' targeted system normally works in order to understand how the biopesticides functions. Finally, *M. oleifera* has provided sources for antimicrobial compounds, as plant-based drug and biopesticide that have made large contributions in human health and plant disease management. The researches on active compounds from the plant form a new source of green remedy against fungal disease in food products and had always been of great interest to plant pathologists. To this end *M. oleifera* could be a useful alternative for fungal plant disease management and can be incorporated as active compound to develop plant-based biopesticides.

14.8 Conclusions and Future prospects

Plant-based control of plant diseases is eco-friendly and an essential component of disease management. In this regard, *Moringa oleifera* contains diverse phytochemical constituents that form a valuable source of new and biologically active compounds possessing antimicrobial property. Most of these compounds are available, less expensive, safe to the environment, less risky of developing resistance in pests and pathogens and pest resurgence and less harmful to biodiversity. Scientists today are investigating for plant products that will serve as a novel chemotherapeutants against plant pathogens. It will be more scientific to standardize techniques of extraction, in vitro and in vivo antimicrobial efficacy testing so that the search for new biologically active compounds could be more systematic and results interpretation would be facilitated. Therefore, there is a need for crucial collaboration with plant pathologists and microbiologists to make complete development of an interesting lead compound into an exploitable product. The results from published research studies to date with *M. oleifera* are very promising. All these findings and observations bring further evidences that established the effectiveness and mechanisms of actions of different morphological parts of *M. oleifera* which have the potential of becoming a powerful and safe means of disease management instead of long-standing chemical pesticides.

References

Abdulkadir IS, Nasir IA, Sofowora A, Yahaya F, Ahmad AA (2015) Phytochemical screening and antimicrobial activities of ethanolic extracts of *Moringa oleifera* Lam on isolate of some pathogens. J App Pharm 7:203–209

Abhishek T, Sharma N, Sharma V, Afroz AA (2013) Review on conventional and non-conventional methods to manage postharvest diseases of perishables. Researcher 5:6–19

Adandonon A, Aveling TAS, Labuschagne N, Tamo M (2006) Biocontrol agents in combination with *Moringa oleifera* extract for integrated control of *Sclerotium* caused cowpea damping-off and stem rot. Eur J Plant Pathol 115:409–418

Aisha SB, Kutama AS, Kabir S, Paul AT (2016) Phytochemical screening and antifungal activity of *Moringa Oleifera* on some selected fungi in Dutse, Jigawa State. Global Adv Res J Agric Sci 5:243–248

Ajibade TO, Arowolo R, Olayemi FO (2013) Phytochemical screening and toxicity studies on the methanol extract of the seeds of *Moringa oleifera*. J Compl Integr Med 10:11–16

Alhakmani F, Kumar S, Khan SA (2013) Estimation of total phenolic content, *in–vitro* antioxidant and anti–inflammatory activity of flowers of *Moringa oleifera*. Asian Pac J Trop Biomed 3:623–627

Angajala G, Ramya R, Subashini R (2014) *In-vitro* antiinflammatory and mosquito larvicidal efficacy of nickel nanoparticles phytofabricated from aqueous leaf extracts of *Aegle marmelos* Correa. Acta Tropica 135:19–26

Anwar F, Latif S, Ashraf M, Gilani AH (2007) *Moringa oleifera*: a food plant with multiple medicinal uses. Phytother Res 21:17–25

Araujo LCC, Aguiar JS, Napoleao TH, Mota FVB, Barros ALS Moura MC (2013) Evaluation of cytotoxic and anti-inflammatory activities of extracts and lectins from *Moringa oleifera* seeds. PLoS One 8:e0081973

Arora DS, Onsare JG (2014) *In vitro* antimicrobial evaluation and phytoconstituents of *Moringa oleifera* pod husks. Ind Crops Prod 52:125–135

Arora DS, Onsare JM, Kuar H (2013) Bioprospecting of *Moringa* (Moringaceae): microbiological perspective. J Pharmacog Phytochem 1:193–215

Arowora KA, Adetunji CO (2014) Antifungal effects of crude extracts of *Moringa oleifera* on *Aspergillus niger* V. *Tieghem* associated with post-harvest rot of onion bulb. SMU Med J 1:214–223

Asiedu-Gyekye IJ, Frimpong-Manso S, Awortwe C, Antwi DA, Nyarko AK (2014) Micro-and macroelemental composition and safety evaluation of the nutraceutical *Moringa oleifera* leaves. J Toxicol 2014:786979. https://doi.org/10.1155/2014/786979

Atawodi SE, Atawodi JC, Idakwo GA (2010) Evaluation of the polyphenol content and antioxidant properties of methanol extracts of the leaves, stem, and root barks of *Moringa oleifera* Lam. J Med Food 13:710–716

Baris O, Gulluce M, Sahin F, Ozer H, Kilic H, Ozkan H, Sokmen M, Ozbek T (2006) Biological activities of the essential oil and methanol extract of *Achillea Biebersteinii* Afan. (Asteraceae). Turk J Biol 30:65–73

Bastiaans L, Paolini R, Baumann DT (2008) Focus on ecological weed management: what is hindering adoption. Weed Res 48:481–491

Batista AB, Oliveira JT, Gifoni JM, Pereira ML, Almeida MG, Gomes VM, Lopes JLS (2014) New insights into the structure and mode of action of Mo-CBP3, an antifungal chitin-binding protein of *Moringa oleifera* seeds. PloS One 9:0111427

Belay K, Sisay M (2014) Phytochemical constituents and physicochemical properties of medicinal plant (*Moringa Oleifera*) around bule hora. Int J Innov Sci Res 8:88–98

Bhattacharjee R, Dey U (2014) An overview of fungal and bacterial biopesticides to control plant pathogens/diseases. Afr J Microbiol Res 8:1749–1762

Broekaert WF, Marien W, Terras FR, De Bolle MF, Proost P, Van Damme J, Rees SB (1992) Antimicrobial peptides from *Amaranthus caudatus* seeds with sequence homology to the cysteine/glycine-rich domain of chitin-binding proteins. Biochemistry 31:4308–4314

Caceres A, Saravia A, Rizzo S, Zabala L, De Leon E, Nave F (1992) Pharmacologic properties of *Moringa oleifera*. 2: screening for antispasmodic, antiinflammatory and diuretic activity. J Ethnopharmacol 36:233–237

Chaudhary K, Chaurasia S (2017) Neutraceutical properties of *Moringa oleifera*: a review. Eur J Pharma Med Res 4:646–655

Chetia B, Gogoi S (2011) Antibacterial activity of the methanolic extract of stem bark of *Spondias pinnata*, *Moringa oleifera* and *Alstonia scholaris*. Asian Trad Med 6:163–167

Choi H, Cho J, Jin Q, Woo ER, Lee DG (2012) Antifungal property of dihydrodehydrodiconiferyl alcohol 9′-O-β-d-glucoside and its pore-forming action in plasma membrane of *Candida albicans*. BBA-Biomembrane 1818:1648–1655

Chuang PH, Lee CW, Chou JY Murugan M, Shieh BJ, Chen HM (2007) Anti-fungal activity of crude extracts and essential oil of *Moringa oleifera* Lam. Bioresour Technol 98:232–236

Cordell GA, Colvard MD (2005) Some thoughts on the future of ethnopharmacology. J Ethnopharmacol 100:5–14

Cos P, Hermans N, De Bruyne T, Apers S, Sindambiwe JB, Berghe DV, Vlietinck AJ (2002) Further evaluation of Rwandan medicinal plant extracts for their antimicrobial and antiviral activities. J Ethnopharmacol 79:155–163

Das K, Tiwari RKS, Shrivastava DK (2010) Techniques for evaluation of medicinal plant products as antimicrobial agent: current methods and future trends. J Med Plants Res 4:104–111

Dewangan G, Koley KM, Vadlamudi VP, Mishra A, Poddar A, Hirpurkar SD (2010) Antibacterial activity of *Moringa oleifera* (drumstick) root bark. J Chem Pharma Res 2:424–428

Duan XH, Jiang R, Wen YJ, Bin JH (2013) Some 2S albumin from peanut seeds exhibits inhibitory activity against *Aspergillus flavus*. Plant Physiol Biochem 66:84–90

Dwivedi SK, Enespa A (2012) Effectiveness of extract of some medical plants against soil borne fusaria causing diseases on *Lycopersicon esculentum* and *Solanum melongena*. Int J Pharm Bio Sci 3:1171–1180

EL-Mohamedy RSR, Abdalla AM (2014) Evaluation of antifungal activity of *Moringa oleifera* extracts as natural fungicide against some plant pathogenic fungi *in-vitro*. J Agric Technol 10:963–982

Ergene A, Guler P, Tan S, Mirici S, Hamzaoglu E, Duran A (2006) Antimicrobial and antifungal activity of *Heracleum sphondylium* subsp. artivinense. Afr J Biotechnol 5:1087–1089

Fahey JW (2005) *Moringa oleifera*: a review of the medical evidence for its nutritional therapeutic, and prophylactic properties. Part 1. Trees for Life J 1:1–15

Faizi S, Siddiqui BS, Saleem R, Siddiqui S, Aftab K, Gilani AUH (1994) Isolation and structure elucidation of new nitrile and mustard oil glycosides from *Moringa oleifera* and their effect on blood pressure. J Nat Prod 57:1256–1261

Freire JEC, Vasconcelos IM, Moreno FBMB, Batista AB, Lobo MDP, Pereira ML (2015) *Mo*-CBP$_3$, an antifungal chitin-binding protein from *Moringa oleifera* seeds, is a member of the 2S albumin family. PLoS One 10:e0119871

Fuglie LJ (1999) The miracle tree: *Moringa oleifera,* natural nutrition for the tropics. Church World Service, Dhakar, p 68

Gifoni JM, Oliveira JT, Oliveira HD, Batista AB, Pereira ML, Gomes AS, Vasconcelos IM (2012) A novel chitin-binding protein from *Moringa oleifera* seed with potential for plant disease control. Peptide Sci 98:406–415

Green RJ (2004) Antioxidant activity of Peanut plant tissues. Masters Thesis. North Carolina State University, USA

Green JM, Beestman GB (2007) Recently patented and commercialized formulation and adjuvant technology. Crop Prot 26:320–327

Gulluce M, Aslan A, Sokmen M, Sahin F, Adiguzel A, Agar G, Sokmen A (2006) Screening the antioxidant and antimicrobial properties of the lichens *Parmelia saxatilis*, *Platismatia glauca, Ramalina pollinaria, Ramalina polymorpha* and *Umbilicaria nylanderiana*. Phytomedicine 13:515–521

Gurjar MS, Ali S, Akhtar M, Singh KS (2012) Efficacy of plant extracts in plant disease management. Agr Sci 3:425–433

Huang HC, Wang HF, Yih KH, Chang LZ, Chang TM (2012) Dual bioactivities of essential oil extracted from the leaves of Artemisia argyi as an antimelanogenic versus antioxidant agent and chemical composition analysis by GC/MS. Int J Mol Sci 13:14679–14697

Iqbal MA (2015) Improving germination and seedling vigour of cowpea (*Vigna unguiculata* L. Walp) with priming techniques. Am-Eur J Agric Environ Sci 15:265–270

Jabeen R, Shahid M, Jamil A, Ashraf M (2008) Microscopic evaluation of the antimicrobial activity of seed extracts of *Moringa oleifera*. Pak J Bot 40:1349–1358

Jamil A, Shahid M, Khan MM, Ashraf M (2007) Screening of some medicinal plants for isolation of antifungal proteins and peptides. Pak J Bot 39:211–221

Janmeda P, Sharma V, Singh L, Paliwal R, Sharma S, Yadav S, Sharma S (2011) Chemopreventive effect of hydroethanolic extract of *Euphorbia neriifolia* leaves against DENA-induced renal carcinogenesis in mice. Asian Pac J Cancer Prev 12:677–683

Jenssen H, Hamill P, Hancock RE (2006) Peptide antimicrobial agents. Clin Microbiol Rev 19:491–511

Kadhim EJ, AL-Shammaa DA (2014) Phytochemical characterisation using GC-MS analysis of methanolic extract of *Moringa oleifera* (family *Moringaceae*) plant cultivated in Iraq. Chem Mater Res 6:9–26

Krishnamurthy PT, Vardarajalu A, Wadhwani A, Patel V (2015) Identification and characterization of a potent anticancer fraction from the leaf extracts of *Moringa oleifera Lam*. Indian J Exp Biol 53:98–103

Kumar PS, Mishra D, Ghosh G, Panda CS (2010) Medicinal uses and pharmacological properties of *Moringa oleifera*. Int J Phytomed 2:210–216

Lacerda AF, Vasconcelos EAR, Pelegrini PB, Grossi-de-Sa MF (2014) Antifungal defensins and their role in plant defense. Front Microbiol 5:116. https://doi.org/10.3389/fmicb.2014.00116

Luqman S, Srivastava S, Kumar R, Maurya AK, Chanda D (2012) Experimental assessment of *Moringa oleifera* Leaf and fruit for its antistress, antioxidant, and scavenging potential using *in vitro* and *in vivo* assays. Evidence-Based Compl Alternat Med 2012:519084. https://doi.org/10.1155/2012/519084

Marei GIK, Rasoul MAA, Abdelgaleil SA (2012) Comparative antifungal activities and biochemical effects of monoterpenes on plant pathogenic fungi. Pestic Biochem Physiol 103:56–61

Marino M, Bersani CCG (2001) Impedance measurement to study antimicrobial activity of essential oils from Lamiaceae and Compositae. Int J Food Microbiol 67:187–195

Matshediso PG, Cukrowska E, Chimuka L (2015) Development of pressurised hot water extraction (PHWE) for essential compounds from *Moringa oleifera* leaf extracts. Food Chem 172:423–427

Mazid S, Kalita JC, Rajkhowa RC (2011) A review on the use of biopesticides in insect pest management. Int J Sci Adv Technol 1:169–178

Mbikay M (2012) Therapeutic potential of *Moringa oleifera* leaves in chronic hyperglycemia and dyslipidemia: a review. Front Pharmacol 3:24. https://doi.org/10.3389/fphar.2012.00024

McClements DJ, Decker EA (2000) Lipid oxidation in oil-in-water emulsions: impact of molecular environment on chemical reactions in heterogeneous food systems. J Food Sci 65:1270–1280

Mehmood T (2015) Optimization of the canola oil based vitamin E nanoemulsions stabilized by food grade mixed surfactants using response surface methodology. Food Chem 183:1–7

Meng X, Yang L, Kennedy JF, Tian S (2010) Effects of chitosan and oligochitosan on growth of two fungal pathogens and physiological properties in pear fruit. Carbohyd Polym 81:70–75

Mishra RK, Soni GC, Mishra RP (2014) A review article: on nanoemulsion. World J Pharm Pharm Sci 3:258–274

Moghimipour E, Handali S (2015) Saponin: properties, methods of evaluation and applications. Ann Res Rev Biol 5:207–220

Mohammed A, Mohammad SH, Mohammad EHC, Mohsinul H (2012) *In vitro* antimicrobial activity of methanolic extract of *Moringa oleifera* Lam. fruits. J Pharmacog Phytochem 4:94–98

Moraes MN, Zabot GL, Prado JM, Meireles MAA (2013) Obtaining antioxidants from botanic matrices applying novel extraction techniques. Food Pub Health 3:195–214

Morais JKS, Gomes VM, Oliveira JTA, Santos IS, Da Cunha M, Oliveira HD, Vasconcelos IM (2010) Soybean toxin (SBTX), a protein from soybeans that inhibits the life cycle of plant and human pathogenic fungi. J Agric Food Chem 58:10356–10363

Moyo B, Masika PJ, Muchenge V (2012) Antimicrobial activities of *Moringa oleifera* Lam leaf extracts. Afr J Biotechnol 11:2797–2802

Mustafa MA, Ali A, Manickam S, Siddiqui Y (2014) Ultrasound-assisted chitosan–surfactant nanostructure assemblies: towards maintaining postharvest quality of tomatoes. Food Biopro Technol 7:2102–2111

Myers D (2005) Surfactant science and technology, 3rd edn. Wiley, New Jersey, p 400

Ncube N, Afolayan SAJ, Okoh AI (2008) Assessment techniques of antimicrobial properties of natural compounds of plant origin: current methods and future trends. Afr J Biotechnol 7:1797–1806

Ocsoy I, Paret ML, Ocsoy MA, Kunwar S, Chen T, You M, Tan W (2013) Nanotechnology in plant disease management: DNA-directed silver nanoparticles on graphene oxide as an antibacterial against *Xanthomonas perforans*. ACS Nano 7:8972–8980

Orole OO, Adejumo TO, Orole RT (2016) Antifungal activity of nine medicinal plants against *Aspergillus* species from cocoa beans (*Theobroma cacao*). Int J Agr Ecol Res 7:1–11

Panda S, Kar A, Sharma P, Sharma A (2013) Cardioprotective potential of N, α-l-rhamnopyranosyl vincosamide, an indole alkaloid, isolated from the leaves of *Moringa oleifera* in isoproterenol induced cardiotoxic rats: *In vivo* and *in vitro* studies. Bioorg Medi Chem Lett 23:959–962

Parekh J, Jadeja D, Chanda S (2005) Efficacy of aqueous and methanol extracts of some medicinal plants for potential antibacterial activity. Turk J Biol 29:203–210

Patwardhan B, Warude D, Pushpangadan P, Bhatt N (2005) Ayurveda and traditional Chinese medicine: a comparative overview. Evidence-Based Compl Altern Med 2:465–473. https://doi.org/10.1093/ecam/neh140

Peixoto JRO, Silva GC, Costa RA, Vieira GHF, Fonteles Filho AA, dos Fernandes Vieira RHS (2011) *In vitro* antibacterial effect of aqueous and ethanolic *Moringa* leaf extracts. Asian Pac J Trop Med 4:201–204

Pereira FSG, da Silva AMRB, Galvao CC, de Lima VF, de Assunçao Montenegro LGL, de Lima-Filho N, da Silva VL (2015) *Moringa oleifera* as sustainable source for energetic biomass. Int J Chem 7:177–185

Pinto CE, Farias DF, Carvalho AF, Oliveira JT, Pereira ML, Grangeiro TB, Vasconcelos IM (2015) Food safety assessment of an antifungal protein from *Moringa oleifera* seeds in an agricultural biotechnology perspective. Food Chem Toxicol 83:1–9

Polychniatou V, Tzia C (2014) Study of formulation and stability of co-surfactant free water-in-olive oil nano-and submicron emulsions with food grade non-ionic surfactants. J Am Oil Chem Soc 91:79–88

Prapassom P, Piyarat N, Paweena R, Chandrapatya A (2012) Effect of crude leaf extracts on *Colletotrichum gloeosporioides* (Penz.) Sacc. Pysche 2012:309046. https://doi.org/10.1155/2012/309046

Prashith KTR, Mallikarjun N, Swathi D, Nayana KV, Meera BA, Rohini TR (2010) Antibacterial and antifungal efficacy of steam distillate of *Moringa oleifera* Lam. J Pharm Sci Res 2:34–37

Qaim M (2010) Benefits of genetically modified crops for the poor: household income, nutrition, and health. New Biotechnol 27:552–557

Rahman MM, Sheikh MMI, Sharmin SA, Islam MS, Rahman MA, Rahman MM, Alam MF (2009) Antibacterial activity of leaf juice and extracts of *Moringa oleifera* Lam. against some human pathogenic bacteria. CMU J Nat Sci 8:219–227

Rai M, Jogee PS, Ingle AP (2015) Emerging nanotechnology for detection of mycotoxins in food and feed. Int J Food Sci Nutr 66:363–370

Raj AJ, Gopalakrishnan VK, Yadav SA, Dorairaj S (2014) Antimicrobial activity of *Moringa oleifera* (Lam.) root extract. J Pharm Res 4:1426–1430

Razis AFA, Ibrahim MD, Kntayya SB (2014) Health benefits of *Moringa oleifera*. Asian Pac J Cancer P15:8571–8576

Rhouma A, Ben Daoud H, Ghanmi S, Ben Salah H, Romdhane M, Demak M (2009) Antimicrobial activities of leaf extracts of *Pistacia* and *Schinus* species against some plant pathogenic fungi and bacteria. J Plant Pathol 91:339–345

Richards A, Veses V, Gow NA (2010) Vacuole dynamics in fungi. Fungal Biol Rev 24:93–105

Rocha MFG, Aguiar FLND, Brilhante RSN, Cordeiro RDA, Teixeira CEC, Castelo-Branco DDSC, Barbosa FG (2011) *Moringa oleifera* and *Vernonia* sp. extracts against *Candida albicans* and *Microsporum canis* isolates from dogs and cats and analysis of toxicity to *Artemia* sp. Cienc Rural 41:1807–1812

Rodríguez-Perez C, Gilbert-Lopez B, Mendiola JA, Quirantes-Pine R, Segura-Carretero A, Ibanez E (2016) Optimization of microwave-assisted extraction and pressurized liquid extraction of phenolic compounds from *Moringa oleifera* leaves by multiresponse surface methodology. Electrophoresis 37:1938–1946

Rodríguez-Rojo S, Varona S, Nunez M, Cocero MJ (2012) Characterization of rosemary essential oil for biodegradable emulsions. Ind Crops Prod 37:137–140

Saadabi AM, Zaid IA (2011) An *in vitro* antimicrobial activity of *Moringa oleifera* L. seed extracts against different groups of microorganisms. Aust J Basic Appl Sci 5:129–134

Sahab AF, Nawar LS (2015) Chemical and phytochemical composition of *Moringa* seed powder and Antifungal activity of seed extracts against seed borne fungi. Int J Chem Tech Res 8:686–695

Salem HB, Makkar HPS (2009) Defatted *Moringa oleifera* seed meal as a feed additive for sheep. Anim Feed Sci Technol 150:27–33

Satish S, Mohana DC, Ranhavendra MP, Raveesha KA (2007) Antifungal activity of some plant extracts against important seed borne pathogens of *Aspergillus sp*. Int J Agr Technol 3:109–119

Satish A, Kumar RP, Rakshith D, Satish S, Ahmed F (2013) Antimutagenic and antioxidant activity of *Ficus benghalensis* stem bark and *Moringa oleifera* root extract. Int J Chem Anal Sci 4:45–48

Satish A, Reddy PV, Sairam S, Ahmed F, Urooj A (2014) Antioxidative effect and DNA protecting property of *Moringa oleifera* root extracts. J Herb Spice Med Plants 20:209–220

Satyanarayana U (2005) Chromatography, 2nd edn, Biochemistry. Books and Allied Pvt Ltd, Kolkata, p 369

Schmourlo G, Mendonça-Filho RR, Alviano CS, Costa SS (2005) Screening of antifungal agents using ethanol precipitation and bioautography of medicinal and food plants. J Ethnopharmacol 96:563–568

Shafiq S, Shakeel F, Talegaonkar S, Ahmad FJ, Khar RK, Ali M (2007) Development and bioavailability assessment of ramipril nanoemulsion formulations. Eur J Pharm Biopharm 66:227–243

Sharma V, Paliwal R (2013) Isolation and characterization of saponins from *Moringa oleifera* (Moringaceae) pods. Int J Pharm Pharm Sci 5:179–183

Sharma V, Paliwal R, Janmeda P, Sharma S (2012) Chemopreventive efficacy of *Moringa oleifera* pods against 7, 12-dimethylbenz [a] anthracene induced hepatic carcinogenesis in mice. Asian Pac J Cancer P 13:2563–2569

Singh A, Van Hamme JD, Ward OP (2007) Surfactants in microbiology and biotechnology: part 2. Application aspects. Biotechnol Adv 25:99–121

Slawecki RA, Ryan EP, Young DH (2002) Novel fungitoxicity assays for inhibition of germination-associated adhesion of *Botrytis cinerea* and *Puccinia recondita* spores. Appl Environ Microbiol 68:597–660

Snyder LR, Kirkland JJ, Glajch JL (2012) Practical HPLC method development, 2nd edn. Wiley, New York, p 542

Sogvar OB, Saba MK, Emamifar A (2016) *Aloe vera* and ascorbic acid coatings maintain postharvest quality and reduce microbial load of strawberry fruit. Postharvest Biol Technol 114:29–35

Spadaro D, Ciavorella A, Dianpeng Z, Garibaldi A, Gullino ML (2010) Effect of culture media and pH on the biomass production and biocontrol efficacy of a Metschnikowia pulcherrima strain to be used as a biofungicide for postharvest disease control. Can J Microbiol 56:128–137

Stohs SJ, Hartman MJ (2015) Review of the safety and efficacy of *Moringa oleifera*. Phytother Res 29:796–804

Tadros TF, Izquierdo P, Esquena J, Solans C (2004) Formation and stability of nano-emulsions. Adv Colloid Interfac 108:303–318

Teixeira EMB, Carvalho MRB, Neves VA, Silva MA, Arantes-Pereira L (2014) Chemical characteristics and fractionation of proteins from *Moringa oleifera* Lam. leaves. Food Chem 147:51–54

Thompson MD, Thompson HJ (2010) Botanical diversity in vegetable and fruit intake: potential health benefits. In: Watson RR, Preddy VR (eds) Bioactice foods in promoting health fruits and vegetables. Academic, San Francisco, pp 3–17

Tijjani A (2018) Control of gray Mold rot disease of tomato caused by Botrytis cinerea with emulsion formulations from *Moringa oleifera* crude extracts. Ph.D thesis. University Putra Malaysia, Selangor, Malaysia

Tijjani A, Adebitan SA, Gurama AU, Aliyu M, Dawakiji AY, Haruna SG, Mohammed NA (2013) Efficacy of some botanicals for control of wet rot disease of mechanically injured sweet potato caused by *Rhizopus stolonifer* in Bauchi State. Int J Sci Res Pub 3:1–10

Tijjani A, Adebitan SA, Gurama AU, Haruna SG, Safiya T (2014) Effect of some selected plant extracts on *Aspergillus flavus*, a causal agent of fruit rot disease of tomato (*Solanum lycopersicum*) in Bauchi State. Int J Biosci 4:244–252

Van Hamme JD, Singh A, Ward OP (2006) Physiological aspects: part 1 in a series of papers devoted to surfactants in microbiology and biotechnology. Biotechnol Adv 24:604–620

Verma AR, Vijayakumar M, Mathela CS, Rao CV (2009) *In vitro* and *in vivo* antioxidant properties of different fractions of *Moringa oleifera* leaves. Food Chem Toxicol 47:2196–2201

Verma VC, Kharwar RN, Gange AC (2010) Biosynthesis of antimicrobial silver nanoparticles by the endophytic fungus *Aspergillus clavatus*. Nanomed 5:33–40

Villas-Boas SG, Mas S, Akesson M, Smedsgaard J, Nielsen J (2005) Mass spectrometry in metabolome analysis. Mass Spectrom Rev 24:613–646

Vinoth B, Manivasagaperumal R, Balamurugan S (2012) Phytochemical analysis and antibacterial activity of *Moringa oleifera* Lam. Int J Res Biol Sci 2:98–102

Vongsak B, Sithisam P, Gritsanapan W (2014) Simultaneous HPLC quantitative analysis of active compounds in leaves of *Moringa oleifera* Lam. J Chromatogr Sci 52:641–645

Wang CJ, Liu ZQ (2007) Foliar uptake of pesticides-present status and future challenge. Pest Biochem Phys 87:1–8

Wang Y, Chang L, Zhao X, Meng X, Liu Y (2012) Gas chromatography-mass spectrometry analysis on compounds in volatile oils extracted from Yuan Zhi (*Radix Polygalae*) and Shi Chang Pu (*Acorus Tatarinowii*) by supercritical CO_2. J Tradit Chin Med 32:459–464

Wang B, Wang RR, Cui ZH, Bi WL, Li JW, Li BQ, Wang QC (2014) Potential applications of cryogenic technologies to plant genetic improvement and pathogen eradication. Biotechnol Adv 32:583–595

Waterman C, Cheng DM, Rojas-Silva P, Poulev A, Dreifus J, Lila MA, Raskin I (2014) Stable, water extractable isothiocyanates from *Moringa oleifera* leaves attenuate inflammation *in vitro*. Phytochem 103:114–122

Xu YX, Hanna MA, Josiah SJ (2007) Hybrid hazelnut oil characteristics and its potential oleochemical application. Indust Crops Prod 26:69–76

Xu L, Zhu H, Ozkan HE, Bagley WE, Krause CR (2011) Droplet evaporation and spread on waxy and hairy leaves associated with type and concentration of adjuvants. Pest Manage Sci 67:842–851

Yadav S, Tomar AK, Yadav RN, Yadav S (2013) Screening of antifungal proteins from plants of Cucurbitaceae family against *Fusarium oxysporum*: potential as biofungicides. Int Res J Environ Sci 2:91–96

Yao Q, Wu CF, Luo P, Xiang XC, Liu JJ, Mou L, Bao JK (2010) A new chitin-binding lectin from rhizome of *Setcreasea purpurea* with antifungal, antiviral and apoptosis-inducing activities. Process Biochem 45:1477–1485

Yassa HD, Tohamy AF (2014) Extract of *Moringa oleifera* leaves ameliorates streptozotocin-induced *Diabetes mellitus* in adult rats. Acta Histochem 116:844–854

Zahid N, Ali A, Manickam S, Siddiqui Y, Maqbool M (2012) Potential of chitosan-loaded nanoemulsions to control different *Colletotrichum* spp. and maintain quality of tropical fruits during cold storage. J Appl Microbiol 113:925–939

Zhang Z, Duan Z, Xu P, Li G (2010) Food security of China: the past, present and future. Plant omics 3:183–189

Zhao S, Zhang D (2013) Supercritical fluid extraction and characterisation of *Moringa oleifera* leaves oil. Sep Purif Technol 118:497–502

Zottich U, Da Cunha M, Carvalho AO, Dias GB, Silva NC (2011) Purification, biochemical characterization and antifungal activity of a new lipid transfer protein (LTP) from *Coffea canephora* seeds with alfa-amylase inhibitor properties. BBA-Gen Subj 1810:375–383

Zvinorova PI, Lekhanya L, Erlwanger K, Chivandi E (2015) Dietary effects of *Moringa oleifera* leaf powder on growth, gastrointestinal morphometry and blood and liver metabolites in Sprague Dawley rats. J Anim Physiol Anim Nutri 99:21–28

Natural Compound from Genus *Brassica* and Their Therapeutic Activities

15

Nida Idrees, Baby Tabassum, Robeena Sarah, and Mohd Kamil Hussain

Contents

Abstract

The genus *Brassica* occupies an important position in the family Brassicaceae because of its oil, food, fibres, minerals, vitamins, soluble sugars, phytochemicals like carotenoids, glucosinolates and phenolic compounds. Phenolic compounds have the characteristics of protection against various diseases, such as cancer and cardiovascular problems. Several preclinical studies have described flavonoids, such as quercetin, kaempferol and glucosinolate possess multiple pharmacological and biological activities, including antioxidant, anti-inflammatory, anticancer, cardioprotective, neuroprotective, anti-osteoporotic estrogenic/antiestrogenic,

Nida Idrees and Baby Tabassum have equally contributed for this chapter.

N. Idrees (✉) · B. Tabassum · R. Sarah
Toxicology Laboratory, Department of Zoology, Government Raza P.G. College, Rampur, Uttar Pradesh, India
e-mail: nidaidrees999@gmail.com; dr.btabassum@gmail.com

M. K. Hussain
Department of Chemistry, Government Raza P.G. College, Rampur, Uttar Pradesh, India

M. S. Akhtar et al. (eds.), *Natural Bio-active Compounds*,
https://doi.org/10.1007/978-981-13-7154-7_15

antidiabetic, antiallergic and antimicrobial activities. The presence of high concentration of these phytochemicals showed several epidemiological benefits, prevented the human body against reactive oxygen species damages and decelerated the progress of many chronic diseases. They are also effective in the oxidation of low-density lipoproteins (LDL), which play an important role in atherosclerosis. Thus, the aim of this chapter is to focus on the therapeutic activities of natural bio-active compounds extracted from genus *Brassica*.

Keywords
Brassica · Human health · Lipoproteins · Phytochemicals · Phytoremediation

15.1 Introduction

Plant-based food contains high rich nutritional value and provides desirable health benefits. Besides nutritional effects, they accommodate the significant amount of bio-active compounds which are beneficial for animals as well as human beings. In the past few decades, edible plants have grabbed attention towards researchers due to the presence of secondary metabolites or phytochemicals. In present days, there is an increase of interest in such type of healthy diet because of their beneficial effects of such phytochemicals. Several epidemiological studies have showed that consumption of plant products in maximum quantity is accomplice with the reduced risk of various chronic diseases, such as atherosclerosis and cancer (Law and Morris 1998; Hashimoto et al. 2002; Kris-Etherton et al. 2002; Gundgaard et al. 2003; Gosslau and Chen 2004). These reports indicated that cruciferous vegetables have natural antioxidants in considerable amount and the presence of high levels of tocopherols, ascorbic acid and carotenoids make the herb more beneficial to health. Brassica crops gained popularity, and its consumption is increasing due to the presence of these beneficial secondary metabolites which is directly related to the reduction of the risk of various chronic diseases including cardiovascular diseases and cancer.

Genus *Brassica* belongs to cruciferous family which is known for a variety of vegetables and oilseed plants such as *Brassica oleracea*, *Brassica napus*, *Brassica rapa*, *Brassica campestris* etc. These vegetables are an important part of the human diet worldwide, are consumed by people all over the world and are considered important food and oilseed crops in China, Japan, India and European countries. Besides consumed as food, they exhibit several pharmacological properties such as anticancer and antioxidant activities (Verhoeven et al. 1997; Cohen et al. 2000; Chu et al. 2002). These herbs carry a large group of glucosinolates which have low antioxidant activity (Plumb et al. 1996), but the hydrolysis of these products resulted in the protection against cancer (Paolini 1998; Keum et al. 2004). In Ayurvedic tradition, mustard is considered as a valuable herb which has therapeutic benefits. Mustard and its oil have been used in relief of joint pain, swelling, fever, cough colds and in the cleaning of the cranial cavity. It is also used in the treatment of various skin diseases and wounds (Manohar et al. 2009). Consumption of *Brassica* vegetables induces glutathione S-transferase in humans and increase the capacity for protection against various cancers (Barcelo et al. 1996). Besides therapeutic uses, a *Brassica* herb is used as phytoremediator against heavy metals. It can be used as a natural, non-toxic and cheap source against heavy metal pollution mainly for cadmium toxicity. Thus, the aim of this chapter is to focus on the therapeutic and pharmacological activities of natural bio-active compounds extracted from genus *Brassica*.

15.2 Economically Important *Brassica* Species

The family Brassicaceae is currently composed of more than 3700 species and 330 genera (Warwick et al. 2006) and is among the ten most economically important plant families. Some important members of this family are cultivated for vegetables, oils, condiments, fodder and are listed in tabular form (Table 15.1).

15.3 Phytochemicals Found in *Brassica*

The considerable attention has been directed towards the identification of natural products that may be used for human consumption regarding health promotion and disease prevention. Glucosinolates, phenolic compounds, vitamins, carotenoids, fibres, soluble sugars and minerals are phytochemicals found in genus *Brassica*. Phytochemicals or secondary metabolites present in the members of Brassicaceae family are directly related to curative or preventive properties against various diseases. Due to the presence of potential antioxidant phenolic compounds such as quercetin, kaempferol and vitamin C, they are associated with health improvements. These vegetables also contain a high amount of lipid-soluble antioxidants such as vitamin A, and carotenoids are assumed to be 20% of the total antioxidative properties (Podsedek 2007).

Table 15.1 Economic importance of *Brassica* species

Common name	Botanical name	Usages
Turnip rape	*B. rapa* spp. *oleifera*	Oilseed
Turnip	*B. rapa* spp. *rapifera*	Fodder, vegetable (root)
Black mustard	*B. nigra*	Oilseed, condiment
Yellow mustard	*B. juncea*, *B. campestris*	Oilseed
White mustard	*Sinapis alba/B. alba*	Oilseed
Rapeseed	*B. napus* var. *toria*	Oilseed
Kale	*B. oleracea* var. *acephala*	Vegetable, fodder (leaves)
Cabbage	*B. oleracea* var. *capitata*	Vegetable (head)
Savoy cabbage	*B. oleracea* var. *sabauda*	Vegetable (terminal buds)
Brussels sprouts	*B. oleracea* var. *gemmifera*	Vegetable (head)
Kohlrabi	*B. oleracea* var. *gongylodes*	Vegetable, fodder (stem)
Cauliflower	*B. oleracea* var. *botrytis*	Vegetable (inflorescence)
Broccoli	*B. oleracea* var. *italic*	Vegetable (inflorescence)
Branching bush kale	*B. oleracea* var. *fruticosa*	Fodder (leaves)

15.3.1 Glucosinolates

Glucosinolates (1) are a class of amino acid-derived, sulfur-rich secondary metabolites found in the order Brassicales, which includes the scientifically and economically important genera of *Brassica*. More than 100 glucosinolates have been reported from 16 plant families. They broke down enzymatically by myrosinase, mainly into isothiocyanates, cyanides and thiocyanates which are the main bio-active components responsible for pharmacological effects (Chew 1988). Glucosinolates contain a β-D-thioglucose group linked to a sulfonated aldoxime moiety and a variable side chain which may be aliphatic, aromatic or indolyl. They are water-soluble compounds as a result of their ionised sulphate, and hydrophilic thioglucose moieties, these compounds cannot be easily separated and purified (Tianxin et al. 2012). Glucosinolates have many pharmacological properties like antibacterial, antifungal, anti-inflammatory, antimutagenic, anticancer, bioherbicidal, etc. Various glucosinolates having aliphatic (Fig. 15.1), benzyl, phenethyl and indolyl chain (Fig. 15.2) have been reported in mustard oilseeds of *Brassica* and related species (Hansen et al. 1995; Bellostas et al. 2007).

15.3.1.1 Aliphatic Glucosinolates

These glucosinolates are derived from methionine, isoleucine, leucine or valine and are mostly present in *Brassica* genus. Important glucosinolates are sinigrin (1a), gluconapin (1b), glucobrassicanapin (1c), progoitrin (1d), epi-progoitrin (1e), napoleiferin (1f), glucoibervirin (1g), glucoiberin (1h), glucoerucin (1i), glucoalyssin (1j) and glucoraphanin (1k).

15.3.1.2 Aromatic Glucosinolates

These glucosinolates are derived from phenylalanine or tyrosine glucotropaeolin (1l), gluconasturtiin (1m), glucosinalbin (1n) and glucolimnanthin (1o).

15.3.1.3 Indole Glucosinolates

Indole glucosinolates are derived from tryptophan; these are exclusively found in the vegetative parts of *Brassica* species, but they have not been reported in oilseeds. These glucosinolates include glucobrassicin (1p), neoglucobrassicin (1q), 4-hydroxy glucobrassicin (1r) and 4-methoxyglucobrassicin (1s) (Fig. 15.2) (Bergmann 1970; Josefsson 1970).

15.3.1.4 Sinigrin

Various glucosinolates such as sinigrin, progoitrin, glucoerucin, glucosinalbin, epi-progoitrin, gluconasturtiin and glucolimnanthin have been reported in different species of oilseeds *Brassica* (Ugolini et al. 2008; Table 15.2). Here, the biological and pharmacological properties of sinigrin have been highlighted only. Sinigrin (1a) is a major glucosinolate, which is extensively found in the members of Brassicaceae family, such as yellow mustard (*Brassica juncea* and *Brassica campestris*), black mustard (*Brassica nigra*), broccoli and brussels sprouts (Fig. 15.3; Table 15.3). Myrosinase-mediated enzymatic degradation of sinigrin leads to the formation of

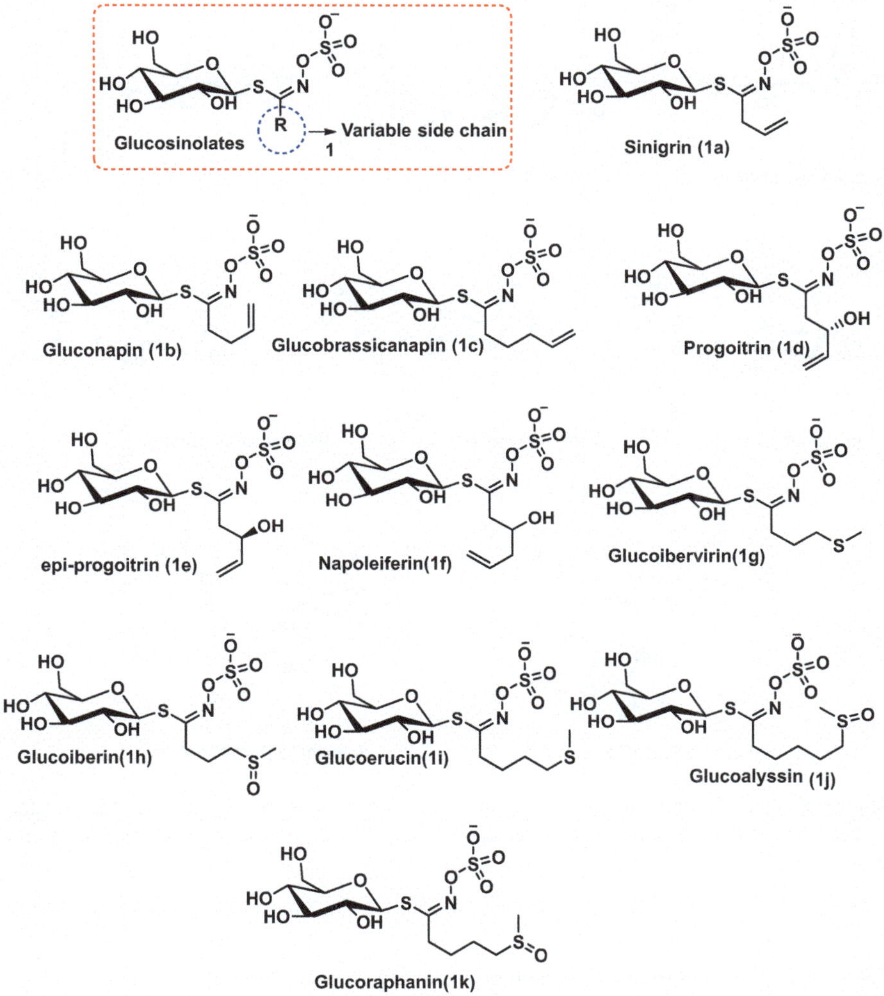

Fig. 15.1 Structure of some important glucosinolates containing aliphatic side chain, isolated from oilseed *Brassica* and related species. (Adopted from Hansen et al. 1995; Bellostas et al. 2007)

allyl isothiocyanate which is an important cancer chemopreventive agent; it prevents DNA damage caused by carcinogens; in addition to chemopreventive properties, sinigrin exhibits multiple biological effects and therapeutic properties such as antioxidant, anti-inflammatory, antifungal and antibacterial activities (Nomura et al. 2005; Mazumder et al. 2015, 2016). In an orthotopic rat bladder cancer model, allyl isothiocyanate (AITC) produced from sinigrin through myrosinase catalysed reaction inhibited bladder cancer growth and blocked muscle invasion (Bhattacharya et al. 2010). Allyl isothiocyanate generated from sinigrin exhibited an antioxidative effect in vivo through suppressing nitric oxide (NO) production (Ippoushi et al. 2010).

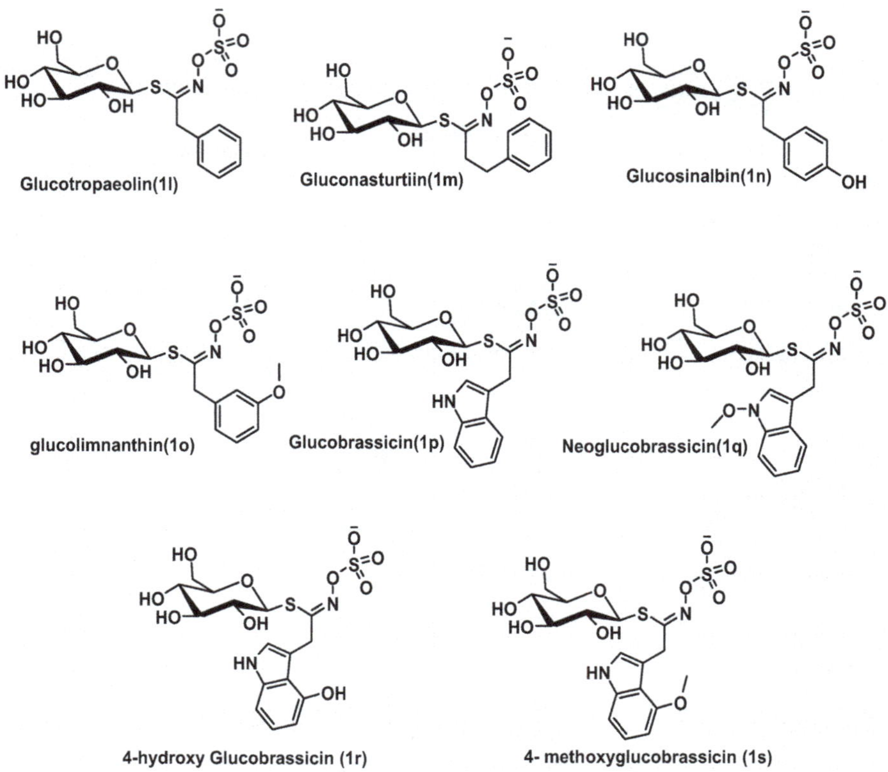

Fig. 15.2 Structure of some important glucosinolates having aromatic and indolyl side chain, isolated from oilseed *Brassica* and related species. (Adopted from Hansen et al. 1995; Bellostas et al. 2007)

Table 15.2 Major glucosinolates present in oilseed *Brassica* and related species

Common name	Botanical name	Major glucosinolates
Indian mustard	*Brassica juncea*	Sinigrin (**1a**)
Karan rai	*Brassica carinata*	Sinigrin
Black mustard	*Brassica nigra*	Sinigrin
Gobhi sarson	*Brassica napus*	Progoitrin (**1d**)
Taramira	*Eruca sativa*	Glucoerucin (**1i**)
Yellow/white mustard	*Sinapis alba*	Glucosinalbin (**1n**)
Cabbage	*Brassica oleracea*	Progoitrin (**1d**)
Crambe/Abyssinian kale	*Crambe abyssinica*	Epi-progoitrin (**1e**)
Land cress	*Barbarea verna*	Gluconasturtiin(**1 m**)
White meadowfoam	*Limnanthes alba*	Glucolimnanthin (**1o**)

Adopted from Ugolini et al. (2008)

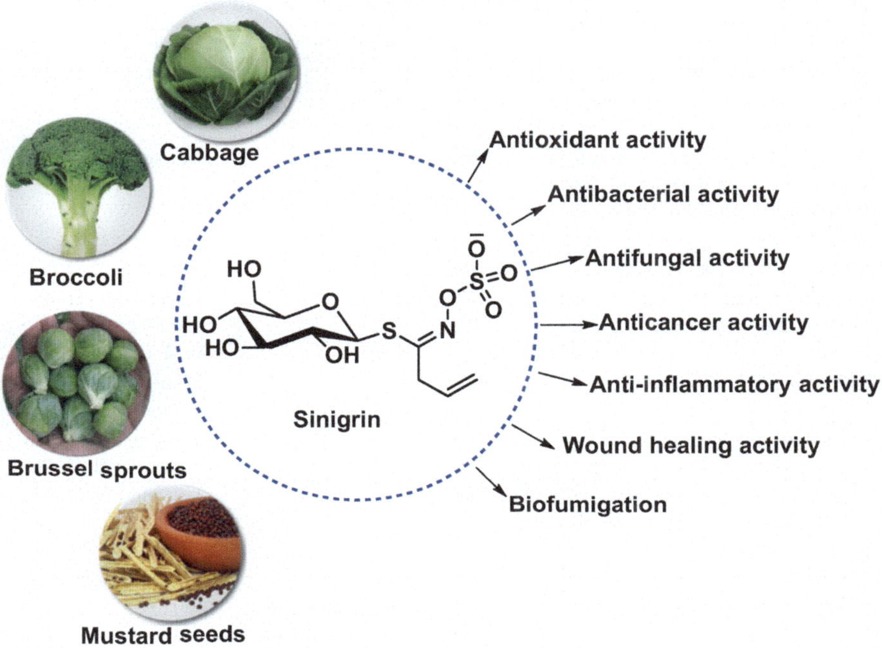

Fig. 15.3 Structure and therapeutic properties of sinigrin

15.3.2 Phenolic Compounds

The phenolic compounds of *Brassica* vegetables have been investigated due to their beneficial effect on human health. On the basis of their molecular structure, these phenolic compounds can be classified into simple phenols, phenolic acids, hydroxycinnamic acid derivatives and flavonoids. Phenolic phytocompounds have received significant attention for being potentially preventive factors against cancer, cardiovascular and several diseases. Foods of plant origin are commonly consumed due to the potential antioxidative effect of these phenolic compounds. On the basis of the number and arrangement of their carbon atoms and position of hydroxyl groups, they can be classified as flavonoids (flavonols, flavones, flavan-3-ols, anthocyanidins, flavanones, isoflavones and others) and non-flavonoids (phenolic acids, hydroxycinnamates, stilbenes etc., were discussed by Crozier et al. (2006) and mostly found conjugated to sugars and organic acids). They also exhibit antioxidant activities, hence, beneficial in different biological activities. They are indulged in capillary protective effects and hindered the growth of various stages of a tumour (Cartea et al. 2011).

15.3.2.1 Flavonoids

Flavonoids are aromatic heterocyclic organic compounds and serve as versatile building blocks for the construction of many biologically active natural products,

Table 15.3 Therapeutic properties of sinigrin (extracted from *Brassica* sp.)

Therapeutic properties of sinigrin	Model animal/cell line	References
Increase in apoptosis in colonic crypts was displayed, when exposed to the carcinogen	Rat	Smith et al. (1998)
Increased 7-methylguanine levels in hepatic DNA, but decreased DNA methylation in the lung and nasal mucosa	Rat	Morse et al. (1988)
Inhibited tongue carcinogenesis	Rat	Tanaka et al. (1992)
Antimicrobial activity against food spoilage and pathogenic organisms	*E. coli*	Luciano and Holley (2009) and Gamage et al. (2009)
Inhibition of bladder cancer	Rat	Bhattacharya et al. (2010)
Anti-hepatocarcinogenesis	Rat	Jie et al. (2014) and Tanaka et al. (1990)
Suppressed the nuclear translocation of NF-$_{\kappa}$B induced by TNF-α. Inhibited the TNF-α-stimulated VCAM-1 expression	Rat	Lee and Lee (2015)
Anti-proliferative, anti-genotoxicity, antimutagenic activities (*B. carinata*)	HL60 (human promyelocytic leukaemia cell line) and *Drosophila melanogaster*	Lozano-Baena et al. (2015)
Wound-healing activity	Human keratinocytes (HaCaT)	Mazumder et al. (2015)

drugs and therapeutic leads. These phytochemicals are extensively present in different parts of crucifers and other plants of various families. Chemically they have 15 carbon skeletons consisting of 3 rings A, B and C. Ring A (benzene ring) and B (pyran ring) are fused to each other through carbon-carbon and carbon-oxygen bonds, while ring B is substituted by a phenyl substituent (benzene ring C). These flavonoids can be classified into different classes such as flavones, flavanones, flavonols, isoflavones, flavanonol and flavan-3-ols (Fig. 15.4). These flavonoids vary in the level of oxygenation and pattern of substitution of the B ring, while individual flavonoid molecules of each class differ in the pattern of substitution of the A and C benzene rings (Middleton 1998).

Their higher concentrations are found in the epidermis of leaves and fruits of brassica herbs as reported by Crozier et al. (2006) and Pereira et al. (2009). Flavonols are the most abundant flavonoids found in *Brassica* crops such as quercetin (**2**), kaempferol (**3**) and isorhamnetin (4) as *O*-glycosides which are extensively derived (Fig. 15.5). Anthocyanins are naturally occurring flavonoid pigments responsible for red, blue and purple colour in plants, possess antioxidative properties and protect plants against excessive light and also have an important role in attracting the pollinating insects. Most common anthocyanins found in *Brassica* are cyanidin (5), pelargonidin (6), delphinidin (7), petunidin (8), malvidin (9) and peonidin (10) (Fig. 15.5). Although flavonoids are beneficial for organisms but at high doses, they

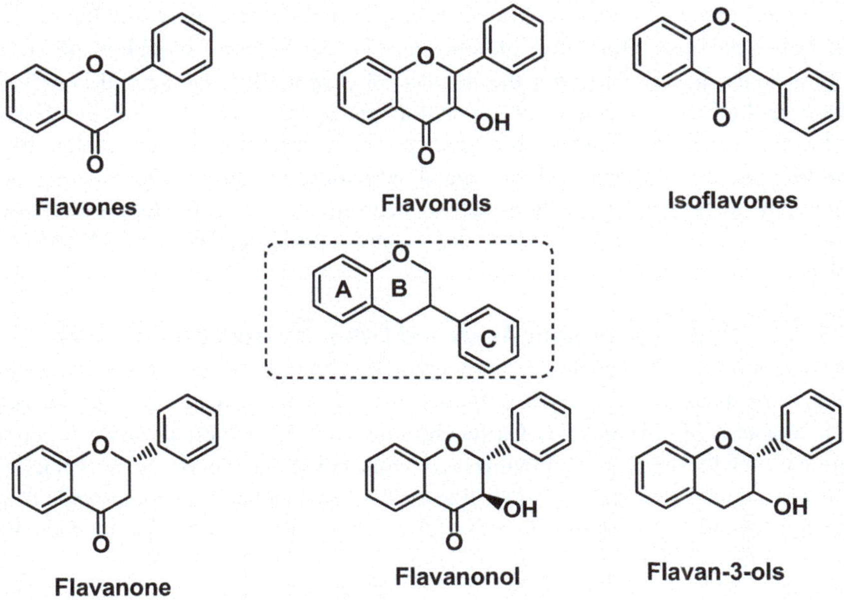

Fig. 15.4 Basic skeleton of different classes of flavonoids

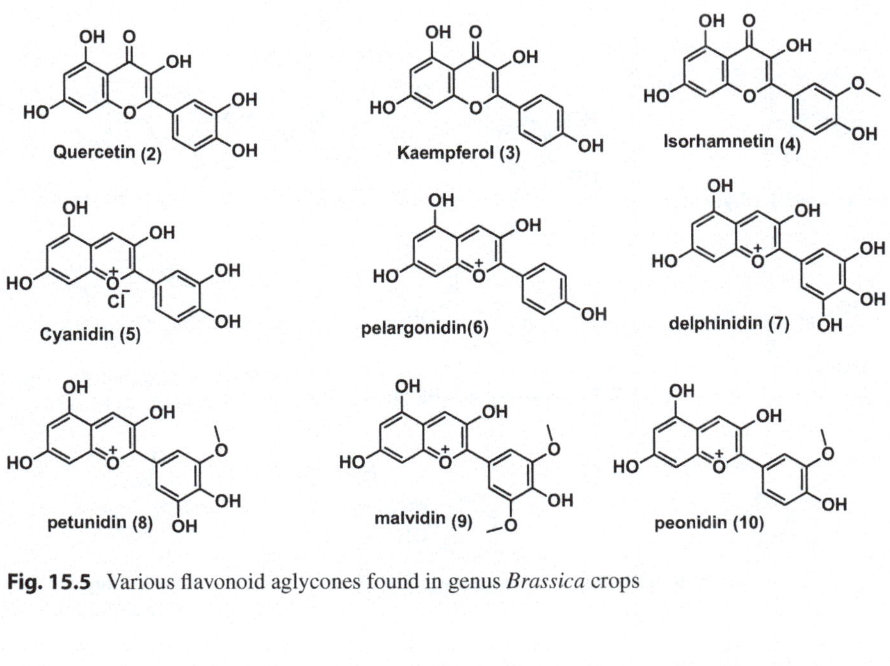

Fig. 15.5 Various flavonoid aglycones found in genus *Brassica* crops

may act as mutagens, and, therefore, their unfavourable effects may well balance up their beneficial ones. More investigations must be led about the toxicological effects of flavonoids, hence clarifying the balance of potentially adverse and beneficial effects including its mechanisms of action inside the body. Several preclinical studies described flavonoids such as quercetin, kaempferol and glucosinolate have multiple pharmacological and biological activities, including antioxidant, anti-inflammatory, anticancer, cardioprotective, neuroprotective, anti-osteoporotic estrogenic/antiestrogenic, antidiabetic antiallergic and antimicrobial activities (Middleton 1998).

15.3.2.2 Hydroxycinnamic Acids and Other Isolated Compounds

They are non-flavonoid phenolic compounds (Fig. 15.6) that are indulged in various defence mechanisms. The most common are *p*-Coumaric acid (11), ferulic acid (12), sinapic acid (13) and 3-O-Caffeoylquinic acid (15) which are often found in conjugation with sugar or other hydroxycinnamic acids found in *Brassica* vegetables as discussed by Cartea et al. (2011). Jing et al. (2014) isolated 14 compounds from the ethyl acetate extract of the oilseeds of *Brassica campestris* and identified sinapic

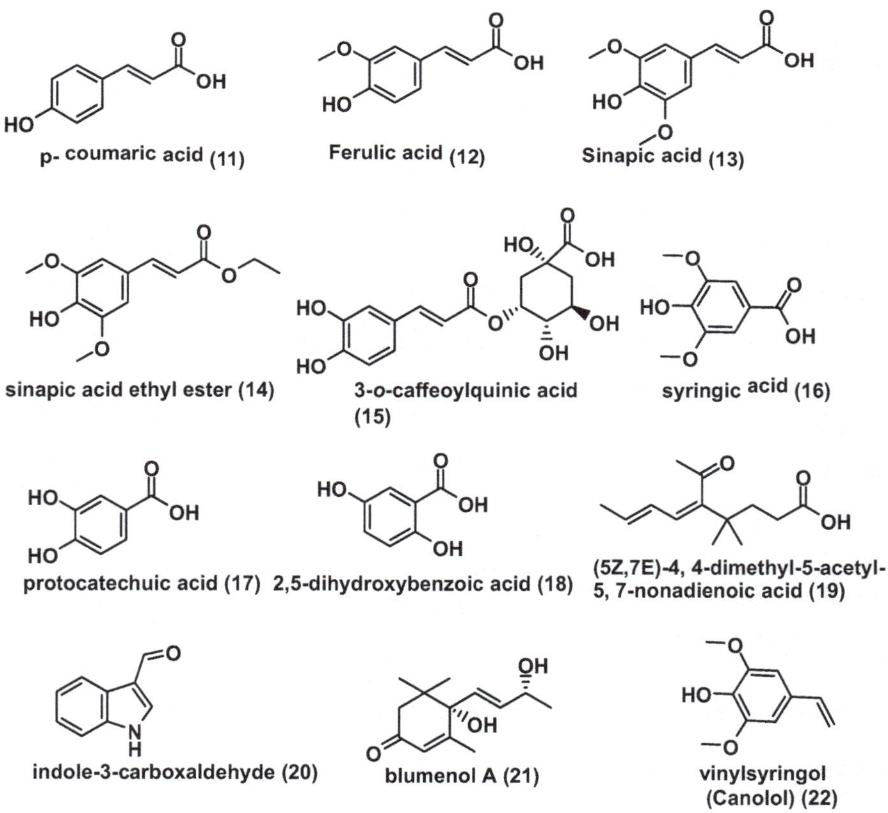

Fig. 15.6 Hydroxycinnamic acids and other compounds found in genus *Brassica* crops

acid (13), sinapic acid ethyl ester (14), syringic acid (16), protocatechuic acid (17), 2,5-dihydroxybenzoic acid (18), (5Z,7E)-4, 4-dimethyl-5-acetyl-5, 7-nonadienoic acid) (19), indole-3-carboxaldehyde (20), blumenol A (21), vinylsyringol (22), crinosterol (23), campesterol (24), 7-oxo-stigmasterol (25) and daucosterol (26) (Fig. 15.7).

15.4 Biological Activities and Their Bioavailability

Phenolic compounds have multiple additional properties including anti-inflammatory, antimicrobial, enzyme inhibition, antiallergic, vascular and cytotoxic antitumour activities. Their antioxidant activity depends on its chemical structure, i.e. the number and position of hydroxyl groups in the molecule; an increase in the number of hydroxyl groups leads to a higher antioxidant activity and confers them redox properties. In the human body, they reduce the oxidative damage from reactive oxygen species (ROS) and decelerate the progress of several chronic diseases. They take part in the oxidation of low-density lipoproteins (LDL), which play a significant role in atherosclerosis. The members of the Brassicaceae family have been shown to produce considerable quantities of antifungal compounds in their roots which were noticed by Schreiner and Koide (1993).

Several studies have been carried out to determine the bioavailability of different phenolic compounds in the diet by using animal models and human assays (Crozier

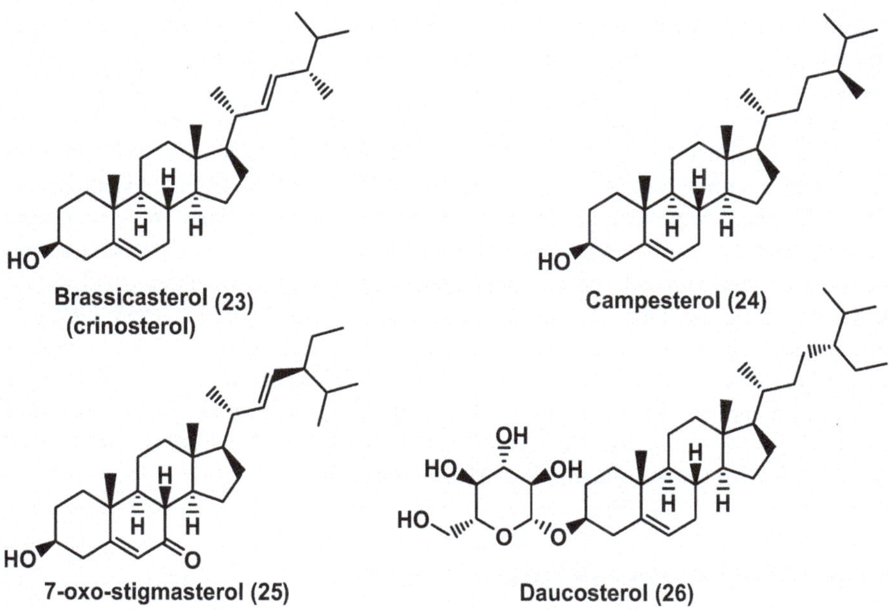

Fig. 15.7 Structures of various sterols isolated from *Brassica campestris*. (Adopted from Jing et al. 2014)

et al. 2006; D'Archivio et al. 2010). These stated that most of the phenolic compounds have a low bioavailability since they are detected in very small amounts both in tissues and plasma. The structural properties of polyphenols affect the rate and extent of their absorption in the colon and the small intestine of humans, as well as the development and appearance of metabolites in plasma (Crozier et al. 2006). The substitution position and amount of substitution of sugar moiety play a key role in absorption and bioavailability of flavonoids. Therefore, more detailed studies are desired on the mechanisms of bioavailability of phenolic compounds, which may be profitable to correlate phenolic intake with one or several absolute amplifications of bioavailability including the concentrations of key bio-active metabolites in plasma and tissues along with potential health effects in epidemiological studies (D'Archivio et al. 2010). Phenolic compounds found in *Brassica* crops also possess several other properties such as the production of hydrogen peroxide in the presence of certain metals and the ability to inhibit nitrosation reactions, scavenge electrophiles and chelate metals; therefore, they perform the mechanism of blocking in the initiation stage of several human diseases.

15.5 Role in Phytoremediation

Environmental pollutants such as heavy metals and pesticides deteriorate the quality of the soil. Cadmium is one of the toxic heavy metal, hence, its remediation is highly desired. The members of Brassicaceae family differentially tolerate cadmium toxicity and show its bioaccumulation with variable magnitude (Qadir et al. 2004). For instance, *Brassica campestris* is a good accumulator of cadmium, which may accumulate 3.5–4.0 times more Cd than in soil (Lai and Chen 2013). However, Thakur and Tiwari (2012) showed that the *B. campestris* had the maximum concentration of Cd (54.68 μg/g) at 50 ppm in 40 days after sowing (DAS), when EDTA was applied to the plants. EDTA is effective only when the plant is in 40 DAS. However, the minimum concentration of Cd was reported in seeds at harvest 95 DAS. Thus, it could be a good accumulator of Cd. Similarly, *B. juncea* also procure special attention because of its pertinence to the process of phytoextraction of heavy metals from soil. It accumulated huge capacity of Cd in the shoots (1450 μg Cd/g dry weight). This was three times higher than *B. napus* (555 μg/g) as investigated by Nouairi et al. (2006). Moreover, Turan and Esringu (2007) reported that the application of EDTA at 3, 6 and 12 mmol/kg increased the concentration of Cu, Cd, Pb and Zn in both shoots and roots, and the dose of 6 mmol/kg was found most effective for uptake of Cd, Cu, Zn and Pb. Therefore, these eco-friendly practices are desired for the remediation of these pollutants.

15.6 Conclusions and Future Prospects

The family, Brassicaceae possesses glucosinolates and phenolic compounds, which may be responsible for the anticancer, anti-inflammatory, antibacterial, antifungal, antioxidant and wound-healing effects and biofumigation. These phytochemicals manifest their benificial effects through various mechanisms. *Brassica campestris*, *B. juncea* and many other members are used in heavy metal remediation especially against cadmium toxicity. Moreover, the consumption of *Brassica* vegetables as diet provides the protection against many chronic diseases as well as serves as a biomarker of heavy metal pollution. In the future, more in-depth studies are required to explore these phytochemicals.

Acknowledgement Authors are thankful to UGC's Maulana Azad National Fellowship.

References

Barcelo S, Gardiner JM, Gescher A, Chipman JK (1996) CYP2E1-mediated mechanism of anti-genotoxicity of the broccoli constituent sulforaphane. Carcinogenesis 17:277–282

Bellostas N, Kachlicki P, Sorensen JC, Sorensen H (2007) Glucosinolate profiling of seed and sprouts of *B. oleracea* varieties used for food. Sci Hortic 114:234–242

Bergmann F (1970) The biosynthesis of glucosinolates in the course of ontogenesis of *Sinapis alba* L. Z Pflanzenphysiol 62:362–375

Bhattacharya A, Li Y, Wade KL, Paonessa JD, Fahey JW, Zhang Y (2010) Allyl isothiocyanate-rich mustard seed powder inhibits bladder cancer growth and muscle invasion. Carcinogenesis 3(12):2105–2110

Cartea ME, Francisco M, Soengas P, Velasco P (2011) Phenolic compounds in *Brassica* vegetables. Molecules 16:251–280

Chew FS (1988) Biological effects of glucosinolates. In: Cutler HG (ed) Biologically active natural products: potential use in agriculture. American Chemical Society, Washington, DC, pp 155–181

Chu YF, Sun J, Wu X, Liu RH (2002) Antioxidant and antiproliferative activities of common vegetables. J Agric Food Chem 50:6910–6916

Cohen J, Kristal R, Stanford J (2000) Fruit and vegetable intakes and prostate cancer. J Natl Cancer Inst 9:61–68

Crozier A, Jaganath IB, Clifford MN (2006) Phenols, polyphenols and tannins: an overview. In: Crozier A, Clifford M, Ashihara H (eds) Plant secondary metabolites: occurrence, structure and role in the human diet. Blackwell, Oxford, pp 1–24

D'Archivio M, Filesi C, Vari R, Scazzocchio B, Masella R (2010) Bioavailability of the polyphenols: status and controversies. Int J Mol Sci 11:1321–1342

Gamage GR, Park HJ, Kim KM (2009) Effectiveness of antimicrobial coated oriented polypropylene/polyethylene films in sprout packaging. Food Res Int 42:832–839

Gosslau A, Chen KY (2004) Nutraceuticals, apoptosis, and disease prevention. Nutrition 20:95–102

Gundgaard J, Nielsen JN, Olsen J, Sorensen J (2003) Increased intake of fruit and vegetables: estimation of impact in terms of life expectancy and healthcare costs. Pub Health Nutr 6:25–30

Hansen M, Moller P, Sorensen H, Trejo MC (1995) Glucosinolate in Broccoli stored under controlled atmosphere. J Am Soc Hortic 120:1069–1074

Hashimoto K, Kawamata S, Usui N, Tanaka A, Uda Y (2002) *In vitro* induction of the anticarcinogenic marker enzyme, quinine reductase, in human hepatoma cells by food extracts. Cancer Lett 180:1–5

Ippoushi K, Takeuchi A, Azuma K (2010) Sinigrin suppresses nitric oxide production in rats administered intraperitoneally with lipopolysaccharide. Food Chem 120:1119–1121

Jie M, Cheung WM, Yu V, Zhou Y, Tong PH, Ho JWS (2014) Anti-proliferative activities of sinigrin on carcinogen-induced hepatotoxicity in rats. PLoS One 9:110–145

Jing WG, Wang ZM, Zhao Y, Fu J, Zhao XL, Liu A (2014) Chemical constituents from seeds of *Brassica campestris*. China J Chinese Mat Med 39:2521–2525

Josefsson E (1970) Pattern, content and biosynthesis of glucosinolates in some cultivated Cruciferae. Swedish Seed Association Publication, Lund, p 42

Keum YS, Jeong WS, Kong ANT (2004) Chemoprevention by isothiocyanates and their underlying molecular signaling mechanisms. Mutat Res 555:191–202

Kris-Etherton PM, Etherton TD, Carlson J, Gardner C (2002) Recent discoveries in inclusive food-based approaches and dietary patterns for reduction in risk for cardiovascular disease. Curr Opin Lipidol 13:397–407

Lai HY, Chen BC (2013) The dynamic growth exhibition and accumulation of Cadmium of Pak Choi (*Brassica campestris* L. ssp. *chinensis*) grown in contaminated soils. Int J Environ Res Public Health 10:5284–5298

Law MR, Morris JK (1998) By how much does fruit and vegetable consumption reduce the risk of ischemic heart disease. Eur J Clin Nutr 52:549–556

Lee HW, Lee KR (2015) Effect of sinigrin on vascular cell adhesion molecule-1 expression in TNF-α-stimulated mouse vascular smooth muscle cells via downregulation of $NF\text{-}_{k}B$ signalling pathways. FASEB J 29:593.15

Lozano-Baena MD, Tasset I, Obregon-Cano S, de Haro-Bailon A, Munoz-Serrano A, Alonso-Moraga A (2015) Antigenotoxicity and tumor growing inhibition by leafy *Brassica carinata* and Sinigrin. Molecules 20:15748–15765

Luciano FB, Holley RA (2009) Enzymatic inhibition by allyl isothiocyanate and factors affecting its antimicrobial action against *Escherichia coli* O157:H7. Int J Food Microbiol 131:240–245

Manohar RP, Pushpan R, Rohini S (2009) Mustard and its uses in Ayurveda. Indian J Tradit Knowl 8:400–404

Mazumder A, Dwivedi A, Du Preez J, Du Plessis J (2015) *In vitro* wound healing and cytotoxic effects of sinigrin–phytosome complex. Int J Pharm 498:283–293

Mazumder A, Dwivedi A, Du Preez J, Du Plessis J (2016) Sinigrin and its therapeutic benefits. Molecules 21:416

Middleton EJ (1998) Effect of plant flavonoids on immune and inflammatory cell function. Adv Exp Med Biol 439:175–182

Morse MA, Wang CX, Amin SG, Hecht SS, Chung FL (1988) Effects of dietary sinigrin or indole-3-carbinol on O6-methylguanine-DNA-transmethylase activity and 4-(methyl nitrosamino)-1-(3-pyridyl)-1-butanone-induced DNA methylation and tumorigenicity in F344 rats. Carcinogenesis 9:891–1895

Nomura T, Shinoda S, Yamori T, Sawaki S, Nagata I, Ryoyama K, Fuke Y (2005) Selective sensitivity to wasabi-derived 6-(methylsulfinyl) hexyl isothiocyanate of human breast cancer and melanoma cell lines studied *in vitro*. Cancer Detect Prev 29:155–160

Nouairi I, Ben Ammar W, Youssef N, Miled Daoud DB, Habib Ghorbal M, Zarrouk M (2006) Comparative study of cadmium effects on membrane lipid composition of *Brassica juncea* and *Brassica napus* leaves. Plant Sci 170:511–519

Paolini M (1998) Brussels sprouts: an exceptionally rich source of ambiguity for anticancer strategies. Toxicol Appl Pharmacol 152:293–294

Pereira DM, Valentao P, Pereira JA, Andrade PB (2009) Phenolics: from chemistry to biology. Molecules 14:2202–2211

Plumb GW, Lambert N, Chambers SJ, Wanigatunga S, Heaney RK, Plumb JA (1996) Are whole extracts and purified glucosinolates from cruciferous vegetables antioxidants. Free Radic Res 25(1):75–86

Podsedek A (2007) Natural antioxidants and antioxidant capacity of brassica vegetables: a review. LWT 40:1–11

Qadir S, Qureshi MI, Javed S, Abdin MZ (2004) Genotypic variation in phytoremediation potential of *Brassica juncea* cultivars exposed to Cd stress. Plant Sci 167:1171–1181

Schreiner RP, Koide RT (1993) Antifungal compounds from the roots of mycotrophic and non-mycotrophic plant species. New Phytol 123:99–105

Smith TK, Lund EK, Johnson IT (1998) Inhibition of dimethylhydrazine-induced aberrant crypt-foci and induction of apoptosis in rat colon following oral administration of the glucosinolate sinigrin. Carcinogenesis 19:267–273

Tanaka T, Mori Y, Morishita Y, Hara A, Ohno T, Kojima T, Mori H (1990) Inhibitory effect of sinigrin and indole-3-carbinol on diethylnitrosamine-induced hepatocarcinogenesis in male ACI/N rats. Carcinogenesis 11:1403–1406

Tanaka T, Kojima T, Morishta Y, Moori H (1992) Inhibitory effects of the natural products indole-3-carbinol and Sinigrin during initiation and promotion phases of 4-nitroquinoline 1-oxide-induced rat tongue carcinogenesis. Jpn J Cancer Res 83:835–842

Thakur A, Tiwari A (2012) Study of concentration of cadmium in *Brassica campestris* during timely sowing and late sowing season in Gwalior region, M.P. Int J Innov Biosci 2:121–125

Tianxin W, Liang H, Qipeng Y (2012) Separation of sinigrin from Indian mustard (*Brassica juncea* L.) seed using macroporous ion-exchange resin. Korean J Chem Eng 29:396–403

Turan M, Esringu A (2007) Phytoremediation based on canola (*Brassica napus* L.) and Indian mustard (*Brassica juncea* L.) planted on spiked soil by aliquot amount of Cd, Cu, Pb, and Zn. Plant Soil Environ 53:7–15

Ugolini L, Nicola GD, Palmieri S (2008) Use of reverse micelles for the simultaneous extraction of oil, proteins, and glucosinolates from cruciferous oilseeds. J Agric Food Chem 56:1595–1601

Verhoeven DT, Verhagen H, Goldbohm RA, van den Brandt PA, van Poppel G (1997) A review of mechanisms underlying anticarcinogenecity by *Brassica* vegetables. Chem Biol Interact 103:79–129

Warwick SI, Francis A, Al-Shehbaz IA (2006) Brassicaceae: species checklist and database on CD-Rom. Plant Syst Evol 259:249–258

Antibacterial and Antifungal Agents of Higher Plants

16

Balasupramaniam Kirubakari, Shanmugapriya, Thiagarajan Sangeetha, Soundararajan Vijayarathna, Yeng Chen, Jagat R. Kanwar, Chiuan Herng Leow, Lai Ngit Shin, Mallappa Kumara Swamy, Sreeramanan Subramaniam, and Sreenivasan Sasidharan

Contents

B. Kirubakari · Shanmugapriya · T. Sangeetha · S. Vijayarathna · C. H. Leow · L. N. Shin
S. Sasidharan (✉)
Institute for Research in Molecular Medicine (INFORMM), Universiti Sains Malaysia, Gelugor, Pulau Pinang, Malaysia
e-mail: srisasidharan@yahoo.com

Y. Chen
Dental Research & Training Unit, and Oral Cancer Research and Coordinating Centre (OCRCC), Faculty of Dentistry, University of Malaya, Kuala Lumpur, Malaysia

J. R. Kanwar
Nanomedicine-Laboratory of Immunology and Molecular Biomedical Research (LIMBR), School of Medicine (SoM), Faculty of Health, Deakin University, Geelong, VIC, Australia

M. K. Swamy
Department of Biotechnology, East West First Grade College of Science, Bengaluru, Karnataka, India

S. Subramaniam
School of Biological Sciences, Universiti Sains Malaysia, Gelugor, Pulau Pinang, Malaysia

M. S. Akhtar et al. (eds.), *Natural Bio-active Compounds*,
https://doi.org/10.1007/978-981-13-7154-7_16

Abstract

The extensive search for therapeutic agents derived from higher plants has recently accelerated owing to the antibiotic resistance crisis, which is a serious warning to human healthcare worldwide. It is estimated that antibiotic-resistant infections will lead to nearly ten million yearly deaths at the global level by the year 2050. Thus, it is wise to search for new alternative antibiotics sources derived from higher plants to tackle this alarming antibiotic resistance crisis issue. Amongst the 250,000–500,000 plant species available in the World, only a small percentage (1–10%) of them have been explored by man. This data shows that there is a great opportunity for exploring these higher plants, which are rich in a wide variety of phytochemicals, and could be used as a good source to develop new antibacterial and antifungal agents. Moreover, the World Health Organization (WHO) suspects about four billion humans (about 80% of the world population) currently use herbal medicine as their primary healthcare. Therefore, this chapter is designated to discuss in detail the prospects of higher plants as antimicrobial agents. Also, this chapter attempts to summarize the antibiotic resistance crisis, current status of higher plants as antimicrobial agents, the extraction process of phytochemicals, in vitro and in vivo antimicrobial activity evaluations and the contribution of important phytochemicals, such as polyphenols, as antimicrobial agents. The methods presented in this chapter are illustrated for the bacterial and fungal model.

Keywords

Animal model · Antibiotic-resistant · Antimicrobial · Higher plants · Phytochemical

16.1 Introduction

The higher plants, characterized by vascular tissue and reproducing either by spores, cones, or flowers, dominate the world's flora and vegetation (Akeroyd and Synge 1992). The higher plants are flora comprising various phytochemicals that could be utilized for healing activities or are precursors for the new synthesis of beneficial remedies (Sofowora 2008; Swamy et al. 2016; Arumugam et al. 2016; Swamy et al. 2017). Therefore, since antiquity, Mother Nature has been a treasure of natural remedies for providing relief from various illnesses affecting human beings. In addition, the oldest written records on Indian, Chinese, Egyptian, Greek and Roman folk medicine have recorded numerous medicinal plants with various therapeutic values and prescriptions used in treating numerous illnesses (Pandey and Kumar

2013; Mohanty et al. 2017). Moreover, plant-based remedies have been reported to be inexpensive, safe and without any adverse side effects, particularly when compared with synthetic chemical drugs, and they can serve patients from low income countries. Higher plants also usually yield numerous secondary metabolites, which comprise a significant source of new antimicrobials against pathogenic bacterial and fungal infections. It is projected that only a small percentage (1–10%) of plant species have been explored by man and animals out of the 250,000 to 500,000 in the world. This data shows the great opportunity of exploring the higher plants, which are rich in a wide variety of phytochemicals, as a good source for developing antibacterial and antifungal agents. Moreover, the World Health Organization (WHO) suspects about four billion humans (about 80% of the World population) currently use herbal remedies as their main healthcare (Kumara Swamy et al. 2011; Kumara et al. 2012; Ekor 2013; Mohanty et al. 2017). Thus, exploration for novel drugs with improved therapeutic values and cheaper phytochemicals from higher plants is a natural choice for developing innovative next-generation therapeutics, especially against pathogenic microbial infection. Consequently, the extensive search for therapeutic agents derived from higher plants has recently accelerated owing to the antibiotic resistance crisis, which is a serious warning to human health worldwide (Rudramurthy et al. 2016). Thus, this chapter is intended to deliver insights into the antibiotic resistance crisis, current status of higher plants as antimicrobial agents, important antimicrobial phytochemicals from higher plants, extraction methods of higher plants and in detail in vitro and in vivo evaluations of higher plants' antimicrobial activities through clinical trials.

16.2 Antimicrobial Resistance Crisis

Recently, a rapid rise has been noticed in the cases of multidrug-resistant microorganisms worldwide. This has become a great threat to the use of antibiotics for treating several microbial diseases (Michael et al. 2014; Rudramurthy et al. 2016). Antibiotics are commonly used against many bacterial infections effectively. Since the discovery of antibiotics, the scenario of medicine has significantly transformed towards safeguarding millions of animal lives (Rossolini et al. 2014; Ventola 2015). Because the antibiotics were effective, they became the only option, and were continuously prescribed to patients suffering from microbial diseases. This practice is still in progress owing to various reasons and has resulted in the crisis of antibiotic resistance. In particular, the resistance catastrophe is mainly attributed to the misuse and overuse of antibiotics/drugs. Also, a complete failure in the development of alternative, but clinically effective drugs, led to this crisis. There are a number of microbes that have become resistant to multi-drugs and threaten the medical world. Therefore, there is an urgent need to invent novel and effective antimicrobial agents to overcome this situation and treat patients against multi-drug resistant microbial infections (Michael et al. 2014; Rudramurthy et al. 2016).

Microorganisms, such as bacteria, fungi, viruses, or protozoa, recurrently interact with humans in several ways. The type of interactions may lead to beneficial or damaging effects in their host organisms. The damaging microbes, also known as pathogens, cause many contagious diseases that can spread from one individual to another. These pathogenic microorganisms are predominantly responsible for human deaths throughout the world. After invading a host body, pathogens multiply quickly by overpowering the immunity or defense lines. The design and function of the microbial genome is responsible for their ability to overcome the defense barriers of the host cell. A wide range of molecular mechanisms, such as enzyme inactivation, target site modification, biofilms formation, evading repair mechanisms and intra-cellular localization, assist in the survival of pathogens in their host cells (Swamy and Rudramurthy 2016). This makes it necessary to use effective antimicrobial agents that can kill these pathogens and control diseases.

Antimicrobial agents are natural, synthetic or semi-synthetic chemical substances with the capability to inhibit or destroy the growth of microbial pathogens. For instance, antibiotics are used against bacteria, while antifungal agents are used against fungi. It is noteworthy that antimicrobial agents cause no harm or very little damage to the host. The era of antimicrobial agents has been in progress since Sir Alexander Fleming discovered the antibiotic penicillin in 1928 (Sengupta et al. 2013; Ventola 2015; Swamy and Rudramurthy 2016). The discovery of antibiotics, such as penicillin, vancomycin, amoxicillin, levofloxacin, tetracycline, rifamycins, etc., in the twentieth century has undoubtedly boosted the state of human health in myriad ways (Laxminarayan et al. 2013; Rudramurthy et al. 2016). Antimicrobial agents destroy pathogenic microbes through various ways, such as the inhibition of DNA replication, inhibiting RNA and protein synthesis, preventing cell wall formation and modifying the other metabolic activities (Swamy and Rudramurthy 2016). Antimicrobial agents are extensively used for treating and preventing many infectious diseases. With the discovery of several new antimicrobial drugs, their application further increased drastically, and this has prompted the development of antibiotic resistance properties in microorganisms. This change can be attributed to the fact that microbes exhibit an unlimited flexibility or adaptableness to different environments because of their incredible genomic manipulability and capability to interchange their genetic material with other species (Rodriguez-Rojas et al. 2013; Ventola 2015). The drug resistance may also occur unexpectedly via mutations (Read and Woods 2014). In addition, incorrect prescription of antimicrobial agents, availability of only a few new antibiotics, extensive agricultural uses and regulatory barriers are some of the other reasons towards the development of microbial resistance. Regardless of warnings on the overuse of antibiotics, even today, they are overprescribed throughout the world leading to serious threats and extensive clinical and economic burden on the healthcare system and patients (Ventola 2015).

Antimicrobial-resistant microbial infections are already prevalent across the globe, and they pose a great challenge to treatment approaches. It is reported that nearly 99,000 deaths occur amongst the two million Americans who develop health associated infections per year, mostly because of antibiotic-resistant pathogens (Ventola 2015). Further, antibiotic-resistant infections are overburdening the nation's healthcare systems. It is estimated that the average medical expense per each patient suffering with a drug-resistant infection ranges between 18,588 and

29,069 US$ (Golkar et al. 2014; Ventola 2015). Some of the drug-resistant strains of bacteria include methicillin-resistant *Staphylococcus aureus* (MRSA), vancomycin-resistant *Enterococci* (VRE), penicillin-resistant *Streptococcus pneumoniae* (PRSP), fluoroquinolone-resistant *Escherichia coli*, carbapenem-resistant *Acinetobacter baumannii*, cephalosporin-resistant Enterobacteriaceae, etc. (Swamy and Rudramurthy 2016).

A rapid emergence of drug-resistant microbes is becoming a great threat to the health benefits offered by the use of antibiotics. This global crisis is mainly caused by the overuse or misuse of antimicrobial agents in addition to the failure of pharmaceutical firms to develop novel antimicrobials to address this challenge. In this direction, many efforts are being carried out to mitigate the problem of drug resistance. A variety of natural compounds isolated from plants, microbes, algae, lichens, etc., still remain the major sources of antimicrobials (Swamy et al. 2016). As plants possess many compounds with chemodiversity and high antimicrobial properties, they are highly preferred and widely exploited in the drug discovery research. However, the antimicrobial activity in relation to chemical structure for many phytocompounds is yet to be clearly understood (Swamy et al. 2016; Swamy and Rudramurthy 2016). The class of polyphenolic compounds obtained from plants exhibit a great diversity in their structures; therefore, their antimicrobial effectiveness against pathogens also differs significantly (Gyawali and Ibrahim 2014; Swamy et al. 2016). In addition, various types of nanoparticles can be effectively employed as alternatives against multidrug-resistant pathogenic microbes (Rudramurthy et al. 2016). It is essential to encourage scientists to keep trying towards the discovery of new lead molecules against the drug-resistant pathogens. Moreover, well-coordinated efforts in implementing new policies and refurbishing investigations are imperative to manage the drug-resistance crisis and other clinical challenges.

16.3 Current Status of Higher Plants as Antimicrobial Agents

Infections caused by pathogenic fungi and bacteria affect the lives of millions of people worldwide. Since time immemorial, infectious diseases caused by pathogenic microbes have been a major cause of death throughout the world. Nowadays, clinically significant pathogenic fungi and bacteria are categorized not only by single drug resistance but also by multiple antibiotic resistant microbes caused by the misuse of antibiotics (Levy 2002). Thus, it is necessary to explore newer drugs to overcome multiple antibiotic resistance issues (Sarkar et al. 2003). Nevertheless, past data of the rapid, extensive appearance of resistance to newly introduced antimicrobial agents show that even new families of antibiotics will have a short life expectancy, and new antibiotics are urgently needed (Coates et al. 2002).

Therefore, the use of plant extracts or novel natural compounds in mixture with conventional antibiotics might hold better potential for promptly providing inexpensive treatment choices against multiple antibiotic resistant microbes (Cheesman et al. 2017). Moreover, about 80% of the world population is reported by WHO to rely on traditional medicine, such as medicinal plants, for their primary healthcare requirements (Bannerman 1983). Furthermore, this view was revised later by WHO and states that the majority of the population of most developing countries now regularly use traditional medicine for their healthcare requirements (WHO 2003). Even in developed countries, complementary or alternative medicine is gaining popularity among the peoples. More importantly, along with the increased incidence of bacterial resistance to antibiotics, there has also been a corresponding decrease in novel antimicrobial discovery. Consequently, there is a great need to search for new classes of antibacterial substances, especially from natural sources such as plant-based medicines in combination with conventional antibiotics. Scientific research shows that plants either contain antimicrobials that can function in synergy with conventional antibiotics or hold compounds that have no inherent antimicrobial activity but are capable of sensitizing the pathogen to a previously unproductive antibiotic (Betoni et al. 2006; Aiyegoro et al. 2009; Aiyegoro et al. 2010). Therefore, a combinational method that allows synergistic interaction between plant extracts and antibiotics is possibly the most effective approach to battle the antimicrobial resistance problem (Inui et al. 2007). Moreover, the drug combination therapy with higher plant extracts can be used to enhance the spectrum of antimicrobial activity, to avoid the development of resistant strains, to reduce toxicity and to achieve synergistic antimicrobial activity (Pankey et al. 2005). Therefore, this line of research of higher plants as antimicrobial agents may eventually prove to be very beneficial.

16.4 Important Antimicrobial Phytochemicals from Higher Plants

Higher plants typically contain numerous biologically active, structurally varied phytochemicals that are valuable as medicines, lead structures or fresh materials and are employed mainly for curing various ailments (Kumar and Pandey 2013). Many higher plants have been laboratory tested as antimicrobial agents. Polyphenols or phenolic compounds are a group of secondary metabolites that are major contributors to the observed antimicrobial activity exhibited by the higher plants. Polyphenols are secondary metabolites formed by higher plants that play numerous important functions in plant physiology and have great possible health properties for human beings, mainly as antimicrobial agents against pathogenic bacterial and fungal infections. Polyphenols can be categorized as flavonoids and non-flavonoids based on their chemical structures (Manach et al. 2004). Flavonoids have C6-C3-C6 carbon builds comprising double phenyl rings (A and B) and a heterocyclic ring (C). On the basis of the hydrogenation level of the heterocyclic ring and the bond site of ring B, flavonoids can be further categorized into numerous subclasses, such as flavones, flavonols, flavanones and isoflavonoids (Xiao 2013). Moreover, stilbenes, chalcones, anthraquinones, ellagitannins, ellagic acids and phenolic acids are categorized as non-flavonoids (Xiao and Kai 2012).

Higher floras are rich in a broad diversity of secondary metabolites, namely polyphenols, which hold antimicrobial properties and could act as alternative, efficient, inexpensive and harmless antimicrobials for the cure of microbial infections. Several attractive results have been established using a combination of natural products to cure illnesses; particularly, the synergistic properties and polypharmacological application of higher plant preparations (Gibbons 2003). Furthermore, the antimicrobial activities of polyphenols have attracted great attention among scientists worldwide due to the prospect of dealing with the drug-resistant microbes that are unresponsive to conventional antibiotics (Tangney and Rasmussen 2013). Polypenols have been recommended for their antimicrobial activities in three modes of action, including direct killing of microbes, synergistic activation of conventional antibiotics and attenuation of microbial pathogenicity (Cushnie and Lamb 2011), in addition to their ability to inactivate efflux pumps, destabilize cytoplasmic membranes and inhibit β-lactamases and topoisomerase to stop the progress of antibiotic resistance in microbes (Daglia 2012). Therefore, polyphenols found in the higher plants can be important phytochemical sources for future and current antimicrobial studies against resistant pathogenic bacterial and fungal infections.

16.5 Extraction Methods of Higher Plants

Plants are highly utilized pharmaceutically due to the rich medicinal values of their phytochemicals, such as phenolics and flavonoids. Various plant extraction methods are widely practiced in galenical development, such as the maceration method, soxhlet extraction method and ultrasound extraction method.

16.5.1 Maceration

Maceration is a well-established plant extraction method which involves soaking coarse powdered plant material in a closed vessel with an appropriate solvent (Jones and Kinghorn 2006; Handa et al. 2008). The selection of solvent mainly depends on the bio-active compounds of interest from the plant material, and thus the solubility of the compounds must be taken into account to choose an appropriate solvent. In addition, chemical characterization of the solvent and extraction yield is also equally imperative to consider while choosing a solvent for plant extraction by the maceration method. Commonly used solvents for the maceration process include hexane, chloroform, ethyl acetate, methanol or ethanol (Yan et al. 2008). The maceration process is usually carried out at room temperature for at least 3 days with occasional agitation to allow the solubilization of phytochemicals from the plant material. The mixture is then filtered, and the final marc is pressed out to completely extract the dissolved bio-active compounds (Pandey and Tripathi 2014).

16.5.2 Soxhlet Extraction

Soxhlet extraction, also known as hot continuous extraction, was developed by van Soxhlet in 1879 (Soxhlet 1879). The finely powdered plant sample is located in a "thimble", which is usually made from a strong filter paper or cellulose. The sample containing "thimble" is placed in the thimble-holder of the Soxhlet extractor, and the distillation flask at the bottom is filled with the extraction solvent. When the solvent is heated, the vapourized solvent is condensed and consecutively fills the thimble containing the plant material. When the solvent level rises, a siphon tube aspirates it from the thimble-holder and discharges it back into the distillation flask. This procedure is continued until the solvent from the siphon tube does not leave residue when evaporated, and thus it is a continuous–discrete plant extraction method (Luque de Castro and Priego-Capote 2010).

16.5.3 Ultrasound Extraction or Sonication Extraction

This extraction method comprises the utilization of ultrasound with frequencies ranging from 20 kHz to 2000 kHz (Handa et al. 2008). The acoustic result from the ultrasound causes cavitation that in turn upsurges the penetrability of the cell wall promoting the release of phytochemicals from the plant material into the solvent. Although this extraction method is easy and cost-effective, higher ultrasound energy is known to cause undesirable alterations to the bio-active compounds.

16.6 In vitro and In vivo Antimicrobial Activities Evaluation

16.6.1 Test Microorganisms and Growth Media

Various Gram-positive and Gram-negative bacteria, yeasts and molds for antimicrobial activity studies can be obtained from the American Type Culture Collection (ATCC). The bacterial strains are cultured in Mueller–Hinton agar (MHA) plates at 37 °C, while the yeasts and fungal are cultured in Sabouraud dextrose agar (SDA) plates and potato dextrose agar (PDA) plates media, respectively, at 28 °C. The stock culture is preserved on agar slants at 4 °C.

16.6.2 Inoculum Preparation for Bacterial and Fungal

One of the most crucial parts in microbiological testing procedures is the preparation of inoculum. This involves a series of steps, including the selection of appropriate colonies, suspension preparation and standardization. There are two approaches for inoculum preparation; direct colony suspension and log phase growth method. For the direct colony method, colonies from cultures not longer than 18–24 h can be used (Cavalieri et al. 2005). Inoculum is standardized immediately after preparation of suspension unlike the log phase method where inoculum is standardized following inoculation and incubation of bacteria/fungi up to log phase growth.

16.6.2.1 Inoculum Suspension Preparation

To prepare the inoculum, pick 3 to 5 well-isolated colonies from an appropriate bacterial culture (grown at correct temperature on correct media) using a sterile loop or cotton swab. Use morphologically similar colonies to avoid an atypical variant (Schwalbe et al. 2007). Make a suspension by suspending the colonies in 5 mL of sterile saline (0.85% NaCl) or media. Next, homogenize the suspension by using a vortex for about 15 s. If a log phase method is carried out, incubate this suspension up to log phase growth first, prior to the standardization step.

16.6.2.2 Inoculum Suspension Standardization

Set up the spectrophotometer and adjust the wavelength to 600 nm. Add 1 mL of un-inoculated sterile saline or media to a clean cuvette and blank the machine. Transfer the same volume of bacterial suspension into another cuvette and read the optical density. Adjust the cell density of the suspension measured to match that of the standard (0.08–0.13). However, this is subjective as different bacteria give a 10^8 CFU/mL at different absorbance value (Schwalbe et al. 2007). If the turbidity of the suspension is higher than the standard, add more diluent, or if lower than the standard, add more bacteria. Proper adjustment of the inoculum cell density is important to ensure that the resulting growth is confluent or almost confluent (Rao 2011).

16.6.2.3 Principle of Spectrophotometric Method

A spectrophotometer is used to measure the turbidity of bacterial suspension as an indirect measure of cell density (Lumen Learning 2018). This instrument functions by transmitting a light beam through a suspension and measuring the amount of light passing through it. The light intensity is detected and converted to a logarithmic value called absorbance (optical density) (Lumen Learning 2018). The turbidity (optical density) of a sample highly depends on the choice of light wavelength used for measurement. OD600, which refers to "optical density of sample at 600 nm", is often used to estimate the growth phase of bacteria or other cells in suspension (London BioHackspace 2018). As the number of bacteria in a suspension increases, the turbidity increases, causing less light to reach the detector. The decrease in light intensity is associated with the increase in absorbance measured by the spectrophotometer. The basic idea is to compare a sample of plain media and a sample of media inoculated with bacteria (London BioHackspace 2018).

16.6.2.4 Bacterial Inoculum Preparation

The bacterial inoculum can be standardized according to the CLSI procedures for aerobic bacteria (CLSI 2012). The bacteria is cultured in Mueller Hinton Broth (MHB) for 18–24 h, followed by comparison of the bacterial suspension in the MHB to the turbidity equal to 0.5 McFarland standard solution ($1–2 \times 10^8$ CFU/mL) with the addition of germ-free saline water.

16.6.2.5 Fungal Inoculum Preparation

Sabouraud Dextrose Agar (SDA) or Potato Dextrose Agar (PDA) slants are prepared in order to cultivate the selected fungal strain (Maghsoodi and Yaghmaei 2010.) The fungal species can be cultured in SDA or PDA at 28 °C until the maximum amount of conidia form (3- to 5-day-old). The conidia are recovered with 5 ml of distilled sterile water. Subsequently, the isolates are added to a Vortex mixer for 15 s and are then moved to a sterilized tube. Afterwards, the inoculum is moved to a sterilized syringe fixed to a sterilized filter with a micro-holes diameter of 11 μm. The suspension is filtered and collected in a germ-free tube. This process filters most of the hyphae, generating an inoculum predominantly composed of spores. Finally, the inoculum size is adjusted to 2.0×10^5 spores/ml by microscopic calculations through a cell-counting hematocytometer slide (Neubauer chamber).

16.6.3 Antimicrobial Disk Diffusion Assay

Antimicrobial activities of the medicinal plant extracts can be determined by the disk diffusion technique (Bauer et al. 1966; Alzoreky and Nakahara 2003). The SDA and PDA plates are seeded with an inoculum size of 10^6 colony-forming units CFU/mL of bacteria or 2×10^5 CFU/mL of yeast cells or fungal spores, respectively, by spreading the inoculum using an L-shaped glass rod. Subsequently, the disks (6.0-mm diameter) are saturated with 25 μL of crude extract at a concentration of 10.0 mg/mL before being placed on the microbe seeded plates. Likewise, each plate also has a blank disk impregnated with solvent alone as a vehicle control at the center of the plate and standard antibiotic disks (6.0-mm diameter) impregnated with 50 mg/mL ampicillin, streptomycin, kanamycin sulfate (for bacteria) and 100 mg/mL nystatin (for fungi) as positive controls. Subsequently, the plates are incubated at 37 °C for 18 h for bacteria and at 28 °C for 48 h for fungi. The inhibition zone around the disk is measured after 18 h of incubation at 37 °C for bacteria and 48 h for fungi at 28 °C, respectively. The antimicrobial activity of the plant crude extracts is measured by determining the diameter of the inhibitory zones (including the diameter of disk) on the agar surface, and values <8 mm are taken as no antimicrobial activity against the tested microbes (Zhu et al. 2005). The antimicrobial activity evaluation should be conducted in triplicate. The qualitative results are reported as the average of three experiments.

16.6.4 Minimum Inhibitory Concentration (MIC) Determination

Normally, the minimum inhibitory concentration (MIC) determinations are employed on plant extracts that show positive results against tested microorganisms by the disk diffusion assay. The maximum dilution of a plant extract that still inhibits the growth of a microorganism is known as the MIC (Misra and Dixit 1978). The comprehensive procedure of the MIC determination is found in the M7-T2 publication of the National Committee for Clinical Laboratory Standards (NCCLS 2002). In short, plant crude extract preparations are serially diluted by using sterile nutrient broth or Sabouraud dextrose broth medium as diluents to produce final plant crude extract concentrations between 1.275 and 200.000 mg/mL. Subsequently, the tubes with various concentrations of plant extract are inoculated with 20 μL/mL of bacterial or yeast suspension, homogenized and incubated at 37 °C for 24 h. The tube with the lowest dilution of the plant extract that showed an inhibitory effect against the tested microbe, resulting in no visible growth of microorganism or absence of turbidity, is verified as the MIC value for the plant extract. Usually, the microorganism growth is shown by the turbidity exhibited in the test tube. The experiment can be performed in triplicate and repeated twice to verify the result. A control experiment can be conducted in parallel to evaluate the influence of the solvent alone (without the crude plant extract) on growth of the microorganisms. The solvent (5% DMSO) is diluted by the same approach with sterile nutrient or Sabouraud dextrose broth, as mentioned above, and inoculation by microbial suspensions accompanied by incubation is done similarly to evaluate the influence of the solvent.

16.6.5 In Vivo Antimicrobial Activity in Animal Model

16.6.5.1 Animals Used in the Experiment

The animal experiments should be conducted following the guidelines of the Animal Care Council of the particular country. Protocol of animal experiments should also be specifically approved by the institutional ethics committee on animal experimentation to perform the animal studies. Pathogen-free adult female laboratory-mice *Musmus culus* (White albino) weighing 30–35 g, 10–12 weeks in age can be used for this study. The mice are familiarized with the room temperature (23 ± 2 °C) and a standard 12 h light/dark cycles in cages (1 mice/cage) for 7 days before the beginning of the research. Briefly, mice are immunosuppressed via intraperitoneal injection with cyclophosphamid as demonstrated by Shah et al. (2008) to facilitate the infection. At the third day of immunosuppression, the mice are fasted overnight.

16.6.5.2 In vivo Assay Using Mice

An example of using an in vivo antimicrobial assay to validate the in vivo antimicrobial activity of medicinal plants on the example of candidias is instigated by *Candida albicans* infection, which is an induced yeast infection model discussed in the following sections. Candidiasis infection is triggered by intravenous (i.v.) inoculation of 0.1 mL of a 10^6 UFC/mL inoculum in germ-free saline water from a fresh 48 h *Candida albicans* culture to mice. Twenty-four hours after infection, 3 mice are euthanized to verify the success of the infection by evaluating the fungal burden in the blood and kidneys samples. Infected mice are divided into 3 groups of 10 animals each, are housed in cages and have access to food and water ad libitum. Extract at the dosage determined from the toxicity study can be administered to treat the infected animals. As for treatment, the intraperitoneal injection (i.p.) technique is used over 3 consecutive days, starting 24 h after infection. Two experimental control groups can be designed; an untreated negative control group is administrated with distilled water alone and a positive control group is administrated with standard antifungal drug nystatin at 10 mg/kg of bw. On day four, the mice are killed by cervical dislocation; blood and kidney are collected from each mouse. Kidney tissues are homogenized in 5 mL of germ-free saline water and then serially diluted. Blood samples are also serially diluted; 0.1 mL of each dilution is plated onto SDA and incubated for 24 h at 37 °C. The colonies are calculated and the colony forming units (CFU) are enumerated per gram of organs (CFU/g) and per milliliter of the blood sample (CFU/ml) (Sasidharan et al. 2008). The kidney samples for histopathology evaluation are stained using the Periodic Acid Schiff (PAS) reagent and also haematoxylin to identify the *C. albicans* infection in kidney cells. Differences in mean CFU in kidneys and blood samples compared to the negative and positive control can be analysed by using a one-way analysis of variance (ANOVA) with a post hoc Tukey test. A P value of <0.05 is considered statistically significant for all comparisons.

16.7 Clinical Trials

Medicinal plants have become an essential and crucial part of public healthcare around the globe. Numerous reports have emphasized the usage of therapeutic plants in traditional and alternative remedies (Eisenberg et al. 1993). Many laboratory level scientific research studies have also proven the efficacy of the medicinal plant to treat various diseases, including issues related to pathogenic microbial infection. Nevertheless, in order to further extend their acceptance among the patient, clinical trials of these medicinal plant products should be fortified. Even though traditional medicine practitioners do not necessitate clinical trials, for its authorization and existence at the global pharmaceutical industry together with modern drugs, it has become a crucial need of the time (Mills 2003). Therefore, to prove the effectiveness of medicinal plant-based products in clinical trials, it is advised to use pharmacological formulations of these products (US Food and Drug Administration 2004). A clinical trial of medicinal plant-based products poses several challenges that need to be addressed, including issues such as those connected to the monetary, moral, product standardization through quality control, the strategy of the study and the law requirements before filing an investigational novel medicine for assessing large phase III trials (Parveen et al. 2015). However, in 2005, the World Health Organization (WHO) distributed working guidelines concerning relevant law requirements needed to support clinical trials of herbal remedies (WHO 2005); hence, clinical trials on medicinal plant-based products based on these WHO working procedures should be encouraged.

16.8 Conclusions and Future Prospects

The potential use of brand new plant-based bio-active products, such as crude extract, active fraction and isolated compound(s), from higher plant origins, is still a very fruitful activity for the production of novel therapeutic agents to advance healthcare, especially against pathogenic microbial infections. It is important to highlight that widespread in vitro, in vivo and clinical trials need to be carried out frequently to identify the active and nontoxic antimicrobial phytochemicals such as polyphenol from higher plants as plant-derived antimicrobial agents. It is beneficial to standardize the procedures of extraction, in vitro and in vivo testing so that the exploration to developed antimicrobial agents from higher plants is more orderly and explanation of results would be eased. This chapter also sheds light on the future prospects of combination therapy comprising plant-derived agents, which is certainly very promising. Future research might offer other new or innovative ways of attaining combinational therapy between plant-derived agents with known antibiotics. Therefore, there is great necessity to continue the exploration for antimicrobial agents from higher plants to combat the antibiotic resistance crisis.

References

Aiyegoro OA, Afolayan AJ, Okoh AI (2009) Synergistic interaction of *Helichrysumpedunculatum* leaf extracts with antibiotics against wound infection associated bacteria. Biol Res 42:327–338

Aiyegoro OA, Afolayan AJ, Okoh AI (2010) Interactions of antibiotics and extracts of *Helichrysumpedunculatum* against bacteria implicated in wound infections. Folia Microbiol 55:176–180

Akeroyd J, Synge H (1992) Higher plant diversity. In: Groombridge B (ed) Global biodiversity. Springer, Dordrecht

Alzoreky NS, Nakahara K (2003) Antibacterial activity of extracts from some edible plants commonly consumed in Asia. Int J Food Microbiol 80:223–230

Arumugam G, Swamy MK, Sinniah UR (2016) *Plectranthus amboinicus* (Lour.) Spreng: botanical, phytochemical, pharmacological and nutritional significance. Molecules 21:369

Bannerman RH (1983) Traditional medicine and healthcare coverage. World Health Organization, Geneva

Bauer AW, Kirby WMM, Sherris JC, Turck M (1966) Antibiotic susceptibility testing by standardized single disk method. Am J Clin Pathol 36:493–496

Betoni JEC, Mantovani RP, Barbosa LN, Di Stasi LC, Fernandes JA (2006) Synergism between plant extract and antimicrobial drugs used on *Staphylococcus aureus* diseases. Mem Inst Oswaldo Cruz 101:387–390

Cavalieri SJ, Rankin ID, Harbeck RJ, Sautter RL, McCarter YS, Sharp SE, Spiegel CA (2005) Manual of antimicrobial susceptibility testing. American Society for Microbiology. https://www.asm.org/ccLibraryFiles/FILENAME/000000002484/Manual%20of%20Antimicrobial%20Susceptibility%20Testing.pdf. Accessed 15 May 2018

Cheesman MJ, Ilanko A, Blonk B, Cock IE (2017) Developing new antimicrobial therapies: are synergistic combinations of plant extracts/compounds with conventional antibiotics the solution. Pharmacogn Rev 11:57–72

CLSI (2012) Methods for dilution antimicrobial susceptibility tests for bacteria that grows aerobically, approved standard-ninth edition," CLSI Document MO7-A9, Clinical and Laboratory Standards Institute, Wayne, PA, USA

Coates A, Hu YM, Bax R, Page C (2002) The future challenges facing the development of new antimicrobial drugs. Nat Rev Drug Discov 1:895–910

Cushnie TP, Lamb AJ (2011) Recent advances in understanding the antibacterial properties of flavonoids. Int J Antimicrob Agents 38:99–107

Daglia M (2012) Polyphenols as antimicrobial agents. Curr Opin Biotechnol 23(2):174–181

Eisenberg DM, Kessler RC, Foster C, Norlock FE, Calkins DR, Delbanco TL (1993) Unconventional medicine in the United States. Prevalence, costs, and patterns of use. N Engl J Med 328:246–252

Ekor M (2013) The growing use of herbal medicines: issues relating to adverse reactions and challenges in monitoring safety. Front Pharmacol 4:177

Geneva, Switzerland: World Health Organization (2005) Operational guidance: information needed to support clinical trials of herbal products. (Document reference who/TDR/GEN/Guidance/051)

Gibbons S (2003) An overview of plant extracts as potential therapeutics. Expert Opin Ther Pat 13(4):489–497

Golkar Z, Bagasra O, Pace DG (2014) Bacteriophage therapy: a potential solution for the antibiotic resistance crisis. J Infect Dev Ctries 8:129–136

Gyawali R, Ibrahim SA (2014) Natural products as antimicrobial agents. Food Control 46:412–429

Handa SS, Khanuja SPS, Longo G, Rakesh DD (2008) Extraction technologies for medicinal and aromatic plants, 1st edn. United Nations Industrial Development Organization and the International Centre for Science and High Technology, Trieste

Inui T, Wang Y, Deng S, Smith DC, Franzblau SG, Pauli GF (2007) Counter-current chromatography based analysis of synergy in an anti-tuberculosis ethnobotanical. J Chromatogr A 1151:211–215

Jones WP, Kinghorn AD (2006) Extraction of plant secondary metabolites. In: Sarker SD, Latif Z, Gray AI (eds) Natural products isolation, 2nd edn. Human Press Inc., New Jersey, pp 323–351

Kumar S, Pandey AK (2013) Chemistry and biological activities of flavonoids: an overview, Sci World J 2013: 162750, 16

Kumara Swamy M, Pokharen N, Dahal S, Anuradha M (2011) Phytochemical and antimicrobial studies of leaf extract of *Euphorbia nerifolia*. J Med Plant Res 5:5785–5788

Kumara SM, Sudipta KM, Lokesh P, Neeki A, Rashmi W, Bhaumik H, Darshil H, Vijay R, Kashyap SSN (2012) Phytochemical screening and *in vitro* antimicrobial activity of *Bougainvillea spectabilis* flower extracts. Int J Phytomed 4:375–379

Laxminarayan R, Duse A, Wattal C, Zaidi AK, Wertheim HF, Sumpradit N, Vlieghe E, Hara GL, Gould IM, Goossens H, Greko C (2013) Antibiotic resistance the need for global solutions. Lancet Infect Dis 13:1057–1098

Levy SB (2002) The antibiotic paradox: how the misuse of antibiotics destroys their curative powers. Perseus Publishing, Cambridge, MA

London BioHackspace (2018) OD600 Spectrophotometer https://biohackspace.org/projects/od600-spectrophotometer/. Accessed on 17 May 2018

Lumen Learning (2018) How microbes grow. https://courses.lumenlearning.com/microbiology/chapter/how-microbes-grow. Accessed on 16 May 2018

Luque de Castro MD, Priego-Capote F (2010) Soxhlet extraction: past and present panacea. J Chromatogr A 1217:2383–2389

Maghsoodi V, Yaghmaei S (2010) Comparison of solid substrate and submerged fermentation for chitosan production by *Aspergillus niger*. Trans C: Chem Chem Eng 17:153–157

Manach C, Scalbert A, Morand C, Remesy C, Jimenez L (2004) Polyphenols: food sources and bioavailability. Am J Clin Nutr 79:727–747

Michael CA, Dominey-Howes D, Labbate M (2014) The antimicrobial resistance crisis: causes, consequences, and management. Front Public Health 2:145

Mills S (2003) Clinical research in complementary therapies: principles, problems and solutions. In: Lewith GT, Jonas WB, Walach H (eds) Elsevier Science: Churchill Livingstone, pp 211–212

Misra SB, Dixit SN (1978) Antifungal properties of leaf extract of *Ranunculus sceleratus* L. Experientia 34:1442–1443

Mohanty SK, Swamy MK, Sinniah UR, Anuradha M (2017) *Leptadenia reticulata* (Retz.) Wight & Arn. (Jivanti): Botanical, agronomical, phytochemical, pharmacological, and biotechnological aspects. Molecules 22:1019. https://doi.org/10.3390/molecules22061019

NCCLS (2002) Methods for dilution antimicrobial susceptibility tests for Bacteria that grow aerobically; approved standard, 3rd edn. NCCLS document M100-S12. Wayne, National Committee for Clinical Laboratory Standards

Pandey AK, Kumar S (2013) Perspective on plant products as antimicrobials agents: a review. Pharmacologia 4:469–480

Pandey A, Tripathi S (2014) Concept of standardization, extraction and pre-phytochemical screening strategies for herbal drug. J Pharmacogn Phytochem 2:115–119

Pankey G, Ashcraft D, Patel N (2005) *In vitro* synergy of daptomycin plus rifampin against *Enterococcus faecium* resistant to both linezolid and vancomycin. Antimicrob Agents Chemother 49:5166–5168

Parveen A, Parveen B, Parveen R, Ahmad S (2015) Challenges and guidelines for clinical trial of herbal drugs. J Pharm Bioallied Sci 7:329–333

Rao TV (2011) Kirby-bauer disc diffusion method, antibiotic susceptibility testing, skill based learning. https://www.slideshare.net/doctorrao/kirbybauer-disc-diffusion-methodantibiotic-susceptibility-testing-skill-based-learning. Accessed on 16 May 2018

Read AF, Woods RJ (2014) Antibiotic resistance management. Evol Med Public Health 2014:147

Rodriguez-Rojas A, Rodriguez-Beltran J, Couce A, Blazquez J (2013) Antibiotics and antibiotic resistance: a bitter fight against evolution. Int J Med Microbiol 303:293–297

Rossolini GM, Arena F, Pecile P, Pollini S (2014) Update on the antibiotic resistance crisis. Curr Opin Pharmacol 18:56–60

Rudramurthy GR, Swamy MK, Sinniah UR, Ghasemzadeh A (2016) Nanoparticles: alternatives against drug-resistant pathogenic microbes. Molecules 21:836

Sarkar A, Kumar KA, Dutta NK, Chakraborty P, Dastidar SG (2003) Evaluation of *in vitro* and *in vivo* antibacterial activity of dobutamine hydrochloride. Indian J Med Microbiol 21:172–178

Sasidharan S, Darah I, Jain K (2008) *In vitro* and *in vivo* antifungal activity of the methanol extract from *Gracilaria changii*. Internet J Pharmacol 6:1–10

Schwalbe R, Steele-Moore L, Goodwin AC (2007) Antimicrobial susceptibility testing protocols, e-book, Taylor and Francis. https://books.google.com.my/books?id=FukzUi3sLvEC&printsec=frontcove&dq=Antimicrobial+Susceptibility+Testing+Protocols&hl=en&sa=X&ved=ahUKEwiDm7qVhJvbAhXMTrwKHbQ0CRwQ6AEIKDAA#v=onepage&qAntimicrobial%20Susceptibility%20Testing%20Protocols&f=false. Accessed on 15 May 2018

Sengupta S, Chattopadhyay MK, Grossart HP (2013) The multifaceted roles of antibiotics and antibiotic resistance in nature. Front Microbiol 4:47

Shah AS, Wakade AS, Juvekar AR (2008) Immunomodulatory activity of methanolic extract of *Murraya koenigii* spreng leaves. Indian J Exp Biol 46:505–509

Sofowora EA (2008) Medicinal plants and traditional medicine in Africa. Wiley, pp 1–10

Soxhlet F (1879) Dinglers' Polyt 232:461

Swamy MK, Rudramurthy GR (2016) Antimicrobial agents: current status and future challenges. Austin Pharmacol Pharm 1:1004

Swamy MK, Sinniah UR, Akhtar MS (2016) Antimicrobial properties of plant essential oils against human pathogens and their mode of action: an updated review. Evid-Based Complement Alternat Med 2016:21. https://doi.org/10.1155/2016/3012462

Swamy MK, Arumugam G, Kaur R, Ghasemzadeh A, Yusoff MM, Sinniah UR (2017) GC-MS based metabolite profiling, antioxidant and antimicrobial properties of different solvent extracts of Malaysian *Plectranthus amboinicus* leaves. Evid-Based Complement Alternat Med 2017:1517683. https://doi.org/10.1155/2017/1517683

Tangney CC, Rasmussen HE (2013) Polyphenols, inflammation, and cardiovascular disease. Curr Atheroscler Rep 15:324

US Food and Drug Administration (2004) Guidance for industry: botanical drug products. http://www.fda.gov/downloads/Drugs/GuidanceComplianceRegulatoryInformation/Guidances/ucm070491.pdf. Accessed on 22 Mar 2014

Ventola CL (2015) The antibiotic resistance crisis: part 1: causes and threats. Pharmacol Ther 40:277

WHO (2003) Traditional medicine. WHO Fact Sheet No. 134, Geneva. http://tinyurl.com/5mrd5. Accessed on 22 May 2018

Xiao JB (2013) Polyphenol-plasma proteins interaction: its nature, analytical techniques, and influence on bioactivities of polyphenols. Curr Drug Metab 14:367–368

Xiao J, Kai G (2012) A review of dietary polyphenol-plasma protein interactions: characterization, influence on the bioactivity, and structure-affinity relationship. Crit Rev Food Sci Nutr 52:85–101

Yan X, Rana J, Chandra A, Vredeveld D, Ware H, Rebhun J, Mulder T, Persons K, Zemaitis D, Li Y (2008) Medicinal herb extraction strategy-a solvent selection and extraction method study. AIChE Annual Meeting, Conference Proceedings, 16–21st November, 2008, Philadelphia, PA, United States

Zhu X, Lo HZR, Lu Y (2005) Antimicrobial activities of *Cynaras colymus* L. leaf, head, and stem extracts. J Food Sci 70:149–152

Bio-active Compounds Isolated from Neem Tree and Their Applications

17

Robeena Sarah, Baby Tabassum, Nida Idrees, and Mohd Kamil Hussain

Contents

Robeena Sarah and Baby Tabassum have equally contributed for this chapter.

R. Sarah (✉) · B. Tabassum · N. Idrees
Toxicology Laboratory, Department of Zoology, Government Raza P.G. College, Rampur, Uttar Pradesh, India
e-mail: sarah.sarah102@gmail.com; dr.btabassum@gmail.com

M. K. Hussain
Department of Chemistry, Government Raza P.G. College, Rampur, Uttar Pradesh, India

M. S. Akhtar et al. (eds.), *Natural Bio-active Compounds*,
https://doi.org/10.1007/978-981-13-7154-7_17

Abstract

The extracts of different parts of neem tree (*Azadirachta indica*) have been well documented for its pharmacological or medicinal properties and their wide applications by indigenous healthcare practices. Pharmacological properties exhibited by plant parts could be further explored for development of new herbal formulations and therapeutic agents. Pharmacological properties of neem tree are due to the presence of active phytochemicals like flavonoids, terpenoids, coumarins, alkaloids, tannins, sulphurous compounds, carbohydrates, proteins and minerals. Various medicinal properties and applications of neem tree have been well documented in ancient Indian system of medicine and scripts such as *Susruta Samhita* and *Charak Samhita*. Over 700 herbal preparations based on *A. indica* have been recognized in traditional system of medicine such as Unani, homoeopathy, Ayurveda and Siddha, and more than 160 local practices are known in different countries of the world in which neem contributes as a main or the sole constituent for curing various diseases. Neem displays various medicinal properties such as antioxidant, anti-inflammatory, antidiabetic, anticancer, antiviral, antibacterial, antigingivitis, antifungal, antiulcer, hepatoprotective, neuroprotective, antipyretic and wound healing activities. All the parts of neem tree have been used as traditional medicines. In addition to its therapeutic potential, neem is being extensively used as eco-friendly commercial agrochemicals and pesticides. The present chapter provides the critical description of phytochemistry and pharmacological properties of different parts of neem tree and its important natural bio-active compounds.

Keywords

Anticancer · Flavonoids · Pharmacology · Phytochemistry · Triterpenoids

17.1 Introduction

In recent era, haphazard use of synthetic chemicals to increase the soil fertility and plant vigour has been associated with the various side effects on human beings and environment. Thus, there is a need to screen and analyse the plants for their natural bio-active compounds as the drugs of eco-friendly and environment-loving nature. Phytochemicals from medicinal plants serve as lead compounds in drug invention and are recognized for their medicinal value as the possible source of bio-active compounds (Prusti et al. 2008). In this regard, *Azadirachta indica* A. Juss (syn. *Melia azadirachta*) is well known in India and its neighbouring countries for more than 2000 years as one of the most important multipurpose, medicinal plants having wide variety of pharmacological activities. It belongs to family Meliaceae and is popularly known as Indian neem (margosa tree) or Indian lilac (Girish and Shankara 2008).

Neem is an evergreen tree, cultivated in various parts of the Indian subcontinent. Every part of the tree has been used as traditional medicine against various human ailments from ancient times (Biswas et al. 2002). The neem tree is an extraordinary plant that has been declared the "tree of the twenty-first century" by the United Nations (UNEP 2012). In 1992, the US National Academy of Science published a

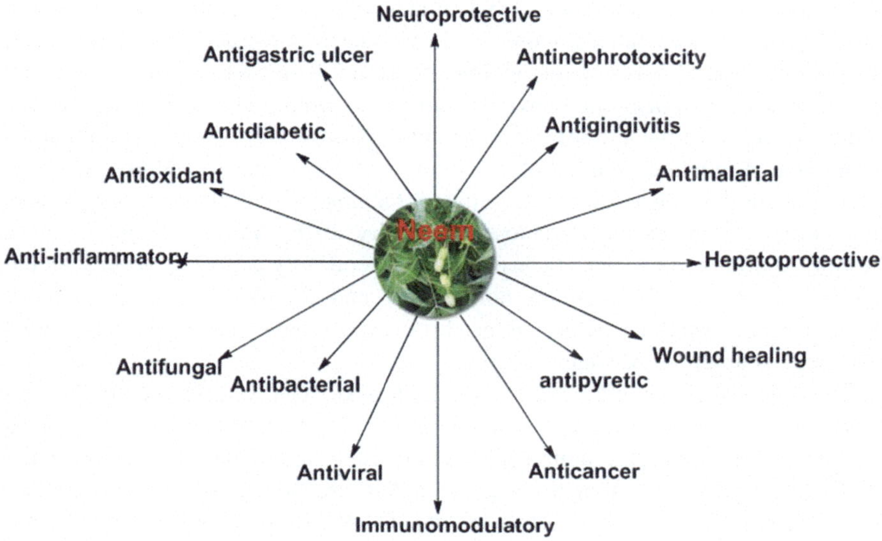

Fig. 17.1 Therapeutic properties of *Azadirachta indica*

report entitled *Neem: A Tree for Solving Global Problems* (NAS 1992; Hashmat et al. 2012). Every part of the plant such as leaves, fruits, seeds, bark and roots contains compounds with proven antioxidant, anti-inflammatory (Chattopadhyay et al. 1993), antidiabetic, anticancer, antiviral, antibacterial, antigingivitis, antifungal, antiulcer, hepatoprotective, nephroprotective, neuroprotective, antipyretic and wound healing activities. It has great prospective in the field of pest management, medicine and environment protection. Neem is a natural source of eco-friendly insecticides, pesticides and agrochemicals (Brahmachari 2004; Fig. 17.1). The present chapter provides the critical description of phytochemistry and pharmacological properties of different parts and important natural bio-active compounds isolated from neem tree.

17.2 Morphological Description and Chemical Compositions

Neem (*A. indica* A. Juss.) is a member of family Meliaceae. It is a tall evergreen tree with hard and scaly bark and alternate leaves, and the small white flowers. It reaches up to 15–20 m and sometimes 35–40 m (Bhowmik et al. 2010). The fruits are green drupes which on ripening turn into golden yellow in colour. The neem tree has long been documented for its unique and broad spectrum medicinal properties for improving human health and activity against insects. Due to its diverse and unique pharmacological properties, this plant is also referred as 'wonder plant', 'village dispensary' or 'living pharmacy'.

Neem has therapeutic implication in management, prevention and treatment of various diseases (Alzohairy 2016; Akhila and Rani 1999). The principal constituent of neem leaves includes protein, carbohydrates, minerals, vitamin C, carotene, etc. In addition, it also contains glutamic acid, tyrosine, aspartic acid, amino acids and several fatty acids. Leaves mainly yield nimbosterol (β-sitosterol) and polyphenolic flavonoid such as quercetin (Govindachari et al. 1998). The bark of neem tree contains chief constituents like nimbidin, nimbin, β-sitosterol, 6-desacetylnimbinene, nimbinone, nimbolicin, nimbidiol, nimbione, margocin, etc. In addition to above terpenoids, the bark also yields an essential oil, tannins and important polysaccharides like glucose, arabinose and fructose (Alzohairy 2016; Akhila and Rani 1999). Flowers yield a waxy material consisting of several fatty acids and several amino acids. The flower also contain nimbosterol and flavonoids like kaempferol, melicitrin, etc., which are helpful in the elimination of phlegm and intestinal worms (Akazawa et al. 2014; Alzohairy 2016).

The tree exudes a gum, which on hydrolysis yields different sugars like L-arabinose, L-fucose, D-galactose and D-glucuronic acid. The older tree exudes a sap containing free sugars, amino acids and organic acids. The sap is useful in the treatment of weakness, skin diseases, stimulant and tonic. Neem seeds contain 29.27% of lipids, 12.10% of proteins and 43.28% of parietal constituents and important source of terpenoids. The azadirachtin is the main phytoconstituent obtained from seeds of neem tree (Akhila and Rani 1999). Nimbidin is another important compound isolated from the neem seed oil which was found to be effective against many skin diseases. Neem cake is the byproduct obtained from fruits and kernels; it is an important source of sulphur, nitrogen, phosphorus and potassium and also used as green manure and cattle feed. Various important parts and their biological activities of different parts of neem tree have been summarized in tabular form (Table 17.1; Fig. 17.2).

17.3 Important Phytochemicals Isolated from Neem Tree

Approximately 250 natural products including diterpenoids, triterpenoids, steroids, flavonoids, coumarins, hydrocarbons, fatty acids, etc. have been isolated from different parts of the neem tree. The tetranortriterpenoids were isolated from the seeds, kernels, oils and leaves, while the other bio-active compounds include alkaloids, flavonoids, phenolic compounds, carotenoids, steroids, ketones and limonoids (aladucin, valassin, meliacin, nimbin, nimbidin, geducin and azadirachtin) (Koul et al. 1989; Uko and Kamalu 2001; Lale 2002). Azadirachtin was first isolated by Butterworth and Morgan (1968). It is a complex tetranortriterpenoid limonoid present in seeds and the key constituent responsible for both antifeedant and toxic effects in insects (Mordue and Nisbet 2000).

17.3.1 Triterpenoid Constitutes of *A. Indica*

Neem tree is one of the richest natural sources of biologically active phytochemicals or secondary metabolites. After the isolation of nimbin in 1942 by Siddiqui, various

Table 17.1 Phytochemical constituents of *A. Indica* and their pharmaceutical properties

Compounds	Pharmaceutical property	References
Nimbidin	Anti-inflammatory, antifungal hypoglycaemic, antibacterial	Mitra et al. (1971) and Biswas et al. (2002)
Nimbin	Anti-inflammatory, antipyretic, fungicidal, antihistamine and antiseptic	Siddiqui (1945) and Kraus (1995)
Nimbolide	Pancreatic cancer, anti-breast cancer, anti-malarial, antibacterial, oral cancer	Elumalai et al. (2014), Liu et al. (2015), and Subramani et al. (2016)
Azadirachtin	Cervical cancer, oral cancer	Kumar et al. (2010) and Priyadarsini et al. (2010)
Gedunin	Impaired allergic responses, anticancer	Kamath et al. (2009), Ferraris et al. (2012), and Tharmarajah et al. (2017)
Mahmoodin	Antibacterial	Atawodi and Atawodi (2009)
Nimocinol	Anti-breast cancer	Suman et al. (2014)
Meliantriol	Antifeedant	Jacobson (1995)
Azadirone	Induces death receptors and sensitizes human cancer cells	Gupta et al. (2013)
Azadiradione/ nimolicin	Neurodegenerative diseases	Nelson et al. (2016)
Epoxyazadiradione	Anti-inflammatory, anticancer	Alam et al. (2012) and Kumar et al. 2018
7-Deacetyl dedunin	Anti-malarial	Pereira et al. (2014)
Salannol	Inhibition of larval growth	Koul et al. (2004)
Nimbinene	Insecticide	Champagne et al. (1992)
β-Sitosterol (nimbosterol)	Antifungal	Govindachari et al. (1998)
Nimbolin A	Anti-termite	Srevino et al. (2007)
Azadiramide A	Breast cancer	Zhu et al. (2018)
Magolone	Antibacterial	Lakshmi et al. (2015)
Margolonone		
Isomargolonone		
Catechin	Anti-inflammatory, immunomodulatory and antioxidant	Atawodi and Atawodi (2009)
Epicatechin		
Quercetin	Antioxidant activity	Sultana et al. (2007)

other natural compounds such as diterpenoids, triterpenoids, steroids, flavonoids, coumarins, hydrocarbons, fatty acids, amino acids and polysaccharides have been isolated and characterized from different parts of the neem tree. Among them, triterpenoid constitutes approximately 150 compounds of diverse oxygenated compounds which are of prime importance and are mainly responsible for the diverse biological activities. Based on the skeletal diversity, these can be classified into different groups (Akhila and Rani 1999).

Fig. 17.2 Medicinally important parts of the neem tree: (**a**) neem tree showing barks; (**b**) flowers; (**c**) fruits; (**d**) neem cake; (**e**) neem oil; (**f**) twigs and leaves

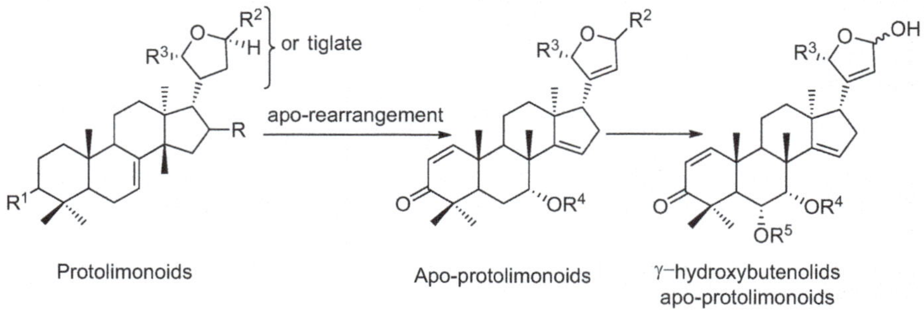

Fig. 17.3 General structures of protolimonoids or protomeliacins and apo-protolimonoids and γ-hydroxybutenolids apo-protolimonoids

17.3.1.1 Protolimonoids or Protomeliacins and Apo-Protolimonoids

Protolimonoids are considered as a precursor of the limonoids and contain a C_8 side chain on C-17 of butyrospermol skeleton, while apo-protolimonoids are characterized by the presence of 4,4,8-trimethylsteroidal skeleton and supposed to be the instant products formed from protolimonoids through an apo-rearrangement. Meliantriol (Jacobson 1995), kulactone (Siddiqui et al. 1991), limocinol (Siddiqui et al. 1991), limocinone, melianone Lavie et al. (1967), nimbocinone (Siddiqui et al. 1986a, b, c, d), nimolinone (Siddiqui et al. 1986a, b, c, d) and odoratone (Siddiqui et al. 2003a, b) are well-known protolimonoids. Azadirachtol (Siddiqui et al. 1985), azadirachnol/naheedin (Siddiqui et al. 1992a, b) and azadirol (Siddiqui et al. 1991), liocin A-B and limocinin are important apo-protolimonoids. γ-hydroxybutenolids apo-protolimonoids are formed by the modification of furan ring of apo-protolimonoids to the γ-hydroxybutenolids. Nimocinolide, isonimocinolide and nimbocinolide are known γ-hydroxybutenolids apo-protolimonoids (Akhila and Rani 1999; Fig. 17.3).

17.3.1.2 Tetranortriterpenoids or Limonoids

Tetranortriterpenoids have 4,4,8-trimethyl-17-furanyl steroid skeleton, and they are the major class of terpenoids isolated from neem. Limonoids can be sub-divided into the following categories:

(A) Basic Limonoids

All four rings (A, B, C and D) of the triterpenoid skeleton are intact, and these limonoids can further be classified according to the skeletal modifications as follows (Fig. 17.4):

(i) Azadirone Skeleton

Azadirone, nimbinin/nimolin, azadiradione/nimolicin, 17β-hydroxy azadiradione (Lavie et al. 1971), neeflone (Nanduri and Banstola 1995), 17β-hydroxy nimbocinol (Lee et al. 1988), nimbocinol (Siddiqui et al. 1986a, b, c, d), 17-epiazadiradione (Kraus and Cramer 1978), meldenin (Connolly et al. 1968), isomeldenin, nimocinol (Suresh et al. 1997), nimocin (Siddiqui et al. 1986a, b, c, d), meldenindiol (Tan and Luo 2011), 1α,2α-epoxy nimocinol (Hallur et al. 2002), 7α-benzoylazadiradione, 7α-benzoylepoxyazadiradione (Kraus and Cramer 1978), 17-epi-nimbocinol (Gaikwad et al. 1990), trichilinone acetate (Siddiqui et al. 2003a, b), 6α-O-acetyl-7-deacetyl nimocinol (Siddiqui et al. 2000), 14,15epoxy nimocinol (Govindachari et al. 1999) and 1α,2α-epoxy-17β-hydroxyazadiradione (Hallur et al. 2002).

(ii) Gedunin Skeleton

Gedunin (Akihisa et al. 2009), 6β-hydroxygedunin (Koul et al. 2003), 7-deacetylgedunin, 7-deacetyl nimolicinol, 7-deacetyl-7α-benzoylgedunin (Kraus et al. 1981), nimolicinol (Hallur et al. 2002), mahmoodin (Siddiqui et al. 1992a, b), azadirinin (Ara et al. 1992) and 1α,2α-epoxynimolicinol (Hallur et al. 2002).

(iii) Vilasinin Skeleton

Vilasinin, vilasinin triacetate (Pachapurkar et al. 1974), nimbidinin (Mitra et al. 1971), meliacinol (Siddiqui et al. 2000), nimbolin A (Ekong et al. 1969), 1,3-diacetyl vilasinin (Kraus et al. 1981), 1-acetyl-7-tigloyl vilasinin (Kraus 1984) and 1,3-diacetyl-7-tigloyl-12α-hydroxyvilasinin (Kumar et al. 1996).

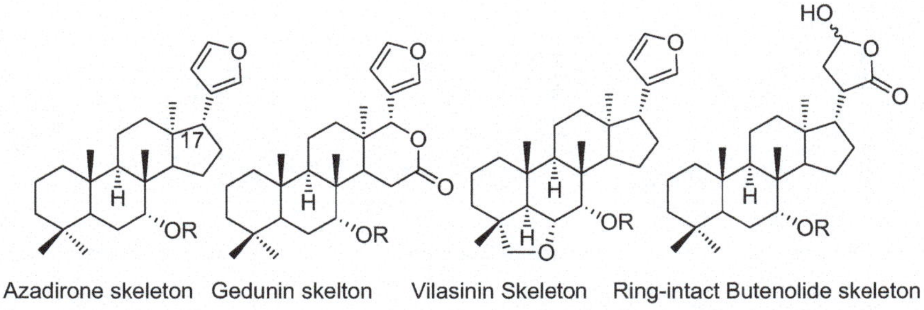

Fig. 17.4 Skeletons of basic limonoids

(iv) Ring-Intact Butenolide Skeleton
Azadironolide, azadiradionolide, isoazadironolide (Siddiqui and Ghiasuddin 1998), nimocinolide, isonimbocinolide (Siddiqui et al. 1986a, b, c, d), isonimbolide, isonimolicinolide, isonimolide (Siddiqui et al. 1987a, b), nimbocinolide, 23-o-methyl nimocinolide, 7α-Senecioyl-7-deacetyl-23-o-methyl nimocinolide (Siddiqui et al. 1999a, b) and 23-o-methylazadironolide (Siddiqui et al. 2003a, b).

(B) C-Seco Limonoids
C-ring of ring tetracyclic core of intact limonoids rearranges further to form C-seco limonoids. On the basis of skeletal diversity, these tetranortriterpenoids can be classified as (Fig. 17.5):

(i) Azadirachtin Skeleton
Azadirachtin A (Morgan and Thornton 1973), azadirachtin B (Govindachari and Gopalakrishnan 1997), azadirachtin D (Rojatkar et al. 1989), azadirachtin E (Rembold 1989), azadirachtin F, azadirachtin H, azadirachtin I (Govindachari et al. 1992a, b), azadirachtin G (Rojatkar and Nagasampagi 1995), azadirachtin K (Govindachari et al. 1992a, b), azadirachtin M (Luo et al. 1999), azadirachtol (Ley et al. 1993), vepaol, isovepaol (Akihisa et al. 2009), 13,14-desepoxy-azadirachtin-A (Govindachari and Gopalakrishnan 1997), 3-deacetyl-11-desoxyazadirachtin (Bilton et al. 1987), 3-deacetyl-3-cinnamoyl-azadirachtin (Kraus et al. 1987) and 3α-acetoxy-1α-hydroxy azadirachtol (Rojatkar and Nagasampagi 1995).

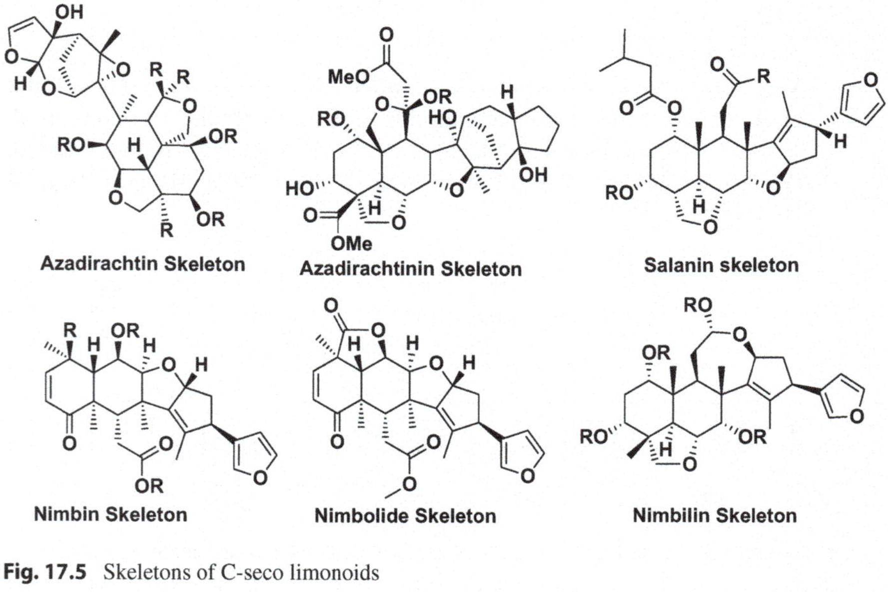

Fig. 17.5 Skeletons of C-seco limonoids

(ii) Azadirachtinin Skeleton
Azadirachtin N (Luo et al. 1999), 1-tigloyl-3-acetyl azadirachtinin (Luo et al. 2001), 1-tigloyl-3-acetyl-methoxy azadirachtinin (Kraus et al. 1987) and 3-tigloyl azadirachtinin (Luo et al. 2001).

(iii) Salannin Skeleton
Salannin (Rajab et al. 1988), salannol, salannol acetate (Rojatkar et al. 1989), 6-deacetylsalannin, 2′,3′-dehydrosalannol (Kraus et al. 1981) and nimbidic acid (Mitra et al. 1971).

(iv) Nimbin Skeleton
Nimbin nimbanal, nimbinol, 6-deacetylnimbin, 6-deacetylnimbinal (Bokel et al. 1990) and 4-epi-nimbin (Devakumar and Mukerjee 1985).

(v) Nimbolide Skeleton
Nimbolide and 28-deoxonimbolide (Bokel et al. 1990).

(vi) Nimbilin Skeleton
Nimbilin, nimbolicin, nimbolin B (Ara et al. 1989a, b) and ochinolide B (Govindachari et al. 1992a, b).

17.3.2 Pentanortriterpenoids

Pentanortriterpenoids are derived from proto or apo-proto skeleton through the collective degradation of five carbon units from eight carbon side chain at C-17 or additionally from another skeletal region. Nimbandiol, 6-acetyl nimbandiol, 6-deacetyl nimbinene and nimbinene are important pentanortriterpenoids obtained from the neem tree (Kraus and Cramer 1981).

17.3.3 Octanortriterpenoids

Octanortriterpenoids are formed by the degradation of complete eight carbon side chain from protolimonoid skeleton resulting in un-substituted C-17. Azadiron, desfurano-azadiradione (Siddiqui et al. 1992a, b), Desfurano-desacetylnimbin-17-one (Siddiqui et al. 2001) and desfurano-6α-hydroxyazadiradion (Siddiqui et al. 2002) octanortriterpenoids were isolated from neem tree.

17.3.4 Nonanortriterpenoids

They are the nine-carbon degraded (‘nonanor’) structure of triterpenoid. Eight-carbon side chain along with one additional carbon from ring D is removed to form this specific class of terpenoids. Two nonanortriterpenoids β-nimolactone and α-nimolactone were isolated from the neem tree (Siddiqui et al. 1992a, b; Fig. 17.6).

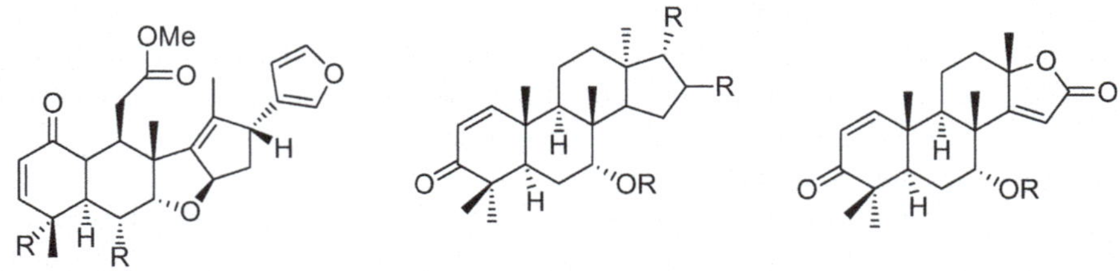

Fig. 17.6 General structures of pentanortriterpenoids, octanortriterpenoids and nonanortriterpenoids

17.4 Therapeutic Applications of some Important Phytochemicals of Neem Tree

Different parts of the neem tree and isolated compounds display diverse pharmacological properties such as neuroprotective, anti-gastric ulcer, antinephrotoxicity, antigingivitis, antidiabetic, antioxidant, anti-malarial, anti-inflammatory, hepatoprotective, antifungal, wound healing, antibacterial, antipyretic, antiviral, anticancer and immunomodulatory activities (Alzohairy 2016). Various constituents from neem tree such as nimbolide, azadirachtin and gedunin target multiple cellular and molecular signalling pathways which are responsible for cancer, such as cell proliferation, apoptosis inflammation, evasion, invasion, and angiogenesis. However, the effectiveness of these limonoids has been investigated only at the preclinical level, and the effects of these phytochemicals on human beings are mostly unexplored (Nagini 2014; Paul et al. 2011) (Figs. 17.7 and 17.8).

17.5 Other Therapeutic Applications of Neem Tree

Almost every part of the neem tree has been known to possess a wide range of therapeutic properties (Biswas et al. 2002; Table 17.2). Some of them are mentioned below.

17.5.1 Neem Twigs in Dental Care

India's rural population starts their day with the chewing neem twigs, while in the urban areas neem toothpaste is preferred. It is quite effective in reducing plaque and gingival inflammation. Chava et al. (2012) reported that dried chewing twigs of *neem* showed maximum antibacterial activity against *S. mutans* as compared to *S. salivarius*, *S. mitis*, and *S. sanguis*, while Chatterjee et al. (2011) determined the effectiveness of neem-based mouth wash for its antigingivitis effect and confirmed that it is equally effective in reducing periodontal indices as chlorhexidine.

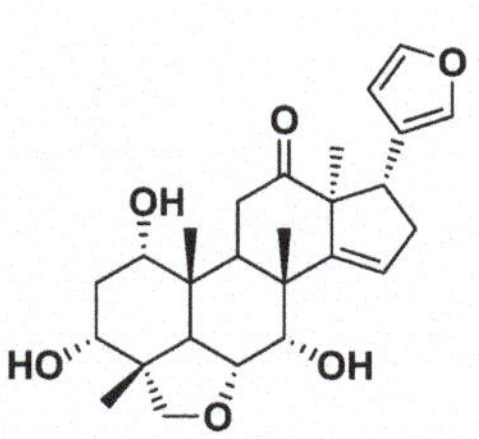
Nimbidin (1)

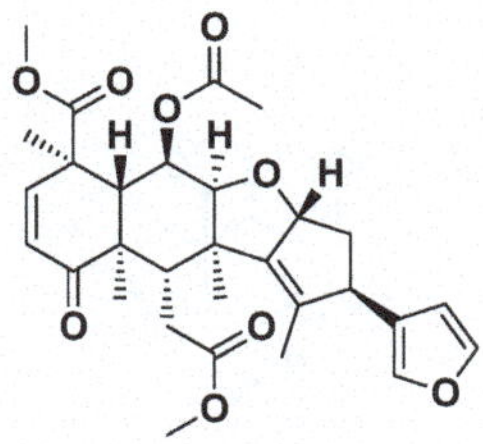
Nimbin (2)

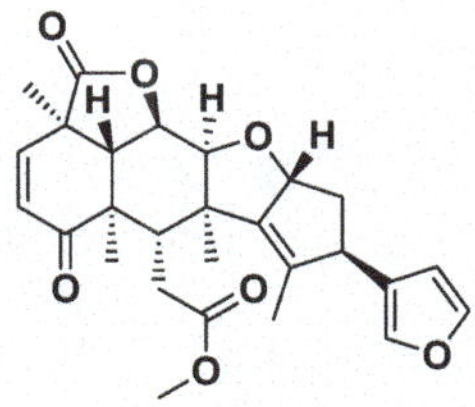
Nimbolide (3)

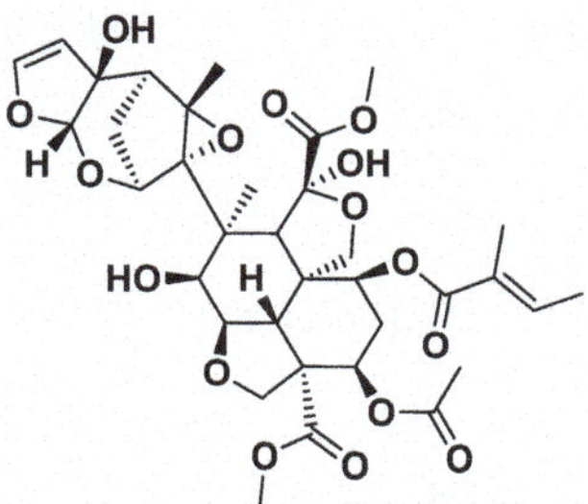
Azadirachtin (4)

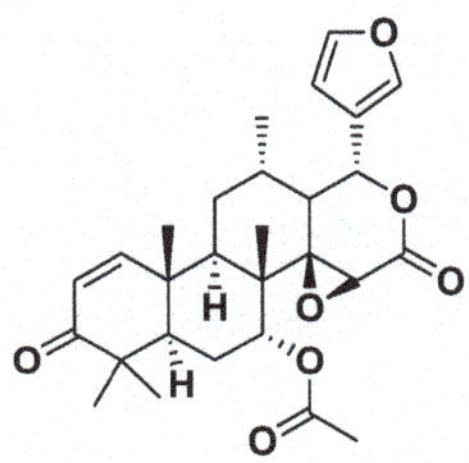
Gedunin (5)

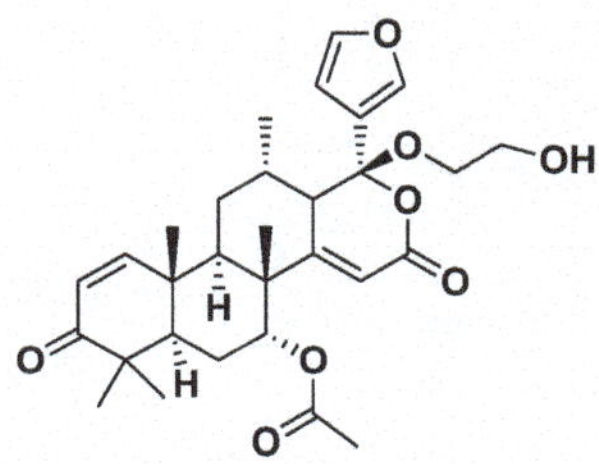
Mahmoodin (6)

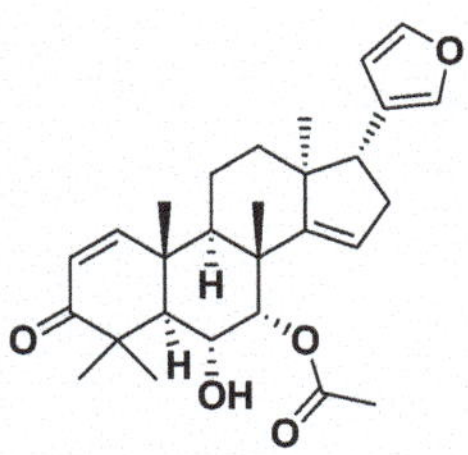
Nimocinol (7)

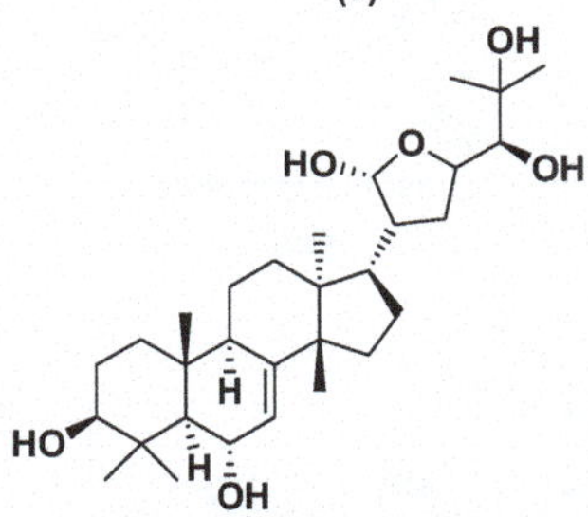
Meliantriol (8)

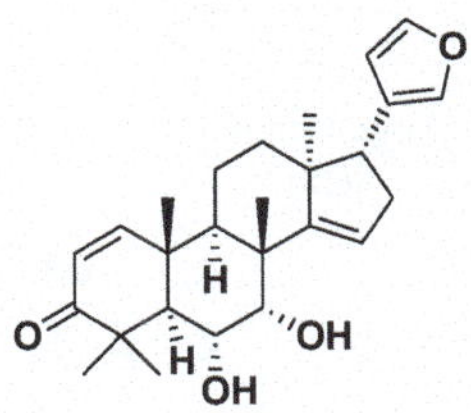
Azadirone (9)

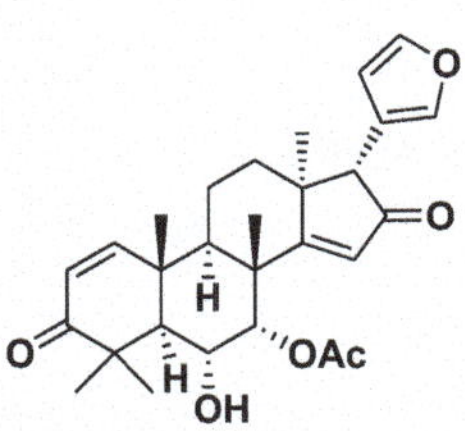
Azadiradione (10)

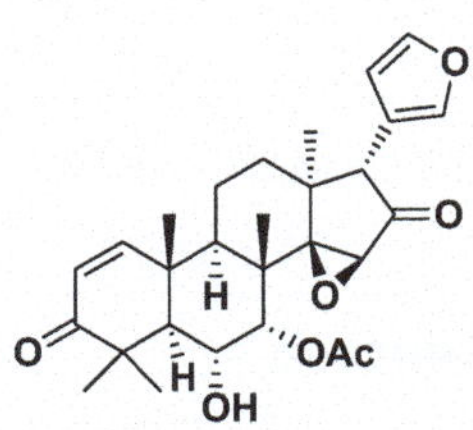
Epoxyazadiradione (11)

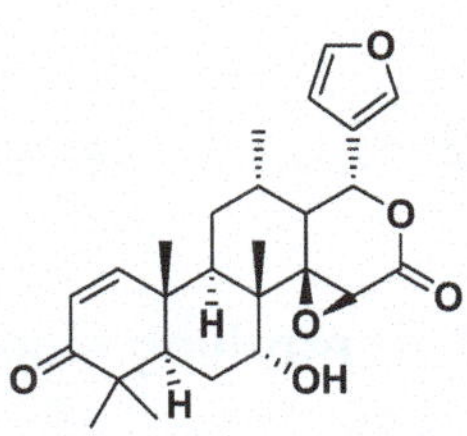
7-Deacetyl Gedunin (12)

Fig. 17.7 Chemical structures of active phytochemical constituents of *Azadirachta indica*

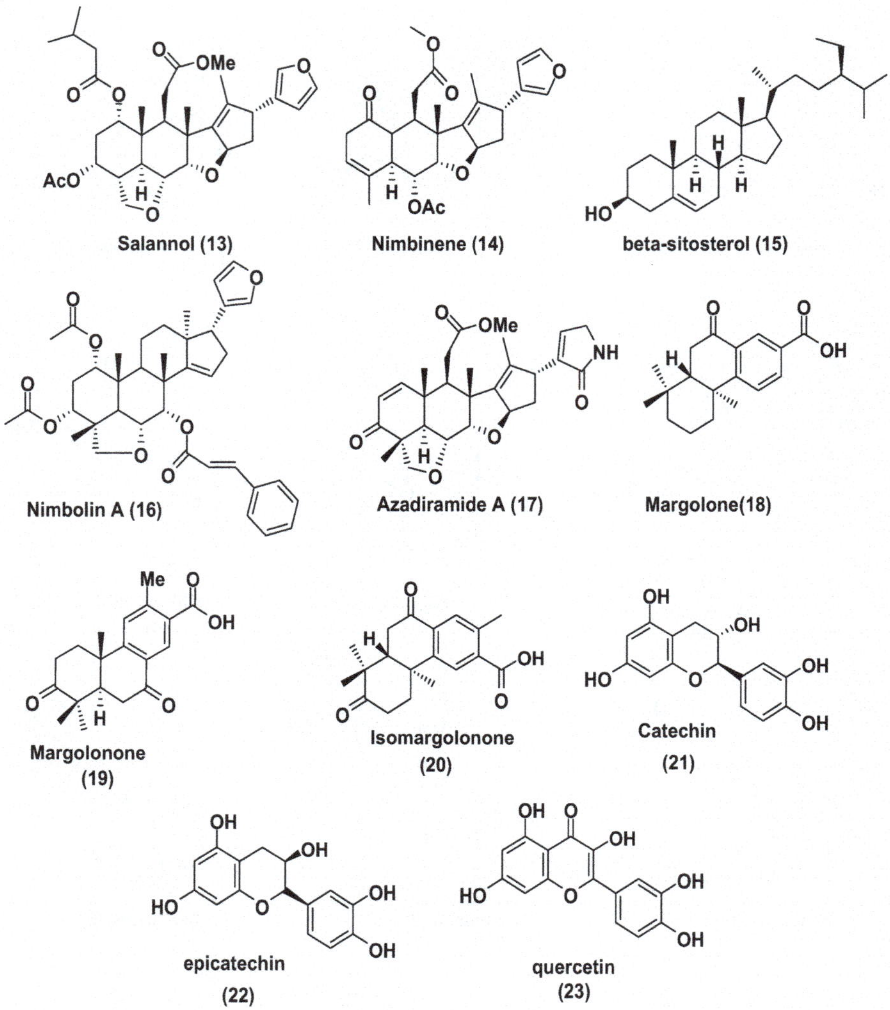

Fig. 17.8 Chemical structures of active phytochemical constituents of *Azadirachta indica*

17.5.2 Controlling Blood Glucose

Diabetes is one of the most prevalent epidemic diseases in most parts of the world. Conventional medicines are good for controlling diabetes but also have some side effects. Consumption of the neem leaves and its extract has been proved to be beneficial for keeping blood sugar under control. Aqueous extract of green leaves of the neem tree considerably decreases blood sugar level and checks adrenaline and glucose-induced hyperglycaemia (Murty et al. (1978). In normal as well as alloxan-induced diabetic rabbits, leaf extract and seed oil exhibited hypoglycaemic effect (Khosla et al. 2000).

Table 17.2 Various parts of *A. indica* with therapeutic applications

Plant part	Therapeutic applications
Leaves	Antioxidant, leprosy, eye problem, intestinal worms and cancer
Flower	Elimination of intestinal parasites and phlegm
Fruit	Diabetes, piles, intestinal parasites, urinary problems, phlegm, wounds healing and leprosy
Seed	Cancer, leprosy, pesticide and elimination of intestinal parasites
Twig	Diabetes, cough, asthma, piles, dental hygiene and urinary disorder
Bark	Acts as an analgesic, curative of fever and skin diseases
Oil	Leprosy, the killing of intestinal worms and pesticide
Gum	Effective against skin diseases, wounds healing and ulcers
Cake	Used as manure and pest repellent

17.5.3 Hepatoprotective Role of Neem

Co-administration of *Azadirachta indica* leaf extract with arsenic to rats effectively reduced the extent of liver damage as levels of serum enzymes and hepatic antioxidants were modulated close to normal (Oyewole 2011). Baligar et al. (2014) evaluated the protective effect of nimbolide against carbon tetrachloride (CCl_4)-induced liver toxicity in rats. They suggested that the nimbolide possesses hepatoprotective effect against CCl_4-induced liver damage. The aqueous extract of neem leaf was found to offer protection against paracetamol-induced liver necrosis in rats and significantly reduce the enhanced levels of serum aspartate aminotransferase (AST), alanine aminotransferase (ALT) and gamma-glutamyl transpeptidase (GGT) (Bhanwra et al. 2000).

17.5.4 Nephroprotective and Neuroprotective Effects

An experiment was made to investigate the effects of methanolic leaf extract of *A. indica* on cisplatin-induced nephrotoxicity and oxidative stress in rats, and results confirmed that extract effectively protects the kidney from cisplatin-mediated oxidative damage (Dhar et al. 1996). Similarly, Abdel (2014) investigated the neuroprotective effects of *Azadirachta indica* leaves against cisplatin (CP)-induced neurotoxicity, and results showed that morphological findings of neem before and after CP injection implied a well-preserved brain tissue.

17.5.5 Anticancer Activity

Alzohairy (2016) stated that some ingredients of neem tree show effective role in the management of cancer through the regulation of cell signalling pathways. Neem modulates the activity of various tumour suppressor genes (e.g. p53, pTEN), angiogenesis (VEGF), transcription factors (e.g. NF-κB) and apoptosis (e.g. bcl2, bax).

17.5.6 Improving Water Quality and Health of Fish

Neem leaf powder efficiently removes cadmium from water and also decrease its concentration in the fish tissues, while neem leaf water extract removes cadmium in low degree but affects significantly the haematological, physiological and immunological state of *O. niloticus*, improving the health status of fish (Osman and Hegazy 2013). Moreover, it is quite helpful in the purification of water and kills harmful microbes and makes the water potable and safe for fish as well as human consumption.

17.5.7 Antimicrobial Activity

Neem shows antimicrobial role through inhibitory effect on microbial growth/potentiality of cell wall breakdown. The Neem nanoemulsion (o/w) exhibited antibacterial activity against strains of bacterial pathogen by disrupting the integrity of the bacterial cell membrane. Different parts of the plant are shown to exhibit antimicrobial effects against a broad range of microorganisms (Jerobin et al. 2015).

17.5.8 Toxicology of Neem

Studies based on animal model and clinical trials confirmed that neem is safe at a certain doses, while its ingredients also showed toxic/adverse effect. Several studies reported that neem oil poisoning causes vomiting, hepatic toxicity, metabolic acidosis and encephalopathy in children (Sinniah and Baskaran 1981; Sundaravalli et al. 1982; Lai et al. 1990). Aqueous extract of seed kernel displayed toxicity to freshwater fish *Oreochromis niloticus* (tilapia) and *Cyprinus carpio* (carp). Retardation of spermatogenesis was observed by feeding neem seed cake to rats. Calves fed with neem seed cake resulted in reduced haemoglobin content in the blood along with depression (Jacobson 1995).

17.6 Conclusions and Future Prospects

The extracts isolated from various parts of neem tree exhibited broad therapeutic potentials as antioxidant, anti-inflammatory, antidiabetic, anticancer, antiviral, antibacterial, antigingivitis, antifungal, antiulcer, hepatoprotective, neuroprotective, antipyretic and wound healing activities. In addition, it is also used as eco-friendly commercial agrochemicals and pesticides. Various studies have supported the traditional uses of neem tree, but still more research work is desired for understanding the behaviour and mode of action of the bio-active compounds of neem for validating its clinical exploitation in the herbal formulations for humans uses. Further, investigations to evaluate the potential toxicity of neem bio-actives using clinical research need to be encouraged.

References

Abdel Moneim AE (2014) *Azadirachta indica* attenuates cisplatin induced neurotoxicity in rats. Indian J Pharm 46:316–321

Akazawa H, Manosroi A, Manosroi J, Akihisa T (2014) Limonoids and flavonoids from the flowers of *Azadirachta indica* var. siamensis, and their melanogenesis-inhibitory and cytotoxic activities. Chem Biodivers 11:73–84

Akhila A, Rani K (1999) Chemistry of the neem tree (*Azadirachta indica* A. Juss.). Fortschr Chem Org Naturst 78:47–149

Akihisa T, Noto T, Takahashi A, Fujita Y, Banno N, Tokuda H, Koike K, Suzuki T, Yasukawa K, Kimura Y (2009) Melanogenesis inhibitory, anti-inflammatory, and chemopreventive effects of limonoids from the seeds of *Azadirachta indica* A. Juss (neem). J Oleo Sci 58:581–594

Alam A, Haldar S, Thulasiram HV, Kumar R, Goyal M, Iqbal MS, Pal C, Dey S, Bindu S, Sarkar S, Pal U, Maitali NC, Bandyopadhyay U (2012) The novel antiinflammatory activity of epoxyazadiradione against macrophage migration inhibitory factor inhibition of tautomerase and proinflammatory activities of macrophage migration inhibitory factor. J Biol Chem 287:24844–24861

Alzohairy MA (2016) Therapeutics role of *Azadirachta indica* (neem) and their active constituents in diseases prevention and treatment. Evid-Based Complement Alternat Med 2016:11, 738250611. https://doi.org/10.1155/2016/7382506

Ara I, Siddiqui BS, Faizi S, Siddiqui S (1989a) Isolation of Meliacin cinnamates from the root bark of *Azadirachta indica* A. Juss (*Meliaceae*). Heterocycles 29:729–735

Ara I, Siddiqui BS, Faizi S, Siddiqui S (1989b) Two new terpenoids from root bark of *Azadirachta indica*. J Nat Prod 52:1209–1213

Ara I, Siddiqui BS, Faizi S, Siddiqui S (1992) Isolation and structure elucidation of the triterpene azadirinin from the root of *Azadirachta indica*. Fitoterapia 63:118–121

Atawodi SE, Atawodi JC (2009) *Azadirachta indica* (neem): a plant of multiple biological and pharmacological activities. Phytochem Rev 8:601–620

Baligar NS, Aladakatti RH, Ahmed M, Hiremath MB (2014) Evaluation of acute toxicity of neem active constituent, nimbolide and its hepatoprotective activity against acute dose of carbon tetrachloride treated albino rats. Int J Pharm Sci Res 5:3455–3466

Bhanwra S, Singh J, Khosla P (2000) Effect of *Azadirachta indica* (neem) leaf aqueous extract on paracetamol-induced liver damage in rats. Indian J Physiol Pharmacol 44:64–68

Bhowmik D, Yadav J, Tripathi KK, Sampath KK (2010) Herbal remedies of *A. indica* and its medicinal application. J Chem Pharm Res 2:62–72

Bilton JN, Broughton HB, Jones PS, Ley SV, Lidert Z, Morgan ED, Rzepa HS, Sheppard RN, Slawin AMZ, Williams DJ (1987) An X-ray crystallographic, mass spectroscopic, and NMR study of the limonoid insect antifeedant azadirachtin and related derivatives. Tetrahedron 43:2805–2815

Biswas K, Chattopadhyay I, Banerjee RK, Bandyopadhyay U (2002) Biological activities and medicinal properties of neem (*Azadirachta indica*). Curr Sci 82:1336–1345

Bokel M, Cramer R, Gutzeit H, Reeb S, Kraus W (1990) Tetranortriterpenoids related to nimbin and nimbolide from *Azadirachta indica* A. Juss (Meliaceae). Tetrahedron 46:775–782

Brahmachari G (2004) Neem-an omnipotent plant: a retrospection. Chem Biochem 5:408–421

Butterworth JH, Morgan E (1968) Isolation of a substance that suppresses feeding in locusts. Chem Commun (1):23–24

Champagne DE, Koul O, Isman MB, Scudder GGE, Towers GHN (1992) Biological activity of limonoids from the rutales. Phytochemistry 31:377–394

Chatterjee A, Saluja M, Singh N, Kandwal A (2011) To evaluate the antigingivitis and antiplaque effect of an *Azadirachta indica* (neem) mouthrinse on plaque induced gingivitis: a doubleblind, randomized, controlled trial. J Indian Soc Periodontol 15:398–401

Chattopadhyay RR, Chattopadhyay RN, Maitra SN (1993) Possible mechanism of anti-inflammatory activity of *Azadirachta indica* leaf extract. Indian J Pharmacol 25:99–100

Chava VR, Manjunath SM, Rajanikanth AV, Sridevi N (2012) The efficacy of neem extract on four microorganisms responsible for causing dental caries viz *Streptococcus mutans, Streptococcus salivarius, Streptococcus mitis* and *Streptococcus sanguis*: an *in vitro* study. J Contemp Dent Pract 13:769–772

Connolly JD, Handa KL, McCrindl R (1968) Further constituents of nim oil: the constitution of meldenin. Tetrahedron Lett 4:437–440

Devakumar C, Mukerjee SK (1985) 4-epinmbin, a new meliacin from from *Azadirachta indica* A. Juss. Indian J Chem B 24:1105–1106

Dhar R, Dawar H, Garg S, Basir SF, Talwar GP (1996) Effect of volatiles from neem and other natural products on gonotrophic cycle and oviposition of *Anopheles stephensi* and *An. Culicifacies* (Diptera: Culicidae). J Med Entomol 33:195–201

Ekong DEU, Fakunle CO, Fasina AK, Okogun JI (1969) The meliacins (limonoids). Nimbolin A and B, two new meliacin cinnamates from *Azadirachta indica* L. and *Melia azedarach* L. J Chem Soc Chem Commun 20:1166–1167

Elumalai P, Arunkumar R, Benson CS, Sharmila G, Arunakaran J (2014) Nimbolide inhibits IGF-I-mediated PI3K/Akt and MAPK signalling in human breast cancer cell lines (MCF-7 and MDA-MB-231). Cell Biochem Funct 32:476–484

Ferraris FK, Moret KH, Figueiredo AB, Penido C, Henriques M (2012) Gedunin, a natural tetranortriterpenoid, modulates T lymphocyte responses and ameliorates allergic inflammation. Int Immunopharmacol 14:82–93

Gaikwad BR, Mayelvaganan T, Vyas BA, Bhat SV (1990) Nimbocinol and 17-epinimbocinol from the nimbidin fraction of neem oil. Phytochemistry 29:3963–3965

Girish K, Shankara BS (2008) Neem–a green treasure. Electron J Biol 4:102–111

Govindachari TR, Gopalakrishnan G (1997) 13,14-desepoxyazadirachtin-A, a tetranortri-terpenoid from *Azadirachta indica*. Phytochemistry 45:397–399

Govindachari TR, Sandhya G, Raj SP (1992a) Azadirachtins H and I: two new tetranortriterpenoids from *Azadirachta indica*. J Nat Prod 55:596–601

Govindachari TR, Sandhya G, Raj SP (1992b) Structure of Azadirachtin K, a new tetranortriterpenoid from Azadirachtins H and I: two new tetranortriterpenoids from *Azadirachta indica*. Indian J Chem B 31:295–298

Govindachari TR, Gopalakrishnan G, Suresh G (1997) Purification of Azadirachtin-B(3-tigloylazadirachtol) by preparative high performance liquid chromatography, using the recycling mode. J Liq Chromatogr 20:1633–1636

Govindachari TR, Suresh G, Gopalakrishnan G, Banumathy B, Masilamani S (1998) Identification of antifungal compound from the seed oil of *Azadirachta indica*. Phytoparasitica 26:109–116

Govindachari TR, Malathi R, Gopalakrishnan G, Suresh G, Rajan SS (1999) Isolation of a new tetranortriterpenoid from the uncrushed green leaves of *Azadirachta indica*. Phytochemistry 52:1117–1119

Gupta SC, Francis SK, Nair MS, Mo YY, Aggarwal BB (2013) Azadirone, a limonoid tetranortriterpene, induces death receptors and sensitizes human cancer cells to tumor necrosis factor-related apoptosis-inducing ligand (TRAIL) through a p53 protein-independent mechanism: evidence for the role of the ROS-ERK-CHOP-death receptor pathway. J Biol Chem 288:32343–32356

Hallur G, Sivaramakrishnan A, Bhat SV (2002) Three new tetranortriterpenoids from neem seed oil. J Nat Prod 65:1177–1179

Hashmat I, Azad H, Ahmed A (2012) Neem (*Azadirachta indica* A. Juss)-a nature's drugstore: an overview. Int Res J Biol Sci 1:76–79

Jacobson M (1995) In: Schmutterer H (ed) The neem tree: source of unique natural products for integrated pest management, medicine, industry and other purposes. VCH, New York, pp 484–495

Jerobin J, Makwana P, Kumar RSS, Sundaramoorthy R, Mukharjee A, Chandrasekaran N (2015) Antibacterial activity of neem nanoemulsion and its toxicity assessment on human lymphocytes in vitro. Int J Nanomedicine 10:77–86

Kamath SG, Chen N, Xiong Y, Wenham R, Apte S, Humphrey M, Cragun J, Lancaster JM (2009) Gedunin, a novel natural substance, inhibits ovarian cancer cell proliferation. Int J Gynecol Cancer 19:1564–1569

Khosla P, Bhanwra S, Singh J, Seth S, Srivastava RK (2000) A study of hypoglycaemic effects of Azadirachta indica (Neem) in normaland alloxan diabetic rabbits. Indian J Physiol Pharmacol 44:69–74

Koul O, Isman M, Katter C (1989) Properties and uses of neem (*Azadirachita indica*). Can J Bot 68:1–11

Koul O, Multani JS, Singh G, Daniewski WM, Berlozecki S (2003) 6β-Hydroxygedunin from *Azadirachta indica*. its potentiation effects with some non-azadirachtin limonoids in neem against lepidopteran larvae. J Agric Food Chem 51:2937–2942

Koul O, Singh G, Singh R, Singh J, Singh J, Daniewski WM, Berlozecki S (2004) Bioefficacy and mode-of-action of some limonoids of salannin group from *Azadirachta indica* A. Juss and their role in a multicomponent system against lepidopteran larvae. J Biosci 29:409. https://doi.org/10.1007/BF02712112

Kraus W (1984) Biologically active compounds from Meliaceae. Stud Org Chem 17:331

Kraus W (1995) Biologically active ingredients-azadirachtin and other triterpenoids. In: Schutterer H (ed) The neem tree *Azadirachta indica* A. Juss and other meliaceous plants. Weinheim, New York, pp 35–88

Kraus W, Cramer R (1978) 17-EPI-azadiradion uno 17-β-hydroxy-azadiradion, zwei neue inhaltsstoffe aus *Azadirachta indica* A. Juss. Tetrahedron Lett 19:2395–2398

Kraus W, Cramer R (1981) Pentanortriterpenoide aus *Azadirachta indica* A. Juss (*Meliaceae*). Chem Ber 114:2375–2381

Kraus W, Cramer R, Sawitzki G (1981) Tetranortriterpenoids from the seeds of *Azadirachta indica*. Phytochemistry 20:117–120

Kraus W, Bokel M, Bruhn A, Cramer R, Klaiber I, Klenk A, Nagl G, Pohnl H, Sadlo H, Vogler B (1987) Structure determination by nmr of azadirachtin and related compounds from *Azadirachta indica* a. Juss (Meliaceae). Tetrahedron 43:2817–2830

Kumar C, Srinivas M, Yakkundi S (1996) Limonoids from the seeds of *Azadirachta indica*. Phytochemistry 43:451–455

Kumar GH, Vidya Priyadarsini R, Vinothini G, Vidjaya Letchoumy P, Nagini S (2010) The neem limonoids azadirachtin and nimbolide inhibit cell proliferation and induce apoptosis in an animal model of oral oncogenesis. Investig New Drugs 28:392–401

Kumar D, Haldar S, Gorain M, Kumar S, Mulani FA, Yadav AS, Miele L, Thulasiram HV, Kundu GC (2018) Epoxyazadiradione suppresses breast tumor growth through mitochondrial depolarization and caspase-dependent apoptosis by targeting PI3K/Akt pathway. BMC Cancer 18:52. https://doi.org/10.1186/s12885-017-3876-2

Lai SM, Lim KW, Cheng HK (1990) Margosa oil poisoning as a cause of toxic encephalopathy. Singap Med J 31:463–465

Lakshmi T, Krishnan V, Rajendran R, Mahusudhanan N (2015) *Azadirachta indica*: a herbal panacea in dentistry - an update. Pharmacogn Rev 9:41–44

Lale NES (2002) Bio-activity and limitation against wide spread use of neem products for the management of insect pests. Nigerian J Appl Biol 3:115–125

Lavie D, Jain MK, Kirson I (1967) Terpenoids. Part VI. The complete structure of melianone. J Chem Soc Commun:1347–1351

Lavie D, Levy EC, Jain MK (1971) Limonoids of biogenetic interest from *Melia Azadirachta* L. Tetrahedron 27:3927–3939

Lee SM, Olsen JI, Schweizer MP, Klocke JA (1988) 7-Deacetyl-17β-hydroxyazadiradione, a new limonoid insect growth inhibitor from *Azadirachta indica*. Phytochemistry 27:2773–2775

Ley SV, Denholm AA, Wood A (1993) The chemistry of azadirachtin. Nat Prod Rep 10:109–157

Liu JF, Hou CH, Lin FL, Tsao YT, Hou SM (2015) Nimbolide induces ROS-regulated apoptosis and inhibits cell migration in osteosarcoma. Int J Mol Sci 16:23405–23424

Luo XD, Ma YB, Wu SH, Wu DG (1999) Two novel Azadirachtin derivatives from *Azadirachta indica*. J Nat Prod 62:1022–1024

Luo XD, Wu SH, Ma YB, Wu DG (2001) Tetranortriterpenoids from insecticidal fraction of *Azadirachta indica*. Nat Prod Res Dev 13:9–13
Mitra CR, Garg HS, Pandey GN (1971) Identification of nimbidic acid and nimbidinin from *Azadirachta indica*. Phytochemistry 10:857–864
Mordue AJ, Nisbet AJ (2000) Azadirachtin from the neem tree *Azadirachta indica*: its action against insects. An Soc Entomol Brasil 29:615–632
Morgan ED, Thornton MD (1973) Azadirachtin in the fruit of melia azedaiuch. Phytochemistry 12:391–392
Murty KS, Rao DN, Rao DK, Murty LBG (1978) A preliminary study of hypoglycaemic and antihyperglycemic effect of *Azadirachta indica*. Indian J Pharm 10:247–250
Nagini S (2014) Neem limonoids as anticancer agents: modulation of cancer hallmarks and oncogenic signalling. In: Machinda Y, Lin C, Tamanoi F (eds) Enzymes: natural products and cancer signaling: isoprenoids, polyphenols and flavonoids. Elsevier, New York, pp 131–147
Nanduri S, Banstola KP (1995) Neeflone, a new tetranortriterpenoid from the flowers of *Azadirachta indica* A. Juss. Meliaceae. Indian J Chem B 34:1019–1020
National Academy of Science (1992) Neem, a tree for solving global problems. NAS, Washington, DC
Nelson VK, Ali A, Dutta N, Ghosh S, Jana M, Ganguli A, Komarov A, Paul S, Dwivedi V, Chatterjee S, Jana NR, Lakhotia SC, Chakrabarti G, Misra AK, Mandal SC, Pal M (2016) Azadiradione ameliorates polyglutamine expansion disease in *Drosophila* by potentiating DNA binding activity of heat shock factor 1. Oncotarget 7:78281–78296
Osman HAM, Hegazy AM (2013) Removal of cadmium from fresh water cultured Nile tilapia *Oreochromis niloticus* using neem leave water extract (NLWE) and neem leave powder (NLP). Nat Sci 11:12–20
Oyewole IO (2011) Ameliorating effect of methanolic leaf extract of *Azadirachta indica* (neem) on Arsenic-induced oxidative damage in rat liver. Int J Toxicol Appl Pharmacol 1:25–28
Pachapurkar RV, Kornule PM, Narayanan CR (1974) A new hexacyclic tetranortriterpenoid. Chem Lett 4:357–358
Paul R, Prasad M, Shah NK (2011) Anticancer biology of *Azadirachta indica* L. (neem): a mini review. Cancer Biol Ther 12:467–476
Pereira TB, Rocha e Silva LF, Amorim RCN, Melo MRS, de Souza RCZ, Eberlin MN, Lima ES, Vasconcellos MC, Pohlit AM (2014) *In-vitro* and *in-vivo* anti-malarial activity of limonoids isolated from the residual seed biomass from *Carapa guianensis* (andiroba) oil production. Malar J 13:317. https://doi.org/10.1186/1475-2875-13-317
Priyadarsini RV, Murugan RS, Sripriya P, Karunagaran D, Nagini S (2010) The neem limonoids azadirachtin and nimbolide induce cell cycle arrest and mitochondria mediated apoptosis in human cervical cancer (HeLa) cells. Free Radic Res 44:624–634
Prusti A, Mishra SR, Sahoo S, Mishra SK (2008) Antibacterial activity of some indian medicinal plants. Ethnobotanical Leaflets 12:227–230
Rajab MS, Bentley MD, Fort RC (1988) Biomimetic formation of a nimbin class limonoid. J Nat Prod 51:1291–1293
Rembold H (1989) The azadirachtins. Their structure and mode of action. ACS Symp Ser 387:150–163
Rojatkar SR, Nagasampagi BA (1995) 3-alpha-acetoxy-1-alpha-hydroxyazadirachtol, a new constituent from *Azadirachta indica* (A-juss) seeds. Indian J Chem B 34:1016–1018
Rojatkar SR, Bhat VS, Kulkarni MM, Joshi VS, Nagasampagi B (1989) Tetranortriterpenoids from *Azadirachta indica*. Phytochemistry 28:203–205
Severino RP, Fernandes JB, Vieira PC, Silva MFGF, Batista Pereira LG, Camargo-Dietrich CGC, Pereira DA, Costa-Leonardo AM, Bueno OC (2007) Biological activity of limonoids from meliaceae against a subterranean termite (*Heterotermes tenuis*) (Isoptera: Rhinotermitidae). Sociobiology 50:947–957
Siddiqui S (1945) Utilization of nim oil and its bitter constituents (nimbidin series) in the pharmaceutical industry. J Sci Ind Res 4:5–10

Siddiqui BS, Ghiasuddin FS (1998) Tetracyclic triterpenoids of the fruit coats of *Azadirachta indica*. Phytochemistry 47:1631–1636

Siddiqui S, Siddiqui BS, Faizi S (1985) Studies in the chemical constituents of *Azadirachta indica* part II: isolation and structure of the new triterpenoid Azadirachtol. Planta Med 51:478–480

Siddiqui S, Faizi S, Mahmood T, Siddiqui BS (1986a) Two new insect growth regulator meliacins from *Azadirachta indica* A. Juss (meliaceae). J Chem Soc Perkin Trans 1:1021–1025

Siddiqui S, Faizi S, Mahmood T, Siddiqui BS (1986b) Isolation of a new tetranortrietrpenoid from *Azadirachta indica* A. Juss (Meliaceae). Heterocycles 24:1319–1324

Siddiqui S, Faizi S, Siddiqui BS (1986c) Studies on the chemical constituents of *Azadirachta indica* A. Juss (Meliaceae), Part VII [1]. Z Naturforsch B 41:922–924

Siddiqui S, Mahmood T, Siddiqui BS, Faizi S (1986d) Isolation of a triterpenoid from *Azadirachta indica*. Phytochemistry 25:2183–2185

Siddiqui S, Mahmood T, Faizi S, Siddiqui BS (1987a) Studies in the chemical constituents of *Azadirachta indica* A. Juss (meliaceae). Part 10. Isolation and structure elucidation of isonimolicinolide, the first 17-acetoxy tetranortriterpenoid and nimolicinoic acid, the first hexanortriterpenoid with an apoeuphane (apotirucallane) skeleton. J Chem Soc Perkin Trans 1:1429–1432

Siddiqui S, Mahmood T, Siddiqui BS, Faizi S (1987b) Isonimolide and isolimbolide, two new tetranortriterpenoids from the twigs of *Azadirachta indica* A. Juss (Meliaceae). Heterocycles 26:1827–1833

Siddiqui S, Siddiqui BS, Ghiasuddin, Faizi S (1991) Terpenoids from fruit coatings of *Azadirachta indica*. Phytochemistry 30:1615–1619

Siddiqui S, Faizi S, Siddiqui BS, Ghiasuddin (1992a) Constituents of *Azadirachta indica*: isolation and structure elucidation of a new antibacterial tetranortriterpenoid, mahmoodin, and a new protolimonoid, naheedin. J Nat Prod 55:303–310

Siddiqui BS, Ghiasuddin, Faizi S, Siddiqui S (1992b) Triterpenoids from the fresh fruit coats of *Azadirachta indica*. Phytochemistry 31:4275–4278

Siddiqui BS, Afshan F, Ghiasuddin, Faizi S, Naqvi SN, Tariq RM (1999a) New insect-growth-regulator meliacin butenolides from the leaves of *Azadirachta indica* A. Juss. J Chem Soc Perkin Trans 1:2367–2370

Siddiqui BS, Ghiasuddin G, Faizi S, Rasheed M (1999b) Triterpenoids of the fruit coats of *Azadirachta indica*. J Nat Prod 62:1006–1009

Siddiqui BS, Afshan F, Ghiasuddin, Faizi S, Naqvi SNH, Tariq RM (2000) Two insecticidal tetranortriterpenoids from *Azadirachta indica*. Phytochemistry 53:371–376

Siddiqui S, Mahmood Siddiqui BS, Afshan F, Faizi S (2001) Three novel tetracyclic triterpenoids of biogenetic interest from the leaves of *Azadirachta indica*. Tetrahedron 57:10281–10286

Siddiqui BS, Afshan F, Faizi S, Naqvi SNH, Tariq RM (2002) Two new triterpenoids from *Azadirachta indica* and their insecticidal activity. J Nat Prod 65:1216–1218

Siddiqui BS, Afshan F, Gulzar T, Sultana R, Naqvi SNH, Tariq RM (2003a) Tetracyclic triterpenoids from the leaves of *Azadirachta indica* and their insecticidal activities. Chem Pharm Bull 51:415–417

Siddiqui BS, Ali ST, Rasheed M, Kardar MN (2003b) Chemical constituents of the flowers of *Azadirachta indica*. Helv Chim Acta 86:2787–2796

Sinniah D, Baskaran G (1981) Margosa oil poisoning as a cause of Reye's syndrome. Lancet 317:487–489

Subramani R, Gonzalez E, Arumugam A, Nandy S, Gonzalez V, Medel J (2016) Nimbolide inhibits pancreatic cancer growth and metastasis through ROS-mediated apoptosis and inhibition of epithelial-to-mesenchymal transition. Sci Rep 6:19819

Sultana B, Anwar F, Przybylski RI (2007) Antioxidant activity of phenolic components present in barks of *Azadirachta indica*, *Terminalia arjuna*, *Acacia nilotica*, and *Eugenia jambolana* Lam. trees. Food Chem 104:1106–1114

Suman S, Das TP, Reddy R, Nyakeriga AM, Luevano JE, Konwar D, Pahari P, Damodaran C (2014) The pro-apoptotic role of autophagy in breast cancer. Br J Cancer 111:309–317

Sundaravalli N, Bhaskar Raju B, Krishnamoorthy KA (1982) Neem oil poisoning. Indian J Pediatr 49:357–359

Suresh G, Narasimhan NS, Palani N (1997) Structure of nimonol from fresh whole green leaves of *Azadirachta indica*. Phytochemistry 45:807–810

Tan QG, Luo XD (2011) Meliaceous limonoids: chemistry and biological activities. Chem Rev 111:7437–7522

Tharmarajah L, Samarakoon SR, Ediriweera MK, Piyathilaka P, Tennekoon KH, Senathilake KS, Rajagopalan U, Galhena PB, Thabrew I (2017) *In vitro* anticancer effect of gedunin on human teratocarcinomal (NTERA-2) cancer stem-like cells. Biomed Res Int 2017:2413197. https://doi.org/10.1155/2017/2413197

Uko OJ, Kamalu NT (2001) The neem tree-uses and potentials. Niger J Exp Appl Biol 2:223–229

UNEP (2012) Neem: the UN's tree of the 21st century. United Nations Environment Programme, Nairobi

Zhu J, Lu X, Fan X, Wu R, Diao H, Yu R, Xu H, Zi J (2018) A new cytotoxic salannin-class limonoid alkaloid from seeds of *Azadirachta indica* A. Juss. Chin Chem Lett 29:1261–1263

Role of Plant Secondary Metabolites as Antidiabetic Agents

18

Varsha Vasantrao Sonkamble, Nilesh Shirish Wagh, and Sandeep Ramchandra Pai

Contents

Abstract

Plant kingdom is considered to be a convenient source for potential therapeutic drugs. It is a preferred choice due to their easy availability, affordability and considered safe with minimal side effects. Owing to these advantages, enormous efforts have been routed toward search for effective plant-derived drugs against life-threatening diseases like cancer, diabetes, and other disorders in cardiovascu-

Varsha Vasantrao Sonkamble and Nilesh Shirish Wagh have equally contributed for this chapter.

V. V. Sonkamble (✉)
School of Life Sciences, Swami Ramanand Teerth Marathwada University, Nanded, Maharashtra, India
e-mail: vvsonkamble@gmail.com

N. S. Wagh · S. R. Pai
Amity Institute of Biotechnology, Amity University, Mumbai, Maharashtra, India
e-mail: waghnil@gmail.com

M. S. Akhtar et al. (eds.), *Natural Bio-active Compounds*,
https://doi.org/10.1007/978-981-13-7154-7_18

lar, neurological, respiratory systems, etc. Nowadays, diabetes is one of the most complex metabolic disorders affecting the pathophysiology of individuals of almost all age groups worldwide. Currently, antidiabetic drugs used for diabetes management are majorly derived from synthetic source. However, these synthetic drugs come with serious complications like hypoglycemia, hyperinsulinemia, fatigue, nausea, toxicity in kidney and liver, etc. Whereas, plants offer a quiver of secondary metabolites which can act as effective, safe antidiabetic drugs with minimal or no side effects. Considering these advantages, this chapter is focused on reporting the uses of plants, or plant parts, and/or plant extracts in diabetes treatment. It is also centered towards assembling the recent studies on uses of plants acting on different targets for controlling diabetes. The present chapter highlights the role of plant secondary metabolites as an acting molecule on different targets for diabetes treatment.

Keywords
Antidiabetic agents · Diabetes · Natural drugs · Secondary metabolites

18.1 Introduction to Secondary Metabolites

Secondary metabolites are organic compounds derived from primary metabolites that support the growth and development of plants without a fundamental role in the maintenance of life processes in plants (Rungsung et al. 2015). A huge number of secondary metabolites with chemically differing structures have been evolved during the plants development stages to ward off herbivores, pathogens, and to benefit for plant's survival. Pharmacologically, these secondary metabolites have been utilized by people for different purposes like for the treatment of health disorders and infections; as flavors, perfume, toxins, arrow poisons, and pesticides; etc. (Wink 2015). Majorly, these secondary metabolites fall into classes comprising terpenoids, phenolics, nitrogen-containing alkaloids, and sulfur-containing compounds (Rungsung et al. 2015). They can be unique to specific species or genera (Kennedy and Wightman 2011).

In most of the developing countries, around 75–80% of the population still uses plant-based medicines as a primary healthcare (Akhtar et al. 2014; Gezahegn et al. 2015; Swamy et al. 2016; Manisha and Kumar 2018). These plant-based medicines are nothing but the secondary metabolites in the form of phytomedicines (Briskin 2016; Akhtar and Swamy 2018a, b). These are the most established known healthcare, accessible to humanity, and used widely in the naturopathic, ayurvedic, homeopathic, and other traditional medical systems that involve natural sources (Vinayagam et al. 2015). They have been proven to possess many fascinating applications far beyond their notable medical uses like antimicrobials, antitumor, cholesterol-lowering, immunosuppressant, antidiabetic, antiprotozoal, anthelminthic, antiviral, antimutagenic, scavenging activity on free radicals, prevention of cardiovascular heart disorder activities and cancer, and many more (Vaishnav and Demain 2011; Ozcan et al. 2014; Akhtar and Swamy 2018a, b).

Among the life-threatening diseases, diabetes is one of the most complex metabolic disorders. It leads to adversities in cerebrovascular, sexual, renal, neurological, and pathophysiological functioning leading to sublethal to lethal conditions in an individual (El-Abhar and Schaalan 2014). Enormous efforts are being carried out towards finding antidiabetic agents restoring the normal metabolic functioning in a diabetic patient. In this regard, antidiabetic agents in the form of plant-derived drugs provide cost-effective and safe remedy for diabetes. Plant-derived drugs in the form of secondary metabolites show potential of regulating impaired glucose metabolism, β-cell functioning, GLP-1 homeostasis, and restoration of insulin levels. The present chapter highlights the role of plant secondary metabolites as an acting molecule on different targets for diabetes treatment.

18.2 Diabetes and Its Cure through Plant Secondary Metabolites

18.2.1 Diabetes

Diabetes mellitus is a worldwide perpetual metabolic issue that results from the unavailability of reduced discharge as well as hampered function of insulin (El-Abhar and Schaalan 2014). Earlier it was described in two forms, one as a genetic disorder and other as a dietary affiliated disorder. Furthermore, it is portrayed as type 1 (insulin-dependent diabetes mellitus), considered as heritable and which can be cured by insulin, while type 2 (non-insulin-dependent diabetes mellitus) is found in elderly persons and can be controlled through proper diet and oral hypoglycemic drugs (Patel et al. 2012). Another category of diabetes is diagnosed in pregnant ladies (devoid of any diabetic history), known as gestational diabetes mellitus (Choudhury et al. 2017). Indications for diabetic circumstances may comprise high-intensity sugar in blood, infrequent thirst, extreme hunger, frequent urination, weight loss, distorted vision, extreme weakness and tiredness, nausea and vomiting, mood changes, irritability, and so on (Tayyab et al. 2012). Considering the adversity and pathophysiology of this complication, it can be managed by dietary limitation, by physical workout, as well as by consuming various synthetic oral hypoglycemic agents and insulin. Since diabetes is a multifactorial disorder, different targets could be used for treatment (El-Abhar and Schaalan 2014). But, synthetic oral hypoglycemic medications which are at present utilized for diabetes treatment have undesirable side effects with high secondary failure rates. These include complications like hypoglycemia, hyperinsulinemia, fatigue, nausea, and toxicity in the kidney and liver (Lacroix and Li-Chan 2014). So, herbs and/or herbal compounds with an alternative to synthetic drugs are recommended for the treatment of diabetes mellitus (Coman et al. 2012).

18.2.2 Herbs and Diabetes

Diet plays a key role in diabetes management. Since decades, ample studies have been explored to establish the correlation between intake of food products or nutrients and incidence of diabetes. Additionally, their impacts on metabolic disorders have also been studied (Lacroix and Li-Chan 2014). However, other than the daily dietary food, several medicinal plants have been reported to show antidiabetic activity by multiple means. In many countries, herbs have been an apt part of traditional medicine for diabetes (Assad and Morse 2013). Thus, herbal medicines are administered nowadays for the well-being, prevention, and treatment of diabetes (Mazumder and Choudhury 2010).

Methanolic extract of *Gymnema sylvestre* and *Andrographis paniculata* leaves possesses significant antihyperglycemic and antioxidative effect along with anti-inflammatory properties as evident from the first week of treatment in streptozotocin-induced diabetes in Sprague Dawley rats (Kumar et al. 2017). *Aloe vera* gel, a very common traditional healthcare product, has a therapeutic effect on diabetes (Jerine Peter and Sabina 2016). The gel comprises of minerals, amino acids, polyphenols, and polysaccharides as active ingredients. Continuous administration of *Aloe vera* juice or gel twice a day reduces blood sugar and triglyceride levels (Jerine Peter and Sabina 2016). *Ocimum sanctum* is also one of the Indian traditional therapeutic plants whose whole plant extract effectively improves diabetes mellitus through its insulin-potentiating and antioxidant activities. The methanol extract of *Ocimum sanctum* (root, stem, leaves, and flowers) was assessed for hypoglycemic activities in streptozotocin-induced mice diabetic models which revealed that daily administration of the extract through oral route for 28 days produced a gentle but sustained reduction in blood glucose levels in diabetic-treated mice (Kumar and Kumar 2016). *Zygophyllum album* is conventionally used in Tunisia diabetes management, digestive tract spasm, and hypertension. To ratify the study, the leaves of *Z. album* were extracted which showed the presence of essential oils. Extracted essential oils were examined for antidiabetic, antidiarrheal, and antihypertensive activities in alloxan-induced diabetic rats. The outcomes revealed a significant decrease in α-amylase activity in the pancreas along with reduced serum glucose level and lower glycated hemoglobin rate post administration of oil. Furthermore, the oil treatment also showed inhibition of key-digestive enzymes (related to diabetes like angiotensin-converting enzyme and pancreatic lipases), hypertension and reduced the symptoms of diarrhea in alloxan-induced diabetic rats (Mnafgui et al. 2016).

Reports are available stating the antidiabetic potential of ethanolic extracts of *Coccinia grandis* leaves. Significantly increased serum insulin levels and body weight along with reduced blood glucose level was observed in rats on treatment of this extract. Their kidney and liver functions along with the lipid profiles were also optimally maintained (Mohammed et al. 2016). *Prunella vulgaris* is another conventionally used herb in diabetes management. Studies were carried out to elucidate the active components of *P. vulgaris* and their acute and sub-chronic antidiabetic, antioxidant, and antinociceptive properties in mice (Raafat et al. 2016). Rosmarinic acid, p-coumaric acid, and caffeic acid were found to be among the active components of

Prunella. Further, the caffeic acid-rich fraction was able to inhibit α-amylase and α-glucosidase enzymes along with improving the oxidative stress and reducing blood glucose levels in mice. It also increased ameliorated thermal hyperalgesia, tactile allodynia, and serum insulin with a potential of reinstating the lipid peroxide levels (Raafat et al. 2016). Similarly, *Catharanthus roseus* is considered as one of the traditional medicines and is used in India, South Africa, China, and Malaysia as a therapeutic agent for diabetes mellitus. A study was conducted by feeding ethanolic extract of *C. roseus* in streptozotocin-induced diabetic Wistar rats. The study demonstrated improved levels of insulin released in islets of Langerhans and showed good correlation with intracellular calcium and hence *C. roseus* assets as an herbal drug to cure deadly diabetes (Al-Shaqha et al. 2015).

Coptis chinensis Franch is also a Chinese medicine traditionally used with multiple pharmacological effects like antidiabetic, anti-inflammation, and anticancer activity. The antidiabetic activity investigations on male Wistar rats using *Coptis* showed a good response against fasting blood glucose, triglyceride, total cholesterol, glutathione, glutathione peroxidases, superoxide dismutases, catalase, malondialdehyde, c-jun *n*-terminal kinase, phosphorylated insulin receptor substrate 1, phosphorylated phosphatidylinositol 3 kinase, and glucose transporter-4 (GLUT-4) (Jiang et al. 2015). *Emblica officinalis* is another traditional Indian medicine of which all plant parts including fruit, seed, leaves, root, bark, and flower are used for various herbal preparations. The leaves showed antipyretic, anti-inflammatory, antioxidant potential, and antidiabetic effects (Jerine Peter and Sabina 2016). The fruit extract showed the potential of reducing cardiac complications, hyperglycemia, and diabetic neuropathy as well as nephropathy. The contents of *E. officinalis* fruit extract revealed the presence of gallic acids, along with gallotanins, ellagic acids, as well as corilagins (D'souza et al. 2014).

Marrubium vulgare is traditionally used in Algeria for diabetes treatment. The analysis of aqueous extracts of leaves was conducted which showed the presence of 15 metabolites belonging to the class of polyphenols. Using the same extracts, in vivo trials were carried out on albinos Wistar rats for diabetic studies which revealed a notable decrease in blood glucose as well as a significant lowering of total lipids, triglycerides, and total cholesterol levels (Boudjelal et al. 2012). *Ophiopogon japonicus* is one of the traditional Chinese medicinal herbs whose roots are used for diabetes treatment. The root contains water-soluble polysaccharide which can be extracted via hot water treatment, shows increased insulin and reduced blood glucose levels, and helps in remediating pancreatic islets destruction when studied in rats (Chen et al. 2011). People of Southern Nigeria use *Axonopus compressus* for diabetes management. Its antidiabetic potential was investigated by studying the efficacy of methanolic leaf extract of the plant in rats which confirmed a decrease in blood glucose levels (Ibeh and Ezeaja 2011). Fruits of *Vaccinium arctostaphylos* are traditionally used in Iran for diabetes treatment. In a study, the efficacy of these fruit extracts was evaluated in the treatment of type 2 diabetic patients. The outcome revealed that the fruit extracts lowered the glucose levels in blood during fasting conditions without any side effects on liver or kidney functions (Kianbakht et al. 2013). Along with the potential of being an antidiabetic agent, these fruit extracts were also reported for antihyperglycemic, antioxidant, and triglyceride-lowering effect activities (Feshani et al. 2011). Conventionally, fruit, barks, and leaves of *Dillenia indica* Linn. are usually used for various purposes. Reports are available stating that methanolic extract of *D. indica* leaves has an antidiabetic and antihyperlipidemic activity which can reduce triglyceride, cholesterol, and transaminase levels in serum (Kumar et al. 2011a, b). Apart from this, alcoholic extracts of *Alangium lamarckii* leaves showed remarkable antidiabetic activity when administered in the streptozotocin-nicotinamide-induced diabetic rats (Kumar et al. 2011a, b), whereas the methanol extract of *A. salvifolium* fruit showed significant hypoglycemic, antidiabetic activity along with a reduced rate of body weight loss in alloxan-induced diabetic and normal animals (Mishra and Garg 2011).

Ginseng, a traditional medicine, is used for over centuries to treat both type 1 and type 2 diabetic patients (Hui et al. 2009). Stimulation of insulin secretion, removal of blockages of intestinal glucose absorption, as well as enhancement in the utilization of glucose periphery were observed after usage of complete ginseng extract. In addition to its antidiabetic properties, pharmacological studies affirm that ginseng possesses multiple actions like neuroprotective, immune modulatory, and anticancerous and has effects on the central nervous system. Moreover, anti-inflammatory, antioxidant, immunostimulating, and anti-apoptosis properties were observed in presence of ginsenosides, one of the active ingredients from *Ginseng* (Hui et al. 2009). Besides this, *Artemis sphaerocephala* Krasch seed powder, one of the traditional Chinese medicines, is also used for the treatment of diabetes. Gum extracted from the seed powder when administered to type 2 diabetic rats via food showed significant lowering in serum cholesterol and triglyceride levels along with low blood glucose levels in fasting condition. The gum also surpassed hyperglycemia and hyperlipemia in a diabetic patient (Xing et al. 2009). Ethanolic extract of *Setaria italica* seed was explored for antimicrobial, hypoglycemic, and antidiarrheal activities which revealed that the extracts showed good antimicrobial activity against Gram-negative bacteria and a promising hypoglycemic property of lowering blood glucose level (Raafat et al. 2016).

As per a recent report, *Flacourtia indica*, *Galium aparine*, *Ipomoea purpurea*, *Mallotus philippensis*, *Mentha longifolia*, and *Fumaria indica* are used as antioxidant, antimicrobial, and antidiabetic agents (Jerine Peter and Sabina 2016). Studies on *Morus alba*, a natural antidiabetic plant, showed the presence of 15 bio-active molecules with α-glucosidase and tyrosinase inhibitory effects including flavanes, prenylated stilbenes, and iminosugars in its leaf extract (Coman et al. 2012). Similarly, studies on *Brassica juncea* showed presence of isothiocyanate, glycoside, singrin, protein, and fixed oil which act as food adjuvants for diabetic patients (Chawla et al. 2013), and specifically its leaves showed the potential of hypoglycemic and antioxidant activity (Najmi et al. 2012). It was also recently reported that administration of *Galega officinalis* extract in diabetic condition upholds the restoration of neutrophils bone marrow pool in addition to the reduction of lymphoblasts number and inhibiting lymphocytes apoptosis process. It was speculated in the study that stabilization of neutrophil functional capability using *G. officinalis* extract can improve the disease through its hypoglycemic action and may further prevent the development and progression of diabetes complications (Nagalievska et al. 2018). Researchers are now investigating the antidiabetic potential of several traditional medicines in the form of extracts procured from different herbs or their plant parts. Few of such plants like *Cinnamomum cassia*, *Scoparia dulcis*, *Ficus racemosa* bark, and *Portulaca oleracea* seeds have reached up to clinical studies in human patients for the elucidation of their antidiabetic potential. Furthermore, laboratory research on few herbal products has reached the diabetic patients by the brand names Diabecon®, Glyoherb® and Diabeta Plus® (Choudhury et al. 2017).

18.3 Herbs and Different Targets in Diabetes Treatment

Potential of herbs in diabetes management through different possible targets has been elaborated in the below sections (Fig. 18.1).

18.3.1 In the Regulation of Insulin Resistance

Insulin resistance is an unusual state of function in which pancreatic β-cells producing insulin are unable to generate a signal transduction pathway in the respective organs like the liver and muscles including adipose tissues. Subsequently, this results in hyperglycemia, hyperinsulinemia, and fatty acid and lipid dysregulation which may be dominant under obesity conditions. Therefore, regulation of insulin resistance is one of the treatments for diabetes (Belwal et al. 2017). Various herbs, as well as herbal compounds, can target or regulate insulin resistance.

Roots of *Helicteres angustifolia* have been used as a folk herbal drug to treat cancer, bacterial infections, inflammations and flu in China. However, the recent

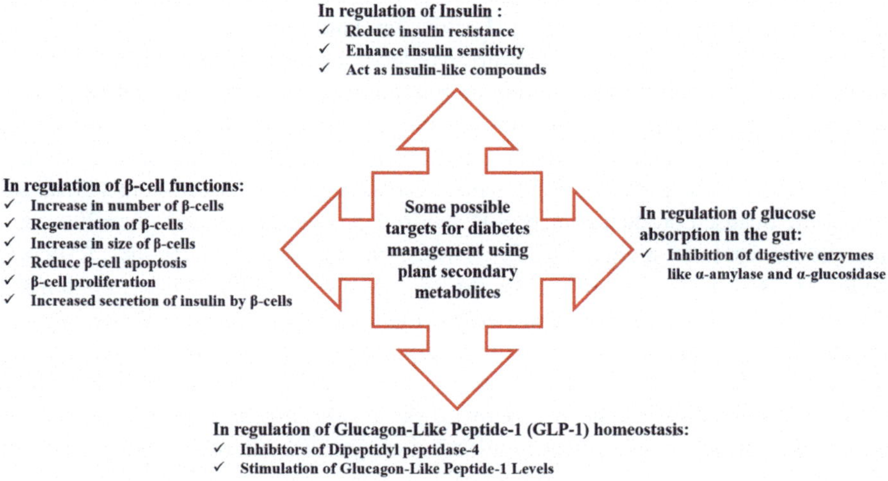

Fig. 18.1 Some possible targets for diabetes management using plant secondary metabolites

studies also report for its antidiabetic potential (Hu et al. 2016). Ethanolic extract of *H. angustifolia* roots exhibited the antidiabetic effect on diabetic rats and could also decrease the insulin resistance (Hu et al. 2016). Similarly, the administration of a commercial blend of *Fucus vesiculosus* and *Ascophyllum nodosum* was associated with improved insulin sensitivity and regulation (Sharifuddin et al. 2015). Reports on the hydroalcoholic extract of *Urtica dioica* leaves showed hypoglycemic activities in male Wistar rats with fructose-induced insulin resistance and a significant reduction in plasma glucose level and fasting insulin resistance index (Choudhury et al. 2017). A recent study on the efficacy of the ethanolic extract of *Anacardium occidentale* Linn. leaves as antidiabetic agent on neonatal streptozotocin-diabetic rats revealed dropped levels of glycated hemoglobin, fasting blood glucose, serum insulin and lipid parameters, as well as renal function biomarkers (Jaiswal et al. 2017). Similarly, *Cinnamomum verum* one of the medicinal plants used worldwide in daily consumption is found to have health-benefiting properties like anti-obesity, dropping blood pressure and plasma glucose levels along with ameliorating dyslipidemia. Owing to its health benefits, it is thus regarded as a probable metabolic syndrome decreasing agent (Mollazadeh and Hosseinzadeh 2016). Beside this, aqueous extract of *Afzelia africana* stem bark was studied for antidiabetic potential and its hematological parameters which revealed its antihyperglycemic properties and aid in avoiding several diabetic complications and improvement in hematological parameters (Oyedemi et al. 2011). Available reports on *Uvaria chamae* extract affirm its potential to act as an antidiabetic and hypolipidemic agent by reducing the levels of plasma glucose, cholesterol, and low-density lipoproteins and further weight loss (Emordi et al. 2016), while the studies on the ethanol extract of *Symplocos cochinchinensis* disclosed the enhanced sensitivity of insulin via downregulation of lipogenesis along with high fructose insulin resistance (Antu et al. 2016). The rhizome of *Gastrodia elata* Blume is widely used in traditional Chinese herbal medicine to treat brain-related diseases like a headache, migraine, dizziness, epilepsy, infantile convulsion, Alzheimer's disease, and stroke. Apart from these health stimulatory properties, it also has some known anti-obesity properties; precisely, it reduces insulin resistance in diet-induced obese rats by decreasing fat accumulation in adipocytes (Kim et al. 2017). As per the available literature, *Biophytum sensitivum* is said to have an insulin-like compound which made it superior for diabetic treatment (Manisha and Kumar 2018). Using *B. sensitivum* along with a combination of 15 different herbs, an ayurvedic licensed product named as Diabedrink has been marketed since 2002. Its ingredients include plants like *B. sensitivum*, *Strychnos potatorum*, *Cyperus rotundus*, *Vetiveria zizanioides*, *Centella asiatica*, *Cyclea peltata*, *Salacia oblonga*, *Coscinium fenestratum*, *Syzygium cumini*, *Embelia ribes*, *Curcuma longa*, *Aerva lanata*, *E. officinalis*, *Mangifera indica*, *Terminalia chebula*, and *Ziziphus jujuba*. The product, however, showed inadequately controlled blood sugar levels, but when studied individually, methanolic extract of *B. sensitivum* leaves disclosed a remarkable insulinotropic effect in alloxan-induced diabetic rabbit which may be regulated through instigating the beta cells of Langerhans to synthesize or release insulin (Manisha and Kumar 2018). From a similar report, it was found that amorfrutin, a natural product present in eatable parts of *Glycyrrhiza foetida* and *Amorpha fruticosa* legumes, could be used as a novel and influential antidiabetic agent. The report also states that treating with amorfrutin prevents resistance to insulin and improves other metabolic and inflammatory parameters without any adverse effects (Weidner et al. 2012). A similar study on total polysaccharides extracted from *Pleurotus ostreatus* assessed for antidiabetic potential revealed the reduction of hyperglycemia and hyperlipidemia levels, improvement in insulin resistance condition, and an increase in glycogen storage. Moreover, it also increased the activity of superoxide dismutase, catalase, as well as glutathione peroxidase and reduced oxidative damage risk and malondialdehyde level (Zhang et al. 2016). Reports on oligonol, a flavanol-rich lychee fruit extract, revealed anti-obesity property in vitro and in vivo. The study revealed that oligonol reduced intracellular lipid contents in the liver and skeletal muscle along with suppressed inflammatory markers and showed insulinotropic effects (Nazrul Islam 2016).

It has also been reported in the literature that blueberries which are a part of our diet consist of anthocyanin as bio-active compounds which epidemiologically, through the preclinical and clinical studies, proved to reduce the risk of diabetes via improving insulin resistance condition (Stull 2016). Similarly, in vivo and ex vivo reports on garlic oil showed its potential of improving insulin and oral glucose tolerance, along with the insulin-stimulated consumption of glucose for producing glycogen in skeletal muscles (Liu et al. 2012). It is also reported that garlic improved glycemic control through increased insulin secretion and enhanced insulin sensitivity (Xie et al. 2011). Similarly, resveratrol improves insulin action in animals, by stimulating intracellular glucose transport and increasing glucose uptake by various cells, while, in the absence of insulin, resveratrol showed its stimulating action on glucose transport (Choudhury et al. 2017).

18.3.2 In the Regulation of β-Cell Functions

β-Cell dysfunction with progressive loss of pancreatic *β*-cell insulin secretion consequently develops insulin resistance, which are the key defects related to alteration from a healthy glycemic state to hyperglycemia (Cersosimo et al. 2014). Since β-cells of the pancreas are responsible for secretion of insulin, many models have been developed to study the pancreatic secretion of insulin. These comprise intact and detached pancreatic islets, perfused pancreas, purified β-cells, and insulin-discharging β-cell lines (Kazeem and Davies 2016). So, this section highlights the recent reports of herbs or herbal compositions which help in regulating β-cell functions for management of diabetes. In this endeavor, a study on diabetic rabbits treated with aqueous extract of *Abroma augusta* leaves states that, on histological analysis of pancreas, an increased number of β-cells were observed (Hussain Mir et al. 2013). A similar study involving aqueous ethanolic extract of *Anacardium occidentale* bark also showed regeneration in pancreatic β-cells (Bassey et al. 2011), while ethanolic extract of *Tecoma stans* stem showed antidiabetic activity which was found to be due to raised insulin secretion from β-cells of the pancreas (Arif et al. 2014). Similarly, the antidiabetic study using methanolic leaf extract of *Anastatica hierochuntica* in rats revealed regeneration of β-cells, weight loss, and synergetic hypoglycemic and hypolipidemic effect and antioxidant activity (Shaban et al. 2011). Another study comprising a treatment of hydroalcoholic extract of *Trigonella foenum-graceum* seed to neonatal streptozotocin-induced type 2 diabetic rats showed a substantial potentiation in glycosylated hemoglobin levels and serum glucose levels with an appropriate weight loss followed by increased number and size of pancreatic β-cells islet (Kulkarni et al. 2012). Similarly, a study comprising of ethanol extracts of *Amaranthus caudatus*, *A. spinosus*, and *A. viridis* leaves for potent antidiabetic and anti-cholesterolemic activity showed their potential to regenerate β-cells and decrease serum lipid levels along with significant weight loss (Girija et al. 2011). Available reports on *Allium sativum* state that it is traditionally used as a hypoglycemic agent against type 1 and type 2 diabetes mellitus wherein its oil when used in rats shows significant effects like reduced body weight, increased serum insulin levels, as well as regeneration of pancreatic islets (Alashkham et al. 2013). Besides this, mangiferin, a xanthonoid found in plants like *Mangifera indica* and *Iris unguicularis*, is suggested to have antioxidant, antitumor, antiviral, and immune-modulatory activities. Mangiferin has been also reported for having antidiabetic activity by reducing β-cell apoptosis and improving β-cell proliferation and hyperplasia, as well as increasing the serum insulin levels, glucose tolerance, and glycemia (Wang et al. 2014). Reports available on curcumin, isolated from rhizomes of *Curcuma longa*, state its potential of possessing antioxidant, antitumor, anti-inflammatory, and antidiabetic activity. Antidiabetic studies using curcumin in diabetic rats showed a gradual decrease in glucose and increase in serum insulin levels along with an enhanced number of pancreatic β-cells (Abdel Aziz et al. 2013). Whereas, conophylline, one of the alkaloids present in plants like *Ervatamia microphylla*, can differentiate and generate pancreatic β-cells in vitro and in vivo (Kodera et al. 2009). One of the polyynes cytopiloyne, isolated from *Bidens pilosa*, showed enhanced glycemic control wherein it regulates β-cells for increased secretion or expression of insulin and protection of islets (Li et al. 2012). Besides this, studies conducted on *Anoectochilus roxburghii* showed the presence of kinsenoside as a major constituent which can control hypoglycemic activity in mice via β-cell repair or regeneration (Qi et al. 2018).

18.3.3 In the Regulation of Glucagon-like Peptide-1 (GLP-1) Homeostasis

Glucagon-like peptide-1 (GLP-1) is an intestinally secreted endocrine L cell-derived peptide, whose receptors are found in β-cells of islets, the brain, cardiovascular system, and lungs (Wang et al. 2015). During hyperglycemia, this GLP-1 helps in decreasing blood glucose levels by stimulating insulin secretion as well as by reducing glucose-dependent glucagon secretion (Wang et al. 2015). GLP-1 have an intense effect on stimulation of insulin release from the pancreas, however, for a very short period. This happens due to breakdown by an enzyme, dipeptidyl peptidase-4 (DPP-4). Thus, it is said that herbal drug that could inhibit the action of DPP-4 would extend the insulin-releasing effect of GLP-1 (Bethala et al. 2013). Hence, DPP-4 inhibitors providing beneficial effects of mimicking or enhancing GLP-1 activity would make them a promising drug candidate for the treatment of type 2 diabetes (Lacroix and Li-Chan 2014; Velásquez et al. 2010). In this perspective, a study conducted on the root, bark, and rhizome of the *Berberis vulgaris* showed antidiabetic potency by means of insulin secretion, glycolysis stimulation, inhibiting DPP-4 leading to enhanced hyperglycemic activity, and increasing GLUT-4, and GLP-1in a rat model (Singh et al. 2015). Additionally, they were clinically used as dysentery, antimicrobial, anti-inflammatory, and cholesterol level-lowering agents (Rungsung et al. 2015). Similar effects were produced by *Momordica charantia* fruit in mice, wherein high serum GLP-1 and lower glucose levels were observed. Besides this, ingestion of bark extract of *Cinnamomum zeylanicum* to a diabetic patient reduced postprandial serum insulin with an increase in the concentration of GLP-1 devoid of disturbing blood glucose (Singh et al. 2015). Moreover, fruits of *Mangifera indica* and *Gardenia jasminoides*, roots of *Glycine max* and *Smallanthus sonchifolius*, leaves of *Ilex paraguariensis* and *Artemisia dracunculus* L., seeds of *Pinus koraiensis*, fibers of *Triticum aestivum*, and bark of *Prunus africana* have been reported to show potential activity of GLP-1 secretion when studied under in vitro and in vivo conditions (Singh et al. 2015). Another study involving water extracts of some plants like *Padina sulcata, Halimeda macroloba, Turbinaria conoides, and Sargassum binderi* showed α-glucosidase and DPP-4 inhibitory potentials, while butanol fraction of *Padina sulcata* and *Sargassum binderi* helped to stimulate GLP-1 secretion (Sharifuddin et al. 2015; Chin et al. 2015). Similarly, ethanolic extract of *Ipomoea batatas* L is found to be rich in derivatives of caffeoylquinic acid and can increase the GLP-1 secretion in a diabetic patient (Domínguez Avila et al. 2017). Apart from this, metformin one of the standard antidiabetic drugs have been suggested to help preserve pancreatic β-cell functions and to increase the concentration of GLP-1 (Lacroix and Li-Chan 2014). Monounsaturated fatty acids from olive oils and α-linolenic acids help to increase GLP-1 blood concentrations, glucose-insulinotropic peptide blood concentrations, as well as high-density lipoprotein cholesterol blood concentrations, while saccharides like glucose, fructose, chitosan, and sucrose significantly increased GLP-1 secretion when studied in rats and humans (Bodnaruc et al. 2016), while saccharides like glucose, fructose, chitosan, and sucrose significantly increased GLP-1 secretion. Similarly, dietary fibers (fermentable), short-chain fatty acids, meat hydrolysate, as well as corn and milk proteins also help in increased secretion of GLP-1 and thereby regulate diabetes (Wang et al. 2015; Lacroix and Li-Chan 2014).

18.3.4 In the Regulation of Glucose Absorption in the Gut

Role of the gut in glucose homeostasis is considered as one of the recent developments in diabetes management. The gut is the primary absorptive site, and it triggers neurohumoral feedback response which can regulate glucose metabolism (Thazhath et al. 2014). So, one of the ways to regulate glucose absorption in the gut is to inhibit the digestive enzymes like α-amylase and α-glucosidase (Williamson 2013).

Different plants and plant constituents have the potential to inhibit α-amylase and α-glucosidase activity which is depicted here from recent reports.

Corchorus olitorius leaves rich in polyphenolic compounds and flavonoids have been used historically to treat certain degenerative conditions. Under in vitro conditions, these have been reported to have α-glucosidase inhibitory activity, making it a potential source of antidiabetic agent for the management of postprandial hyperglycemia and diabetic complications caused due to oxidative stress (Choudhury et al. 2017). *Sclerocarya birrea*, traditionally used in South Africa for various medicinal purposes including diabetes, exerted both α-amylase and α-glucosidase inhibitory action in vitro and significantly prevented a rise in postprandial blood glucose levels (Mogale et al. 2011). Reports available on extracts derived from peel and pulps of several apple varieties showed in vitro inhibitory potential against α-glucosidase and α-amylase activity (Vinayagam et al. 2015). From a recent study, it was found that brown seaweeds like *Laminaria japonica* and *Hizikia fusiforme* possess several health-supporting properties including the potential to inhibit α-amylase activity and α-glucosidase activity when studied in in vitro and in vivo models (Kang et al. 2018). Whereas aqueous extract of *Morinda lucida* exhibited α-amylase and α-glucosidase inhibitory activity speculated due to the presence of phytochemicals like saponins, tannins, and flavonoids (Kazeem et al. 2013). Similarly, methanolic fruit extract of *Chaenomeles sinensis* demonstrated antidiabetic potential owing to its significant activity as α-glucosidase and β-glucosidase inhibitor. Furthermore, they inhibit lower α- and β-galactosidase also (Sancheti et al. 2009). Studies on acetone extract of *Rosa canina* fruit showed the presence of daucosterol and D-glucono-1,4-lactone which were found to act as good α-glucosidase inhibitors as concluded from in vivo studies (Asghari et al. 2015). Likewise, polyphenol extracts from the burs of *Castanea mollissima* Blume exhibited potential antioxidant and hypoglycemic activities. It was found that the extract rich in ellagitannins could act as an efficacious dietary supplement for diabetes management through the inhibition of α-glucosidase as evident from in vitro and in vivo studies (Zhang et al. 2014). In a similar study, naringenin has been reported for various pharmacological effects like antioxidant, anticancer, antiatherogenic, hepatoprotective, nephroprotective, and antimutagenic properties. Moreover, naringenin also possesses a postprandial hyperglycemic property by delaying the release of glucose leading to inhibition of intestinal α-glucosidase (Priscilla et al. 2014). In another study on aqueous and methanolic seed extracts of *Linum usitatissimum*, it was revealed that the extracts showed significantly more inhibition potential of murine pancreatic glucosidase as compared to acarbose, a commonly used α-glucosidase inhibitor (Bhat et al. 2011), while *Murraya koenigii* and *Ocimum tenufloru* also showed good pancreatic and intestinal glucosidase inhibitory potential. Two phlorotannins, fucofuroeckol A and dioxinodehydroeckol, isolated from *Eisenia bicyclis* demonstrated significant inhibitory activity against α-glucosidase and α-amylase (Eom et al. 2012). In another study using ethanol extracts and boiling water extracts from white onion, red onion, and commercial dry onion powder, bitter melon, pumpkin, yam, and medlar, it was demonstrated that the extracts exhibited inhibitory activities against α-glucosidase (Wu and Xu 2014). In a similar study, it was found that fenugreek seeds containing fiber are thought to slow gastric emptying and thereby carbohydrate digestion and absorption (Campbell 2010).

18.4 Plant Secondary Metabolites and Different Targets in Diabetes Treatment

Plants are considered as a "biosynthetic laboratory" for a multitude of chemical compounds known as active constituents that exert physiological effects and give plants their therapeutic properties. These compounds or products play a significant role in drug discovery and development process. Some of the drugs which are prepared from plant products like morphine, cocaine, codeine, digitoxin, quinine, and pilocarpine are still in use for different purposes (Rungsung et al. 2015). So, plant-based drugs (mainly their secondary metabolites) have provided an outstanding contribution to modern therapeutics. While in case of diabetes, many herbal constituents are claimed to reduce the blood glucose level by various means, some of them are listed here.

Myrcia bella is traditionally used in Brazil for diabetes treatment. Therefore, in vivo studies of leaves extract of *M. bella* were carried out which showed hypoglycemic properties and regulation of glucose uptake by the liver. Further, extracts were characterized for bio-active compounds and found the presence of flavonoid aglycones, flavonoid-O-glycosides, and acylated flavonoid-O-glycosides derivatives of quercetin and myricetin (Vareda et al. 2014). Similarly, myricetin found in aerial parts of *Abelmoschus moschatus* subsequently enhanced utilization of glucose to low down the plasma glucose in a deficient insulin level condition (Ahmed 2012). In a similar study conducted in Brazil, it was reported that *Bauhinia forficate*, a herbal drug for the treatment of diabetes, consists of kaempferol and quercetin O-glycosides as active ingredients (Jerine Peter and Sabina 2016). In another report, it was found that the consumption of supplementary tocopherol (vitamin C)- and ascorbic acid (vitamin E)-rich plant extract could decrease induced inflammatory response in patients with diabetes mellitus type 2. It was speculated that vitamin C and vitamin E supplementation can attenuate the incidence of some proposed pathological effects of diabetes mellitus (Osadebe et al. 2014). Reports available on *Cassia tora* showed the presence of tocopherol, ascorbic acid, and maltodextrin as an active ingredient which help to improve serum lipid levels in type 2 diabetic subjects without serious adverse effects (Ahmed 2012). Fruit pulp of *Eugenia jambolana* shows presence of cyanidin 3,5-diglucoside, dolphinidin-3,5-diglucoside, malvidin-3,5-diglucoside, petunidin-3,5-diglucoside, and peonidin-3,5-diglucoside which help in the treatment and management of diabetes (Sharma et al. 2016). Similarly, (6RS)-22-hydroxy-23,24,25,26,27-pentanor-vitamin-D3–6,19-sulfurdioxide-adduct, manoalide, and 5β-cholestane-3α,7α,12α,24,25,26-hexol isolated from *Helianthus annuus* seeds, when studied via in silico approach, showed potent antidiabetic property (Sonkamble et al. 2018). A new saponin, elatoside E isolated from roots of *Aralia elata* was found to have efficient hypoglycemic potential when tested in rats. While catharanthine, an alkaloid isolated from *C. roseus*, was also reported to lower blood sugar levels.

Similarly, trigonelline isolated from seed water extract of *T. foenum-graecum* reduced blood glucose in the glucose tolerance test in sub-diabetic and moderately diabetic rabbits. Whereas, betulin isolated from root bark acetone extract of *Euclea undulata* var. reported depleted blood glucose level and inhibited α-glucosidase activity in comparison to acarbose (Arif et al. 2014). Various reports have speculated that the hydroxycinnamic acids like caffeic acid, chlorogenic acid, and caffeoylquinic acid have α-glucosidase inhibitory potential. Investigations reported that methyl caffeate isolated from the *Solanum torvum* fruit has α-glucosidase inhibitory potential. Similarly, 3-O-caffeoylquinic acid (chlorogenic acid) and its structural isomer, 5-O-caffeoylquinic acid, isolated from the leaves of *Nerium indicum* have shown to inhibit α-glucosidases in a noncompetitive manner. However, o-hydroxycinnamic acid, m-hydroxycinnamic acid, p-hydroxycinnamic acid, ferulic acid, and gallic acid hardly inhibited α-glucosidases activity. Ferulic acid, however, helped to decrease blood glucose levels, glucose-6-phosphatase, and phosphoenolpyruvate carboxykinase activities along with higher glycogen and insulin, while gallic acid improves β-cell regeneration, insulin secretion, and lipid profile and could be used as a drug to bring about insulin secretagogue and hypolipidemic effect. Apart from this, 5-caffeoyl quinic acid and 5-caffeoyl-4-methoxy quinic acid isolated from the fruit of *Viburnum dilatatum* Thunb showed the strongest inhibitory activity on isomaltase and glucoamylase. Moreover, caffeic acids also help to improve cerebral insulin resistance and leptin resistance, promote brain carbohydrate metabolism, and protect brain neural cells (Punithavathi et al. 2011; Patel and Goyal 2011; Xiao et al. 2013). Isoorientin and 3-caffeoylquinic acid isolated from *Cecropia obtusifolia Bertol* when administered to type 2 diabetic patients showing steady response to conventional medication showed health-promoting effects on lipid and carbohydrate metabolisms. Whereas, pyrrolidine alkaloid like 2,5-dihydroxymethyl-3,4-dihydroxypyrrolidine and four piperidine alkaloids, i.e., 1-deoxymannojirimycin, 1-deoxynojirimycin, α-homonojirimycin, and 7-O-β-Dglucopyranosyl α-homonojirimycin isolated from the methanolic extract of *Commelina communis* showed inhibitory activity against α-glucosidase (Ahmed 2012). Bergenin, a major constituent isolated from roots of *Caesalpinia digyna* Rottler, showed a significant evidence for being antidiabetic in action when administered in streptozotocinnicotinamide-induced diabetic rats (Kumar et al. 2012). Whereas, Hesperetin-7-rhamnoglucoside, verbascoside, 3-(1-naphthoyl)-benzoate, (4-cinnamoyl-3,5-dihydroxy-phenoxy)-acetate, and 4-(2,4-dimethoxy-3,6-dimethyl-benzoyl)-oxy-2-hydroxy-3,6-dimethyl-benzoate isolated from leaves of *Tinospora cordifolia* showed α-amylase and α-glucosidase inhibitory potential via in silico studies (Sonkamble et al. 2015). Triterpenoids isolated from *Momordica charantia* has also been described for showing antidiabetic activity (Jerine Peter and Sabina 2016). While a new prenylated dibenzofuran, achyrofuran, derived from the plant *Achyrocline satureioides* significantly lowers the blood glucose levels when administered orally (Osadebe et al. 2014). Similarly, isoorientin, vitexin, ursolic acid, and protocatechuic acid found in *Fagopyrum tataricum* are reported to improve high glucose induced insulin resistance in mouse hepatocytes (Kazeem and Davies 2016). Huge amounts of protocatechuic acid present in raspberry, blueberry, mulberry, strawberry, cranberry, gooseberry, loquat fruit, wine, honey, and soybean have the potential to normalize hyperglycemia, reverse dyslipidemia generally associated with diabetes, improve high-density lipoprotein and low-density lipoprotein levels in diabetic rats, as well as increase glutathione peroxidase, superoxide dismutase, and catalase along with decrease in lipid peroxidation level in tissues of diabetic rats (Vinayagam et al. 2015). Moreover, contents of decoction prepared from seeds of *Coriandrum sativum* were found to have polyacetylenes, 3-O-β-D-fructofuranoside, 1-O-β-D-glucopyranoside, 4-O-β-D-fructofuranoside, 1-O-β-D-(6-O-4-hydroxybenzoyl)-glucopyranoside, 3-O-β-D-glucopyranoside, 1-O-β-D-fructofuranoside, 4-O-β-D-glucopyranoside, and 1-O-β-D-(6-O-4-methoxybenzoyl)-glucopyranoside of 2-C-methyl-D-erythritol along with monoterpenoids, glycosides, monoterpenoid glucoside sulfates, and aromatic compound glycosides. These contents were responsible for antihyperglycemic, insulin-releasing, and insulin-like activity (Afifi-Yazar et al. 2011). Pongamol and karanjin found in the fruit of *Pongamia pinnata* exhibited antihyperglycemic activity in streptozotocin-induced diabetic rats by lowering the glucose levels (Hung et al. 2012), whereas shamimin, a C-flavonol glucoside sequestered from *Bombax ceiba*, produced a notable blood glucose reduction in rats (Osadebe et al. 2014).

Administration of *Swertia punicea* extract and themethyl-swertianin and bellidifolin fractions of the extract to mice with type 2 diabetes improved insulin sensitivity and insulin signaling processes and depleted the glucokinase activity including increased activity of glucose-6-phosphatase, enzymes that are involved in the secretion of insulin from pancreatic β-cells (Coman et al. 2012). Isoflavones like medicarpin, formononetin, mucronulatol, tectorigenin, biochanin A, calycosin, daidzein, and genistein were isolated from *Dalbergia odorifera* heartwood extract and *Pueraria thunbergiana* root extract inhibited yeast α-glucosidase (Choi et al. 2010). Similarly, α-glycosidase inhibitory activity was shown by corosolic acid isolated from the ethyl acetate leaf extract of *Lagerstroemia speciosa* (Hung et al. 2012). *Cinnamomum cassia* is known in literature to produce a variety of health-promoting properties including antidyspepsia, antiflatulent, anti-influenza, antiarthritis, against cold and rheumatism, antidiarrheal, antimicrobial, and antiemetic. Its bark is known to be beneficial in type 2 diabetes and antioxidant activity, and their major constituents include eugenol and coumarins (Singh et al. 2012). Reports are also available stating that ellagic acid and 2′-(2,3,6-trihydroxy-4-carboxyphenyl) ellagic acid isolated from *Caesalpinia ferrea* Mart helps to contribute to the relief of long-term diabetic complications (Ahmed 2012). *Salacia oblonga*, *S. prinoides*, and *S. reticulata* are used for several years in traditional medications for the treatment of diabetes and corpulence. Methanolic and aqueous extracts of *Salacia* stem and roots independently inhibited the α-glucosidase and α-amylase activities. Furthermore, the active compounds identified in these extracts were found to be mangiferin, salacinol, kotalanol, and kotalagenin 16-acetate. Report further states that under in vitro conditions, mangiferin causes concentration-dependent α-glucosidase inhibition, while higher inhibition of increased serum glucose levels in maltose- and sucrose-loaded rats was caused by salacinol, kotalanol, and kotalagenin as compared to acarbose (El-Abhar and Schaalan 2014). Bellidifolin is a xanthone isolated from *Swertia japonica* and is identified as a hypoglycemic agent (Osadebe et al.

2014). Similarly, different classes of phytochemical constituents were isolated from roots of *Helicteres angustifolia* L. which includes triterpenoids like betulinic acid, oleanolic acid, helicteric acid, and methyl heliceate; flavonoids like kaempferol 3-O-b-D-glucopyranoside, 5,8-dihydroxy-7,40-dimethoxyflavone, takakin 8-O-b-D-glucuronide 600-methyl ester, and takakin 8-O-b-D-glucuronide 200-sodium sulfate; phenolic acids like rosmarinic acid; quinines like mansonone E, F, H, and H methyl ester; lignans like lariciresinol, lirioresinol-B, and (+)-pinoresinol; and cucurbitacins like cucurbitacin D and J. Among these phytochemical constituents, betulinic acid, oleanolic acid, and rosmarinic acid have been shown to exhibit antidiabetic activity (Hu et al. 2016). *Heliotropium strigosum* has been reported to have antidiabetic and antioxidant activities. The characterization of methanol extract of *Heliotropium strigosum* revealed several compounds like chromotropic acid, quercetin, trans-4-hydroxy-3-methoxy cinnamic acid, vanillic acid, gallic acid, caffeic acid, m-coumaric acid, p-coumaric acid, syringic acid, sinapic acid, and ferulic acid (Qayyum et al. 2016). In a further study, it was found that syringic acid normalized glycemia and insulinemia, when administered to diabetic Wistar rats (Domínguez Avila et al. 2017).

18.5 Conclusion and Future Prospects

Plants provide a plethora of secondary metabolites which have tremendous potential for use as antidiabetic, antimicrobial, anti-inflammatory, antioxidant, etc. agents. These plants are used traditionally for various medications; however, recent scientific explorations are providing insights about the active ingredients in those plants which are responsible for various therapeutic or pharmacological activities. As per the available evidences, plant secondary metabolites have the potential of regulating impaired glucose metabolism, β-cell functioning, GLP-1 homeostasis, and restoration of insulin levels. Through use of plant secondary metabolites in diabetes management, one can reduce the adverse effects caused due to administration of synthetic drugs. Although plants have been used historically for diabetes management in the form of herbal drugs/formulations, they have demerits like differences in plant active ingredients due to seasonal, geographical, and physiological variations in plant source, unoptimized dosage (in case of crude extracts and decoctions), etc. Hence to overcome these demerits, active ingredients in herbal drugs/formulations having antidiabetic potential need to be characterized for the presence of active ingredients and their dosage; ADMET (absorption, distribution, metabolism, excretion, and toxicity) properties should be optimized before considering them as a drug. Thus this is an attempt to highlight some of the endeavors on use of plant secondary metabolites having antidiabetic potential.

Acknowledgments Authors are thankful to Dr. Neetin S. Desai, Director, Amity Institute of Biotechnology, Amity University Mumbai for providing necessary resources and guidance. Moreover, authors are also thankful to Dr. Laxmikant H. Kamble, School of Life Sciences, Swami Ramanand Teerth Marathwada University, Nanded for his valuable editorial assistance.

References

Abdel Aziz MT, El-Asmar MF, Rezq AM, Mahfouz SM, Wassef MA, Fouad HH, Ahmed HH, Taha FM (2013) The effect of a novel curcumin derivative on pancreatic islet regeneration in experimental type-1 diabetes in rats (long term study). Diabetol Metab Syndr 5:75. https://doi.org/10.1186/1758-5996-5-75

Afifi-Yazar FU, Kasabri V, Abu-Dahab R (2011) Medicinal plants from Jordan in the treatment of cancer: traditional uses vs. *in vitro* and *in vivo* evaluations-part 1. Planta Med 77:1203–1209

Ahmed AM (2012) Medical plant and human health. In: Ghista DN (ed) Biomedical science, engineering and technology. Intech, pp 165–190. http://www.intechopen.com/books/biomedical-science-engineering-and-technology/medical-plant-and-humanhealth. Accessed 28 Sept 2018

Akhtar MS, Swamy MK (2018a) Anticancer plants: properties and application-volume 1. Springer, Singapore, pp 1–582

Akhtar MS, Swamy MK (2018b) Anticancer plants: natural products and biotechnological implements-volume 2. Springer, Singapore, pp 1–561

Akhtar MS, Birhanu G, Demisse S (2014) Antimicrobial activity of *Piper nigrum* L. and *Cassia didymobotyra* L. leaf extract on selected food borne pathogens. Asian Pac J Trop Dis 4:S911–S919

Alashkham FA, Osman MT, Adnan A, Bakar NS (2013) Histopathological and biochemical effects of *Allium sativum* oil administration on type 1 diabetic rats. Res J Pharm Biol Chem Sci 4:1045–1053

Al-Shaqha WM, Khan M, Salam N, Azzi A, Chaudhary AA (2015) Anti-diabetic potential of *Catharanthus roseus* Linn. And its effect on the glucose transport gene (GLUT-2 and GLUT-4) in streptozotocin induced diabetic wistar rats. BMC Complement Altern Med 15:379

Antu KA, Riya MP, Nair A, Mishra A, Srivastava AK, Raghu KG (2016) *Symplocoscochin chinensis* enhances insulin sensitivity via the down regulation of lipogenesis and insulin resistance in high energy diet rat model. J Ethnopharmacol 193:500–509

Arif T, Sharma B, Gahlaut A, Kumar V, Danbur R (2014) Anti-diabetic agents from medicinal plants: a review. Chem Biol Lett 1:1–13

Asghari B, Peyman S, Farimani MM, Ebrahimi SN (2015) α-Glucosidase inhibitors from fruits of *Rosacanina* L. Rec Nat Prod 9:276–283

Assad S, Morse J (2013) A review on the efficacy in humans of four popular herbs used for the treatment of diabetes. Spatula DD 3:13–22

Bassey T, Eliakim-Ikechukwu C, Ihentuge C (2011) Effect of ethanolic stem-bark extract of *Anacardium occidentale* (cashew) on the histology of the pancreas of diabetic wistar rats. J Biol Agric Healthcare 2:153–156

Belwal T, Nabavi SF, Nabavi SM, Habtemariam S (2017) Dietary anthocyanins and insulin resistance: when food becomes a medicine. Nutrients 9:1–22

Bethala K, Nalamolu KR, Sisinthy SP (2013) Herbal remedies for diabetes- a review on conventionally widely available plants. Int J Invent Pharm Sci 1:149–155

Bhat M, Zinjarde SS, Bhargava SY, Kumar AR, Joshi BN (2011) Antidiabetic indian plants: a good source of potent amylase inhibitors. Evid-Based Complement Alternat Med 2011:810207. https://doi.org/10.1093/ecam/nen040

Bodnaruc AM, Prud'homme D, Blanchet R, Giroux I (2016) Nutritional modulation of endogenous glucagon-like peptide-1 secretion: a review. Nutr Metab 13:92

Boudjelal A, Henchiri C, Siracusa L, Sari M, Ruberto G (2012) Compositional analysis and *in vivo* anti-diabetic activity of wild Algerian *Marrubium vulgare* L. infusion. Fitoterapia 83:286–292

Briskin DP (2016) Medicinal plants and phytomedicines. Linking plant biochemistry and physiology to human health. Plant Physiol 124:507–514

Campbell AP (2010) Diabetes and dietary supplements. Clin Diabetes 28:35–39

Cersosimo E, Solis-Herrera C, Trautmann ME, Malloy J, Triplitt CL (2014) Assessment of pancreatic β-cell function: review of methods and clinical applications. Curr Diabetes Rev 10:2–42

Chawla R, Thakur P, Chowdhry A, Jaiswal S, Sharma A, Goel R, Sharma J, Priyadarshi SS, Kumar V, Sharma RK, Arora R (2013) Evidence based herbal drug standardization approach in coping with challenges of holistic management of diabetes: a dreadful lifestyle disorder of 21st century. J Diabetes Metab Disord 12:1–16

Chen X, Jin J, Tang J, Wang Z, Wang J, Jin L, Lu J (2011) Extraction, purification, characterization and hypoglycemic activity of a polysaccharide isolated from the root of *Ophiopogon japonicus*. Carbohydr Polym 83:749–754

Chin YX, Lim PE, Maggs CA, Phang SM, Sharifuddin Y, Green BD (2015) Anti-diabetic potential of selected Malaysian seaweeds. J Appl Phycol 27:2137–2148

Choi CW, Choi YH, Cha MR, Yoo DS, Kim YS, Yon GH, Hong KS, Kim YH, Ryu SY (2010) Yeast α-glucosidase inhibition by isoflavones from plants of leguminosae as an *in-vitro* alternative to acarbose. J Agric Food Chem 58:9988–9993

Choudhury H, Pandey M, Hua CK, Mun CS, Jing JK, Kong L, Ern LY, Ashraf NA, Kit SW, Yee TS, Pichika MR, Gorain B, Kesharwani P (2017) An update on natural compounds in the remedy of *Diabetes mellitus*: a systematic review. J Tradit Complement Med 8:361–376

Coman C, Rugină OD, Socaciu C (2012) Plants and natural compounds with antidiabetic action. Not Bot Horti Agrobot Cluj-Napoca 40:314–325

D'souza JJ, D'souza PP, Fazal F, Kumar A, Bhat HP, Baliga MS (2014) Anti-diabetic effects of the Indian indigenous fruit *Emblica officinalis* Gaertn: active constituents and modes of action. Food Funct 5:635

Domínguez Avila J, Rodrigo García J, González Aguilar G, de la Rosa L (2017) The antidiabetic mechanisms of polyphenols related to increased glucagon-like Peptide-1 (GLP1) and insulin signaling. Molecules 22:E903. https://doi.org/10.3390/molecules22060903

El-Abhar HS, Schaalan MF (2014) Phytotherapy in diabetes: review on potential mechanistic perspectives. World J Diabetes 5:176–197

Emordi JE, Agbaje EO, Oreagba IA, Iribhogbe OI (2016) Antidiabetic and hypolipidemic activities of hydroethanolic root extract of *Uvaria chamae* in streptozotocin induced diabetic albino rats. BMC Complement Altern Med 16:468

Eom SH, Lee SH, Yoon NY, Jung WK, Jeon YJ, Kim SK, Lee MS, Kim YM (2012) α-Glucosidase and α-amylase-inhibitory activities of phlorotannins from *Eisenia bicyclis*. J Sci Food Agric 92:2084–2090

Feshani AM, Kouhsari SM, Mohammadi S (2011) *Vaccinium arctostaphylos*, a common herbal medicine in Iran: molecular and biochemical study of its antidiabetic effects on alloxan-diabetic Wistar rats. J Ethnopharmacol 133:67–74

Gezahegn Z, Akhtar MS, Woyessa D, Tariku Y (2015) Antibacterial potential of *Thevetia peruviana* leaf extracts against food associated pathogens. J Coast Life Med 3:150–157

Girija K, Lakshman K, Udaya C, Sabhya Sachi G, Divya T (2011) Anti–diabetic and anti–cholesterolemic activity of methanol extracts of three species of Amaranthus. Asian Pac J Trop Biomed 1:133–138

Hu X, Cheng D, Zhang Z (2016) Antidiabetic activity of *Helicteres angustifolia* root. Pharm Biol 54:938–944

Hui H, Tang G, Go VLW (2009) Hypoglycemic herbs and their action mechanisms. Chin Med 4:11

Hung H-Y, Qian K, Morris-Natschke SL, Hsu CS, Lee KH (2012) Recent discovery of plant derived anti-diabetic natural products. Nat Prod Rep 29:580

Hussain Mir S, Darzi MM, Mir MS (2013) Efficacy of *Abroma augusta* on biochemical and histomorphological features of alloxan-induced diabetic rabbits. Iran J Pathol 8:153–158

Ibeh BO, Ezeaja MI (2011) Preliminary study of antidiabetic activity of the methanolic leaf extract of *Axonopus compressus* (*P. beauv*) in alloxan-induced diabetic rats. J Ethnopharmacol 138:713–716

Jaiswal YS, Tatke PA, Gabhe SY, Vaidya AB (2017) Antidiabetic activity of extracts of *Anacardium occidentale* Linn. Leaves on n-streptozotocin diabetic rats. J Tradit Complement Med 7:421–427

Jerine Peter S, Sabina EP (2016) Global current trends in natural products for diabetes management: a review. Int J Pharm Pharm Sci 8:20–28

Jiang S, Wang Y, Ren D, Li J, Yuan G, An L, Du P, Ma J (2015) Antidiabetic mechanism of *Coptischinensis* polysaccharide through its antioxidant property involving the JNK pathway. Pharm Biol 53:1022–1029

Kang S, Kim E, Kang I, Lee M, Lee Y (2018) Anti-diabetic effects and anti-inflammatory effects of *Laminaria japonica* and *Hizikia fusiforme* in skeletal muscle: *in vitro* and *in vivo* model. Nutrients 10:491

Kazeem MI, Davies TC (2016) Anti-diabetic functional foods as sources of insulin secreting, insulin sensitizing and insulin mimetic agents. J Funct Foods 20:122–138

Kazeem MI, Adamson JO, Ogunwande IA (2013) Modes of inhibition of α-amylase and α-glucosidase by aqueous extract of *Morinda lucida* Benth leaf. Biomed Res Int 2013:527570. https://doi.org/10.1155/2013/527570

Kennedy DO, Wightman EL (2011) Herbal extracts and phytochemicals: plant secondary metabolites and the enhancement of human brain function. Adv Nutr 2:32–50

Kianbakht S, Abasi B, Dabaghian FH (2013) Anti-hyperglycemic effect of *Vaccinium arctostaphylos* in type 2 diabetic patients: a randomized controlled trial. Forsch Komplementmed 20:17–22

Kim MJ, Yang HJ, Moon BR, Kim JE, Kim KS, Park S (2017) *Gastrodia elata* Blume rhizome aqueous extract improves arterial thrombosis, dyslipidemia, and insulin response in testosterone-deficient rats. Evid-Based Complement Alternat Med 2017:2848570

Kodera T, Yamada S, Yamamoto Y, Hara A, Tanaka Y, Seno M, Umezawa K, Takei I, Kojima I (2009) Administration of conophylline and betacellulin-delta4 increases the beta-cell mass in neonatal streptozotocin-treated rats. Endocr J 56:799–806

Kulkarni CP, Bodhankar SL, Ghule AE, Mohan V, Thakurdesai PA (2012) Antidiabetic activity of *Trigonella foenum-graecum* L. seeds extract (IND01) in neonatal streptozotocin-induced (n-stz) rats. Diabetol Croat 41:29–40

Kumar B, Kumar A (2016) Diabetes mellitus and its control by *Ocimum sanctum* extract in mice diabetic model. Int J Curr Microbiol App Sci 5:795–810

Kumar R, Pate DK, Prasad SK, Sairam K, Hemalatha S (2011a) Antidiabetic activity of alcoholic leaves extract of *Alangium lamarckii* Thwaites on streptozotocin–nicotinamide induced type 2 diabetic rats. Asian Pac J Trop Med 4:904–909

Kumar S, Kumar V, Prakash O (2011b) Antidiabetic and antihyperlipidemic effects of *Dillenia indica* (L.) leaves extract. Braz J Pharm Sci 47:373–378

Kumar R, Patel DK, Prasad SK, Laloo D, Krishnamurthy S, Hemalatha S (2012) Type 2 antidiabetic activity of bergenin from the roots of *Caesalpinia digyna* Rottler. Fitoterapia 83:395–401

Kumar P, Rani S, Arunjyothi B, Chakrapani P, Rojarani A (2017) Evaluation of antidiabetic activity of *Gymnema sylvestres* and *Andrographis paniculata* in streptozotocin induced diabetic rats. Int J Pharmacogn Phytochem Res 9:22–25

Lacroix IME, Li-Chan ECY (2014) Overview of food products and dietary constituents with antidiabetic properties and their putative mechanisms of action: a natural approach to complement pharmacotherapy in the management of diabetes. Mol Nutr Food Res 58:61–78

Li J, Liu T, Wang L, Guo X, Xu T, Wu L, Qin L, Sun W (2012) Antihyperglycemic and antihyperlipidemic action of cinnamaldehyde in C57BLKS/J db/db mice. J Tradit Chin Med 32:446–452

Liu CT, Hsu TW, Chen KM, Tan Y-P, Lii CK, Sheen LY (2012) The antidiabetic effect of garlic oil is associated with ameliorated oxidative stress but not ameliorated level of pro-inflammatory cytokines in skeletal muscle of streptozotocin-induced diabetic rats. J Tradit Complement Med 2:135–144

Manisha, Kumar S (2018) *Biophytum sensitivum* DC: a review. Int J Pharm Sci Res 9:27–36

Mazumder PB, Choudhury MD (2010) A review on antidiabetic plants used by the people of Manipur characterized by hypoglycemic activity. Asian Univ J Sci Technol Biol Environ Sci 6:167–175

Mishra A, Garg GP (2011) Antidiabetic activity of *Alangium salvifolium* in alloxan induced diabetic rats. Int Res J Pharm 2:101–105

Mnafgui K, Kchaou M, Ben Salah H, Hajji R, Khabbabi G, Elfeki A, Allouche N, Gharsallah N (2016) Essential oil of *Zygophyllum album* inhibits key-digestive enzymes related to diabetes and hypertension and attenuates symptoms of diarrhea in alloxan-induced diabetic rats. Pharm Biol 54:1326–1333

Mogale M, Lebelo S, Thovhogi N, De Freitas AN, Shai LJ (2011) α-Amylase and α-glucosidase inhibitory effects of *Sclerocarya birrea* [(a. Rich.) Hochst.] subspecies caffra (Sond) Kokwaro (Anacardiaceae) stem-bark extracts. Afr J Biotechnol 10:15033–15039

Mohammed SI, Chopda MZ, Patil RH, Vishwakarma KS, Maheshwari VL (2016) *In vivo* antidiabetic and antioxidant activities of *Coccinia grandis* leaf extract against streptozotocin induced diabetes in experimental rats. Asian Pac J Trop Dis 6:298–304

Mollazadeh H, Hosseinzadeh H (2016) Cinnamon effects on metabolic syndrome: a review based on its mechanisms. Iran J Basic Med Sci 19:1258–1270

Nagalievska M, Sabadashka M, Hachkova H, Sybirna N (2018) *Galega officinalis* extract regulate the diabetes mellitus related violations of proliferation, functions and apoptosis of leukocytes. BMC Complement Altern Med 18:1–13

Najmi A, Akhtar M, Aqil M, Mujeeb M, Pillai KK, Khan V (2012) A pharmacological appraisal of medicinal plants with antidiabetic potential. J Pharm Bioallied Sci 4:27

Nazrul Islam M (2016) Insulinotropic effect of herbal drugs for management of diabetes mellitus: a congregational approach. Biosens J 05:2–4

Osadebe P, Odoh E, Uzor P (2014) Natural products as potential sources of antidiabetic drugs. Br J Pharm Res 4:2075–2095

Oyedemi SO, Adewusi EA, Aiyegoro OA, Akinpelu DA (2011) Antidiabetic and haematological effect of aqueous extract of stem bark of *Afzelia africana* (smith) on streptozotocin–induced diabetic Wistar rats. Asian Pac J Trop Biomed 1:353–358

Ozcan T, Akpinar-Bayizit A, Yilmaz-Ersan L, Delikanli B (2014) Phenolics in human health. Int J Chem Eng Appl 5:393–396

Patel SS, Goyal RK (2011) Cardioprotective effects of gallic acid in diabetes-induced myocardial dysfunction in rats. Pharm Res 3:239–245

Patel DK, Kumar R, Laloo D, Hemalatha S (2012) Natural medicines from plant source used for therapy of diabetes mellitus: an overview of its pharmacological aspects. Asian Pac J Trop Dis 2:239–250

Priscilla DH, Roy D, Suresh A, Kumar V, Thirumurugan K (2014) Naringenin inhibits α-glucosidase activity: a promising strategy for the regulation of postprandial hyperglycemia in high fat diet fed streptozotocin induced diabetic rats. Chem Biol Interact 210:77–85

Punithavathi VR, Prince PSM, Kumar R, Selvakumari J (2011) Antihyperglycaemic, antilipid peroxidative and antioxidant effects of gallic acid on streptozotocin induced diabetic Wistar rats. Eur J Pharmacol 650:465–471

Qayyum A, Sarfraz RA, Ashraf A, Adil S (2016) Phenolic composition and biological (antidiabetic and antioxidant) activities of different solvent extracts of an endemic plant (*Heliotropium strigosum*). J Chil Chem Soc 61:2828–2831

Qi C, Zhou Q, Yuan Z, Luo Z, Dai C, Zhu H, Chen C, Xue Y, Wang J, Wang Y, Liu Y, Xiang M, Sun W, Zhang J, Zhang Y (2018) Kinsenoside: a promising bioactive compound from *Anoectochilus* species. Curr Med Sci 38:11–18

Raafat K, Wurglics M, Schubert-Zsilavecz M (2016) *Prunella vulgaris* L. active components and their hypoglycemic and antinociceptive effects in alloxan-induced diabetic mice. Biomed Pharmacother 84:1008–1018

Rungsung W, Ratha KK, Dutta S, Dixit AK, Hazra J (2015) Secondary metabolites of plants in drugs discovery. World J Pharm Res 4:604–613

Sancheti S, Sancheti S, Seo S-Y (2009) *Chaenomeles sinensis*: a potent α-and β-glucosidase inhibitor. Am J Pharmacol Toxicol 4:8–11

Shaban F, Al-Azzawie HF, Mohammed AS (2011) Effect of alcoholic *Anastatica hierochuntica* extract on some biochemical and histological parameters in alloxan induced diabetic rats. Iraqi J Sci 52:445–455

Sharifuddin Y, Chin YX, Lim PE, Phang SM (2015) Potential bioactive compounds from seaweed for diabetes management. Mar Drugs 13:5447–5491

Sharma RJ, Gupta RC, Singh S, Bansal AK, Singh IP (2016) Stability of anthocyanins- and anthocyanidins-enriched extracts, and formulations of fruit pulp of *Eugenia jambolana* ('jamun'). Food Chem 190:808–817

Singh RK, Pandey KB, Rizvi SI (2012) Medicinal properties of some Indian spices. Ann Phytomed 1:29–33

Singh R, Bhat GA, Sharma P (2015) GLP-1 secretagogues potential of medicinal plants in management of diabetes. J Pharmacogn Phytochem 4:197–202

Sonkamble VV, Wagh NS, Kamble LH (2015) An in silico exploration for novel inhibitors of α-amylase and α-glucosidase extracted from *Tinospora cordifolia*. Pharmacophore 6:281–298

Sonkamble VV, Wagh NS, Kamble LH (2018) Inhibition of α-amylase and α-glucosidase by (6RS)-22-hydroxy-23,24,25,26,27-cholestane-3α,7α,12α,24,25,26-hexol isolated from acetone extract of *Helianthus annuus* L. seeds. Int J Pharm Pharm Sci 10:39–49

Stull AJ (2016) Blueberries' impact on insulin resistance and glucose intolerance. Antioxidants 5:E44

Swamy MK, Akthar MS, Sinniah UR (2016) Antimicrobial properties of plant essential oils against human pathogens and their mode of action: an updated review. Evid-Based Complement Alternat Med 2016:3012462. https://doi.org/10.1155/2016/3012462

Tayyab F, Lal SS, Mishra M, Kumar U (2012) A review: medicinal plants and its impact on diabetes. World J Pharm Res 1:1019–1046

Thazhath SS, Wu T, Young RL, Horowitz M, Rayner CK (2014) Glucose absorption in small intestinal diseases. Expert Rev Gastroenterol Hepatol 8:301–312

Vaishnav P, Demain AL (2011) Unexpected applications of secondary metabolites. Biotechnol Adv 29:223–229

Vareda PMP, Saldanha LL, Camaforte NADP, Violato NM, Dokkedal AL, Bosqueiro JR (2014) *Myrcia bella* leaf extract presents hypoglycemic activity via PI3k/Akt insulin signaling pathway. Evid-Based Complement Alternat Med 2014:543606. https://doi.org/10.1155/2014/543606

Velásquez DA, Beiroa D, Vázquez MJ, Romero A, López M, Diéguez C, Nogueiras R (2010) Central GLP-1 actions on energy metabolism. Vitam Horm 84:303–317

Vinayagam R, Jayachandran M, Xu B (2015) Antidiabetic effects of simple phenolic acids: a comprehensive review. Phytother Res 30:184–199

Wang HL, Li CY, Zhang B, Liu YD, Lu BM, Shi Z, An N, Zhao LK, Zhang JJ, Bao JK, Wang Y (2014) Mangiferin facilitates islet regeneration and β-cell proliferation through upregulation of cell cycle and β-cell regeneration regulators. Int J Mol Sci 15:9016–9035

Wang X, Liu H, Chen J, Li Y, Qu S (2015) Multiple factors related to the secretion of glucagon-like peptide-1. Int J Endocrinol 2015:651757. https://doi.org/10.1155/2015/65175

Weidner C, de Groot JC, Prasad A, Freiwald A, Quedenau C, Kliem M, Witzke A, Kodelja V, Han CT, Giegold S, Baumann M, Klebl B, Siems K, Müller-Kuhrt L, Schürmann A, Schüler R, Pfeiffer AFH, Schroeder FC, Büssow K, Sauer S (2012) Amorfrutins are potent antidiabetic dietary natural products. Proc Natl Acad Sci USA 109:7257–7262

Williamson G (2013) Possible effects of dietary polyphenols on sugar absorption and digestion. Mol Nutr Food Res 57:48–57

Wink M (2015) Modes of action of herbal medicines and plant secondary metabolites. Medicine 2:251–286

Wu H, Xu B (2014) Inhibitory effects of onion against α-glucosidase activity and its correlation with phenolic antioxidants. Int J Food Prop 17:599–609

Xiao J, Kai G, Yamamoto K, Chen X (2013) Advance in dietary polyphenols as α-glucosidases inhibitors: a review on structure-activity relationship aspect. Crit Rev Food Sci Nutr 53:818–836

Xie W, Zhao Y, Zhang Y (2011) Traditional Chinese medicines in treatment of patients with type 2 diabetes mellitus. Evid-Based Complement Alternat Med 2011:726723. https://doi.org/10.1155/2011/726723

Xing XH, Zhang ZM, Hu XZ, Wu RQ, Xu C (2009) Antidiabetic effects of *Artemisia sphaerocephala* Krasch. Gum, a novel food additive in China, on streptozotocin-induced type 2 diabetic rats. J Ethnopharmacol 125:410–416
Zhang J, Zhao S, Yin P, Yan L, Han J, Shi L, Zhou X, Liu Y, Ma C (2014) A-glucosidase inhibitory activity of polyphenols from the burs of *Castanea mollissima* blume. Molecules 19:8373–8386
Zhang Y, Hu T, Zhou H, Zhang Y, Jin G, Yang Y (2016) Antidiabetic effect of polysaccharides from *Pleurotus ostreatus* in streptozotocin-induced diabetic rats. Int J Biol Macromol 83:126–132

Plant Metabolites and Pharmacological Activities of *Leptadenia pyrotechnica* (Forssk.) Decne

19

Sabahat Javid, Sunbal Khalil Chaudhari, Iqra Munir, Muhammad Shoaib Amjad, Khalid Farooq Akbar, Farhat Yasmeen, and Mohd Sayeed Akhtar

Contents

Sunbal Khalil Chaudhari and Mohd Sayeed Akhtar have equally contributed for this chapter.

S. Javid · S. K. Chaudhari (✉) · I. Munir · K. F. Akbar
Department of Botany, The University of Lahore, Sargodha, Pakistan
e-mail: sunbal.khalil@imbb.uol.edu.pk; khalilsunbal@yahoo.com

M. S. Amjad
Department of Botany, Women Univeristy of Azad Jammu & Kashmir, Bagh, Pakistan

F. Yasmeen
Department of Botany, Women University Swabi, Khyber Pakhtunkhwa, Pakistan

M. S. Akhtar
Department of Botany, Gandhi Faiz-e-Aam College, Shahjahanpur, Uttar Pradesh, India
e-mail: sayeedbot@gmail.com

M. S. Akhtar et al. (eds.), *Natural Bio-active Compounds*,
https://doi.org/10.1007/978-981-13-7154-7_19

Abstract

Leptadenia pyrotechnica is an important medicinal plant of the family, Asclepiadaceae. It is usually found in the desert habitat, and generally used for the treatment of various diseases. The green parts of the plant species are used as fodder and vegetable, and are a very important source of protein, calcium, dietary fiber, phosphorus, vitamin C, and iron. Good quality fibers and many active pharmacological constituents are produced from this plant. The crude extracts of *L. pyrotechnica* contains five polyphenolic compounds, i.e., caffeic acid, epicatechin, gallic acid, vanillic acid, and quercetin-3-β-D-glucoside. The aim of the present chapter is to entail the details of nutritional, economic, and medicinal values, phytochemistry, and conservation status of this important plant.

Keywords

Crude extract, Conservation, Pharmacological constituents, Medicinal plants, Phytochemistry

19.1 Introduction

Leptadenia pyrotechnica (Forsk) Decne. belongs to Asclepiadaceae and is commonly known as Khimp/khipp in India and Pakistan (Qureshi 2004; Bahaduri and Mojumder 2008). It is an ascending, erect, and leafless shrub up to 0.5–2.6 m in height with green-colored stem. The flowers are yellowish green, bisexual, pentamerous, and actinomorphic with joined sepals at the base and free at above. The blooming and fruiting period is from August to January, and the fruits are cooked as vegetable (Verma et al. 2014a, b). It is commonly used in traditional system of medicine (Shetty and Singh 1991) and is the major species of the Middle East specifically the United Arab Emirates. It has numerous nutritional, economic, and medicinal importances. It is a xerophytic plant and can grow well under adverse climatic conditions of the arid region having very low rainfall (Migahid et al. 1972). This plant has great soil-binding properties because of its long and extensive root system which makes the plant as a main choice in desert afforestations and sand dune fixation programs (Alyemeni 2000; Qureshi et al. 2012). In India, the threatened Budha Pushkar in Rajasthan is successfully restored by planting *Leptadenia pyrotechnica* on the edges of surrounding sand dunes (Sharma and Chouhan 2008). The aim of the present chapter is to entail the details of nutritional, economic, and medicinal values, phytochemistry, and conservation status of this important plant.

19.2 Economic, Nutritional, and Medicinal Importance

It has immense nutritional, medicinal, and economic significance and is used carefully by the individuals that live in the desert areas (Singh et al. 2007). The green parts of the plant species are used as fodder and vegetable. It is considered an important source of amino acids, proteins, calcium, dietary fiber, vitamin C, iron, and phosphorus (Goyal and Sharma 2009). Good quality fibers are produced from this plant and are used for making ropes, carpets, bags, and mats and even in textile by mixing with the polyester and cotton. However, the raw material was used in paper and pulp industries (Goyal and Sharma 2006). The plant species are generally used as a fuel and also used in making excellent firewood and long-burning matchsticks. In addition, several bio-active pharmacological constituents such as alkaloids, cardenolides, flavonoids, triterpenes, sterols, and polyoxypregnane derivatives are produced from *L. pyrotechnica*. The natural bio-active compound has anti-inflammatory, antitumor, diuretic, and analgesic activities (Moustafa et al. 2007, 2009a; Chaudhary et al. 2011; Khasawneh et al. 2011; Soliman et al. 2012).

19.2.1 Antimicrobial Activity

Antifungal activity of the *L. pyrotechnica* leaves extracts against four fungi, viz., *Aspergillus niger*, *A. flavus*, *Fusarium oxysporum*, and *F. moniliformis*, was evaluated by Rekha et al. (2013). Among the various leaves extracts, methanol leaves extract showed pronounced results against *A. flavus*, while the aqueous leaves extract was more effective against *F. moniliformis*. Similarly, hexane and ethyl ether leaves extracts showed more resistance against *A. niger*. The finding of this research is in the line of agreements of several past researchers findings (Fabry et al. 1998; Ahmad et al. 2000; Boer et al. 2005; Nair et al. 2005). Similarly, Mehmooda et al. (2012) used the various solvent extracts of fruit and root in the disc diffusion assay against *Staphylococcus aureus* and *S. epidermidis*. The result showed that the growth of both pathogens was efficiently inhibited by these plant parts, particularly root extracts. However, the methanolic extract of these plant parts produced pronounced results by hindering growth and development of both pathogens.

19.2.2 Wound Healing Activity

The wound healing activity of *L. pyrotechnica* was evaluated by Shaw and Singh (2014). They prepared ethanolic extract by macerating root and aerial parts powder into petroleum ether and ethanol and observed that 4% (w/w) aerial parts showed more wound healing action on Wistar albino rats compared to root extracts due to the presence of flavonoids and terpenoides contents in the roots and aerial parts.

19.2.3 Anthelmintic activity

Kumar et al. (2011) evaluated the methanolic extract of the plant *L. pyrotechnica* against the *Pheretima posthuma* and stated that standard drug took more time to cause death and paralysis of *P. posthuma* as compared to the methanolic extract in a dose-dependent manner.

19.2.4 Antitumor Activity

Moustafa et al. (2009a) evaluated the antitumor activity on the brine shrimps. They have taken alcoholic extracts of aerial parts of the plants and proved its antitumor activities due to the presence of the alkaloids in the aerial parts.

19.2.5 Antidiabetic Activity

Chaudhary et al. (2011) conducted an experiment on antidiabetic activity of *L. pyrotechnica* in streptozotocin-induced diabetic rats. They administered the methanolic extract of *L. pyrotechnica* (MELP) to rats at 100, 200, and 300 mg/kg for 21 days. The results showed that the MELP-administered rats exhibited the dosage-dependent reduction of blood glucose level along with reduction of glycogen in the liver, cholesterol, and triglycerides in serum.

19.2.6 Antioxidant, Anti-lipoxygenase, and Cytotoxic Activity

Khasawneh et al. (2011) investigated the antioxidant, anti-lipoxygenase, and cytotoxic activity of n-butanol, ethyl acetate, and water partitioning fractions of aerial parts of the *L. pyrotechnica* by using ABTS, DPPH, FRAP, and β-carotene bleaching assay against MCF-7 (human breast cancer cells line). The results showed good antioxidant, anti-lipoxygenase, and cytotoxic potentials. However, Watafua and Geidam (2014) examined the subacute toxicity of the ethanolic extract of *L. pyrotechnica* on albino rats for 21 days at the concentrations of 50, 100, and 150 mg/kg of their weight. After 21 days, the serum and liver of rats were taken to check the toxicity. The results showed that the administration of ethanolic extract of *L. pyrotechnica* was slightly toxic to the liver.

19.3 Phytochemistry

Dhawan and Singh (1976) reported free pools of amino acids and sugars from the *L. pyrotechnica*. They characterized the l-lysine-alanine, l-arginine, l-threonine, l-methionine, and l-isoleucine, dl-alanyl-l-alanine and glycyl-l-alanine, and also the sugars (raffinose, sucrose, glucose, and fructose) from the stems. Manavalan and

Mittal (1976) reported the isolation of the triterpenoid (i.e., taraxerol, fernenol, β-sitosterol) from the dried aerial parts. Similarly, Ali et al. (2001) reported a new glycerol-oleanolic acid conjugate, namely, pyrotechnoic acid from the ethanolic extract of L. pyrotechnica. However, Afifi et al. (2002) reported sterols, triterpenes, and five flavonoid compounds from the *L. pyrotechnica*. The sterols and triterpenes were β-amyrin, β-sitosterol, lupeol, and betulin, and five flavonoids were quercetin, kaempferol-3-o-β-D-glucoside, isorhamnetin-3-o-rutinoside, quercetin-3′-o-β-D-glucoside, and rutin.

Cioffi et al. (2006) reported 18 new pregnane glycosides from the plants of the *L. pyrotechnica* with sarcostin, 11-hydroxy sarcostin, and deacetylmetaplexigenin as the aglycon moieties; acetyl, benzoyl, cinnamoyl, p-coumaroyl, and nicotinoyl ester moieties linked at C-12 and/or C-20 of the aglycon; and hexapyranose, 6-deoxy-3-o-methylhexapyranose, and 2,6-dideoxy-3- o-methyhexapyranose sugars linked at C-3 of their aglycons. Similarly, Moustafa et al. (2007) reported various lipid constituents from the aerial parts of the *L. pyrotechnica*. The compounds were characterized as 3 terpenes (i.e., taraxerol, phytol, and squalene); 5 sterols (cholesterol, fucosterol, campesterol, stigmasterol, and β-sitosterol); and 15 fatty acids, 11 n-alkanol, 1 n-alkene (3-tetradecene), and 18 aromatic hydrocarbons with 5-phenyl-undecanes and 6-phenyl-tridecane as main constituents. Rastogi and Mehrotra (2008) reported the presence of cetyl alcohol, β-sitosterol, n-triacontane, β-amyrin acetate, and lupanol-3-O-diglucoside. However, the uncharacterized compound, namely, leptidine glucoside, was isolated from whole plant of *L. pyrotechnica*.

Moustafa et al. (2009a, b, c) reported the separation of 24 alkaloids from the aboveground parts of the *L. pyrotechnica* belonging to pyrrole, pyridine, pyrazine, and indole. They also reported the presence of glycosides such as 14,19-dihydroxycard-20(22)-enolide-3-o-[β-dglucopyranosyl-β-d-ucopyranoside], 14,19-dihydroxycard-20(22)-enolide-3-o-[β-d-gluco-pyranosyl-β-ddigitoxoside], and19-dihy-droxycard20(22)-enolide-3-o-β-d-digitoxo-side and six flavonoids, such as kaempferol-3-o-α-1-rhamnopyranosyl(1‴→6″)-o-β-dglucopyranoside, kaempferol-3-o-β-rhamnopyranosyl (1‴→6″)-o-β-d-glucopyranoside, texasin-7-o-β-d-glucopyranoside, kaempferol-3-o-β-d-glucopyranoside, kaempferol, and kaempferide-3-o-αrhamnopyranosyl(1‴→6″)-o-β-d-glucopyranoside. Similarly, Amal et al. (2009) isolated the six flavonoids, namely, kaempferol 3-O-α-L-rhamnopyranosyl(1‴→6″)-O-β-D-glucopyranoside, kaempferol 4′-methyl ether 3-O-β-D-rutinoside (kaempferide 3-O-β-D-rutinoside), kaempferol-3-O-β-D-glucopyranosyl (1‴→6″)-O-β-D-glucopyranoside, kaem-pferol-3-O-β-D-glucopyranoside, texasin 3-O-glucoside, and quercetin 3-Ogalactoside. Sherwani et al. (2009) isolated 12,13-epoxy octadec-cis-9-enoic acid (vernolic acid) (32%) from the seeds of *L. pyrotechnica* by direct acetolysis. However, Mohammad et al. (2011) studied the aqueous ethanolic extracts of aerial part of *L. pyrotechnica* and found that the crude extract contains five polyphenolic compounds such as caffeic acid, epicatechin, gallic acid, vanillic acid, and quercetin-3-β-D-glucoside, where vanillic acid, quercetin-3-β-D-glucoside, and epicatechin were present in the highest amount.

19.4 Ethnobotanical Uses of *Leptadenia pyrotechnica*

19.4.1 Uses in Pakistan and India

In Pakistan, Ahmed et al. (2012) suggested the use of leaf and shoot powder of *L. pyrotechnica* for constipation, obesity, and dysmenorrheal. Qasim et al. (2014) reported the use of this plant for upper gastrointestinal track disorder, spermatorrhea, and impotency. *Leptadenia pyrotechnica* is traditionally used for different purposes, while in India, a stem decoction is used to cure gout and rheumatism.

19.4.2 Uses in Yemen

In Yemen, wounds are treated by applying crushed stem, and the kidney disorders are treated by making infusions of the aboveground parts. Twigs are softened, and the liquid is taken to cure urinary retention. Butter milk is mixed with the infusion of the whole plant and used for the handling of various stomach disorders and uterine prolapses. All plant parts of *L. pyrotechnica* are used in folk remedies to treat gynecological, muscular, and skeleton problems. Sap obtained from plant is applied on the skin to cure dermatitis, psoriasis, smallpox, eczema, rheumatism, infantile diarrhea, and asthma (Gulshan et al. 2012; Schmeizer and Gurib-fakim 2013; Verma et al. 2014a, b). Leaf paste is used to remove thorns by applying over the injury (Gulshan et al. 2012; Verma et al. 2014a, b). The stem and fruit decoction is use to cure chronic renal problems and ear ache and also as purgative and carminative (Ahmad et al. 2014), while the roots are used for the cure of constipation and stomach complaints. The plant decoction is also used to treat sterility and venereal diseases and to avoid spontaneous abortions. However, the bark of root is mixed with cow's milk and used as purgative (Gulshan et al. 2012).

19.4.3 Uses in Saudi Arabia

In Saudi Arabia, the decoction of seeds and whole plant is used to treat flu and tissue disorder (Randa and Youssef 2013). Watery juice of this plant is used against ringworm, while the young branches are used to make ropes (Bhatti et al. 2001). Similarly, the stem juice was used for constipation, obesity, dysmenorrhea flu, kidney pain, cold, ringworm, skin diseases, diabetes, tuberculosis, and hepatitis (Diallo et al. 1999).

19.4.4 Uses in Senegal

In Senegal, it is used for infant children as laxative and also for muscle and kidney pain. The leaf paste or latex is used against the thorn injury and ringworm (Qureshi 2002). Its fiber has expectorant activity (Al-Yahiya 1986), and the plant sap is used

Table 19.1 Traditional uses of different plant parts of *L. pyrotechnica*

Plant material	Uses	References
Leaf paste	To remove the throne/throne injury	Upadhyay et al. (2010)
Stem juice	To cure the cough, flu	Diallo et al. (1999)
Plant fiber	Used as anti-expectorant and antihistaminic	Al-Yahiya (1986)
Plant sap	For skin diseases and diabetes, smallpox, psoriasis	Kateva and Galav (2006), Maydell (1990)
Leaves and shoots	Used for fever, hepatitis, constipation, and obesity	Ahmad et al. (2014)
Whole plant	It's warmed juice taken orally to remedy the jaundice	Sharma et al. (2012)
Seed	Macerated seed lotion is used as eye lotion	Maydell (1990)
Yong twigs	Used as toothbrush	Maydell (1990)

as remedy of skin diseases and diabetes (Praveen et al. 2007), while the stem is used for the removal of kidney stones. The *L. pyrotechnica* extract remarkably lowers the rate of plaque formation in aorta (Saleh et al. 2012). The roots are used as vegetables (Ali et al. 2001), and its boiled filtrate is used to cure tuberculosis (Patal et al. 2014). Some of the significant traditional uses are summarized in tabular form (Table 19.1).

19.5 Challenges in the Conservation of *L. pyrotechnica*

L. pyrotechnica is a frequently distributed in the desert areas. Nowadays, the plant species is under great pressure, which appeals the researchers to take the action. In order to prevent this situation, in time and effective conservation measures are required for the benefit of mankind. In this regard, Vinod et al. (2003) focused on the need of sustainable utilization and conservation of biodiversity. To protect biodiversity, different research groups across the globe have made efforts for plants conservation (Malik et al. 2005; Parabia et al. 2007; Ray and Bhattacharya 2008). Various methods and biotechnological approaches such as plant tissue culture, organ culture, etc. have been used for the conservation and protection of this significant plant species (Parabia et al. 2007).

19.6 Conclusions and Future Prospects

L. pyrotechnica is the most significant plant of desert plants with multiple uses. The plant species possesses antimicrobial, antifungal, anticancer, anthelmintic, antioxidant, wound healing, antiatherosclerotic, antidiabetic activities, etc. In the future, more focused researches are needed to explore and conserve this valuable medicinal plant.

References

Afifi MS, Hassan MA, Ali EA, Badria FA, Mahmood EF (2002) Phytochemical and biological investigation of *Leptadenia pyrotechnica* decne growing in Egypt. J Pharm Sci 18:122–138

Ahmad I, Mehmood Z, Mohammad P, Ahmed S (2000) Antimicrobial potency and synergistic activity of five traditionally used Indian medicinal plants. J Med Aroma Plant Sci 22:173–176

Ahmad N, Mahmood A, Tahir SS, Mahmood A, Bano A, Malik RN, Hassan S, Ishtiaq M (2014) Relative importance of indigenous medicinal plants from Layyah district, Punjab province. Pak J Ethnopharmacol 155:509–523

Ahmed H, Wariss AK, Anjum S, Mukhtar M (2012) Ethnobotanical studies of plant resources of Cholistan desert. Pak Int J Sci Res 3:1782–1788

Ali MS, Kausar F, Malik M (2001) Pyrotechnoic acid: a glycol-oleanolic acid conjugate from *Leptadenia pyrotechnica*. J Chem Soc Pak 23:180–182

Al-Yahiya MA (1986) Triterpenoid from *Leptadenia pyrotechnica*. Fitoterapia 57:179–185

Alyemeni NM (2000) Ecological studies on sand dunes vegetation in Al-Kharj region, Saudi Arabia. Saudi J Bio Sci 7:64–87

Amal YM, Khodair AI, Saleh M (2009) Isolation, structure elucidation of flavonoid constituents from *Leptadenia pyrotechnica* and evaluation of their toxicity and antitumor activity. Pharma Biol 47:539–552

Bahaduri SK, Mojumder P (2008) Medium density particle board from khimp plant. Nat Prod Rad 7:106–110

Bhatti GR, Qureshi R, Shah M (2001) Ethnobotany of Qadan Wari of Nara desert. Pak J Bot 33:801–812

Boer HJ, Kool A, Broberg A (2005) Antifungal and antibacterial activity of some herbal remedies from Tanzania. J. Ethnopharmacol 96:461–469

Chaudhary S, Khosa RL, Jha KK, Kumar S (2011) Evaluation of antidiabetic activity of whole plant of *Leptadenia pyrotechnica* (Forssk.) Decne against streptozotocin induced diabetes in rats. Pharmacol Online 2:1196–1204

Cioffi G, Sanogo R, Vassallo A, Dal Piaz F, Autore G, Marzocco S, De Tommasi N (2006) Pregnane glycoside from *Leptadenia pyrotechnica*. J Nat Prod 69:625–635

Dhawan AK, Singh H (1976) Free pools of amino acids and sugars in *Leptadenia pyrotechnica*. Curr Sci 45:198

Diallo D, Hveem B, Mahmoud MA, Berge G, Paulsen BS, Maiga S (1999) An Ethnobotanical study of herbal drugs of Gourma district, Mali. Pharm Biol 37:80–91

Fabry W, Okemo PO, Ausorg R (1998) Antibacterial activity of East African medicinal plants. J Ethnopharmacol 60:19–24

Goyal M, Sharma SK (2006) Prospects and dimensions for utilization arid foods. Yash Publishing House, Bikaner

Goyal M, Sharma SK (2009) Traditional wisdom and value addition prospects of and foods of desert region of North West India. Indian J Trad Knowl 8:581–585

Gulshan AB, Dasti AA, Hussain S, Atta MI, Amin-uddin M (2012) Indigenous uses of medicinal plants in rural areas of Dera ghazi khan, Punjab, Pakistan. ARPN J Agric Biol Sci 7:750–762

Kateva SS, Galav PK (2006) Addition to the traditional folk herbal medicines from Shekhwati region of Rajasthan. Indian J Tradit Knowl 5:494–500

Khasawneh MA, Elwy HM, Hamza AA, Fawzi NM, Hassan AH (2011) Antioxidant, anti-lipoxygenase and cytotoxic activity of *Leptadenia pyrotechnica* (Forssk.) Decne polyphenolic constituents. Molecules 16:7510–7521

Kumar S, Chaudhary S, Jha KK (2011) Anthelminthic activity on the *Leptadenia pyrotechnica* (forsk.) decne. J Nat Prod Plant Resour 1:56–59

Malik SK, Chaudhury R, Kalia RK (2005) Rapid *in vitro* multiplication and conservation of *Garcinia indica*: a tropical medicinal tree species. Sci Horti 106:539–553

Manavalan R, Mittal BM (1976) Constituents of the aerial parts of *Leptadenia pyrotechnica*. Planta Med 39:25–30

Mehmooda M, Rahmatullah Q, Muhammad A, Muhammad G (2012) Antibacterial activity of root and fruit extracts of *Leptadenia pyrotechnica* (Asclepiadaceae) from Pakistan. Pak J Bot 44:1209–1213

Migahid AM, Batanoun K, Abdelwah A (1972) Eco-physiological studies on desert plants. 7. Water relations of *Leptadenia pyrotechnica* (Forsk.) Decne. Growing in Egyptian desert. Oecologia 10:79–91

Mohammad AK, Hanan ME, Alaaeldin AH, Nael MF, Ahmed HH (2011) Antioxidant, antilipoxygenase and cytotoxic activity of *Leptadenia pyrotechnica* (Forsk.) decne polyphenolic constituents. Molecules 16:7510–7521

Moustafa AMY, Khodair AI, Saleh MA (2007) Phytochemical investigation and toxicological studies of lipid constituents isolated from *Leptadenia pyrotechnica*. J Pharmacol Toxicol 2:681–697

Moustafa AMY, Khodair AI, Mahmoud AS (2009a) GC-MS investigation and toxicological evaluation of alkaloids from *Leptadenia pyrotechnica*. Pharm Biol 47:994–1003

Moustafa AMY, Khodair AI, Mahmoud AS (2009b) Structural elucidation and evaluation of toxicity and antitumour activity of cardiac glycosides isolated from *Leptadenia pyrotechnica*. Pharm Biol 47:826–834

Moustafa AMY, Khodair AI, Saleh MA (2009c) Isolation, structural elucidation of flavonoid constituents from *Leptadenia pyrotechnica* and evaluation of their toxicity and antitumor activity. Pharm Biol 47:539–552

Nair R, Kalariya T, Chanda S (2005) Antibacterial activity of some selected Indian medicinal flora. Turk J Biol 29:41–47

Parabia FM, Gami B, Kothari IL, Mohan JSS, Parabia MH (2007) Effect of plant growth regulators on *in vitro* morphogenesis of *Leptadenia reticulata* (Retz.) W. and A. from nodal explants. Curr Sci 92:1290–1293

Patel SK, Desai PR, Pandey VB (2014) Ethnomedicinal plants used by the tribals in Bhiloda taluka of Sabarkantha district, Gujarat. Indian J Adv Plant Res 1:33–36

Praveen B, Upadhayay RS, Kumar A (2007) Traditional use of medicinal plants among rural communities of churu district in the Thar Desert. Indian J Ethnopharmacol 113:387–399

Qasim M, Abideen Z, Adnan YM, Ansari R, Gull B, Khan MA (2014) Traditional ethno-botanical uses of medicinal plants from coastal areas of Pakistan. J Coast Life Med 2:22–30

Qureshi R (2002) Ethnobotany of Rohri Hill, Sindh, Pakistan. Hamdard Med 45:86–94

Qureshi R (2004) Floristic and ethnobotanical study of Desert-Nara Region, Sindh. PhD thesis. Shah Abdul Latif University, Pakistan, pp 317–454

Qureshi R, Munazir M, Abul-Soad AA, Jatoi MA, Shabbir G (2012) *In vitro* callus induction protocol for *Leptadenia pyrotechnica* using various explants. J Med Plants Res 6:379–382

Randa S, Youssef A (2013) Medicinal and non-medicinal uses of some plants found in the middle region of Saudi Arabia. J Med Plants Res 7:2501–2513

Rastogi RP, Mehrotra BN (2008) Compendium of Indian Medicinal Plant. 1st edn, vol 4, (1985-1989). CDRI, Lucknow and National Institute of Science Communication and Information Resources, New Delhi, India, p 430

Ray A, Bhattacharya S (2008) Storage and plant regeneration from encapsulated shoot tips of *Rauvolfia serpentina*-an effective way of conservation and mass propagation. South Afr J Bot 74:776–779

Rekha K, Kumar, Singh M (2013) Antifungal Potential of *Leptadenia pyrotechnica* against some pathogenic fungi. Int J Eng Sci Invent 2:27–29

Saleh I, Aehs G, Amani SA, Abd ER, Mohammed D (2012) Anti-inflammatory activity, safety and protective effects of *Leptadenia pyrotechnica, haloxylon salicornicum* and *Ochradenus baccatus* in ulcerative colitis. Phytopharmacol 2:58–71

Schmeizer G, Gurib-fakim A (2013) Plant resources of tropical Africa 11(2). Medicinal Plants 2. PROTA foundation, Wageningen, Netherlands/CTA, Wageningen, The Netherlands, p 155

Sharma KC, Chouhan CS (2008) Ecology and restoration of lake Budha Pushkar-a threatened water body of Ajmer, Rajasthan. In: Sengupta M, Dalwani R (eds) The 12th world Lake conference, pp 1757–1764

Sharma J, Gairola S, Gaur RD, Painuli RM (2012) The treatment of jaundice with medicinal plants in indigenous communities of the sub-himalayan region of Uttarakhand. Indian J Ethnopharmacol 143:262–291

Shaw NR, Singh GK (2014) Wound healing activity of ethanolic extract of aerial and root part of *Leptadenia pyrotechnica* (Fork.) Decne. Int J Pharm Biol Sci 5:671–675

Sherwani MRK, Mathur A, Afreena SR (2009) *Leptadenia pyrotechnica* seed oil: a good source of epoxy acid. J Lipid Sci Technol 41:95–97

Shetty BV, Singh V (1991) Flora of India (Series-II), Flora of Rajasthan. Bot Surv India 2:453–480

Singh JP, Rathore VS, Beniwal RK (2007) Kheep (*Leptadenia pyrotechnica*): Potential rangeland shrub of Western Rajasthan, India. Indian J Plant Genet Resour 20:199–203

Soliman GA, Donia AERM, Awaad AS, Alqasoumi SI, Yusufoglu H (2012) Effect of *Emex spinosa, Leptadenia pyrotechnica, Haloxylon salicornicum* and *Ochradenus baccatus* extracts on the reproductive organs of adult male rats. Pharm Biol 50:105–112

Upadhyay B, Praveen DAK, Kumar A (2010) Ethnomedicinal and Ethnopharmacological - statistical studies of eastern Rajasthan, India. J Ethnopharmacol 129:64–86

Verma N, Jha KK, Chaudhary S, Singh O, Kumar A (2014a) Phytochemistry, pharmacology and traditional uses of *Leptadenia pyrotechnica*- an important medicinal plant. Indian J Pharma Biol Res 2:128–134

Verma N, Jha KK, Chaudhary S, Singh O, Kumar A (2014b) Phytochemistry, pharmacology and traditional uses of *Leptadenia pyrotechnica*. An important medicinal plant. Indian J Pharm Bio Res 2:128–134

Vinod A, Shekhawat NS, Singh RP (2003) Micropropagation of *Leptadenia reticulata*: a medicinal plant. In Vitro Cell Dev Boil 39:180–185

von Maydell HJ (1990) Trees and shrubs of the Sahel: their characteristics and uses. Deutsche Gesellschaft fur Technische Zusammenarbeit, Germany

Watafua M, Geidam MA (2014) Effects of the ethanolic extract of aerials of *Leptadenia pyrotechnica* (Forsk.) Decnse. on the liver function of Wister strain Albino Rats. Int J Sci Res Manag 2:605–610

Functioning of Organosulfur Compounds from Garlic (*Allium sativum* Linn) in Targeting Risk Factor-Mediated Atherosclerosis: A Cross Talk Between Alternative and Modern Medicine

20

Parvej Ahmad, Sahir Sultan Alvi, and M. Salman Khan

Contents

P. Ahmad · S. S. Alvi · M. S. Khan (✉)
Clinical Biochemistry and Natural Product Research Lab, Department of Biosciences, Integral University, Lucknow, Uttar Pradesh, India
e-mail: contactskhan@gmail.com

M. S. Akhtar et al. (eds.), *Natural Bio-active Compounds*,
https://doi.org/10.1007/978-981-13-7154-7_20

Abstract

Garlic (*Allium sativum* Linn), a member of Liliaceae family has been used in different cultures as foodstuffs and medicine. It possesses potent biological activities, i.e., anti-inflammatory, antidiabetic, anticancer, and anti-atherosclerotic properties that are accredited to the organosulfur compound (OSCs) as well as other phytochemicals content of garlic, viz., phenolic acids, flavonoids, thiosulfinates, and anthocyanins that either work alone or synergistically by complex mechanisms. Among all the dietary phytochemicals from *A. sativum* Linn, OSCs have been found to show greater effects against various metabolic disorders. Major OSCs from garlic owing to its high therapeutic and pharmacological properties are allicin, diallyl trisulfide (DAT), allyl methyl trisulfide, diallyl disulfide (DAD), ajoene, and many others. The current chapter summarizes the therapeutic efficacy of garlic and its bio-active OSCs against risk factor-mediated atherosclerotic cardiovascular diseases (ASCVD) via modulation of distinct molecular mechanisms, i.e., antioxidant potential, lipid and lipoprotein metabolism, HDL functionality through paraoxonase-1 (PON-1) activity, HMG-CoA reductase (HMG-R) expression and functionality, LDL-receptor (LDL-R), LDL oxidation status, effects on macrophage, foam cell formation and plaque stabilization, inflammatory signalling pathways and hypertension, etc. We also hypothesized that garlic and its OSCs could be beneficial in targeting proprotein convertase subtilisin/kexin type-9 (PCSK-9) and LDL-R pathway in order to treat and manage hypercholesterolemia especially for the patients facing inadequate lipid lowering with classical HMG-R inhibitors (statins) and statin intolerance. Finally to sum up the whole, we conclude that the garlic and its OSCs may be promoted from alternative to mainstream medicine in targeting risk factor-mediated ASCVD.

Keywords

Atherosclerosis · Cholesterol · Homeostasis · Inflammation · Lipoprotein metabolism

20.1 Introduction

Garlic (*Allium sativum* Linn.) has long been used as a dietary spice and food additive and is considered as one of the important medicinal plants with novel therapeutic sources (Mikaili et al. 2013; Santhosha et al. 2013; Ansari et al. 2018; Czompa et al. 2018). It has numerous biological activities, i.e., lipid-lowering, anti-inflammatory, anti-asthmatic, antimutagenic, antidiabetic, anticancer, and anti-atherosclerotic properties that are accredited to its rich content of various volatile OSCs and other phytochemicals, viz., phenolic acids, flavonoids, thiosulfinates, and anthocyanins that either work alone or synergistically by complex mechanisms (Lanzotti et al. 2014; Petropoulos et al. 2018). Among all the dietary phytochemicals from *A. sativum* L., OSCs have been found to show greater effects of therapeutic modulations against various metabolic disorders (Ahmed 2018; Kimura et al. 2017; Zhai et al. 2018). The major OSCs, i.e., allicin, alliin, diallyl trisulfide (DATS), allyl methyl trisulfide (AMTS), diallyl disulfide (DADS), ajoene, and others, have been found to show anticancer, antioxidant, anti-inflammatory, immunomodulatory, antimicrobial, hypoglycemic, and cardiovascular protections (Viswanathan et al. 2014; Schäfer and Kaschula 2014).

In the age of medicines and methodologies, when we are floating through the advancements and achievements, we are being hijacked by various metabolic as well as infectious and inflammatory diseases. Among these, atherosclerotic cardiovascular disease (ASCVD) is the most prominent mortality cause worldwide (Heron 2013). ASCVD, a class of complex multifactorial arterial disease, results due to various stimuli and characterized by aberrations of lipoproteins and cholesterols levels (Mendis et al. 2011). The two foremost drug targets, i.e., PCSK-9 and HMG-R, play very crucial role in the cholesterol homeostasis (Abifadel et al. 2009; Goldstein and Brown 2015; Alvi et al. 2016). When considering different cholesterols, low-density lipoprotein-cholesterol (LDL-C) is the supreme risk factor for ASCVD onset, thus considered as a promising target for ASCVD management (Besseling et al. 2014). Animals get rid of this atherogenic LDL-C through hepatic LDL receptors (LDL-R). PCSK-9, a serine protease, combines to LDL-R at epidermal growth factor-like repeat-A (EGF-A) domain and directs its lysosomal degradation and thus increases the circulatory LDL-C level (Horton et al. 2007) and thus has arisen as a novel therapeutic target for the treatment of hypercholesterolemia.

On the other hand, HMG-CoA reductase (HMG-R), the rate-limiting enzyme of the biosynthetic pathway of cholesterol, also considered as central player in the cholesterol production, was therefore considered as attractive target for cholesterol homeostasis (Goldstein and Brown 2015; Alvi et al. 2016). HMG-R inhibitors, statins, diminish the CVD risk; however, the use of these medications has been associated with drug intolerance and uneven lipid-lowering efficiency (Davidson et al. 2007; Keyamura et al. 2014; Reiner 2014). Moreover, statins also upregulate the PCSK-9 via upregulating the expression of SREBP-2 which coordinately regulates the PCSK-9 expression along with LDL-R expression, thus limiting its beneficial effects (Abifadel et al. 2009; Alvi et al. 2017b). In this order, PCSK-9 has been established as a novel therapeutic target against elevated cholesterol levels. Among all the currently used strategies for PCSK-9 targeting, human mAbs, i.e., alirocumab and evolocumab, have been used intensely in human clinical trials (Yadav et al. 2016). These mAbs have shown various limitations regardless of their potential to decrease LDL-C levels such as repeated injections and high expense (Alvi et al. 2017a).

To sum up the whole, HMG-R inhibitors, particularly statins, display various adverse effects, i.e., cytotoxicity, whereas none of the inhibitors of PCSK-9 have been approved for long-term clinical administration due to antidrug antibody (ADA) generation and not preferred due to cost-related consequences. Therefore, discovery of dual-action inhibitors targeting PCSK-9 and HMG-R activity from natural sources may greatly reflect forthcoming era of lipid-lowering therapy, particularly in the subjects with familial hypercholesterolemia that are experiencing uneven

LDL-C lowering effects with statins. In this order, an array of recent studies demonstrated the role of various medicinal plants and their therapeutic natural products against HMG-R and PCSK-9 activity to combat hypercholesterolemia (Iqbal et al. 2014a, b, 2015; Alvi et al. 2016, 2017a, b). But combinatorial information at one place regarding the role of garlic and OSCs in targeting risk factor-mediated ASCVD is still lacking. Therefore, the present chapter summarizes the therapeutic efficacy of garlic and its bio-active OSCs against risk factor-mediated ASCVD via modulation of distinct molecular mechanisms, i.e., antioxidant potential, lipid and lipoprotein metabolism, HMG-R expression and functionality, LDL-R, LDL oxidation status, effects on macrophage, foam cell formation and plaque stabilization, inflammatory signalling pathways, and hypertension.

20.2 Garlic and Its Bio-active Principals: The Organosulfur Compounds

Garlic has been used as an impressive dietary agent and belongs to the Liliaceae family, the most prominent family known to have beneficial effects against atherosclerosis, cancer, diabetes, and aging in various in vivo studies (Choudhary et al. 2011; Santhosha et al. 2013; Chiu et al. 2016). Most of the health beneficial impacts of garlic are accredited to its rich content in secondary metabolites, i.e., steroidal saponins, OSCs, flavonoids, and so on (Fig. 20.1). Among these metabolites, OSCs have been largely investigated for their therapeutic efficacy. These compounds

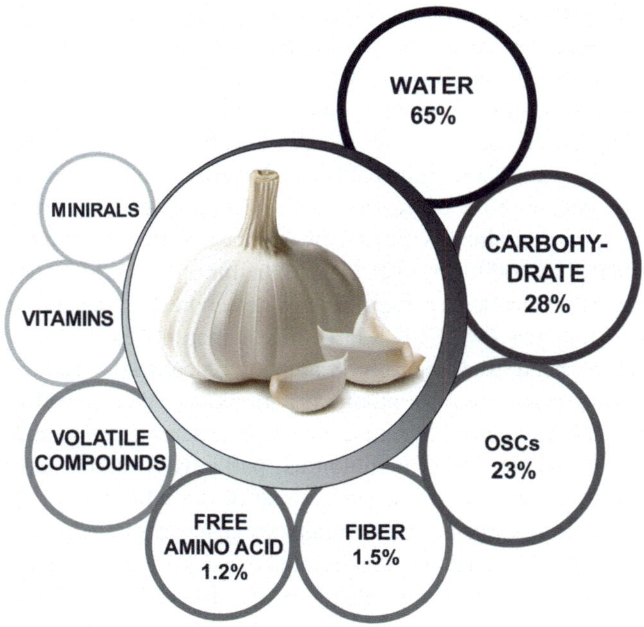

Fig. 20.1 Various types of secondary metabolite content of garlic

include γ-glutamyl peptides: γ-l-glutamyl-S-allyl-l-cysteine (GSAC), γ-l-glutamyl-S-(trans-1-propenyl)-l-cysteine (GSPC), and γ-l-glutamyl-S-methyl-l-cysteine (GSMC). Moreover, biosynthesis of S-alk(en)yl-l-cysteine sulfoxides (ACSOs) generates intermediates, i.e., S-alk(en)yl-cysteines such as (+)-S-allyl-l-cysteine (SAC) and (+)-S-(trans-1-propenyl)-l-cysteine (SPC) (Yoshimoto et al. 2015). The corresponding ACSOs produced are (+)-S-allyl-l-cysteine sulfoxide (alliin), (+)-S-(trans-1-propenyl)-l-cysteine sulfoxide (isoalliin), and (+)-S-methyl-l-cysteine sulfoxide (methiin) (Iciek et al. 2009; Beato et al. 2012). Garlic also possess OSCs like diallyl thiosulfinate (allicin), DATS, AMTS, DADS, ajoene, and others, which show anticancer, antioxidant, anti-inflammatory, immunomodulatory, hypoglycemic, and cardiovascular protections (Schäfer and Kaschula 2014; Viswanathan et al. 2014). Some of the major OSCs have been structurally represented in Fig. 20.2. Whole garlic typically contains ~1% alliin, (+)-S-methyl-L-cysteine sulfoxide (SMCS), S-propargyl-cysteine (SPC), and (+)-S-(trans-1-propenyl)-L-cysteine sulfoxide. Researchers also reported the presence of γ-glutamyl-S-(trans-1-propenyl)-L-cysteine and γ-glutamyl-S-allyl-mercapto-L-cysteine in garlic cloves (Amagase et al. 2001). Alliin or SACS and allicin are important OSCs present in the garlic and are responsible for their higher medicinal values (Amagase et al. 2001; Vigrinia 2006; Schäfer and Kaschula 2014).

20.3 Bioavailability, Metabolism, and Pharmacokinetics of OSCs

The garlic and its preparations have been used as a remedy since a long ago without the knowledge of molecular rationale behind their potent health benefits (Rivlin 2001; Chauhan 2005; Ray et al. 2011). Bioavailability, an important factor in drug discovery, is defined as "the fraction of administered drug that can reach plasma and body tissues without undergoing any change in its structure and function" (Abourashed 2013). Gastrointestinal (GI) lining presents challenges to the absorption of oral medications in the active form as it acts as lipid barrier (Abourashed 2013). Optimum physical and chemical properties enable an oral drug candidate to persist in GI conditions as well as penetrate the GI barrier and provide it desired bioavailability. Therefore, the detailed information and rationale for the pharmacokinetics and metabolism of OSCs must be clearly illustrated before their pharmacological implications. Till date very little amount of data is present regarding bioavailability, absorption, metabolism, and pharmacokinetics of these OSCs. Novel drug delivery systems such as microemulsion, liposomes, and nanoparticles have also developed to increase the stability, bioavailability, and systemic circulation time of relatively stable OSCs such as DADS and DATS. The biopharmaceutical evaluations of these novel drug delivery systems have been recently focused in the development of potential pharmaceutical products based on these organosulfur compounds (Gao et al. 2013). After ingestion of raw garlic, allyl methyl sulfide (AMS), allyl methyl disulfide (AMDS), diallyl sulfide (DAS), DADS, DATS, and dimethyl sulfide were discovered in the breath of the tested volunteers. AMDS,

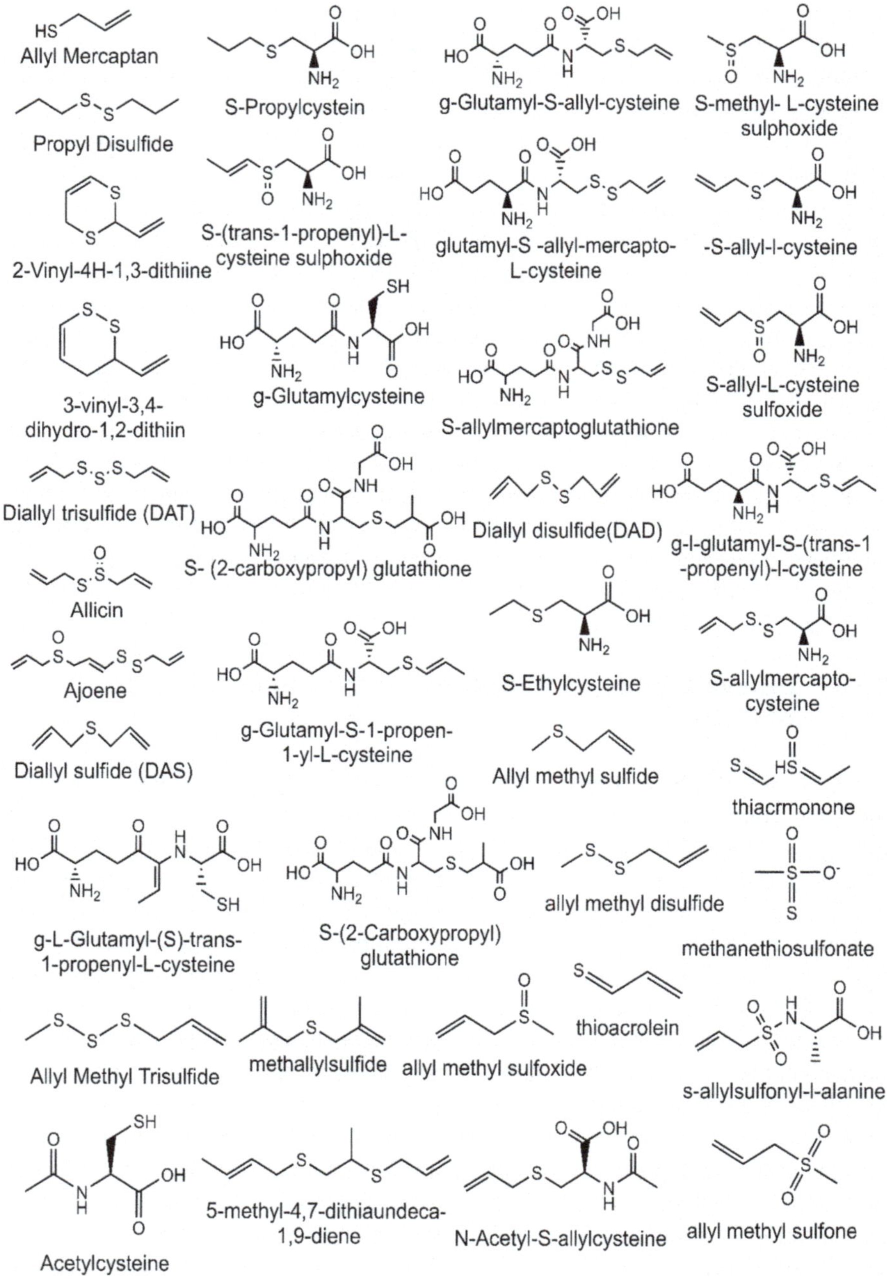

Fig. 20.2 Structural illustration of some major OSCs isolated from garlic

DAS, DADS, and DATS reached the maxima shortly within the 2–3 h, while the concentrations of others increased much more slowly (Taucher et al. 1996).

A pharmacokinetic study of DADS demonstrated that in addition to allyl mercaptan (AM) and AMS, allyl methyl sulfoxide (AMSO) and allyl methyl sulfone ($AMSO_2$) were detected as metabolites of DADS in various organs of rats (Germain et al. 2002). DADS was detected only during the first 2 h after administration in the liver and then transiently detected in the plasma but undetected in the urine. The level of DADS in the liver and blood was less than 0.5% of that in the stomach. The apparent halftime ($T_{1/2}$s) of DADS was found to be <1 h in the isolated rat liver, which was too short for the assessment of its in vivo pharmacokinetic factors (Gao et al. 2013). $AMSO_2$ and AMSO appeared to be the oxidative products of AMS. The $T_{1/2}$'s of four DADS metabolites were found to be 4.39, 6.78, 7.16, and 8.64 h for AM, AMS, AMSO, and $AMSO_2$, respectively. The peak concentrations (Cmax's) of four DADS metabolites were determined to be 8, 8, 376, and 1440 μM for AM, AMS, AMSO, and $AMSO_2$, respectively, indicating that the effective therapeutic concentrations of these active metabolite(s) may be potentially achievable (Gao et al. 2013).

Allicin and alliin are garlic-derived OSCs and are responsible for its high pharmacological properties (Schäfer and Kaschula 2014). Naturally allicin is produced enzymatically from its precursor, alliin, by damaging the plant tissue. The lipid-soluble cysteine sulfoxide, i.e., alliin, is converted to water-soluble products (such as dehydroalanine and allyl sulfenic acid) upon hydrolysis by the enzyme alliinase, which makes them more stable and more bioavailable (Abourashed 2013; Borlinghaus et al. 2014). Allyl sulfenic acid molecules condense freely to form one molecule of allicin (Ilić et al. 2011). A flavin-containing monooxygenase in garlic produces alliin through the conversion and biosynthesis of γ-glutamyl-S-allyl-L-cysteine (Yoshimoto et al. 2015). S-allyl-mercapto-glutathione and S-allylmercaptocysteine were revealed as derivatives of allicin in in vitro studies (Horev-Azaria et al. 2009). Ajoene exists in trans E-isomer and cis(Z)-isomeric forms, in which Z-isomer showed strong bioactivities when compared to E-isomer, whereas the E-isomer showed higher stability during storage (Hassan 2004; Yoo et al. 2012).

20.4 Mechanistic Insights into Anti-atherosclerotic Potential of Garlic and Its OSCs

20.4.1 Modulatory Effects of Garlic and Its OSCs Against Oxidative Stress/Redox Status

Oxidative stress, a condition of disproportionality between the level of free radicals/oxidants and antioxidant defense system, is an important factor in health. Aqueous garlic extract exerts antioxidant potential via scavenging reactive oxygen species (ROS) and improving the cellular antioxidant enzymes activity, i.e., superoxide dismutase (SOD), catalase (CAT), and glutathione peroxidase (GPx). Furthermore, garlic also inhibited the oxidative injury in liver through AMP-activated protein kinase (Han et al. 2011). Several OSCs from crushed garlic including alliin, DAS, and DATS have been shown to have specific antioxidant potential (Lee et al. 2012). These compounds are well known for their antioxidant nature as they diminish the rate of nitric oxide (NO) generation via inhibiting nuclear factor-κB (NF-κB), downregulating inducible nitric oxide synthase (iNOS) expression and inhibiting NF-κB nuclear translocation (Ryu et al. 2015). Fresh garlic-derived γ-glutamyl-S-allyl-cysteine peptide also possesses the radical scavenging and metal-chelating capacities (Tan et al. 2015).

The ratio of reduced glutathione (GSH) (an abundant endogenous antioxidant) and oxidized glutathione is a good predictor of intracellular redox status (Izigov et al. 2011). Allicin, another bio-active OSC from *A. sativum* Linn, shields the cells against oxidative stress by stimulating the production of antioxidant molecules and enzymes as well as scavenging the free radicals (Chung 2006; Ansari et al. 2018). Allicin copes with imbalanced redox status via production of glutathione from its metabolic derivatives. Moreover, it is also believed to quench hydroxyl radicals, 1,1-diphenyl-2-picrylhydrazyl (DPPH), and singlet oxygen due to its spin trapping and spin resonance abilities (Okada et al. 2005). Other evidences have shown that allicin is effective in suppressing xanthine oxidase (XO) which results in the production of toxic superoxides (Chung 2006).

Garlic is used as an important dietary source of natural antioxidants because of its high content OSCs like DADS and S-allylmercaptocystein (SAMC) (Sener et al. 2005; Ansari et al. 2018). DATS is able to release H_2S through interactions with biological thiols, including GSH within red blood cells (Benavides et al. 2007). SAMC, major OSC from aged garlic extract (AGE), showed strong free radical scavenging activity in DPPH assays, advocating the potent antioxidant activity of AGE (Numagami et al. 1996). AMS and DAS have been shown to improve hypertension-induced structural changes (Castro et al. 2010). Moreover, Cruz et al. (2007) showed that SAC reduced the hypertension and renal damage in nephrectomized rats due to its antioxidant nature. Another study by Dirsch et al. (1998) reported that allicin and ajoene decreased the accumulation of NO by blocking the activity of iNOS in LPS-stimulated RAW 264.7 macrophages. Similar to garlic extract, SACS is also reported to be antioxidative in nature, and due to this antioxidant property, it protects against lipid peroxidation and subsequent atherosclerosis (Helen et al. 2003; Zhai et al. 2018).

20.4.2 Effects of Garlic and Its Extract on Lipid and Lipoprotein Metabolism

ASCVD is caused by various risk factors, i.e., raised serum TC, LDL and an increase in LDL oxidation, infection and inflammation, hypertension, and smoking (Alvi et al. 2015, 2017a; Kodai et al. 2015). The report of Mikaili and co-workers (2013) indicated that garlic could protect against atherosclerosis via its ability to reduce lipid content into the arteries and also inhibits intracellular accumulation of lipid (Mathew and Biju 2008). Other studies showed that garlic extracts and fractions protect high-fat diet-induced hyperlipidemia in rats and mice via modulation of serum LDL to HDL ratio (Mohammadi and Oshaghi 2014). Garlic powder was also reported to protect against hypercholesterolemia by reducing the level of TC and LDL-cholesterol (Sobenin et al. 2010). However, black garlic has been studied intensely for its antioxidative, anti-inflammatory, and hypocholesterolemic effects (Lee et al. 2011; Colin-Gonzalez et al. 2012). Garlic oil is also known to cause a significant decrease in LDL and VLDL levels along with an increase in HDL levels (Bordia 1981). Intake of 400 mg garlic and 1 mg allicin twice daily in hyperlipidemic patients has significantly reduced TC, LDL-C, and TGs and increased the HDL-C levels (Kojuri et al. 2007). The level of cholesterol, TGs, phospholipids, and ß-lipoproteins were significantly declined in the individuals consuming 10–50 g of garlic/week (Sainani et al. 1979). These results indicate that dietary consumption of garlic has a beneficial role in lipid homeostasis.

20.4.3 Hypolipidemic Activity of OSCs

In particular, aforementioned beneficial therapeutic effects of garlic were achieved due to the presence of its OSCs (Yun et al. 2014). Allicin and SAC from AGE and DADS from garlic oil are the major active constituents responsible for anti-atherosclerotic effects (Yu-Yan and Liu 2001). The supplementation with single OSCs (alliin, allitride, and SMCS) reduced the CVD risk in obese rats via decreasing the body weight and lipid content (Kook et al. 2009). Ajoene, a garlic-derived highly stable OSC, is conventionally well known for its lipid-lowering and antithrombotic properties (Miroddi et al. 2011). Furthermore, the lipid-lowering effects of these OSCs are attributed to their ability to inhibit the activity of cholesterogenic enzyme, i.e., HMG-R and stimulation of hepatocellular antioxidants, such as CAT and glutathione peroxidase (GPx) (Lin and Yin 2008). SPC has also shown cardioprotection in ischemic heart disease, and garlic-derived polysulfides may be convenient in the treatment of myocardial ischemic disease (Lavu et al. 2011; Wen and Zhu 2015).

20.4.4 Effect of Garlic and Its OSCs on HMG-R Expression and Its Functionality

In hepatocytes, HMG-R acts the central cholesterogenic enzyme which converts acetyl-CoA into mevalonate (Goldstein and Brown 2015; Alvi et al. 2016). A large set of studies have proved the cardioprotective ability of garlic as it inhibits the hepatic cholesterol synthesis via the inhibition of HMG-R activity (Rahman and Lowe 2006; Gorinstein et al. 2007; Mathew and Biju 2008; Saud et al. 2016). In a study, Zahid-Ashraf et al. (2005) investigated the type 2 diabetic patients that were taking garlic tablets along with allicin and found that this garlic preparation resulted in reduced levels of blood TC and LDL-C while the level of HDL-C was increased. These desirable effects of allicin were achieved through the inhibition of HMG-R activity. Hypolipidemic activities of garlic by the inhibition of HMG-R activity were also supported by other in vivo experiments, and this HMG-R inhibitory activity of garlic was due to the presence of its OSCs such as allicin, ajoene, SAC, DADS, and SMCS (Augusti et al. 2005; Kojuri et al. 2007; Gorinstein et al. 2006).

In one study, two lipophilic OSCs, DAS and DADS, and two hydrophilic OSCs, s-ethyl cysteine and n-acetyl cysteine, reduced lipid peroxidation by the activation of antioxidant enzymes, i.e., glutathione-s-transferase (GST) and CAT and modulation of lipogenic enzymes such as HMG-R (Yin and Cheng 2003; Tsai et al. 2005). The ajoene has also been reported to reduce the HMG-R in order to protect hypercholesterolemia (Banerjee and maulik 2002). Apart from the HMG-R inhibitory activity, garlic and various OSCs have shown profound inhibitory effects against human squalene monooxygenase, an enzyme in cholesterol biosynthetic pathway (Gupta and Porter 2001). Investigations also established that water-soluble OSCs from garlic, i.e., SAC, are less cytotoxic and known to inhibit the HMG-R activity more efficiently than the lipid-soluble OSC, i.e., DAS (Yeh and Liu 2001).

20.4.5 Effects of Garlic and Its OSCs on Thrombosis and Platelet Aggregation

The recruitment of blood constituents within the vascular system of a living animal leads to the formation of an abnormal mass termed thrombosis and leads to the atherosclerotic events (Kesieme et al. 2011). On the other hand, platelets are megakaryocyte-derived small anucleated cells and can readily adhere and accumulate into the vascular injury which subsequently leads to the hemostasis and blood clotting (Yun et al. 2016). This process may also lead to thrombosis and vessel obstruction, the most common way leading to heart attacks (Hou et al. 2015). Therefore, inhibition of the thrombosis and platelet aggregation could be an ideal approach to limit the progression of atherosclerosis. Garlic and its OSCs, especially allicin and its thiosulfinates, have been found to possess antithrombotic activities in various studies (Cavagnaro et al. 2007). Allicin also exhibits antimyocardial fibrosis effect and the mechanism related to TGF-β/Smads signal transduction (Li et al. 2016). DATS has also been reported to possess antithrombotic activity, whereas methyl allyl trisulfide (MATS) is well known for its antiplatelet effects as it inhibits the arachidonic acid cascade (Ariga et al. 2000; Choi and Park 2012). Ajoene also exerts anti-aggregatory and antithrombotic potential that results due to its interaction with fibrinogen receptors (Taylor et al. 2006), and it also inhibits arachidonic acid-stimulated platelet aggregation irreversibly (Srivastava and Tyagi 1993). These reports from various in vitro and in vivo experiments are suggesting the antithrombotic and anti-aggregatory potential of garlic extract and its OSCs.

Other athero-protective modulations of AGE are protection against platelet aggregation and inhibition of prostanoid synthesis. Garlic reduces the risks of ASCVD by inhibiting platelet aggregation and lowering the levels of cholesterol and blood pressure. These protective effects are attributed to the fact that allicin is degraded into diallyl polysulfides by H_2S that ultimately prevents myocardial injury and dysfunction (Bradley et al. 2016). Garlic-derived ajoene has been shown to modulate in vitro platelet aggregation (Teranishi et al. 2003). Sodium 2-propenyl thiosulfate from garlic prevents aggregation of canine platelets via modulation of cyclooxygenase activity (Chang et al. 2005a). Allicin and aromatic thiosulfonate have also been as the major antiplatelet ingredients from garlic (MacDonald et al. 2004).

20.4.6 Effect on Smooth Muscle Cell, Macrophages, Foam Cell, and Plaque Stabilization

Irregular proliferation of vascular smooth muscle cells (VSMCs), major cells in the atherosclerotic plaques, is one of the most important factors playing a central role in the atherosclerotic events (Faries et al. 2002). Moreover, atherosclerosis is also stimulated by the activation of endothelial cells (ECs) that invite monocytes and other immune cells from the blood into the arterial wall. Macrophages tend to transform into foam cells after they engulf/phagocytize cholesterol and other fatty materials (Faries et al. 2002). AMS and DAS have been shown to inhibit angiotensin II-induced migration in aortic SMCs via modulation of ROS generation, p27 downregulation, and MAPK activation (Castro et al. 2010). Generation of H_2S is believed to be the major component responsible for the cardioprotective effects of garlic and its OSCs (Mikaili et al. 2013). In the erythrocytes, the garlic-derived OSCs are converted into H_2S which results in vasodilation of the blood and thereby reducing blood pressure and vascular injury that may lead to the plaque formation (Ginter and Simko 2010). Thus AMS and DAS could be established as an effective antioxidant against hypertension-induced arteriostructural changes. Other reports also showed that allicin could protect against atherosclerotic plaque formation either by preventing the uptake of LDL into the intima or by breaking down of macrophages (Ali et al. 2000; Gonen et al. 2005).

20.4.7 Effect on LDL Oxidation

Uptake of oxidized LDL (ox-LDL) by vascular ECs is considered to be the most crucial phase in the initiation of atherosclerotic plaques. Expression of various chemokines like monocyte chemotactic protein (MCP-1) and adhesion molecules like intercellular adhesion molecule-1 (ICAM-1), vascular cell adhesion molecule-1 (VCAM-1), and E- and P-selectins have been shown to be stimulated by ox-LDL (Li et al. 2008, 2003). In response to inflammatory stimuli, formation of foam cells (lipid-laden macrophages) takes place which play key role in the atherosclerosis (Angelovich et al. 2015). AGE has shown the preventive effects against oxidative damage of LDL, thereby reducing the level of circulatory ox-LDL that is more prone to be accumulated into macrophages, SMCs, and blood vessel walls to develop fatty streaks (Efendy et al. 1997). A more recent study by Saud et al. (2016) demonstrated the instrumental impact of garlic in the treatment of hyperlipidemia via restricting hepatic HMG-R activity and also by inhibiting the LDL-C oxidation. In addition, it also decreases the aggregation of platelets and prevents lipid peroxidation in erythrocytes and LDL which is attributed to its profound antioxidant potential (Rahman and Lowe 2006).

Another study by Akash et al. (2014) evident that garlic-derived SMC decreases serum lipids maintains the redox conditions and lipid peroxidation by increasing the antioxidant enzyme activity. Allicin was shown to suppress the generation of lipid peroxidation products, i.e., conjugated dienes and malondialdehyde (Okada et al. 2005). An in vitro study by Naidu (2002) reported the protective role of allyl sulfide on oxidation of human LDL. Lei et al. (2008) reported that DADS and DATS suppress LDL oxidation events and ox-LDL-induced vascular cell adhesion. Other researchers also reported that garlic oil-derived DAS, DADS, and DATS reduce ox-LDL-induced adhesion molecule expression and can effectively suppress in vitro LDL oxidation (Ou et al. 2003; Huang et al. 2004). These findings are strongly advocating the athero-protective efficacy of garlic and its bio-active OSCs.

20.4.8 Effect of Garlic on HDL-Associated PON-1 Expression and Activity

The progression of ASCVD is triggered by the oxidative modification of LDL in the arterial wall cells which in turn activates the macrophages leading to foam cell formation. HDL has been shown an inverse association with the risk of developing atherosclerosis as it removes excess arteries cholesterol and reduces its serum levels (Goswami et al. 2009). The activity of serum paraoxonase1 (PON-1), namely, arylesterase, a calcium-dependent esterase and closely associated to HDL, is attributed to the anti-atherogenic properties of HDL (Mackness and Mackness 2004). PON-1 functions as HDL-associated antioxidative enzyme and diminishes the LDL oxidation in order to lower the risk of ASCVD (Aviram and Rosenblat 2004). Further, the studies also showed that the expression and activity of PON-1 can also be modulated by dietary antioxidants (Leckey et al. 2010; Alvi et al. 2017a, b). In the same vein, Jaiswal and Rizvi (2014) showed that extract from *Allium* spp. can protect against $HgCl_2$-induced oxidative insult by upregulating the PON-1 expression as well as enhancing its activity and can also protect LDL oxidative modifications. Another study reported that AGE stimulates PON-1 activity and diminishes CCl_4-induced liver and cardiac dysfunction. These defensive properties of AGE against CCl_4 toxicity may be attributed to its high contents of OSCs with potent antioxidant and free radical quenching activities (Abdel-Wahhab et al. 2012).

20.4.9 Effect on Inflammation and Signalling

Recently, infection and inflammation have been found to critically power the pathogenesis of ASCVD (Weiner et al. 2014). In particular, lipopolysaccharide (LPS), an immune response-triggering molecule of the Gram-negative bacteria, may lead to a pro-inflammatory and systemic inflammatory response (Weiner et al. 2014). Various herbal products have been investigated recently for their ability to reduce inflammation, i.e., flavonoids, carotenoids, and plants such as turmeric, tomato, and garlic (Dong et al. 2015; Alvi et al. 2017a, b). The studies have reported previously that garlic also shows anti-inflammatory and immunomodulatory activity in several disease conditions including ASCVD (Lee et al. 2012; Schäfer and Kaschula 2014; Shukla and Kalra 2007). Preventive mechanism of garlic in ASCVD and obesity treatment is attributed to the modulatory effects of OSCs which downregulate the cytokine secretion with immunomodulation and anti-inflammatory effects (Arreola et al. 2015). Furthermore, Quintero-Fabián et al. (2013) reported that alliin possesses anti-inflammatory activity by reducing TNF-α, IL-6, and iNOS level in LPS-stimulated 3T3-L1 cells.

A huge set of reports suggested that the signalling pathways of MAPKs are critical regulators for LPS-stimulated inflammatory responses (Hsu et al. 2010; Chung 2011; Fisk et al. 2014). It has been demonstrated that DATS inhibits the inflammatory cascades by the downregulation of the AKT/TGFβ-mediated NF-κB and MAPK signalling (Chang and Karin 2001). Moreover, DATS has been used for the treatment of inflammation-associated neurodegenerative disorders and periodontal inflammation by downregulation of NFκB and MAPK signalling pathways (You et al. 2013; Ho and Su 2014; Fu et al. 2015). SAC and S-propargyl-cysteine have also been found to possess antioxidant and anti-inflammatory activities (Kim et al. 2013; Colin-Gonzalez et al. 2015; Wen and Zhu 2015).

SAMC and DADS have been found to inhibit NF-κB-mediated inflammatory cascades (Liu et al. 2015; Saud et al. 2016). Allicin displayed a significant protective effect against PM2.5-induced EA.hy926 endothelial cell injury via ERK1/2 pathway-induced inflammation and oxidative stress (Wan et al. 2016). Other studies also advocate the protective role of allicin against oxidative stress, inflammatory cascades, and vascular dysfunction (El-Sheakh et al. 2015; Panyod et al. 2016). However, allicin was found to stimulate the secretion of pro-inflammatory mediators such as IFN-γ, TNF-α, and IL-12*p*70 in BALB/c mice when administered orally (Feng et al. 2012). In vitro studies revealed that DAS, DADS, DATS, and SAMC downregulated the iNOS expression/activity and generation of NO and also inhibited production of inflammatory cytokines and NF-κB activity (Chang et al. 2005b; Liu et al. 2006; Kim et al. 2016). Alliin downregulated the mRNA and protein expression of pro-inflammatory genes IL-6 and MCP-1 in LPS-induced 3T3-L1 adipocytes (Quintero-Fabián et al. 2013).

20.4.10 Effect on Hypertension: OSCs vs H_2S-Mediated Vasodilation

Hypertension is one of the most frequently encountered risk factors among metabolic syndrome patients and thought to be the predecessor to ASCVD development (Patel et al. 2016). Garlic extracts and oils have been reported to have antihypertensive effects in various animal studies via modulations of NO synthesis, blood pressure, H_2S-mediated vasodilation, LDL oxidation events, or by inhibiting the activity of angiotensin-converting enzyme (Dhawan and Jain 2004; Zahid-Ashraf et al. 2005; Ginter and Simko 2010). SAC, the most prominent OSCs from garlic and AGE, have shown profound antihypertensive and renoprotective effects suggesting their beneficial role in prevention of hypertension or delaying the renal damage (Cruz et al. 2007). Banerjee and Maulik (2002) reported that oral administration of allicin to hypertensive rats could also exhibit hypotensive properties. It is also considered as a systematic vasodilator prostacyclin synthase and mediates the dilation of mesenteric circulation via the mechanisms other than secretion of prostaglandin and β- adrenergic pathway (Sobenin et al. 2009). Allicin from garlic powder also targets neuroeffector junction in the eyes of the rabbit in order to attain its hypotensive impacts (Chu et al. 1993).

SAC has been shown to be beneficial in maintaining the peripheral as well as central blood pressure in patients facing uncontrolled hypertension via improving arterial stiffness, inflammation, and lipid abnormalities (Ried et al. 2016; Cruz et al. 2007). The beneficial impact of OSCs in the modulation of CVD includes various mechanisms, i.e., the production of H_2S, free radical scavenging, and gene regulation in cholesterol homeostatic pathway (Tocmo et al. 2015). Cystathionine γ-lyase helps to produce H_2S that protects against hypertension through vasodilation in CVD, diabetes, and other metabolic syndromes (Ju et al. 2015). H_2S is enzymatically synthesized from L-cysteine and homocysteine, whereas, isoforms of NOS mediate the production of NO (Nagpure and Bian 2016; Wu et al. 2016). The potent therapeutic efficacy of these two gases advocates the importance of garlic and its OSCs in the treatment and management of risk factor-mediated atherosclerosis. Moreover, organopolysulfides (DAS, DADS, and DATS) have been shown to possess antihypertensive actions via attenuation in the level of angiotensin I-converting enzyme type-I and angiotensin II concentrations (Al-Qattan et al. 2006; Al-Malki 2016).

20.5 OSCs vs PCSK-9-LDL-R Pathway: Current Perspectives and Future Horizons

In the last decade, the circulating liver-derived protein, PCSK-9, has been established as a foremost drug target in ASCVD treatment. Secreted PCSK-9 interacts with hepatic LDL-R on its epidermal growth factor-like repeat-A (EGF-A) domain (Alvi et al. 2017a) and triggers their lysosomal degradation leading to the diminished clearance and enhanced accumulation of LDL-C in the circulation (Zhang et al. 2007). PCSK-9 expression is transcriptionally controlled by hepatic SREBPs and hepatocyte nuclear factor-1 (HNF-1). Human molecular genomic analysis has established a linkage between PCSK-9 functionality and circulating LDL-C levels (Abifadel et al. 2009). LDL-R and PCSK-9 are transcriptionally regulated by a common transcription factor, SREBP-2, which limits the desired LDL-C lowering action of statins as they coordinately upregulate PCSK-9 and LDL-R expression (Careskey et al. 2008; Welder et al. 2010). There are distinct approaches to target PCSK-9, i.e., protein inhibitors, siRNA against EGF-A-LDL-R, peptide inhibitors, and monoclonal antibodies (mAbs) (Tail et al. 2014; Dong et al. 2015). Among all these approaches, human mAbs alone or in combination with statins have been extensively used in the major human trials (Yadav et al. 2016). Regardless of the desired LDL-C lowering efficacy, various adverse effects are associated with these PCSK-9 mAbs such as repeated self-administered injection and development of antidrug Abs as well as the huge cost over a long-lasting prescription. Therefore, discovery of novel natural products targeting PCSK-9 may be considered as breakthrough in the forthcoming era of lipid-lowering therapy, particularly for the subjects with familial hypercholesterolemia that are facing inadequate LDL-C lowering with statins.

In this order, recently, various researchers across the globe paid their attention on the discovery of natural PCSK-9 inhibitors to combat hypercholesterolemia (Tail et al. 2014; Dong et al. 2015; Alvi et al. 2017a, b). Considering the beneficial therapeutic impact of the garlic and its bio-active OSCs against risk factor-induced hypercholesterolemia and atherosclerosis, we hypothesized that garlic and its bio-active OSCs may possess potent hypercholesterolemic properties via targeting both the PCSK-9 expression and its functionality at protein levels, thereby decreasing the lysosomal degradation of LDL-R in order to enhance the recycled number of hepatic LDL-R that ultimately may lead to accelerated rate of LDL-C clearance from circulation. On the other hand, to the best of our knowledge, there is no report demonstrating the effect of garlic and its OSCs on SREBP-2-mediated LDL-R expression and functionality. Therefore, these therapeutic agents may also have some effects on the expression and functionality of LDL-R, thereby increasing the functional number of hepatic LDL-R available for LDL-C metabolism. Thus, reduced atherogenic load (LDL-C levels) in the circulation can protect the individuals from developing atherosclerotic plaques. The possible beneficial mechanisms of garlic and its OSCs in targeting risk factor-mediated atherosclerosis have been schematically illustrated in Fig. 20.3. Garlic and its OSCs may protect against risk factor-mediated atherosclerosis via downregulation of HMG-R expression (1), inhibition of HMG-R activity (2), upregulation of LDL-R (3), enhanced recycling of LDL-R to the hepatocyte surfaces making them available to upcoming LDL-C (4), downregulation of PCSK-9 expression (5), reduction of PCSK-9 affinity for EGF-A portion of LDL-R (6), increased processing of SREBP-2 during high cholesterol present in the circulation (7), reduction in LDL oxidation events (8), plaque formation (9), lipid and lipoprotein metabolism (10), and PON-1 activity (11) (*HL* hepatic lipase, *FFA* free fatty acids, *CETP* cholesteryl ester transfer protein, *PLTP* plasma phospholipid transfer protein, *ER* endoplasmic reticulum).

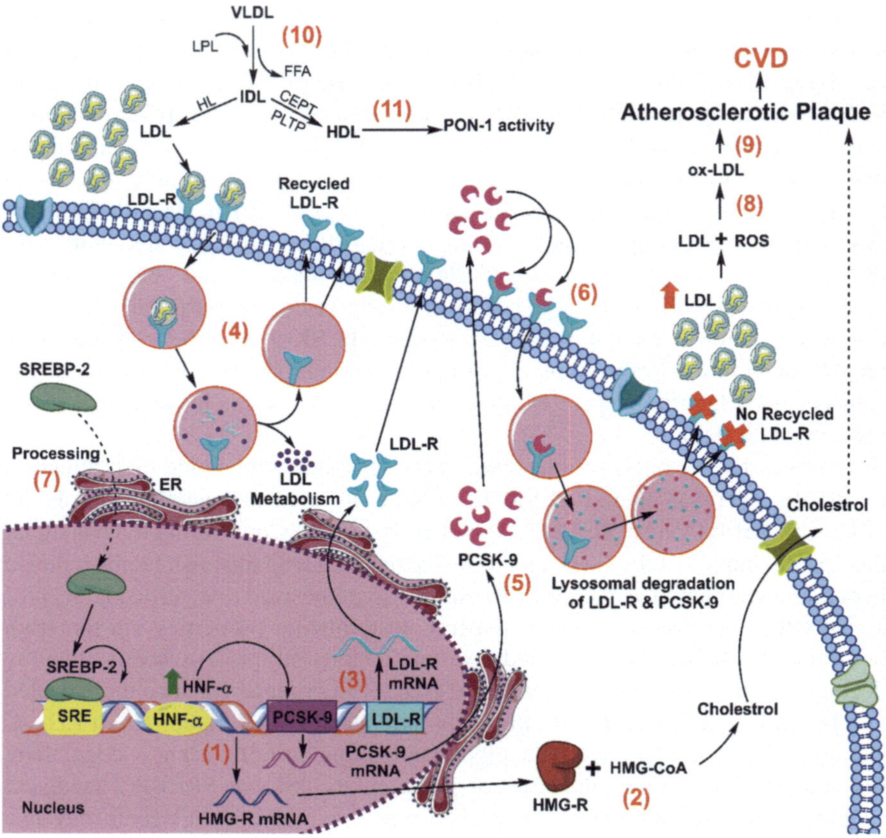

Fig. 20.3 Proposed regulatory mechanisms of garlic and its OSCs against cholesterol homeostasis

20.6 Conclusions and Future Prospects

Hypercholesterolemia and atherosclerosis are the leading cause of deaths worldwide, and high levels of blood lipids, including cholesterol, LDL-C, and TG, are the key risk factors associated with ASCVD. Dietary phytochemicals from garlic and OSCs have been found to show profound effects against various metabolic disorders. Major OSCs from garlic owing to its high therapeutic and pharmacological properties are allicin, DAS, DAT, AMTS, DAD, SACS, SAC, ajoene, allicin, and alliin, apart from many others. Based on above studies, we concluded that the therapeutic efficacy of garlic and its bio-active OSCs against risk factor-mediated ASCVD is due to the modulation of distinct biochemical and molecular mechanisms, i.e., antioxidant potential, amelioration of lipid, and lipoprotein metabolism, enhancing the HDL functionality through paraoxonase-1 (PON-1) activity, regulation of HMG-R and LDL-R expression and functionality, reduction in LDL oxidation events, effects on macrophage activation, foam cell formation and plaque

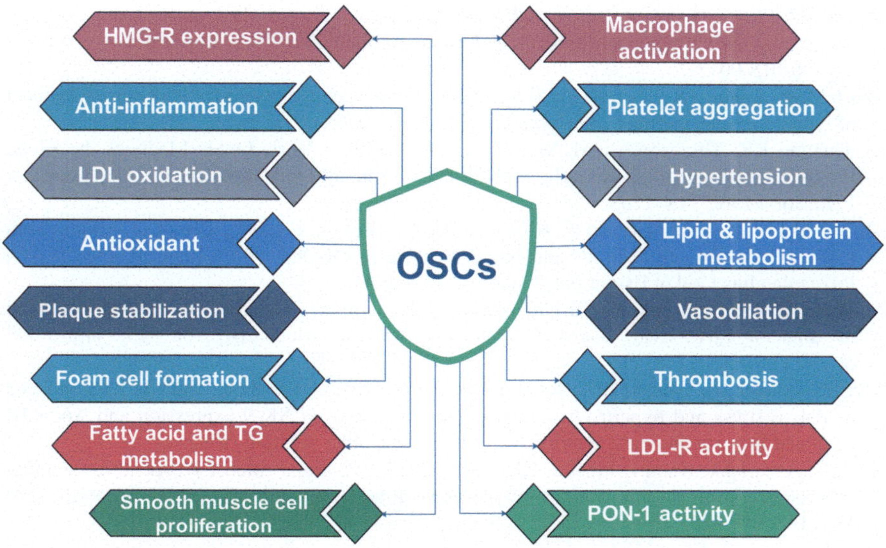

Fig. 20.4 Mechanisms of garlic-derived OSCs in protecting various risk factor-induced hypercholesterolemia and atherosclerosis

stabilization, suppression of inflammatory signalling pathways, and cure against hypertension (Fig. 20.4).

We also hypothesized that garlic and its OSCs could have beneficial effects on PCSK-9 and LDL-R pathway in order to treat and manage hypercholesterolemia especially for the patients facing inadequate lipid lowering with classical HMG-R inhibitors (statins) and statin intolerance. More studies are still needed in order to decipher the molecular pathways of range of OSCs in modulating atherosclerosis and associated conditions. Finally to sum up the whole, we concluded that the garlic and its OSCs may be promoted from alternative to mainstream medicine in targeting risk factor-mediated ASCVD.

Acknowledgments We are highly indebted to Prof. S. W. Akhtar, Chancellor, for providing us the state-of the-art research laboratory. This study was partly supported by University Internal Grant (BRTF, 2017-18). This chapter has manuscript communication number IU/R&D/2018-MCN000383.

References

Abdel-Wahhab KGE-D, Khadrawy YA, Mannaa FAE (2012) Aged garlic extract enhances paraoxonase 1 activity and suppress oxidative stress in CCl_4- intoxicated rats. Comun Sci 3:55–63

Abifadel M, Rabes JP, Devillers M, Munnich A, Erlich D, Junien C, Varret M, Boileau C (2009) Mutations and polymorphisms in the proprotein convertase subtilisin kexin 9 (PCSK9) gene in cholesterol metabolism and disease. Hum Mutat 30:520–529

Abourashed EA (2013) Bioavailability of plant-derived antioxidants. Antioxidants 2:309–325

Ahmed RA (2018) Hepatoprotective and antiapoptotic role of aged black garlic against hepatotoxicity induced by cyclophosphamide. J Basic Applied Zool 79:8. https://doi.org/10.1186/s41936-018-0017-7

Akash MSH, Rehman K, Chen S (2014) Spice plant *Allium cepa*: dietary supplement for treatment of type 2 diabetes mellitus. Nutrition 30:1128–1137

AL-Quattan KK, Thomson M, AL-Musawa'a S et al (2006) Nitric Oxide Mediates the Blood Pressure Lowering Effect of Garlic in the Rat Two-Kidney, One-Clip Model of Hypertension. J Nutrition 136:774S–776S.

Ali M, Al-Qattan KK, Al-Enezi F, Khanafer RMA, Mustafa T (2000) Effect of Allicin from garlic powder on serum lipids and blood pressure in rats fed with a high cholesterol diet. Prostaglandins Leukot Essent Fatty Acids 62:253–259

Al-Malki AL (2016) Inhibition of α-glucosidase by thiosulfinate as a target for glucosemodulation in diabetic rats. Evid Based Complement Alternat Med 2016:7687915. https://doi.org/10.1155/2016/7687915

Alvi SS, Ansari IA, Ahmad MK, Iqbal J, Khan MS (2017b) Lycopene amends LPS induced oxidative stress and hypertriglyceridemia via modulating PCSK-9 expression and Apo-CIII mediated lipoprotein lipase activity. Biomed Pharmacother 96:1082–1093

Alvi SS, Ansari IA, Khan I, Iqbal J, Khan MS (2017a) Potential role of lycopene in targeting proprotein convertase subtilisin/kexin type-9 to combat hypercholesterolemia. Free Radic Biol Med 108:394–403

Alvi SS, Ansari IA, Khan MS (2015) Pleiotropic role of lycopene in protecting various risk factors mediated atherosclerosis. Ann Phytomed 4:54–60

Alvi SS, Iqbal D, Ahmad S, Khan MS (2016) Molecular rationale delineating the role of lycopene as a potent HMG-CoA reductase inhibitor: *in vitro* and *in silico* study. Nat Prod Res 30:2111–2114

Amagase H, Petesch BL, Matsuura H, Kasuga S, Itakura Y (2001) Intake of garlic and its bioactive components. J Nutr 131:S955–S962

Angelovich TA, Hearps AC, Jaworowski A (2015) Inflammation- induced foam cell formation in chronic inflammatory disease. Immunol Cell Biol 93:683–693

Ansari IA, Khan I, Khan MS, Akhtar MS (2018) Organosulfur compounds of garlic as potent chemotherapeutic agents against cancer. In: Akhtar MS, Swamy MK (eds) Anticancer plants: clinical trials and nanotechnology. Springer, Singapore, pp 253–272

Ariga T, Tsuj K, Seki T, Moritomo T, Yamamoto JI (2000) Antithrombotic and antineoplastic effects of phyto-organo sulfur compounds. Bio Factors 13:251–255

Arreola R, Quintero-Fabian S, Lopez-Roa RI, Flores-Gutiérrez EO, Reyes-Grajeda JP, Carrera-Quintanar L, Ortuño-Sahagún D (2015) Immunomodulation and anti-inflammatory effects of garlic compounds. J Immunol Res 2015:401630. https://doi.org/10.1155/2015/401630

Augusti KT, Chackery J, Jacob J, Kuriakose S, George S, Nair SS (2005) Beneficial effects of a polar fraction of garlic (*Allium sativum* Linn) oil in rats fed with two different high fat diets. Indian J Exp Biol 43:76–83

Aviram M, Rosenblat M (2004) Paraoxonases 1, 2, and 3, oxidative stress, and macrophage foam cell formation during atherosclerosis development. Free Radic Biol Med 37:1304–1316

Banerjee SK, Maulik SK (2002) Effect of garlic on cardiovascular disorders: a review. J Nutr 1:1–4

Beato VM, Sánchez AH, de Castro A, Montaño A (2012) Effect of processing and storage time on the contents of organosulfur compounds in pickled blanched garlic. J Agric Food Chem 60:3485–3491

Benavides GA, Squadrito GL, Mills RW, Patel HD, Isbell TS, Patel RP, Darley-Usmar VM, Doeller JE, Kraus DW (2007) Hydrogen sulphide mediates the vasoactivity of garlic. Proc Natl Acad Sci USA 104:17977–17982

Besseling J, Kindt I, Hof M, Kastelein JJ, Hutten BA, Hovingh GK (2014) Severe heterozygous familial hypercholesterolemia and risk for cardiovascular disease: a study of a cohort of 14000 mutation carriers. Atherosclerosis 233:219–223

Bordia A (1981) Effect of garlic and blood lipids in patients with coronary heart disease. Am J Clin Nutr 34:2100–2103

Borlinghaus J, Albrecht F, Gruhlke MC, Nwachukwu ID, Slusarenko AJ (2014) Allicin: Chemistry and biological properties. Molecules 19:12591–12618

Bradley JM, Organ CL, Lefer DJ (2016) Garlic-derived organic polysulfides and myocardial protection. J Nutr 146:403–409

Careskey HE, Davis RA, Alborn WE, Troutt JS, Cao G, Konrad RJ (2008) Atorvastatin increases human serum levels of proprotein convertase subtilisin/ kexin type 9. J Lipid Res 49:394–398

Castro C, Lorenzo AG, Gonzalez A, Cruzado M (2010) Garlic components inhibit Angiotensin II-induced cell-cycle progression and migration: Involvement of cell-cycle inhibitor p27Kip1 and mitogen-activated protein kinase. Mol Nutr Food Res 54:781–787

Cavagnaro PF, Camargo A, Galmarini CR, Simon PW (2007) Effect of cooking on garlic (*Allium sativum* L.) antiplatelet activity and thiosulfinates content. J Agric Food Chem 55:1280–1288

Chang HP, Hung SY, Chen YH (2005b) Modulation of cytokine secretion by garlic oil derivatives is associated with suppressed nitric oxide production in stimulated macrophages. J Agric Food Chem 53:2530–2534

Chang HS, Yamato O, Yamasaki M, Maede Y (2005a) Modulatory influence of sodium 2-propenyl thiosulfate from garlic on cyclooxygenase activity in canine platelets: Possible mechanism for the anti-aggregatory effect. Prostaglandins Leukot Essent Fatty Acids 72:351–355

Chang L, Karin M (2001) Mammalian MAP kinase signaling cascades. Nature 410:37–40

Chauhan NB (2005) Multiplicity of garlic health effects and Alzheimer's disease. J Nutr Health Aging 9:421–432

Chiu HF, Shen YC, Huang TY, Venkatakrishnan K, Wang CK (2016) Cardioprotective efficacy of red wine extract of onion in healthy hypercholesterolemic subjects. Phytother Res 30:380–385

Choi YH, Park HS (2012) Apoptosis induction of U937 human leukemia cells by diallyl trisulfide induces through generation of reactive oxygen species. J Biomed Sci 11:19–50

Choudhary K, Choudhary R, Choudhary PR, Chawla VK (2011) Beneficial effect associated with use of watery and alcoholic extract of garlic as a supplement in hyperlipidemic Guniea pigs. J Bang Soc Physiol 6:22–26

Chu TC, Ogidigben M, Han JC, Potter DE (1993) Allicin-induced hypotension in rabbit eyes. J Ocul Pharmacol Ther 9:201–209

Chung KF (2011) p38 mitogen-activated protein kinase pathways in asthma and copd. Chest J 139:1470–1479

Chung LY (2006) The antioxidant properties of garlic compounds: allyl cysteine, alliin, allicin, and allyl disulfide. J Med Food 9:205–213

Colin-Gonzalez AL, Ali SF, Túnez I, Santamaría A (2015) On the antioxidant, neuroprotective and anti-inflammatory properties of S-allyl cysteine: an update. Neurochem Int 89:83–91

Colin-Gonzalez AL, Santana RA, Silva-Islas CA, Chánez-Cárdenas ME, Santamaría A, Maldonado PD (2012) The antioxidant mechanisms underlying the aged garlic extract- and S-allylcysteineinduced protection. Oxidative Med Cell Longev 2012:907162. https://doi.org/10.1155/2012/907162

Cruz C, Correa-Rotter R, Sanchez-Gonzalez DJ, Hernández-Pando R, Maldonado PD, Martínez-Martínez CM, Medina-Campos ON, Diana Aguilar ET, Chirino YI, Pedraza-Chaverri J (2007) Renoprotective and antihypertensive effects of S-allylcysteine in 5/6 nephrectomized rats. Am J Physiol Renal Physiol 293:F1691–F1698

Czompa A, Szoke K, Prokisch J, Gyongyosi A, Bak I, Balla G, Tosaki A, Lekli I (2018) Aged (Black) versus raw garlic against ischemia/reperfusion-induced cardiac complications. Int J Mol Sci 19:E1017. https://doi.org/10.3390/ijms19041017

Davidson MH, Armani A, McKenney JM, Jacobson TA (2007) Safety considerations with fibrate therapy. Am J Cardiol 99:S3–S18

Dhawan V, Jain S (2004) Effect of garlic supplementation on oxidized low-density lipoproteins and lipid peroxidation in patients of essential hypertension. Mol Cell Biochem 266:109–115

Dirsch VM, Kiemer AK, Wagner H (1998) Effect of allicin and ajoene, two compounds of garlic, on inducible nitric oxide synthase. Atherosclerosis 139:333–339

Dong B, Li H, Singh AB, Cao A, Liu J (2015) Inhibition of PCSK9 transcription by Berberine involves down-regulation of hepatic HNF1-α protein expression through the ubiquitin-proteasome degradation pathway. J Biol Chem 290:4047–4058

Efendy JL, Simmons DL, Campbell GR, Campbell JH (1997) The effect of the aged garlic extract, "Kyolic", on the development of experimental atherosclerosis. Atherosclerosis 132:37–42

El-Sheakh AR, Ghoneim HA, Suddek GM, Ammar ESM (2015) Attenuation of oxidative stress, inflammation, and endothelial dysfunction in hypercholesterolemic rabbits by allicin. Can J Physiol Pharmacol 94:216–224

Faries PL, Rohan DI, Wyers MC (2002) Vascular smooth muscle cells derived from atherosclerotic human arteries exhibit greater adhesion, migration, and proliferation than venous cells. J Surg Res 104:22–28

Feng Y, Zhu X, Wang Q, Jiang Y, Shang H, Cui L, Cao Y (2012) Allicin enhances host pro–inflammatory immune responses and protects against acute murine malaria infection. Malar J 11:268. https://doi.org/10.1186/1475-2875-11-268

Fisk M, Gajendragadkar PR, Mäki-Petäjä KM, Wilkinson IB, Cheriyan J (2014) Therapeutic potential of p38 MAP kinase inhibition in the management of cardiovascular disease. Am J Cardiovasc Drugs 14:155–165

Fu E, Tsai MC, Chin YT, Tu HP, Fu MM, Chiang CY, Chiu HC (2015) The effects of diallyl sulphide upon *Porphyromonas gingivalis* lipopolysaccharide stimulated pro-inflammatory cytokine expressions and nuclear factor kappa B activation in human gingival fibroblasts. J Periodontal Res 50:380–388

Gao C, Jiang X, Wang H, Zhao Z, Wang W (2013) Drug metabolism and pharmacokinetics of organosulfur compounds from garlic. J Drug Metab Toxicol 4:159. https://doi.org/10.4172/2157-7609.1000159

Germain E, Auger J, Ginies C, Siess MH, Teyssier C (2002) *In vivo* metabolism of diallyl disulphide in the rat: identification of two new metabolites. Xenobiotica 32:1127–1138

Ginter E, Simko V (2010) Garlic (*Allium sativum* L) and cardiovascular diseases. Bratisl Lek Listy 111:452–456

Goldstein JL, Brown MS (2015) A century of cholesterol and coronaries: from plaques to genes to statins. Cell 161:161–172

Gonen A, Harat D, Rabinkov A, Miron T, Mirelman D, Wilchek M, Weiner L, Ulman E, Levkovitz H, Ben-Shushan D, Shaish A (2005) The antiatherogenic effect of Allicin: possible mode of action. Pathobiology 72:325–334

Gorinstein S, Jastrzebski Z, Namiesnik J, Leontowicz H, Leontowicz M, Trakhtenberg S (2007) The Atherosclerotic heart disease and protecting properties of garlic: Contemporary data. Mol Nutr Food Res 51:1365–1381

Gorinstein S, Leontowicz H, Leontowicz M, Drzewiecki J, Najman K, Katrich E, Barasch D, Yamamoto K, Trakhtenberg S (2006) Raw and boiled garlic enhances plasma antioxidant activity and improves plasma lipid metabolism in cholesterol-fed rats. Life Sci 78:655–663

Goswami B, Tayal D, Gupta N, Mallika V (2009) Paraoxonase: a multifaceted biomolecule. Clin Chim Acta 410:1–12

Gupta N, Porter TD (2001) Garlic and garlic derived compounds inhibit human squalene monooxygenase. J Nutr 131:1662–1667

Han CY, Ki SH, Kim YW, Noh K, Lee DY, Kang B, Ryu JH, Jeon R, Kim EH, Hwang SJ, Kim SG (2011) Ajoene, a stable garlic by-product, inhibits high fat diet-induced hepatic steatosis and oxidative injury through LKB1-dependent AMPK activation. Antioxid Redox Signal 14:187–202

Hassan HT (2004) Ajoene (natural garlic compound): a new anti-leukaemia agent for AML therapy. Leuk Res 28:667–671

Helen A, Krishnakumar K, Vijayammal PL, Augusti KT (2003) A comparative study of antioxidants S-Allyl cysteine sulfoxide and vitamin E on the damages induced by nicotine in rats. Pharmacology 67:113–117

Heron M (2013) Deaths: leading causes for 2010. Natl Vital Stat Rep 62:1–97

Ho SC, Su MS (2014) Evaluating the anti-neuroinflammatory capacity of raw and steamed garlic as well as five organosulfur compounds. Molecules 19:17697–17714

Horev-Azaria L, Eliav S, Izigov N, Pri-Chen S, Mirelman D, Miron T, Rabinkov A, Wilchek M, Jacob-Hirsch J, Amariglio N, Savion N (2009) Allicin up-regulates cellular glutathione level in vascular endothelial cells. Eur J Nutr 48:67–74

Horton JD, Cohen JC, Hobbs HH (2007) Molecular biology of PCSK9: its role in LDL metabolism. Trends Biochem Sci 32:71–77

Hou Y, Carrim N, Wang Y, Gallant RC, Marshall A, Ni H (2015) Platelets in hemostasis and thrombosis: Novel mechanisms of fibrinogen-independent platelet aggregation and fibronectinmediated protein wave of hemostasis. J Biomed Res 29:437–444

Hsu MJ, Chang CK, Chen MC, Chen BC, Ma HP, Hong CY, Lin CH (2010) Apoptosis signal-regulating kinase 1 in peptidoglycan-induced COX-2 expression in macrophages. J Leukoc Biol 87:1069–1082

Huang CN, Horng JS, Yin MC (2004) Antioxidative and antiglycative effect of six organosulfur compounds in low- density lipoprotein and plasma. J Agric Food Chem 52:3674–3678

Iciek M, Kwiecien I, Włodek L (2009) Biological properties of garlic and garlic derived organosulfur compounds. Environ Mol Mutagen 50(3):247–265

Ilić D, Nikolić VD, Nikolić LB, Stanković MZ, Stanojević LP, Cakić MD (2011) Allicin and related compounds: Biosynthesis, synthesis and pharmacological activity. Facta Univ Phys Chem Technol 9:9–20

Iqbal D, Khan MS, Khan A, Khan M, Ahmad S, Srivastava AK, Bagga P (2014b) *In vitro* screening for β-hydroxy-β-methylglutarylcoa reductase inhibitory and antioxidant activity of sequentially extracted fractions of Ficus palmate Forsk. Bio Med Res Int 2014:762620. https://doi.org/10.1155/2014/762620

Iqbal D, Khan MS, Khan MS, Ahmad S, Srivastava AK (2014a) An *in vitro* and molecular informatics study to evaluate the antioxidative and β-hydroxy-β-methylglutaryl-CoA reductase inhibitory property of Ficusvirens Ait. Phytother Res 28:899–908

Iqbal D, Khan MS, Khan MS, Ahmad S, Hussain MS, Ali M (2015) Bioactivity guided fractionation and hypolipidemic property of a novel HMG-CoA reductase inhibitor from Ficusvirens Ait. Lipids Health Dis 14:15. https://doi.org/10.1186/s12944-015-0013-6

Izigov N, Farzam N, Savion N (2011) S-allylmercapto-N-acetylcysteine up-regulates cellular glutathione and protects vascular endothelial cells from oxidative stress. Free Radic Biol Med 50:1131–1139

Jaiswal N, Rizvi SI (2014) Onion extract (*Allium cepa* L.), quercetin and catechin up-regulate paraoxonase 1 activity with concomitant protection against low-density lipoprotein oxidation in male Wistar rats subjected to oxidative stress. J Sci Food Agric 94:2752–2757

Ju Y, Untereiner A, Wu L, Yang G (2015) H_2S-induced sulfhydration of pyruvate carboxylase contributes to gluconeogenesis in liver cells. Biochimica Biophysica Acta Gene Sub 1850:2293–2303

Kesieme E, Kesieme C, Jebbin N, Irekpita E, Dongo A (2011) Deep vein thrombosis: a clinical review. J Blood Med 2:59–69

Keyamura Y, Nagano C, Kohashi M, Niimi M, Nozako M, Koyama T, Yasufuku R, Imaizumi A, Itabe H, Yoshikawa T (2014) Add-on effect of probucol in atherosclerotic, cholesterol-fed rabbits treated with atorvastatin. PLoS One 9:e96929. https://doi.org/10.1371/journal.pone.0096929

Kim D, Kang MJ, Hong SS, Choi YH, Shin JH (2016) Antiinflammatory effects of functionally active compounds isolated from aged black garlic. Phytother Res 31:53–61

Kim SR, Jung YR, An HJ, Kim DH, Jang EJ, Choi YJ, Moon KM, Park MH, Park CH, Chung KW, Bae HR (2013) Anti-wrinkle and anti-inflammatory effects of active garlic components and the inhibition of MMPs via NF-κB signalling. PLoS One 8:e0073877. https://doi.org/10.1371/journal.pone.0073877

Kimura S, Tung YC, Pan MH, Su NW, Lai YJ, Cheng KC (2017) Black garlic: a critical review of its production, bioactivity, and application. J Food Drug Anal 25:62–70

Kodai S, Takemura S, Kubo H, Azuma H, Minamiyama Y (2015) Therapeutic administration of an ingredient of aged garlic extracts, S-allyl cysteine resolves liver fibrosis established by carbon tetrachloride in rats. J Clin Biochem Nutr 56:179–185

Kojuri J, Vosoughi AR, Akrami M (2007) Effects of *Anethum graveolens* and garlic on lipid profile in hyperlipidemic patients. Lipids Health Dis 6:5. https://doi.org/10.1186/1476-511X-6-5

Kook S, Kim GH, Choi K (2009) The antidiabetic effect of onion and garlic in experimental diabetic rats: meta-analysis. J Med Food 12:552–560

Lanzotti V, Scala F, Bonanomi G (2014) Compounds from *Allium* species with cytotoxic and antimicrobial activity. Phytochem Rev 13:769–791

Lavu M, Bhushan S, Lefer DJ (2011) Hydrogen sulphide mediated cardioprotection: mechanisms and therapeutic potential. Clin Sci 120:219–229

Leckey LC, Garige M, Varatharajalu R, Gong M, Nagata T, Spurney CF, Lakshman RM (2010) Quercetin and ethanol attenuate the progression of atherosclerotic plaques with concomitant up regulation of paraoxonase1 (PON1) gene expression and PON1 activity in LDLR−/−mice. Alcohol Clin Exp Res 34:1535–1542

Lee DY, Li H, Lim HJ, Jeon R, Ryu JH (2012) Anti-inflammatory activity of sulfur-containing compounds from garlic. J Med Food 15:992–999

Lee EN, Choi YW, Kim HK, Park JK, Kim HJ, Kim MJ, Lee HW, Kim KH, Bae SS, Kim BS, Yoon S (2011) Chloroform extract of aged black garlic attenuates TNF-alphainduced ROS generation, VCAM-1 expression, NF-kappa B activation and adhesiveness for monocytes in human umbilical vein endothelial cells. Phytother Res 25:92–100

Lei YP, Chen HW, Sheen LY, Lii CK (2008) Diallyl disulfide and diallyl trisulfide suppress oxidized LDL-induced vascular cell adhesion molecule and E-selectin expression through protein kinase A- and B-dependent signaling pathways. J Nutr 138:996–1003

Li D, Liu L, Chen H, Sawamura T, Mehta JL (2003) LOX-1, an oxidized LDL endothelial receptor, induces CD40/CD40L signaling in human coronary artery endothelial cells. Arteriosclerosis Throm Vas Bio 23:816–821

Li SC, Ma LN, Chen J, Li YK (2016) Effect of allicin on myocardial fibrosis after myocardial infarction in rats and its relationship with TGF*β*/Smads signal transduction. China J Chin Mater Medica 41:2517–2521

Lin CC, Yin MC (2008) Effects of cysteine-containing compounds on biosynthesis of triacylglycerol and cholesterol and anti-oxidative protection in liver from mice consuming a high-fat diet. Br J Nutr 99:37–43

Liu CT, Wong PL, Lii CK, Hse H, Sheen LY (2006) Antidiabetic effect of garlic oil but not diallyl disulfide in rats with streptozotoc in induced diabetes. Food Chem Toxicol 44:1377–1384

Liu Y, Yan J, Han X, Hu W (2015) Garlic-derived compound S-allylmercaptocysteine (SAMC) is active against anaplastic thyroid cancer cell line 8305C (HPACC). Technol Health Care 23:S89–S93

MacDonald JA, Marchand ME, Langler RF (2004) Improving upon the *in vitro* biological activity of antithrombotic disulfides. Blood Coagul Fibrinolysis 15:447–450

Mackness M, Mackness B (2004) Paraoxonase 1 and atherosclerosis: is the gene or the protein more important. Free Radic Biol Med 37:1317–1323

Mathew BC, Biju RS (2008) Neuroprotective effects of garlic-A review. Libyan J Med 3:23–33

Mendis S, Puska P, Norrving B (2011) Global Atlas on cardiovascular disease prevention and control. WHO in Collaboration with the WHF and the WSO 3-18. http://www.who.int/cardiovascular_diseases/publications/atlas_cvd/en/. Accessed 13 Sept 2018

Mikaili P, Maadirad S, Moloudizargari M, Aghajanshakeri S, Sarahroodi S (2013) Therapeutic uses and pharmacological properties of garlic, shallot, and their biologically active compounds. Iran J Basic Med Sci 16:1031–1048

Miroddi M, Calapai F, Calapai G (2011) Potential beneficial effects of garlic in oncohematology. Mini Rev Med Chem 11:461–472

Mohammadi A, Oshaghi EB (2014) Effect of garlic on lipid profile and expression of LXR alpa in intestine and liver of hypercholesterolemic mice. J Diabetes Metab Disord 13:20. https://doi.org/10.1186/2251-6581-13-20

Nagpure BV, Bian JS (2016) Interaction of hydrogen sulphide with nitric oxide in the cardiovascular system. Oxidative Med Cell Longev. https://doi.org/10.1155/2016/6904327

Naidu A, Thippeswamy N (2002) Inhibition of human low density lipoprotein oxidation by active principles from spices. Mol Cell Biochem 229:19

Numagami Y, Sato S, Ohnishi T (1996) Attenuation of rat ischemic brain damage by aged garlic extracts: a possible protecting mechanism as antioxidants. Neurochem Int 29:135–143

Okada Y, Tanaka K, Fujita I, Sato E, Okajima H (2005) Antioxidant activity of thiosulfinates derived from garlic. Redox Rep 10:96–102

Ou CC, Tsao SM, Lin MC, Yin MC (2003) Protective action on human LDL against oxidation and glycation by four organosulfur compounds derived fron garlic. Lipids 38:219–224

Panyod S, Wu WK, Ho CT, Lu KH, Liu CT, Chu YL, Lai YS, Chen WC, Lin YE, Lin SH, Sheen LY (2016) Diet supplementation with allicin protects against alcoholic fatty liver disease in mice by improving anti-inflammation and antioxidative functions. J Agric Food Chem 64:7104–7113

Patel P, Ordunez P, Di Pette D, Escobar MC, Hassell T, Wyss F, Hennis A, Asma S, Angell S (2016) Improved blood pressure control to reduce cardiovascular disease morbidity and mortality: The standardized hypertension treatment and prevention project. J Clin Hypertens 18:1284–1294

Petropoulos SA, Fernandes Â, Ntatsi G (2018) Konstantinos Petrotos, Lillian Barros, Isabel CFR Ferreira. Nutritional value, chemical characterization and bulb morphology of greek garlic landraces. Molecules 23:E319. https://doi.org/10.3390/molecules23020319

Quintero-Fabián S, Ortuño-Sahagún D, Vázquez-Carrera M, López-Roa RI (2013) Alliin, a garlic (*Allium sativum*) compound, prevents LPS-induced inflammation in 3T3-L1 adipocytes. Mediat Inflamm 2013:381815. https://doi.org/10.1155/2013/381815

Rahman K, Lowe GM (2006) Garlic and cardiovascular disease: a critical review. J Nutr 136:S736–S740

Ray B, Chauhan NB, Lahiri DK (2011) The "aged garlic extract": (AGE) and one of its active ingredients S-allyl-L-cysteine (SAC) as potential preventive and therapeutic agents for Alzheimer's disease (AD). Curr Med Chem 18:3306–3313

Reiner Z (2014) Resistance and intolerance to statins. Nutr Metab Cardiovasc Dis 24:1057–1066

Ried K, Travica N, Sali A (2016) The effect of aged garlic extract on blood pressure and other cardiovascular risk factors in uncontrolled hypertensives: the AGE at heart trial. Integr Blood Press Control 9:9–21

Rivlin R (2001) Historical perspective on the use of garlic. J Nutr 131:S951–S954

Ryu JH, Park HJ, Jeong YY, Han S, Shin JH, Lee SJ, Kang MJ, Sung NJ, Kang D (2015) Aged red garlic extract suppresses nitric oxide production in lipopolysaccharide-treated RAW 264.7 macrophages through inhibition of NF-B. J Med Food 18:439–445

Sainani GS, Desai DB, Gorhe NH, Natu SM, Pise DV, Sainani PG (1979) Effect of dietary garlic and onion on serum lipid profile in jain community. Indian J Med Res 69:776–780

Santhosha SG, Jamuna P, Prabhavathi SN (2013) Bioactive components of garlic and their physiological role in health maintenance: a review. Food Biosci 3:59–74

Saud SM, Li W, Gray Z, Matter MS, Colburn NH, Young MR, Kim YS (2016) Diallyl disulfide (DADS), a constituent of garlic, inactivates NF- B and prevents colitisinduced colorectal cancer by inhibiting GSK-3. Cancer Prev Res 9:607–615

Schäfer G, Kaschula CH (2014) The immunomodulation and anti-inflammatory effects of garlic organosulfur compounds in cancer chemoprevention. Anti Cancer Agents Med Chem 14:233–240

Sener G, Sehirli O, Ipci Y, Ercan F, Şirvanci S, Gedik N, Yeĝen BÇ (2005) Aqueous garlic extract alleviates ischaemia-reperfusion-induced oxidative hepatic injury in rats. J Pharm Pharmacol 57:145–150

Shukla Y, Kalra N (2007) Cancer chemoprevention with garlic and its constituents. Cancer Lett 247:167–181

Sobenin IA, Andrianova IV, Fomchenkov IV, Gorchakova TV, Orekhov AN (2009) Time-Released garlic powder tablets lower systolic and diastolic blood pressure in men with mild and moderate arterial hypertension. Hypertens Res 32:433–437

Sobenin IA, Pryanishnikov VV, Kunnova LM, Rabinovich YA, Martirosyan DM, Orekhov AN (2010) The effects of time-released garlic powder tablets on multifunctional cardiovascular risk in patients with coronary artery disease. Lipids Health Dis 9:119. https://doi.org/10.1186/1476-511X-9-119

Srivastava KC, Tyagi OD (1993) Effects of a garlic-derived principle (Ajoene) on aggregation and arachidonic acid metabolism in human blood platelets. Prostaglandins Leukot Essent Fatty Acids 49:587–595

Tail MH, Chen PK, Chen PY, Wu MJ, Ho CT, Yen JH (2014) Curcumin enhances cell-surface LDLR level and promotes LDL uptake through down-regulation of PCSK9 gene expression in HepG2 cells. Mol Nutr Food Res 58:2133–2145

Tan D, Zhang Y, Chen L, Liu L, Zhang X, Wu Z, Bai B, Ji S (2015) Decreased glycation and structural protection properties of γ-glutamyl-S-allyl-cysteine peptide isolated from fresh garlic scales (*Allium sativum* L.). Nat Prod Res 29:2219–2222

Taucher J, Hansel A, Jordan A, Lindinger W (1996) Analysis of compounds in human breath after ingestion of garlic using proton-transfer-reaction mass spectrometry. J Agric Food Chem 44:3778–3782

Taylor P, Noriega R, Farah C, Abad MJ, Arsenak M, Apitz R (2006) Ajoene inhibits both primary tumor growth and metastasis of B16/BL6 melanoma cells in C57BL/6 mice. Cancer Lett 239:298–304

Teranishi K, Apitz-Castro R, Robson SC, Romano E, Cooper DK (2003) Inhibition of baboon platelet aggregation *in vitro* and *in vivo* by the garlic derivative, ajoene. Xenotransplantation 10:374–379

Tocmo R, Liang D, Lin Y, Huang D (2015) Chemical and biochemical mechanisms underlying the cardioprotective roles of dietary organopolysulfides. Front Nutr 2:1. https://doi.org/10.3389/fnut.2015.00001

Tsai TH, Tsai PJ, Ho SC (2005) Antioxidant and anti-inflammatory activities of several commonly used spices. J Food Sci 70:C93–C97

Vigrinia L (2006) The analysis of onion and garlic. J Chromatogr A 1112:3–22

Viswanathan V, Phadatare AG, Mukne A (2014) Antimycobacterial and antibacterial activity of *Allium sativum* bulbs. Ind J Pharma Sci 76:256–261

Wan Q, Yang YP, Liu ZY (2016) Allicin prevents EA.hy926 endothelial cell injury induced by PM2.5 via inhibiting ERK1/2 pathway. Chin Pharmacol Bull 32:692–687

Weiner SD, Ahmed HN, Jin Z, Cushman M, Herrington DM, Nelson JC, Di Tullio MR, Homma S (2014) Systemic inflammation and brachial artery endothelial function in the multi-ethnic study of atherosclerosis (MESA). Heart 100:862–866

Welder G, Zineh I, Pacanowski MA, Troutt JS, Cao G, Konrad RJ (2010) High-dose atorvastatin causes a rapid sustained increase in human serum PCSK9 and disrupts its correlation with LDL cholesterol. J Lipid Res 51:2714–2721

Wen YD, Zhu YZ (2015) The Pharmacological effects of S propargyl- cysteine, a novel endogenous H2S- producing compound. Handbook Exp Pharm 230:325–336

Wu D, Hu Q, Ma F, Zhu YZ (2016) Vasorelaxant effect of a new hydrogen sulfide-nitric oxide conjugated donor in isolated rat aortic rings through cGMP pathway. Oxidative Med Cell Longev 2016:10. https://doi.org/10.1155/2016/7075682

Yadav K, Sharma M, Ferdinand KC (2016) Proprotein convertase subtilisin/kexin type 9 (PCSK9) Inhibitors: present perspectives and future horizons. Nutr Metab Cardiovasc Dis 26:853–862

Yeh Y, Liu L (2001) Cholesterol effect of garlic extracts and organosulfur compounds: human and animal studies. J Nutr 131:S989–S993

Yin MC, Cheng WS (2003) Antioxidant and antimicrobial effects of four garlic-derived organosulfur compounds in ground beef. Meat Sci 63:23–28

Yoo M, Lee S, Kim S, Shin D (2012) Validation of HPLC method for determination of E- and Z-ajoene in oil-macerated garlic juice. Food Sci Biotechnol 21:107–111

Yoshimoto N, Onuma M, Mizuno S, Sugino Y, Nakabayashi R, Imai S, Tsuneyoshi T, Sumi SI, Saito K (2015) Identification of a flavin-containing S-oxygenating monooxygenase involved in alliin biosynthesis in garlic. Plant J 83:941–951

You S, Nakanishi E, Kuwata H, Chen J, Nakasone Y, He X, He J, Liu X, Zhang S, Zhang B, Hou DX (2013) Inhibitory effects and molecular mechanisms of garlic organosulfur compounds on the production of inflammatory mediators. Mol Nutr Food Res 57:2049–2060

Yun HM, Ban JO, Park KR, Lee CK, Jeong HS, Han SB, Hong JT (2014) Potential therapeutic effects of functionally active compounds isolated from garlic. Pharmacol Ther 142:183–195

Yun SH, Sim EH, Goh RY, Park JI, Han JY (2016) Platelet activation: the mechanisms and potential biomarkers. BioMed Rese Int'l 2016:9060143. https://doi.org/10.1155/2016/9060143

Yu-Yan Y, Liu L (2001) Cholesterol lowering effect of garlic extracts and organosulfur compounds: Human and animal studies. J Nutr 131:989–993

Zahid-Ashraf M, Hussain ME, Fahim M (2005) Antiatherosclerotic effects of dietary supplementations of garlic and turmeric restoration of endothelial function in rats. Life Sci 77:837–857

Zhai B, Zhang C, Sheng Y (2018) Hypoglycemic and hypolipidemic effect of S-allyl-cysteine sulfoxide (alliin) in DIO mice. Sci Rep 8:3527. https://doi.org/10.1038/s41598-018-21421-x

Zhang DW, Lagace TA, Garuti R, Zhao Z, McDonald M, Horton JD, Cohen JC, Hobbs HH (2007) Binding of proprotein convertase subtilisin/kexin type 9 to epidermal growth factor-like repeat A of low density lipoprotein receptor decreases receptor recycling and increases degradation. J Biol Chem 282:18602–18612

Biological Activities and Nutritional Value of *Physalis peruviana* L.

21

Nosheen Akhtar, Sunbal Khalil Chaudhari, Sohaib Arshad, Hira Fatima, Muhammad Shoaib Amjad, and Saboon

Contents

N. Akhtar · Saboon
Department of Botany, PMAS Arid Agriculture University, Rawalpindi, Pakistan

S. K. Chaudhari (✉) · H. Fatima
Department of Botany, The University of Lahore, Sargodha, Pakistan
e-mail: sunbal.khalil@imbb.uol.edu.pk; khalilsunbal@yahoo.com

S. Arshad
Department of Plant Sciences, Quaid-i-Azam University, Islamabad, Pakistan

M. S. Amjad
Department of Botany, Women Univeristy of Azad Jammu & Kashmir, Bagh, Pakistan

M. S. Akhtar et al. (eds.), *Natural Bio-active Compounds*,
https://doi.org/10.1007/978-981-13-7154-7_21

Abstract

The ever increasing population urges the demand for alternative crops of high economic values with medicinal importance. In this regard, *Physalis peruviana* L. is explored for its minerals, vitamins, oils, carotenoids, ascorbic acid, and metabolite contents. It possesses antioxidant, antidiabetic, anticancerous, antihepatotoxic, anti-inflammatory, and antibacterial activities due to its natural bio-active constituents, particularly, physalin and withanolide. The present chapter provides an overview on the natural bio-active contents and its biological activities of *P. peruviana*.

Keywords

Antioxidant · Biological activities · Metabolites contents · Physalin · Withanolide

21.1 Introduction

Physalis peruviana L. is the member of Solanaceae family, and commonly known as "cape gooseberry," rasbhari, poha, peruviana cherry, giant ground cherry, golden berry, aguaymanto, uvilla, and uchuva (Raquira et al. 2014). It is the native of South Africa and is distributed to Asia, Europe, and Pacific regions of the world as a fruit and ornamental plant. This plant was introduced in Australia in the end of the eighteenth century. Later on, it was planted in Hawaii, Israel, and Jamaica. Nowadays, it is cultivated in China, England, Florida, India, Malaya, New Zealand, and Tasmania (Morton 1987). The plant has the ability to grow in a diverse habitat and even in poor sandy soil, which shows its adoptive behavior toward hoarse environmental conditions. It occurs in plantations, gardens, and forests up to 2000 m in subalpine woodland. It grows up to 0.6–0.9 m in height, but sometimes it can grow up to the height of 1.8 m. The hairs exist on the plant surface, and the leaves are ovate, opposite, slender, pointed, and heart shape. It bears yellow flowers which contain five large purple spots near their base, and these flowers are pollinated by insects or wind; sometimes self-pollination also occurs. Usually cross-pollination is favored as its stigmas are receptive 2 days before the release of pollens (Lagos et al. 2008). It reproduces naturally by seeds. In Columbia, it is propagated by cutting (Moreno et al. 2009). It is a short-day plant as its flowering occurs under 8 h day length (Lagos et al. 2008). Its fruit is orange, globose, smooth, and juicy and contains numerous small yellowish seeds and takes 70–80 days to mature and 4–5 g in weight (Graca et al. 1985). The fruit is protected inside the calyx which is termed as capacho; it facilitates the transport of fruits and also keeps it fresh, so it is easily sold as a fresh fruit product. Husk of fruit protects it from insects and birds. The calyx increases in size until the fruit is fully grown. In intact calyx, fruit remains fresh for 1 month (Raquira et al. 2014). Thus, the present chapter provides an overview on the natural bio-active contents and its biological activities of *P. peruviana*.

21.2 Chemical Constituents and Properties

The fruit is highly nutritious and contains minerals; vitamins A, B, and C; iron; phosphorus; carotenoids; glycosides; physalins; pseudo-steroid; phytosterols; and antioxidants. Pectin is also its component, which helps to sooth the gastrointestinal tracts and prevents constipation (Nunes et al. 2008). Moreover, it is well known for its natural bio-active compounds and also its uses as diuretic, antispasmodic, sedative, antiseptic, and analgesic (Hassanien 2011).

21.2.1 Aroma and Flavor

These are the important attributes which affect the consumption of fruits. The aroma is due to the presence of different compounds: esters, alcohols, terpenes, ketones, lactones aldehydes, and oxides. The concentration of these compounds varies at different developmental stages (Othman et al. 2014). The most abundant constituent was alcohols (39.27%), and the alcohols found in highest concentration were 1-hexanol followed by eucalyptol and 4-terpineol, which may play important roles in fruit flavor and in the scent of flowers. Esters (38.52%) also give a specific flavor, while terpenes and its derivatives (7.31%) are responsible for varietal characters of the fruits (Nunes et al. 2008). Amino acids and fatty acids are the precursors of volatile aldehydes (7.05%). Other volatile compounds like ketones (3.97%), lactones (2.09%), acids (0.09%), and oxides (0.89%) exist in less concentration (Yilmaztakin 2014).

21.2.2 Oil Content

Fruit and seeds of the plant contain high concentration of oils, which is edible and a rich source of natural antioxidants, fatty acids, and phytosterols. It is more stable to oxidation and contains 28.7% crude fiber, 24.5% carbohydrate, 17.8% protein, 6.6% moisture, and 3.1% ash (Ramadan et al. 2008). The major fatty acids of oils are palmitoleic acid, γ-linolenic acid, and palmitic acid; however, the godoleic, erucic, nervonic, lignoceric, and dihomo-γ-linolenic acids are also present in trace amount. Fruits have 2.0% lipid (1.8% lipid is resent in the seeds and 0.2% is in the pulp). Higher amount of saturates such as trienes and monoenes exist in pulp oil, where neutral lipids are present in the seed oil (Ramadan and Moersel 2007). Moreover, the oils also contain high level of vitamin K, and the skin of the fruit contains high level of that vitamin (Wong 1978). The lipid plays a very crucial role in the production of many proteins. It enhances skin health, kills cancer cells, acts as antioxidants, and also reduces the risk of heart attack (Ward et al. 2010). Its oil is also nutritionally valuable due to high tocopherol, phylloquinone, and linoleic acids (Otles and Cagindi 2007).

21.2.3 Mineral Contents

Minerals are vital for the proper functioning of human body. Some minerals are required in small amount called as micronutrients, while others are required in large amount called as macronutrients. These nutrients are important because they are present in all stages of growth and reproduction. *P. peruviana* fruit contains calcium, sodium, phosphorus, iron, potassium, and zinc. Phosphorus and calcium are the major components of human skeleton, so they are required for growth and strength of bones. The major function of iron is to carry oxygen from lungs to whole body. Zinc acts as a nonenzymatic antioxidant and scavenge free radicals which facilitates the immune system. Sodium and potassium are important for physiological processes. Fruit juice contains higher content of copper and manganese which act as cofactor for enzymes. Potassium performs homeostasis and maintains heart health (Rodrigues et al. 2009).

21.2.4 Carotenoid Contents

P. peruviana fruit contains carotenoids, which is responsible for its orange color. They are fat-soluble pigments. The main carotenoids are β-carotene and lycopene. Carotenoids have significantly high antioxidant potential (Moreno et al. 2009). The deficiency of vitamin A is the chief nutritional problem of population in developing countries, as human body cannot synthesize vitamin A and depend on the intake of food having provitamin A. Carotene particularly β-carotene is important for its provitamin activity, so, during human intake, provitamin A carotenoid will be metabolized to form vitamin A (Ahmed 2014).

21.2.5 Ascorbic Acid Contents

Ascorbic acid exists in its fruit in high amount. It is a water-soluble vitamin, so its amount exceeds than 50% in fruit. In many enzymatic reactions, it acts as antioxidants so scavenge oxygen radicals but at elevated temperature form carboxylic compound which cause nonenzymatic browning (Dinam et al. 1997).

21.2.6 Vitamin Contents

Many benefits related with the consumption of *P. peruviana* fruits are due to its nutritional importance as it contains high level of vitamins. In order to meet the Recommended Daily Allowance (RDA) of vitamin C in the United States, only 20 units of that fruit is required as 10g fruit contains 4mg vitamin C (Ahmed 2014). Vitamin A is important for the proper immune functioning and good eyesight, and it also improves gene transcription. It also acts as antioxidant, and it reduces the risk of cancer and degenerative diseases like rheumatoid arthritis and cystic fibrosis. Vitamin B is important for vitality and good mood; it consists of niacin, riboflavin, and thiamine. Niacin maintains energy level and repairs DNA; riboflavin facilitates cell growth and repair, whereas thiamine keeps the central nervous system healthy (Valdenegro et al. 2010).

21.2.7 Saponin Contents

Saponins exist in the leaf extract of *P. peruviana*. It is used as a mild detergent and in intracellular histochemical staining of intercellular proteins. Moreover, it also possesses antioxidant, antifungal, anticancer, and anti-inflammatory activities and is used in hyperglycemia and hypercholesterolemia for weight reduction (Deb 1979).

21.2.8 Physalin Contents

Many drugs are used to suppress the immune responses in autoimmune diseases like organ transplantation and allergies and possess negative effects. *P. peruviana* contains physalin A, physalin B, physalin D, and physalin F. All of these are pseudosteroids and are quite helpful in treating immune-mediated diseases like cancer and inflammation (Chiang et al. 1992).

21.2.9 Withanolide Contents

Members of Solanaceae contain withanolide as its name indicates it was first isolated from *Withania somnifera*. They are actually a group of steroidal lactones obtained from *P. peruviana*. Withanolides show broad spectrum of biological and pharmacological properties and also possess anti-inflammatory, antitumor, immunomodulatory insect antifeedant, hepatoprotective, antibacterial, and cytotoxic activities (Lu et al. 2010). Withanolide E and 4β-hydroxywithanolide showed insect antifeedant activity against *Spodoptera littoralis* (Ascher et al. 1980). Calderon et al. (2012) reported that the concentration of 4β-hydroxywithanolide varies in different plant organs, and it is correlated to its developmental stages. In calyx its concentration is too high; thus, it provides the defense against the herbivory. Different types of withanolide glycosides such as blumenol A, perulactone, perulactone B, and (P)-(S)-dehydrovomifolial were extracted from *P. peruviana*. Dinam et al. (1997) extracted the presence of two withanolides: (20S, 22R)-5β,6β-epoxy-4β-14β,15α-trihydroxy-1-oxowith-2, 24-dienolide and (20R, 22R)-5α, 6β, 14α, 20, 27-pentahydroxyl-1-oxowith-24-enolide. Similarly, Lan et al. (2009) observed the presence of 17 withanolide in *P. peruviana*. Out of 17, 10 withanolide (withanolid E, 4β-hydroxywithanolide E, withanolid C, withanolid S, withaperuvin, physalactone, withaperuvin D, withaphysanolid) are well known, while 7 withanolide (phyperunolid A, phyperunolid B, phyperunolid C, phyperunolid D, phyperunolid E, phyperunolid F, peruvianoxid) were recently discovered.

21.3 Biological Activities

21.3.1 Antioxidant Activity

Production of free radicals is one of the causes for different types of diseases like aging, stroke, inflammation, coronary, heart diseases, cancer, renal failure, and rheumatism in humans. These radicals are produced inside the body of humans as a result of different metabolic processes (Ramdan et al. 2008); these radicals are also known as reactive oxygen species (ROS), and they exist in the form of $^1O_2$ (singlet oxygen), H_2O_2 (hydrogen peroxide), ·OH (hydroxyl radical), and $\cdot O^{2-}$ (superoxide anion). These ROS cause tissue injury, oxidation of enzymes, protein damage, DNA damage, and lipid peroxidation. Antioxidants mitigate the effect of ROS by scavenging activities (Moeinian et al. 2011). Antioxidant defense systems are of two types, i.e., enzymatic antioxidants and nonenzymatic antioxidants. Enzymatic antioxidants include peroxidase, superoxide dismutase, catalase, glutathione reductase, and ascorbate peroxidase, whereas nonenzymatic antioxidants are glutathione, ascorbic acid, and tocopherol (Lu et al. 2010). Intake of antioxidants in food is recommended for healthy life. Fruit of *Physalis peruviana* shows antioxidant properties (Wu et al. 2005). Its fruit contains flavonoids, carotene, and polyphenols rutin and myricetin which scavenge free radicals. Its antioxidant activity was analyzed by DPPH method which shows that the content of flavonols, rutin and myricetin, is positively correlated with the antioxidant activity (Licodiedoff et al. 2013). The antioxidant activities of *P. peruviana* aqueous extract and its protective effect against acetaminophen (APAP)-induced hepatotoxicity in rats showed potent hepatoprotective effect against APAP-induced liver injury in rats (Chang et al. 2008).

21.3.2 Antibacterial Activity

Physalis peruviana was effective against both gram-positive and gram-negative bacteria, but it showed more inhibitory action against gram-positive strains. It showed antibacterial activity against *Staphylococcus epidermidis* 14990, *S. aureus* A950277, *Escherichia coli* DH5-a, *Lactococcus lactis* ATCC 11454, and *Erwinia herbicola* pv. *Gypsopholia* 824 due to the presence of withanolide contents (Jaca and Kambizi 2011). The ethanolic extracts of *Physalis peruviana* showed the maximum inhibitory actions compared to other solvents. Moreover, medicinal properties of *P. peruviana* in the treatment of constipation, diarrhea, common flu, and sore throats were also demonstrated by Eiras et al. (2012).

21.3.3 Antihepatotoxic Activity

Roots of *Physalis peruviana* were seen to be effective for eliminating fibrosis in rats, and it was checked by kidney and liver histopathological investigation that they are protected from fibrosis. Extract of *Physalis peruviana* shows antihepatotoxic activity against CCl_4-induced hepatotoxicity (Arun and Asha 2007).

21.3.4 Anti-inflammatory Activity

Juice of *P. peruviana* fruit showed anti-inflammatory activity. Jaun and his co-worker applied fruit juice in rabbit eyes for the treatment of *pterygium,* and the results confirmed and validated its use in folk medicines. Its anti-inflammatory activity is due to phytosterols and withanolide (Pardo et al. 2008).

21.3.5 Anti-diabetic Activity

Diabetes mellitus is caused by defects in insulin production, and it results in high blood glucose level (Kadima et al. 2016). *Physalis peruviana* contains hypoglycemic activity. The dry powder of *P. peruviana* was used to make aqueous decoction, which decreases the glucose amount in guinea pigs, but the high doses may cause intoxication (Kasali et al. 2013). The presence of cardiac glycoside is also reported, which showed antidiabetic activity. Cardiac glycosides increase Na^+ ion concentration in myocytes by Na^+/K^+ pumps which in turn enhance the level of Ca^{2+} ions. So, more calcium is available for contraction of heart muscles, which reduces the distention of the heart (Ramadan et al. 2008).

21.3.6 Anti-cancerous Activity

Puente et al. (2011) verified the anticancer action of *P. peruviana*. The methanolic leaf extract showed the inhibition of both the aberrantly active stat3 in human tumor cells and tumor necrosis factor-α (TNF-α)-induced NF-kappa B activity. It has been concluded that physalin F is an effective anticancer agent and it targets the NF-kappa B cells and also induces cell apoptosis in human renal carcinoma; thus, it could be used for further clinical trials (Wu et al. 2012). Physalins B and F have cancer suppressive activity by inhibiting the formation of pro-inflammatory cytokines and stimulation of macrophage. It also inhibited the proliferation of lymphocytes. Physalins increase the enzymatic activity of catalase and superoxide dismutase (SOD) to prevent free radical damaging effect to pancreatic B cells (Kadima et al. 2016).

Lung cancer is a major disease in United States, and many therapies are done for its treatment; however, the development of drugs for lung cancer is still a challenge. Many epidemiological studies evaluated that intake of fruits and vegetables reduced the threat of many cancers (Arun and Asha 2007). *P. peruviana* contains C28 steroidal lactones, withanolides, and physalins. In human hepatocellular carcinoma (HepG2) cell line, extract of *P. peruviana* brings on apoptosis. So, it could act as anticancer and anti-inflammatory agents, and the 4β-hydroxywithanolide is the main component, which showed the potential anticancer activity (Dinam et al. 1997).

21.3.7 Anti-hepatoma Activity

Apoptosis or programmed cell death generally participates significantly in the regulation of homeostasis of cell. In order to treat cancer, apoptosis is induced in cells. The ethanolic extract of *P. peruviana* induced apoptosis in human HepG2 cells in a time-dependent manner and increases the number apoptosis cells in percentage in cell cultures (Wu et al. 2004).

21.3.8 Renoprotective Activity

The extract of *P. peruviana* also showed renoprotective activity against acute renal injury in rats. It contains polyphenol (quercetin, myricetin, and kaempferol) and flavonoids and scavenges the reactive oxygen species and also possesses metal-chelating properties, which directly or indirectly alleviate the biological and biochemical parameters of the kidney and its histology (Ahmed 2014).

21.3.9 Ethnomedicinal Uses

It is sweet in taste and is used in jam, ice creams, chocolate, and juices. Dried berries are used for garnishing. Its fruit is decorative in appearance, so it is famous in restaurants for garnishing purpose. Traditional South American people make infusion from roots and leaves of *P. peruviana*, which is used to improve respiratory diseases such as laryngitis (Jaun et al. 2007). It shows phytotherapeutic properties so used in folk medicines. In Europe it is used in herbal medicines to treat gout, kidney stones, and urinary tract disorders. It is widely used as anti-mycobacterial, antipyretic, antileukemic, immunomodulatory, and anticancer in folk medicines (Cakir et al. 2014). Its leaves are heated and used as poultice. Infusion of its leaves is used to treat abdominal ailments in children. Juice of fresh leaves is used for stomach problems; it is diuretic and also shows antibiotic activity against *Staphylococcus* (Lan et al. 2009). In Chinese medicines, *P. peruviana* is used for the treatment of sore throat, cough, and fever. Its roots and leaves are diuretic used against worms, bowel complaints, stomachache, anemia, and abdominal disorders (Holm et al. 1979).

21.3.10 Other Uses

21.3.10.1 Weight Loss

Fruit of *Physalis peruviana* L. contains low amount of fats and calories, so it assists in weight loss. 100 g of fresh berries contains 44k calories.

21.3.10.2 Treatment of Asthma

Leaf extract contains alkaloids due to which it is used to cure asthma (Ascher et al. 1980).

21.3.10.3 Prevention of Diseases

There are many natural enemies of *P. peruviana*, which may disturb its normal growth and functioning. There are some bacteria which attack on it and cause bacterial leaf spot (*Xanthomonas* spp.) in India and Queensland (Kishun et al. 1977), bacterial wilt *Candidatus liberibacter* species in New Zealand, and *Pseudomonas solanacearum* in Hawaii (Liefting et al. 2009). In Colombia, *Fusarium oxysporum* caused vascular wilting in *P. peruviana*. The plants are also prone to attack by *Phytophthora infestans*, *Peronospora hyoscyami*, *Drechslera rostrata*, and *Asteridiella acervata*. Powdery mildew and soft brown scale were the most troublesome diseases in South Africa (Cardenas et al. 2011). There are some insect pests of *P. peruviana* which include the dipteran *Bactrocera latifrons* in Hawaii; *Helicoverpa armigera* in India; lepidopteran *Tuta absoluta* in Italy and cutworms; and *Agrotis* species in South Africa. Red spider also attacks *P. peruviana*, and if they were grown near potato fields, they also got attacked by potato tuber moth (Badon et al. 2001). It also acts as a host for many viruses including tobacco mosaic virus in India; tomato chlorotic spot virus (Eiras et al. 2012); cucumber mosaic virus in India and tomato spotted wilt virus in South Africa (Graca et al. 1985); and potato spindle tuber viroid in Slovenia, New Zealand, Germany, Turkey, and the Netherlands (Ward et al. 2010).

21.4 Breeding Strategies

Fruit size is controlled genetically, which is also related to environmental conditions. Climatic factors such as light intensity and temperature have strong influence on the nutritional quality of fruit. Growth, quality, and rate of production are strongly influenced by tropical altitude at lower elevations; fruits grow faster and have greater diameter as compared to those at high elevation (Fischer et al. 2007). Plant growth and yield could be improved by the application of mulch as it suppresses weeds and increases temperature of the soil about 5 °C, and it also prevents fruit contact with the ground and increases moisture retention. So, by the application of mulch, 13 t/ha yield of *P. peruviana* could be expected (Klinas 1986). Evans and Poorter (2001) studied the rate of photosynthesis, specific leaf area, and nitrogen partitioning in the leaves at two different light levels and concluded that *P. peruviana* has C_3 metabolism. It grows as heliophytes in full sunlight. The sowing is done in spring season (from February to April), germinates quickly, and is recognized as frost-sensitive and drought-tolerant plant.

21.5 Conclusions and Future Prospects

Physalis peruviana attract the world due to its nutritional value and unique storage properties. It may be act as a potential new candidate for foods and drinks as its juice is a rich source of sugars as well as water- and fat-soluble bio-active compounds. Moreover, it could act as a new source of bio-active phytochemicals for the pharmaceutical industry.

References

Ahmed LA (2014) Renoprotective effect of Egyptian cape gooseberry fruit (*Physalis peruviana* L.) against acute renal injury in rats. Sci World J 2014:273870. https://doi.org/10.1155/2014/273870

Arun M, Asha VV (2007) Preliminary studies on antihepatotoxic effect of *Physalis peruviana* Linn. (Solanaceae) against carbon tetrachloride induced acute liver injury in rats. J Ethnopharmacol 111:110–114

Ascher KRS, Nemny NE, Eliyahu M, Kirson I, Abraham A, Glotter E (1980) Insect anti-feedant properties of withanolides and related steroids from Solanaceae. Experientia 36:998–999

Badon SG, Folcia AM, Cerri AM, Vilella F (2001) Population parameters of *Macrosiphum euphorbiae* (Thomas) (Homoptera: Aphididae) in *Physalis peruviana* (L.) (Solanaceae) crops. Rev Fac Agron 21:265–269

Cakir O, Pekmez M, Cepni E, Candar B, Fidan K (2014) Evaluation of biological activities of *Physalis peruviana* ethanol extracts and expression of *Bcl-2* genes in *HeLa* cells. Food Sci Technol 34:422–430

Calderon JM, Ruiz V, Castellanos L (2012) Within and between plant variation of 4b-hydroxiwithanolide E in cape gooseberry (*Physalis peruviana*; Solanaceae). Biochem System Ecol 41:21–25

Cardenas M, Grajales V, Sierra R, Rojas A, Gonzalez Almario V, Vargas A, Marín M, Fermín G, Lagos LE, Grunwald NJ, Bernal A, Salazar C, Restrepo S (2011) Genetic diversity of *Phytophthora infestans* in the Northern Andean region. BMC Genet 12:12–23

Chang JC, Lin CC, Wu SJ, Lin DL, Wang SS, Miaw CL, Ng LT (2008) Antioxidative and hepatoprotective effects of *Physalis peruviana* extract against acetaminophen-induced liver injury in rats. Pharma Biol 46:724–731

Chiang HC, Jaw S, Chen C, Kan W (1992) Antitumor agent, physalin F from *Physalis angulata* L. Anticancer Res 12:837–843

Deb DB (1979) Solanaceae in India. In: Hawkes JG, Lester RN, Skelding AD (eds) The biology and taxonomy of the solanaceae. Academic, London, pp 87–112

Dinam LS, Sarker V, Sik V (1997) 28-hydroxywithnolide E from *Physalis peruviana*. Photochem 44:509–512

Eiras M, Costa IFD, Chaves ALR, Colariccio A, Harakava R, Tanaka FAO, Garcez RM, Silva LA (2012) First report of a *Tospo*virus in a commercial crop of Cape gooseberry in Brazil. New Dis Rep 25:1–25

Evans JR, Poorter H (2001) Photosynthetic acclimation of plants to growth irradiance: the relative importance of specific leaf area and nitrogen partitioning in maximizing carbon gain. Plant Cell Environ 24:755–767

Fischer G, Ebert G, Ludders P (2007) Production, seeds and carbohydrate contents of Cape gooseberry (*Physalis peruviana* L.) fruits grown at two contrasting Colombian altitudes. J Appl Bot Food Qual 81:29–35

Graca JV, Trench TN, Martin MM (1985) Tomato spotted wilt virus in commercial Cape gooseberry (*Physalis peruviana*) in Transkei. Plant Pathol 34:451–453

Hassanien MFR (2011) *Physalis peruviana*. A rich source of bio-active phytochemicals for functional foods and pharmaceuticals. Food Rev Intl 27:259–273

Holm LG, Pancho JV, Herberger JP, Plucknett DL (1979) A geographical atlas of world weeds. Wiley, New York

Jaca TP, Kambizi L (2011) Antibacterial properties of some wild leafy vegetables of the Eastern Cape province. South Afr J Med Plants Res 5:2624–2628

Kadima JN, Kasali FMB, Bavhure AO, Mahano FM, Bwironde FM (2016) Comparative antidiabetic potential and survival function of *Harungana madagascariensis*, *Physalis peruviana*, *Solanum americanum* and *Tithonia diversifolia* extracts on alloxan induced diabetes in guinea-pigs. Human J 5:196–206

Kasali F, Kadima J, Mpiana P, Ngbolua K, Tshibangu D (2013) Assessment of antidiabetic activity and acute toxicity of leaf extracts from *Physalis peruviana* L. in guinea-pig. Asian Pac J Trop Biomed 3:841–846

Kishun R, Pandey R, Banerjee AK (1977) Occurrence of *Xanthomonas vesicatoria* on Physalis peruviana L. A new host record for India. Indian J Farm Sci 4:122–123

Klinas DJ (1986) Cape gooseberry (*Physalis peruviana*) production systems. N Z J Exp Agric 14:425–430

Lagos BTC, Cabrera FAV, Escobar HC, Florez JEM (2008) Sexual reproduction of the Cape gooseberry. Acta Agron 57:81–87

Lan YH, Chang FR, Pan MJ, Wu CC, Wu SJ, Chen SL, Wang SS, Wu MJ, Wu YC (2009) New cytotoxic withanolides from *Physalis peruviana*. Food Chem 116:462–469

Licodiedoff S, Koslowski LAD, Ribani RH (2013) Flavonols and antioxidant activity of *Physalis peruviana* L. fruit at two maturity stages Acta Scient. Technol Mari 35:393–399

Liefting LW, Weir BS, Pennycook SR, GRG C (2009) *Candidatus liberibacter solanacearum* associated with plants in the family Solanaceae. Intl J Syst Evol Microbiol 59:2274–2276

Lu TM, Lin PH, Yao Q, Chen C (2010) Chemical and molecular mechanisms of antioxidants, experimental approaches and model system. J Cell Mol Med 14:840–860

Moeinian MR, Zargari K, Hasannour J (2011) Effect of boron foliar spray application on quality characteristics and growth parameters of wheat grains under drought stress. Am-Eur J Agric Environ Sci 10:593–599

Moreno NH, Alvarez-Herrera JG, Balaguera-López HE, Fischer G (2009) Asexual propagation of cape gooseberry (*Physalis peruviana* L.) using different substrates and auxin levels. Agron Colomb 27:341–348

Morton J (1987) Cape gooseberry. In: Fruits of warm climates, pp 430–434

Nunes C, Coimbra MA, Saraiva J, Rocha SM (2008) Study of the volatile components of a candied plum and estimation of their contribution to the aroma. Food Chem 111:897–905

Othman MS, Nada AHS, Zaki E, Ahmad A, Moneim (2014) Effect of *Physalis peruviana* L. on cadmium-induced testicular toxicity in rats. Biol Trace Element Res 159:278–287

Otles S, Cagindi O (2007) Determination of vitamin K1 content in olive oil, chard and human plasma by RP-HPLC method with UV–Vis detection. Food Chem 100:1220–1222

Pardo JM, Fontanilla MR, Ospina LF, Espinosa L (2008) Determining the pharmacological activity of *Physalis peruviana* fruit juice on rabbit eyes and fibroblast primary cultures. Investig Ophthalmol Vis Sci 49:3074–3079

Puente LA, Pinto-Muñoz CA, Castro ES, Cortés M (2011) *Physalis peruviana* Linnaeus, the multiple properties of a highly functional fruit: a review. Food Res Int 44:1733–1740

Ramadan MF, Moersel JT (2007) Impact of enzymatic treatment on chemical composition, physicochemical properties and radical scavenging activity of golden berry (*Physalis peruviana* L.) juice. J Sci Food Agric 87:452–460

Ramadan MF, Sitohy MZ, Moersel JT (2008) Solvent and enzyme-aided aqueous extraction of goldenberry (*Physalis peruviana* L.) pomace oil impact of processing on composition and quality of oil and meal. Eur Food Res Technol 226:1445

Raquira WA, Borda D, Carvajal LH (2014) *Potyvirus* affecting uchuva (*Physalis peruviana* L.) in centro agropecuario marengo. Colombia Agric Sci 5:897–905

Rodrigues E, Rockenbach I, Cataneo C, Gonzaga V, Chaves E, Fett R (2009) Minerals and essential fatty acids of the exotic fruit *Physalis peruviana* L. Food Sci Technol 29:642–645

Valdenegro M, Henriquez C, Lutz M, Almonacid S, Simpson R (2010) Drum dried, liophylized dried and traditional drying of golden berry (*Physalis peruviana* L.): effects in nutritional and healthy quality. In: Proceedings of the international conference on food innovation, Food Innova, Valencia, California, USA

Ward LI, Tang J, Veerakone S, Quinn BD, Harper SJ, Delmiglio C, Clover GRG (2010) First report of Potato spindle tuber viroid in cape gooseberry (*Physalis peruviana*) in New Zealand. Plant Dis 94:479

Wong W (1978) A hypothesis on the pathogenesis of *Pterygiums*. Ann Ophthalmol 10:303–308

Wu SJ, Ng L, Lin DL, Huang SN, Wang SS, Lin CC (2004) *Physalis peruviana* extract induces apoptosis in human Hep G2 cells through CD95/CD95L system and the mitochondrial signaling transduction pathway. Cancer Lett 215:199–208

Wu SJ, Ng LT, Huang YM, Lin DL, Wang SS, Huang SN, Lin CC (2005) Antioxidant activities of *Physalis peruviana*. Biol Pharma Bull 28:963–966

Wu SY, Leu YL, Chang YL, Wu TS, Kuo PC, Liao YR, Teng CM, Pan SL (2012) Physalin F induces cell apoptosis in human renal carcinoma cells by targeting NF-kappaB and generating reactive oxygen species. PLoS One 7:e40727

Yilmaztekin M (2014) Analysis of volatile components of Cape gooseberry (*Physalis peruviana* L.) grown in Turkey by HS-SPME and GC-MS. Sci World J 2014:796097. https://doi.org/10.1155/2014/796097

Essential Oil of *Baccharis milleflora* in the Atlantic Rain Forest of the Paraná State in Brazil: Chemical Composition and Biological Evaluation

22

Wanderlei do Amaral, Cícero Deschamps, Humberto R. Bizzo, Marco Antonio S. Pinto, Luiz Everson da Silva, Luiz A. Biasi, Thiago S. Freitas, Janaína E. Rocha, Fábia F. Campina, Maria S. Costa, Camila F. Bezerra, and Henrique D. M. Coutinho

Contents

Abstract

Essential oils are broadly used in the food, perfume, and pharmaceutical industries. These are associated with ecological functions of defense and attraction of pollinators, and usually undergo quantitative and qualitative variation in response to the environment. Thus, the aim of this chapter is to highlight the essential oil content, chemical composition, and biological activities of *Baccharis milleflora*.

Wanderlei do Amaral and Luiz Everson da Silva have equally contributed for this chapter.

W. do Amaral (✉) · C. Deschamps · H. R. Bizzo · M. A. S. Pinto · L. E. da Silva
L. A. Biasi · T. S. Freitas · J. E. Rocha · F. F. Campina · M. S. Costa
C. F. Bezerra · H. D. M. Coutinho
Universidade Federal do Paraná, Matinhos, Brazil
e-mail: wdoamaral@hotmail.com; luiever@gmail.com; luiz_everson@yahoo.de

M. S. Akhtar et al. (eds.), *Natural Bio-active Compounds*,
https://doi.org/10.1007/978-981-13-7154-7_22

Keywords
Atlantic Forest · Carqueja · Essential oil · Medicinal plants · Terpenes

22.1 Introduction

Asteraceae is a well-known family of valuable medicinal plants with about 250 genera and 2000 species (Di Stasi et al. 2002; Souza and Lorenzi 2008). The genus *Baccharis* is represented by more than 500 species, distributed predominantly in South America (Verdi et al. 2005). About 120 species are described, distributed in greater concentration in the Southern region of Brazil (Barroso 1976; Verdi et al. 2005). This genus produces secondary metabolites of the flavonoids and terpenoids groups (Davies 2004; Verdi et al. 2005). These are associated with ecological functions of defense and attraction of pollinators and usually undergo quantitative and qualitative variations in response to the environment (Sanchez et al. 2006; Simoes et al. 2007; Dicke and Baldwin 2010; Bedoya-Perez et al. 2014). As a result of the fact that they are complex mixtures of volatile substances, lipophilic, usually odoriferous, and liquid (Simoes et al. 2007), they are widely used in the food, pharmaceutical, and cosmetic industries (Gobbo-Netto and Lopes 2007; Biasi and Deschamps 2009; Yunes 2012). Recent works have demonstrated several biological activities of the essential oils of the genus *Baccharis*, such as leishmanicidal and trypanocidal (Grecco et al. 2010a, b; Santos et al. 2010), antinociceptive and inflammatory processes (Santos et al. 2010), an antioxidant (Guimarães et al. 2012), and insecticidal and antifungal (Kurdelas et al. 2012).

The production of essential oils in Brazil does not meet the demand; besides this the national and international market has been showing interest in new essences (Bizzo et al. 2009; Souza et al. 2010). According to an estimate, approximately 65% of the market of essences comes from cultivated species and 1% from wild species (Bandoni and Czepak 2008). Yunes (2012) advocated that the intensification of studies of the Brazilian flora, in an interdisciplinary way, aiming at the identification of promising species for the production of volatile oils, is used as inputs to obtain the assets to be included in new available to the national health system.

It is the largest biodiversity in the world with over 40,000 plant species described (Brandao et al. 2006; Ferro et al. 2006; Forzza et al. 2012). According to Forzza et al. (2012), Brazil ranks first among megadiverse countries, with 56% of endemic plant species, with the Atlantic Forest biome having the highest number of plant species among Brazilian biomes, with more than 19,000 species, of which more than 7600 are endemic species. The Campos Gerais is considered as biodiversity hotspots (Bilenca and Minarro 2004). They have a very particular structure, function, and dynamics and represent highly interactive ecosystems (IBGE 1992) existence by abiotic factors, by anthropogenic action, and by natural disturbances such as frost, drought, and especially fire (Pillar 2006). In this sense, post-harvest factors, such as drying, and variations in environmental factors affect the yield and composition of the essential oils (Goebbo-Netto and Lopes 2007; Bezerra et al. 2008).

In addition, the indiscriminate use of antibiotics has contributed to select resistant bacterial strains (Silveira et al. 2006). In this context, many efforts have been carried out in the discovery of new drugs to face these resistant pathogens. Thus, biologically active natural compounds can be an alternative in antibiotic modulation research to combat infectious agents, especially multiresistant ones. Although there are several floristic and structural studied for the State of Paraná, the aromatic flora of the Atlantic Forest, is still a little known. In this context, we believe the bioprospection of aromatic species program in the Atlantic rain forest can contribute to the identification of essential oils produced by the species in interaction with the environment and can be applied in different sectors of the industry. Thus, the content and chemical composition of the essential oil of fresh and dried leaves of *Baccharis milleflora* was evaluated as well as the biological activities in order to contribute to the knowledge of the aromatic flora of Paraná and to identify potential species for cultivation and propagation. Thus, the aim of this chapter is to highlight the essential oil content, chemical composition, and biological activities of *Baccharis milleflora*.

22.1.1 Botanical Materials and Extraction of Essential Oils

The terminal branches with cladodes were collected for extracting the essential oil. The oil was extracted by hydro-distillation method from the dried materials, and the total mass of the oil was determined. Identification of the chemical constituents of the essential oil was performed by gas chromatography coupled to mass spectrometry. To identify and quantify the essential oil, components of the samples were subjected to gas chromatography coupled to the mass spectrometer. The chromatograph used was of the brand Agilent 7890A, with a detector of ionization (FID), operating at 250 °C on an HP% column (30 m length, 25 mm internal diameter, and 0.25 μm film thickness), using hydrogen as the entraining gas (1.0 ml, min^{1}). For each sample, it was injected with 1.0 μl in a nozzle heated between 250 °C and 280 °C, operating at mode with split-flow (1:5). The temperature setting of the oven was 60 °C at 240 °C and a heating rate of 3 °C/min. Coupled to a chromatograph, the mass spectra were obtained using the same column chromatography under the same conditions as above using helium as the carrier gas 91 ml/min. Ionization was used at 70 eV. Ionization source (70 eV) was maintained at 220 °C, the analyzer at 150 °C, and the transfer line at 260 °C. The linear retention indices were calculated from the retention periods of the components of essential oils and those of a homologous series of n-alkanes injected in the same column and under the same analysis conditions as above. The identification of the chemical constituents was obtained by comparing their mass spectra with those of the libraries (WILEY 1994) and also their linear retention indexes, calculated from the injection of a homologous series of hydrocarbons (C7–C26), and it was compared with literature data (ADAMS 2007).

22.1.2 Preparation of the Substances

About 10 mg (10,000 μg) of the oil was weighed and placed in individualized Eppendorfs, diluted in 0.5 ml of DMSO. This first solution was placed in a falcon tube and added another 9265 ml of water, making a total of 9765 ml of 1024 μg/ml solution for each substance. This solution was used for the MIC and modulation tests.

22.1.3 Preparations of Bacterial Inoculums and Antibiotics

Bacterial cultures were seeded in Petri dishes containing HIA and placed in the oven at 37 °C for growth for 24 h. After this period, a sample of each microbial culture and diluted in identified test tubes was carried out in triplicate. After this procedure, the turbidity of the solution was tested with a McFarland control. The test Eppendorfs were prepared in triplicate for each bacterium and for each substance, containing 1350 μl of 10% BHI with 150 μl of the inoculum (corresponding to 10% of the total solution) for the MIC. The antibiotics used in the test were norfloxacin, gentamicin, and erythromycin at initial concentration of 1024 μg/ml).

22.1.4 Determination of Minimum Inhibitory Concentration (MIC)

About 100 μl of the final inoculums solution was added to each well of the microdilution plate, serially, with 100 μl solution of each oil column, varying at concentrations of 512 μg/ml in the first well and 0.5 μg/ml in the last well. Micro-dilutions were performed in triplicate. Plates were taken to the incubator for 24 h at 37 °C. The determination of bacterial MIC was done using the addition of 20 μl of resazurin in each well and ocular observation after 1 h.

22.1.5 Modulation of Antibiotic Activity by Direct Contact

To verify the modulation of the antibacterial effect of the antibiotics against the tested strains, the method proposed by Coutinho (2008). Eppendorf tubes contain the subinhibitory concentration (MIC/8), 10% BHI content according to volume of subinhibitory concentration, and 150 μl of bacterial suspension (corresponding to 10 % of the solution). For the control, Eppendorf tubes were prepared with 1.5 ml of solution containing 1350 μl of BHI (10%) and 150 μl of microorganisms' suspension. The plate was filled numerically by adding 100 μL of this solution into each well. Subsequently, serial micro-dilution with 100 μl of the antibiotic was done.

22.2 Chemical Compositions of Essential Oils

A total of 20 and 25 chemical compounds were identified in the essential oil composition of the fresh and dried samples of the species in a total of 49.7 and 63.2% of identified compounds, respectively (Table 22.1). Evaluating different samples of *Baccharis milleflora*, Agostini et al. (2005) managed, on average, to identify 51.8% of the essential oil compounds of the species, while Lago et al. (2008) reported the essential oils of 6 species of the genus *Baccharis* and identified 67 chemical compounds in the essential oils of fresh samples. Similarly, Agostini et al. (2005)

Table 22.1 Relative percentage of chemical components of the essential oil of the fresh and dried samples of *B. milleflora*

Compounds	RI[a]	RI[b]	Fresh sample	Dry sample
α-Pinene	937	932	–	1,6
β-Pinene	979	974	–	0,4
Myrcene	991	988	0,9	2,9
β-Phellandrene	1031	1075	0,2	2,9
(E)-β-Ocimene	1050	1044	0,8	1,9
Linalool	1100	1095	–	0,2
β-Elemene	1389	1389	0,2	0,3
(E)-β-Caryophyllene	1416	1417	2,8	5,6
α-Humulene	1450	1452	0,8	2,2
Sesquisabinene	1457	1443	0,9	1,3
γ-Muurolene	1478	1478	3,0	3,8
Bicyclogermacrene	1492	1494	4,7	4,4
α-Muurolene	1496	1500	–	0,2
β-Bisabolene	1505	1505	0,5	1,0
Cubebol	1511	1514	0,8	0,7
δ-Cadineno	1520	1522	2,1	2,4
Spathulenol	1574	1577	0,7	2,4
Caryophyllene oxide	1578	1582	3,1	4,5
Viridiflorol	1586	1592	18,6	16,8
Rosifoliol	1602	1600	1,8	1,1
1-Epi-Cubenol	1623	1627	1,2	1,8
Eremoligenol	1629	1629	–	1,9
Epi-α-muurolol	1639	1640	3,3	1,7
β-Eudesmol	1645	1649	1,7	1,8
α-Cadinol	1650	1652	1,6	1,8
Monoterpenes (%)			1,7	6,8
Oxygenated monoterpenes (%)			–	0,2
Sesquiterpenes (%)			15,1	20,4
Oxygenated sesquiterpenes (%)			30,2	28,4
Others (%)			2,7	7,4
Total compounds identified (%)			49,7	63,2

[a]RI = retention index calculated
[b]RI = literature retention index

analyzed the essential oils of 6 species and identified 36 chemical compounds in the dry samples with an average of 13 compounds in each essential oil samples. Our study showed the recovery of sesquiterpenes (45.3 and 48.8%) from fresh and dry leaves, respectively, with a low concentration of oxygenated monoterpenes (0.2%) after drying. However, the average chemical composition identified was predominantly sesquiterpenes (34.73%), also with a low concentration of oxygenated monoterpenes on average less than 0.1% (Agostini et al. 2005). The species *B. milleflora* has viridiflorol (18.6 and 16.8%) in the dry and fresh samples. Previous studies have shown that changes in the composition and content of the essential oil for the species evaluated in different regions may be related to environmental conditions, chemo-types, phases of the vegetative cycle, soil nutrient contents, time, and harvesting time, among other factors (Sangwan et al. 2001), and showed a relationship between the two variables. The drying influenced the composition of the essential oil that has the main components as α-pinene, β-pinene, linalool, alpha-muurolene, and eremoligenol. After drying, there was an increase in the percentage of myrcene, β-felandren, (E)-β-caryophyllene, alpha-humulene, and caryophyllene oxide in the essential oil composition of the species, and there was also a decrease in viridiflorol and epi-α-muurolol after drying (Table 22.1). The changes in the composition of the essential oil after drying are due to oxidation and volatilization processes and consequently formation of other compounds (Yunes 2012), thus altering the composition and concentrations of the constituents of the oil essential.

22.3 Biological Activity of *B. milleflora*

The minimum inhibitory concentration (MIC) assay showed the antimicrobial activity of *B. milleflora* essential oil against the *S. aureus* and *E. coli* strains (Table 22.2). The minimum MIC was observed against Gram-positive *S. aureus* 10 strain (102 μg/ml), a multidrug-resistant strain isolated from the clinics. The modulatory activity of the essential oils of *B. milleflora* was observed against *S. aureus* (Fig. 22.1), whereas the essential oil demonstrated synergism when associated with norfloxacin and with the gentamicin, reducing the MIC when associated with these antibiotics. Against the Gram-negative strain *E. coli* 06, the essential oil demonstrated indifference when associated with all drugs (Fig. 22.2).

Table 22.2 Minimum inhibitory concentration (MIC) of the essential oil of *B. milleflora* against Gram-positive and Gram-negative bacteria (μg/ml)

Bacteria	MIC values
Gram-positive strains of *S. aureus* (ATCC 6538)	≥1024
Gram-positive strains of *E. coli* (ATCC 25922)	≥1024
Gram-negative strains of *S. aureus* (10)	102
Gram-negative strains of *E. coli* (06)	≥1024

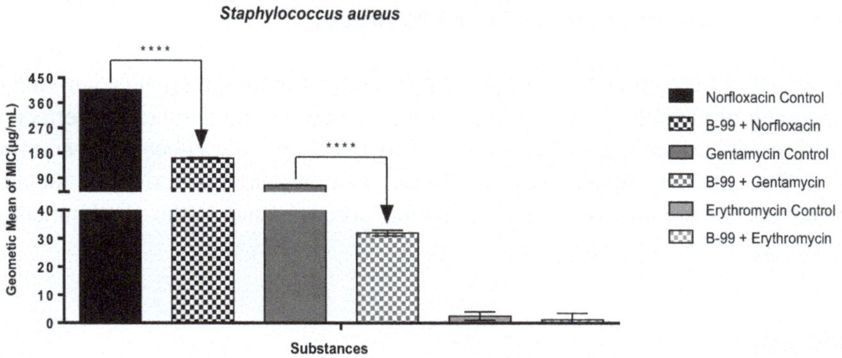

Fig. 22.1 Modulatory of the antibiotic activity effect of the essential oil of *B. milleflora* associated with antibiotics against multidrug-resistant bacterial strains of *S. aureus*

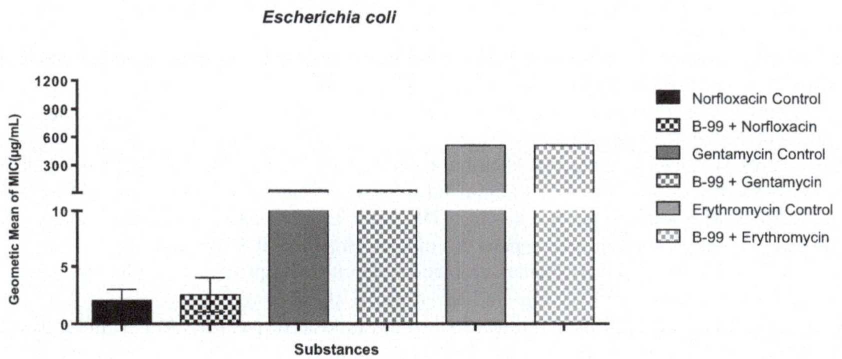

Fig. 22.2 Modulatory of the antibiotic activity effect of the essential oil of *B. milleflora* associated with antibiotics against a multidrug-resistant bacterial strain of *E. coli*

The biological and modulatory activities of essential oils have been associated with the presence of terpenes. Due to their toxic effects on the structure and function of the cell membranes, by their lipophilic characteristics, terpenes displace from the aqueous phase toward membrane structures (Sikkema et al. 1994) causing Na expansion of the cell membrane, enhancing the fluidity and the permeability. These alterations affect the structure of the transmembrane proteins, the respiratory chain, and the ion transport processes (Trombetta 2005). Similarly, results were also obtained by past researchers against different strains of bacteria and fungi (Chaves et al. 2018; Oliveira-Tintino et al. 2018; Santos et al. 2018).

22.4 Conclusions and Future Prospects

The essential oil of *Baccharis milleflora* is enriched in sesquiterpenes with major component of viridiflorol. The essential oil possesses strong antibacterial effect on Gram-positive bacteria and also has capability to enhance the effect of the antibiotics norfloxacin and gentamicin. In the future, more research is desired on the anatomical characterization of the lysing cellular structures and on the development of a drying protocol of the species.

References

Adams RP (2007) Identification of essential oil components by gas chromatography/mass spectroscopy. Allured Publishing, Carol Stream

Agostini F, Santos ACA, Rossato M, Pansera MR, Zaterra F, Wasum R, Serafini LA (2005) Estudo do óleo essencial de algumas espécies do gênero *Baccharis* (Asteraceae) do sul do Brasil. Rev Brasil de Farmacogn 15:215–220

Bandoni AL, Czepak MP (2008) Os recursos aromáticos no Brasil: seu aproveitamento industrial para a produção de aromas e sabores. Edufes, Vitória, ES, p 623

Barroso GM (1976) Compositae-Subtribo Baccharidinae Hoffmann: Estudo das espécies ocorrentes no Brasil. Rodriguésia 40:28–31

Bedoya-Perez MA, Isler I, Banks PB, Mcarthur C (2014) Roles of the volatile terpene, 1,8-cineole, in plant–herbivore interactions: a foraging odor cue as well as a toxin. Oecologia 174:827–837

Bezerra AME, Mdedeiros-Filho S, Oliveira LDM, Silveira ER (2008) Produção e composição química da macela em função da época de colheita. Horti Brasil 3:780–784

Biasi LA, Deschamps C (2009) Plantas aromáticas do cultivo à produção de óleo essencial. 1ª edição. Layer Studio Gráfico e Editora Ltda, Curitiba, p 160

Bilenca DN, Minarro FO (2004) Identificação de áreas valiosas de pastizal (AVPs) em lãs pampas y campos de Argentina, Uruguay y sur de Brasil. Fundación Vida Silvestre, Argentina

Bizzo HR, Hovell AM, Rezende CM (2009) Óleos essenciais no Brasil: Aspectos gerais, desenvolvimento e perspectivas. Quím Nova 32:588–594

Brandao MGL, Cosenza GP, Moreira R, Monte-Mor R (2006) Medicinal plants and other products from the Brazilian official pharmacopoeia. Rev Brasil Farmacogn 16:408–420

Chaves TP, Pinheiro REE, Melo ES, Soares MJS, Souza JSN, Andrade TB, Lemos TLG, Coutinho HDM (2018) Essential oil of *Eucalyptus camaldulensis* Dehn potentiates β-lactam activity against *Staphylococcus aureus* and *Escherichia coli* resistant strains. Indust Crops Prod 112:70–74

Coutinho HDM (2008) *In vitro* anti-staphylococcal activity of *Hyptis martiusii* Benth against methicillin-resistant *Staphylococcus aureus*: MRSA strains. Rev Brasil Farmacogn 18:670–675

Davies P (2004) Fichas técnicas de cultivo. In: Estudios em domesticación y cultivo de espécies medicinales y aromáticas nativas. Publicação Técnica do Instituto Nacional de Investigacion Agropecuária (INIA), Projeto FPTA 137, Uruguai

Di Stasi LC, Hiruma-Lima CA, Santos CM, Guimarães EM (2002) Asterales medicinais. In: Di Stasi LC (ed) Plantas medicinais na amazônia e na mata atlântica. UNESP, São Paulo

Dicke M, Baldwin IT (2010) The evolutionary context for herbivore induced plant volatiles: Beyond the 'cry for help'. Trends Plant Sci 15:167–175

Ferro AFP, Bonacelli MBM, Assad ALD (2006) Oportunidades tecnológicas e estratégias concorrenciais de gestão ambiental: o uso sustentável da biodiversidade brasileira. Gestão & Produção 13:489–501

Forzza RC, Baumgratz JFA, Bicudo CEM, Canhos DAL, Carvalho AA Jr, Coelho MAN, Costa AF, Costa DP, Hopkins MG, Leitman PM, Lohmann LG, Lughadha EN, Maia LC, Martinelli

G, Menezes M, Morim MP, Peixoto AL, Pirani JR, Prado J, Queiroz LP, Souza S, Souza VC, Stehmann JR, Sylvestre LS, Walter BMT, Zappi DC (2012) New Brazilian floristic list highlights conservation challenges. Bioscience 62:39–45

Gobbo-Netto L, Lopes N (2007) Plantas medicinais: Fatores de influência no conteúdo de metabólitos secundário. Quím Nova 30:374–381

Grecco SS, Gimenes L, Ferreira MJP, Romoff P, Favero OA, Zalewski CA, Lago JHG (2010a) Triterpenoids and phenolic derivatives from *Baccharis uncinella* DC. (Asteraceae). Biochem Syst Ecol 38:1234–1237

Grecco SS, Reimão JQ, Tempone AG, Sartorelli P, Romoff P, Ferreira MJP, Fávero OA, Lago JHG (2010b) Isolation of an antileishmanial and antitrypanosomal flavanone from the leaves of *Baccharis retusa* DC. (Asteraceae). Parasitol Res 106:1245–1248

Guimarães NSS, Mello JC, Paiva JS, Bueno PCP, Berretta AA, Torquato RJ, Nantes IL, Rodrigues T (2012) *Baccharis dracunculifolia*, the main source of green propolis, exhibits potent antioxidant activity and prevents oxidative mitochondrial damage. Food Chem Toxicol 50:1091–1097

IBGE – Instituto Brasileiro de Geografia e Estatística – Manual Técnico de vegetação Brasileira – manuais técnicos em geociências (1992) 1-92/Mallard WG, Linstrom PJ (1992) NIST Chemistry Webbook. National Institute os Standards and Technology, Gaithersburg

Kurdelas RR, López S, Lima B, Feresin GE, Zygadlo J, Zacchino S, López ML, Tapia A, Freile ML (2012) Chemical composition, anti-insect and antimicrobial activity of *Baccharis darwinii* essential oil from Argentina, Patagonia. Indust Crop Prod 40:261–267

Lago LHG, Romoff P, Fávero OA (2008) Composição química dos óleos essenciais das folhas de seis espécies do gênero *Baccharis* de "campos de altitude" da mata atlântica paulista. Quím Nova 31:727–730

Oliveira-Tintino CDM, Tintino SR, Limaverde PW, Figueredo FG, Campina FF, Cunha FAB, Costa RHS, Pereira PS, Lima LF, Matos YMLS, Coutinho HDM, Siqueira-Júnior JP, Balbino VQ, Silva TG (2018) Inhibition of the essential oil from *Chenopodium ambrosioides* L. and α- terpinene on the NorA efflux-pump of *Staphylococcus aureus*. Food Chem 262:72–77

Pillar VP (2006) Estado atual e desafios para a conservação dos campos. Workshop. Instituto de Biociências da Universidade Federal do Rio Grande do Sul, Brazil

Sanchez F, Korine C, Steeghs M, Laarhoven LJ, Cristescu S, Harrem FM, Dudley R, Pinshow B (2006) Ethanol and methanol as possible odor cues for Egyptian fruit bats (*Rousettus aegyptiacus*). J Chem Ecol 32:1289–1300

Sangwan NS, Farooqi AHA, Shabih F, Sangwan RS (2001) Regulation of essential oil production in plants. Plant Growth Regul 34:3–21

Santos DA, Fukui MJ, Nanayakkara NPD, Khan SI, Sousa JPB, Bastos JK, Andrade SF, Silva Filho AA, Quintão NLM (2010) Anti-inflammatory and antinociceptive effects of *Baccharis dracunculifolia* DC (Asteraceae) in different experimental models. J Ethnopharmacol 127:543–550

Santos JFS, Rocha JE, Bezerra CF, Silva MKN, Matos YMLS, Freitas TS, Santos ATL, Cruz RP, Machado AJT, Rodrigues THS, Brito ES, Sales DL, Almeida WO, Costa JGM, Coutinho HDM, Morais-Braga MFB (2018) Chemical composition, antifungal activity and potential anti-virulence evaluation of the *Eugenia uniflora* essential oil against *Candida* spp. Food Chem 261:233–239

Sikkema J, De Bont JAM, Poolman B (1994) Interactions of cyclic hydrocarbons with biological membranes. J Biol Chem 269:8022–8028

Silveira GP, Nome F, Gesser JC (2006) Estratégias Utilizadas no Combate a Resistência Bacteriana. Quím Nova 29:844–855

Simoes CMO, Schenkel EP, Gosman G, Mello JCP, Mentz LA, Petrovick PR (2007) Farmacognosia: da planta ao medicamento, 6th edn. Simões CMO, Spitzer V. Porto Alegre/Florianópolis, Brazil

Souza CV, Lorenzi H (2008) Botânica Sistemática: Guia ilustrado para identificação das famílias de fanerógamas nativas e exóticas no Brasil, baseado em APG II. 2ª edição, Editora Plantarum, Nova Odessa, SP, Brazil

Souza SAM, Meira MR, Figueiredo LS, Martins ER (2010) Óleos essências: aspectos econômicos e sustentáveis. Enciclopéd Biosfer 6:1–10

Trombetta D (2005) Mechanisms of antibacterial action of three monoterpenes. Antimicrob Agents Chemother 49:2474–2478

Verdi LG, Brighente IMC, Pizzolatti MG (2005) Gênero *Baccharis* (Asteraceae): Aspectos químicos, econômicos e biológicos. Quím Nova 28:85–94

Wiley Registry of Mass Spectral Data (1994) 6th edn. Wiley Inter-science, New York

Yunes RA (2012) Química de produtos naturais: novos fármacos e a moderna farmacognosia; Yunes RA, Cechinel Filho C, Org; 3ª Edition. Univali, Itajaí, Brazil, p 384

The manufacturer's authorised representative in the EU is Springer
Nature Customer Service Centre GmbH, Europaplatz 3, 69115 Heidelberg,
Germany. If you have any concerns regarding our products, please
contact ProductSafety@springernature.com

Printed and bound by CPI Group (UK) Ltd, Croydon, CR0 4YY
25/09/2026
02222886-0011